An Introduction to
Mixed-Signal IC Test and Measurement

THE OXFORD SERIES IN ELECRICAL AND COMPUTER ENGINEERING

ADEL S. SEDRA, Series Editor

Allen and Holberg, *CMOS Analog Circuit Design, 2nd Edition*
Bobrow, *Elementary Linear Circuit Analysis, 2nd Edition*
Bobrow, *Fundamentals of Electrical Engineering, 2nd Edition*
Burns and Roberts, *An Introduction to Mixed-Signal IC Test and Measurement*
Campbell, *The Science and Engineering of Microelectronic Fabrication, 2nd Edition*
Chen, *Analog & Digital Control System Design*
Chen, *Linear System Theory and Design, 3rd Edition*
Chen, *System and Signal Analysis, 2nd Edition*
Chen, *Digital Signal Processing*
Comer, *Digital Logic and State Machine Design, 3rd Edition*
Cooper and McGillem, *Probabilistic Methods of Signal and System Analysis, 3rd Edition*
DeCarlo and Lin, *Linear Circuit Analysis, 2nd Edition*
Dimitrijev, *Understanding Semiconductor Devices*
Fortney, *Principles of Electronics: Analog & Digital*
Franco, *Electric Circuits Fundamentals*
Granzow, *Digital Transmission Lines*
Guru and Hiziroğlu, *Electric Machinery and Transformers, 3rd Edition*
Hoole and Hoole, *A Modern Short Course in Engineering Electromagnetics*
Jones, *Introduction to Optical Fiber Communication Systems*
Krein, *Elements of Power Electronics*
Kuo, *Digital Control Systems, 3rd Edition*
Lathi, *Modern Digital and Analog Communications Systems, 3rd Edition*
Lathi, *Signal Processing and Linear Systems*
Lathi, *Linear Systems and Signals*
Martin, *Digital Integrated Circuit Design*
McGillem and Cooper, *Continuous and Discrete Signal and System Analysis, 3rd Edition*
Miner, *Lines and Electromagnetic Fields for Engineers*
Parhami, *Computer Arithmetic*
Roberts and Sedra, *SPICE, 2nd Edition*
Roulston, *An Introduction to the Physics of Semiconductor Devices*
Sadiku, *Elements of Electromagnetics, 3rd Edition*
Santina, Stubberud and Hostetter, *Digital Control System Design, 2nd Edition*
Sarma, *Introduction to Electrical Engineering*
Schaumann and Van Valkenburg, *Design of Analog Filters*
Schwarz, *Electromagnetics for Engineers*
Schwarz and Oldham, *Electrical Engineering: An Introduction, 2nd Edition*
Sedra and Smith, *Microelectronic Circuits, 4th Edition*
Stefani, Savant, Shahian, and Hostetter, *Design of Feedback Control Systems, 4th Edition*
Van Valkenburg, *Analog Filter Design*
Warner and Grung, *Semiconductor Device Electronics*
Warner and Grung, *MOSFET Theory and Design*
Wolovich, *Automatic Control Systems*
Yariv, *Optical Electronics in Modern Communications, 5th Edition*

An Introduction to
Mixed-Signal IC Test and Measurement

Mark Burns
Texas Instruments, Incorporated

Gordon W. Roberts
McGill University

New York   Oxford
OXFORD UNIVERSITY PRESS
2001

Oxford University Press

Oxford New York
Athens Auckland Bangkok Bogotá Buenos Aires Calcutta
Cape Town Chennai Dar es Salaam Delhi Florence Hong Kong Istanbul
Karachi Kuala Lumpur Madrid Melbourne Mexico City Mumbai
Nairobi Paris São Paulo Shanghai Singapore Taipei Tokyo Toronto Warsaw

and associated companies in
Berlin Ibadan

Copyright © 2001 by Texas Instruments, Incorporated

Published by Oxford University Press, Inc.
198 Madison Avenue, New York, New York, 10016
http://www.oup-usa.org

Oxford is a registered trademark of Oxford University Press

All rights reserved. No part of this publication may be reproduced,
stored in a retrieval system, or transmitted, in any form or by any means,
electronic, mechanical, photocopying, recording, or otherwise,
without the prior permission of Oxford University Press.

Library of Congress Cataloging-in-Publication Data

Burns, Mark, 1962–
  An introduction to mixed-signal IC test and measurement / Mark Burns, Gordon Roberts
    p.   cm.   — (Oxford series in electrical and computer engineering)
  Includes bibliographical references and index.
  ISBN-13 978-0-19-514016-3
  ISBN-10 0-19-514016-8

    1. Integrated circuits—Testing. 2. Mixed signal circuits—Testing. I. Roberts, Gordon
W., 1959– II. Title. III. Series
  TK7874 .B825 2000
  621.3815—dc21
                                                                    00-042770

Printing number: 9 8

Printed in the United States of America
on acid-free paper

Contents

PREFACE xvii

Chapter 1: Overview of Mixed-Signal Testing

1.1 MIXED-SIGNAL CIRCUITS 1
    1.1.1 Analog, Digital, or Mixed-Signal? 1
    1.1.2 Common Types of Analog and Mixed-Signal Circuits 2
    1.1.3 Applications of Mixed-Signal Circuits 3
1.2 WHY TEST MIXED-SIGNAL DEVICES? 5
    1.2.1 The CMOS Fabrication Process 5
    1.2.2 Real-World Circuits 5
    1.2.3 What Is a Test Engineer? 8
1.3 POST-SILICON PRODUCTION FLOW 10
    1.3.1 Test and Packaging 10
    1.3.2 Characterization versus Production Testing 11
1.4 TEST AND DIAGNOSTIC EQUIPMENT 11
    1.4.1 Automated Test Equipment 11
    1.4.2 Wafer Probers 13
    1.4.3 Handlers 13
    1.4.4 E-Beam Probers 14
    1.4.5 Focused Ion Beam Equipment 15
    1.4.6 Forced-Temperature Systems 15
1.5 NEW PRODUCT DEVELOPMENT 16
    1.5.1 Concurrent Engineering 16
1.6 MIXED-SIGNAL TESTING CHALLENGES 17
    1.6.1 Time to Market 18
    1.6.2 Accuracy, Repeatability, and Correlation 18
    1.6.3 Electromechanical Fixturing Challenges 18
    1.6.4 Economics of Production Testing 19

Chapter 2: The Test Specification Process

2.1 DEVICE DATA SHEETS 23
    2.1.1 Purpose of a Data Sheet 23
    2.1.2 Structure of a Data Sheet 24
    2.1.3 Electrical Characteristics 27
2.2 GENERATING THE TEST PLAN 31
    2.2.1 To Plan or Not to Plan 31

2.2.2 Structure of a Test Plan   35
2.2.3 Design Specifications versus Production Test Specifications   36
2.2.4 Converting the Data Sheet into a Test Plan   37
2.3 COMPONENTS OF A TEST PROGRAM   38
2.3.1 Test Program Structure   38
2.3.2 Test Code and Digital Patterns   38
2.3.3 Binning   40
2.3.4 Test Sequence Control   40
2.3.5 Waveform Calculations and Other Initializations   41
2.3.6 Focused Calibrations and DIB Checkers   41
2.3.7 Characterization Code   42
2.3.8 Simulation Code   42
2.3.9 "Debuggability"   42
2.4 SUMMARY   43

Chapter 3: DC and Parametric Measurements

3.1 CONTINUITY   45
3.1.1 Purpose of Continuity Testing   45
3.1.2 Continuity Test Technique   46
3.1.3 Serial versus Parallel Continuity Testing   48
3.2 LEAKAGE CURRENTS   50
3.2.1 Purpose of Leakage Testing   50
3.2.2 Leakage Test Technique   50
3.2.3 Serial versus Parallel Leakage Testing   51
3.3 POWER SUPPLY CURRENTS   51
3.3.1 Importance of Supply Current Tests   51
3.3.2 Test Techniques   51
3.4 DC REFERENCES AND REGULATORS   52
3.4.1 Voltage Regulators   52
3.4.2 Voltage References   55
3.4.3 Trimmable References   55
3.5 IMPEDANCE MEASUREMENTS   56
3.5.1 Input Impedance   56
3.5.2 Output Impedance   58
3.5.3 Differential Impedance Measurements   59
3.6 DC OFFSET MEASUREMENTS   60
3.6.1 V_{MID} and Analog Ground   60
3.6.2 DC Transfer Characteristics (Gain and Offset)   60
3.6.3 Output Offset Voltage (V_O)   61
3.6.4 Single-Ended, Differential, and Common-Mode Offsets   62
3.6.5 Input Offset Voltage (V_{OS})   64
3.7 DC GAIN MEASUREMENTS   65
3.7.1 Closed-Loop Gain   65
3.7.2 Open-Loop Gain   68
3.8 DC POWER SUPPLY REJECTION RATIO   71
3.8.1 DC Power Supply Sensitivity   71
3.8.2 DC Power Supply Rejection Ratio   72

3.9 DC COMMON-MODE REJECTION RATIO 72
3.9.1 CMRR of Op Amps 72
3.9.2 CMRR of Differential Gain Stages 75
3.10 COMPARATOR DC TESTS 77
3.10.1 Input Offset Voltage 77
3.10.2 Threshold Voltage 78
3.10.3 Hysteresis 78
3.11 VOLTAGE SEARCH TECHNIQUES 79
3.11.1 Binary Searches versus Step Searches 79
3.11.2 Linear Searches 80
3.12 DC TESTS FOR DIGITAL CIRCUITS 82
3.12.1 I_{IH}/I_{IL} 82
3.12.2 V_{IH}/V_{IL} 82
3.12.3 V_{OH}/V_{OL} 82
3.12.4 I_{OH}/I_{OL} 82
3.12.5 I_{OSH} and I_{OSL} Short Circuit Current 82
3.13 SUMMARY 83

Chapter 4: Measurement Accuracy

4.1 TERMINOLOGY 87
4.1.1 Accuracy and Precision 87
4.1.2 Systematic Errors 88
4.1.3 Random Errors 88
4.1.4 Resolution (Quantization Error) 88
4.1.5 Repeatability 89
4.1.6 Stability 90
4.1.7 Correlation 91
4.1.8 Reproducibility 92
4.2 CALIBRATIONS AND CHECKERS 93
4.2.1 Traceability to Standards 93
4.2.2 Hardware Calibration 93
4.2.3 Software Calibration 93
4.2.4 System Calibrations and Checkers 96
4.2.5 Focused Instrument Calibrations 97
4.2.6 Focused DIB Circuit Calibrations 101
4.2.7 DIB Checkers 102
4.2.8 Tester Specifications 103
4.3 DEALING WITH MEASUREMENT ERROR 106
4.3.1 Filtering 106
4.3.2 Averaging 111
4.3.3 Guardbanding 113
4.4 BASIC DATA ANALYSIS 114
4.4.1 Datalogs 114
4.4.2 Histograms 115
4.4.3 Noise, Test Time, and Yield 118
4.5 SUMMARY 120

Chapter 5: Tester Hardware

5.1 MIXED-SIGNAL TESTER OVERVIEW 123
    5.1.1 General-Purpose Testers versus Focused Bench Equipment 123
    5.1.2 Generic Tester Architecture 123
5.2 DC RESOURCES 125
    5.2.1 General-Purpose Multimeters 125
    5.2.2 General-Purpose Voltage/Current Sources 127
    5.2.3 Precision Voltage References and User Supplies 128
    5.2.4 Calibration Source 128
    5.2.5 Relay Matrices 128
    5.2.6 Relay Control Lines 130
5.3 DIGITAL SUBSYSTEM 131
    5.3.1 Digital Vectors 131
    5.3.2 Digital Signals 131
    5.3.3 Source Memory 132
    5.3.4 Capture Memory 132
    5.3.5 Pin Card Electronics 134
    5.3.6 Timing and Formatting Electronics 136
5.4 AC SOURCE AND MEASUREMENT 139
    5.4.1 AC Continuous Wave Source and AC Meter 139
    5.4.2 Arbitrary Waveform Generators 139
    5.4.3 Waveform Digitizers 140
    5.4.4 Clocking and Synchronization 141
5.5 TIME MEASUREMENT SYSTEM 141
    5.5.1 Time Measurements 141
    5.5.2 Time Measurement Interconnects 142
5.6 COMPUTING HARDWARE 143
    5.6.1 User Computer 143
    5.6.2 Tester Computer 144
    5.6.3 Array Processors and Distributed Digital Signal Processors 144
    5.6.4 Network Connectivity 144
5.7 SUMMARY 144

Chapter 6: Sampling Theory

6.1 ANALOG MEASUREMENTS USING DSP 147
    6.1.1 Traditional versus DSP-Based Testing of AC Parameters 147
6.2 SAMPLING AND RECONSTRUCTION 148
    6.2.1 Use of Sampling and Reconstruction in Mixed-Signal Testing 148
    6.2.2 Sampling: Continuous-Time and Discrete-Time Representation 149
    6.2.3 Reconstruction 152
    6.2.4 The Sampling Theorem and Aliasing 159
    6.2.5 Quantization Effects 161
    6.2.6 Sampling Jitter 166
6.3 REPETITIVE SAMPLE SETS 170
    6.3.1 Finite and Infinite Sample Sets 170
    6.3.2 Coherent Signals and Noncoherent Signals 171

      6.3.3 Peak-to-RMS Control in Coherent Multitones   173
      6.3.4 Spectral Bin Selection   175
6.4 SYNCHRONIZATION OF SAMPLING SYSTEMS   179
      6.4.1 Simultaneous Testing of Multiple Sampling Systems   179
      6.4.2 ATE Clock Sources   181
      6.4.3 The Challenge of Synchronization   183
6.5 SUMMARY   184

Chapter 7: DSP-Based Testing

7.1 ADVANTAGES OF DSP-BASED TESTING   189
      7.1.1 Reduced Test Time   189
      7.1.2 Separation of Signal Components   189
      7.1.3 Advanced Signal Manipulations   190
7.2 DIGITAL SIGNAL PROCESSING   190
      7.2.1 DSP and Array Processing   190
      7.2.2 Fourier Analysis of Periodic Signals   191
      7.2.3 The Trigonometric Fourier Series   192
      7.2.4 The Discrete-Time Fourier Series   195
      7.2.5 Complete Frequency Spectrum   205
      7.2.6 Time and Frequency Denormalization   210
      7.2.7 Complex Form of the DTFS   211
7.3 DISCRETE-TIME TRANSFORMS   213
      7.3.1 The Discrete Fourier Transform   213
      7.3.2 The Fast Fourier Transform   216
      7.3.3 Interpreting the FFT Output   218
7.4 THE INVERSE FFT   230
      7.4.1 Equivalence of Time- and Frequency-Domain Information   230
      7.4.2 Parseval's Theorem   232
      7.4.3 Applications of the Inverse FFT   233
      7.4.4 Frequency-Domain Filtering   234
      7.4.5 Noise Weighting   239
7.5 SUMMARY   240
APPENDIX A.7.1   241

Chapter 8: Analog Channel Testing

8.1 OVERVIEW   249
      8.1.1 Types of Analog Channels   249
      8.1.2 Types of AC Parametric Tests   250
      8.1.3 Review of Logarithmic Operations   250
8.2 GAIN AND LEVEL TESTS   251
      8.2.1 Absolute Voltage Levels   251
      8.2.2 Absolute Gain and Gain Error   256
      8.2.3 Gain Tracking Error   258
      8.2.4 PGA Gain Tests   260
      8.2.5 Frequency Response   265

8.3 PHASE TESTS 273
    8.3.1 Phase Response 273
    8.3.2 Group Delay and Group Delay Distortion 278
8.4 DISTORTION TESTS 280
    8.4.1 Signal to Harmonic Distortion 280
    8.4.2 Intermodulation Distortion 283
8.5 SIGNAL REJECTION TESTS 284
    8.5.1 Common-Mode Rejection Ratio 284
    8.5.2 Power Supply Rejection and Power Supply Rejection Ratio 287
    8.5.3 Channel-to-Channel Crosstalk 289
    8.5.4 Clock and Data Feedthrough 293
8.6 NOISE TESTS 293
    8.6.1 Noise 293
    8.6.2 Idle Channel Noise 294
    8.6.3 Signal to Noise, Signal to Noise and Distortion 296
    8.6.4 Spurious Free Dynamic Range 298
    8.6.5 Weighting Filters 300
8.7 SIMULATION OF ANALOG CHANNEL TESTS 304
    8.7.1 MATLAB Model of an Analog Channel 304
8.8 SUMMARY 308

Chapter 9: Sampled Channel Testing

9.1 OVERVIEW 315
    9.1.1 What Are Sampled Channels? 315
    9.1.2 Examples of Sampled Channels 315
    9.1.3 Types of Sampled Channels 318
9.2 SAMPLING CONSIDERATIONS 320
    9.2.1 DUT Sampling Rate Constraints 320
    9.2.2 Digital Signal Source and Capture 321
    9.2.3 Simultaneous DAC and ADC Channel Testing 326
    9.2.4 Mismatched Fundamental Frequencies 330
    9.2.5 Undersampling 333
    9.2.6 Reconstruction Effects in AWGs, DACs, and Other Sampled-Data Circuits 335
9.3 ENCODING AND DECODING 338
    9.3.1 Signal Creation and Analysis 338
    9.3.2 Data Formats 339
    9.3.3 Intrinsic Errors 344
9.4 SAMPLED CHANNEL TESTS 350
    9.4.1 Similarity to Analog Channel Tests 350
    9.4.2 Absolute Level, Absolute Gain, Gain Error, and Gain Tracking 351
    9.4.3 Frequency Response 356
    9.4.4 Phase Response (Absolute Phase Shift) 359
    9.4.5 Group Delay and Group Delay Distortion 360
    9.4.6 Signal to Harmonic Distortion, Intermodulation Distortion 360
    9.4.7 Crosstalk 361
    9.4.8 CMRR 362

        9.4.9   PSR and PSRR   362
        9.4.10 Signal-to-Noise Ratio and ENOB   363
        9.4.11 Idle Channel Noise   363
   9.5 SUMMARY   364

Chapter 10: Focused Calibrations

   10.1 OVERVIEW   369
        10.1.1 Traceability to National Standards   369
        10.1.2 Why Are Focused Calibrations Needed?   370
        10.1.3 Types of Focused Calibrations   372
        10.1.4 Mechanics of Focused Calibration   372
        10.1.5 Program Structure   375
   10.2 DC CALIBRATIONS   376
        10.2.1 DC Offset Calibration   376
        10.2.2 DC Gain and Offset Calibrations   378
        10.2.3 Cascading DC Offset and Gain Calibrations   380
   10.3 AC AMPLITUDE CALIBRATIONS   382
        10.3.1 Calibrating AWGs and Digitizers   382
        10.3.2 Low-Level AWG and Digitizer Amplitude Calibrations   389
        10.3.3 Amplitude Calibrations for ADC and DAC Tests   390
   10.4 OTHER AC CALIBRATIONS   392
        10.4.1 Phase Shifts   392
        10.4.2 Digitizer and AWG Synchronization   396
        10.4.3 DAC and ADC Phase Shifts   396
        10.4.4 Distortion Tests   396
        10.4.5 Noise Tests   397
   10.5 ERROR CANCELLATION TECHNIQUES   397
        10.5.1 Avoiding Absolute Calibration   397
        10.5.2 Gain and Phase Matching   397
        10.5.3 Differential Gain and Differential Phase   399
   10.6 SUMMARY   400

Chapter 11: DAC Testing

   11.1 BASICS OF CONVERTER TESTING   403
        11.1.1 Intrinsic Parameters versus Transmission Parameters   403
        11.1.2 Comparison of DACs and ADCs   404
        11.1.3 DAC Failure Mechanisms   405
   11.2 BASIC DC TESTS   405
        11.2.1 Code-Specific Parameters   405
        11.2.2 Full-Scale Range   406
        11.2.3 DC Gain, Gain Error, Offset, and Offset Error   406
        11.2.4 LSB Step Size   409
        11.2.5 DC PSS   410
   11.3 TRANSFER CURVE TESTS   410
        11.3.1 Absolute Error   410
        11.3.2 Monotonicity   412

11.3.3 Differential Nonlinearity   412
11.3.4 Integral Nonlinearity   416
11.3.5 Partial Transfer Curves   419
11.3.6 Major Carrier Testing   420
11.3.7 Other Selected-Code Techniques   423
11.4 DYNAMIC DAC TESTS   424
11.4.1 Conversion Time (Settling Time)   424
11.4.2 Overshoot and Undershoot   426
11.4.3 Rise Time and Fall Time   426
11.4.4 DAC-to-DAC Skew   426
11.4.5 Glitch Energy (Glitch Impulse)   427
11.4.6 Clock and Data Feedthrough   428
11.5 DAC ARCHITECTURES   428
11.5.1 Resistive Divider DACs   428
11.5.2 Binary-Weighted DACs   430
11.5.3 PWM DACs   431
11.5.4 Sigma-Delta DACs   433
11.5.5 Companded DACs   434
11.5.6 Hybrid DAC Architectures   435
11.6 TESTS FOR COMMON DAC APPLICATIONS   435
11.6.1 DC References   435
11.6.2 Audio Reconstruction   436
11.6.3 Data Modulation   436
11.6.4 Video Signal Generators   436
11.7 SUMMARY   437
APPENDIX A.11.1   437

Chapter 12: ADC Testing

12.1 ADC TESTING VERSUS DAC TESTING   447
12.1.1 Comparison of DACs and ADCs   447
12.1.2 Statistical Behavior of ADCs   448
12.2 ADC CODE EDGE MEASUREMENTS   454
12.2.1 Edge Code Testing versus Center Code Testing   454
12.2.2 Step Search and Binary Search Methods   455
12.2.3 Servo Method   455
12.2.4 Linear Ramp Histogram Method   456
12.2.5 Conversion from Histograms to Code Edge Transfer Curves   457
12.2.6 Accuracy Limitations of Histogram Testing   460
12.2.7 Rising Ramps versus Falling Ramps   461
12.2.8 Sinusoidal Histogram Method   462
12.3 DC TESTS AND TRANSFER CURVE TESTS   467
12.3.1 DC Gain and Offset   467
12.3.2 INL and DNL   468
12.3.3 Monotonicity and Missing Codes   469
12.4 DYNAMIC ADC TESTS   470
12.4.1 Conversion Time, Recovery Time, and Sampling Frequency   470
12.4.2 Aperture Jitter   472
12.4.3 Sparkling   472

12.5 ADC ARCHITECTURES 473
- 12.5.1 Successive Approximation Architectures 473
- 12.5.2 Integrating ADCs (Dual-Slope and Single-Slope) 474
- 12.5.3 Flash ADCs 475
- 12.5.4 Semiflash ADCs 476
- 12.5.5 PDM (Sigma-Delta) ADCs 477

12.6 TESTS FOR COMMON ADC APPLICATIONS 479
- 12.6.1 DC Measurements 479
- 12.6.2 Audio Digitization 479
- 12.6.3 Data Transmission 479
- 12.6.4 Video Digitization 480

12.7 SUMMARY 480

Chapter 13: DIB Design

13.1 DIB BASICS 483
- 13.1.1 Purpose of a Device Interface Board 483
- 13.1.2 DIB Configurations 484
- 13.1.3 Importance of Good DIB Design 486

13.2 PRINTED CIRCUIT BOARDS 486
- 13.2.1 Prototype DIBs versus PCB DIBs 486
- 13.2.2 PCB CAD Tools 487
- 13.2.3 Multilayer PCBs 488
- 13.2.4 PCB Materials 489

13.3 DIB TRACES, SHIELDS, AND GUARDS 490
- 13.3.1 Trace Parasitics 490
- 13.3.2 Trace Resistance 490
- 13.3.3 Trace Inductance 491
- 13.3.4 Trace Capacitance 496
- 13.3.5 Shielding 502
- 13.3.6 Driven Guards 503

13.4 TRANSMISSION LINES 504
- 13.4.1 Lumped- and Distributed-Element Models 504
- 13.4.2 Transmission Line Termination 508
- 13.4.3 Parasitic Lumped Elements 514

13.5 GROUNDING AND POWER DISTRIBUTION 514
- 13.5.1 Grounding 514
- 13.5.2 Power Distribution 516
- 13.5.3 Power and Ground Planes 517
- 13.5.4 Ground Loops 518

13.6 DIB COMPONENTS 519
- 13.6.1 DUT Sockets and Contactor Assemblies 519
- 13.6.2 Contact Pads, Pogo Pins, and Socket Pins 520
- 13.6.3 Electromechanical Relays 521
- 13.6.4 Socket Pins 524
- 13.6.5 Resistors 525
- 13.6.6 Capacitors 526
- 13.6.7 Inductors and Ferrite Beads 528
- 13.6.8 Transformers and Power Splitters 528

13.7 COMMON DIB CIRCUITS 530
- 13.7.1 Local Relay Connections 530
- 13.7.2 Relay Multiplexers 532
- 13.7.3 Selectable Loads 533
- 13.7.4 Analog Buffers (Voltage Followers) 533
- 13.7.5 Instrumentation Amplifiers 534
- 13.7.6 V_{MID} Reference Adder 535
- 13.7.7 Current-to-Voltage and Voltage-to-Current Conversions 536
- 13.7.8 Power Supply Ripple Circuits 536

13.8 COMMON DIB MISTAKES 540
- 13.8.1 Poor Power Supply and Ground Layout 540
- 13.8.2 Crosstalk 541
- 13.8.3 Transmission Line Discontinuities 541
- 13.8.4 Resistive Drops in Circuit Traces 541
- 13.8.5 Tester Instrument Parasitics 541
- 13.8.6 Oscillations in Active Circuits 542
- 13.8.7 Poor DIB Component Placement and PCB Layout 542

13.9 SUMMARY 543
APPENDIX A.13.1 543

Chapter 14: Design for Test (DfT)

14.1 OVERVIEW 549
- 14.1.1 What Is DfT? 549
- 14.1.2 Built-In Self-Test 550
- 14.1.3 Differences between Digital DfT and Analog DfT 550
- 14.1.4 Why Should We Use DfT? 551

14.2 ADVANTAGES OF DfT 551
- 14.2.1 Lower Cost of Test 551
- 14.2.2 Increased Fault Coverage and Improved Process Control 553
- 14.2.3 Diagnostics and Characterization 553
- 14.2.4 Ease of Test Program Development 554
- 14.2.5 System-Level Diagnostics 555
- 14.2.6 Economics of DfT 555

14.3 DIGITAL SCAN 556
- 14.3.1 Scan Basics 556
- 14.3.2 IEEE Std. 1149.1 Standard Test Access Port and Boundary Scan 557
- 14.3.3 Full Scan and Partial Scan 559

14.4 DIGITAL BIST 562
- 14.4.1 Pseudorandom BILBO Circuits 562
- 14.4.2 Memory BIST 563
- 14.4.3 Microcode BIST 564

14.5 DIGITAL DfT FOR MIXED-SIGNAL CIRCUITS 565
- 14.5.1 Partitioning 565
- 14.5.2 Digital Resets and Presets 566
- 14.5.3 Device-Driven Timing 567
- 14.5.4 Lengthy Preambles 569

14.6 MIXED-SIGNAL BOUNDARY SCAN AND BIST 569
- 14.6.1 Mixed-Signal Boundary Scan (IEEE Std. 1149.4) 569

14.6.2 Analog and Mixed-Signal BIST   571
14.7 AD HOC MIXED-SIGNAL DfT   573
   14.7.1 Common Concepts   573
   14.7.2 Accessibility of Analog Signals   573
   14.7.3 Analog Test Buses, T-Switches, and Bypass Modes   575
   14.7.4 Separation of Analog and Digital Blocks   577
   14.7.5 Loopback Modes   579
   14.7.6 Precharging Circuits and AC Coupling Shorts   580
   14.7.7 On-Chip Sampling Circuits   581
   14.7.8 PLL Testability Circuits   583
   14.7.9 DAC and ADC Converters   584
   14.7.10 Oscillation BIST   585
   14.7.11 Physical Test Pads   585
14.8 SUBTLE FORMS OF ANALOG DfT   585
   14.8.1 Robust Circuits   585
   14.8.2 Design Margin as DfT   586
   14.8.3 Avoiding Overspecification   586
   14.8.4 Predictability of Failure Mechanisms   586
   14.8.5 Conversion of Analog Functions to Digital   587
   14.8.6 Reduced Tester Performance Requirements   587
   14.8.7 Avoidance of Trim Requirements   587
14.9 I_{DDQ} 587
   14.9.1 Digital I_{DDQ} 587
   14.9.2 Analog and Mixed-Signal I_{DDQ} 588
14.10 SUMMARY   589
APPENDIX A.14.1   589

Chapter 15: Data Analysis

15.1 INTRODUCTION TO DATA ANALYSIS   597
   15.1.1 The Role of Data Analysis in Test and Product Engineering   597
   15.1.2 Visualizing Test Results   597
15.2 DATA VISUALIZATION TOOLS   598
   15.2.1 Datalogs (Data Lists)   598
   15.2.2 Lot Summaries   599
   15.2.3 Wafer Maps   600
   15.2.4 Shmoo Plots   601
   15.2.5 Histograms   604
15.3 STATISTICAL ANALYSIS   606
   15.3.1 Mean (Average) and Standard Deviation (Variance)   606
   15.3.2 Probabilites and Probability Density Functions   607
   15.3.3 The Standard Gaussian Cumulative Distribution Function $\Phi(z)$ 611
   15.3.4 Non-Gaussian Distributions   615
   15.3.5 Guardbanding and Gaussian Statistics   618
   15.3.6 Effects of Measurement Variability on Test Yield   620
   15.3.7 Effects of Reproducibilty and Process Variation on Yield   623
15.4 STATISTICAL PROCESS CONTROL   627
   15.4.1 Goals of SPC   627
   15.4.2 Six-Sigma Quality   628

15.4.3 Process Capability, C_p, and C_{pk} 628
15.4.4 Gauge Repeatability and Reproducibility   630
15.4.5 Pareto Charts   631
15.4.6 Scatter Plots   631
15.4.7 Control Charts   633
15.5 SUMMARY   634

Chapter 16: Test Economics

16.1 PROFITABILITY FACTORS   641
16.1.1 What Is Meant by Test Economics?   641
16.1.2 Time to Market   641
16.1.3 Testing Costs   642
16.1.4 Yield Enhancement   642
16.2 DIRECT TESTING COSTS   643
16.2.1 Cost Models   643
16.2.2 Cost of Test versus Cost of Tester   643
16.2.3 Throughput   645
16.3 DEBUGGING SKILLS   649
16.3.1 Sources of Error   649
16.3.2 The Scientific Method   649
16.3.3 Practical Debugging Skills   651
16.3.4 Importance of Bench Instrumentation   652
16.3.5 Test Program Structure   652
16.3.6 Common Bugs and Techniques to Find Them   653
16.4 EMERGING TRENDS   655
16.4.1 Test Language Standards   655
16.4.2 Test Simulation   656
16.4.3 Noncoherent Sampling   658
16.4.4 Built-In Self-Test   658
16.4.5 Defect-Oriented Testing   658
16.5 SUMMARY   659

ANSWERS TO SELECTED PROBLEMS   663

INDEX   677

Preface

Integrated circuits incorporating both digital and analog functions have become increasingly prevalent in the semiconductor industry. Complex digital circuits are now commonly combined with analog circuits as part of the continuing drive toward higher levels of electronic system integration. For example, complex microprocessors are frequently combined with high-performance analog and mixed-signal circuits to form so-called "system-on-a-chip" devices. An example of this is a single chip modem combining a digital signal processor with precision analog-to-digital and digital-to-analog functions on a single silicon die. Such devices offer the semiconductor customer significant savings in manufacturing costs due to the resulting reduction of chip-to-chip interconnections.

Mixed-signal IC test and measurement has grown into a highly specialized field of electrical engineering. However, test engineering is still a relatively unknown profession compared with IC design engineering. It has become harder to hire and train new engineers to become skilled mixed-signal test engineers. It may take one to two years for a mixed-signal test engineer to develop enough knowledge and experience to develop adequate test solutions. The slow learning curve for mixed-signal test engineers is largely due to the shortage of written materials and university-level courses on the subject of mixed-signal testing. While many books have been devoted to the subject of digital test and testability, the same cannot be said for analog and mixed-signal automated test and measurement.

Training for mixed-signal test engineers has historically started with a sink-or-swim training course covering the use of the test equipment itself, with little or no training on the basics of mixed-signal test and measurement. This equipment-centric approach to training is analogous to teaching a student how to drive by simply explaining the mechanics of the automobile itself (pull this knob, push that pedal, etc.). It would be unwise to assign such an inadequately trained student to drive from L.A. to Pittsburgh without a roadmap and without a working knowledge of trivialities such as stop lights and police sirens. Similarly, a new test engineer is often assigned to develop tests for a complex circuit without training in basic test definitions and common test techniques.

The test engineer is also expected to contribute to the definition of testability circuits that are incorporated into the design of the device to be tested. Again, there is little formal reference material or training on the subject of basic mixed-signal design for test (DfT). As a result, new test engineers often overlook basic deficiencies in the circuit architecture that prevent the device from being tested thoroughly and economically.

This book was written in response to the shortage of basic course material for mixed-signal test and measurement. The book assumes a solid background in analog and digital circuits as well as a working knowledge of computers and computer programming. A background in digital signal processing and statistical analysis is also helpful, though not absolutely necessary. This material is designed to be useful as both a university textbook and as a reference manual for the beginning professional test engineer. Like many specialized technical materials, this book will most likely become partially outdated before publication. Hopefully, it will at least serve as an amusing historical record of how things were done back in the twentieth century.

The prerequisite for this book is a junior-level course in linear continuous-time and discrete-time systems, as well as exposure to elementary probability and statistical concepts. Fortunately, these two courses are usually required at most universities.

The book is divided into 16 chapters. Chapter 1 presents an introduction to the context in which mixed-signal testing is performed and why it is necessary. Chapter 2 examines the process by which test programs are generated, from device data sheet to test plan to test code. Test program structure and functionality are also discussed in Chapter 2. Chapter 3 introduces basic DC measurement definitions, including continuity, leakage, offset, gain, DC power supply rejection ratio, and many other types of fundamental DC measurements.

Chapter 4 covers the basics of absolute accuracy, resolution, software calibration, standards traceability, and measurement repeatability. In addition, basic data analysis is presented in Chapter 4. A more thorough treatment of data analysis and statistical analysis is delayed until Chapter 15.

Chapter 5 takes a closer look at the architecture of a generic mixed-signal ATE tester. The generic tester includes instruments such as DC sources, meters, waveform digitizers, arbitrary waveform generators, and digital pattern generators with source and capture functionality.

Chapter 6 presents an introduction to both ADC and DAC sampling theory. DAC sampling theory is applicable to both DAC circuits in the device under test and to the arbitrary waveform generators in a mixed-signal tester. ADC sampling theory is applicable to both ADC circuits in the device under test and to waveform digitizers in a mixed-signal tester. Coherent multi-tone sample sets are also introduced as an introduction to DSP based testing. Chapter 7 further develops sampling theory concepts and DSP-based testing methodologies, which are at the core of many mixed-signal test and measurement techniques. FFT fundamentals, windowing, frequency domain filtering, and other DSP-based testing fundamentals are covered in Chapters 6 and 7.

Chapter 8 shows how basic AC channel tests can be performed economically using DSP-based testing. This chapter covers only nonsampled channels, consisting of combinations of op amps, analog filters, PGAs and other continuous-time circuits. Chapter 9 explores many of these same tests as they are applied to sampled channels, which include DACs, ADCs, sample and hold (S/H) amplifiers, etc.

Chapter 10 explains how the basic accuracy of ATE test equipment can be extended using specialized software routines. This subject is not necessarily taught in formal ATE tester classes, yet it is critical in the accurate measurement of many DUT performance parameters.

Testing of DACs is covered in Chapter 11. Several kinds of DACs are studied, including traditional binary-weighted, resistive ladder, pulse-width modulation (PWM), and sigma-delta architectures. Traditional measurements like INL, DNL, and absolute error are discussed. Several kinds of DAC architectures are explored, with an emphasis on their respective weaknesses and common testing methodologies. Chapter 12 builds upon the concepts in Chapter 11 to show how ADCs are commonly tested. Again, several different kinds of ADCs are studied, including binary-weighted, dual-slope, flash, semiflash, and sigma-delta architectures. The weaknesses of each design are explained, as well as the common methodologies used to probe their weaknesses.

Chapter 13 explores the gray art of mixed-signal DIB design. Topics of interest include component selection, power and ground layout, crosstalk, shielding, transmission lines, and tester loading. Chapter 13 also illustrates several common DIB circuits and their use in mixed-signal testing.

Chapter 14 gives a brief introduction to some of the techniques for analog and mixed-signal design for test. There are fewer structured approaches for mixed-signal DfT than for purely digital DfT. The more common ad hoc methods are explained, as well as some of the industry standards such as IEEE Std. 1149.1 and 1149.4.

A brief review of statistical analysis and Gaussian distributions is presented in Chapter 15. This chapter also shows how measurement results can be analyzed and viewed using a variety of software tools and display formats. Datalogs, shmoo plots, and histograms are discussed. Also, statistical process control (SPC) is explained, including a discussion of process control metrics such as C_p and C_{pk}.

Chapter 16 examines the economics of production testing. The economics of test are affected by many factors such as equipment purchase price, test floor overhead costs, test time, dual-head testing, multisite testing, and time to market. A test engineer's debugging skills heavily impacts time to market. Chapter 16 examines the test debugging process to attempt to set down some general guidelines for debugging mixed-signal test programs. Finally, emerging trends that affect test economics and test development time are presented in Chapter 16. Some or all of these trends will shape the future course of mixed-signal test and measurement.

The preliminary versions of this complete manuscript were reviewed by a number of students and practicing test engineers. We would like to thank those professionals and students who gave us extensive corrections and feedback to improve this textbook: Steve Lyons (Lucent Technologies/Teradyne, Inc.), Jim Larson and Gary Moraes (Teradyne, Inc.), Justin Ewing (Texas A&M University/Texas Instruments, Inc.) Pramodchandran Variyam (Georgia Tech/Texas Instruments, Inc.), and Geoffrey Zhang (Texas Instruments, Inc.). We also thank Juli Boman (Teradyne, Inc.) and Ted Lundquist (Schlumberger Test Equipment) for providing photographs for Chapter 1.

We would also like to extend our sincere appreciation to Dr. Rainer Fink and Dr. Jay Porter of Texas A&M University, Dr. Cajetan Akujuobi of Prairieview A&M University, and Dr. Simon Ang of the University of Arkansas for their help in developing this textbook. Their early adoption of this work at their respective universities has helped to shape the book's content and expose its many weaknesses.

We are extremely grateful to the staff at Oxford University Press, who have helped guide us through the process of writing an enjoyable book. First, we would like to acknowledge the help and constructive feedback of the publishing editor, Peter Gordon. The editorial development help of Karen Shapiro was greatly appreciated.

Finally, on behalf of the test engineering profession, Mark Burns would like to extend his gratitude to Del Whittaker, David VanWinkle, Bob Schwartz, Ming Chiang, and Brian Evans, all of Texas Instruments, Inc., for allowing him to develop this book as part of his engineering duties for the past three years. It takes great courage and vision for corporate management to expend resources on the production of a work that may ultimately help the competition. Mark also extends his appreciation to his parents, Burt and Shirley Burns, whose financial and emotional support helped him through four years at the Massachusetts Institute of Technology.

On behalf of Gordon Roberts, he would like to extend his sincere appreciation to all the dedicated staff members and graduate students associated with the Microelectronics and Computer Systems (MACS) Laboratory at McGill University. Professors Nicholas Rumin and David Lowther, past and present chairmen of the department of electrical and computer engineering, deserve special mention for initially believing in this project and allowing it to take root and flourish at McGill University. He would also like to note the enormous contribution made by his friend, past graduate thesis supervisor and present-day mentor, Professor Adel Sedra of the University of Toronto, for his invaluable advice over the past two decades. Professor Sedra taught him more about the world of microelectronics than anyone else. Finally, and, most important, Gordon Roberts would like to express his sincere gratitude to his best friend and partner, Eileen O'Reilly, for her constant support and encouragement during this project. Her dedication to their two children, Brigid Maureen and Sean Gordon, gave him the peace of mind needed to work on this book with Mark. For this, he will be forever in debt.

Mark Burns	Gordon W. Roberts
Texas Instruments, Inc.	McGill University
Dallas, Texas	Montreal, Quebec, Canada

CHAPTER 1

Overview of Mixed-Signal Testing

1.1 MIXED-SIGNAL CIRCUITS

1.1.1 Analog, Digital, or Mixed-Signal?

Before delving into the details of mixed-signal IC test and measurement, one might first ask a few good questions. Exactly what are mixed-signal circuits? How are they used in typical applications? Why do we have to test mixed-signal circuits in the first place? What is the role of a test engineer, and how does it differ from that of a design engineer or product engineer? Most training classes offered by mixed-signal tester companies assume that the students already know the answers to these questions. For instance, a typical automated test equipment (ATE) training class shows the students how to program the per-pin current leakage measurement instruments in the tester before the students even know why leakage current is an important parameter to measure. This book will answer many of the what's, when's, and why's of mixed-signal testing, as well as the usual how's. Let's start with a very basic question: what is a mixed-signal circuit?

A mixed-signal circuit can be defined as a circuit consisting of both digital and analog elements. By this definition, a comparator is one of the simplest mixed-signal circuits. It compares two analog voltages and determines if the first voltage is greater than or less than the second voltage. Its digital output changes to one of two states depending on the outcome of the comparison. In effect, a comparator is a one-bit analog-to-digital converter (ADC). It might also be argued that a simple digital inverter is a mixed-signal circuit, since its digital input controls an "analog" output which swings between two fixed voltages, rising, falling, overshooting, and undershooting according to the laws of analog circuits. In fact, in certain extremely high-frequency applications the outputs of digital circuits have been tested using mixed-signal testing methodologies.[1]

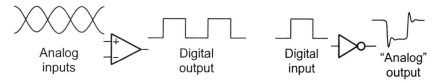

Figure 1.1. Comparator and inverter - analog, digital, or mixed-signal?

Some mixed-signal experts might argue that a comparator and an inverter are not mixed-signal devices at all. The comparator is typically considered an analog circuit, while an inverter is considered a digital circuit (Figure 1.1). Other examples of borderline mixed-signal devices are analog switches and programmable gain amplifiers. The purist might argue that mixed-signal circuits are those that involve some sort of nontrivial interaction between digital signals and analog signals. Otherwise, the device is simply a combination of digital logic and separate analog circuitry coexisting on the same die or circuit board. The line between mixed-signal circuits and analog or digital circuits is blurry if one wants to be pedantic.

Fortunately, the blurry lines between digital, analog, and mixed-signal are completely irrelevant in the context of mixed-signal test and measurement. Most complex mixed-signal devices include at least some stand-alone analog circuits that do not interact with digital logic at all. Thus, the testing of op amps, comparators, voltage references, and other purely analog circuits must be included in a comprehensive study of mixed-signal testing. This book encompasses the testing of both analog and mixed-signal circuits, including many of the borderline examples. Digital testing will only be covered superficially, since testing of purely digital circuits has been extensively documented elsewhere.[2-4]

1.1.2 Common Types of Analog and Mixed-Signal Circuits

Analog circuits (also known as *linear circuits*) include operational amplifiers, active or passive filters, comparators, voltage regulators, analog mixers, analog switches, and other specialized functions such as Hall effect transistors. One of the very simplest circuits that can be considered to fall into the mixed-signal realm is the CMOS analog switch. In this circuit, the resistance of a CMOS transistor is varied between high impedance and low impedance under control of a digital signal. The off-resistance may be as high as one megaohm or more, while the on-resistance may be 100 Ω or less. Banks of analog switches can be interconnected in a variety of configurations, forming more complex circuits such as analog multiplexers and demultiplexers and analog switch matrices.

Another simple type of mixed-signal circuit is the programmable gain amplifier (PGA). The PGA is often used in the front end of a mixed-signal circuit to allow a wider range of input signal amplitudes. Operating as a digitally adjusted volume control, the PGA is set to high gains for low-amplitude input signals and low gains for high-amplitude input signals. The next circuit following a PGA is thus provided with a consistent signal level. Many circuits require a consistent signal level to achieve optimum performance. These circuits therefore benefit from the use of PGAs.

PGAs and analog switches involve a trivial interaction between the analog and digital circuits. This is why they are not always considered to be mixed-signal circuits at all. The most common circuits that can truly be considered mixed-signal devices are analog to digital converters (A/Ds or ADCs) and digital to analog converters (D/As or DACs). While the abbreviations A/D and ADC are used interchangeably in the electronics industry, this book will always use the term ADC for consistency. Similarly, the term DAC will be used throughout the book rather than D/A. An ADC is a circuit that samples a continuous analog signal at specific points in time and converts the sampled voltages (or currents) into a digital representation. Each digital representation is called a *sample*. Conversely, a DAC is a circuit that converts digital samples into analog voltages (or currents). ADCs and DACs are the most common mixed-signal components in complex mixed-signal designs, since they form the interface between the physical world and the world of digital logic.

Comprehensive testing of DACs and ADCs is an expansive topic, since there are a wide variety of ADC and DAC designs and a wide variety of techniques to test them. For example, an ADC which is only required to sample once per second may employ a dual slope conversion architecture, whereas a 100-MHz video ADC may have to employ a much faster flash conversion architecture. The weaknesses of these two architectures are totally different. Consequently, the testing of these two converter types is totally different. Similar differences exist between the various types of DACs.

Another common mixed-signal circuit is the phase locked loop, or PLL. PLLs are typically used to generate high-frequency reference clocks or to recover a synchronous clock from an asynchronous data stream. In the former case, the PLL is combined with a digital divider to construct a frequency multiplier. A relatively low-frequency clock, say, 50 MHz, is then multiplied by an integer value to produce a higher-frequency master clock, such as 1 GHz. In the latter case, the recovered clock from the PLL is used to latch the individual bits or bytes of the incoming data stream. Again, depending on the nature of the PLL design and its intended use, the design weaknesses and testing requirements can be very different from one PLL to the next.

1.1.3 Applications of Mixed-Signal Circuits

Many mixed-signal circuits consist of combinations of amplifiers, filters, switches, ADCs, DACs, and other types of specialized analog and digital functions. End-equipment applications such as cellular telephones, hard disk drives, modems, motor controllers, and multimedia audio and video products all employ complex mixed-signal circuits. While it is important to test the individual circuits making up a complex mixed-signal device, it is also important to perform system-level tests. System-level tests guarantee that the circuit as a whole will perform as required in the end-equipment application. Thorough testing of large-scale mixed-signal circuits therefore requires at least a basic understanding of the end-equipment application in which the circuits will be used.

As an example of a mixed-signal application, let us consider a common consumer product using many mixed-signal subcircuits. Figure 1.2 shows a simplified block diagram of a complex mixed-signal application, the digital cellular telephone. It represents an excellent example of a complex mixed-signal system because it employs a variety of mixed-signal components. Since the digital cellular telephone will be used as an example throughout this book, we shall examine its operation in some detail.

A cellular telephone consists of many analog, digital, and mixed-signal circuits working together in a complex fashion. The cellular telephone user interfaces with the keyboard and display to answer incoming calls and to initiate outgoing calls. The control microprocessor handles the interface with the user. It also performs many of the supervisory functions of the telephone, such as helping coordinate the handoff from one base station to the next as the user travels through each cellular area. The control microprocessor selects the incoming and outgoing transmission frequencies by sending control signals to the frequency synthesizer. The synthesizer often consists of several PLLs, which control the mixers in the radio frequency (RF) section of the cellular telephone. The mixers convert the relatively low-frequency signals of the base-band interface to extremely high frequencies that can be transmitted from the cellular telephone's radio antenna. They also convert the very high-frequency incoming signals from the base station into lower-frequency signals that can be processed by the base-band interface.

The voice-band interface, digital signal processor (DSP), and base-band interface perform most of the complex operations. The voice-band interface converts the user's voice into digital samples using an ADC. The volume of the voice signal from the microphone can be adjusted automatically using a programmable gain amplifier (PGA) controlled by either the DSP or the control microprocessor. Alternatively, the PGA may be controlled with a specialized digital circuit built into the voice-band interface itself. Either way, the PGA and automatic adjustment mechanism form an automatic gain control (AGC) circuit. Before the voice signal can be digitized by the voice-band interface ADC, it must first be low-pass filtered to avoid unwanted high-frequency components that might cause aliasing in the transmitted signal. (Aliasing is a type of distortion that can occur in sampled systems, making the speaker's voice difficult to understand.)

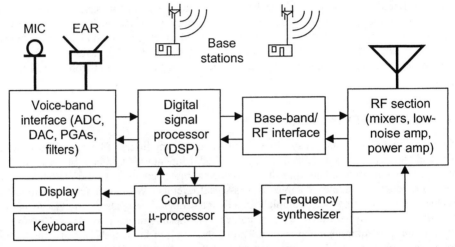

Figure 1.2. Digital cellular telephone.

The digitized samples are sent to the DSP, where they are compressed using a mathematical process called *vocoding*. The vocoding process converts the individual samples of the sound pressure waves into samples that represent the *essence* of the user's speech. The vocoding algorithm calculates a time-varying model of the speaker's vocal tract as each word is spoken. The characteristics of the vocal tract change very slowly compared to the sound pressure waves of the speaker's voice. Therefore, the vocoding algorithm can compress the important characteristics of speech into a much smaller set of data bits than the digitized sound pressure samples. The vocoding process is therefore a type of data compression algorithm that is specifically tailored for speech. The smaller number of transmitted bits frees up airspace for more cellular telephone users. The vocoder's output bits are sent to the base-band interface and RF circuits for modulation and transmission. The base-band interface acts like a modem, converting the digital bits of the vocoder output into modulated analog signals. The RF circuits then transmit the modulated analog waveforms to the base station.

In the receiving direction, the process is reversed. The incoming voice data are received by the RF section and demodulated by the base-band interface to recover the incoming vocoder bit stream. The DSP converts the incoming bit stream back into digitized samples of the incoming speaker's voice. These samples are then passed to the DAC and low pass reconstruction filter of the voice-band interface to reconstruct the voltage samples of the incoming voice. Before the received voice signal is passed to the earpiece, its volume is adjusted using a second PGA. This earpiece PGA is adjusted by signals from the control microprocessor, which monitors the telephone's volume control buttons to determine the user's desired volume setting. Finally, the signal must be passed through a low impedance buffer to provide the current necessary to drive the earpiece.

Several common cellular telephone circuits are not shown in Figure 1.2. These include DC voltage references and voltage regulators that may exist on the voice-band interface or the base-band processor, analog multiplexers to control the selection of multiple voice inputs, and power-on reset circuits. In addition, a watchdog timer is often included to periodically wake the control microprocessor from its battery-saving idle mode. This allows the microprocessor to receive information such as incoming call notifications from the base station. Clearly, the digital cellular telephone represents a good example of a complex mixed-signal system. The various circuit blocks of a cellular telephone may be grouped into a small number of individual integrated circuits, called a *chipset*, or they may all be combined into a single chip. The test engineer must be ready to test the individual pieces of the cellular telephone and/or to test the cellular telephone as a whole. The increasing integration of circuits into a single semiconductor die is one of the most challenging aspects of mixed-signal test engineering.

1.2 WHY TEST MIXED-SIGNAL DEVICES?

1.2.1 The CMOS Fabrication Process

Integrated circuits (ICs) are fabricated using a series of photolithographic printing, etching, and doping steps. Using a digital CMOS fabrication process as an example, let us look at the idealized IC fabrication process. Some of the steps involved in printing a CMOS transistor pair are illustrated in Figure 1.3a-f. Starting with a lightly doped P$^-$ wafer, a layer of silicon dioxide (SiO$_2$) is deposited on the surface (Figure 1.3a). Next, a negative photoresist is laid down on top of the silicon dioxide. A pattern of ultraviolet light is then projected onto the photoresist using a photographic mask. The photoresist becomes insoluble in the areas where the mask allows the ultraviolet light to pass (Figure 1.3b). An organic solvent is used to dissolve the nonexposed areas of the photoresist (Figure 1.3c). After baking the remaining photoresist, the exposed areas of oxide are removed using an etching process (Figure 1.3d). Next, the exposed areas of silicon are doped to form an N-well using either diffusion or ion implantation (Figure 1.3e).

After many additional steps of printing, masking, etching, implanting, and chemical vapor deposition,[5] a complete integrated circuit can be fabricated as illustrated in Figure 1.3f. The uneven surfaces are exaggerated in the diagram to show that the various layers of oxide, polysilicon, and metal are not at all flat. Even with these exaggerations, this diagram only represents an idealized approximation of actual fabricated circuit structures. The actual circuit structures are not nearly as well defined as textbook diagrams would lead us to believe. Cross sections of actual integrated circuits reveal a variety of nonideal physical characteristics that are not entirely under the semiconductor manufacturer's control. Certain characteristics, such as doping profiles that define the boundaries between P and N regions, are not even visible in a cross-section view. Nevertheless, they can have a profound effect on many important analog and mixed-signal circuit characteristics.

1.2.2 Real-World Circuits

Like any photographic printing process, the IC printing process is subject to blemishes and imperfections. These imperfections may cause catastrophic failures in the operation of any individual IC, or they may cause minor variations in performance from one IC to the next. Mixed-signal ICs are often extremely sensitive to tiny imperfections or variations in the printing and doping processes. Many of the fabrication defects that cause problems in mixed-signal devices are difficult to photograph, even with a powerful scanning electron microscope (SEM).

For example, a doping error may or may not cause an observable physical defect. However, doping errors can introduce large DC offsets, distortions, and other problems that result in IC performance failures.

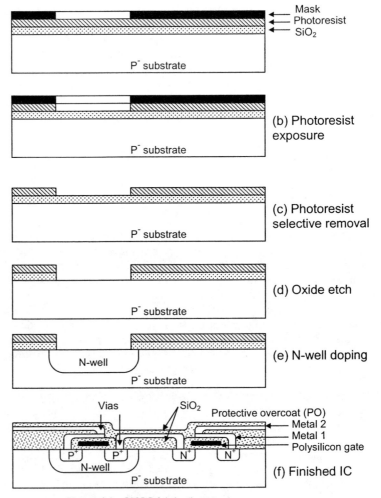

Figure 1.3. CMOS fabrication steps.

Certain types of defects can be photographed quite easily. Figure 1.4 shows a nondefective circuit as photographed using a FIB machine (a device similar to a scanning electron microscope). Compared to the idealized textbook circuit representation, the metal traces are rounded and imperfect.

In digital circuits, such imperfections in shape may be largely unimportant. However, in mixed-signal circuits, the parasitic capacitance between these traces and surrounding structures may represent significant circuit elements. The exact three-dimensional shape of a metal line and its spacing to adjacent layers may therefore affect the performance of the circuit under test. As circuit geometries continue to shrink, these performance sensitivities will only become more exaggerated. Although a mixed-signal circuit may be essentially functional in the presence of these minor imperfections, it may not meet all its required specifications. For this reason, mixed-signal devices are often tested exhaustively to guard against defects that are not necessarily catastrophic.

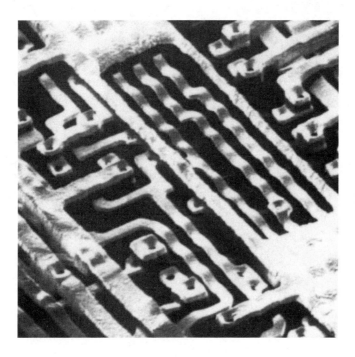

Figure 1.4. FIB micrograph of metal traces on an integrated circuit obtained using a Schlumberger AMS 3000 (photo courtesy Schlumberger Test Equipment).

Catastrophic defects such as short circuits and open circuits are often easier to detect with test equipment than the subtler ones common in mixed-signal devices. Not surprisingly, the catastrophic defects are often much easier to photograph as well. Several typical defect types are shown in Figures 1.5–1.8. Figure 1.5 shows a defective metal contact, or via, caused by underetching. Figure 1.6 shows a defective via caused by photomask misalignment. A completely defective via usually results in a totally defective circuit, since it represents a complete open circuit. A more subtle problem is a partially connected via, which may exhibit an abnormally high contact resistance. Depending on the amount of excess resistance, the results of a partially connected via can range from minor DC offset problems to catastrophic distortion problems.

Figure 1.7 shows incomplete etching of the metal surrounding a circuit trace. Incomplete etching can result in catastrophic shorts between circuit nodes. Finally, Figure 1.8 shows a surface defect caused by particulate matter landing on the surface of the wafer or on a photographic mask during one of the processing steps. Again, this type of defect results in a short between circuit nodes. Other catastrophic defects include surface scratches, broken bond wires, and surface explosions caused by electrostatic discharge in a mishandled device. Defects such as these are the reason each semiconductor device must be tested before it can be shipped to the customer.

Figure 1.5. Underetched via.

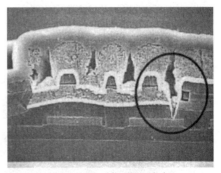
Figure 1.6. Misaligned via.

Figure 1.7. Incomplete metal etch.

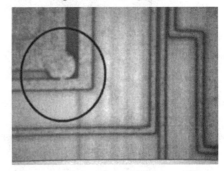
Figure 1.8. Blocked etch (particulate defect).

It has been said that production testing adds no value to the final product. Testing is an expensive process that drives up the cost of integrated circuits without adding any new functionality. Testing cannot change the quality of the individual ICs; it can only *measure* quality if it already exists. However, semiconductor companies would not spend money to test products if the testing process did not add value. This apparent discrepancy is easily explained if we recognize that the product is actually the entire shipment of devices, not just the individual ICs. The quality of the product is certainly improved by testing, since defective devices are not shipped. Therefore, testing does add value to the product, as long as we define the product correctly.

1.2.3 What Is a Test Engineer?

We have mentioned the term *test engineer* several times without actually defining what test engineering is. Perhaps this would be a good time to discuss the traditional roles of test engineers, design engineers, product engineers, and systems engineers. Although each of these engineering professions is involved in the development and production of semiconductor devices, each profession entails its own set of tasks and responsibilities. The various engineering professions are easiest to define if we examine the process by which a new semiconductor product is developed and manufactured.

A new semiconductor product typically begins in one of two ways. Either a customer requests a particular type of product to fill a specific requirement, or a marketing organization realizes an opportunity to produce a product that the market needs. In either case, systems engineers help define the technical requirements of the new product so that it will operate correctly in the end-equipment application. The systems engineers are responsible for defining and documenting the customer's requirements so that the rest of the engineering team can design the product and successfully release it to production.

After the systems engineers have defined the product's technical requirements, design engineers develop the corresponding integrated circuit. Hopefully, the new design meets the technical requirements of the customer's application. Unfortunately, integrated circuits sometimes fail to meet the customer's needs. The failure may be due to a fabrication defect or it may be due to a flaw or weakness in the circuit's design. These failures must be detected before the product is shipped to the customer.

The test engineer's role is to generate hardware and software that will be used by automated test equipment (ATE) to guarantee the performance of each device after it is fabricated. The test software directs the ATE tester to apply a variety of electrical stimuli (such as digital signals and sine waves) to the device under test (DUT). The ATE tester then observes the DUT's response to the various test stimuli to determine whether the device is good or bad (Figure 1.9). A typical mixed-signal DUT must pass hundreds or even thousands of stimulus/response tests before it can be shipped to the customer.

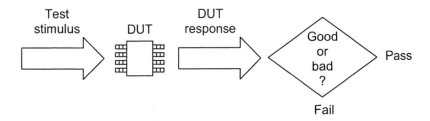

Figure 1.9. Test stimulus and DUT response verification.

Sometimes the test engineer is also responsible for developing hardware and software that modifies the structure of the semiconductor die to adjust parameters like DC offset and AC gain, or to compensate for grotesque manufacturing defects. Despite claims that production testing adds no value, this is one way in which the testing process can actually enhance the quality of the individual ICs. Circuit modifications can be made in a number of ways, including laser trimming, fuse blowing, and writing to nonvolatile memory cells.

The test engineer is also responsible for reducing the cost of testing through test time reductions and other cost-saving measures. The test cost reduction responsibility is shared with the product engineer. The product engineer's primary role is to support the production of the new device as it matures and proceeds to profitable volume production. The product engineer helps identify and correct process defects, design defects, and tester hardware and software defects.

Sometimes the product engineering function is combined with the test engineering function, forming a single test/product engineering position. The advantage of the combined job function is that the product engineering portion of the job can be performed with a much more thorough understanding of the device and test program details. The disadvantage is that the product engineering responsibilities may interfere with the ability of the engineer to become an expert on the use of the complex test equipment. The choice of combined versus divided job functions is highly dependent on the needs of each organization.

1.3 POST-SILICON PRODUCTION FLOW

1.3.1 Test and Packaging

After silicon wafers have been fabricated, many additional production steps remain before a final packaged device is ready for shipment to the customer. The untested wafers (Figure 1.10) must first be probed using automated test equipment to prevent bad dies from passing on to further production steps. The bad dies can be identified using ink dots, which are applied either after each die is tested or after the whole wafer has been tested. Offline inking is a method used to electronically track bad dies using a computer database. Using pass/fail information from the database, bad dies are inked after the wafer has been removed from the test equipment.

Figure 1.10. Untested wafer.

The wafers are then sawed into individual dies and the good ones are attached to lead frames. Lead frames are punched metal holders that eventually become the individual leads of the packaged device. Bond wires are attached from each die's bond pads to the appropriate lead of the lead frame. Then plastic is injection-molded around the dies and lead frame to form packaged devices. Finally, the individual packaged devices are separated from one another by trimming them from the lead frame.

After the leads have been trimmed and formed, the devices are ready for final testing on a second ATE tester. Final testing guarantees that the performance of the device did not shift during the packaging process. For example, the insertion of plastic over the surface of the die changes the electrical permitivity near the surface of the die. Consequently, trace-to-trace capacitances are increased, which may affect sensitive nodes in the circuit. In addition, the injection-molded plastic introduces mechanical stresses in the silicon, which may consequently introduce DC voltage shifts. Final testing also guarantees that the bond pads are all connected and that the die was not cracked, scratched, or otherwise damaged in the packaging process. After final testing, the devices are ready for shipment to the end-equipment manufacturer. Figure 1.11 shows a tray of tested quad flat pack (QFP) devices in a plastic carrier tray.

Figure 1.11. Tested QFP devices in a plastic carrier tray.

1.3.2 Characterization versus Production Testing

When prototype devices are first characterized, the ATE test program is usually very extensive. Tests are performed under many different conditions to evaluate worst-case conditions. For instance, the distortion of an amplifier output may be worse under one loading condition than another. All loading conditions must be tested to identify which one represents the worst-case test. Other examples of exhaustive characterization testing would be DC offset testing using multiple power supply voltages and harmonic distortion testing at multiple signal levels. Characterization testing must be performed over a large number of devices and over several production lots of material before the results can be considered statistically valid and trustworthy.

Characterization testing can be quite time consuming due to the large number of tests involved. Extensive characterization is therefore economically unacceptable in high-volume production testing of mixed-signal devices. Once worst-case test conditions have been established and the design engineers are confident that their circuits meet the required specifications, a more streamlined production test program is needed. The production test program is created from a subset of the characterization tests. The subset must be carefully chosen to guarantee that no bad devices are shipped. Product and test engineers must work very closely to make sure that the reduced test list still catches all manufacturing defects.

1.4 TEST AND DIAGNOSTIC EQUIPMENT

1.4.1 Automated Test Equipment

Automated test equipment is available from a number of commercial vendors, such as Teradyne, LTX, Agilent Technologies, and Schlumberger, to name a few. The Teradyne, Inc. Catalyst mixed-signal tester is shown in Figure 1.12. High-end ATE testers often consist of three major components: a test head, a workstation, and the mainframe.

Figure 1.12. Teradyne Catalyst mixed-signal tester (photo courtesy Teradyne, Inc.).

The computer workstation serves as the user interface to the tester. The test engineer can debug test programs from the workstation using a variety of software tools from the ATE vendor. Manufacturing personnel can also use the workstation to control the day-to-day operation of the tester as it tests devices in production.

The mainframe contains power supplies, measurement instruments, and one or more computers that control the instruments as the test program is executed. The mainframe may also contain a manipulator to position the test head precisely. It may also contain a refrigeration unit to provide cooled liquid to regulate the temperature of the test head electronics.

Although much of the tester's electronics are contained in the mainframe section, the test head contains the most sensitive measurement electronics. These circuits are the ones which require close proximity to the device under test. For example, high-speed digital signals benefit from short electrical paths between the tester's digital drivers and the pins of the DUT. Therefore, the ATE tester's digital drivers and receivers are located in the test head close to the DUT.

Figure 1.13. Device interface board (DIB) showing local circuits (left) and DUT socket (right).

A device interface board (DIB) forms the electrical interface between the ATE tester and the DUT. The DIB is also known as a *performance board*, *swap block*, or *family board*, depending on the ATE vendor's terminology. DIBs come in many shapes and sizes, but their main function is to provide a temporary (socketed) electrical connection between the DUT and the electrical instruments in the tester. The DIB also provides space for DUT-specific local circuits such as load circuits and buffer amplifiers that are often required for mixed-signal device testing.

1.4.2 Wafer Probers

Wafer probers are robotic machines that manipulate wafers as the individual dies are tested by the ATE equipment. The prober moves the wafer underneath a set of tiny electrical probes attached to a probe card. The probes are connected to the electrical resources of the ATE tester through a probe interface board (PIB). The PIB is a specialized type of DIB board that may be connected to the probe card through coaxial cables and/or spring-loaded contacts called *pogo pins*. The PIB and probe card serve the same purpose for the wafer that the DIB board serves for the packaged device. They provide a means of temporarily connecting the DUT to the ATE tester's electrical instrumentation while testing is performed.

The prober informs the tester when it has placed each new die against the probes of the probe card. The ATE tester then executes a series of electrical tests on the die before instructing the prober to move to the next die. The handshaking between tester and prober insures that the tester only begins testing when a die is in position and that the prober does not move the wafer in midtest. Figure 1.14 shows a wafer prober manufactured by Electroglas, Inc., and closeup views of a probe card and its probe tips.

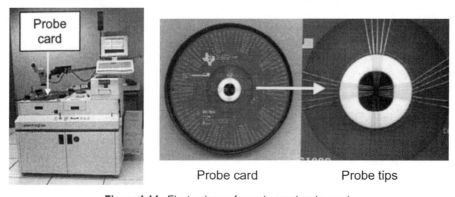

Figure 1.14. Electroglas wafer prober and probe card.

1.4.3 Handlers

Handlers are used to manipulate packaged devices in much the same way that probers are used to manipulate wafers. Handlers fall into two categories: gravity-fed and robotic. Robotic handlers are also known as *pick-and-place* handlers. Gravity-fed handlers are normally used with dual inline packages, while robotic handlers are used with devices having pins on all four sides or pins on the underside (ball grid array packages, for example). Figure 1.15 shows a gravity-fed handler. A robotic handler is shown in Figure 1.16.

Either type of handler has one main purpose: to make a temporary electrical connection between the DUT pins and the DIB board. Gravity-fed handlers often perform this task using a contactor assembly that grabs the device pins from either side with metallic contacts that are in turn connected to the DIB board. Robotic handlers usually pick up each device with a suction arm and then plunge the device into a socket on the DIB board.

In addition to providing a temporary connection to the DUT, handlers are also responsible for sorting the good DUTs from the bad ones based on test results from the ATE tester. Some handlers also provide a controlled thermal chamber where devices are allowed to "soak" for a few minutes so they can either be cooled or heated before testing. Since many electrical parameters shift with temperature, this is an important handler feature.

Figure 1.15. MultiTest gravity-fed handler.

Figure 1.16. Delta robotic handler.

1.4.4 E-Beam Probers

Electron beam probers, or e-beam probers as they are often called, are used to probe internal device signals while the device is being stimulated by the tester. These machines are very similar to scanning electron microscopes (SEMs). Unlike an SEM, an e-beam prober is designed to display variations in circuit voltage as the electron beam is swept across the surface of an operating DUT. Variations in the voltage levels on the metal traces in the IC appear as different shades of gray in the e-beam display (Figure 1.17). e-beam probers are extremely powerful diagnostic tools, since they provide measurement access to internal circuit nodes.

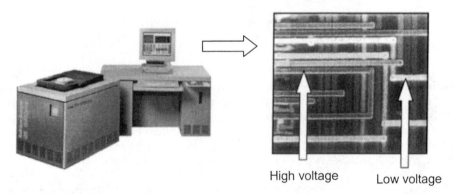

Figure 1.17. Schlumberger IDS-10000 electron beam prober
(photo courtesy Schlumberger Test Equipment).

1.4.5 Focused Ion Beam Equipment

Focused ion beam (FIB) equipment is used in conjunction with e-beam probers to modify the device's metal traces and other physical structures. A FIB machine can cut holes in oxide and metal traces and can also lay down new metallic traces on the surface of the device (Figure 1.18). Experimental design changes can be implemented without waiting for a complete semiconductor fabrication cycle. The results of the experimental changes can then be observed on the ATE tester to determine the success or failure of the experimental circuit modifications.

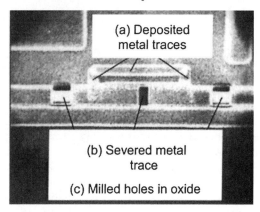

Figure 1.18. Circuit modifications implemented using FIB equipment.

1.4.6 Forced-Temperature Systems

As previously mentioned, a handler's thermal chamber allows characterization and testing of large numbers of DUTs at a controlled temperature. When characterizing a small number of DUTs at a variety of temperatures, a less expensive and cumbersome method of temperature control is needed. Portable forced-temperature systems allow DUT performance characterization

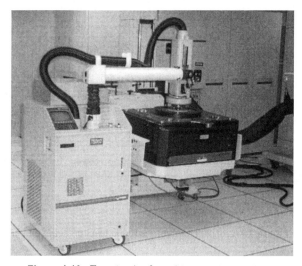

Figure 1.19. Temptronics forced-temperature system.

under a variety of controlled thermal conditions (Figure 1.19). The nozzle of a forced-temperature system can be seated against the DIB board or bench characterization board, forming a small thermal chamber for the DUT. Many forced-temperature systems are able to raise or lower the DUT's ambient temperature across the full military range (–55 to +125°C).

1.5 NEW PRODUCT DEVELOPMENT

1.5.1 Concurrent Engineering

On a poorly managed project, the test engineer might not see the specifications for a device to be tested until after the first prototype devices arrive. The devices must be screened as soon as possible to ship good prototypes to the customer even if they were never designed with testability in mind. In this case, the test engineer's role is completely reactive.

By contrast, the test engineer's role on a well-managed project is proactive. The design engineers and test engineers work together to add testability functions to the design that make the device easier and less expensive to test. The test engineer presents a test plan to the design engineers, explaining all the tests that are to be performed once the device is in production. The design engineers can catch mistakes in the test engineer's understanding of the device operation. They can help eliminate unnecessary tests or point out shortfalls in the proposed test list. This proactive approach is commonly called *concurrent engineering*. True concurrent engineering involves not only design and test engineering personnel, but also systems engineering, product engineering, and manufacturing personnel. Figure 1.20 shows a simplified concurrent engineering flow for the development and production release of a new semiconductor product.

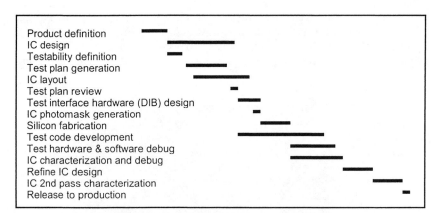

Figure 1.20. Concurrent engineering project flow.

The flow begins with a definition of the device requirements. These include product features, electrical specifications, power consumption requirements, die area estimates, etc. Once the device requirements are understood, the design team begins to design the individual circuits. In the initial design meetings, test and product engineers work with the design engineers to define the testability features that will make the device less expensive to test and manufacture. Test modes are added to the design to allow access to internal circuit nodes that otherwise would be unobservable in production testing. These observability test modes can be very useful in diagnosing device design flaws.

After the test modes are defined, the test engineer begins working on a test plan while the design process continues. Initially, the main purpose of a test plan is to allow design engineers and test engineers to agree upon a set of tests that will guarantee the quality of a product once it is in production. Eventually, the test plan will serve as documentation for future test and product engineers that may inherit the test program once it is complete. A well-written test plan contains brief background information about the product to be tested, the purpose of each test as it relates to the device specification, setup conditions for each test, and a hardware setup diagram for each test. Once the test plan is complete, all engineers working on the project meet to review the proposed test plan. Last-minute corrections and additions are added at this time. Design engineers point out deficiencies in the proposed test coverage while product engineers point out any problems that may arise on the production floor.

Once the test plan has been approved, the test engineer begins to design the necessary test interface hardware that will connect the automated test equipment to the device under test. Once the initial test hardware has been designed, the test engineer begins writing a test program that will run on the ATE tester. In modern ATE equipment, the test engineer can also debug many of the software routines in the test program before silicon arrives, using an offline simulation environment running on a stand-alone computer workstation.

After the design and layout of the device is complete, the fabrication masks are created from the design database. The database release process is known by various names, such as tape-out or pattern generation. Until pattern generation is complete, the test engineer cannot be certain that the pinout or functionality of the design will not undergo last-minute modifications. The test interface hardware is often fabricated only after the pattern generation step has been completed.

While the silicon wafers and the DIB board are fabricated, the test engineer continues developing the test program. Once the first silicon wafers arrive, the test engineer begins debugging the device, DIB hardware, and software on the ATE tester. Any design problems are reported to the design engineers, who then begin evaluating possible design errors. A second design pass is often required to correct errors and to align the actual circuit performance with specification requirements. Finally, the corrected design is released to production by the product engineer, who then supports the day-to-day manufacturing of the new product.

Of course, the idealized concurrent engineering flow is a simplification of what happens in a typical company doing business in the real world. Concurrent engineering is based on the assumption that adequate personnel and other resources are available to write test plans and generate test hardware and software before the first silicon wafers arrive. It also assumes that only one additional design pass is required to release a device to production. In reality, a high-performance device may require several design passes before it can be successfully manufactured at a profit. This flow also assumes that the market does not demand a change in the device specifications in midstream - a poor assumption in a dynamic world. Nevertheless, concurrent engineering is consistently much more effective than a disjointed development process with poor communication between the various engineering groups.

1.6 MIXED-SIGNAL TESTING CHALLENGES

1.6.1 Time to Market

Time to market is a pressing issue for semiconductor manufacturers. Profit margins for a new product are highest shortly after it has been released to the market. Margins begin to shrink as competitors introduce similar products at lower prices. The lack of a complete, cost-effective test program is often the main bottleneck preventing the release a new product to profitable volume production.

Mixed-signal test programs are particularly difficult to produce in a short period of time. Surprisingly, the time spent writing test code is often significantly less than the time spent learning about the device under test, defining the test plan, designing the test hardware, and debugging the ATE test solution once silicon is available. Much of the time spent in the debugging phase of test development is actually spent debugging device problems. Mixed-signal test engineers often spend as much time running experiments for design engineers to isolate design errors as they spend debugging their own test code. Perhaps the most aggravating debug time of all is the time spent tracking down problems with the tester itself or the tester's software.

1.6.2 Accuracy, Repeatability, and Correlation

Accuracy is a major concern for mixed-signal test engineers. It is very easy to get an answer from a mixed-signal ATE tester that is simply incorrect. Inaccurate answers are caused by a bewildering number of problems. Electromagnetic interference, improperly calibrated instruments, improperly ranged instruments, and measurements made under incorrect test conditions can all lead to inaccurate test results.

Repeatability is the ability of the test equipment and test program to give the same answer multiple times. Actually, a measurement that never changes at all is suspicious. It sometimes indicates that the tester is improperly configured, giving the same incorrect answer repeatedly. A good measurement typically shows some variability from one test program execution to the next, since electrical noise is present in all electronic circuits. Electrical noise is the source of many repeatability problems.

Another problem facing mixed-signal test engineers is correlation between the answers given by different pieces of measurement hardware. The customer or design engineer often finds that the test program results do not agree with measurements taken using bench equipment in their lab. The test engineer must determine which answer is correct and why there is a discrepancy. It is also common to find that two supposedly identical testers or DIB boards give different answers or that the same tester gives different answers from day to day. These problems frequently result from obscure hardware or software errors that may take days to isolate. Correlation efforts can represent a major portion of the time spent debugging a test program.

1.6.3 Electromechanical Fixturing Challenges

The test head and DIB board must ultimately make contact to the DUT through the handler or prober. There are few mechanical standards in the ATE industry to specify how a tester should be docked to a handler or prober. The test engineer has to design a DIB board that not only meets electrical requirements, but also meets the mechanical docking requirements. These requirements include board thickness, connector locations, DUT socket mechanical holes, and various alignment pins and holes.

Handlers and probers must make a reliable electrical connection between the DUT and the tester. Unfortunately, the metallic contacts between DUT and DIB board are often very inductive and/or capacitive. Stray inductance and capacitance of the contacts can represent a major problem, especially when testing high-impedance or high-frequency circuits. Although several companies have marketed test sockets that reduce these problems, a socketed device will often not perform quite as well as a device soldered directly to a printed circuit board. Performance differences due to sockets are yet another potential source of correlation error and extended time to market.

1.6.4 Economics of Production Testing

Time is money, especially when it comes to production test programs. A high-performance tester may cost two million dollars or more, depending on its configuration. Probers and handlers may cost five hundred thousand dollars or more. If we also include the cost of providing floor space, electricity, and production personnel, it is easy to understand why testing is an expensive business.

One second of test time can cost a semiconductor manufacturer three to five cents. This may not seem expensive at first glance, but when test costs are multiplied by millions of devices a year the numbers add up quickly. For example, a five-second test program costing four cents per second times one million devices per quarter costs a company $800,000 per year in bottom-line profit. Testing is perhaps the fastest-growing portion of the cost of manufacturing a mixed-signal device. Continuous process improvements and better photolithography allow the design engineers to add more functions on a single semiconductor chip at little or no additional cost. Unfortunately, test time (especially data collection time) cannot be similarly reduced by simple photolithography. A 100-Hz sine wave takes 10 ms per cycle no matter how small we shrink a transistor. The only hope of salvation from photolithography is the addition of test features into the design itself that aid in the testing of the DUT.

Mainframe ATE equipment is designed to minimize test time and maximize overall product throughput. For example, many testers can be equipped with two test heads that share the mainframe instruments in a multiplexing fashion (Figure 1.21). The purpose of the second head is to allow the tester to simultaneously test a device on one head while a second handler or prober is moving and sorting devices on the other head. Dual-head testing is especially important when the handler index time (the time it takes to remove one DUT from the tester and insert the next one) is significant compared to the test time. When a handler or prober is docked to each test head, the tester can run almost continuously. Thus dual-head testing allows a more efficient use of the expensive tester hardware.

Another feature common in mainframe testers is multisite capability. Multisite testing is a process in which multiple devices are tested on the same test head simultaneously with obvious savings in test cost. The word "site" refers to each socketed DUT. For example, site 0 corresponds to the first DUT; site 1 corresponds to the second DUT, etc. Multisite testing is primarily a tester operating system feature, although duplicate tester instruments must be added to the tester to allow simultaneous testing on multiple DUT sites.

Clearly, production test economics is an extremely important issue in the field of mixed-signal test engineering. Not only must the test engineer perform accurate measurements of mixed-signal parameters, but the measurements must be performed as quickly as possible to reduce production costs. Since a mixed-signal test program may perform hundreds or even thousands of measurements on each DUT, each measurment must be performed in a small fraction of a second. The conflicting requirements of low test time and high accuracy will be a recurring theme throughout this book.

Figure 1.21. Teradyne A575 with dual test heads.

Problems

1.1. List four examples of analog circuits.

1.2. List four examples of mixed-signal circuits.

1.3. Questions 1.3–1.6 relate to the cellular telephone in Figure 1.2. Which type of mixed-signal circuit acts as a volume control for the cellular telephone earpiece?

1.4. Which type of mixed-signal circuit converts the speaker's voice into digital samples?

1.5. Which type of mixed-signal circuit converts incoming modulated voice data into digital samples?

1.6. Which type of digital circuit vocodes the speaker's voice samples before they are passed to the base-band interface?

1.7. When a PGA is combined with a digital logic block to keep a signal at a constant level, what is the combined circuit called?

1.8. Assume a particle of dust lands on a photomask during the photolithographic printing process of a metal layer. List at least one possible defect that might occur in the printed IC.

1.9. Why does the cleanliness of the air in a semiconductor fabrication area affect the number of defects in IC manufacturing?

1.10. List at least four production steps after wafers have been fabricated.

1.11. Why would it be improper to draw conclusions about a design based on characterization data from one or two devices?

1.12. List three main components of an ATE tester.

1.13. What is the purpose of a DIB board?

1.14. What type of equipment is used to handle wafers as they are tested by an ATE tester?

1.15. List three advantages of concurrent engineering.

1.16. What is the purpose of a test plan?

1.17. List at least four challenges faced by the mixed-signal test engineer.

1.18. Assume a test program runs on a tester that costs the company 3 cents per second to operate. This test cost includes tester depreciation, handler depreciation, electricity, floor space, personnel, etc. How much money can be saved per year by reducing a 5-s test program to 3.5 s, assuming 5 million devices per year are to be shipped. Assume that only 90% of devices tested are good, and that the average time to find a bad device drops to 0.5 s.

1.19. Assume the profit margin on the device in problem 1.18 is 20% (i.e., for each $1 worth of devices shipped to the customer, the company makes a profit of 20 cents). How many dollars worth of product would have to be shipped to make a profit equal to the savings offered by the streamlined test program in Problem 1.18? If each device sells for $1.80, how many devices does this represent? What obvious conclusion can we draw about the importance of test time reduction versus the importance of selling and shipping additional devices?

References

1. Mark Burns, *High Speed Measurements Using Undersampled Delta Modulation*, 1997 Teradyne User's Group proceedings, Teradyne, Inc., 321 Harrison Ave., Boston, MA 02118

2. Miron Abramovici, Melvin A. Breuer, Arthur D. Friedman, *Digital Systems Testing and Testable Design*, Revised Printing, IEEE Press, New York, NY, January, 1998, ISBN: 0780310624

3. Parag K. Lala, *Practical Digital Logic Design and Testing*, Prentice Hall, Upper Saddle River, New Jersey, 1996, ISBN: 0023671718

4. J. Max Cortner, *Digital Test Engineering*, John Wiley & Sons, 605 Third Ave., New York, NY 10158-0012, 1987, ISBN: 0471851353

5. David A. Johns, Ken Martin, *Analog Integrated Circuit Design*, John Wiley & Sons, 605 Third Ave., New York, NY 10158-0012, 1996, ISBN: 0471144487

CHAPTER 2

The Test Specification Process

2.1 DEVICE DATA SHEETS

2.1.1 Purpose of a Data Sheet

Beginning test engineers often spend a great deal of time learning how to generate test plans for mixed-signal devices. A test plan is a written list of tests and test procedures that will be used to verify the quality of a particular device under test (DUT). The definition of a production test plan usually begins with a device *data sheet* or *specification sheet*, as it is often called. Unfortunately, the data sheet does not directly translate into a finite list of all required production tests. For example, a low-pass filter ripple specification of ±1.0 dB states that gain variation is guaranteed at each and every frequency in the passband of the filter. Of course, semiconductor manufacturers do not test every possible frequency in production. Test plan generation sometimes seems like more of an art than a science, especially when one tries to define exactly how a data sheet is translated into a test plan. Before exploring the art and science of mixed-signal test plan generation, let us look at the function and structure of a data sheet in detail.

The data sheet serves many purposes. When development of a new device begins, the data sheet serves as the design specification. Design engineers refer to the data sheet as a blueprint to make sure they design the functions that the marketing and systems engineering organizations have specified. As the project progresses, the test and product engineers refer to the data sheet to define the test list. The test list must be comprehensive enough to guarantee that the manufactured devices meet the data sheet specifications. Throughout the design process, the customer refers to the data sheet while designing the device into the system level end application. The data sheet thus serves as the formal communication channel between the marketing and engineering personnel engaged in a project.

The test engineer often detects data sheet mistakes and ambiguities while writing the test plan or developing the test program. In effect, the test engineer is the first customer for a new design. Likewise, the tester and device interface board (DIB) can be considered the new device's first application. Data sheet errors should be promptly corrected to prevent further mistakes. For instance, if an inappropriate measurement is specified in the initial data sheet, it is the test engineer's responsibility to make sure the error gets corrected or clarified so that a sensible test plan can be defined. For this reason it is important to know which organization is responsible for controlling the data sheet's contents.

The answer to the ownership question depends somewhat on the type of device being developed. There are two kinds of devices: catalog and custom. A catalog device is one that is defined by the semiconductor manufacturer or by an IC design house. Once defined, a catalog device is offered to multiple customers for use in their end applications. A custom device, by contrast, is defined by a specific customer. It must meet that customer's exact requirements.

In the case of a catalog device, the systems engineering or marketing organization controls the data sheet. The test engineer only needs to get agreement from the design and systems engineers to make a data sheet change. In the case of a custom device, the customer and systems engineer share responsibility for the contents of the data sheet. In addition to approvals from the marketing or systems engineering team, the customer's approval is also required before the data sheet can be modified.

Depending on the customer's requirements, data sheet changes may be very easy to implement or they may be impossibly difficult. Regardless of the customer's needs, though, specification changes requested at the last minute give the appearance of a poorly run organization. For this reason, it is a good idea for the test engineer to get involved very early in the definition of a device so that specification changes can be suggested in a timely manner. Suggestions made early in the new product development cycle give a customer more confidence that the testing process is under control.

2.1.2 Structure of a Data Sheet

Data sheets may contain any of the following sections: a feature summary and description, principles of operation, absolute maximum ratings, electrical characteristics, timing diagrams, application information, characterization data, circuit schematic, and die layout. The sections that are most pertinent to test engineering are the device description, principles of operation, electrical characteristics, timing diagrams, and package/pinout information. Before we can understand the process by which the production test list is developed, we must first understand the purpose of each of these data sheet sections.

Figure 2.1 shows an example data sheet for a digital-to-analog converter (DAC). This data sheet is taken from a Texas Instruments data acquisition circuits data book.[1] The first page of the data sheet provides a quick device summary. The feature summary allows the customer to quickly gauge the device's fit to a particular application. The test engineer can generally ignore this section since the same information is typically called out in subsequent sections of the data sheet. The pinout and package information is much more relevant to test engineering. The test engineer refers to the pinout and package information to design the DIB for each package type.

The device description gives a quick overview of the device's functionality. Together with the principles of operation (Figure 2.2), the device description defines the various operations of the device in detail. The test program must guarantee all these functions, though not necessarily in a straightforward manner. For instance, the device description may depict a circuit that divides an externally generated 1-MHz reference clock by one million, producing a 1-s timebase. Since straightforward testing would represent an unacceptably long test time of 1 s, this function might be tested in an indirect manner.

There are many indirect ways to guarantee the operation of a 1-s timebase counter without spending 1 s of test time. For example, a special test mode might split the divider into two separate stages that each count to one thousand in only 1 ms. The two divider stages could then be tested simultaneously to guarantee the functionality of the whole. Total test time would be only 1 ms (plus overhead introduced by the tester). This is an example of a design for test (DfT) test mode. It serves no purpose in the system-level end application. The customer does not need to split the divider into two halves and may not even need to know that it can be placed in this test mode at all. Therefore, test modes may or may not be documented in the data sheet.

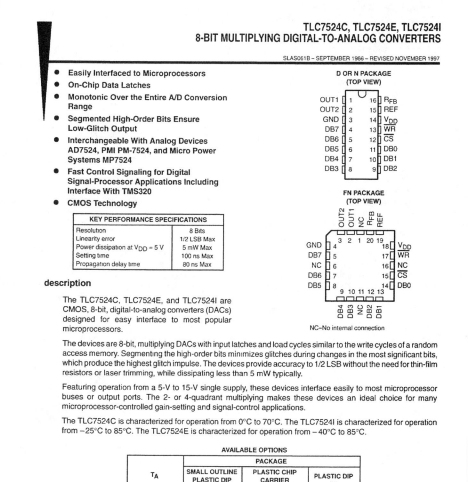

Figure 2.1. Data sheet for 8-bit multiplying DAC: features, description, and pinout.

TLC7524C, TLC7524E, TLC7524I
8-BIT MULTIPLYING DIGITAL-TO-ANALOG CONVERTERS

SLAS061B – SEPTEMBER 1986 – REVISED NOVEMBER 1997

PRINCIPLES OF OPERATION

The TLC7524C, TLC7524E, and TLC7524I are 8-bit multiplying DACs consisting of an inverted R-2R ladder, analog switches, and data input latches. Binary-weighted currents are switched between the OUT1 and OUT2 bus lines, thus maintaining a constant current in each ladder leg independent of the switch state. The high-order bits are decoded. These decoded bits, through a modification in the R-2R ladder, control three equally-weighted current sources. Most applications only require the addition of an external operational amplifier and a voltage reference.

The equivalent circuit for all digital inputs low is seen in Figure 2. With all digital inputs low, the entire reference current, I_{ref}, is switched to OUT2. The current source I/256 represents the constant current flowing through the termination resistor of the R-2R ladder, while the current source I_{lkg} represents leakage currents to the substrate. The capacitances appearing at OUT1 and OUT2 are dependent upon the digital input code. With all digital inputs high, the off-state switch capacitance (30 pF maximum) appears at OUT2 and the on-state switch capacitance (120 pF maximum) appears at OUT1. With all digital inputs low, the situation is reversed as shown in Figure 2. Analysis of the circuit for all digital inputs high is similar to Figure 2; however, in this case, I_{ref} would be switched to OUT1.

The DAC on these devices interfaces to a microprocessor through the data bus and the $\overline{CS}$ and $\overline{WR}$ control signals. When $\overline{CS}$ and $\overline{WR}$ are both low, analog output on these devices responds to the data activity on the DB0–DB7 data bus inputs. In this mode, the input latches are transparent and input data directly affects the analog output. When either the $\overline{CS}$ signal or $\overline{WR}$ signal goes high, the data on the DB0–DB7 inputs are latched until the $\overline{CS}$ and $\overline{WR}$ signals go low again. When $\overline{CS}$ is high, the data inputs are disabled regardless of the state of the $\overline{WR}$ signal.

These devices are capable of performing 2-quadrant or full 4-quadrant multiplication. Circuit configurations for 2-quadrant or 4-quadrant multiplication are shown in Figures 3 and 4. Tables 1 and 2 summarize input coding for unipolar and bipolar operation respectively.

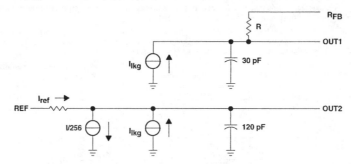

Figure 2. TLC7524 Equivalent Circuit With All Digital Inputs Low

POST OFFICE BOX 655303 • DALLAS, TEXAS 75265

Figure 2.2. Eight-bit multiplying DAC: principles of operation.

Consider a second example of indirect testing which is more applicable to mixed-signal circuits. A data sheet states that a programmable gain amplifier (PGA) can be set to gains from 0 to 30 dB in 2-dB steps. It may or may not be necessary to test each and every gain step. If the PGA is designed with a binary weighted resistor structure, it might be possible to measure only four of the sixteen gain steps, corresponding to the four gain setting paths of the binary architecture. The other twelve gain steps might be calculated mathematically depending on the accuracy requirements of the test and the robustness of the design.

It is up to the test engineer and design engineer to work through all the required functionality in the principles of operation and determine what series of tests and test modes constitute an acceptable balance between test thoroughness and costly test time. The astute design engineer will make architectural decisions based not only on circuit performance but also on test efficiency. The experienced test engineer serves a critical role in helping to define what kinds of circuits can be most efficiently tested.

Many of the features listed in the device description and principles of operation do not result in measurements of electrical parameters. These features are verified using what is often referred to as a go/no-go test or functional test. Functional tests result in a simple pass/fail result with no numerical reading. Parametric tests, by comparison, are those that return a value that must be compared against one or more test limits to determine pass/fail results.

The 1-s timer is a good example of a circuit that can be tested with a functional test. It is not necessary to measure the exact countdown period in seconds and fractions of a second. The digital counter circuit either divides by one million or it does not. This type of digital logic verification is known as a *functional pattern test*.

The only way the 1-s period of time could be in error is if the divider circuits are not functional or if the 1-MHz external reference clock is not set to the correct frequency. An incorrect external frequency setting does not need to be tested during IC production, since it is not a function of device performance. Only a functional pattern test is required to guarantee the 1-s interval. An automated software process is often used to generate functional pattern tests. The test engineer should verify that all digital functionality has been guaranteed by either automatically generated patterns or by hand-coded functional pattern tests.

In highly customized mixed-signal devices, the test engineer needs to understand the end application of the device. Otherwise, the concurrent engineering process described in Chapter 1 will be impeded and the test engineer will not be able to contribute to the design definition and debug. The device description and principles of operation provide the test engineer with much of the information needed to understand the system into which the device will be placed.

2.1.3 Electrical Characteristics

Electrical characteristics (or electrical specifications) provide the test limits and test conditions for many of the parametric tests in a mixed-signal test program. Figures 2.3 and 2.4 show the electrical characteristics for the 8-bit multiplying DAC. While the format of the electrical characteristics section may vary widely from one manufacturer to another, there are some common features. There are generally parameter names to the left side of the chart, followed by test conditions. Often, a series of notes are listed below the electrical characteristics that give more complete background information for some of the specifications.

TLC7524C, TLC7524E, TLC7524I
8-BIT MULTIPLYING DIGITAL-TO-ANALOG CONVERTERS

SLAS061B – SEPTEMBER 1986 – REVISED NOVEMBER 1997

recommended operating conditions

		$V_{DD} = 5$ V			$V_{DD} = 15$ V			UNIT
		MIN	NOM	MAX	MIN	NOM	MAX	
Supply voltage, V_{DD}		4.75	5	5.25	14.5	15	15.5	V
Reference voltage, V_{ref}			±10			±10		V
High-level input voltage, V_{IH}		2.4			13.5			V
Low-level input voltage, V_{IL}				0.8			1.5	V
CS setup time, $t_{su(CS)}$		40			40			ns
CS hold time, $t_{h(CS)}$		0			0			ns
Data bus input setup time, $t_{su(D)}$		25			25			ns
Data bus input hold time, $t_{h(D)}$		10			10			ns
Pulse duration, $\overline{WR}$ low, $t_{w(WR)}$		40			40			ns
Operating free-air temperature, T_A	TLC7524C	0		70	0		70	°C
	TLC7524I	–25		85	–25		85	
	TLC7524E	–40		85	–40		85	

electrical characteristics over recommended operating free-air temperature range, $V_{ref} = \pm 10$ V, OUT1 and OUT2 at GND (unless otherwise noted)

PARAMETER			TEST CONDITIONS	$V_{DD} = 5$ V			$V_{DD} = 15$ V			UNIT
				MIN	TYP	MAX	MIN	TYP	MAX	
I_{IH}	High-level input current		$V_I = V_{DD}$			10			10	µA
I_{IL}	Low-level input current		$V_I = 0$			–10			–10	µA
I_{lkg}	Output leakage current	OUT1	DB0–DB7 at 0 V, $\overline{WR}$, $\overline{CS}$ at 0 V, $V_{ref} = \pm 10$ V			±400			±200	nA
		OUT2	DB0–DB7 at V_{DD}, $\overline{WR}$, $\overline{CS}$ at 0 V, $V_{ref} = \pm 10$ V			±400			±200	
I_{DD}	Supply current	Quiescent	DB0–DB7 at V_{IH}min or V_{IL}max			1			2	mA
		Standby	DB0–DB7 at 0 V or V_{DD}			500			500	µA
k_{SVS}	Supply voltage sensitivity, Δgain/ΔV_{DD}		$\Delta V_{DD} = \pm 10\%$		0.01	0.16		0.005	0.04	%FSR/%
C_i	Input capacitance, DB0–DB7, $\overline{WR}$, $\overline{CS}$		$V_I = 0$			5			5	pF
C_o	Output capacitance	OUT1	DB0–DB7 at 0 V, $\overline{WR}$, $\overline{CS}$ at 0 V			30			30	pF
		OUT2				120			120	
		OUT1	DB0–DB7 at V_{DD}, $\overline{WR}$, $\overline{CS}$ at 0 V			120			120	
		OUT2				30			30	
Reference input impedance (REF to GND)				5		20	5		20	kΩ

TEXAS INSTRUMENTS
POST OFFICE BOX 655303 • DALLAS, TEXAS 75265

Figure 2.3. Recommended operating conditions and electrical characteristics.

TLC7524C, TLC7524E, TLC7524I
8-BIT MULTIPLYING DIGITAL-TO-ANALOG CONVERTERS

SLAS061B – SEPTEMBER 1986 – REVISED NOVEMBER 1997

operating characteristics over recommended operating free-air temperature range, $V_{ref} = \pm 10$ V, OUT1 and OUT2 at GND (unless otherwise noted)

PARAMETER	TEST CONDITIONS	$V_{DD} = 5$ V			$V_{DD} = 15$ V			UNIT
		MIN	TYP	MAX	MIN	TYP	MAX	
Linearity error				±0.5			±0.5	LSB
Gain error	See Note 1			±2.5			±2.5	LSB
Settling time (to 1/2 LSB)	See Note 2			100			100	ns
Propagation delay from digital input to 90% of final analog output current	See Note 2			80			80	ns
Feedthrough at OUT1 or OUT2	$V_{ref} = \pm 10$ V (100-kHz sinewave) $\overline{WR}$ and $\overline{CS}$ at 0 V, DB0–DB7 at 0 V			0.5			0.5	%FSR
Temperature coefficient of gain	$T_A = 25°C$ to MAX		±0.004			±0.001		%FSR/°C

NOTES: 1. Gain error is measured using the internal feedback resistor. Nominal full scale range (FSR) = V_{ref} – 1 LSB.
2. OUT1 load = 100 Ω, C_{ext} = 13 pF, $\overline{WR}$ at 0 V, $\overline{CS}$ at 0 V, DB0 – DB7 at 0 V to V_{DD} or V_{DD} to 0 V.

operating sequence

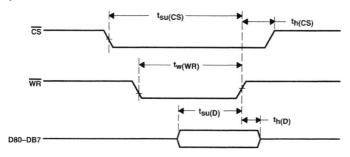

POST OFFICE BOX 655303 • DALLAS, TEXAS 75265

Figure 2.4. Electrical specifications and timing diagram.

In the example in Figure 2.4, Note 1 states that the "Gain error is measured using the internal feedback resistor..." This piece of information is vital, since the gain error specification is ambiguous without it. Data sheet ambiguities can lead to frustrating correlation efforts. For instance, if Note 1 was missing, the test engineer might use the internal resistor to measure gain error while the customer uses an external resistor. The two engineers might waste days trying to agree upon the correct value of gain error for a particular group of DUTs. Unfortunately, the data sheet seldom lists all possible test conditions for each measurement in complete detail. The test plan must fill in the gaps as needed. The test engineer should also suggest that clarifications be added to the data sheet wherever serious ambiguities might cause the customer problems at a later time.

On the right side of the electrical specification table are the test limits. These are divided into three categories: MIN, TYP, and MAX. The MIN column and MAX column represent the minimum and maximum values allowed for a passing device. These may or may not all be tested in production on every single device. Nevertheless, all specifications should be tested in an extended characterization version of the test program. The extended test program verifies that the device design meets all of the specifications listed in the electrical characteristics section of the data sheet.

The TYP column represents the typical reading expected from a good device. If a TYP value is specified at all, it is often just the average of the MIN and MAX test limits. The production test program does not generally guarantee the TYP value. For example, it is not necessary to verify that the average reading for a large number of tested devices is equal to the TYP value. Since the TYP value has no guaranteed correlation to the devices tested, it has much less value to the customer than the guaranteed MIN and MAX specifications.

The TYP column is sometimes used to specify parameters that are guaranteed by design and/or process. For example, an 8-bit DAC has 8 bits of resolution by definition. Resolution is sometimes listed as a typical specification, but only as a means of formally communicating the number of DAC input bits. The TYP column is also used to list characterization data for parameters that are difficult or impossible to measure in a cost-effective manner. For example, input capacitance is often listed as a typical specification because it is largely dominated by the design layout and by the device package. In cases such as this, characterization data can be collected from many devices to prove that the parameter never fails and therefore does not need to be tested in production.

Sometimes the data sheet lists a parameter with a note stating that it is "guaranteed, not tested" or "guaranteed by design." This is a formal way to notify the customer that this specification has been characterized and shown to be good by design, and is therefore not tested in production. However, the lack of such a notice should not be taken as a guarantee that the parameter is tested in production. Most data books contain a notification that parameters may or may not be tested in production, but that they are nevertheless guaranteed by the manufacturer to meet minimum and maximum specifications.

In addition to electrical characteristics, Figure 2.4 also shows the timing diagram for the example 8-bit DAC. Timing diagrams are critical to test program development. The digital patterns used in mixed-signal tests are sometimes generated manually due to frequency synchronization issues that will become more apparent in subsequent chapters. At present there are few if any good automation schemes that allow the design engineer to specify mixed-signal tests in a way that allows automatic translation into a debugged test program. The mixed-signal test generation process is still largely manual. Thus timing diagrams are still very pertinent to mixed-signal test engineers.

Application information is often added to the data sheet to aid the customer in designing the end application. Figure 2.5 shows the application diagram for the example 8-bit DAC. This particular application diagram shows the customer how to use the DAC in voltage mode rather than current mode. Application information is often very helpful to the test engineer, as well as the customer. Often the application information helps the test engineer understand the intended application for the device or helps in designing a thorough test list. Application information can also be helpful in the design of circuitry located on the automated test equipment's device interface board (DIB).

Figure 2.6 shows a functional block diagram for the example DAC. The functional block diagram is extremely important on complex devices since it provides a top-level representation of all the device functions in a single diagram. Like the application information section, the functional block diagram helps the customer (and the test engineer) understand the overall functionality of a complex mixed-signal device. Figure 2.6 also shows the absolute maximum and recommended operating conditions for the example 8-bit DAC. Absolute maximum ratings are not intended for production testing. These are specified limits beyond which device damage may occur. The recommended operating conditions, by contrast, list production test conditions such as minimum and maximum supply voltage under which all test limits must be met. The recommended operating conditions are therefore quite important to the test engineer, since they define the permutations of test conditions under which all the specifications must be met.

Figure 2.7 shows characterization data for a low-offset JFET op amp. Characterization data may or may not be included in a data sheet. If it is included, it does not necessarily represent guaranteed data. It is analogous to the TYP data column and is not necessarily guaranteed by the production test program. Certain characterization plots such as statistical histograms may be collected using the production tester simply because it is the easiest way to generate the data. However, characterization plots are more often generated using bench equipment such as oscilloscopes and spectrum analyzers.

2.2 GENERATING THE TEST PLAN

2.2.1 To Plan or Not to Plan

Strictly speaking, test plans are not absolutely necessary. A test engineer can certainly generate a test program by simply sitting down at the tester computer and entering code based on the device data sheet. There are several problems with this type of undisciplined approach. First, device testability will probably not be identified early enough to allow the addition of test features to the design. Test plans force the design engineers and test engineers to work through all the details of testing at an early stage in the design cycle. Second, the test engineer may create test-to-test compatibility problems if the details of all tests are not known up front. For example, a clocking scheme that works well for one test may be incompatible with the clocking scheme required for a subsequent test. The first test may then need to be rewritten from scratch so that the clocking schemes mesh properly.

If a test plan is not clearly documented before coding begins, then the test engineer lacks the necessary overview of the test program that allows all the tests to fit together efficiently.

TLC7524C, TLC7524E, TLC7524I
8-BIT MULTIPLYING DIGITAL-TO-ANALOG CONVERTERS

SLAS061B – SEPTEMBER 1986 – REVISED NOVEMBER 1997

APPLICATION INFORMATION

voltage-mode operation

It is possible to operate the current-multiplying DAC in these devices in a voltage mode. In the voltage mode, a fixed voltage is placed on the current output terminal. The analog output voltage is then available at the reference voltage terminal. Figure 1 is an example of a current-multiplying DAC, which is operated in voltage mode.

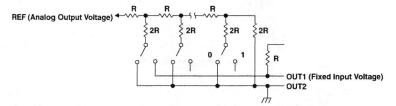

Figure 1. Voltage Mode Operation

The relationship between the fixed-input voltage and the analog-output voltage is given by the following equation:

$$V_O = V_I (D/256)$$

where

V_O = analog output voltage
V_I = fixed input voltage
D = digital input code converted to decimal

In voltage-mode operation, these devices meet the following specification:

PARAMETER	TEST CONDITIONS	MIN	MAX	UNIT
Linearity error at REF	V_{DD} = 5 V, OUT1 = 2.5 V, OUT2 at GND, T_A = 25°C		1	LSB

POST OFFICE BOX 655303 • DALLAS, TEXAS 75265

Figure 2.5. Eight-bit multiplying DAC: application information.

TLC7524C, TLC7524E, TLC7524I
8-BIT MULTIPLYING DIGITAL-TO-ANALOG CONVERTERS

SLAS061B – SEPTEMBER 1986 – REVISED NOVEMBER 1997

functional block diagram

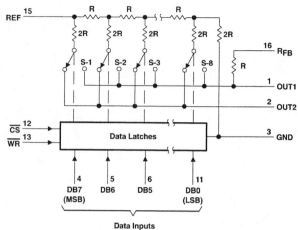

Terminal numbers shown are for the D or N package.

absolute maximum ratings over operating free-air temperature range (unless otherwise noted)

Supply voltage range, V_{DD} .... −0.3 V to 16.5 V
Digital input voltage range, V_I .... −0.3 V to V_{DD} + 0.3 V
Reference voltage, V_{ref} .... ±25 V
Peak digital input current, I_I .... 10 μA
Operating free-air temperature range, T_A: TLC7524C .... 0°C to 70°C
 TLC7524I .... −25°C to 85°C
 TLC7524E .... −40°C to 85°C
Storage temperature range, T_{stg} .... −65°C to 150°C
Case temperature for 10 seconds, T_C: FN package .... 260°C
Lead temperature 1,6 mm (1/16 inch) from case for 10 seconds: D or N package .... 260°C

POST OFFICE BOX 655303 • DALLAS, TEXAS 75265

2

Figure 2.6. Eight-bit multiplying DAC: functional block diagram and absolute maximum ratings.

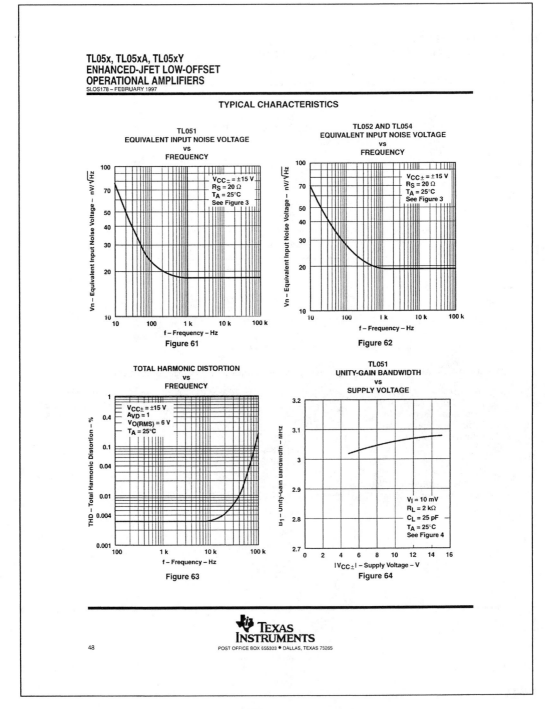

Figure 2.7. Low-offset JFET op amp: characterization data.

Similarly, test hardware such as DIBs and probe interface hardware cannot be properly designed until all test details are known. Finally, the test plan helps to identify shortfalls in the target tester's capabilities. Early identification of tester deficiencies allows the test engineer time to find acceptable work-arounds. Sometimes, the design engineer can even modify the IC design to accommodate tester deficiencies. This is another example of design for test (DfT).

2.2.2 Structure of a Test Plan

The structure of a mixed-signal test plan varies from one engineer to the next and from company to company. Since test plans are not generally published outside a company, they tend to be less formal and less structured than device data sheets. The primary purpose of a test plan is to serve as a roadmap for the test engineer while the test program is being generated. However, it is also used as the official communication channel between test engineers, design engineers, product engineers, and even customers. Depending on the needs and tastes of the person or organization generating a test plan, it might include any of a number of sections. The following are some suggestions for a well-written test plan.

A thorough test plan includes device background information that cannot be found in the data sheet. For example, a device may have test modes that are not documented in the data sheet. The test plan is an ideal place to list all test modes and how they are utilized. The test plan is also a good place to explain special test requirements that relate to the end application of the device. For example, a data sheet might list a parameter called error vector magnitude that is documented more completely in a separate telecommunication standard. It is a good idea to explain some of the details of a test like this so that a new engineer does not have to spend hours researching the purpose and meaning of the test.

When reading another person's test program, it can sometimes be difficult to understand why a particular test is being performed. A test plan should explain the purpose of each test and how it relates to a data sheet specification. It should also explain any assumptions the test engineer is making about a particular test. For example, an amplifier's absolute gain may be specified from 0 to 10 kHz. If the test engineer plans to test absolute gain at only three frequencies, then the test plan should explain why those frequencies were chosen. Was it an arbitrary choice, or does it relate to expected weaknesses in the design? It should also answer other questions: What was the input signal level? What typical level is expected from the output of the device? Are other tests like signal-to-noise ratio tested at the same time as absolute gain?

Another very useful addition to a test plan is hardware setup diagrams for each test. A hardware diagram can include as much or as little detail as needed to explain the test. If too much detail is included, then the diagram becomes too cluttered to clearly explain the test conditions. It would probably be unnecessary, for example, to show the full schematic for an op amp gain stage having a gain of 10. It would be perfectly acceptable to simply draw a triangular buffer with the label "x10 gain." Likewise, it is unnecessary to draw an entire 256-pin DUT if only three pins of the DUT are relevant to a given test.

Consider the simple diagram in Figure 2.8, showing a test diagram for a DAC full-scale output test. This diagram looks simple, but it may actually represent a test for a small portion of a much more complicated device. Since most of the other pins of the DUT are irrelevant during this test, they can be eliminated from the diagram for clarity. Simplified block diagrams such as this are often more useful in test plan diagrams than fully detailed schematic representations. Another way to eliminate clutter is to document default conditions. Default conditions such as

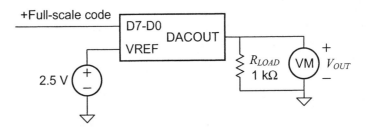

Figure 2.8. Test plan hardware setup diagram.

nominal power supply settings and nominal digital timing can be documented in a "defaults" section at the beginning of the test plan. Unless otherwise specified, these defaults are assumed to apply to all remaining test descriptions.

Test descriptions should be documented in a tester-independent manner if possible. This makes it easier for all engineering team members to understand the purpose of the test and how it should ideally be implemented. If the test plan describes each test in a tester-independent manner, it also makes it easier to convert test programs from one tester to another. Tester-specific information such as digitizer sampling rates and meter range settings can be added to the test plan, but they should perhaps be set aside in a separate subsection after the generic tester-independent description.

2.2.3 Design Specifications versus Production Test Specifications

Since the data sheet is initially used as a design specification, it may contain parameters that either cannot be tested or do not need to be tested in production. These internal specifications are meant for the design engineers only, and should probably be removed from the final data sheet. Consider a very simple example of an on-chip resistor specification of 100 kΩ plus or minus 10%. A resistance test would normally be performed by forcing current across the resistor, measuring the voltage drop, and applying Ohm's law to calculate resistance. However, if the resistor is permanently connected into the feedback path of an op amp gain stage, then resistance may not be directly measurable.

Beginning test engineers sometimes agonize over how to measure an internal parameter such as this without realizing that it is only an internal design specification. Such parameters do not necessarily need to be tested in production. The experienced test engineer knows to verify with the design engineer whether or not the specification is a design specification or a production specification. If it is indeed a production specification and the test engineer has no access to measure the parameter, then something must be done to make the parameter measurable. A design for test (DfT) structure might be added to the design early in the design cycle to allow the necessary test access.

In the on-chip resistor example the resistance and its tolerance is specified only to remind the design engineer of the requirements necessary to support another parameter such as absolute gain. Absolute gain of a simple op amp inverting gain stage is the result of several parameters including two resistance values. The op amp gain stage will probably require an absolute gain specification, which can be measured directly without pulling the resistor out of the gain circuit. If absolute gain performance is the driving force behind the resistance specification, there is no reason to measure the resistance values in production. Measuring the important parameter, absolute gain in this example, is sufficient. No DfT structure would be required to test the resistors' values in production.

2.2.4 Converting the Data Sheet into a Test Plan

One of the most difficult things to teach a new test engineer is how to convert a data sheet into a corresponding test plan. The difficulty arises from the infinite permutations of possible tests implied by the data sheet. Unlike many digital devices, mixed-signal circuits have an obnoxious habit of interacting with one another in unexpected ways. For example, the signal-to-noise ratio of an amplifier may change depending on the operating mode of a completely separate digital circuit on the opposite side of the die.

This effect may worsen with varying power supply voltage setting, die temperature, etc. How does a test engineer know which operating modes should be tested when there are so many possible permutations of test conditions? Obviously, the production test program must consist of a small subset of possible test conditions but it is difficult to define the pruning process in a scientific manner. There is really no fixed series of steps that can reliably convert a data sheet into a test plan. Nevertheless, a few suggestions are listed in the following paragraphs as a starting point.

For each sentence in the device description and principles of operation, make sure a test is defined that guarantees the device can perform the described operation. For example, if the description states that a DAC can operate in either of two digital interface modes, make sure there is a test that verifies both interface modes. Perhaps the DAC has to be tested in both modes, or perhaps the DAC can be tested in only one mode and a much faster digital test can be executed to verify the other mode. These details must be discussed with the design engineers and systems engineers.

The central function of a mixed-signal test program is the measurement of each of the electrical specifications listed in the device data sheet. For each electrical parameter, make sure that a test is defined to measure the parameter in all modes of operation. For example, a DAC might have a particular linearity specification, but it may operate in two different modes with three different voltage ranges and two possible power supply voltages.

Unfortunately, the permutations of possible test conditions will often grow to an unrealistic test list. If the permutations of modes and test conditions are too large, then discuss the problem with the design engineers and try to identify which permutations are likely to cause the most problems. These so-called *worst-case conditions* are commonly used to prune an infinite test list down to a more manageable subset of critical tests.

Each electrical specification should raise a series of questions. If the device includes internal control registers, how should they be set during each test? What loading should be applied to the output pins of the device? Should an external voltage reference be supplied to the device or should an internal voltage reference be enabled? Can the device drive the capacitive load presented by the tester instruments or will a voltage follower need to be added to the device interface hardware? Are there important electrical specifications that have not been included in the data sheet? Asking detailed questions like this can save the test engineer many headaches later in the project.

In the early stages of device development, some parameters may be listed as TBD (to be determined). Make sure that the innocent-looking TBD placeholder for a simple DC offset test does not become an unexpected nightmare like "offset = 1 pV MAX" at the last minute. There is a huge difference in test methodology between a 100-mV DC offset test and 1-pV offset test. Asking for rough estimates of expected performance is a good idea, even if the exact specification is not known.

After considering all these questions, a limited set of tests must be specified. In the end, it may be impossible to predict what modes of operation represent worst-case test conditions. To be safe, the test engineer usually specifies as many permutations as is reasonable. The initial test program has to be written so that the test list and test conditions can be easily modified. In this manner, worst-case conditions can be determined through empirical characterization of the device performance. After thorough characterization of many lots of production devices, the test list can be pruned down to an optimum set of tests that most efficiently probe the DUT's weaknesses.

2.3 COMPONENTS OF A TEST PROGRAM

2.3.1 Test Program Structure

The test program is a detailed, tester-specific version of the test plan, written in the target tester's native language. It may at times deviate from the test plan if the target tester is incapable of performing tests exactly as specified by the test plan. In these cases, comments should be added to the program to make this discrepancy clear. Major deviations should be approved by the other members of the engineering team.

Tester languages vary from low-level C routines to very sophisticated graphical user interface environments. Despite wide differences in their details, tester languages often share some basic structural components. Test programs typically consist of all or most of the following sections: waveform creation and other tester initializations, calibrations, continuity, DC parametric tests, AC parametric tests, digital patterns (also known as *functional tests*), digital timing tests, test sequence control, test limits, and binning control.

Well-written test programs often contain extensive characterization code to perform tests not specifically required by the data sheet. These characterization tests allow the design engineers to better evaluate the quality and robustness of the IC design. A thorough test program may also contain code that allows offline simulation of the tests so that certain portions of the program can be debugged without a tester or device. Let us take a more detailed look at some of the structural components commonly found in a mixed-signal test program.

2.3.2 Test Code and Digital Patterns

Test code and digital patterns make up the bulk of mixed-signal test programs. Test code controls the order and timing of instrument settings, signal generation, and signal measurements that make up each measurement in the test program. Test code typically does not control the real-time details of each instrument, though.

For example, the data generated from the digital subsystem of a tester are not clocked out one bit at a time by the test code. Instead, the test code simply calls for the tester's digital subsystem to begin exercising the desired digital pattern at the appropriate time. The digital subsystem then takes care of the details of generating the individual ones and zeros. Thus the test code for a DAC full-scale output test might look something like this in pseudocode:

```
dac_full_scale_voltage( )
{
    set VI1 = 2.5 V; /* Set the DAC's voltage reference to 2.5 volts */
    start digital pattern = "dac_full_scale"; /* Set the DAC output to +full scale (2.5 V) */
    connect meter: DAC_OUT /* Connect the DAC voltmeter to the DAC output */
    fsout = read_meter( ); /* Read the voltage level at the DAC_OUT pin */
    test fsout; /* Compare the DAC full scale output to the data sheet limits */
}
```

Digital patterns consist of groups of data bits called *vectors*. Each vector represents the drive and expect data that are to be sent out on each of the tester's digital pins at a specific time. Drive data specifies the desired state at the input to the DUT (HI, LOW, or HIZ). Compare data (also called *expect data*) specifies the required digital output from the device. Vectors are usually sent out at a regular rate, called the *bit cell rate*. Digital patterns usually contain not only the 1/0 drive and expect data, but also the sequencing information for the vectors. The digital pattern sequencing commands allow branching, looping, and other vector sequencing operations that make the pattern more compact. To generate a pair of clocks at two different frequencies from digital pins CLK1 and CLK2, one might write the following pseudocode pattern:

```
label           pattern control     CLK1    CLK2
-------------------------------------------------
START                               0       0   /* Vector one */
                                    1       0   /* Vector two, etc. */
                                    0       1
            Jump START              1       1   /* Infinite loop */
```

This pattern would continue in an infinite loop, producing two frequencies. The CLK1 frequency would be twice that of the CLK2 frequency.

The test code and the digital pattern must operate in stepped synchronization for mixed-signal tests. It would be unfortunate in the DAC test above if the digital pattern "dac_full_scale" did not execute until 50 ms after the meter measurement had already been performed. For this reason, mixed-signal testers include handshaking functions in both the test code and digital pattern control that allow the tester computer and digital pattern subsystem to keep in step with one another.

Another pattern issue unique to mixed-signal testing is that the pattern often must be executed at a very precise frequency. It is not acceptable to round off the period of the vector rate to the nearest nanosecond as is often done in purely digital test programs. The reason for this will become more apparent in Chapters 6 and 7, "Sampling Theory" and "DSP-Based Testing."

2.3.3 Binning

One of the functions of a test program is to sort each device into one of several categories, called *bins*, depending on the outcome of the various tests. The most obvious bins are "pass" and "fail," but there are several others that might be added. For example, a continuity test is usually inserted at the beginning of the test program.

The purpose of the continuity test is to verify that all the electrical contacts between the tester and the DUT have been successfully connected. If a large percentage of devices fail the continuity test, this indicates a probable error in the tester hardware. It is therefore a good idea to use separate bins for continuity failures and data sheet failures so that the production staff can more easily recognize tester hardware problems.

Binning is not always a pass/fail operation. Sometimes there are different grades of passing devices. If a device is designed to operate at 100 MHz but some of the manufactured devices are actually able to operate at 120 MHz, then the test program might be set up to split these devices into two quality grades, "good" and "great." Bin 1 might represent the 120-MHz devices, while Bin 2 might represent the devices that could only operate up to 100 MHz. The-120 MHz devices would be labeled differently than the 100-MHz devices. They would also be priced differently, of course. We are all familiar with higher prices for faster PC microprocessors and memory chips.

Fast binning is a term used to describe a tester's ability to bin a bad device as soon as it fails any test. This is done to prevent a bad device from wasting valuable tester time after it has already produced a failing result. The test and product engineers should work together to ensure that the most commonly failed tests are placed near the beginning of the test program. This allows the tester to sort bad devices as quickly as possible.

The tester generates a binning signal that tells the handler or prober what to do with the various categories of devices. Until recent years, bad die on a wafer were often squirted with red ink dots to designate them as failures. Now this inking is commonly performed offline or is done in a purely virtual manner using pass/fail databases and production lot ID numbers. At final test, different grades of packaged devices are sorted into separate plastic tubes or trays by the handler.

2.3.4 Test Sequence Control

Test sequence may be controlled in a number of ways, depending on the sophistication of the tester's software environment. In older testers, the order of the various tests was simply determined by the order of test routine execution. Comparisons of measured results against test limits were performed after each measurement. Test limits were therefore scattered all through the test program along with the instrument setups and measurement code. Such a scattered arrangement made it difficult to identify which test limits were applied to a device in a given version of the test program. This made it difficult to verify that the test program limits matched the data sheet limits.

As tester software environments matured, a new type of test code module evolved to allow a more convenient summary of test flow, test limits, and binning information. The new code module, called a *sequencer* by some vendors, contains the test routine function calls, the test limits, and the binning information for each test result. The sequencer code allows the programmer to order the tests and group all the test limits into a central location in the test code, separate from the test routines themselves. The sequencer code thus provides a convenient summary of the test list, test order, and pass/fail limits for each test. This makes it easier to audit the program for compliance with data sheet test limits. Depending on the tester's software environment, the sequencer modules may be coded as text or they may consist of graphical interface objects linked together with arrows to indicate program flow and binning decisions.

2.3.5 Waveform Calculations and Other Initializations

Mixed-signal test programs use many precomputed waveforms. A 1-kHz gain test requires a sinusoidal waveform that does not change from one program execution to the next. Waveforms that do not need to change are precomputed and stored either in arrays or directly into memory banks in the tester instruments themselves. Digital waveforms are also precomputed and stored in the digital subsystem of the tester. Many of the required initializations such as waveform computations are performed only once when the test program is first loaded. Performing these initializations only once saves a large amount of test execution time.

Other operations, such as resetting tester instruments to a default state, must be performed each time the program is run. The details of initializations are very specific to each tester, but most testers involve some type of first-run initialization code. One major class of first-run code is focused calibrations and checkers.

2.3.6 Focused Calibrations and DIB Checkers

Sometimes the instrumentation in a tester does not have sufficient accuracy for a given test. If not, a special routine called a *focused calibration* is required when the program first runs. The focused calibration routine determines the inaccuracy of the instrument using slower, more accurate instrumentation as a reference. The inaccuracies of the faster instrument can then be corrected in a process known as *software calibration*.

Software calibrations must also be performed on circuitry placed on the device interface board. Assume an op amp voltage follower is placed on the DIB to buffer a weak device output. The gain and offset of the voltage follower adds errors into any measured results. The test engineer must calibrate the gain and offset of the voltage follower using focused calibrations to achieve maximum accuracy. Sometimes focused calibrations can be as difficult to develop as the device measurements themselves, especially when extreme accuracy is required in the final test result.

Fortunately, many software calibrations are hidden from the user in the tester's operating system. These calibrations are performed automatically when the program is first loaded. Other calibrations are performed on a regular basis, such as once per week. Software calibration is discussed in greater detail in Chapter 10, "Focused Calibrations."

Electromechanical relays, op amp circuits, comparators, and other active circuits are commonly placed on the DIB to extend the tester's functionality and accuracy. These circuits are subject to failure. The test program should include DIB checker code to verify the functionality of any circuitry placed on the device interface board. This allows production personnel to avoid running thousands of good devices through a bad DIB before discovering the error. DIB checker routines are usually run along with focused calibrations when the program is first loaded.

2.3.7 Characterization Code

Characterization tests are often added to a test program to allow thorough evaluation of the first few lots of production material. Thorough characterization of a new device is critical, since it allows the design engineers to identify and correct the marginal portions of the design. An example of a characterization test would be a filter response test implemented at each frequency from 100 Hz to 10 kHz in 100-Hz increments. Such a test would never be cost-effective in a production test program, but it would provide thorough information about the filter's gain versus frequency characteristics.

2.3.8 Simulation Code

Simulation code is sometimes added to a mixed-signal test program to allow some of the mathematical routines to be verified. For example, the ideal output of a DAC might be simulated and stored into an array for use by a DAC linearity calculation routine. Offline code debugging techniques like this are a good way to reduce debug time and avoid wasting valuable tester time. However, such simulations are not entirely effective in uncovering errors such as incorrect DUT register settings or improper tester instrument range settings.

A more advanced type of simulation, known as *test simulation* or *virtual test* allows true closed-loop simulation of the tester and device. Using test simulation, a software model of the tester stimulates a model of the DUT according to the instructions in the test program. The tester model and tester operating system then capture the responses from the DUT model and compare them to test limits. Test simulation is explained in more detail in Chapter 16, "Test Economics."

2.3.9 "Debuggability"

It is said that the three most important things in real estate are location, location, and location. It might be said of test program structure that the three most important things are debuggability, debuggability, and debuggability (despite the fact that "debuggability" is not a real word). A study at Texas Instruments showed that the test program debugging process takes about 20% of an average test engineer's week. The debugging time was found to be roughly twice the time spent writing test code.

Debugging is not only a matter of finding and fixing test code bugs. It is also a matter of locating measurement correlation errors, intermittent failures, and hardware problems including bad DIB layout and broken tester modules. More important, test debugging often turns into design debugging.

Design debugging activities account for a large portion of the test program debugging time. One of a mixed-signal test engineer's most valuable roles is to help the design engineers isolate design problems. A good test engineer with a well-structured test program can quickly modify the program and run experiments for design engineers or customers. These experiments are critical to reducing the time it takes to get the problems worked out of a new mixed-signal design. The success or failure of a mixed-signal product often depends on how well the design engineers, test engineers, and product engineers work together to resolve design problems.

2.4 Summary

In this chapter, we have reviewed the basic structure of a data sheet, and we have seen how tabular entries and comments in a data sheet translate first into a test plan and then into a test program. The translation is not a simple one-to-one process, with each data sheet entry corresponding to one clearly defined test. Rather, the process requires a great deal of thought, experience, and common sense to guarantee the intent of the data sheet specifications without literally performing millions of possible tests in a production test program.

In Chapter 3, we will begin looking at the test development process, starting with the definitions of some of the most basic tests in a mixed-signal test program, the analog DC tests. Since these tests require very simple hardware and software, we will study DC tests first to gain some familiarity with the language and methodology of analog and mixed-signal testing. Our study of true mixed-signal tests, involving a mixture of analog and digital signals, will be delayed until later chapters so that we can first develop a fundamental understanding of issues such as accuracy and repeatability.

Problems

2.1. (a) List at least three purposes of a data sheet. (b) List at least six types of information that can be found in a data sheet. (c) In which section of the data sheet are the maximum, minimum, and typical specification limits listed?

2.2. What is a test plan? Is there a rigorous method to convert any given data sheet into a test plan? Do the electrical characteristics listed in the tables of a data sheet correspond one-to-one with individual DUT measurements?

2.3. Do the absolute maximum ratings need to be verified during production testing?

2.4. Problems 2.4–2.8 refer to the TLC7524C data sheet in Figures 2.1–2.6. What output load resistance should be attached to the DAC output (OUT1) during the settling time test? What capacitance should be attached in parallel with the load resistance?

2.5. What is the ideal relationship between the output voltage and the input voltage and digital input code when the DAC is operated in voltage mode?

2.6. What state must be applied to the $\overline{WR}$ and $\overline{CS}$ signals to allow the DAC output voltage to change according to the data at DB0-DB7? If the $\overline{CS}$ signal is high and the $\overline{WR}$ signal is low, what will happen to the DAC output when the data signals DB0-DB7 change from 00000000 to 11111111?

2.7. When the TLC7524C is packaged in the FN package, what device signal is attached to pin 16? What signal is attached to pin 9? What pin is connected to the positive power supply?

2.8. When the TLC7524C is powered with a 5-V supply, what is the maximum power dissipated by the device? If a TLC7524C draws 1.5 mA from the V_{DD} supply, would it pass the power dissipation specification?

2.9. At what frequency does the total harmonic distortion for a TL051 JFET op amp exceed 0.004%? Is the distortion at this frequency guaranteed to be less than 0.004% on every device?

2.10. Can a DUT's specifications be measured under all possible test conditions?

2.11. Which section of a test program tells the handler or prober whether a device is good or bad?

2.12. What is the purpose of the DIB checker section of a test program?

2.13. What is the purpose of focused calibrations?

References

1. Data Book, *Data Acquisition Circuits - Data Conversion and DSP Analog Interfaces*, Texas Instruments, Inc., P.O. Box 809066, Dallas, TX 75380-9066

CHAPTER 3

DC and Parametric Measurements

3.1 CONTINUITY

3.1.1 Purpose of Continuity Testing

Before a test program can evaluate the quality of a device under test (DUT), the DUT must be connected to the ATE tester using a test fixture such as a device interface board (DIB). A typical interconnection scheme is shown in Figure 3.1. When packaged devices are tested, a socket or handler contactor assembly provides the contact between the DUT and the DIB. When testing a bare die on a wafer, the contact is made through the probe needles of a probe card. The tester's instruments are connected to the DIB through one or more layers of connectors such as spring-loaded pogo pins or edge connectors. The exact connection scheme varies from tester to tester, depending on the mechanical/electrical performance tradeoffs made by the ATE vendor.

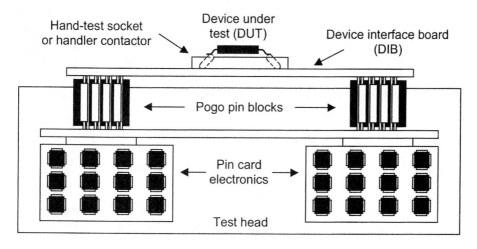

Figure 3.1. ATE test head to DUT interconnections.

In addition to pogo pins and other connectors, electromechanical relays are often used to route signals from the tester electronics to the DUT. A relay is an electrical switch whose position is controlled by an electromagnetic field. The field is created by a current forced through a coil of wire inside the relay (Figure 3.2). Relays are used extensively in mixed-signal testing to modify the electrical connections to and from the DUT as the test program progresses from test to test.

45

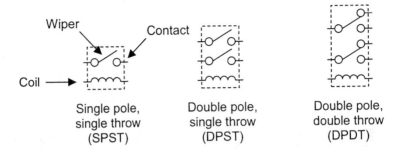

Figure 3.2. Electromechanical relays.

Any of the electrical connections between a DUT and the tester can be defective, resulting in open circuits or shorts between electrical signals. For example, the wiper of a relay can become stuck in either the open or closed position after millions of open/close cycles. While interconnect problems may not pose a serious problem in a lab environment, defective connections can be a major source of tester down time on the production floor. Continuity tests (also known as *contact tests*) are performed on a device to verify that all the electrical connections are sound. If continuity testing is not performed, then the production floor personnel cannot distinguish between bad lots of silicon and defective test hardware connections. Without continuity testing, thousands of good devices could be rejected simply because a pogo pin was bent or because a relay was defective.

3.1.2 Continuity Test Technique

Continuity testing is usually performed by detecting the presence of on-chip protection circuits. These circuits protect each input and output of the device from electrostatic discharge (ESD) and other excessive voltage conditions. The ESD protection circuits prevent the input and output pins from exceeding a small voltage above or below the power supply voltage or ground. Diodes and silicon-controlled rectifiers (SCRs) can be used to short the excess currents from the protected pin to ground or to a power terminal.

An ESD protection diode conducts the excess ESD current to ground or power any time the pin's voltage exceeds one diode drop above (or below) the power or ground voltage. SCRs are similar to ESD protection diodes, but they are triggered by a separate detection circuit. Any of a variety of detection circuits can be used to trigger the SCR when the protected pin's voltage exceeds a safe voltage range. Once triggered, an SCR behaves like a forward-biased diode from the protected pin to power or ground (Figure 3.3). The SCR remains in its triggered state until the excessive voltage is removed. Since an SCR behaves much like a diode when triggered, the term "protection diode" is used to describe ESD protection circuits whether they employ a simple diode or a more elaborate SCR structure. We will use the term "protection diode" throughout the remainder of this book with the understanding that a more complex circuit may actually be employed.

DUT pins may be configured with either one or two protection diodes, connected as shown in Figure 3.4. Notice that the diodes are reverse-biased when the device is powered up, assuming normal input and output voltage levels. This effectively makes them "invisible" to the DUT circuits during normal operation.

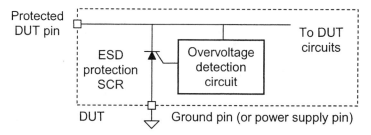

Figure 3.3. SCR-based ESD protection circuit.

To verify that each pin can be connected to the tester without electrical shorts or open circuits, the ATE tester forces a small current across each protection diode in the forward-biased direction. The DUT's power supply pins are set to zero volts to disable all on-chip circuits and to connect the far end of each diode to ground. ESD protection diodes connected to the positive supply are tested by forcing a current I_{CONT} into the pin as shown in Figure 3.5 and measuring the voltage, V_{CONT}, that appears at the pin with respect to ground. If the tester does not see the expected diode drops on each pin, then the continuity test fails and the device is not tested further. Protection diodes connected to the negative supply or ground are tested by reversing the direction of the forced current.

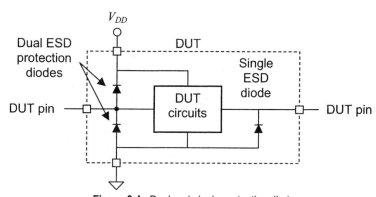

Figure 3.4. Dual and single protection diodes.

In the case of an SCR-based protection circuit, the current source initially sees an open circuit. Because the current source output tries to force current into an open circuit, its output voltage rises rapidly. The rising voltage soon triggers the SCR's detection circuit. Once triggered, the SCR accepts current from the current source and the voltage returns to one diode drop above ground. Thus the difference between a diode-based ESD protection circuit and an SCR-based circuit is hardly noticeable during a continuity test.

The amount of current chosen is typically between 100 µA and 1 mA, but the ideal value depends on the characteristics of the protection diodes. Too much current may damage the diodes, while too little current may not fully bias them. The voltage drop across a good protection diode usually measures between 550 and 750 mV. For the purpose of illustration, we shall assume that a conducting diode has voltage drop of 0.7 V. A dead short to ground will

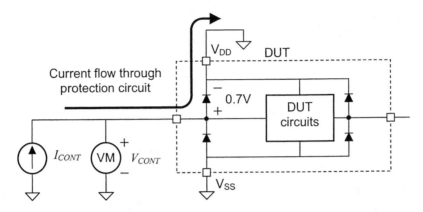

Figure 3.5. Checking the continuity of the diode connected to the positive supply. The other diode is tested by reversing the direction of the forced current.

result in a reading of 0 V, while an open circuit will cause the tester's current source to reach a programmed clamp voltage.

Many mixed-signal devices have multiple power supply and ground pins. Continuity to these power and ground connections may or may not be testable. If all supply pins or all ground pins are not properly connected to ground, then continuity to some or all of the nonsupply pins will fail. However, if only some of the supply or ground pins are not grounded, the others will provide a continuity path to zero volts. Therefore, the unconnected power supply or ground pins may not be detected. One way to test the power and ground pins individually is to connect them to ground one at a time, using relays to break the connections to the other power and ground pins. Continuity to the power or ground pin can then be verified by looking for the protection diode between it and another DUT pin.

Occasionally, a device pin may not include any protection diodes at all. Continuity to these unprotected pins must be verified by an alternative method, perhaps by detecting a small amount of current leaking into the pin or by detecting the presence of an on-chip component such as a capacitor or resistor. Since unprotected pins are highly vulnerable to ESD damage, they are used only in special cases.

One such example is a high-frequency input requiring very low parasitic capacitance. The space-charge layer in a reverse-biased protection diode might add several picofarads of parasitic capacitance to a device pin. Since even a small amount of stray capacitance presents a low impedance to very high-frequency signals, the protection diode must sometimes be omitted to enhance electrical performance of the DUT.

3.1.3 Serial versus Parallel Continuity Testing

Continuity can be tested one pin at a time, an approach known as *serial continuity testing*. Unfortunately, serial testing is a time-consuming and costly approach. Modern ATE testers are capable of measuring continuity on all or most pins in parallel rather than measuring the protection diode drops one at a time. These testers accomplish parallel testing using so-called *per-pin measurement instruments* as shown in Figure 3.6(a).

Clearly it is more economical to test all pins at once using many current sources and voltage meters. Unfortunately, there are a few potential problems to consider. First, a fully parallel test of pins may not detect pin-to-pin shorts. If two device pins are shorted together for some reason, the net current through each diode does not change. Twice as much current is forced through the parallel combination of two diodes. The shorted circuit configuration will therefore result in the expected voltage drop across each diode, resulting in both pins passing the continuity test. Obviously, the problem can be solved by performing a continuity test on each pin in a serial manner at the cost of extra test time. However, a more economical approach is to test every other pin for continuity on one test pass while grounding the remaining pins. Then the remaining pins can be tested during a second pass while the previously tested pins are grounded. Shorts between adjacent pins would be detected using this dual-pass approach, as illustrated in Figure 3.6(b).

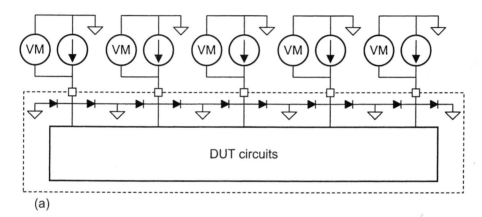

(a)

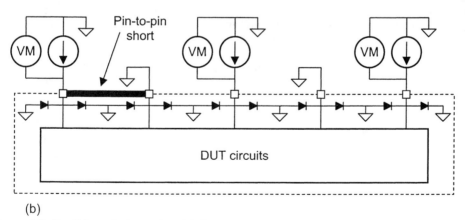

(b)

Figure 3.6. Parallel continuity testing: (a) full parallel testing with possible adjacent fault masking; (b) minimizing potential adjacent fault masking by exciting every second pin.

A second, subtler problem with parallel continuity testing is related to analog measurement performance. Both analog pins and digital pins must be tested for continuity. On some testers the per-pin continuity test circuitry is limited to digital pins only. The analog pins of the tester may not include per-pin continuity measurement capability. On these testers, continuity testing on analog pins can be performed one pin at a time using a single current source and voltmeter. These two instruments can be connected to each device pin one at a time to measure protection diode drops. Of course, this is a very time-consuming serial test method, which should be avoided if possible.

Alternatively, the analog pins can be connected to the per-pin measurement electronics of digital pins. This allows completely parallel testing of continuity. Unfortunately, the digital per-pin electronics may inject noise into sensitive analog signals. Also, the signal trace connecting the DUT to the per-pin continuity electronics adds a complex capacitive and inductive load to the analog pin, which may be unacceptable. The signal trace can also behave as a parasitic radio antenna into which unwanted signals can couple into analog inputs. Clearly, full parallel testing of analog pins should be treated with care. One solution to the noise and parasitic loading problems is to isolate each analog pin from its per-pin continuity circuit using a relay. This complicates the DIB design but gives high performance with minimal test time. Of course, a tester having per-pin continuity measurement circuits on both analog and digital pins represents a superior solution.

3.2 LEAKAGE CURRENTS

3.2.1 Purpose of Leakage Testing

Each input pin and output pin of a DUT exhibits a phenomenon called *leakage*. When a voltage is applied to a high-impedance analog or digital input pin, a small amount of current will typically leak into or out of the pin. This current is called *leakage current*, or simply *leakage*. Leakage can also be measured on output pins that are placed into a nondriving high-impedance mode. A good design and manufacturing process should result in very low leakage currents. Typically the leakage is less than 1 µA, although this can vary from one device design to the next.

One of the main reasons to measure leakage is to detect improperly processed integrated circuits. Leakage can be caused by many physical defects such as metal filaments and particulate matter that forms shorts and leakage paths between layers in the IC. Another reason to measure leakage is that excessive leakage currents can cause improper operation of the customer's end application. Leakage currents can cause DC offsets and other parametric shifts. A third reason to test leakage is that excessive leakage currents can indicate a poorly processed device that initially appears to be functional but which eventually fails after a few days or weeks in the customer's product.[1] This type of early failure is known as *infant mortality*.

3.2.2 Leakage Test Technique

Leakage is measured by simply forcing a DC voltage on the input or output pin of the device under test and measuring the small current flowing into or out of the pin. Unless otherwise specified in the data sheet, leakage is typically measured twice. It is measured once with an input voltage near the positive power supply voltage and again with the input near ground (or negative supply). These two currents are referred to as I_{IH} (input current, logic high) and I_{IL} (input current, logic low), respectively.

Digital inputs are typically tested at the valid input threshold voltages, V_{IH} and V_{IL}. Analog input leakage is typically tested at specific voltage levels listed in the data sheet. If no particular input voltage is specified, then the leakage specification applies to the entire allowable input voltage range. Since leakage is usually highest at one or both input voltage extremes, it is often measured at the maximum and minimum allowable input voltages. Output leakage (I_{OZ}) is measured in a manner similar to input leakage, though the output pin must be placed into a high-impedance (HIZ) state using a test mode or other control mechanism.

3.2.3 Serial versus Parallel Leakage Testing

Leakage, like continuity, can be tested one pin at a time (serial testing) or all pins at once (parallel testing). Since leakage currents can flow from one pin to another, serial testing is superior to parallel testing from a defect detection perspective. However, from a test time perspective, parallel testing is desired. As in continuity testing, a compromise can be achieved by testing every other pin in a dual-pass approach.

Continuity tests are usually implemented by forcing DC current and measuring voltage. By contrast, leakage tests are implemented by forcing DC voltage and measuring current. Since the tests are similar in nature, tester vendors generally design both capabilities into the per-pin measurement circuits of the ATE tester's pin cards. Analog leakage, like analog continuity, is often measured using the per-pin resources of digital pin cards. Again, a tester with per-pin continuity measurement circuits on both analog and digital pins represents a superior solution, assuming the extra per-pin circuits are not prohibitively expensive.

3.3 POWER SUPPLY CURRENTS

3.3.1 Importance of Supply Current Tests

One of the fastest ways to detect a device with catastrophic defects is to measure the amount of current it draws from each of its power supplies. Many gross defects such as those illustrated in Figures 1.5–1.8 result in a low-impedance path from one of the power supplies to ground. Supply currents are often tested near the beginning of a test program to screen out completely defective devices quickly and cost effectively.

Of course, the main reason to measure power supply current is to guarantee limited power consumption in the customer's end application. Supply current is an important electrical parameter for the customer who needs to design a system that consumes as little power as possible. Low power consumption is especially important to manufacturers of battery operated equipment like cellular telephones. Even devices that draw large amounts of current by design should draw only as much power as necessary. Therefore, power supply current tests are performed on most if not all devices.

3.3.2 Test Techniques

Most ATE testers are able to measure the current flowing from each voltage source connected to the DUT. Supply currents are therefore very easy to measure in most cases. The power supply is simply set to the desired voltage and the current from its output is measured using one of the tester's ammeters.

When measuring supply currents, the only difficulties arise out of ambiguities in the data sheet. For example, are the analog outputs loaded or unloaded during the supply current test? Is digital block XYZ operating in mode A, mode B, or idle mode? In general, it is safe to assume that the supply currents are to be tested under worst-case conditions.

The test engineer should work with the design engineers to attempt to specify the test conditions that are likely to result in worst-case test conditions. These test conditions should be spelled out clearly in the test plan so that everyone understands the exact conditions used during production testing. Often the actual worst-case conditions are not known until the device has been thoroughly characterized. In these cases, the test program and test plan have to be updated to reflect the characterized worst-case conditions.

Supply currents are often specified under several test conditions, such as power-down mode, standby mode, and normal operational mode. In addition, the digital supply currents are specified separately from the analog supply currents. I_{DD} (CMOS) and I_{CC} (bipolar) are commonly used designations for supply current. I_{DDA}, I_{DDD}, I_{CCA}, and I_{CCD} are the terms used when analog and digital supplies are measured separately.

Many devices have multiple power supply pins that are connected to a common power supply in normal operation. Design engineers often need to know how much current is flowing into each individual power supply pin. Sometimes the test engineer can accommodate this requirement by connecting each power supply pin to its own supply. Other times there are too many DUT supply pins to provide each with its own separate power supply. In these cases, relays can be used to temporarily connect a dedicated power supply to the pin under test.

Another problem that can plague power supply current tests is settling time. The supply current flowing into a DUT must settle to a stable value before it can be measured. The tester and DIB circuits must also settle to a stable value. This normally takes 5–10 ms in normal modes of DUT operation. But in power-down modes, the specified supply current is often less than 100 μA. Since the DIB usually includes bypass capacitors for the DUT, each capacitor must be allowed to charge until the average current into or out of the capacitor is stable.

The charging process can take hundreds of milliseconds if the current must stabilize within microamps. Some types of bypass capacitors may even exhibit leakage current greater than the current to be measured. A typical solution to this problem is to connect only a small bypass capacitor (say 0.1 μF) directly to the DUT and then connect a larger capacitor (say 10 μF) through a relay as shown in Figure 3.7. The large bypass capacitor can be disconnected temporarily while the power-down current is measured.

3.4 DC REFERENCES AND REGULATORS

3.1.1 Voltage Regulators

A voltage regulator is one of the most basic analog circuits. The function of a voltage regulator is to provide a well-specified and constant output voltage level from a poorly specified and sometimes fluctuating input voltage. The output of the voltage regulator would then be used as the supply voltage for other circuits in the system. Figure 3.8 illustrates the conversion of a 6- to 12-V ranging power supply to a fixed 5-V output level.

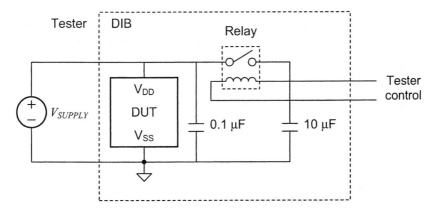

Figure 3.7. Arranging different-sized bypass capacitors to minimize power supply current settling behavior.

Voltage regulators can be tested using a fairly small number of DC tests. Some of the important parameters for a regulator are output no-load voltage, output voltage or load regulation, input or line regulation, input or ripple rejection, and dropout voltage.

Output no-load voltage is measured by simply connecting a voltmeter to the regulator output with no load current and measuring the output voltage V_O.

Load regulation measures the ability of the regulator to maintain the specified output voltage V_O under different load current conditions I_L. As the output voltage changes with increasing load current, one defines the output voltage regulation as the percentage change in the output voltage (relative to the ideal output voltage, $V_{O\text{-}NOM}$) for a specified change in the load current. Load regulation is measured under minimum input voltage conditions

$$\text{load regulation} \equiv 100\% \times \frac{\Delta V_O}{V_{O\text{-}NOM}} \bigg|_{\max\{\Delta I_L\},\, \text{minimum } V_I} \quad (3.1)$$

The largest load current change, max{ΔI_L}, is created by varying the load current from the minimum rated load current (typically 0 mA) to the maximum rated load current.

Load regulation is sometimes specified as the absolute change in voltage, ΔV_O, rather than as a percentage change in V_O. The test definition will be obvious from the specification units (i.e. volts or percentage).

Line regulation or *input regulation* measures the ability of the regulator to maintain a steady output voltage over a range of input voltages. Line regulation is specified as the percentage change in the output voltage as the input line voltage changes over its largest allowable range. Like the load regulation test, line regulation is sometimes specified as an absolute voltage change rather than a percentage. Line regulation is measured under maximum load conditions

$$\text{line regulation} \equiv 100\% \times \frac{\Delta V_O}{V_{O\text{-}NOM}} \bigg|_{\max\{\Delta V_I\},\, \text{maximum } I_L} \quad (3.2)$$

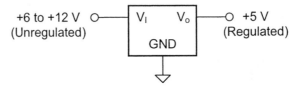

Figure 3.8. 5-V DC voltage regulator.

For the regulator shown in Figure 3.8, with the appropriate load connected to the regulator output, the line regulation would be computed by first setting the input voltage to 6 V, measuring the output voltage, then readjusting the input voltage to 12 V, and again measuring the output voltage to calculate ΔV_O. The line regulation would then be computed using Eq. (3.2).

Input rejection or *ripple rejection* is the ratio of the maximum input voltage variation to the output voltage swing, measured at a particular frequency (commonly 120 Hz) or a range of frequencies. It is a measure of the circuit's ability to reject periodic fluctuations of rectified AC voltage signals applied to the input of the regulator. Input rejection can also be measured at DC using the input voltage range and output voltage swing measured during the line regulation test.

Dropout voltage is the lowest voltage that can be applied between the input and output pins without causing the output to drop below its specified minimum output voltage level. Dropout voltage is tested under maximum current loading conditions. It is possible to search for the exact dropout voltage by adjusting the input voltage until the output reaches its minimum acceptable voltage, but this is a time-consuming test method. In production testing, the input can simply be set to the specified dropout voltage plus the minimum acceptable output voltage. The output is then measured to guarantee that it is equal to or above the minimum acceptable output voltage.

Exercises

3.1. The output of a 5-V voltage regulator varies from 5.10 V under no-load condition to 4.85 V under a 5 mA maximum rated load current. What is its load regulation?

Ans. 250 mV or 5%.

3.2. The output of a 5-V voltage regulator varies from 5.05 to 4.95 V when the input voltage is changed from 14 to 6 V under a maximum load condition of 10 mA. What is its line regulation?

Ans. 100 mV or 2%.

3.3. A 9-V voltage regulator is rated to have a load regulation of 3% for a maximum load current of 15 mA. Assuming a no-load output voltage of 9 V, what is the worst-case output voltage at the maximum load current?

Ans. 8.73 V.

3.4.2 Voltage References

Voltage regulators are commonly used to supply a steady voltage while also supplying a relatively large amount of current. However, many of the DC voltages used in a mixed-signal device do not draw a large amount of current. For example, a 1-V DAC reference does not need to supply 500 mA of current. For this reason, low-power voltage references are often incorporated into mixed-signal devices rather than high-power voltage regulators.

The output of on-chip voltage references may or may not be accessible from the external pins of a DUT. It is common for the test engineer to request a set of test modes so that reference voltages can be measured during production testing. This allows the test program to evaluate the quality of the DC references even if they have no explicit specifications in the data sheet. The design and test engineers can then determine whether failures in the more complicated AC tests may be due to a simple DC voltage error in the reference circuits. DC reference test modes also allow the test program to trim the internal DC references for more precise device operation.

3.4.3 Trimmable References

Many high-performance mixed-signal devices require reference voltages that are trimmed to very exact levels by the ATE tester. DC voltage trimming can be accomplished in a variety of ways. The most common way is to use a programmable reference circuit that can be permanently adjusted to the desired level. One such arrangement is shown in Figure 3.9. The desired level is programmed using fuses, or a nonvolatile digital control mechanism such as EEPROM or flash memory bits. Fuses are blown by forcing a controlled current across each fuse that causes it to vaporize. Fuses can be constructed from either metal or polysilicon. If EEPROM or flash memory is added to a mixed-signal device, then this technology may offer a superior alternative to blown fuses, as EEPROM bits can be rewritten if necessary.

There are various algorithms for finding the digital value that minimizes reference voltage error. In the more advanced trimming architectures such as the one in Figure 3.9, the reference can be experimentally adjusted using a bypass trim value rather than permanently blowing the

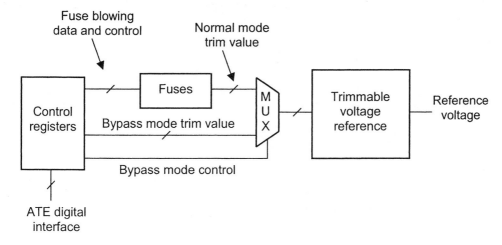

Figure 3.9. Trimmable reference circuit.

fuses. In this example, the bypass trim value is enabled using a special test mode control signal, bypass mode control. Once the best trim value has been determined by experimental trials, the fuses are permanently blown to set the desired trim value. Then, during normal operation, the bypass trim value is disabled and the programmed fuses are used to control the voltage reference.

Trimming can also be accomplished using a laser trimming technique. In this technique, a laser is used to cut through a portion of an on-chip resistor to increase its resistance to the desired value. The resistance value in turn adjusts the DC level of the voltage reference. The laser trimming technique can also be used to trim gains and offsets of analog circuits. Laser trimming is more complex than trimming with fuses or nonvolatile memory. It requires special production equipment linked to the ATE tester.

Laser trimming must be performed while the silicon wafer is still exposed to open air during the probing process. Since metal fuses can produce a conductive sputter when they vaporize, they too are usually trimmed during the wafer probing process. By contrast, polysilicon fuses and EEPROM bits can be blown either before or after the device is packaged.

There is an important advantage to trimming DC levels after the device has been packaged. When plastic is injected around the silicon die, it can place slight mechanical forces on the die. This in turn introduces DC offsets. Because of these DC shifts, a device that was correctly trimmed during the wafer probing process may not remain correctly trimmed after it has been encapsulated in plastic. Another potential DC shift problem relates to the photoelectric effect. Since light shining on a bare die introduces photoelectric DC offsets, a bare die must be trimmed in total darkness. Of course, wafer probers are designed with this requirement in mind. They include a black hood or other mechanism to shield bare die from light sources.

3.5 IMPEDANCE MEASUREMENTS

3.5.1 Input Impedance

Input impedance (Z_{IN}), also referred to as *input resistance*, is a common specification for analog inputs. In general, impedance refers to the behavior of both resistive and reactive (capacitive or inductive) components in the circuit. As the discussion in this chapter is restricted to DC, inductors and capacitors have zero reactance, and as such, make no contribution to impedance. Hence, impedance and resistance refer to the same quantity at DC.

Input impedance is a fairly simple measurement to make. If the input voltage is a linear function of the input current (i.e., if it behaves according to Ohm's law), then one simply forces a

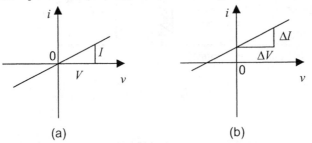

Figure 3.10. Input i-v characteristic curves for (a) linear impedance and (b) nonlinear impedance.

voltage V and measures a current I, or vice versa, and computes the input impedance according to

$$Z_{IN} = \frac{V}{I} \tag{3.3}$$

Figure 3.10(a) illustrates the input i-v relationship of a device satisfying Ohm's law. Here we see that the i-v characteristic is a straight line passing through the origin with a slope equal to Z_{IN}^{-1}. In many instances, the i-v characteristic of an input pin is a straight line but does not pass through the origin as shown in Figure 3.10(b). Such situations typically arise from biasing considerations where the input terminal of a device is biased by a constant current source such as that shown in Figure 3.11 or the input impedance is terminated with an unknown voltage source other than ground.

In cases such as these, one cannot use Eq. (3.3) to compute the input impedance, as it will not lead correctly to the slope of the i-v characteristic. Instead, one measures the change in the input current (ΔI) that results from a change in the input voltage (ΔV) and computes the input impedance using

$$Z_{IN} = \frac{\Delta V}{\Delta I} \tag{3.4}$$

If the input impedance is so low that it would cause excessive currents to flow into the pin, another approach is needed. The alternative method is to force two controlled currents and measure the resulting voltage difference. This is often referred to as a *force-current/measure-voltage* method. Input impedance is again calculated using Eq. (3.4).

Example 3.1

In the input impedance test setup shown in Figure 3.11, voltage source SRC1 is set to 2 V and current flowing into the pin is measured at 0.055 mA. Then SRC1 is set to 1 V and the input current is measured again at 0.021 mA. What is the input impedance?

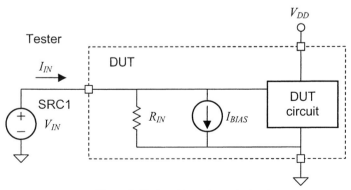

Figure 3.11. Input impedance test setup.

Solution:

Input impedance, Z_{IN}, which is a combination of R_{IN} and the input impedance of the block labeled "DUT Circuit," is calculated using Eq. (3.4) as follows

$$Z_{IN} = \frac{2\text{ V} - 1\text{ V}}{0.055\text{ mA} - 0.021\text{ mA}} = 29.41\text{ k}\Omega$$

Note that the impedance could also have been measured by forcing 0.050 and 0.020 mA and measuring the voltage difference. However, the unpredictable value of I_{BIAS} could cause the input voltage to swing beyond the DUT's supply rails. For this reason, the forced-current measurement technique is reserved for low values of resistance.

In Example 3.1, the values of the excitation consisting of 2 and 1 V are somewhat irrelevant. We could just as easily have used 2.25 and 1.75 V. However, the larger the difference in voltage, the easier it is to make an accurate measurement of current change. This is true throughout many types of tests. Large changes in voltages and currents are easier to measure than small ones. The test engineer should beware of saturating the input of the device with excessive voltages, though. Saturation could lead to extra input current resulting in an inaccurate impedance measurement. The device data sheet should list the acceptable range of input voltages.

3.5.2 Output Impedance

Output impedance (Z_{OUT}) is measured in the same way as input impedance. It is typically much lower than input impedance; so it is usually measured using a force-current/measure-voltage technique. However, in cases where the output impedance is very high, it may be measured using the force-voltage/measure-current method instead.

Example 3.2

In the output impedance test setup shown in Figure 3.12, current source SRC1 is set to 10 mA and the voltage at the pin is measured, yielding 1.61 V. Then SRC1 is set to –10 mA and the output voltage is measured at 1.42 V. What is the total output impedance (R_{OUT} plus the amplifier's output impedance)?

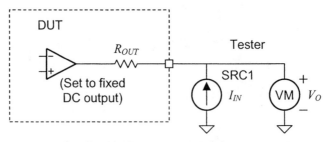

Figure 3.12. Output impedance test setup.

Solution:

Using Eq. (3.4) with Z_{IN} replaced by Z_{OUT}, we write

$$Z_{OUT} = \frac{1.61 \text{ V} - 1.42 \text{ V}}{10 \text{ mA} - (-10 \text{ mA})} = 9.5 \text{ }\Omega$$

3.5.3 Differential Impedance Measurements

Differential impedance is measured by forcing two differential voltages and measuring the differential current change. Example 3.3 illustrates this approach. Differential input impedance would be measured in a similar manner.

Example 3.3

In the differential output impedance test setup shown in Figure 3.13 current source SRC1 is set to 10 mA, SRC2 is set to –10 mA and the differential voltage at the pins is measured at 201 mV. Then SRC1 is set to –10 mA, SRC2 is set to 10 mA, and the output voltage is measured at –199 mV. What is the differential output impedance?

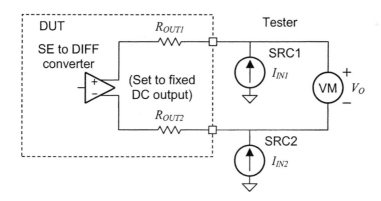

Figure 3.13. Differential output impedance test setup.

Solution:

The output impedance is found using Eq. (3.4) to be

$$Z_{OUT} = \frac{201 \text{ mV} - (-199 \text{ mV})}{20 \text{ mA} - (-20 \text{ mA})} = 10 \text{ }\Omega$$

3.6 DC OFFSET MEASUREMENTS

3.6.1 V_{MID} and Analog Ground

Many analog and mixed-signal integrated circuits are designed to operate on a single power supply voltage (V_{DD} and ground) rather than a more familiar bipolar supply (V_{DD}, V_{SS}, and ground). Often these single-supply circuits generate their own low-impedance voltage between V_{DD} and ground that serves as a reference voltage for the analog circuits. This reference voltage, which we will refer to as V_{MID}, may be placed halfway between V_{DD} and ground or it may be placed at some other fixed voltage such as 1.35 V. In some cases, V_{MID} may be generated off-chip and supplied as an input voltage to the DUT.

To simplify the task of circuit analysis, we can define any circuit node to be 0 V and measure all other voltages relative to this node. Therefore, in a single-supply circuit having a V_{DD} of 3 V, a V_{SS} connected to ground, and an internally generated V_{MID} of 1.5 V, we can redefine all voltages relative to the V_{MID} node. Using this definition of 0 V, we can translate our single-supply circuit into a more familiar bipolar configuration with V_{DD} = +1.5 V, V_{MID} = 0 V, and V_{SS} = –1.5 V (Figure 3.14).

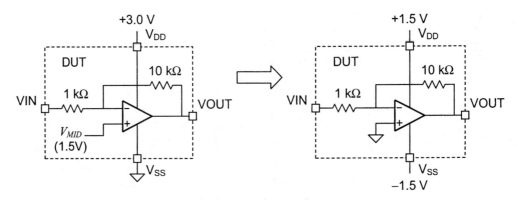

Figure 3.14. Redefining V_{MID} as 0 V to simplify circuit analysis.

Several integrated circuit design textbooks refer to this type of V_{MID} reference voltage as analog ground, since it serves as the ground reference in single-supply analog circuits. This is an unfortunate choice of terminology from a test engineering standpoint. Analog ground is a term used in the test and measurement industry to refer to a high-quality ground that is separated from the noisy ground connected to the DUT's digital circuits. In fact, the term "ground" has a definite meaning when working with measurement equipment since it is actually tied to earth ground for safety reasons. In this textbook, we will use the term analog ground to refer to a quiet 0 V voltage for use by analog circuits and the term V_{MID} to refer to an analog reference voltage (typically generated on-chip) that serves as the IC's analog "ground."

3.6.2 DC Transfer Characteristics (Gain and Offset)

The input-output DC transfer characteristic for an ideal amplifier is shown in Figure 3.15. The input-output variables of interest are voltage, but they could just as easily be replaced by current

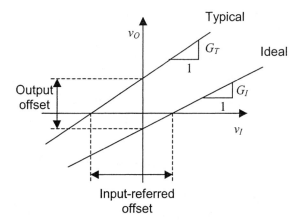

Figure 3.15. Amplifier input-output transfer characteristics in its linear region.

signals. As the real world is rarely accommodating to IC and system design engineers, the actual transfer characteristic for the amplifier would deviate somewhat from the ideal or expected curve. To illustrate the point, we superimpose another curve on the plot in Figure 3.15 and label it "Typical."

In order to maintain correct system operation, design engineers require some assurance that the amplifier transfer characteristic is within acceptable tolerance limits. Of particular interest to the test engineer are the gain and offset voltages shown in the figure. In this section we shall describe the method to measure offset voltages (which is equally applicable to current signals as well) and the next section will describe several methods used to obtain amplifier gain.

3.6.3 Output Offset Voltage (V_O)

The output offset (V_O) of a circuit is simply the difference between its ideal DC output and its actual DC output when the input is set to some fixed reference value, normally analog ground or V_{MID}. Output offset is depicted in Figure 3.15 for an input reference value of 0 V. As long as the output is not noisy and there are no AC signal components riding on the DC level, output offset is a trivial test. If the signal is excessively noisy, the noise component must be removed from the DC level in one of two ways. First, the DC signal can be filtered using a low-pass filter. The output of the filter is measured using a DC voltmeter. ATE testers usually have a low-pass filter built into their DC meter for such applications. The low-pass filter can be bypassed during less demanding measurements in order to minimize the overall settling time. The second method of reducing the effects of noise is to collect multiple readings from the DC meter and then mathematically average the results. This is equivalent to a software low-pass filter.

Sometimes sensitive DUT outputs can be affected by the ATE tester's parasitic loading. Some op amps will become unstable and break into oscillations if their outputs are loaded with the stray capacitance of the tester's meter and its connections to the DUT. An ATE meter may add as much as 200 pF of loading on the output of the DUT depending on the connection scheme chosen by the test engineer. The design engineer and test engineer should evaluate the possible effects of the tester's stray capacitance on each DUT output. It may be necessary to add a buffer amplifier to the DIB to provide isolation between the DUT output and the tester's instruments.

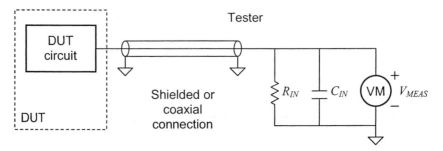

Figure 3.16. Meter impedance loading.

The input impedance of the tester can also shift DC levels when very high-impedance circuit nodes are tested. Consider the circuit in Figure 3.16 where the DUT is assumed to have an output impedance R_{OUT} of 100 kΩ. The DC meter in this example has an input impedance R_{IN} of 1 MΩ. According to the voltage divider principle with two resistors in series, the voltage that appears across the meter V_{MEAS} with respect to the output V_O of the DUT is

$$V_{MEAS} = \frac{R_{IN}}{R_{IN}+R_{OUT}}V_O = \frac{1\text{ M}\Omega}{1\text{ M}\Omega+100\text{ k}\Omega}V_O$$
$$= 0.909 V_O$$

It is readily apparent that a relative error of

$$\text{relative error} = \frac{V_O - V_{MEAS}}{V_O} = \frac{(1-0.909)}{1} = 0.091$$

or 9.1% is introduced into this measurement. A unity gain buffer amplifier may be necessary to provide better isolation between the DUT and tester instrument.

3.6.4 Single-Ended, Differential, and Common-Mode Offsets

Single-ended output offsets are measured relative to some ideal or expected voltage level when the input is set to some specified reference level. Usually these two quantities are the same and are specified on the data sheet. Differential offset is the difference between two outputs of a differential circuit when the input is set to a stated reference level. For simplicity sake, we shall use V_O to denote the output offset for both the single-ended and differential case. It should be clear from the context which offset is being referred to. The output common-mode voltage V_{CM-O} is defined as the average voltage level at the two outputs of a differential circuit. Common-mode offset V_{O-CM} is the difference between the output common-mode voltage and the ideal value under specified input conditions.

Example 3.4

Consider the single-ended to differential converter shown in Figure 3.17. The two outputs of the circuit are labeled OUTP and OUTN. A 1.5-V reference voltage V_{MID} is applied to the input of

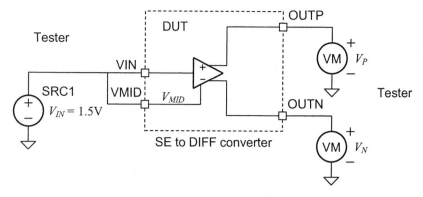

Figure 3.17. Differential output offset test setup.

the circuit and ideally, the outputs should both produce V_{MID}. The voltages at OUTP and OUTN denoted V_P and V_N, respectively, are measured with a meter, producing the following two readings:

$$V_P = 1.507 \text{ V} \quad \text{and} \quad V_N = 1.497 \text{ V}$$

With an expected output reference level of $V_{MID} = 1.50$ V, compute the differential and common-mode offsets.

Solution:

$$\text{OUTP single-ended offset voltage, } V_{O\text{-}P} = V_P - V_{MID} = +7 \text{ mV}$$

$$\text{OUTN single-ended offset voltage, } V_{O\text{-}N} = V_N - V_{MID} = -3 \text{ mV}$$

$$\text{differential offset, } V_O = V_P - V_N = +10 \text{ mV}$$

$$\text{Output common-mode voltage, } V_{CM\text{-}O} = (V_P + V_N) / 2 = 1.502 \text{ V}$$

$$\text{Common-mode offset, } V_{O\text{-}CM} = V_{CM\text{-}O} - V_{MID} = 2 \text{ mV}$$

In the preceding example, V_{MID} is provided to the device from a highly accurate external voltage source. But what happens when the V_{MID} reference is generated from an on-chip reference circuit which itself has a DC offset? Typically there is a separate specification for the V_{MID} voltage in such cases; the input of the DUT should be connected to the V_{MID} voltage, if it is possible to do so and the output offsets are then specified relative to the V_{MID} voltage rather than the ideal value.

Thus the inputs and outputs are treated as if V_{MID} was exactly correct. Any errors in the V_{MID} voltage are evaluated using a separate V_{MID} DC voltage test. In this manner, DC offset errors caused by the single-ended to differential converter can be distinguished from errors in the V_{MID} reference voltage. This extra information may prove to be very useful to design engineers who must decide what needs to be corrected in the design.

3.6.5 Input Offset Voltage (V_{OS})

Input offset voltage (V_{OS}) refers to the negative of the voltage that must be applied to the input of a circuit in order to restore the output voltage to a desired reference level, that is, analog ground or V_{MID}. If an amplifier requires a +10 mV input to be applied to its input to force the output level to analog ground, then V_{OS} = -10 mV. It is common in the literature to find V_{OS} defined as the output offset V_O divided by the measured gain G of the circuit

$$V_{OS} \equiv \frac{V_O}{G} \tag{3.5}$$

If an amplifier has a gain of 10 V/V and its output has an output offset of 100 mV, then its input offset voltage is 10 mV. This will always be true provide the values used in Eq. (3.5) are derived from the circuit in its linear region of operation. In high-gain circuits, such as an open-loop op amp, it is not uncommon to find the amplifer in a saturated state when measuring the output offset voltage. As such, Eq. (3.5) is not applicable.

Exercises

3.4. For a x10 amplifier characterized by $V_{OUT} = 10V_{IN} + 5$, what are its input and output offset voltages?

Ans. +0.5 V (input), 5 V (output).

3.5. For a x10 amplifier characterized by $V_{OUT} = 10V_{IN} - V_{IN}^2 + 5$ over a 10-V range, what is its input and output offset voltages?

Ans. +0.477 V (input), 5 V (output).

3.6. A voltmeter with an input impedance of 100 kΩ is to measure the DC output of an amplifier with an output impedance of 500 kΩ. What is the expected relative error made by this measurement?

Ans. 16.6%.

3.7. A differential amplifier has an output OUTP of 3.3 V and an output OUTN of 2.8 V with its input set to a V_{MID} reference level of 3 V. What are the single-ended and differential offsets? The common-mode offset?

Ans. 0.3 V and –0.2V (SE), 0.5 V (DIFF), 50 mV (CM).

3.8. A perfectly linear amplifier has a measured gain of 5.1 V/V and an output offset of –3.2 V. What is the input offset voltage?

Ans. –0.627 V.

3.7 DC GAIN MEASUREMENTS

3.7.1 Closed-Loop Gain

Closed-loop DC gain is one of the simplest measurements to make, as the input-output signals are roughly comparable in level. Closed-loop gain, denoted G, is defined as the slope of the amplifier input-output transfer characteristic, as illustrated in Figure 3.15. We refer to this gain as closed-loop as it typically contrived from a set of electronic devices configured in a negative feedback loop. It is computed by simply dividing the change in output level of the amplifier or circuit by the change in its input

$$G = \frac{\Delta V_O}{\Delta V_I} \tag{3.6}$$

DC gain is measured using two DC input levels that fall inside the linear region of the amplifier. This latter point is particularly important, as false gain values are often obtained when the amplifier is unknowingly driven into saturation by poorly chosen input levels. The range of linear operation should be included in the test plan.

Gain can also be expressed in decibels (dB). The conversion from volt-per-volt to decibels is simply

$$G(\text{dB}) = 20 \log_{10} |G(\text{V/V})| \tag{3.7}$$

The logarithm function in Eq.(3.7) is a base-10 log as opposed to a natural log.

Example 3.5

An amplifier with an expected gain of -10 V/V is shown in Figure 3.18. Both the input and output levels are referenced to an internally generated voltage V_{MID} of 1.5 V. SRC1 is set to 1.4 V and an output voltage of 2.51 V is measured with a voltmeter. Then SRC1 is set to 1.6 V and an output voltage of 0.47 V is measured. What is the DC gain of this amplifier in V/V? What is the gain in decibels?

Solution:

The gain of the amplifier is computed using Eq. (3.5) as

$$G = \frac{2.51 \text{ V} - 0.47 \text{ V}}{1.4 \text{ V} - 1.6 \text{ V}} = -10.2 \text{ V/V}$$

or, in terms of decibels

$$G = 20 \log_{10} |-10.2| = 20.172 \text{ dB}$$

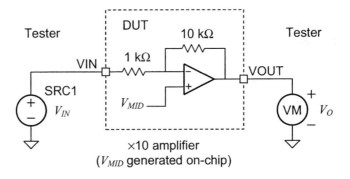

Figure 3.18. A ×10 amplifier gain test setup.

Gain may also be specified for circuits with differential inputs and/or outputs. The measurement is basically the same.

Example 3.6

A fully differential amplifier with an expected gain of +10 V/V is shown in Figure 3.19. SRC1 is set to 1.6 V and SRC2 is set to 1.4 V. This results in a differential input of 200 mV. An output voltage of 2.53 V is measured at OUTP and an output voltage of 0.48 V is measured at OUTN. This results in a differential output of 2.05 V. Then SRC1 is set to 1.4 V and SRC2 is set to 1.6 V. This results in a differential input level of –200 mV. An output voltage of 0.49 V is measured at OUTP and an output voltage of 2.52 V is measured at OUTN. The differential output voltage is thus -2.03 V. Using the measured data provided, compute the differential gain of this circuit.

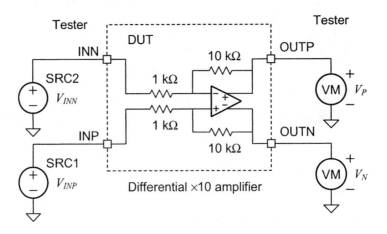

Figure 3.19. Differential ×10 amplifier gain test setup.

Solution:

The differential gain is found using Eq. (3.5) to be

$$G = \frac{2.05\ \text{V} - 2.03\ \text{V}}{200\ \text{mV} - (-200\ \text{mV})} = +10.2\ \text{V/V}$$

Differential measurements can be made by measuring each of the two output voltages individually and then computing the difference mathematically. Alternatively, a differential voltmeter can be used to directly measure differential voltages. Obviously the differential voltmeter approach will work faster than making two separate measurements. Therefore, the use of a differential voltmeter is the preferred technique in production test programs. Sometimes the differential voltage is very small compared to the DC offset of the two DUT outputs. A differential voltmeter can often give more accurate readings in these cases.

In cases requiring extreme accuracy, it may be necessary to measure the input voltages as well as the output voltages. The DC voltage sources in most ATE testers are well calibrated and stable enough to provide a voltage error no greater than 1 mV in most cases. If this level of error is unacceptable, then it may be necessary to use the tester's high-accuracy voltmeter to measure the exact input voltage levels rather than trusting the sources to produce the desired values. The gain equation in the previous example would then be

$$G = \frac{2.05\ \text{V} - 2.03\ \text{V}}{V_1 - V_2}$$

where V_1 and V_2 are the actual input voltages measured using a differential voltmeter.

Exercises

3.9. Voltages of 0.8 and 4.1 V appear at the output of a single-ended amplifier when an input of 1.4 and 1.6 V is applied, respectively. What is the gain of the amplifier in V/V? What is the gain in decibels?

Ans. –16.5 V/V, 24.35 dB.

3.10. An amplifier is characterized by $V_{OUT} = 2.5\ V_{IN} + 1$ over an output voltage range of 0 to 10 V. What is the amplifier output for a 2-V input? Similarly for a 3-V input? What is the corresponding gain of this amplifier in V/V over the 1-V swing? What is the gain in decibels?

Ans. 6 V, 8.5 V, +2.5 V/V, 7.96 dB.

3.11. An amplifier is characterized by $V_{OUT} = 2.5\ V_{IN} + 0.25\ V_{IN}^2 + 1$ over an output voltage range of 0 to 12 V. What is the amplifier output for a 2-V input? Similarly for a 3-V input? What is the corresponding gain of this amplifier in V/V over the 1-V swing? What is the gain in decibels? Would a 4-V input represent a valid test point?

Ans. 7 V, 10.75 V, +3.75 V/V, 11.48 dB, No – the output would exceed 12 V.

The astute reader may have noticed that the gain and impedance measurements are fairly similar, in that they both involve calculating a slope from a DC transfer characteristic pertaining to the DUT. Moreover, they do not depend on any value for the offsets, only that the appropriate slope is obtained from the linear region of the transfer characteristic.

3.7.2 Open-Loop Gain

Open-loop gain (abbreviated G_{ol}) is a basic parameter of op amps. It is defined as the gain of the amplifier with no feedback path from output to input. Since many op amps have G_{ol} values of 10,000 V/V or more, it is difficult to measure open-loop gain with the straightforward techniques of the previous examples. It is difficult to apply a voltage directly to the input of an open loop op amp without causing it to saturate, forcing the output to one power supply rail or the other. For example, if the maximum output level from an op amp is ±5 V and its open-loop gain is equal to 10,000 V/V, then an input-referred offset of only 500 µV will cause the amplifier output to saturate. Since many op amps have input-referred offsets ranging over several millivolts, we cannot predict what input voltage range will result in unsaturated output levels.

We can overcome this problem using a second op amp connected in a feedback path as shown in Figure 3.20. The second amplifier is known as a *nulling amplifier*. The nulling amplifier forces its differential input voltage to zero through a negative feedback loop formed by resistor string R_2 and R_1, together with the DUT op amp. This loop is also known as a servo loop[2]. By doing so, the output of the op amp under test can be forced to a desired output level according to

$$V_{O-DUT} = 2V_{MID} - V_{SRC1} \tag{3.8}$$

where V_{MID} is a DC reference point (grounded in the case of dual-supply op amps, non-grounded for single-supply op amps) and V_{SRC1} is the programmed DC voltage from SRC1. The nulling amplifier and its feedback loop compensate for the input-referred offset of the DUT amplifier. This ensures that the DUT output does not saturate due to its own input-referred offset.

The two matched resistors, R_3, are normally chosen to be around 100 kΩ as a compromise between source loading and op amp bias induced offsets. Since the gain around the loop is extremely large, feedback capacitor C is necessary to stabilize the loop. A capacitance value of 1 to 10 nF is usually sufficient. R_{LOAD} provides the specified load resistance for the G_{ol} test.

Under steady-state conditions, the signal that is fed back to the input of the DUT amplifier denoted V_{IN-DUT} is directly related to the nulling amplifier output V_{O-NULL} according to

$$V_{IN-DUT} = V^+_{DUT} - V^-_{DUT} = \frac{R_1}{R_1 + R_2}(V_{O-NULL} - V_{MID}) \tag{3.9}$$

where V^+_{DUT} and V^-_{DUT} are the positive and negative inputs to the DUT amplifier, respectively. Subsequently, the open-loop voltage gain of the DUT amplifier is found from Eqs. (3.6), (3.8), and (3.9) to be given by

$$G_{ol} = \frac{\Delta V_{O-DUT}}{\Delta V_{IN-DUT}} = -\left(\frac{R_1 + R_2}{R_1}\right) \frac{\Delta V_{SRC1}}{\Delta V_{O-NULL}} \tag{3.10}$$

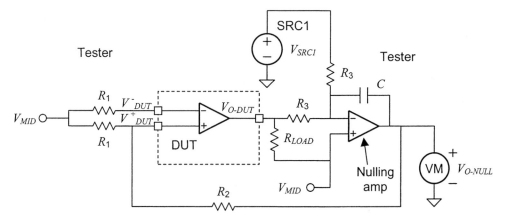

Figure 3.20. Open-loop gain test setup using nulling amplifier.

The nulling loop method allows the test engineer to force two desired outputs and then indirectly measure the tiny inputs that caused those two outputs. In this manner, very large gains can be measured without measuring tiny voltages. Of course the accuracy of this approach depends on accurately knowing the values of R_1 and R_2, and on matching the two resistors, labeled as R_3.

In order to maximize the signal handling capability of the test setup shown in Figure 3.20, and avoid saturating the nulling amplifer, it is a good idea to set the voltage divider ratio to a value approximately equal to the inverse of the expected open-loop gain of the DUT op amp

$$\frac{R_1}{R_1+R_2} \approx \frac{1}{G_{ol}} \qquad (3.11)$$

from which we can write $R_2 \approx G_{ol} R_1$.

Example 3.7

For the nulling amplifier setup shown in Figure 3.20 with R_1=100 Ω, R_2=100 kΩ and R_3=100 kΩ, together with V_{MID} set to a value midway between the two power supply levels (its actual value is not important as all signals will be referenced to it), SRC1 is set to V_{MID} + 1 V and a voltage of V_{MID} + 2.005 V is measured at the nulling amplifier output. SRC1 is set to V_{MID} − 1 V and a voltage of V_{MID} + 4.020 V is measured at the nulling amplifier output. What is the open-loop gain of the amplifier?

Solution:

Open loop gain is calculated using the following procedure. First the change or swing in the nulling amplifier output $\Delta V_{O\text{-}NULL}$ is computed

$$\Delta V_{O-NULL} = 2.005 \text{ V} - 4.020 \text{ V} = -2.015 \text{ V}$$

then, using Eq. (3.9) the voltage swing at the input of the DUT amplifier, $\Delta V_{IN\text{-}DUT}$, is calculated

$$\Delta V_{IN-DUT} = \frac{R_1}{R_1 + R_2} \Delta V_{O-NULL}$$

$$= \frac{100}{100 + 100k}(-2.015 \text{ V})$$

$$= -2.013 \text{ mV}$$

Making use of the fact that ΔV_{SRC1} is 2 V, which forces $\Delta V_{O\text{-}DUT} = -2$ V, the open-loop gain of the amplifier is found to be

$$G_{ol} = \frac{\Delta V_{O-DUT}}{\Delta V_{IN-DUT}} = \frac{-2 \text{ V}}{-2.013 \text{ mV}} = 993.5 \text{ V/V}$$

If the op amp in the preceding example had an open-loop gain closer to 100 V/V instead of 1000 V/V, then the output of the nulling amplifier would have produced a voltage swing of 20 V instead of 2 V. The nulling amplifier would have been dangerously close to clipping against its output voltage rails (assuming ±15-V power supplies). In fact, if a 5-V op amp were used as the nulling amplifier, it would obviously not be able to produce the 20-V swing.

In the example, the nulling amplifier should have produced two voltages centered around V_{MID}. Instead, it had an average or common-mode offset level of approximately 3 V from this value. A detailed circuit analysis reveals that this offset is caused exclusively by the input-referred offset of the DUT. Hence, the offset that appears at the output of the nulling amplifier, denoted $V_{O\text{-}NULL\text{-}Offset}$, can be used to compute the input-referred offset of the DUT, $V_{OS\text{-}DUT}$.

Exercises

3.12. For the nulling amplifier setup shown in Figure 3.20 with $R_1=100$ Ω, $R_2=100$ kΩ, and $R_3=100$ kΩ, an SRC1 voltage swing of 1 V results in a 2.3-V swing at the output of the nulling amplifier. What is the open-loop gain in V/V of the DUT amplifier? What is the gain in decibels?

Ans. 435.2 V/V, 52.77 dB.

3.13. For the nulling amplifier setup shown in Figure 3.20 with $R_1=1$ kΩ, $R_2=100$ kΩ, and $R_3=100$ kΩ, an offset of 2.175 V + V_{MID} appears at the output of the nulling op amp when the SRC1 voltage is set to V_{MID}. What is the input offset of the DUT amplifier?

Ans. 21.5 mV.

3.14. For the nulling amplifier setup shown in Figure 3.20 with $R_1=100$ Ω, $R_2=500$ kΩ and $R_3=100$ kΩ, and the DUT op amp having an open-loop gain of 4,000 V/V, what is the output swing of the nulling amplifier when the SRC1 voltage swings by 1 V?

Ans. 1.25 V.

Input-referred offset would then be calculated using

$$V_{OS-DUT} = \frac{R_1}{R_1 + R_2} V_{O-NULL-Offset} \qquad (3.12)$$

As this method involves the same measured data used to compute the open-loop gain, it is a commonly used method to determine the op amp input-referred offset. For the parameters and measurement values described in Example 3.7, the input-referred offset voltage for the DUT is

$$V_{OS-DUT} = \frac{100}{100 + 100k} \left(\frac{4.020 \text{ V} + 2.005 \text{ V}}{2} \right)$$
$$= 3.0 \text{ mV}$$

3.8 DC Power Supply Rejection Ratio

3.8.1 DC Power Supply Sensitivity

Power supply sensitivity (PSS) is a measure of the circuit's dependence on a constant supply voltage. Normally it is specified separately with respect to the positive or negative power supply voltages and denoted PSS^+ and PSS^-. PSS is defined as the change in the output over the change in either power supply voltage with the input held constant

$$PSS^+ \equiv \left. \frac{\Delta V_O}{\Delta V_{PS^+}} \right|_{V_{in} \text{ constant}} \quad \text{and} \quad PSS^- \equiv \left. \frac{\Delta V_O}{\Delta V_{PS^-}} \right|_{V_{in} \text{ constant}} \qquad (3.13)$$

In effect, PSS is a type of gain test in which the input is one of the power supply levels.

Example 3.8

The input of the ×10 amplifier in Figure 3.21 is connected to its own V_{MID} source forcing 1.5 V. The power supply is set to 3.1 V and a voltage of 1.5011 V is measured at the output of the amplifier. The power supply voltage is then changed to 2.9 V and the output measurement changes to 1.4993 V. What is the PSS of the amplifier in V/V? What is the PSS in decibels?

Solution:

As the positive power supply (V_{DD}) is being changed by SRC1, the positive power supply sensitivity is

$$PSS^+ = \frac{\Delta V_O}{\Delta V_{SRC1}} = \frac{1.5011 \text{ V} - 1.4993 \text{ V}}{3.1 \text{ V} - 2.9 \text{ V}} = 9 \text{ mV/V} = -40.92 \text{ dB}$$

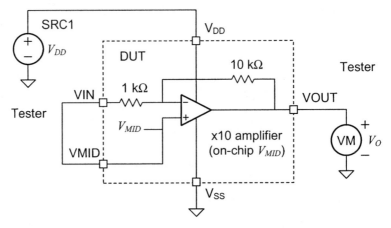

Figure 3.21. Power supply sensitivity test setup.

3.8.2 DC Power Supply Rejection Ratio

Power supply rejection ratio (PSRR) is defined as the power supply sensitivity of a circuit divided by the magnitude of the closed-loop gain of the circuit in its normal mode of operation. Normally it is specified separately with respect to each power supply voltage. Mathematically, we write

$$\text{PSRR}^+ \equiv \frac{\text{PSS}^+}{|G|} \quad \text{and} \quad \text{PSRR}^- \equiv \frac{\text{PSS}^-}{|G|} \tag{3.14}$$

In Example 3.8, we found PSS$^+$=0.009 V/V. In Example 3.5, the DC gain of this same circuit was found to be -10.2 V/V. Hence the PSRR$^+$ would be

$$\text{PSRR}^+ = \frac{\text{PSS}^+}{|G|} = \frac{0.009 \text{ V/V}}{10.2 \text{ V/V}} = 882 \ \mu\text{V/V}$$

Power supply rejection ratio is often converted into decibel units

$$\text{PSRR}^+\big|_{\text{dB}} = 20\log_{10}\left(882 \ \mu\text{V/V}\right) = -61.09 \text{ dB}$$

3.9 DC COMMON-MODE REJECTION RATIO

3.9.1 CMRR of Op Amps

Common-mode rejection ratio (CMRR) is a measurement of a differential circuit's ability to reject a common-mode signal V_{CM} at its inputs. It is defined as the magnitude of the common-mode gain G_{CM} divided by the differential gain G_D, given by

$$\text{CMRR} \equiv \left| \frac{G_{CM}}{G_D} \right| \quad (3.15)$$

This expression can be further simplified by substituting for the common-mode gain $G_{CM} = \Delta V_O / \Delta V_{CM}$, together with the definition for input-referred offset voltage defined in Eq. (3.5), as follows

$$CMRR = \left| \frac{\Delta V_O / \Delta V_{CM}}{G_D} \right| = \left| \frac{\Delta V_O / G_D}{\Delta V_{CM}} \right| = \left| \frac{\Delta V_{OS}}{\Delta V_{CM}} \right| \quad (3.16)$$

The rightmost expression suggests the simplest procedure to measure CMRR; one simply measures ΔV_{OS} subject to a change in the input common-mode level ΔV_{CM}. One can measure ΔV_{OS} directly or indirectly, as the following two examples illustrate.

Example 3.9

Figure 3.22 shows a simple CMRR test fixture for an op amp. The test circuit is basically a difference-amplifier configuration with the two inputs tied together. V_{MID} is set to 1.5 V and an input common-mode voltage of 2.5 V is applied using SRC1. An output voltage of 1.501 V is measured at the output of the op amp. Then SRC1 is changed to 0.5 V and the output changes to 1.498 V. What is the CMRR of the op amp?

Solution:

As the measurement was made at the output of the circuit, we need to infer from these results the ΔV_{OS} for the op amp. This requires a few steps: The first is to find the influence of the op amp input-referred offset voltage V_{OS} on the test circuit output. As in Section 3.7.2, detailed circuit

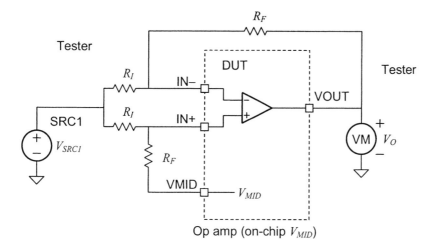

Figure 3.22. Op amp CMRR test setup.

analysis reveals

$$V_O = \frac{R_I + R_F}{R_I} V_{OS}$$

With all resistors equal and perfectly matched, $V_O = 2\,V_{OS}$. Hence, $\Delta V_O = 2\,\Delta V_{OS}$, or when rearranged, $\Delta V_{OS} = 0.5\,\Delta V_O$. Subsequently, substituting measured values $\Delta V_O = 1.501$ V–1.498 V = 3 mV, we find $\Delta V_{OS} = 1.5$ mV. This result can now be substituted into Eq. (3.16), together with $\Delta V_{CM} = \Delta V_{SRC1} = 2.5$ V – 0.5 V = 2.0 V, leading to a CMRR = 750 µV/V or –62.5 dB.

There is one major problem with this technique for measuring op amp CMRR: the resistors must be known precisely and carefully matched. A CMRR value of –100 dB would require resistor matching to 0.0001%, an impractical value to achieve in practice. A better test circuit setup is the nulling amplifier configuration shown in Figure 3.23. This configuration is very similar to the one used previously to measure the open-loop gain and input offsets of Section 3.7. The basic circuit arrangement is identical, only the excitation and the position of the voltmeter are changed. With this test setup, one can vary the common-mode input to the DUT and measure the differential voltage between the input SRC1 and the nulling amplifier output, which we shall denote as $V_{O\text{-}NULL}$. This in turn can then be used to deduce the input-referred offset for the DUT amplifier according to

$$V_{OS\text{-}DUT} = \frac{R_1}{R_1 + R_2} V_{O\text{-}NULL} \tag{3.17}$$

Subsequently, the CMRR of the op amp is given by

$$CMRR = \frac{R_1}{R_1 + R_2} \left| \frac{\Delta V_{O\text{-}NULL}}{\Delta V_{SRC1}} \right| \tag{3.18}$$

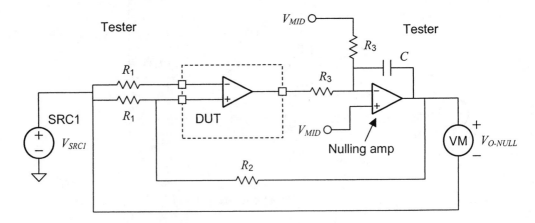

Figure 3.23. Op amp CMRR test setup using nulling amplifier.

Example 3.10

For nulling amplifier setup shown in Figure 3.23 with R_1=100 Ω, R_2=100 kΩ, and R_3=100 kΩ, together with V_{MID} set to a value midway between the two power supply levels, SRC1 is set to +2.5 V and a differential voltage of 10 mV is measured between SRC1 and the output of the nulling amplifier. Then SRC1 is set to 0.5 V and the measured voltage changes to −12 mV. What is the CMRR of the op amp?

Solution:

Using Eq. (3.17), we deduce

$$\Delta V_{OS-DUT} = \frac{R_1}{R_1 + R_2} \Delta V_{O-NULL}$$
$$= \frac{100}{100 + 100k} \left[10 \text{ mV} - (-12 \text{ mV}) \right]$$
$$= 22 \, \mu\text{V}$$

for a corresponding ΔV_{SRC1} = 2.5 V − 0.5 V, or 2.0 V. Thus the CMRR is

$$\text{CMRR} = \frac{22 \, \mu\text{V}}{2.0 \text{ V}} = 11 \frac{\mu\text{V}}{\text{V}} = -99.17 \text{ dB}$$

3.9.2 CMRR of Differential Gain Stages

Integrated circuits often use op amps as part of a larger circuit such as a differential input amplifier. In these cases, the CMRR of the op amp is not as important as the CMRR of the circuit as a whole. For example, a differential amplifier configuration such as the one in Figure 3.22 may have terrible CMRR if the resistors are poorly matched, even if the op amp itself has a CMRR of −100 dB. The differential input amplifier CMRR specifications include not only the effects of the op amp, but also the effects of on-chip resistor mismatch. As such, we determine the CMRR using the original definition given in Eq. (3.15). Our next example will illustrate this.

Example 3.11

Figure 3.24 illustrates the test setup to measure the CMRR of a differential amplifier having a nominal gain of 10. No assumption about resistor matching is made. Both inputs are connected to a common voltage source SRC1 whose output is set to 2.5 V. A voltage of 1.501 V is measured at the output of the DUT. Then SRC1 is set to 0.5 V and a second voltage of 1.498 V is measured at the DUT output. Next the differential gain of the DUT circuit is measured using the technique described in Section 3.7.1. The gain was found to be 10.2 V/V. What is the CMRR?

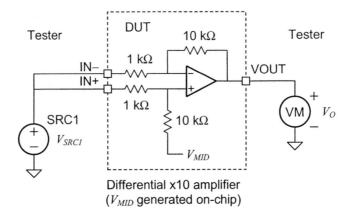

Figure 3.24. A ×10 Differential amplifier CMRR test setup.

Solution:

Since $\Delta V_O = 1.501\text{ V} - 1.498\text{ V} = 3\text{ mV}$ corresponding to a $\Delta V_{CM} = \Delta V_{SRC1} = 2.0\text{ V}$, the common-mode gain G_{CM} is calculated to be equal to 0.0015 V/V. In addition, we are told that the differential gain G_D is 10.2 V/V; thus we find the CMRR from the following

$$CMRR = \left|\frac{G_{CM}}{G_D}\right| = \left|\frac{0.0015\text{ V/V}}{10.2\text{ V/V}}\right| = 0.000147 = -76.65\text{ dB}$$

Exercises

3.15. An amplifier has an expected CMRR of -100 dB. For a 1-V change in the input common-mode level, what is the expected change in the input offset voltage of this amplifier?

Ans. 10 µV.

3.16. For the nulling amplifier CMRR setup in Figure 3.23 with $R_1 = 100\ \Omega$, $R_2 = 500\text{ k}\Omega$, and $R_3 = 100\text{ k}\Omega$, SRC1 is set to +3.5 V and a differential voltage of 210 mV is measured between SRC1 and the output of the nulling amplifier. Then SRC1 is set to 0.5 V and the measured voltage changes to −120 mV. What is the CMRR of the op amp in decibels?

Ans. 21.99 µV/V, -93.15 dB.

3.10 COMPARATOR DC TESTS

3.10.1 Input Offset Voltage

Input offset voltage for a comparator is defined as the differential input voltage that causes the comparator to switch from one output logic state to the other. The differential input voltage can be ramped from one voltage to another to find the point at which the comparator changes state. This switching point is, however, dependent on the input common-mode level. One usually tests for the input offset voltage under worst-case conditions as outlined in the device test plan.

Example 3.12

The comparator in Figure 3.25 has a worst-case input offset voltage of ±50 mV and a midsupply voltage of 1.5V. Describe a test setup and procedure with which to obtain its input offset voltage.

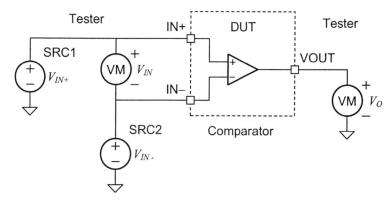

Figure 3.25. Comparator input offset voltage test setup.

Solution:

The comparator in Figure 3.25 is connected to two voltage sources, SRC1 and SRC2. SRC2 is set to 1.5 V and SRC1 is ramped upward from 1.45 to 1.55 V, as the switching point is expected to lie within this range. When the output changes from logic LO to logic HI, the differential input voltage V_{IN} is measured, resulting in an input offset voltage reading of +5 mV. The V_{IN} voltage could be deduced by simply subtracting 1.5 V from the SRC1 voltage, assuming the DC sources force voltages to an accuracy of a few hundred microvolts. This is usually a questionable assumption, though. It is best to measure small voltages using a voltmeter rather than assume the tester's DC sources are set to exact voltages.

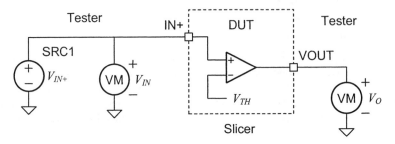

Figure 3.26. Slicer threshold voltage test setup.

3.10.2 Threshold Voltage

Sometimes a fixed reference voltage is supplied to one input of a comparator, forming a circuit known as a *slicer*. The input offset voltage specification is typically replaced by a single-ended specification, called *threshold voltage*.

The slicer in Figure 3.26 is tested in a similar manner as the comparator circuit in the previous example. Assuming the threshold voltage is expected to fall between 1.45 and 1.55 V, the input voltage from SRC1 is ramped upward from 1.45 to 1.55 V. The output switches states when the input is equal to the slicer's threshold voltage.

Notice that threshold voltage will be affected by the accuracy of the on-chip voltage reference, V_{TH}. In theory, the threshold voltage should be equal to the sum of the slicer's reference voltage V_{TH} plus the input offset voltage of the comparator. Threshold voltage error is defined as the difference between the actual and ideal threshold voltages.

3.10.3 Hysteresis

In the comparator input offset voltage example, the output changed when the input voltage reached 5 mV. This occurred on a rising input voltage. On a falling input voltage, the threshold may change to a lower voltage. This characteristic is called *hysteresis*, and it may or may not be an intentional design feature. Hysteresis is defined as the difference in threshold voltage between a rising input test condition and a falling input condition.

Example 3.13

The comparator in Figure 3.25 is connected to two voltage sources, SRC1 and SRC2. SRC2 is set to 1.5 V and SRC1 is ramped upward from 1.45 to 1.55 V in 1-mV steps. When the output changes from logic LO to logic HI, the differential input voltage is measured, resulting in an input offset voltage reading of +5 mV. Then the input is ramped downward from 1.55 to 1.45 V and the output switches when the input voltage reaches −3 mV. What is the hysteresis of this comparator?

Solution:

The hysteresis is equal to the difference of the two input offset voltages

$$5 \text{ mV} - (-3 \text{ mV}) = 8 \text{ mV}$$

It should be noted that input offset voltage and hysteresis may change with different common-mode input voltages. Worst-case test conditions should be determined during the characterization process.

Exercises

3.17. A comparator has an input offset voltage of 50 mV and its positive terminal is connected to a 1-V level, at what voltage on the negative terminal does the comparator change state?

Ans. 0.950 V.

3.18. A slicer circuit is connected to a 1.65 V reference V_{TH} and has a comparator input offset voltage of 11 mV. At what voltage level will the slicer change state?

Ans. 1.76 V.

3.19. A comparator has a measured hysteresis of 9 mV and switches state on a rising input at 2.100 V. At what voltage does the comparator change to a low state on a falling input?

Ans. 2.091 V.

3.11 VOLTAGE SEARCH TECHNIQUES

3.11.1 Binary Searches versus Step Searches

The technique of ramping input voltages until an output condition is met is called a *ramp search*, or *step search*. Step searches are time-consuming and not well suited for production testing. A more efficient binary search technique may be used to reduce test time while maintaining the desired search resolution.

In a binary search, the input is adjusted up or down using a successive approximation algorithm. A binary search can be applied to the comparator input offset voltage test described in the previous section. Instead of ramping the input voltage from 1.45 to 1.55 V, the comparator input is set to 1.5 V and the output is observed. If the output is high, then the input is increased by one quarter of the 100-mV search range (25 mV) to try to make the output go low. If, on the other hand, the output is low, then the input is reduced by 25 mV to try to force the output high. Then the output is observed again. This time, the input is adjusted by one-eighth of the search range (12.5 mV). This process is repeated until the desired input adjustment resolution is reached.

The problem with the binary search technique is that it does not work well in the presence of hysteresis. The binary search algorithm assumes that the input offset voltage is the same whether the input voltage is increased or decreased. If the comparator exhibits hysteresis, then there are two different threshold voltages to be measured. To get around this problem without reverting to the time-consuming ramp search technique, a hybrid approach can be used. A binary search can be used to find the approximate threshold voltage quickly. Then a step search can be used with a much smaller search voltage range.

Another solution to the hysteresis problem is to use a modified binary search algorithm in which the output state of the comparator is returned to a consistent logic state between binary search approximations. The output state is set to a consistent level between approximations by forcing the input either well above or well below the threshold voltage. In this way, steps are always taken in one direction, avoiding hysteresis effects. To measure hysteresis, a binary search is used once with the output state forced high between approximations. Then input offset is measured again with the output state forced low between approximations. The difference in input offset readings is equal to the hysteresis of the comparator.

3.11.2 Linear Searches

Linear circuits can make use of an even faster search technique called a *linear search*. A linear search is similar to the binary search, except that the input approximations are based on a linear interpolation of input-output relationships. For example, if a 0-mV input to a buffer amplifier results in a 10-mV output and a 1-mV input results in a 20-mV output, then a –1-mV input will probably result in a 0-mV output. The linear search algorithm keeps refining its guesses using a simple $V_{OUT} = M \times V_{IN} + B$ algorithm until the desired accuracy is reached. The following example will illustrate the method.

Example 3.14

Using a linear search algorithm find the input offset voltage V_{OS} for a ×10 amplifier.

Solution:

The input to a ×10 amplifier is set to 0 V and the output is measured, yielding a reading of 120 mV. The gain M is known to be approximately 10, since this is supposed to be a ×10 amplifier. The value of offset B can be approximately determined using the $V_{OUT} = M \times V_{IN} + B$ linear equation, that is

$$120 \text{ mV} = M \times 0 \text{ mV} + B = 10 \times 0 \text{ mV} + B$$
$$\Rightarrow B = 120 \text{ mV (first-pass guess)}$$

Since 0 mV is the desired output, the next estimate for V_{OS} can be calculated using the linear equation again

$$0 \text{ mV (desired } V_{OUT}) = M \times V_{IN} + B = 10 \times V_{IN} + 120 \text{ mV}$$

Rewriting this equation to solve for V_{IN}, we get

$$V_{IN} = \frac{(0 \text{ mV-}120 \text{ mV})}{10} = -12 \text{ mV}$$

Applying the best guess of –12 mV to the input, another output measurement is made, resulting in a reading of 8 mV. Now we have two equations in two unknowns

$$120 \text{ mV} = M \times 0 \text{ mV} + B$$
$$8 \text{ mV} = M(-12 \text{ mV}) + B$$

from which a more accurate estimate of M and B can be made. Solving for the two unknowns

$$M = \frac{120 \text{ mV-}8 \text{ mV}}{0 \text{ mV-}(-12 \text{ mV})} = 9.333 \text{ V/V}$$

$$B = 10 \text{ mV} - [M(-12 \text{ mV})] = 122 \text{ mV}$$

The next input approximation should be close enough to the input offset voltage to produce an output of 0 mV, that is

$$V_{OS} = \frac{(0 \text{ mV-}B)}{M} = \frac{(0 \text{ mV-}122 \text{ mV})}{9.333} = -13.1 \text{ mV}$$

The input offset voltage of the ×10 amplifier is therefore –13.1 mV, assuming the circuit is linear. In cases where the input-output relationship is not linear, the linear search technique will still work, but will require more iterations of the above process. During each iteration, the linear interpolations are calculated using the most recent two input-output data points until the input converges to the desired measurement resolution.

Exercises

3.20. For an amplifier characterized by $V_{OUT} = 10V_{IN} - V_{IN}^2 + 5$ over a ±5 V output voltage range, determine the input offset voltage using a binary search process. The input offset voltage is known to fall between 464 and 496 mV. How many search iterations are required for a maximum error of 1 mV? List the input values and corresponding outputs.

Ans. A 32-mV search range with 2-mV resolution is required, requiring four binary iterations: (1) –480 mV, –30.4 mV; (2) –472 mV, +57 mV; (3) –476 mV, +13.4 mV; (4) –478 mV, –8.5 mV. The final estimate is thus -477 mV ($V_{OS} = +477$ mV; true answer is +477.2 mV).

3.21. Repeat Exercise 3.20 using a linear search process starting with two points at $V_{IN} = -250$ mV and –750 mV. How many iterations are required for < 1 mV error in V_{OS}?

Ans. Two iterations produce estimates of $V_{OS} = +471.6$ mV and $V_{OS} = +477.1$ mV.

3.12 DC Tests for Digital Circuits

3.12.1 I_{IH}/I_{IL}

The data sheet for a mixed-signal device usually lists several DC specifications for digital inputs and outputs. Input leakage currents (I_{IH} and I_{IL}) were discussed in Section 3.2.2. Input leakage is also specified for digital output pins that can be set to a high-impedance state.

3.12.2 V_{IH}/V_{IL}

The input high voltage (V_{IH}) and input low voltage (V_{IL}) specify the threshold voltage for digital inputs. It is possible to search for these voltages using a binary search or step search, but it is more common to simply set the tester to force these levels into the device as a go/no-go test. If the device does not have adequate V_{IH} and V_{IL} thresholds, then the test program will fail one of the digital pattern tests that are used to verify the DUT's digital functionality. To allow a distinction between pattern failures caused by V_{IH}/V_{IL} settings and patterns failing for other reasons, the test engineer may add a second identical pattern test that uses more forgiving levels for V_{IH}/V_{IL}. If the digital pattern test fails with the specified V_{IH}/V_{IL} levels and passes with the less demanding settings, then V_{IH}/V_{IL} thresholds are the likely failure mode.

3.12.3 V_{OH}/V_{OL}

V_{OH} and V_{OL} are the output equivalent of V_{IH} and V_{IL}. V_{OH} is the minimum guaranteed voltage for an output when it is in the high state. V_{OL} is the maximum guaranteed voltage when the output is in the low state. These voltages are usually tested in two ways. First, they are measured at DC with the output pin set to static high/low levels. Sometimes a pin cannot be set to a static output level due poor design for test considerations, so only a dynamic test can be performed. Dynamic V_{OH}/V_{OL} testing is performed by setting the tester to expect high voltages above V_{OH} and low voltages below V_{OL}. The tester's digital electronics are able to verify these voltage levels as the outputs toggle during the digital pattern tests. Dynamic V_{OH}/V_{OL} testing is another go/no-go test approach, since the actual V_{OH}/V_{OL} voltages are verified but not measured.

3.12.4 I_{OH}/I_{OL}

V_{OH} and V_{OL} levels are guaranteed while the outputs are loaded with specified load currents, I_{OH} and I_{OL}. The tester must pull current out of the DUT pin when the output is high. This load current is called I_{OH}. Likewise, the tester forces the I_{OL} current into the pin when the pin is low. These currents are intended to force the digital outputs closer to their V_{OH}/V_{OL} specifications, making the V_{OH}/V_{OL} tests more difficult for the DUT to pass. I_{OH} and I_{OL} are forced using a diode bridge circuit in the tester's digital pin card electronics. The diode bridge circuit is discussed in more detail in Chapter 5, "Tester Hardware."

3.12.5 I_{OSH} and I_{OSL} Short Circuit Current

Digital outputs often include a current-limiting feature that protects the output pins from damage during short circuit conditions. If the output pin is shorted directly to ground or to a power supply pin, the protection circuits limit the amount of current flowing into or out of the pin. Short circuit current is measured by setting the output to a low state and forcing a high voltage (usually V_{DD}) into the pin. The current flowing into the pin (I_{OSL}) is measured with one of the tester's current meters. Then the output is set to a high state and 0 V is forced at the pin. The current flowing out of the pin (I_{OSH}) is again measured with a current meter.

3.13 Summary

This chapter has presented only a few of the many DC tests and techniques the mixed-signal test engineer will encounter. Several chapters or perhaps even a whole book could be devoted to highly accurate DC test techniques. However, this book is intended to address mixed-signal testing. Hopefully, the limited examples given in this chapter will serve as a solid foundation from which the test engineer can build a more diversified DC measurement skill set.

DC measurements are trivial to define and understand, but they can sometimes be excruciatingly difficult to implement. A DC offset of 100 mV is very easy to measure if the required accuracy is ±10 mV. On the other hand if 1-µV accuracy is required, the test engineer may find this to be one of the more daunting test challenges in the entire project. The accuracy and repeatability requirements of seemingly simple tests like DC offset can present a far more challenging test problem than much more complicated AC tests.

Accuracy and repeatability of measurements is the subject of the next chapter. This topic pertains to a wide variety of analog and mixed-signal tests. Much of a test engineer's time is consumed by accuracy and repeatability problems. These problems can be one of the most aggravating aspects of mixed-signal testing. The successful resolution of a perplexing accuracy problem can also be one of the most satisfying parts of the test engineer's day.

Problems

3.1. The output of a 10-V voltage regulator varies from 9.95 V under no-load condition to 9.34 V under a 10-mA maximum rated load current. What is its load regulation?

3.2. The output of a 5-V voltage regulator varies from 4.86 to 4.32 V when the input voltage is changed from 14 to 6 V under a maximum load condition of 10 mA. What is its line regulation?

3.3. A 9-V voltage regulator is rated to have a load regulation of 150 mV for a maximum load current of 15 mA. Assuming a no-load output voltage of 9 V, what is the expected output voltage at the maximum load current?

3.4. A 6-V regulator has an output no-load voltage specification of 5.75 V (MIN) to 6.25 V (MAX), a load regulation specification of 150 mV (MAX) and a dropout voltage specification of 1.5 V (MAX). With a 7.5-V input voltage, what is the lowest output voltage that a passing regulator could produce under maximum loading conditions?

3.5. A voltage of 1.2 V is dropped across an input pin when a 100 µA current is forced into the pin. Subsequently, a 1.254-V level occurs when the current is increased to 200 µA. What is the input impedance?

3.6. The input pin of a device is characterized by the i-v relationship: $i = 0.001 v + 100$. What is the impedance seen looking into this pin?

3.7. Voltages of 1.2 and 3.3 V appear at the output of an amplfier when currents of −10 and +10 mA, respectively, are forced into its output. What is the output impedance?

3.8. The no-load output voltage of an amplifier is 4 V. When a 600-Ω load is attached to the output, the voltage drops to 3 V. What is the amplifier's output impedance?

3.9. For a ×10 amplifier characterized by $V_{OUT} = 10V_{IN} - V_{IN}^2 + 5$ over a ±15-V range, what are its input and output offset voltages?

3.10. A voltmeter introduces a measurement error of -5% while measuring a 1-V offset from an amplifier. What is the actual reading captured by the voltmeter?

3.11. A voltmeter with an input impedance of 500 kΩ is used to measure the DC output of an amplifier with an output impedance of 500 kΩ. What is the expected relative error made by this measurement?

3.12. A differential amplifier has outputs of 2.4 V (OUTP) and 2.7 V (OUTN) with its input set to a V_{MID} reference level of 2.5 V. What are the single-ended and differential offsets? The common-mode offset? (All offsets are to be measured with respect to V_{MID}.)

3.13. A perfectly linear amplifier has a measured gain of 5.1 V/V and an output offset of −3.2 V. What is the input offset voltage?

3.14. Voltages of 0.8 and 4.1 V appear at the output of a single-ended amplifier when inputs of 1.4 and 1.6 V are applied, respectively. What is the gain of the amplifier in V/V? What is the gain in decibels?

3.15. An amplifier is characterized by $V_{OUT} = 3.5V_{IN} + 1$ over the input voltage range 0 to 5 V. What is the amplifier output for a 2-V input? Similarly for a 3-V input? What is the corresponding gain of this amplifier in V/V over the 1-V swing? What is the gain in decibels?

3.16. An amplifier is characterized by $V_{OUT} = 1.5V_{IN} + 0.35V_{IN}^2 + 1$ over the input voltage range 0 to 5 V. What is the amplifier output for a 1-V input? Similarly for a 3-V input? What is the corresponding gain of this amplifier in V/V over the 1-V swing? What is the gain in decibels?

3.17. For the nulling amplifier setup shown in Figure 3.20 with R_1=100 Ω, R_2=200 kΩ, and R_3=50 kΩ, an SRC1 input swing of 1 V results in a 130-mV swing at the output of the nulling amplifier. What is the open-loop gain of the DUT amplifier in V/V? What is the gain in decibels?

3.18. For the nulling amplifier setup shown in Figure 3.20 with R_1=200 Ω, R_2=100 kΩ, and R_3=100 kΩ, and a V_{MID} of 2.5 V, an offset of 3.175 V (relative to ground) appears at the output of the nulling op amp when the input is set to V_{MID}. What is the input offset of the DUT amplifier?

3.19. For the nulling amplifier setup shown in Figure 3.20 with R_1=100 Ω, R_2=300 kΩ, and R_3=100 kΩ, and the DUT op amp having an open-loop gain of 1000 V/V, what is the output swing of the nulling amplifier when the input swings by 1 V?

3.20. The input of a ×10 amplifier is connected to a voltage source forcing 1.75 V. The power supply is set to 4.9 V and a voltage of 1.700 V is measured at the output of the amplifier. The power supply voltage is then changed to 5.1 V and the output measurement changes to 1.708 V. What is the PSS? What is the PSRR if the measured gain is 9.8 V/V?

3.21. For nulling amplifier CMRR setup shown in Figure 3.23 with R_1=100 Ω, R_2=300 kΩ, and R_3=100 kΩ, SRC1 is set to +3.5 V and a differential voltage of 130 mV is measured between SRC1 and the output of the nulling amplifier. Then SRC1 is set to 1.0 V and the measured voltage changes to –260 mV. What is the CMRR of the op amp in decibels?

3.22. An amplifier has an expected CMRR of -85 dB. For a 1-V change in the input common-mode level, what is the expected change in the input offset voltage of this amplifier?

3.23. A comparator has an input offset voltage of 6 mV and its negative terminal is connected to a 2.5-V level, at what voltage on the positive terminal does the comparator change state?

3.24. A slicer circuit is connected to a 2-V reference and has a threshold voltage error of 20 mV, at what voltage level will the slicer change state?

3.25. If a slicer's 2.5-V reference has an error of +100 mV and the comparator has an input offset of –10 mV, what threshold voltage should we expect?

3.26. A comparator has a measured hysteresis of 10 mV and switches state on a rising input at 2.5 V. At what voltage does the comparator change to a low state on a falling input?

3.27. For an amplifier characterized by $V_{OUT} = 6V_{IN} + 0.5V_{IN}^2 - 2$ over a ±1-V input voltage range, determine the input offset voltage using a linear search process, starting with two points at ±1 V. After how many iterations did the answer change by less than 1 mV? How many iterations would have been required using a binary search from –1 to +1 V?

References

1. Sreejit Chakravarty, Paul J. Thadikaran, *Introduction to IDDQ Testing*, May, 1997, Kluwer Academic Publishers, Boston, MA, ISBN: 0792399455

2. Analog Devices application note*, *How to Test Basic Operational Amplifier Parameters*, Analog Devices, Inc., Norwood, MA, July, 1982

---

* The nulling amplifier/servo loop methods presented in this chapter were adapted from the referenced application note to allow compatiblity with single-supply op amps having a V_{MID} reference voltage. The technique has been presented with permission from Analog Devices, Inc.

CHAPTER 4

Measurement Accuracy

4.1 TERMINOLOGY

4.1.1 Accuracy and Precision

In conversational English, the terms *accuracy* and *precision* are virtually identical in meaning. Roget's Thesaurus[1] lists these words as synonyms and Webster's Dictionary[2] gives almost identical definitions for them. However, these terms are defined very differently in engineering textbooks[3-5]. Combining the definitions from these and other sources gives us an idea of the accepted technical meaning of the words:

> Accuracy – The difference between the average of measurements and a standard sample for which the "true" value is known. The degree of conformance of a test instrument to absolute standards, usually expressed as a percentage of reading or a percentage of measurement range (full scale).

> Precision – The variation of a measurement system obtained by repeating measurements on the same sample back-to-back using the same measurement conditions.

According to these definitions, precision refers only to the repeatability of a series of measurements. It does not refer to consistent errors in the measurements. A series of measurements can be incorrect by 2 V, but as long as they are consistently wrong by the same amount, then the measurements are considered to be precise.

This definition of precision is somewhat counterintuitive to most people, since the words *precision* and *accuracy* are so often used synonymously. Few of us would be impressed by a "precision" voltmeter exhibiting a consistent 2-V error! Fortunately, the word *repeatability* is far more commonly used in the test engineering field than the word *precision*. This textbook will use the term *accuracy* to refer to the overall closeness of an averaged measurement to the true value and *repeatability* to refer to the consistency with which that measurement can be made. The word *precision* will be avoided.

Unfortunately, the definition of accuracy is also somewhat ambiguous. Many sources of error can affect the accuracy of a given measurement. The accuracy of a measurement should probably refer to all possible sources of error. However, the accuracy of an instrument (as distinguished from the accuracy of a measurement) is often specified in the absence of repeatability fluctuations and instrument resolution limitations. Rather than trying to decide which of the various error sources are included in the definition of accuracy, it is probably more useful to discuss some of the common error components that contribute to measurement inaccuracy. It is incumbent upon the test engineer to make sure all components of error have been accounted for in a given specification of accuracy.

4.1.2 Systematic Errors

Systematic errors are those that show up consistently from measurement to measurement. For example, assume an amplifier's output exhibits an offset of 100 mV from the ideal value of 0 V. Using a digital voltmeter (DVM) we could take multiple readings of the offset over time and record each measurement. A typical measurement series might look like this:

101 mV, 103 mV, 102 mV, 101 mV, 102 mV, 103 mV, 103 mV, 101 mV, 102 mV...

This measurement series shows an average error of about 2 mV from the true value of 100 mV. Errors like this are caused by consistent errors in the measurement instruments. The errors can result from a combination of many things, including DC offsets, gain errors, and nonideal linearity in the DVM's measurement circuits. Systematic errors can often be reduced through a process called *calibration*. Various types of calibration will be discussed in more detail in Section 4.2.

4.1.3 Random Errors

Notice in the preceding example that the measurements are not repeatable. The DVM gives readings from 101 to 103 mV. Such variations do not surprise most engineers because DVMs are relatively inexpensive. On the other hand, when a two million dollar piece of ATE equipment cannot produce the same answer twice in a row, eyebrows may be raised.

Inexperienced test engineers are sometimes surprised to learn that an expensive tester cannot give perfectly repeatable answers. They may be inclined to believe that the tester software is defective when it fails to produce the same result every time the program is executed. However, experienced test engineers recognize that a certain amount of random error is to be expected in analog and mixed-signal measurements.

Random errors are usually caused by thermal noise or other noise sources in either the DUT or the tester hardware. One of the biggest challenges in mixed-signal testing is determining whether the random errors are caused by bad DIB design, by bad DUT design, or by the tester itself. If the source of error is found and cannot be corrected by a design change, then averaging or filtering of measurements may be required. Averaging and filtering are discussed in more detail in Section 4.3.

4.1.4 Resolution (Quantization Error)

In the 100-mV measurement list, notice that the measurements are always rounded off to the nearest millivolt. The measurement may have been rounded off by the person taking the measurements, or perhaps the DVM was only capable of displaying three digits. ATE measurement instruments have similar limitations in measurement resolution. Limited resolution results from the fact that continuous analog signals must first be converted into a digital format before the ATE computer can evaluate the test results. The tester converts analog signals into digital form using analog-to-digital converters (ADCs).

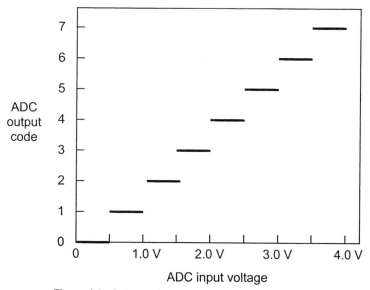

Figure 4.1. Output codes versus input voltages for an ideal 3-bit ADC.

ADCs by nature exhibit a feature called *quantization error*. Quantization error is a result of the conversion from an infinitely variable input voltage (or current) to a finite set of possible digital output results from the ADC. Figure 4.1 shows the relationship between input voltages and output codes for an ideal 3-bit ADC. Notice that an input voltage of 1.2 V results in the same ADC output code as an input voltage of 1.3 V. In fact, any voltage from 1.0 to 1.5 V will produce an output code of 2.

If this ADC were part of a crude DC voltmeter, the meter would produce an output reading of 1.25 V any time the input voltage falls between 1.0 and 1.5 V. This inherent error in ADCs and measurement instruments is caused by quantization error. The resolution of a DC meter is often limited by the quantization error of its ADC circuits.

If a meter has 12 bits of resolution, that means it can resolve a voltage to one part in $2^{12}-1$ (one part in 4095). If the meter's full-scale range is set to ± 2 V, then a resolution of approximately 1 mV can be achieved (4 V / 4095 levels). This does not automatically mean that the meter is accurate to 1 mV, it simply means the meter cannot resolve variations in input voltage smaller than 1 mV. An instrument's resolution can far exceed its accuracy. For example, a 23-bit voltmeter might be able to produce a measurement with a 1-μV resolution, but it may have a systematic error of 2 mV.

4.1.5 Repeatability

Nonrepeatable answers are a fact of life for mixed-signal test engineers. A large portion of the time required to debug a mixed-signal test program can be spent tracking down the various sources of poor repeatability. Since all electrical circuits generate a certain amount of random noise, measurements such as those in the 100-mV offset example are fairly common. In fact, if a test engineer gets the same answer 10 times in a row, it is time to start looking for a problem. Most likely, the tester instrument's full-scale voltage range has been set too high, resulting in a measurement resolution problem. For example, if we configured a meter to a range having a 10-mV resolution, then our measurements from the prior example would be very repeatable (100 mV, 100 mV, 100 mV, 100 mV, etc.). A novice test engineer might think this is a terrific result, but the meter is just rounding off the answer to the nearest 10-mV increment due to an input ranging problem. Unfortunately, a voltage of 104 mV would also have resulted in this same series of perfectly repeatable, perfectly incorrect measurement results. Repeatability is desirable, but it does not in itself guarantee accuracy.

Exercises

4.1. A 5-mV signal is measured with a meter ten times resulting in the following sequence of readings: 5 mV, 6 mV, 9 mV, 8 mV, 4 mV, 7 mV, 5 mV, 7 mV, 8 mV, 11 mV. What is the average measured value? What is the systematic error?

Ans. 7 mV, 2 mV.

4.2. A meter is rated at 8-bits and has a full-scale range of ±5 V. What is the measurement uncertainty of this meter, assuming only quantization errors from an ideal meter ADC?

Ans. ± 19.5 mV.

4.3. A signal is to be measured with a maximum uncertainty of ±0.5 µV. How many bits of resolution are required by an ideal meter having a ±1 V full-scale range?

Ans. 21 bits.

4.1.6 Stability

A measurement instrument's performance may drift with time, temperature, and humidity. The degree to which a series of supposedly identical measurements remains constant over time, temperature, humidity, and all other time-varying factors is referred to as *stability*. Stability is an essential requirement for accurate instrumentation.

Shifts in the electrical performance of measurement circuits can lead to errors in the tested results. Most shifts in performance are caused by temperature variations. Testers are usually equipped with temperature sensors that can automatically determine when a temperature shift has occurred. The tester must be recalibrated anytime the ambient temperature has shifted by a few degrees. The calibration process brings the tester instruments back into alignment with known electrical standards so that measurement accuracy can be maintained at all times.

After the tester is powered up, the tester's circuits must be allowed to stabilize to a constant temperature before calibrations can occur. Otherwise, the measurements will drift over time as the tester heats up. When the tester chassis is opened for maintenance or when the test head is opened up or powered down for an extended period, the temperature of the measurement electronics will typically drop. Calibrations then have to be rerun once the tester recovers to a stable temperature.

Shifts in performance can also be caused by aging electrical components. These changes are typically much slower than shifts due to temperature. The same calibration processes used to account for temperature shifts can easily accommodate shifts of components caused by aging. Shifts caused by humidity are less common, but can also be compensated for by periodic calibrations.

4.1.7 Correlation

Correlation is another activity that consumes a great deal of mixed-signal test program debug time. Correlation is the ability to get the same answer using different pieces of hardware or software. It can be extremely frustrating to try to get the same answer on two different pieces of equipment using two different test programs. It can be even more frustrating when two supposedly identical pieces of test equipment running the same program give two different answers.

Of course correlation is seldom perfect, but how good is good enough? In general, it is a good idea to make sure the correlation errors are less than one-tenth of the full range between the minimum test limit and the maximum test limit. However, this is just a rule of thumb. The exact requirements will differ from one test to the next. Whatever correlation errors exist, they must be considered part of the measurement uncertainty, along with nonrepeatability and systematic errors.

The test engineer must consider several categories of correlation. Test results from a mixed-signal test program cannot be fully trusted until the various types of correlation have been verified. The more common types of correlation include tester-to-bench, tester-to-tester, program-to-program, DIB-to-DIB, and day-to-day correlation.

Tester-to-Bench Correlation

Often, a customer will construct a test fixture using bench instruments to evaluate the quality of the device under test. Bench equipment such as oscilloscopes and spectrum analyzers can help validate the accuracy of the ATE tester's measurements. Bench correlation is a good idea, since ATE testers and test programs often produce incorrect results in the early stages of debug. In addition, IC design engineers often build their own evaluation test setups to allow quick debug of device problems. Each of these test setups must correlate to the answers given by the ATE tester. Often the tester is correct and the bench is not. Other times, test program problems are uncovered when the ATE results do not agree with a bench setup. The test engineer will often need to help debug the bench setup to get to the bottom of correlation errors between the tester and the bench.

Tester-to-Tester Correlation

Sometimes a test program will work on one tester, but not on another presumably identical tester. The differences between testers may be catastrophically different, or they may be very subtle. The test engineer should compare all the test results on one tester to the test results obtained using other testers. Only after all the testers agree on all tests is the test program and test hardware debugged and ready for production.

Similar correlation problems arise when an existing test program is ported from one tester type to another. Often, the testers are neither software compatible nor hardware compatible with one another. In fact, the two testers may not even be manufactured by the same ATE vendor. A myriad of correlation problems can arise because of the vast differences in DIB layout and tester software between different tester types. To some extent, the architecture of each tester will determine the best test methodology for a particular measurement. A given test may have to be executed in a very different manner on one tester versus another. Any difference in the way a measurement is taken can affect the results. For this reason, correlation between two different test approaches can be very difficult to achieve. Conversion of a test program from one type of tester to another can be one of the most daunting tasks a mixed-signal test engineer faces.

Program-to-Program Correlation

When a test program is streamlined to reduce test time, the faster program must be correlated to the original program to make sure no significant shifts in measurement results have occurred. Often, the test reduction techniques cause measurement errors because of reduced DUT settling time and other timing-related issues. These correlation errors must be resolved before the faster program can be released into production.

DIB-to-DIB Correlation

No two DIBs are identical, and sometimes the differences cause correlation errors. The test engineer should always check to make sure that the answers obtained on multiple DIB boards agree. DIB correlation errors can often be corrected by *focused calibration* software written by the test engineer (this will be discussed further in Section 4.2 and in Chapter 10, "Focused Calibrations").

Day-to-Day Correlation

Correlation of the same DIB and tester over a period of time is also important. If the tester and DIB have been properly calibrated, there should be no drift in the answers from one day to the next. Subtle errors in software and hardware often remain hidden until day-to-day correlation is performed. The usual solution to this type of correlation problem is to improve the focused calibration process.

4.1.8 Reproducibility

The term *reproducibility* is often used interchangeably with *repeatability*, but this is not a correct usage of the term. The difference between reproducibility and repeatability relates to the effects of correlation and stability on a series of supposedly identical measurements. Repeatability is most often used to describe the ability of a single tester and DIB board to get the same answer multiple times as the test program is repetitively executed.

Reproducibility, by contrast, is the ability to achieve the same measurement result on a given DUT using any combination of equipment and personnel at any given time. It is defined as the statistical deviation of a series of supposedly identical measurements taken over a period of time. These measurements are taken using various combinations of test conditions that ideally should not change the measurement result. For example, the choice of equipment operator, tester, DIB board, etc., should not affect any measurement result.

Consider the case in which a measurement is highly repeatable, but not reproducible. In such a case, the test program may consistently pass a particular DUT on a given day, and yet consistently fail the same DUT on another day or on another tester. Clearly, measurements must be both repeatable and reproducible to be production-worthy.

4.2 CALIBRATIONS AND CHECKERS

4.2.1 Traceability to Standards

Every tester and bench instrument must ultimately correlate to standards maintained by a central authority, such as the National Institute of Standards and Technology (NIST). In the United States, this government agency is responsible for maintaining the standards for pounds, gallons, inches, and electrical units such as volts, amperes, and ohms. The chain of correlation between the NIST and the tester's measurements involves a series of calibration steps that transfers the "golden" standards of the NIST to the tester's measurement instruments.

Many testers have a centralized standards reference, which is a thermally stabilized instrument in the tester mainframe. The standards reference is periodically replaced by a freshly calibrated reference source. The old one is sent back to a certified calibration laboratory, which recalibrates the reference so that it agrees with NIST standards. Similarly, bench instruments are periodically recalibrated so that they too are traceable to the NIST standards. By periodically refreshing the tester's traceability link to the NIST, all testers and bench instruments can be made to agree with one another.

4.2.2 Hardware Calibration

Hardware calibration is a process of physical "knob tweaking" that brings a piece of measurement instrumentation back into agreement with calibration standards. For instance, oscilloscope probes often include a small screw that can be used to nullify the overshoot in rapidly rising digital edges. This is one common example of hardware calibration.

One major problem with hardware calibration is that it is not a convenient process. It generally requires a manual adjustment of a screw or knob. Robotic screwdrivers might be employed to allow partial automation of the hardware calibration process. However, the use of robotics is an elaborate solution to the calibration problem. Full automation can be achieved using a simpler procedure known as *software calibration*.

4.2.3 Software Calibration

Using software calibration, ATE testers are able to correct hardware errors without adjusting any physical knobs. The basic idea behind software calibration is to separate the instrument's ideal operation from its nonidealities. Then a *model* of the instrument's nonideal operation can be constructed, followed by a *correction* of the nonideal behavior using a mathematical routine written in software. Figure 4.2 illustrates this idea for a voltmeter.

In part (a) a "real" voltmeter is modeled as a cascade of two parts: (1) an ideal voltmeter, and (2) a black box that relates the voltage across its input terminals v_{DUT} to the voltage that is measured by the ideal voltmeter, $v_{measured}$. This relationship can be expressed in more mathematical terms as

$$v_{measured} = f(v_{DUT}) \quad (4.1)$$

where $f(\cdot)$ indicates the functional relationship between $v_{measured}$ and v_{DUT}.

The true functional behavior $f(\cdot)$ is seldom known; so one assumes a particular behavior or model, such as a first-order model given by

$$v_{measured} = G v_{DUT} + \textit{offset} \qquad (4.2)$$

where G and *offset* are the gain and offset of the voltmeter, respectively. These values must be determined from measured data. Subsequently, a mathematical procedure is written in software that performs the inverse mathematical operation

$$v_{calibrated} = f^{-1}(v_{measured}) \qquad (4.3)$$

where $v_{calibrated}$ replaces v_{DUT} as an estimate of the true voltage that appears across the terminals of the voltmeter as depicted in Figure 4.2(b). If $f(\cdot)$ is known precisely, then $v_{calibrated} = v_{DUT}$.

In order to establish an accurate model of an instrument, precise reference levels are necessary. The number of reference levels required to characterize the model fully will depend on its order, that is, the number of parameters used to describe the model. For the linear or first-order model described, it has two parameters, G and *offset*. Hence, two reference levels will be required.

To avoid conflict with the meter's normal operation, relays are used to switch in these reference levels during the calibration phase. For example, the voltmeter in Figure 4.3 includes a pair of calibration relays, which can connect the input to two separate reference levels, V_{ref1} and V_{ref2}. During a system level calibration, the tester closes one relay and connects the voltmeter to

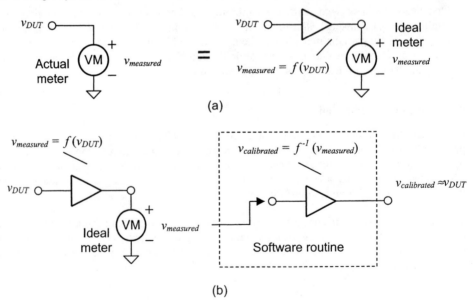

Figure 4.2. (a) Modeling a voltmeter with an ideal voltmeter and a nonideal component in cascade. (b) Calibrating the nonideal effects using a software routine.

V_{ref1} and measures the voltage, which we shall denote as $v_{measured1}$. Subsequently, this process is repeated for the second reference level V_{ref2} and the voltmeter provides a second reading, $v_{measured2}$.

Based on the assumed linear model for the voltmeter, we can write two equations in terms of two unknowns

$$v_{measured1} = GV_{ref1} + offset$$
$$v_{measured2} = GV_{ref2} + offset \quad (4.4)$$

Using linear algebra (Gauss-Jordan elimination method), the two model parameters can then be solved to be

$$G = \frac{v_{measured2} - v_{measured1}}{V_{ref2} - V_{ref1}} \quad (4.5)$$

and

$$offset = \frac{v_{measured1} V_{ref2} - v_{measured2} V_{ref1}}{V_{ref2} - V_{ref1}} \quad (4.6)$$

The parameters of the model, G and $offset$, are also known as *calibration factors*, or cal factors for short.

When subsequent DC measurements are performed, they are corrected using the stored calibration factors according to

$$v_{calibrated} = \frac{v_{measured} - offset}{G} \quad (4.7)$$

This expression is found by isolating v_{DUT} on one side of the expression in Eq. (4.2) and replacing it by $v_{calibrated}$.

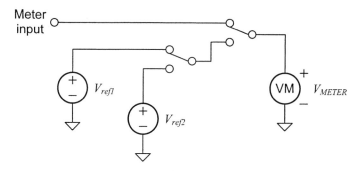

Figure 4.3. DC voltmeter gain and offset calibration paths.

Of course, this example is only for purposes of illustration. Most testers use much more elaborate calibration schemes to account for linearity errors and other nonideal behavior in the meter's ADC and associated circuits.. Also, the meter's input stage can be configured many ways, and each of these possible configurations needs a separate set of calibration factors. For example, if the input stage has ten different input ranges, then each range setting requires a separate set of calibration factors. Fortunately for the test engineer, most instrument calibrations happen behind the scenes. The calibration factors are measured and stored automatically during the tester's periodic system calibration and checker process.

Exercises

4.4. A meter reads 0.5 mV and 1.1 V when connected to two precision reference levels of 0 and 1 V, respectively. What are the offset and gain of this meter? Write the calibration equation for this meter.

Ans. 0.5 mV, 1.0995 V/V, $v_{calibrated} = (v_{measured} - 0.5\ mV)/1.0995$.

4.5. A meter is assumed characterized by a second-order equation of the form: $v_{measured} = \textit{offset} + G_1 v_{calibrated} + G_2 v_{calibrated}^2$. How many precision DC reference levels are required to obtain the parameters of this second-order expression?

Ans. Three.

4.6. A meter is assumed characterized by a second-order equation of the form: $v_{measured} = \textit{offset} + G_1 v_{calibrated} + G_2 v_{calibrated}^2$. Write the calibration equation for this meter in terms of the unknown calibration factors.

Ans. $v_{calibrated} = \dfrac{-G_1 + \sqrt{G_1^2 + 4G_2 v_{measured}}}{2G_2}$ or $v_{calibrated} = \dfrac{-G_1 - \sqrt{G_1^2 + 4G_2 v_{measured}}}{2G_2}$

depending on the data conditions.

4.2.4 System Calibrations and Checkers

Testers are calibrated on a regular basis to maintain traceability of each instrument to the tester's calibration reference source. In addition to calibrations, software is also executed to verify the functionality of hardware and make sure it is production worthy. This software is called a *checker program*, or *checker* for short. Often calibrations and checkers are executed in the same program. If a checker fails, the repair and maintenance (R&M) staff replaces the failing tester module with a good one. After replacement, the new module must be completely recalibrated.

There are several types of calibrations and checkers. These include calibration reference source replacement, performance verification (PV), periodic system calibrations and checkers, instrument calibrations at load time, and focused calibrations. Calibration reference source replacement and recalibration was discussed in Section 4.2.1. A common replacement cycle time for calibration sources is once every six months.

To verify that the tester is in compliance with all its published specifications, a more extensive process called *performance verification* may be performed. Although full performance verification is typically performed at the tester vendor's production floor, it is seldom performed on the production floor. By contrast, periodic system calibrations and checkers are performed on a regular basis in a production environment. These software calibration and checker programs verify that all the system hardware is production worthy.

Since tester instrumentation may drift slightly between system calibrations, the tester may also perform a series of fine-tuning calibrations each time a new test program is loaded. The extra calibrations can be limited to the subset of instruments used in a particular test program. This helps to minimize program load time. To maintain accuracy throughout the day, these calibrations may be repeated on a periodic basis after the program has been loaded. They may also be executed automatically if the tester temperature drifts by more than a few degrees.

Finally, focused calibrations are often required to achieve maximum accuracy and to compensate for nonidealities of DIB board components such as buffer amplifiers and filters. Unlike the ATE tester's built-in system calibrations, focused calibration and checker software is the responsibility of the test engineer. Focused calibrations fall into two categories: (1) focused instrument calibrations and (2) focused DIB calibrations and checkers.

4.2.5 Focused Instrument Calibrations

Testers typically contain a combination of slow, accurate instruments and fast instruments that may be less accurate. The accuracy of the faster instruments can be improved by periodically referencing them back to the slower more accurate instruments through a process called *focused calibration*. Focused calibration is not always necessary. However, it may be required if the test engineer needs higher accuracy than the instrument is able to provide using the built-in calibrations of the tester's operating system.

A simple example of focused instrument calibration is a DC source calibration. The DC sources in a tester are generally quite accurate, but occasionally they need to be set with minimal DC level error. A calibration routine that determines the error in a DC source's output level can be added to the first run of the test program. A high-accuracy DC voltmeter can be used to measure the actual output of the DC source. If the source is in error by 1 mV, for instance, then the requested voltage is reduced by 1 mV and the output is retested. It may take several iterations to achieve the desired value with an acceptable level of accuracy.

A similar approach can be extended to the generation of a sinusoidal signal requiring an accurate RMS value from an arbitrary waveform generator (AWG). A high-accuracy AC voltmeter is used to measure the RMS value from the AWG. The discrepancy between the measured value and the desired value is then used to adjust the programmed AWG signal level. The AWG output level will thus converge toward the desired RMS level as each iteration is executed.

Example 4.1

A 2.500-V signal is required from a DC source as shown in Figure 4.4. Describe a calibration procedure that can be used to ensure that 2.500 V ± 500 µV does indeed appear at the output of the DC source.

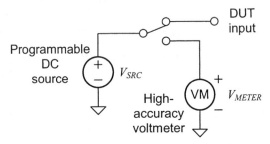

Figure 4.4. DC source focused calibration.

Solution:

The source is set to 2.500 V and a high-accuracy voltmeter is connected to the output of the source using a calibration path internal to the tester. Calibration path connections are made through one or more relays such as the ones in Figure 4.3. Assume the high-accuracy voltmeter reads 2.510 V from the source. The source is then reprogrammed to 2.500 V - 10 mV and the output is remeasured. If the second meter reading is 2.499 V, then the source is reprogrammed to 2.500 V - 10 mV + 1 mV and measured again. This process is repeated until the meter reads 2.500 V (plus or minus 500 µV). Once the exact programmed level is established, it is stored as a calibration factor (e.g., calibration factor = 2.500 V - 10 mV + 1 mV = 2.491 V). When the 2.500-V DC level is required during subsequent program executions, the 2.491-V calibration factor is used as the programmed level rather than 2.500 V. Test time is not wasted searching for the ideal level after the first calibration is performed. However, calibration factors may need to be regenerated every few hours to account for slow drifts in the DC source. This recalibration interval is dependent on the type of tester used.

Another application of focused instrument calibration is spectral leveling of the output of an AWG. An important application of AWGs is to provide a composite signal consisting of N sine waves or *tones* all having equal amplitude at various frequencies and arbitrary phase. Such waveforms are in a class of signals commonly referred to as *multitone* signals. Mathematically a multitone signal $y(t)$ can be written as

$$y(t) = A_0 + A_1 \sin(2\pi f_1 t + \phi_1) + \cdots + A_N \sin(2\pi f_N t + \phi_N)$$
$$= A_0 + \sum_{k=1}^{N} A_k \sin(2\pi f_k t + \phi_k)$$

(4.8)

where A_k, f_k, and ϕ_k denotes the amplitude, frequency, and phase, respectively, of the kth tone. A multitone signal can be viewed in either the time domain or in the frequency domain. Time-domain views are analogous to oscilloscope traces, while frequency-domain views are analogous to spectrum analyzer plots. The frequency-domain graph of a multitone signal contains a series of vertical lines corresponding to each tone frequency and whose length[*] represents the root-

---

[*] Spectral density plots are commonly defined in engineering textbooks with the length of the spectral line representing one-half the amplitude of a tone. In most test engineering work, including spectrum analyzer displays, it is more common to find this length defined as an RMS quantity.

mean-square (RMS) amplitude of the corresponding tone. Each line is referred to as a *spectral line*. Figure 4.5 illustrates the time and frequency plots of a composite signal consisting of three tones of frequencies 1, 2.5 and 4.1 kHz, all having an RMS amplitude of 2 V. Of course, the peak amplitude of each sinusoid in the multitone is simply $\sqrt{2} \times 2$ or 2.82 V, so we could just as easily plot these values as peak amplitudes rather than RMS. This book will consistently display frequency-domain plots using RMS amplitudes.

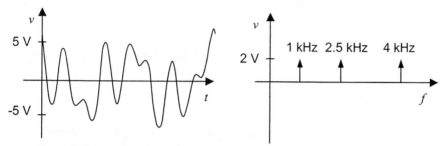

Figure 4.5. Time-domain and frequency-domain views of a three-tone multitone.

The AWG produces its output signal by passing the output of a DAC through a low-pass anti-imaging filter. Due to its frequency behavior, the filter will not have a perfectly flat magnitude response. The DAC may also introduce frequency-dependent errors. Thus the amplitudes of the individual tones may be offset from their desired levels. We can therefore model this AWG multitone situation as illustrated in Figure 4.6. The model consists of an ideal source connected in cascade with a linear block whose gain or magnitude response is described by $G(f)$, where f is the frequency expressed in Hz. To correct for the gain change with frequency, the amplitude of each tone from the AWG is measured individually using a high-accuracy AC voltmeter. The ratio between the actual output and the requested output corresponds to $G(f)$ at that frequency. This gain can then be stored as a calibration factor that can subsequently be retrieved to correct the amplitude error at that frequency. The calibration process is repeated for each tone in the multitone signal. The composite signal can then be generated with corrected amplitudes by dividing the previous requested amplitude at each frequency by the corresponding AWG gain calibration factor. Because the calibration process equalizes the amplitudes of each tone, the process is called *multitone leveling*.

As testers continue to evolve and improve, it may become increasingly unnecessary for the test engineer to perform focused calibrations of the tester instruments. Focused calibrations were once necessary on almost all tests in a test program. Today, they can sometimes be omitted with little degradation in accuracy. Nevertheless, the test engineer must evaluate the need for focused

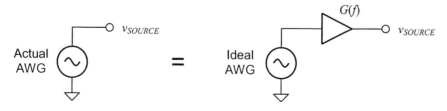

Figure 4.6. Modeling an AWG as a cascaded combination of an ideal source and frequency-dependent gain block.

calibrations on each test. Even if calibrations become unnecessary in the future, the test engineer should still understand the methodology so that test programs on older equipment can be comprehended.

Calibration of circuits on the DIB, on the other hand, will probably always be required. The tester vendor has no way to predict what kind of buffer amplifiers and other circuits will be placed on the DIB board. The tester operating system will never be able to provide automatic calibration of these circuits. The test engineer is fully responsible for understanding the calibration requirements of all DIB circuits.

Example 4.2

A multitone signal consisting of three tones at 1.0, 2.5, and 4.1 kHz is desired from an AWG. Each tone should have exact RMS amplitude of 2.0 V, corresponding to a peak amplitude of $\sqrt{2}\times 2.0$ V. This multitone should have 0 DC offset. Using Eq. (4.8), a three-tone signal is mathematically created with parameters $A_0 = 0$, $A_1 = A_2 = A_3 = \sqrt{2}\times 2$, $f_1 = 1$ kHz, $f_2 = 2.5$ kHz, $f_3 = 4.1$ kHz and is written as

$$y(t) = 2\sqrt{2}\sin(2\pi\times 1\text{ kHz}\times t) + 2\sqrt{2}\sin(2\pi\times 2.5\text{ kHz}\times t) + 2\sqrt{2}\sin(2\pi\times 4.1\text{ kHz}\times t)$$

Sequentially, beginning with the lowest-frequency tone and progressing up in frequency, each tone is loaded into the AWG and the sine wave is passed from the AWG into a high-accuracy AC RMS voltmeter. For each tone, the voltmeter reads: 1.980, 2.023 and 1.950 V. Compute the calibration factors and provide a formula that describes the modified three-tone signal.

Solution:

Three calibration factors are calculated as the ratio of the measured signal to the desired signal

$$cal_1 = G(1\text{ kHz}) = \frac{1.980\text{ V}}{2.0\text{ V}} = 0.99\text{ V/V}$$

$$cal_2 = G(2.5\text{ kHz}) = \frac{2.023\text{ V}}{2.0\text{ V}} = 1.012\text{ V/V}$$

$$cal_3 = G(4.1\text{ kHz}) = \frac{1.950\text{ V}}{2.0\text{ V}} = 0.975\text{ V/V}$$

As long as the AWG is linear, it should be possible to get exactly 2.0 V at each tone by asking for 2.0 V divided by the appropriate calibration factor. The three-tone signal is thus created using the equation

$$y(t) = \frac{2\sqrt{2}}{cal_1}\sin(2\pi\times 1\text{ kHz}\times t) + \frac{2\sqrt{2}}{cal_2}\sin(2\pi\times 2.5\text{ kHz}\times t) + \frac{2\sqrt{2}}{cal_3}\sin(2\pi\times 4.1\text{ kHz}\times t)$$

This waveform is loaded into the AWG and the three-tone signal is produced with equal levels of 2.0 V RMS per tone.

4.2.6 Focused DIB Circuit Calibrations

Often circuits are added to a DIB board to improve the accuracy of a particular test or to buffer the weak output of a device before sending it to the tester electronics. As the signal-conditioning DIB circuitry is added in cascade with the test instrument, a model of the test setup is identical to that given in Figure 4.2(a). The only difference is that functional block $v_{measured} = f(v_{DUT})$ includes both the meter and the DIB's behavior. As a result, the focused instrument calibrations of Section 4.2.3 can be used with no modifications. Conversely, the meter may already have been calibrated so that the functional block $f(\cdot)$ covers the DIB circuitry only. One must keep track of the extent of the calibration to avoid any double counting.

Example 4.3

The op amp in Figure 4.7 has been added to a DIB board to buffer an output of a DUT. The buffer will be used to condition the DC signal from the DUT before sending it to a calibrated DC voltmeter resident in the tester. If the output is not buffered, then we may find that the DUT breaks into oscillations as a result of the stray capacitance arising along the lengthy signal path leading from the DUT to the tester. The buffer prevents these oscillations by substantially reducing stray capacitance at the DUT output. In order to perform an accurate measurement, the behavior of the buffer must be accounted for. Outline the steps to perform a focused DC calibration on the op amp buffer stage.

Solution:

To perform a DC calibration of the output buffer amplifier it is necessary to assume a model for the op amp buffer stage. It is reasonable to assume that the buffer is fairly linear over a wide range of signal levels, so that the following linear model can be used

$$v_{measured} = G v_{DUT} + \textit{offset}$$

Subsequently, following the same procedure as outlined in Section 4.2.3, a pair of known voltages are applied to the input of the buffer from source SRC1 via the relay connection and the output of the buffer is measured with a voltmeter. This temporary connection is called a *calibration path*. As an example, let SRC1 force 2 V and assume that an output voltage of 2.023 V is measured using the voltmeter. Next the input is dropped to 1 V, resulting in an output

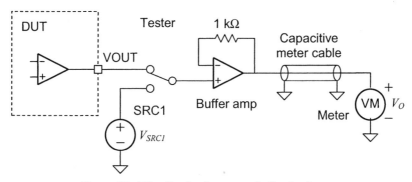

Figure 4.7. DC calibration for op amp buffer circuit.

voltage of 1.012 V. Using Eq. (4.5), we find the buffer has gain given by

$$G = \frac{2.023 \text{ V} - 1.012 \text{ V}}{2 \text{ V} - 1 \text{ V}} = 1.011 \text{ V/V}$$

and the offset is found from Eq. (4.6) to be

$$offset = \frac{1.012 \text{ V} \cdot 2 \text{ V} - 2.023 \text{ V} \times 1 \text{ V}}{2 \text{ V} - 1 \text{ V}} = 1 \text{ mV}$$

Hence, the DUT output v_{DUT} and the voltmeter value $v_{measured}$ are related according to

$$v_{measured} = 1.011 \text{ V/V} \times v_{DUT} + 0.001 \text{ V}$$

The goal of the focused DC calibration procedure is to find an expression that relates the DUT output in terms of the measured value. Hence, by rearranging the expression and replacing $v_{calibrated}$ for v_{DUT}, we obtain

$$v_{calibrated} = \frac{v_{measured} - 0.001 \text{ V}}{1.011 \text{ V/V}}$$

For example, if the voltmeter reads 1.732 V, the actual voltage appearing at its terminals is actually

$$v_{calibrated} = \frac{1.732 \text{ V} - 0.001 \text{ V}}{1.011 \text{ V/V}} = 1.712 \text{ V}$$

If the original uncalibrated answer had been used, there would have been a 20-mV error! This example shows why focused DUT calibrations are so important to accurate measurements.

When buffer amplifiers are used to assist the measurement of AC signals, a similar calibration process must be performed on each frequency that is to be measured. Like the AWG calibration example, the buffer amplifier also has a nonideal frequency response and will affect the reading of the meter. Its gain variation, together with the meter's frequency response, must be measured at each frequency used in the test during a calibration run of the test program. Assuming that the meter has already been calibrated, the frequency response behavior of the DIB circuitry must be correctly accounted for. This is achieved by measuring the gain in the DIB's signal path at each specific test frequency. Once found, it is stored as a calibration factor. If additional circuits such as filters, ADCs, etc., are added on the DIB board and used under multiple configurations, then each unique signal path must be individually calibrated. Chapter 10, "Focused Calibrations," will address these and other issues in greater depth.

4.2.7 DIB Checkers

In addition to focused DIB calibrations, the test program should also include DIB checkers to verify the basic operation of as many DIB circuits and signal paths as possible. DIB failures can be a major source of downtime on a production floor unless thorough checkers are available to quickly diagnose DIB hardware failures. The first run of the test program should not only calibrate the DIB circuits, but it should also perform a go/no-go test on as many of the DIB board components and signal paths as is possible. It is seldom possible to pass signals through every possible relay and every possible trace on the DIB board. However, every path and every component that can be tested with checker code should be verified.

A good example of a circuit in which a checker is useful is a relay path. While the gain through a relay seldom requires focused calibration, relays can become defective with age. They can also be welded shut by high currents or they can become stuck in the open state. The DIB checker code should verify that each accessible relay can be opened and then closed. The easiest way to do this is to apply a 1 V / -1 V voltage pair at the input to the relay and look for a 2-V swing at its output while the relay is closed. To verify the relay can be opened, the program should look for little or no output swing with the 1 V / -1 V input.

Exercises

4.7. A DC source is assumed characterized by a third-order equation of the form $V_{MEASURED} = 0.005 + V_{PROGRAMMED} - 0.003\, V^3_{PROGRAMMED}$ and is required to generate a DC level of 2.6 V. However, when programmed to produce this level, only 2.552 V is measured. Using iteration, determine a value of the programmed source voltage that will establish a measured voltage of 2.6 V to within a ± 1 mV accuracy.

Ans. 2.651 V.

4.8. An AWG has a gain response described by $\left(\sqrt{1+\left(f/10^3\right)^2}\right)^{-1}$ and is to generate three tones at frequencies of 1, 2, and 3 kHz. What are the calibration factors?

Ans. 0.707, 0.447, and 0.316.

4.2.8 Tester Specifications

The test engineer should exercise diligence when evaluating tester instrument specifications. It can be difficult to determine whether or not a particular tester instrument is capable of making a particular measurement with an acceptable level of accuracy. The tester specifications usually do not include the effects of uncertainty caused by instrument repeatability limitations. All the specification conditions must be examined carefully. Consider the following DC meter example.

A DC meter consisting of an analog-to-digital converter and a programmable gain amplifier (PGA) is shown in Figure 4.8. The programmable gain stage is used to set the range of the meter so that it can measure small signals as well as large ones. Small signals are measured with the highest gain setting of the PGA, while large signals are measured with the lowest gain setting. This ranging process effectively changes the resolution of the ADC so that its quantization error is minimized.

Calibration software in the tester compensates for the different PGA gain settings so that the digital output of the meter's ADC can be converted into an accurate voltage reading. The calibration software also compensates for linearity errors in the ADC and offsets in the PGA and

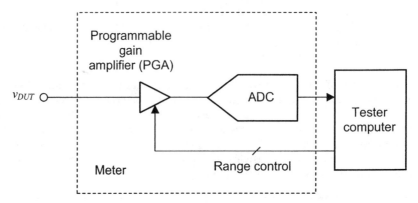

Figure 4.8. Simplified DC voltmeter with input ranging amplifier.

ADC. Fortunately, the test engineer does not have to worry about these calibrations because they happen automatically.

Table 4.1 shows an example of a specification for a fictitious DC meter, the DVM100. This meter has five different input ranges, which can be programmed in software. The different ranges allow small voltages to be measured with greater accuracy than large voltages. The accuracy is specified as a percentage of the measured value, but there is an accuracy limit of 1 mV for the lower ranges and 2.5 mV for the higher ranges.

This accuracy specification probably assumes that the measurement is made 100 or more times and averaged. For a single nonaveraged measurement, there may also be a repeatability error to consider. It is not clear from the table above what assumptions are made about averaging. The test engineer should make sure that all assumptions are understood before relying on the accuracy numbers.

Table 4.1. DVM100 DC Voltmeter Specifications

Range	Resolution	Accuracy (% of Measurement)
±0.5 V	15.25 µV	±0.05 % or 1 mV (whichever is greater)
±1 V	30.5 µV	±0.05 % or 1 mV
±2 V	61.0 µV	±0.05 % or 1 mV
±5 V	152.5 µV	±0.10 % or 2.5 mV
±10 V	305.2 mV	±0.10 % or 2.5 mV

Note: All specs apply with the measurement filter enabled.

Example 4.4

A DUT output is expected to be 100 mV. Our fictitious DC voltmeter, the DVM100, is set to the 0.5 V range to achieve the optimum resolution and accuracy. The reading from the meter (with the meter's input filter enabled) is 102.3 mV. Calculate the accuracy of this reading (excluding possible repeatability errors). What range of outputs could actually exist at the DUT output with this reading?

Solution:

The measurement error would be equal to ±0.05% of 100 mV, or 50 µV, but the specification has a lower limit of 1 mV. The accuracy is therefore ±1 mV. Based on the single reading of 102.3 mV, the actual voltage at the DUT output could be anywhere between 101.3 and 103.3 mV.

Exercises

4.9. A voltmeter is specified to have an accuracy of ±1% of programmed range. If a DC signal is measured on a ±1 V range, what are the minimum and maximum DC levels that might have been present at the meter's input during this measurement?

Ans. 0.5 V ±10 mV (i.e., the input could lie anywhere between 490 and 510 mV).

In addition to the ranging hardware, the meter also has a low-pass filter in series with its input. The filter can be bypassed or enabled, depending on the measurement requirements. Repeatability is enhanced when the low-pass filter is enabled, since the filter reduces electrical noise in the input signal. Without this filter the accuracy would be degraded by nonrepeatability. The filter undoubtedly adds settling time to the measurement, since all low-pass filters require time to stabilize to a final DC value. The test engineer must often choose between slow, repeatable measurements and fast measurements with less repeatability.

It may be possible to empirically determine through experimentation that this DC voltmeter has adequate resolution and accuracy to make a DC offset measurement with less than 100 µV of error. Since this level of accuracy is far better than the instrument's ±1 mV specifications, though, the instrument should probably not be trusted to make such a measurement in production. The accuracy might hold up for 100 days and then drift toward the specification limits of 1 mV on day 101.

Another possible scenario is that multiple testers may be used that do not all have 100-µV performance. Tester companies are often conservative in their published specifications, meaning that the instruments are often better than their specified accuracy limits. This is not a license to use the instruments to more demanding specifications. It is much safer to use the specifications as printed, since the vendor will not take any responsibility for use of instruments beyond their official specifications.

Sometimes the engineer may have to design front-end circuitry such as PGAs and filters onto the DIB board itself. The DIB circuits might be needed if the front-end circuitry of the meter is inadequate for a high-accuracy measurement. Front-end circuits may also be added if the signal from the DUT cannot be delivered cleanly through the signal paths to the tester instruments. Very high-impedance DUT signals might be susceptible to externally coupled noise, for example. Such signals might benefit from local buffering and amplification before passing to the tester instrument. The test engineer must calibrate any such buffering or filtering circuits using a focused DIB calibration.

4.3 DEALING WITH MEASUREMENT ERROR

4.3.1 Filtering

Analog filters are often used in tester hardware to remove unwanted signal components before measurement. A DC voltmeter may include a low-pass filter as part of its front end. The purpose of the filter is to remove all but the lowest-frequency components. It acts as a hardware averaging circuit to improve the repeatability of the measurement. More effective filtering is achieved using a filter with a low cutoff frequency, since a lower cutoff frequency excludes more electrical noise. Consequently, a lower frequency cutoff corresponds to better repeatability in the final measurement.

Unfortunately, it takes longer to measure a series of DC voltages with a low-pass filter in the signal path. Since the filter has a settling time that is inversely proportional to the cutoff frequency, though, a lower cutoff frequency adds extra test time while the filter settles to a stable DC level. Thus, there is an inherent tradeoff between repeatability and test time. The following two examples will quantify this tradeoff for a first-order system.

Example 4.5

The simple *RC* low-pass circuit shown in Figure 4.9 is used to filter the output of a DUT containing a noisy DC signal. For a particular measurement, the signal component is assumed to change from 0 to 1 V, instantaneously. How long does it take the filter to settle to within 1% of its final value? By what factor does the settling time increase when the filter's 3-dB bandwidth is decreased by a factor of 10?

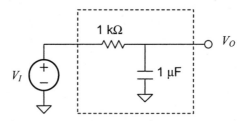

Figure 4.9. *RC* low-pass filter.

Solution:

From the theory of first-order networks, the step response of the circuit starting from rest (i.e., $v_I = 0$) in Figure 4.9 is

$$v_o(t) = S\left(1 - e^{-t/\tau}\right) \tag{4.9}$$

where $S = 1$ V is the magnitude of the step and $\tau = RC = 10^{-3}$ s. Moreover, the 3-dB bandwidth ω_b (expressed in rad/s) of a first-order network is $1/RC$, so we can rewrite the above expression as

$$v_o(t) = S\left(1 - e^{-\omega_b t}\right) \quad (4.10)$$

Clearly, the time $t = t_S$ the output reaches an arbitrary output level of V_O is then

$$t_S = -\frac{\ln\left(\frac{S - V_o}{S}\right)}{\omega_b} \quad (4.11)$$

Further, we recognize that $(S - V_o)/S$ is the settling error ε or the accuracy of the measurement, so we can rewrite Eq. (4.11) as

$$t_S = -\frac{\ln(\varepsilon)}{\omega_b} \quad (4.12)$$

Hence, the time it takes to reach within 1% of 1 V, or 0.99 V, is

$$t_S = -\frac{\ln(0.01)}{1/10^{-3}} = 4.6 \text{ ms}$$

Since settling time and 3-dB bandwidth are inversely related according to Eq. (4.12), a tenfold decrease in bandwidth leads to a tenfold increase in settling time. Specifically, the settling time becomes 46 ms.

Example 4.6

The simple *RC* low-pass circuit shown in Figure 4.9 is used to filter the output of a DUT containing a noisy DC signal. If the noise voltage has a constant spectral density of η V²/Hz, what is the RMS noise voltage that appears at the output of the filter? If the filter bandwidth decreases by a factor of 10, by what factor does the output noise voltage decrease?

Solution:

To answer this question we must rely on our knowledge of noise and linear system theory. We shall make use of frequency-domain techniques. While periodic signals have power at distinct frequency locations (see, for example, Figure 4.5), noise signals have their power spread out over the entire frequency spectrum. As such, noise signals are characterized by a noise spectral density function, which we shall denote as $S(f)$. It represents the average power over a 1-Hz bandwidth centered at each frequency, f. To simplify our discussion, $S(f)$ will have units of volts-squared/hertz or V²/Hz. It can also be expressed in terms of amps-squared/hertz or watts/hertz. (Data sheets often specify noise using volts per root-hertz, which is simply the square root of the noise spectral density as we have defined it here.) The total mean-squared value of the noise is obtained by integrating the spectral density over the entire frequency spectrum. Thus the RMS value of the noise signal can also be obtained in the frequency domain using the following relationship

$$V_{RMS} = \sqrt{\int_0^\infty S(f)\,df} \qquad (4.13)$$

Now to get back to the question at hand, the noise that appears at the output of the filter is related to the input noise voltage according to the following

$$S_{n_O}(f) = S_{n_I}(f)|G(f)|^2 \qquad (4.14)$$

where $S_{n_I}(f)$ and $S_{n_O}(f)$ are the input and output noise voltage spectral densities, respectively, and $G(f)$ is the system input-output transfer function. Hence, the RMS value of the output noise voltage V_{n_O} is

$$V_{n_O} = \sqrt{\int_0^\infty S_{n_O}(f)\,df} \qquad (4.15)$$

or, once we substitute Eq. (4.14), we obtain

$$V_{n_O} = \sqrt{\int_0^\infty S_{n_I}(f)|G(f)|^2\,df} \qquad (4.16)$$

Since we are told

$$S_{n_I}(f) = \eta \; \frac{V^2}{Hz} \qquad (4.17)$$

and that we can calculate the system transfer function for the RC low-pass filter to be

$$G(f) \equiv \frac{V_O}{V_I}(f) = \frac{\omega_b}{\omega_b + j2\pi f} = \frac{1}{1 + j\dfrac{2\pi f}{\omega_b}} \qquad (4.18)$$

where $\omega_b = 1/RC$. We can then write Eq. (4.16) as

$$V_{n_O} = \sqrt{\eta \int_0^\infty \left|\frac{1}{1 + j\dfrac{2\pi f}{\omega_b}}\right|^2 df} \qquad (4.19)$$

Integrating and performing the square root, we obtain the RMS output noise voltage to be

$$V_{n_o} = \frac{1}{2}\sqrt{\eta \omega_b} \quad \text{V} \tag{4.20}$$

Here we clearly see that the output noise voltage depends on two factors: the level of the input noise voltage spectral density and the filter's 3-dB bandwidth expressed in rad/s. If the filter's bandwidth is decreased by a factor of 10, then a $\sqrt{10}$ reduction in output noise RMS voltage will occur. We can also conclude that the repeatability will improve.

However, at this time we can not formally quantify the improvement until a mathematical definition for repeatability is given. For now, we will offer without proof that the total variation in measurement values is proportional to the level of noise. Thus, in the example above, we can expect the variability of measurements to improve by a factor of $\sqrt{10}$. We shall offer a formal analysis of the relationship between noise and repeatability in Chapter 15, "Data Analysis."

In the preceding example, the concept of a noise spectral density was introduced and used to characterize the noise at the input of the filter. Subsequently, it was used to determine the RMS level of the noise at the output of the filter. Often the test engineer knows only the RMS noise value from the output of the DUT, rather than the output noise spectral density. Interestingly enough, using Eq. (4.20) we can work backwards and obtain an estimate of the spectral density level η coming from the DUT

$$\eta = 4\frac{(V_{DUT})^2}{\omega_{DUT}} \quad \frac{\text{V}^2}{\text{Hz}} \tag{4.21}$$

where V_{DUT} is the DUT output noise (RMS volts) and ω_{DUT} is the 3-dB bandwidth of the DUT.

Exercises

4.10. What is the 3-dB bandwidth of the *RC* circuit of Figure 4.9, expressed in Hertz, when $R = 1$ kΩ and $C = 2.2$ nF?

Ans. 72.34 kHz.

4.11. How long does it take a first-order *RC* low-pass circuit with $R = 1$ kΩ and $C = 2.2$ nF to settle to 5% of its final value?

Ans. 6.6 µs.

4.12. A noise signal having a spectral density of 10^{-9} V^2/Hz is applied to the *RC* circuit of Figure 4.9 with $R = 1$ kΩ and $C = 2.2$ nF. What is the RMS-level of the noise voltage that appears at the output?

Ans. 10.7 mV RMS.

Provided $\omega_b \ll \omega_{DUT}$, it is reasonable to assume that the noise has a constant spectral density over the frequencies of interest given by Eq. (4.21). Substituting Eq. (4.21) back into Eq. (4.20), we can write

$$V_{n_o} = V_{DUT} \sqrt{\frac{\omega_b}{\omega_{DUT}}} \quad \text{V} \tag{4.22}$$

This expression clearly illustrates the noise reduction gained by filtering the output. The smaller the ratio ω_b/ω_{DUT}, the greater the noise reduction. Other types of filtering circuits can be placed on the DIB board when needed. For example, a very narrow bandpass filter may be placed on the DIB board to clean up noise components in a sine wave generated by the tester. The filter allows a much more ideal sine wave to the input of the DUT than the tester would otherwise be able to produce.

Exercises

4.13. By what factor should the bandwidth of an *RC* low-pass filter be decreased in order to reduce the variation in a DC measurement from 250 µV-RMS to 100 µV-RMS. By what factor does the settling time increase.

Ans. The bandwidth should be decreased by 6.25 (=2.5^2). Settling time increases by 2.5.

4.14. The variation in the output signal of a DUT is 1 mV RMS. Assume that the DUT's output follows a first-order frequency response and has a 3-dB bandwidth of 100 Hz. Estimate the output noise voltage spectral density.

Ans. 6.37×10^{-9} V^2/Hz.

4.15. The variation in the output RMS signal of a DUT is 1 mV, but it needs to be reduced to a level closer to 500 µV. What filter bandwidth is required to achieve this level of repeatability? Assume that the DUT's output follows a first-order frequency response and has a 3-dB bandwidth of 1000 Hz

Ans.. 250 Hz.

4.16. A DUT output consisting of a noise component having a spectral density of 10^{-6} V^2/Hz is to be measured 100 times and the results averaged. What is RMS value of the noise component in the final result?

Ans. 100 µV RMS.

4.17. The output of a DUT has an uncertainity of 10 mV. How many samples should be combined in order to reduce the uncertainity to 100 µV?

Ans. 10,000.

4.3.2 Averaging

Averaging is a form of discrete-time filtering. Averaging can be used to improve the repeatability of a measurement. For example, we can average the following series of nine voltage measurements

101 mV, 103 mV, 102 mV, 101 mV, 102 mV, 103 mV, 103 mV, 101 mV, 102 mV

and obtain an average of 102 mV. There is a good chance that a second series of nine unique measurements will again result in something close to 102 mV. If the length of the series is increased, the answer will become more repeatable and reliable. But there is a point of diminishing returns. To reduce the effect of noise on the voltage measurement by a factor of two, one has to take four times as many readings and average them. At some point, it becomes prohibitively expensive (i.e., from the point of view of test time) to improve repeatability.

To better understand the above statement, consider representing a sequence of numbers as $x(n)$, where n indicates the order at which the samples appear in the sequence. Further, let $x(n)$ consist of both signal and noise. Now, consider $x(n)$ as input to a discrete-time system whose output is the average value of $x(n)$ and the $N-1$ previous input samples. Mathematically, we can write the input-output relationship as

$$y(n) = \frac{1}{N}\left[x(n) + x(n-1) + \cdots + x(n-N+1)\right]$$
$$= \frac{1}{N}\sum_{k=1}^{N} x(n-k+1) \quad (4.23)$$

This system is called an *N-point running averager* and it can easily be shown that it has a frequency response[*] for $-\tfrac{1}{2} \leq f \leq \tfrac{1}{2}$ given by

$$G(f) = \left(\frac{\sin(2\pi f N/2)}{N \sin(2\pi f/2)}\right)\left[\cos(2\pi f(N-1)/2) - j\sin(2\pi f(N-1)/2)\right]$$
$$= \left(\frac{\sin(2\pi f N/2)}{N \sin(2\pi f/2)}\right) e^{-j2\pi f(N-1)/2} \quad (4.24)$$

Technically, we are really only interested in the output after N samples; that is, the output is downsampled or decimated by N

$$y_D(n) = y(nN) \quad (4.25)$$

To introduce its effect on the system's frequency response will only add more complication to an otherwise sophisticated explanation. Nonetheless, regardless of when the output is obtained, one

---

[*] The frequency response is obtained by first evaluating the z-transform of Eq. (4.23) and then substituting $z = e^{j2\pi f t}$.

would not expect the behavior of the noise to be influenced by the observation window. So we shall work with the running average result only as the conclusions are the same.

If the noise component in the input signal has a constant spectral density of η V²/Hz, the output noise spectral density will then be given by Eq. (4.13). The RMS voltage that appears in the output discrete-time signal is found from an expression very similar to the continuous-time case given in Eq. (4.16). However, in this case the integration is performed from $-\frac{1}{2}$ to $\frac{1}{2}$ to account for the periodicity of $G(f)$ and because the sampling period is 1

$$V_{n_o} = \sqrt{\int_{-1/2}^{1/2} S_{n_I}(f) |G(f)|^2 \, df} \qquad (4.26)$$

Substituting the appropriate relationships, we obtain

$$V_{n_o} = \sqrt{\eta \int_{-1/2}^{1/2} \left| \left[\frac{\sin(2\pi f N/2)}{N \sin(2\pi f/2)} \right] e^{-j2\pi f(N-1)/2} \right|^2 df} \qquad (4.27)$$

Finally, solving the integration and taking the square root, we obtain

$$V_{n_o} = \sqrt{\frac{\eta}{N}} \quad \text{V} \qquad (4.28)$$

If we denote the noise from the DUT before averaging as V_{DUT}, it is easy to show that $V_{n_o} = \sqrt{\eta}$. This can be seen directly from Eq. (4.28) when $N = 1$. However, with no averaging taking place, $V_{n_o} = V_{DUT}$. Hence, we can write the final expression as

$$V_{n_o} = \frac{V_{DUT}}{\sqrt{N}} \quad \text{V} \qquad (4.29)$$

Here we see the output noise voltage reduces the input noise before averaging by the factor $\sqrt{N}$. Hence, to reduce the noise RMS voltage by a factor of two requires an increase in the sequence length, N, by a factor of four.

AC measurements can also be averaged to improve repeatability. A series of sine wave signal level measurements can be averaged to achieve better repeatability. However, one should not try to average readings in decibels. If a series of measurements is expressed in decibels, they should first be converted to linear form using the equation $V = 10^{dB/20}$ before applying averaging. Normally, the voltage or gain measurements are available before they are converted to decibels in the first place; so the conversion from dB to linear units or ratios is not necessary. Once the average voltage level is calculated, it can be converted to decibels using the equation $dB = 20\log_{10}(V)$. To understand why we should not perform averaging on decibels, consider the sequence 0, -20, -40 dBV. The average of these values is –20 dBV. However, the actual voltages are 1 V, 100 mV, and 10 mV. Thus the correct average value is (1 V + 0.1 V + 0.01 V) / 3 = 37 mV, or -8.64 dBV.

4.3.3 Guardbanding

Guardbanding is an important technique for dealing with the uncertainty of each measurement. If a particular measurement is known to be accurate and repeatable with a worst-case uncertainty of $\pm\varepsilon$, then the final test limits should be tightened from the data sheet specification limits by ε to make sure no bad devices are shipped to the customer. In other words

$$\text{guardbanded upper test limit} = \text{upper specification limit} - \varepsilon$$
$$\text{guardbanded lower test limit} = \text{lower specification limit} + \varepsilon \quad (4.30)$$

So, for example, if the data sheet limit for the offset on a buffer output is −100 mV minimum, 100 mV maximum, and an uncertainty of ±10 mV exists in the measurement, the test program limits should be set to −90 mV minimum and 90 mV maximum. This way, if the device output is 101 mV and the error in its measurement is −10 mV, the resulting reading of 91 mV will cause a failure as required. Of course, a reading of 91 mV may also represent a device with an 81-mV output and a +10 mV measurement error.

In such cases, guardbanding has the unfortunate effect of disqualifying good devices. Ideally, we would like all guardbands to be set to 0 so that no good devices will be discarded. To minimize the guardbands we must improve the repeatability and accuracy of each test, but this typically requires longer test times. There is a balance to be struck between repeatability and the number of good devices rejected. At some point, the added test time cost of a more repeatable measurement outweighs the cost of discarding a few good devices.

Example 4.7

Table 4.2 lists a set of output values from a DUT together with their measured values. It is assumed that the upper test limit is 100 mV and the measurement uncertainty is ±6 mV. How many good devices are rejected because of the measurement error? How many good devices are rejected if the measurement uncertainty is increased to ±10 mV?

Table 4.2. DUT Output and Measured Values

DUT Output	Measured Value
105 mV	101 mV
101 mV	107 mV
98 mV	102 mV
96 mV	95 mV
86 mV	92 mV
72 mV	78 mV

Solution:

From the DUT output column on the left, four devices are below the upper test limit of 100 mV and should be accepted. The other two should be rejected. Now with a measurement uncertainty of ±6 mV, according to Eq. (4.30) the guardbanded upper test limit is 94 mV. With the revised test limit, only two devices are acceptable. The others are all rejected. Hence, two otherwise good devices are disqualified.

If the measurement uncertainty increases to ±10 mV, then the guardbanded upper test limit becomes 90 mV. Five devices are rejected and only one is accepted. Consequently, three otherwise good devices are disqualified.

Exercises

4.18. A device is expected to exhibit a worst-case offset voltage of ±50 mV and is to be measured using a voltmeter having an accuracy of only ±5 mV. Where should the guardbanded test limits be set?

Ans. ±45 mV.

4.19. The guardband of a particular measurement is 10 mV and the test limit is set to ±25 mV. What are the original device specification limits?

Ans. ±35 mV.

4.20. The following lists a set of output voltage values from a group of DUTs together with their measured values: {(2.3, 2.1), (2.1, 1.6), (2.2, 2.1), (1.9, 1.6), (1.8, 1.7), (1.7, 2.1), (1.5, 2.0)}. If the upper test limit is 2.0 V and the measurement uncertainty is ±0.5 V, how many good devices are rejected due to the measurement error?

Ans. Four devices (all good devices are rejected by the 1.5-V guardbanded upper test limit).

4.4 BASIC DATA ANALYSIS

4.4.1 Datalogs

A datalog is a concise listing of the test results generated by a test program. Datalogs are the primary means by which test engineers evaluate the quality of a device as it is tested. The format of a datalog typically includes a test category, test description, upper and lower test limits, and a measured result. The exact format of datalogs varies from one tester type to another, but they all convey similar information.

A short datalog from a Teradyne A580 tester is listed in Figure 4.10. Each line of the datalog contains a shorthand description of the test. For example, "DAC Gain Error" is the name given to test number 5000. The gain error is part of the S_VDAC_SNR test group and is executed during a test routine called T_VDAC_SNR. The upper and lower limits for the test are also listed. Using test number 5000 as an example, the lower limit of DAC Gain Error is −1.00 dB, the upper limit is +1.00 dB, and the measured value for this DUT is −0.13 dB.

The datalog lists an easily recognizable fail flag for values that fall outside the specified test limits. Test 7004 shows a test failure in which the measurement reads 1.23 LSBs (least significant bits). The upper limit is 0.9 LSBs, so this test fails. Because the device is not a good one, it is categorized into Bin 10 in this example. Bin 1 usually represents a good device, while other bins usually represent various categories of failure.

Sometimes multiple good bins are defined, allowing devices to be graded into several passing categories. For instance, a microprocessor may fail full speed digital pattern testing at 750 MHz, but may operate perfectly well at 500 MHz. In such a case, the 500-MHz processor might be graded into Bin 2 and sold at a lower price while higher-grade 750-MHz processors are graded into Bin 1 and sold at a premium.

```
Sequencer:   S_continuity
   1000 Neg PPMU Cont       Failing Pins:   0
Sequencer:   S_VDAC_SNR
   5000 DAC Gain Error     T_VDAC_SNR   -1.00   dB <    -0.13  dB      < 1.00   dB
   5001 DAC S/2nd          T_VDAC_SNR    60.0   dB <=    63.4  dB
   5002 DAC S/3rd          T_VDAC_SNR    60.0   dB <=    63.6  dB
   5003 DAC S/THD          T_VDAC_SNR    60.00  dB <=    60.48 dB
   5004 DAC S/N            T_VDAC_SNR    55.0   dB <=    70.8  dB
   5005 DAC S/N+THD        T_VDAC_SNR    55.0   dB <=    60.1  dB
Sequencer:   S_UDAC_SNR
   6000 DAC Gain Error     T_UDAC_SNR   -1.00   dB <    -0.10  dB      < 1.00   dB
   6001 DAC S/2nd          T_UDAC_SNR    60.0   dB <=    86.2  dB
   6002 DAC S/3rd          T_UDAC_SNR    60.0   dB <=    63.5  dB
   6003 DAC S/THD          T_UDAC_SNR    60.00  dB <=    63.43 dB
   6004 DAC S/N            T_UDAC_SNR    55.0   dB <=    61.3  dB
   6005 DAC S/N+THD        T_UDAC_SNR    55.0   dB <=    59.2  dB
Sequencer:   S_UDAC_Linearity
   7000 DAC POS ERR        T_UDAC_Lin  -100.0   mV <     7.2   mV      < 100.0   mV
   7001 DAC NEG ERR        T_UDAC_Lin  -100.0   mV <     3.4   mV      < 100.0   mV
   7002 DAC POS INL        T_UDAC_Lin   -0.90   lsb <    0.84  lsb     < 0.90    lsb
   7003 DAC NEG INL        T_UDAC_Lin   -0.90   lsb <   -0.84  lsb     < 0.90    lsb
   7004 DAC POS DNL        T_UDAC_Lin   -0.90   lsb <    1.23  lsb (F) < 0.90    lsb
   7005 DAC NEG DNL        T_UDAC_Lin   -0.90   lsb <   -0.83  lsb     < 0.90    lsb
   7006 DAC LSB SIZE       T_UDAC_Lin    0.00   mV <     1.95  mV      < 100.00  mV
   7007 DAC Offset V       T_UDAC_Lin  -100.0   mV <     0.0   mV      < 100.0   mV
   7008 Max Code Width     T_UDAC_Lin    0.00   lsb <    1.23  lsb     < 1.50    lsb
   7009 Min Code Width     T_UDAC_Lin    0.00   lsb <    0.17  lsb     < 1.50    lsb
Bin:   10
```

Figure 4.10. Example datalog from a Teradyne A580 tester.

4.4.2 Histograms

When a test program is executed multiple times on a single DUT, it is common to get multiple answers due to imperfect repeatability. If we repeatedly execute the test program corresponding to Figure 4.10 and display only the results from test 5000, the DAC Gain Error test may show slight repeatability errors:

```
   5000 DAC Gain Error     T_VDAC_SNR   -1.000  dB <    -0.127 dB      < 1.000  dB
   5000 DAC Gain Error     T_VDAC_SNR   -1.000  dB <    -0.129 dB      < 1.000  dB
   5000 DAC Gain Error     T_VDAC_SNR   -1.000  dB <    -0.125 dB      < 1.000  dB
   5000 DAC Gain Error     T_VDAC_SNR   -1.000  dB <    -0.131 dB      < 1.000  dB
```

```
5000 DAC Gain Error    T_VDAC_SNR  -1.000 dB <  -0.129 dB  < 1.000 dB
5000 DAC Gain Error    T_VDAC_SNR  -1.000 dB <  -0.128 dB  < 1.000 dB
5000 DAC Gain Error    T_VDAC_SNR  -1.000 dB <  -0.132 dB  < 1.000 dB
5000 DAC Gain Error    T_VDAC_SNR  -1.000 dB <  -0.130 dB  < 1.000 dB
5000 DAC Gain Error    T_VDAC_SNR  -1.000 dB <  -0.134 dB  < 1.000 dB
5000 DAC Gain Error    T_VDAC_SNR  -1.000 dB <  -0.131 dB  < 1.000 dB
...
```

Notice that the resolution of the datalog output has been increased to three digits after the decimal point in this second example. The printout resolution is specified in the test program. The extra resolution in this second example allows the variations in the results to be more easily seen. Otherwise, all the results above would have been rounded off to –0.13 dB. It is important to note that there may be a difference between datalog output resolution and the resolution of the value as it is compared against test limits. Testers generally use the full resolution of the measurement when comparing DUT performance against the specification limits. Only the datalog output values are rounded, not the actual measured values.

We can view the repeatability of a group of readings using a tool called a *histogram*. A histogram for the repeatability example above is shown in Figure 4.11. It shows both numerical results and a plot of the distribution of measured values. The distribution plot is a convenient graphical way to understand the repeatability of the measurement. Ideally, the distribution should be closely packed, as the example in Figure 4.11 shows. However, if the measurement repeatability is poor the histogram spreads out into a larger range of values.

This histogram output also displays a number of useful values. The population size is listed next to the heading "Total Results =". It indicates how many times the measurement was repeated on the same device. Histograms are also used to look at distributions of measurements over many DUTs to determine how much variability there is from one device to another. In the case of multiple DUTs, the total results would be the number of DUTs tested.

The larger the population of results, the more trustworthy a histogram becomes. A histogram with less than 50 results is statistically questionable because of the limited sample size. Ideally a histogram should contain results from at least 100 devices (or 100 test executions on a single device in the case of repeatability studies).

Another useful item in the histogram is the population statistics. The mean μ and standard deviation σ are the most important of these. The mean, or average, represents the most probable value of a measured variable. The best approximation will be made when the number of readings N of the same quantity is very large. The mean value is defined as

$$\mu = \frac{1}{N}\sum_{n=1}^{N} x(n) \qquad (4.31)$$

For the DAC Gain Error example shown in Figure 4.11, the mean value from 110 measurements is -0.1300 dB. The standard deviation is a measure of the dispersion or uncertainty of the measured quantity about the mean value, μ. If the values tend to be concentrated near the mean, the standard deviation is small, while if the values tend to be distributed far from the mean, the standard deviation is large. Standard deviation is defined as

$$\sigma = \sqrt{\frac{1}{N}\sum_{n=1}^{N}[x(n)-\mu]^2} \qquad (4.32)$$

```
Test No    Test Function Name      Test Label                      Units

  5000     T_VDAC_SNR               DAC Gain Error                  dB

Lower Test Limit=         -1dB    Upper Test Limit=                1dB

                  - DISTRIBUTION STATISTICS -
Lower Pop Limit=         -Infinity   Upper Pop Limit=        +Infinity
Total Results=              110      Results Accepted=             110
Underflows=                   0      Overflows=                      0
Mean=                  -0.13003dB    Std Deviation=         0.00292899
Mean - 3 Sigma=        -0.13882dB    Mean + 3 Sigma=        -0.12125dB
Minimum Value=         -0.13594dB    Maximum Value=         -0.12473dB

                     - PLOT STATISTICS -
Lower Plot Limit=         -0.14dB    Upper Plot Limit=         -0.12dB
Cells=                        15     Cell Width=           0.0013333dB
Full Scale Percent=        16.36%    Full Scale Count=              18
```

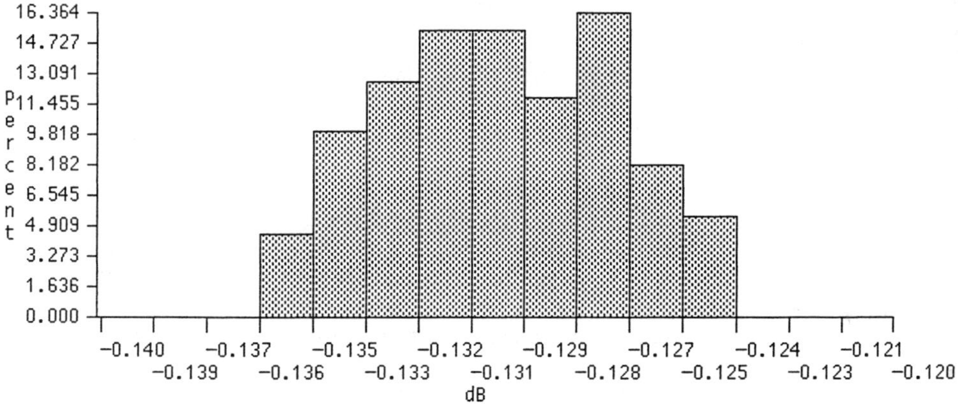

Figure 4.11. Histogram of the DAC gain error test.

Standard deviation and mean are expressed in identical units. In our DAC example, the standard deviation was found to be 0.0029 dB. Notice that we have broken the previously stated rule that we should not calculate averages (and standard deviations for that matter) using logarithmic units such as the decibel. However, we are often able to take this statistical shortcut when all the values are very near one another. The reason for this is that logarithmic curves are approximately linear over a limited span of values. It would not be advisable to take such a shortcut if the range of values covered a 20-dB span. We can still view histograms of logarithmic values to get a general idea of the mean and standard deviation, but we should remember that these values are not entirely accurate. To be perfectly correct we would need to convert the logarithmic values into linear units such as V/V before reading the mean and standard deviation from a histogram.

The histogram in Figure 4.11 exhibits a common feature for analog and mixed-signal measurements. The distribution of values has a shape similar to a bell. The bell curve (also called a *normal distribution* or *Gaussian distribution*) is a common one in the study of statistics. According to the central limit theorem,[6] any summation of a large number ($N > 30$) of statistically independent random variables converges toward a Gaussian distribution. The

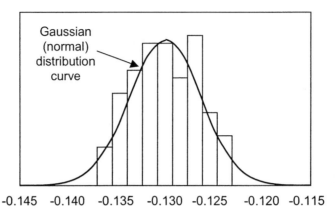

Figure 4.12. Continuous normal (Gaussian) distribution for the DAC gain example.

variations in a typical mixed-signal measurement are caused by a summation of many different sources of noise and crosstalk in both the device and the tester instruments. As a result, many mixed-signal measurements exhibit the familiar Gaussian distribution.

For the DAC Gain Error example, Figure 4.12 illustrates the results using a smooth curve through all the measured points. As is evident, the graph has a bell shape, quite similar to a Gaussian distribution. In theory, a Gaussian distribution extends to infinity. In other words, if you are willing to wait billions of years you should eventually see an answer of +200 dB in the DAC gain example. In reality, measurements are only near-Gaussian. As a result, the answer in the example will probably never stray more than a few tenths of a dB away from –0.13 dB.

Although the real world only approximates the ideal Gaussian distribution, the comparison is close enough to allow some general statements. First, the standard deviation of a Gaussian distribution is roughly equal to one sixth of the total variation from the minimum value to the maximum value. In the distribution of Figure 4.12 the standard deviation is 0.0029 dB; so we would expect to see values ranging from –0.139 to –0.121 dB. These values are displayed in the sample histogram beside the labels "Mean –3 sigma" and "Mean +3 sigma." The actual minimum and maximum values are not as extreme, which indicates that this is not truly a Gaussian distribution.

It is quite common to obtain distributions that are not Gaussian, though the distributions should be more or less bell-shaped. Common deviations from the familiar bell shape are bimodal distributions (Figure 4.13) and outliers (Figure 4.14). When looking for measurement repeatability on a single DUT, these distributions are a warning sign that something is not sufficiently repeatable. When looking for distributions of a parameter across many DUTs, non-Gaussian plots may indicate a poor design or a fabrication process that needs to be fixed. They may also indicate a test program that is yielding incorrect answers.

4.4.3 Noise, Test Time, and Yield

The yield of a given lot of material is defined as the ratio of the total good devices divided by the total devices tested:

$$\text{yield} = \frac{\text{total good devices}}{\text{total devices tested}} \times 100\% \qquad (4.33)$$

If 10,000 parts are tested and only 7,000 devices pass all tests, then the yield on that lot of 10,000 devices is 70%. As explained in Section 4.3.3, there are inherent tradeoffs between repeatability, test time, and production yield.

If a device is well designed and a particular measurement is sufficiently repeatable, then there will be few failures on that measurement. But if the distribution of readings on a production lot of devices is skewed so that it is close to one of the test limits, then production yields are likely to fall. In other words, more good devices will fall beyond the guardband region and be disqualified. Obviously, a measurement with poor accuracy or poor repeatability will just exacerbate the problem.

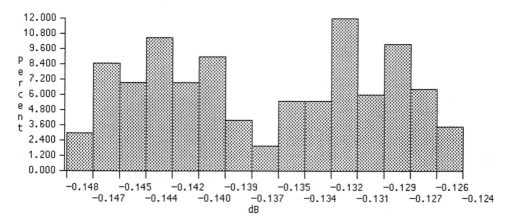

Figure 4.13. Bimodal distribution.

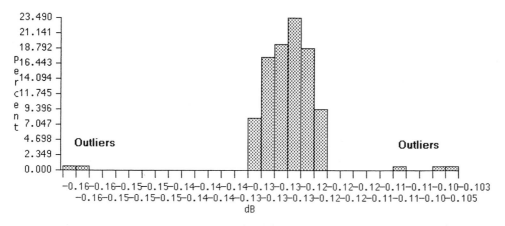

Figure 4.14. Statistical outliers.

In Section 4.3.1 it was shown that averaging or filtering a measurement can make it more repeatable, with the unfortunate consequence of longer test time. On the other hand, centering the design between test limits and making the design insensitive to process variations might make it unnecessary to achieve such accuracy and repeatability. Test costs can be dramatically reduced with well-centered designs with plenty of margin. Unfortunately, well-centered designs with good margin are often power hungry or silicon area intensive. The test engineer and design engineer should work together to achieve an optimum balance between low silicon overhead with minimal power consumption and low test cost with minimum averaging. Only by working as a team can these two engineering disciplines produce a cost-effective product that will succeed in the marketplace.

4.5 Summary

In this chapter we have introduced the concept of accuracy and repeatability and shown how these concepts affect device quality and production test economics. We have examined many contributing factors leading to inaccuracy and nonrepeatability. Using software calibrations, we can eliminate or at least reduce many of the effects leading to measurement inaccuracy. Measurement repeatability can be enhanced through averaging and filtering, at the expense of added test time. The constant balancing act between adequate repeatability and minimum test time represents a large portion of the test engineer's workload. One of fundamental skills that separates good test engineers from average test engineers is the ability to quickly identity and correct problems with measurement accuracy and repeatability. Doing so while maintaining low test times and high yields is the mark of a great test engineer.

In the next chapter, we will take a brief vacation from mathematical equations to examine the measurement instrumentation and other architectural features common to many mixed-signal ATE testers. We will study the various types of instruments such as waveform digitizers, arbitrary waveform generators, digital pattern generators, and other mixed-signal ATE instruments. It might seem that this fundamental topic should have been presented in an earlier chapter. The subject has been delayed until Chapter 5 because many of the architectural features of mixed-signal testers are easier to understand in the context of accuracy, repeatability, and their combined effects on test time and production yield.

Exercises

4.21. A 5-mV signal is measured with a meter ten times, resulting in the following sequence of readings: 7 mV, 6 mV, 9 mV, 8 mV, 4 mV, 7 mV, 5 mV, 7 mV, 8 mV, 11 mV. What is the mean value? What is the standard deviation?

Ans. 7.2 mV, 1.887 mV.

4.22. If 15,000 devices are tested with a yield of 63%, how many devices passed the test?

Ans. 9450 devices.

Problems

4.1. A 55-mV signal is measured with a meter ten times resulting in the following sequence of readings: 57 mV, 60 mV, 49 mV, 58 mV, 54 mV, 57 mV, 55 mV, 57 mV, 48 mV, 61 mV. What is the average measured value? What is the systematic error?

4.2. A DC voltmeter is rated at 14 bits of resolution and has a full-scale input range of ±5 V. Assuming the meter's ADC is ideal, what is the maximum quantization error that we can expect from the meter? What is the error as a percentage of the meter's full-scale range?

4.3. A 100 mV signal is to measured with a worst-case error of ±10 µV. A DC voltmeter is set to a full-scale range of ±1 V. Assuming that quantization error is the only source of inaccuracy in this meter, how many bits of resolution would this meter need to have to make the required measurement? If the meter in our tester only has 14 bits of resolution but has binary-weighted range settings (i.e., ±1 V, ±500 mV, ±250 mV, etc.) how would we make this measurement?

4.4. A voltmeter is specified to have an accuracy error of ±0.1% of full-scale range on a ±1-V scale. If the meter produces a reading of 0.323 V DC, what is the minimum and maximum DC levels that might have been present at the meter's input during this measurement?

4.5. A meter reads −1.039 mV and 1.121 V when connected to two highly accurate reference levels of −1 V and 1 V, respectively. What is the offset and gain of this meter? Write the calibration equation for this meter.

4.6. A DC source is assumed characterized by a third-order equation of the form: $V_{MEASURED} = 0.004 + V_{PROGRAMMED} + 0.001\, V_{PROGRAMMED}^2 - 0.007\, V_{PROGRAMMED}^3$ and is required to generate a DC level of 1.25 V. However, when programmed to produce this level, only 1.242 V is measured. Using iteration, determine a value of the programmed source voltage that will establish a measured voltage of 1.25 V to within a ± 0.5 mV accuracy.

4.7. An AWG has a gain response described by $G(f) = \sqrt{\dfrac{1}{1+\left(\dfrac{f}{4000}\right)^2}}$ and is to generate three tones at frequencies of 1, 2, and 3 kHz. What are the gain calibration factors? What voltage levels would we request if we wanted an output level of 500 mV RMS at each frequency?

4.8. Several DC measurements are made on a signal path that contains a filter and a buffer amplifier. At input levels of 1 and 3 V, the output was found to be 1.02 and 3.33 V, respectively. Assuming linear behavior, what is the gain and offset of this filter-buffer stage?

4.9. Using the setup and results of Problem 4.8, what is the calibrated level when a 2.13 V level is measured at the filter-buffer output? What is the size of the uncalibrated error?

4.10. A simple RC low-pass circuit is constructed using a 1-kΩ resistor and a 10-µF capacitor. This RC circuit is used to filter the output of a DUT containing a noisy DC signal. If the DUT's noise voltage has a constant spectral density of 100 nV $/\sqrt{Hz}$, what is the RMS noise voltage that appears at the output of the RC filter? If the we decrease the capacitor value to 2.2 µF, what is the RMS noise voltage at the RC filter output?

4.11. Assume that we want to allow the *RC* filter in Problem 4.10 to settle to within 0.2% of its final value before making a DC measurement. How much settling time does the first *RC* filter in Problem 4.10 require? Is the settling time of the second *RC* filter greater or less than that of the first filter?

4.12. A DC meter collects a series of repeated offset measurements at the output of a DUT. A first-order low-pass filter such as the one in Problem 4.10 is connected between the DUT output and the meter input. A histogram is produced from the repeated measurements. The histogram shows a Gaussian distribution with a 50-mV difference between the maximum value and minimum value. It can be shown that the standard deviation, σ, of the histogram of a repeated series of identical DC measurements on one DUT is proportional to the RMS noise at the meter's input. Assume the difference between the maximum and minimum measured values is roughly equal to 6σ. How much would we need to reduce the cutoff frequency of the low-pass filter to reduce the nonrepeatability of the measurements from 50 to 10 mV? What would this do to our test time, assuming the test time is dominated by the settling time of the low-pass filter?

4.13. The DUT in Problem 4.12 can be sold for $1.25, assuming it passes all tests. If it does not pass all tests, it cannot be sold at all. Assume that the more repeatable DC offset measurement in Problem 4.12 results in a narrower guardband requirement, causing the production yield to rise from 92% to 98%. Also assume that the cost of testing is known to be 3.5 cents per second and that the more repeatable measurement adds 250 ms to the test time. Does the extra yield obtained with the lower filter cutoff frequency justify the extra cost of testing resulting from the filter's longer settling time?

References

1. Albert H. Moorehead, et.al., *The New American Roget's College Thesaurus*, New American Library, 1633 Broadway, New York, NY, 10019, 1985, p. 6
2. *Webster's New World Dictionary*, Simon and Schuster, Inc. 1230 Avenue of the Americas, New York, NY, 10020, August 1995, pp. 5, 463
3. Bob Metzler, *Audio Measurement Handbook*, Audio Precision, Inc., Beaverton, OR, 97075, August, 1993, p. 147
4. William David Cooper, *Electronic Instrumentation and Measurement Techniques*, 2nd Edition, Prentice Hall, Englewood Cliffs, NJ, 1978, ISBN: 0132517108, pp. 1, 2
5. Rudolf F. Graf, *Modern Dictionary of Electronics*, Newnes Press, Boston, MA, July, 1999, ISBN: 0750698667, pp. 5, 6, 584
6. Mark J. Kiemele, Stephen R. Schmidt, Ronald J. Berdine, *Basic Statistics, Tools for Continuous Improvement*, Fourth Edition, Air Academy Press, 1155 Kelly Johnson Blvd., Suite 105, Colorado Springs, CO, 80920, 1997, ISBN: 1880156067, pp. 9-71

CHAPTER 5

Tester Hardware

5.1 MIXED-SIGNAL TESTER OVERVIEW

5.1.1 General-Purpose Testers versus Focused Bench Equipment

General-purpose mixed-signal testers must be capable of testing a variety of dissimilar devices. On any given day, the same mixed-signal tester may be expected to test video palettes, cellular telephone devices, data transceivers, and general-purpose ADCs and DACs. The test requirements for these various devices are very different from one another. For example, the cellular telephone base-band interface shown in Figure 1.2 may require a phase trajectory error test or an error vector magnitude test. Dedicated bench instruments can be purchased that are specifically designed to measure these application-specific parameters. It would be possible to install one of these stand-alone boxes into the tester and communicate with it through an IEEE-488 GPIB bus. However, if every type of DUT required two or three specialized pieces of bolt-on hardware, the tester would soon resemble Frankenstein's monster and would be prohibitively expensive.

The mixed-signal production tester cannot be focused toward a specific type of device if it is to handle a variety of DUTs. Instead of implementing tests like phase trajectory error and error vector magnitude using a dedicated bench instrument, the tester must emulate this type of equipment using a few general-purpose instruments. The instruments are combined with software routines to simulate the operation of the dedicated bench instruments.

5.1.2 Generic Tester Architecture

Mixed-signal testers come in a variety of "flavors" from a variety of vendors. Unlike the ubiquitous PC, testers are not at all standardized in architecture. Each ATE vendor adds special features that they feel will give them a competitive advantage in the marketplace. Consequently, mixed-signal testers from different vendors do not use a common software platform. Furthermore, a test routine implemented on one type of tester is often difficult to translate to another tester type because of subtle differences in the hardware tradeoffs designed into each tester. Nevertheless, most mixed-signal testers share many common features. In this chapter, we will examine these common features without delving too deeply into specific details for any particular brand of tester.

Figure 5.1 shows a generic mixed-signal tester architecture. It includes system computers, DC sources, DC meters, relay control lines, relay matrix lines, time measurement hardware, arbitrary waveform generators, waveform digitizers, clocking and synchronization sources, and a digital subsystem for generating and evaluating digital patterns and signals. This chapter will briefly examine the operation of each of these tester subsystems.

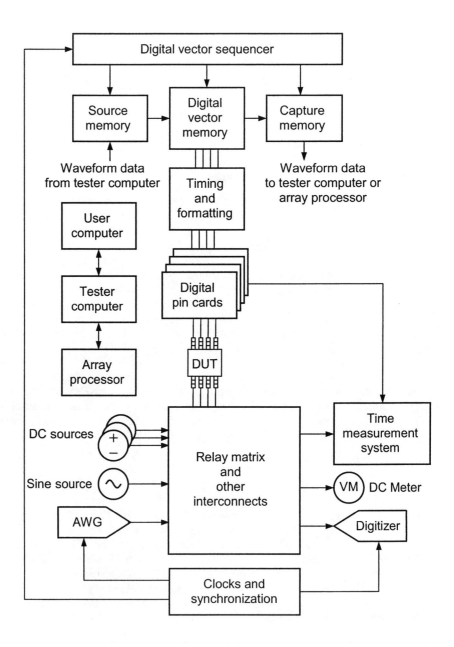

Figure 5.1. Generic mixed-signal tester architecture.

5.2 DC RESOURCES

5.2.1 General-Purpose Multimeters

Most testers incorporate a high-accuracy multimeter that is capable of making fast DC measurements. A tester may also provide a slower, very high-accuracy voltmeter for more demanding measurements such as those needed in focused calibrations. However, this slower instrument may not be usable for production tests because of the longer measurement time. The fast, general-purpose multimeter is used for most of the production tests requiring a nominal level of accuracy.

A very simple DC voltmeter block diagram was presented in Figure 4.8. A more detailed DC multimeter structure is shown in Figure 5.2. This meter can handle either single-ended or differential inputs. Its architecture includes a high-impedance differential to single-ended converter (instrumentation amplifier), a low-pass filter, a programmable gain amplifier (PGA) for input ranging, a high-linearity ADC, integration hardware, and a sample-and-difference stage. It also includes an input multiplexer stage to select one of several input signals for measurement.

The instrumentation amplifier provides a high-impedance differential input. The high impedance avoids potential DC offset errors caused by bias current leaking into the meter. For single-ended measurements, the low end of the meter may be connected to ground through relays in the input selection multiplexer. The multimeter can also be connected to any of the tester's general-purpose DC voltage sources to measure their output voltage. The meter can also measure current flowing from any of the DC sources. This capability is very useful for measuring power supply currents, impedances, leakage currents, and other common DC parametric values. A PGA placed before the meter's ADC allows proper ranging of the instrument to minimize the effects of the ADC's quantization error (see Sections 4.1.4 and 4.2.8).

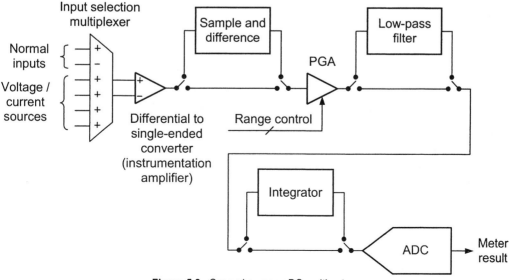

Figure 5.2. General-purpose DC multimeter.

The meter may also include a low-pass filter in its input path. The low-pass filter removes high-frequency noise from the signal under test, improving the repeatability of DC measurements. This filter can be enabled or bypassed using software commands. It may also have a programmable cutoff frequency so that the test engineer can make tradeoffs between measurement repeatability and test time (see Section 4.3.1). In addition, some meters may include an integration stage, which acts as a form of hardware averaging circuit to improve measurement repeatability.

Finally, a sample-and-difference stage is included in the front end of many ATE multimeters. The sample-and-difference stage allows highly accurate measurements of small differences between two large DC voltages. During the first phase of the measurement, a hardware sample-and-hold circuit samples a voltage. This first reference voltage is then subtracted from a second voltage (near the first voltage) using an amplifier-based subtractor. The difference between the two voltages is then amplified and measured by the meter's ADC, resulting in a high-resolution measurement of the difference voltage. This process reduces the quantization error that would otherwise result from a direct measurement of the large voltages using the meter's higher voltage ranges.

Example 5.1

A single-ended DC voltmeter has a resolution of 12 bits. It also features a sample-and-difference front-end circuit. We wish to use this meter to measure the differential offset voltage of a DUT's output buffer. Each of the two outputs is specified to be within a range of 1.35 V ± 10 mV, and the differential offset is specified to be ±5 mV. The meter input can be set to any of the following ranges: ±10 V, ±1 V, ±100 mV, ±10 mV, and ±1 mV. Assuming all components in the meter are perfectly linear (with the exception of the meter's quantization error), compare the accuracy achieved using two simple DC measurements with the accuracy achieved using the sample-and-difference circuit.

Solution:

The simplest way to measure offset using a single-ended DC voltmeter is to connect the meter to the OUTP output, measure its voltage, connect the meter to the OUTN output, measure its voltage, and subtract the second voltage from the first (see Example 3.4). Using this approach, we have to set the meter's input range to ±10 V to accommodate the 1.35 V DUT output signals. Thus each measurement may have a quantization error of as much as $\pm \frac{1}{2}\left(\frac{20 \text{ V}}{2^{12}-1}\right) = \pm 2.44$ mV.

Therefore, our total error might be as high as ±4.88 mV, assuming the quantization error from the first measurement is positive, while the quantization error from the second measurement is negative. Since we have a specification limit of ±5 mV, this will be an unacceptable test method.

Using the sample-and-difference circuitry, we could range the meter input to the worst-case difference between the two outputs, which is 5 mV, assuming a good device. The lowest meter range that will accommodate a 5-mV signal is ±10 mV. However, we also need to be able to collect readings from bad devices for purposes of characterization. Therefore, we will choose a range of ±100 mV, giving us a compromise between accuracy and characterization flexibility.

During the first phase of the sample-and-difference measurement, the voltage at the OUTN pin is sampled onto a holding capacitor internal to the meter. Then the meter is connected to the OUTP pin and the second phase of the measurement amplifies the difference between the OUTP voltage and the sampled OUTN voltage. Since the meter is set to a range of ±100 mV, a 100-mV difference between OUTP and OUTN will produce a full-scale 10 V input to the meter's ADC. This serves to reduce the effects of the meter's quantization error. The maximum error is given by $\pm\frac{1}{2}\left(\frac{100\text{ mV}}{2^{12}-1}\right) = \pm 12.2\ \mu V$. Again, our worst-case error is twice this amount, or ±24.4 μV, which is well within the requirements of our measurement.

5.2.2 General-Purpose Voltage/Current Sources

Most testers include general-purpose DC voltage/current sources, commonly referred to as *V/I sources* or *DC sources*. These programmable power supplies are used to provide the DC voltages and currents necessary to power up the DUT and stimulate its DC inputs. Many general-purpose supplies can force either voltage or current, depending on the testing requirements. On most testers, these supplies can be switched to multiple points on the DIB board using the tester's DC matrix (see Section 5.2.5). As mentioned in the previous section, the system's general-purpose meter can be connected to any DC source to measure its output voltage or its output current.

Figure 5.3 shows a conceptual block diagram of a DC source having a differential Kelvin connection. A differential Kelvin connection consists of four lines (high force, low force, high sense, and low sense) for forcing highly accurate DC voltages. The Kelvin connection forms a feedback loop that allows the DC source to force an accurate differential voltage through the resistive wires between the source and DUT. Without the Kelvin connection, the small resistance in the force line interconnections ($R_{TRACE-H}$ and $R_{TRACE-L}$) would cause a small *IR* voltage drop. The voltage drop would be proportional to the current through the DUT load (R_{LOAD}). The small

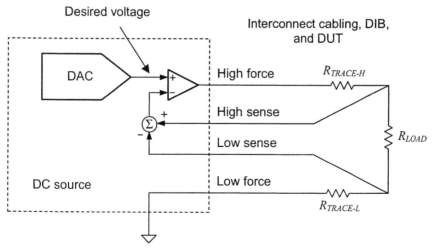

Figure 5.3. General-purpose DC source with Kelvin connections (conceptual diagram).

IR voltage drop would result in errors in the voltage across the DUT load. The sense lines of a Kelvin connection carry no current. Therefore, they are immune to errors caused by *IR* voltage drops.

A sense line is provided on the high side of the DC source and also on the low side of the source. The low-side sense line counteracts the parasitic resistance in the current return path. Since most instruments are referenced to ground, the low sense lines for all the DC instruments in a tester are often lumped into a single ground sense signal called DZ (device zero), DGS (device ground sense), or some other vendor-specific nomenclature. This is one of the most important signals in a mixed-signal tester, since it connects the DUT's ground voltage back to the tester's instruments for use as the entire test system's 0 V "golden zero reference." If any voltage errors are introduced into this ground reference signal relative to the DUT's ground, all the instruments will produce DC voltage offsets.

5.2.3 Precision Voltage References and User Supplies

Mixed-signal testers sometimes include high-accuracy, low-noise voltage references. These voltage sources can be used in place of the general-purpose DC sources when the noise and accuracy characteristics of the standard DC source are inadequate. One common example of a precision voltage reference application is the voltage reference for a high-resolution ADC or DAC. Any noise and DC error on the DC reference of an ADC or DAC translates directly into gain error and increased noise, respectively, in the output of the converter. A precision voltage reference is sometimes used to solve this problem.

Testers may also include nonprogrammable user power supplies with high output current capability. These fixed supplies provide common power supply voltages (±5 V, ±15 V, etc.) for DIB circuits such as op amps and relay coils. This allows DIB circuits to operate from inexpensive fixed power supplies having high current capability instead of tying up the tester's more expensive programmable DC sources.

5.2.4 Calibration Source

The mixed-signal tester's calibration source was discussed in detail in Section 4.2. The purpose of a calibration source is to provide traceability of standards back to a central agency such as the National Institute of Standards and Technology (NIST). The calibration source must be recalibrated on a periodic basis (six months is a common period). Often, the source is removed from the tester and sent to a certified standards lab for recalibration. The old calibration source is replaced by a freshly calibrated one so that the tester can continue to be used in production. On some testers, the high-accuracy multimeter serves as the calibration source. Also, some testers may have multiple instruments that serve as the calibration sources for various parameters such as voltage or frequency. Clearly, this is a highly tester-specific topic. Calibration and standards traceability is discussed in more detail in Chapter 10, "Focused Calibrations."

5.2.5 Relay Matrices

A relay matrix is a bank of electromechanical relays that provides flexible interconnections between many different tester instruments and the DUT. There may be several types of relay matrix in a tester, but they all perform a similar task. At different points in a test program, a particular DUT input may require a DC voltage, an AC waveform, or a connection to a

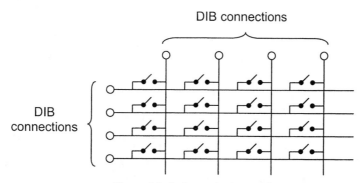

Figure 5.4. Instrument relay matrix.

voltmeter. A relay matrix allows each instrument to be connected to a DUT pin at the appropriate time.

A general-purpose 4 x 4 relay matrix is shown in Figure 5.4. General-purpose relay matrices are used to connect and disconnect various circuit nodes on the DIB board. They have no hardwired connections to tester instruments. Therefore, the purpose and functionality of a general-purpose relay matrix depends on the test engineer's DIB design. A more instrument-specific matrix is shown in Figure 5.5. It allows flexible interconnections between specific tester instruments and pins of the DUT through connections on the DIB board.

In addition to relay matrices, many other relays and signal paths are distributed throughout a mixed-signal tester to allow flexibility in interconnections without adding unnecessary relays to the DIB board. The exact architecture of relays, matrices, and signal paths varies widely from one ATE vendor's tester to the next.

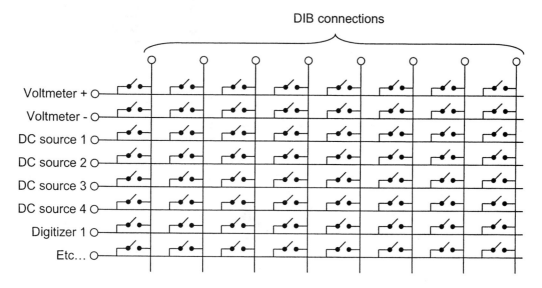

Figure 5.5. General-purpose relay matix.

5.2.6 Relay Control Lines

Despite the high degree of interconnection flexibility provided by the general-purpose relay matrix and other instrument interconnect hardware, there are always cases where a local DIB relay (placed near the DUT) is imperative. Usually the need for a local DIB relay is driven by performance of the DUT. For example, there is no better way to get a low-noise ground signal to the input of a DUT than to provide a local relay placed on the DIB directly between the DUT input and the DUT's local ground plane.

Certainly it is possible to feed the local ground through a DIB trace, through a remote relay matrix, and back through another DIB trace, but this connection scheme invariably leads to poor analog performance. The DIB traces are, after all, radio antennae. Many noise problems can be traced to poor layout of ground connections between the DUT and its ground plane. Local DIB relays minimize the radio antenna effect. Local DIB relays are also used to connect device outputs to various passive loads and other DIB circuits.

The test program controls the local DIB relays, opening and closing them at the appropriate time during each test. The relay coils are driven by the tester's relay control lines. A relay control line driver is shown in Figure 5.6. On some testers, the control line is capable of reading back the state of the voltage on the control line through a readback comparator. The readback comparator allows a low-cost method for determining the state of a digital signal.

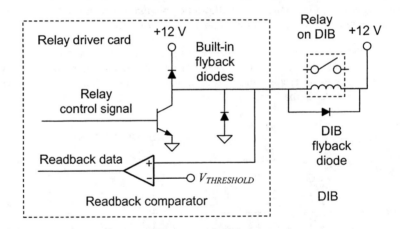

Figure 5.6. Relay coil driver with flyback protection diodes.

Relay coils produce an inductive kickback when the current is suddenly changed between the on and off states. The inductive kickback, or flyback as it is known, is induced according to the inductance formula $v(t) = L\, di/dt$. Since high kickback voltages could potentially damage the output circuits of the relay driver, its output circuits contain flyback protection diodes to shunt the excess voltage to a DC source or to ground. Many test engineers also add flyback diodes across the coils of the relay, as shown in Figure 5.6. The extra diode is probably redundant. However, many engineers consider it good practice to add extra flyback diodes even though they ake up quite a bit of DIB board space. To eliminate the board space issue, the test engineer can choose slightly more expensive relays with built-in flyback diodes.

5.3 DIGITAL SUBSYSTEM

5.3.1 Digital Vectors

A mixed-signal tester must test digital circuits as well as mixed-signal and analog circuits. The mixed-signal and digital-only sections of the DUT are exercised using the tester's digital subsystem. The digital subsystem can present high, low, and high-impedance (HIZ) logic levels to the DUT. It can also compare the outputs from the DUT against expected responses to determine whether the digital logic of the DUT has been manufactured without defects. The tester applies a sequence of drive data to the device and simultaneously compares outputs against expected results. Each drive/compare cycle is called a *digital vector*. A series of digital vectors is called a *digital pattern*. An example digital pattern for testing a simple 3-bit counter is shown in Figure 5.7. The vectors of a digital pattern are usually sourced at a constant frequency, although some testers allow the period of each vector to be set independently. The ability to change digital timing on a vector-by-vector basis is commonly called *timing on the fly*.

```
                                          R C
                                          E L
                                          S O
                                          E C Q Q Q
                                          T K 2 1 0
VECTOR    LABEL          COMMAND     TSET S S S S S COMMENT
0000000   TEST_COUNTER:                1   0 0 X X X
0000001                                1   0 1 X X X Reset Counter
0000002                                1   0 0 L L L Check for all 0's
0000003                                1   0 1 L L L
0000004                                1   1 0 L L L Release Reset
0000005                                1   1 1 L L H Test for 001
0000006                                1   1 0 X X X
0000007                                1   1 1 L H L Test for 010
0000008                                1   1 0 X X X
0000009                                1   1 1 L H H Test for 011
0000010                                1   1 0 X X X
0000011                                1   1 1 H L L Test for 100
0000012                                1   1 0 X X X
0000013                                1   1 1 H L H Test for 101
0000014                                1   1 0 X X X
0000015                                1   1 1 H H L Test for 110
0000016                                1   1 0 X X X
0000017                                1   1 1 H H H Test for 111
0000018                                1   1 0 X X X
0000019                                1   1 1 L L L Test for wrap
0000020                  HALT          1   1 0 X X X
```

Figure 5.7. Digital pattern for 3-bit binary counter.

5.3.2 Digital Signals

In addition to the simple pass/fail digital pattern tests, the tester must also be capable of sourcing and capturing digital signals. Digital signals are digitized representations of continuous waveforms such as sine waves and multitones. Digital signals are distinct from digital vectors in that they typically carry analog signal information rather than purely digital information. Usually, the samples of a digital signal must be applied to a DUT along with a repetitive digital pattern that keeps the device active and initiates DAC and/or ADC conversions. Each cycle of the repeating digital pattern is called a *frame*.

During a mixed-signal test, the repeating frame vectors must be combined with the nonrepeating digital signal sample information to form a repetitive sampling loop. Combining the digital frame vectors with digital signal data, a long sequence of waveform samples can be sent to or captured from the DUT with a very short digital frame pattern. In effect, the sampling frame results in a type of data compression that minimizes the amount of vector memory needed for the tester's digital subsystem.

Looping frames are commonly used when testing DACs and ADCs. A sequence of samples must be loaded into a DAC to produce a continuous sequence of voltages at the DAC's output. In the case of ADC testing, digital signals must be captured and stored into a bank of memory as the looping frame initiates each ADC conversion.

5.3.3 Source Memory

When testing DACs, the digital signal samples representing the desired DAC analog waveform are typically computed in the tester's main test program code. The digital signal samples are stored into a digital subsystem memory block called *source memory* (or *send memory* in some testers). The digital frame data, on the other hand, are stored in vector memory. To generate a repeating frame with a new sample for each loop, the contents of the vector memory and source memory are spliced together in real time as the digital pattern is executed.

An example digital pattern for a DAC sine wave test is shown in Figure 5.8. This pattern shows a combination of a looping frame of ones and zeros combined with digital signal placeholders (W symbols in this example). Ws are placed wherever analog waveform sample data is to be supplied by source memory. Each W may be either high or low during each loop of the frame, depending on the contents of source memory. The address pointer for source memory is incremented by one sample each time through the frame loop so that a series of different samples can be sent to the DAC.

Because its data are generated algorithmically by the main test program, a digital signal can be modified quickly without changing the frame loop pattern. The ability to quickly modify the digital signal data is especially useful during the DUT debug and characterization phase. For example, a DAC may normally be tested using a 1-kHz sine wave digital signal. During the DAC characterization phase, however, the frequency might be swept from 100 Hz to 10 kHz to look for problem areas in the DAC's design. This would be impossibly cumbersome if the digital pattern had to be generated using an expanded, nonlooping sequence of ones and zeros. In fact, some tester architectures attempt to substitute deep, nonlooping vector memory in place of source memory. This may reduce the cost of tester hardware, but it invariably results in frustrated users. One of the main differences between a mixed-signal tester and a digital tester with bolt-on analog instruments is the presence of source and capture memories in the digital subsystem. Other differences will be pointed out throughout this chapter.

5.3.4 Capture Memory

Devices such as ADCs produce a series of digitized waveform samples that must be captured and stored into a bank of memory called *capture memory* (or *receive memory*). Capture memory serves the opposite function of source memory. Each time the sampling frame is repeated, the digital output from the device is stored into the capture memory. The capture memory address pointer is incremented each time a digital sample is captured. Once a complete set of samples have been collected, they are transferred to an array processor or to the tester computer for analysis. A simple ADC sine wave test pattern is shown in Figure 5.9.

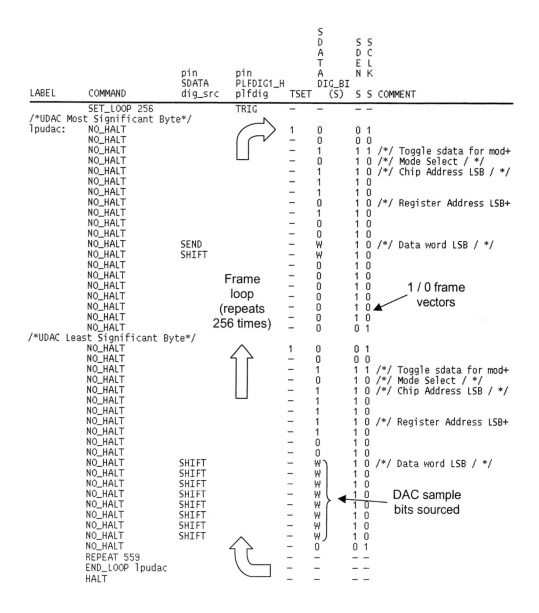

Figure 5.8. DAC data sourcing test pattern.

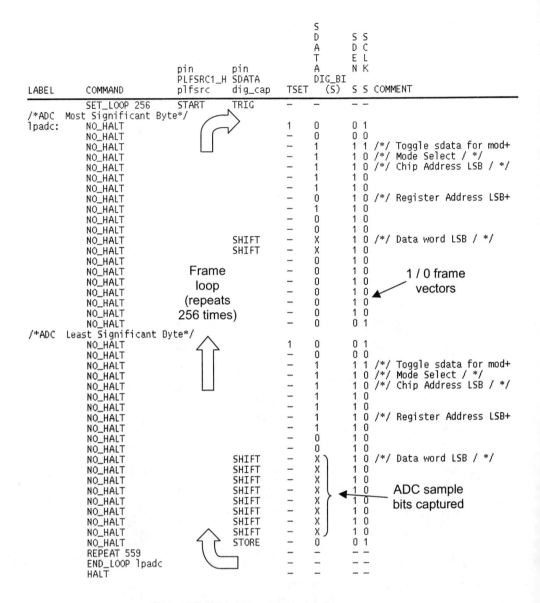

Figure 5.9. ADC data capture test pattern.

5.3.5 Pin Card Electronics

The pin card electronics for each digital channel are located inside the test head on most mixed-signal testers. A pin card electronics board may actually contain multiple channels of identical circuitry. Each channel's circuits consist of a programmable driver, a programmable

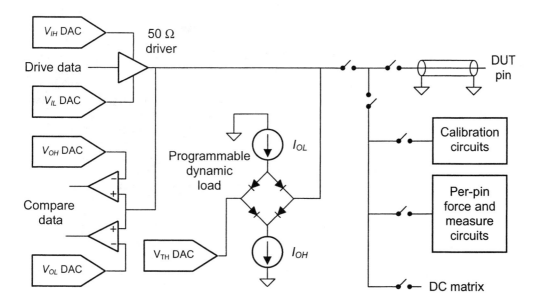

Figure 5.10. Digital pin card circuits.

comparator, various relays, dynamic current load circuits, and other circuits necessary to drive and receive signals to and from the DUT. A generic digital pin card is shown in Figure 5.10.

The driver circuitry consists of a fixed impedance driver (typically 50 Ω) with two programmable logic levels, V_{IH} and V_{IL}. These levels are controlled by a pair of driver level DACs whose voltages are controlled by the test program. The driver can also switch into a high-impedance state (HIZ) at any point in the digital pattern to allow data to come from the DUT into the pin card's comparator. The driver circuits may also include programmable rise and fall times, though fixed rise and fall times are more common. Normally the fixed rise and fall times are designed to be as fast as the ATE vendor can make them. Rise and fall times between 1 and 3 ns are typical in today's testers.

The comparator also has two programmable logic levels, V_{OH} and V_{OL}. These are also controlled by another pair of DACs whose voltages are controlled by the test program. The pin card comparator is actually a pair of comparators, one for the V_{OH} level and one for V_{OL}. If the DUT signal is below V_{OL}, then the signal is considered a logic low. If the DUT is above V_{OH}, then it is considered a logic high. If the DUT output is between these thresholds, then the output state is considered a midpoint voltage. If it is outside these thresholds, then it is considered a valid logic level. Comparator results can also be ignored using a mask. Thus there are typically three drive states (HI, LO, and HIZ) and five compare states (HI, LO, and MID, VALID, and MASK).

The usefulness of the valid comparison is not immediately obvious. If we want to test for valid V_{OH} and V_{OL} voltages from the output of a nondeterministic circuit such as an ADC, we cannot set the tester to expect HI or LO. This is because electrical noise in the ADC and tester will produce somewhat unpredictable results at the ADC output. However, we can set the tester to expect valid logic levels during the appropriate digital vectors without specifying whether the

ADC should produce a HI or a LO. While the pin card tests for valid logic levels, the samples from the ADC are collected into the digital capture memory for later analysis.

In addition to the drive and compare circuits, digital pin cards may also include dynamic load circuits. A dynamic load is a pair of current sources connected to the DUT output with a diode bridge circuit as shown in Figure 5.10. The diode bridge forces a programmable current into the DUT output whenever its voltage is below a programmable threshold voltage, V_{TH}. It forces current out of the DUT output whenever its voltage is above V_{TH}. The sink and source current settings correspond to the DUT's I_{OH} and I_{OL} specifications (see Section 3.12.4).

Another extremely important function that a digital pin card provides is its per-pin measurement capability. The per-pin measurement circuits of a pin card form a low-resolution, low-current DC voltage/current source for each digital pin. The per-pin circuits also include a relatively low-resolution voltage/current meter. The low-resolution and low-current capabilities are usually adequate for performing certain DC tests like continuity and leakage testing. These DC source and measure circuits can also be used for other types of simple DC tasks like input or output impedance testing.

Some testers may also include overshoot suppression circuits that serve to dampen the overshoot and undershoot characteristics in rapidly rising or falling digital signals. The overshoot and undershoot characteristics are the result of a low impedance DUT output driving into the DIB traces and coaxial cables leading to the digital pin card electronics. The ringing is minimized as the signal overshoot is shunted to a DC level through a diode.

Digital pin cards also include relays connected to other tester resources such as calibration standards and system DC meters and sources. These connections can be used for a variety of purposes, including calibration of the pin card electronics during the tester's system calibration process. The exact details of these connections vary widely from one tester type to another.

5.3.6 Timing and Formatting Electronics

When looking at a digital pattern for the first time, it is easy to interpret the ones and zeros very literally, as if they represent all the information needed to create the digital waveforms. However, most ATE testers apply timing and formatting to the ones and zeros to create more complicated digital waveforms while minimizing the number of ones and zeros that must be stored in pattern memory.

Timing and formatting is a type of data compression and decompression. The pattern data are formatted using the ATE tester's formatter hardware, which is typically located inside the tester mainframe or on the pin card electronics in the test head. Figure 5.11 shows how the pattern data are combined with timing and formatting information to create more complex waveforms. Notice that the unformatted data in Figure 5.11 require four times as much 1/0 information and four times the bit cell frequency to achieve the same digital waveform as the formatted data. Another key advantage to formatted waveforms is that the formatting hardware in a high-end mixed-signal tester is capable of placing the rising and falling edges with an accuracy of a few tens of picoseconds. This gives us better control of edge timing than we could expect to achieve using subgigahertz clocked digital logic.

The programmable drive start and stop times illustrated in Figure 5.11 are generated using digital delay circuitry inside the formatter circuits of the tester. Drive and compare timing is

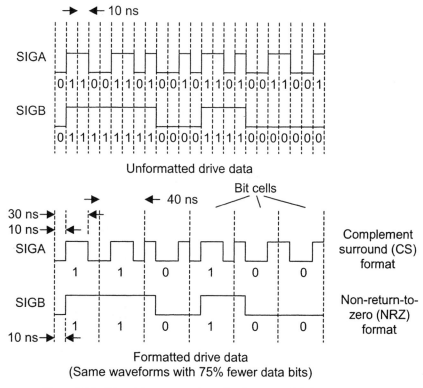

Figure 5.11. Drive data compression using formats and timing.

refined during a calibration process called *deskewing*. This allows subnanosecond accuracy in the placement of driven edges and in the placement of compare times (called *strobes* and *windows*). Strobe comparisons are performed at a particular point in time, while window comparisons are performed throughout a period of time. Window timing is typically used when comparing DUT outputs against expected patterns, while strobe timing is typically used when collecting data into capture memory. Again, this depends on the specific tester.

Figure 5.12 shows examples of several different formatting and timing combinations that create many different waveforms from the same digital data stream. In each case, the drive data sequence is 110X00. The compare data sequence is HHLXLL. Notice that certain formats such as Clock High and Clock Low ignore the pattern data altogether. Since digital pin cards can both drive and expect data, a distinction is made between a driven signal (1 or 0) and an expected signal (H or L). This notation is used for clarity in this book, though it is not universally used in the test industry. In fact, some digital pattern standards define H/L as driven data and 1/0 as expected data.

Example 5.2

Two digital signals, SIGA and SIGB, are generated by an ATE tester's pattern generator. The pattern generator's vector rate (i.e., its bit cell rate) is set to 4 MHz. SIGA is programmed to RO

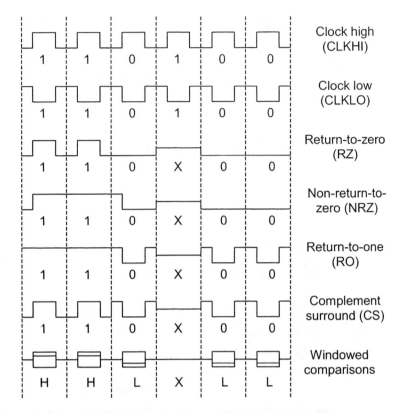

Figure 5.12. Some common digital formats.

format, while SIGB is programmed to NRZ format. The start time for SIGA is programmed to 50 ns and the stop time is programmed to 125 ns. Its initial state is programmed to logic high. The start time for SIGB is programmed to 25 ns and the stop time is programmed to 175 ns. Its initial state is programmed to logic low.

The following digital pattern is executed. Draw a timing diagram for the two signals SIGA and SIGB produced by this pattern. Show the bit cells in the timing diagram and calculate their period. Assume that we want to produce this same pair of signals using a bank of static random access memory (SRAM) whose address is incremented at a fixed rate (i.e., nonformatted ones and zeros). What SRAM depth would be required to produce this same pair of signals?

SIGA	SIGB
0	1
0	0
1	1
0	1
1	0
1	1

Solution:

Figure 5.13 shows the digital waveforms resulting from the specified pattern and timing set. The vector rate is specified to be 4 MHz; so the bit cell period is 250 ns. Also notice that NRZ format does not have a stop time; so the 175-ns stop time setting is irrelevant. In this example, all

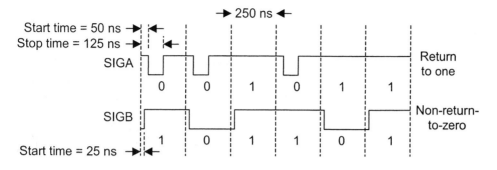

Figure 5.13. Formatted data using return-to-one and non-return-to-zero formats.

timing edges fall on 25-ns boundaries. If we wanted to generate this same pattern using nonformatted data from a bank of SRAM clocked at a fixed frequency, we would have to source a sequence of 6×(250 ns / 25 ns) = 60 bits from SRAM memory at a digital vector rate of 1/(25 ns) = 40 MHz.

5.4 AC SOURCE AND MEASUREMENT

5.4.1 AC Continuous Wave Source and AC Meter

The simplest way to apply and measure single-tone AC waveforms is to use a continuous wave source (CWS) and an RMS voltmeter. The CWS is simply set to the desired frequency and voltage amplitude to stimulate the DUT. The RMS voltmeter is equally simple to use. It is connected to the DUT output and the RMS output is measured with a single test program command.

But the CWS and RMS voltmeter suffer from a few problems. First, they are only able to measure a single frequency during each measurement. This would be acceptable for bench characterization, but in production testing it would lead to unacceptably long test times. As we will see in Chapters 6 through 9, DSP-based multitone testing is a far more efficient way to test AC performance because multiple frequencies can be tested simultaneously.

Another problem that the RMS voltmeter introduces is that it cannot distinguish the DUT's signal from distortion and noise. Using DSP-based testing, these various signal components can easily be separated from one another. This ability makes DSP-based testing more accurate and reliable than simple RMS-based testing. DSP-based testing is made possible with a more advanced stimulus/measurement pair, the arbitrary waveform generator and the waveform digitizer.

5.4.2 Arbitrary Waveform Generators

An arbitrary waveform generator (AWG) consists of a bank of waveform memory, a DAC that converts the waveform data into stepped analog voltages, and a programmable low-pass filter

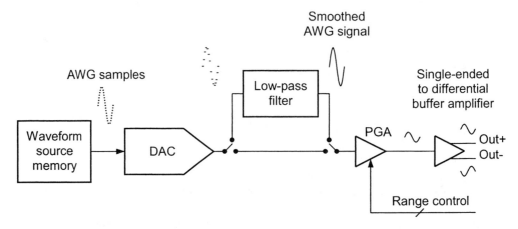

Figure 5.14. Arbitrary waveform generator.

section, which smoothes the stepped signal into a continuous waveform. An AWG usually includes an output scaling circuit (PGA) to adjust the signal level. It may also include differential outputs and DC offset circuits. Figure 5.14 shows a typical AWG and waveforms that might be seen at each stage in its signal path. (Mathematical signal samples are represented as dots to distinguish them from reconstructed voltages.)

An AWG is capable of creating signals with frequency components below the low-pass filter's cutoff frequency. The frequency components must also be less than one-half the AWG's sampling rate. This so-called Nyquist criterion will be explained in the next chapter, "Sampling Theory." An AWG might create the three-tone multitone illustrated in Figure 4.5. It might also be used to source a sine wave for distortion testing or a triangle wave (up ramp / down ramp) for ADC linearity testing (see Chapter 12, "ADC Testing"). Flexibility in signal creation is the main advantage of AWGs compared to simple sine wave or function generators.

5.4.3 Waveform Digitizers

An AWG converts digital samples from a waveform memory into continuous-time waveforms. A digitizer performs the opposite operation, converting continuous-time analog waveforms into digitized representations. The digitized samples of the continuous waveform are collected into a waveform capture memory. The structure of a typical digitizer is shown in Figure 5.15. A digitizer usually includes a programmable low-pass filter to limit the bandwidth of the incoming signal. The purpose of the bandwidth limitation is to reduce noise and prevent signal aliasing, which we will discuss in Chapter 6, "Sampling Theory."

Like the DC meter, the digitizer has a programmable gain stage at its input to adjust the signal level entering the digitizer's ADC stage. This minimizes the noise effects of quantization error from the digitizer's ADC. Waveform digitizers may also include a differential to single-ended conversion stage for measuring differential outputs from the DUT. Digitizers may also include a sample-and-hold circuit at the front end of the ADC to allow undersampled measurements of very high-frequency signals. Undersampling is explained in more detail in Chapter 6, "Sampling Theory."

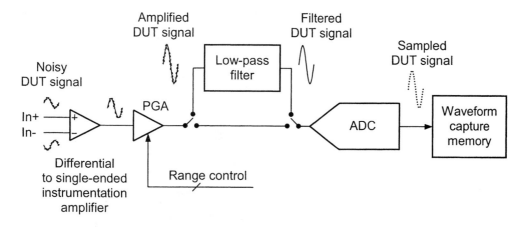

Figure 5.15. Waveform digitizer.

5.4.4 Clocking and Synchronization

Many of the subsections and instruments in a mixed-signal tester derive their timing from a central frequency reference. For example, the digital patterns in the frame loops in Figures 5.8 and 5.9 are generated at a specific frequency. This frequency determines the repetition rate of the sample loop, and therefore sets the frequency of the DAC or ADC sampling rates. The AWG and digitizer also operate from clock sources that must be synchronized to each other and to the digital pattern's frame loop repetition rate.

Figure 5.16 shows a clock distribution scheme that allows synchronized sampling rates between all the DSP-based measurement instruments. Since the clocking frequency for each instrument is derived from a common source, frequency synchronization is possible. Without precise sampling rate synchronization, the accuracy and repeatability of all the DSP-based measurements in a mixed-signal test program would be degraded.

The reason these clocks must all be synchronized will become more apparent in Chapter 6, "Sampling Theory," and Chapter 7, "DSP-Based Testing." Proper synchronization of sample rates between the various AWGs, digitizers, and digital pattern generators is another of the key distinguishing features of a mixed-signal tester. A digital tester with bolt-on analog instruments often lacks a good clocking and synchronization architecture.

5.5 TIME MEASUREMENT SYSTEM

5.5.1 Time Measurements

Digital and mixed-signal devices often require a variety of time measurements, such as frequency, period, duty cycle, rise and fall times, jitter, skew, and propagation delay. These parameters can be measured using the ATE tester's time measurement system (TMS).

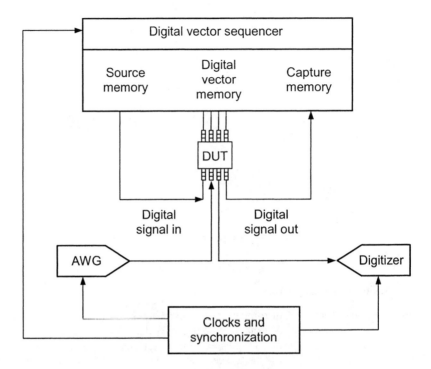

Figure 5.16. Synchronization in a mixed-signal tester.

Figure 5.17 illustrates several of the time measurement capabilities of a typical TMS. Most TMS instruments are capable of measuring these parameters within an accuracy of a few nanoseconds. Some of the more advanced TMS instruments can measure parameters such as jitter to a resolution of less than 1 ps.

Timing parameters that do not change from cycle to cycle (i.e., rise time, fall time, etc.) can sometimes be measured using a very high-bandwidth undersampling waveform digitizer. An undersampling digitizer is similar in nature to the averaging mode of a digitizing oscilloscope. Like digitizing oscilloscopes, undersampling digitizers require a stable, repeating waveform. Thus nonperiodic features such as jitter and random glitches cannot be measured using an undersampling approach. Unfortunately, undersampling digitizers are often considerably slower than dedicated time measurement instruments.

5.5.2 Time Measurement Interconnects

One of the most important questions to consider about a TMS instrument is how its input and interconnection paths affect the shape of the waveform to be measured. It does little good to measure a rise time of 1 ns if the shape of the signal's rising edge has been distorted by a 50-Ω coaxial connection. It is equally futile to try to measure a 100-ps rising edge if the bandwidth of the TMS input is only 300 MHz. Accurate timing measurments require a high-quality signal path between the DUT output and the TMS time measurement circuits.

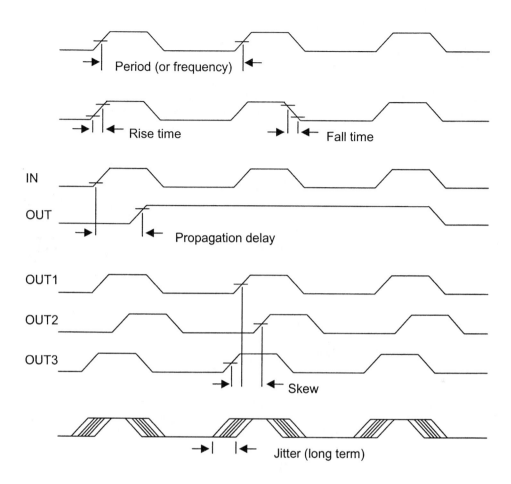

Figure 5.17. Time measurements.

5.6 COMPUTING HARDWARE

5.6.1 User Computer

Mixed-signal testers typically contain several computers and signal processors. The test engineer is most familiar with the user computer, since this is the one which is attached to the keyboard. The user computer is responsible for all the editing and compiling processes necessary to debug a test program. It is also responsible for keeping track of the datalogs and other data collection information. On low-cost testers, the user computer may also drive the measurement electronics as well. On more advanced mainframe testers, the execution of the test program, including I/O functions to the tester's measurement electronics, may be delegated to one or more tester computers located inside the tester's mainframe.

5.6.2 Tester Computer

The tester computer executes the compiled test program and interfaces to all the tester's instruments through a high-speed data backplane. By concentrating most of its processing power on the test program itself, the tester computer can execute a test program more efficiently than the user computer. The tester computer also performs all the mathematical operations on the data collected during each test. In some cases, the more advanced digital signal processing (DSP) operations may be handled by a dedicated array processor to further reduce test time. However, computer workstations have become fast enough in recent years that the DSP operations are often handled by the tester computer itself rather than a dedicated array processor.

5.6.3 Array Processors and Distributed Digital Signal Processors

Many mixed-signal testers include one or more dedicated array processors for performing DSP operations quickly. This is another difference between a mixed-signal tester and a bolted-together digital/analog tester. Some mixed-signal instruments may even include local DSP processors for computing test results before they are transferred to the tester computer. This type of tester architecture and test methodology is called *distributed processing*. Distributed processing can reduce test time by splitting the DSP computation task among several processors throughout the tester. Test time is further reduced by eliminated much of the raw data transfer that would otherwise occur between digitizer instruments and a centralized tester computer or array processor. Unfortunately, distributed processing may have the disadvantage that the resulting test code may be harder to understand and debug.

5.6.4 Network Connectivity

The user computer and/or tester computer are typically connected into a network using ethernet or similar networking hardware. This allows data and programs to be quickly transferred to the test engineer's desk for offline debugging and data analysis. It also allows for large amounts of production data to be stored and analyzed for characterization purposes.

5.7 SUMMARY

In this chapter we have examined many of the common building blocks of a generic mixed-signal tester. Of course, there are many differences between any two ATE vendors' preferred tester architectures. For example, ATE Vendor A may use a sigma-delta-based digitizer and AWG, while ATE Vendor B may choose to use a more conventional sucessive approximation architecture for its AWG and digitizer. Each architecture has advantages and disadvantages, which the test engineer must deal with. The test engineer's approach to measuring a given parameter will often be driven by the vendor's architectural choices. In the end, though, each tester has to test the same variety of mixed-signal parameters regardless of its architectural peculiarities. A test engineer's job often involves testing parameters the tester was simply not designed to measure. This can be one of the more challenging and interesting parts of a test engineer's task.

In the following chapters we will see how digitizers, AWGs, and digital pattern generators, combined with digital signal processing, can provide greater speed and accuracy than conventional measurement techniques. We will also explain why it is so critical to mixed-signal testing that we achieve precise synchronization of sampling frequencies bewteen all the tester's instruments. Although the next two chapters represent some of the most difficult material in the book, they also contain some of the most important material. Most mixed-signal testing involves DSP-based measurements of one type or another; so the student will need to devote special attention to these chapters.

Problems

5.1. Name at least six types of subsystems found in a typical mixed-signal tester.

5.2. What is the purpose of the low-pass filter in a DC multimeter's front end?

5.3. What is the purpose of the PGA in a DC multimeter's front end?

5.4. A single-ended DC voltmeter features a sample-and-difference front-end circuit. We wish to use this meter to measure the differential offset voltage of a DUT's output buffer. Each of the two outputs is specified to be within a range of 3.5 V ± 25 mV, and the differential offset is specified in the device data sheet to be ±15 mV. The meter input can be set to any of the following ranges: ±10 V, ±5 V, ±2 V, and ±1 V. The meter has a maximum error of 0.1% of its programmed range. The error includes all sources of inaccuracy (quantization error, linearity error, gain error, etc.). Compare the accuracy achieved using two simple DC measurements with the accuracy achieved using the sample-and-difference circuit. Assume no errors due to nonrepeatability.

5.5. Why are Kelvin connections used to connect high-current DC power supplies to the DUT?

5.6. Name an instance where a local DIB relay might prove to be a better choice for interconnecting signals than a general-purpose relay matrix.

5.7. What is the purpose of the diodes in the output stage of the relay driver in Figure 5.6?

5.8. What is the difference between a digital pattern and a digital signal?

5.9. Why are the number of vectors in the frame loop and the frequency of the digital vectors in a sampling frame important when developing a digital pattern for a mixed-signal test?

5.10. What is the purpose of source memory?

5.11. What is the purpose of capture memory?

5.12. In Figure 5.8, SDATA is a serial input/output (I/O) interface to a DUT containing a 10-bit DAC. The drive data for SDATA consists of a combination of ones, zeros, and Ws. The ones and zeros represent digital logic states that select the DAC for writing. The 10-bit write is broken into two eight-bit write operations. (The first 8-bit write operation contains only the two most significant bits of DAC data.) The Ws represent digital signal data. The digital vectors are supplied at a constant rate of 6 MHz. This pattern supplies 256 samples to the DAC using a total of 600 vectors (40 + 559 + 1) per frame loop. At what rate are the digital signal samples written to the DAC? How long does it take to supply all 256 samples to the DAC?

5.13. In Figure 5.9, the SDATA interface is used to read samples from an ADC located on the DUT from Problem 5.12. The Xs on SDATA represent the time at which the 10 bits of each ADC sample are captured into capture memory. The Xs represent a high-impedance drive state. Why might Xs be required at this point in the pattern rather than ones and zeros?

5.14. Why is formatting and timing information combined with one/zero information to produce digital waveforms?

5.15. A series of digital bits are driven from a digital pin card at a rate of 1 MHz (1-µs period). The series of bits are 10110X1. The format for this pin is set to return-to-zero (RZ) format. Its initial state is set to logic low. The start time for the drive data is set to 500 ns, and the stop time is set to 900 ns. Draw this waveform using the notation in Figure 5.12. Draw the waveform timing approximately to scale. Next, draw the waveform that would result if we set the format to non-return-to-zero (NRZ). To produce these waveforms using clocked digital logic without timing and formatting circuits, what clock rate would be required? If we wanted to be able to set the start and stop times to 500 and 901 ns, respectively, at what rate would we have to operate the clocked digital logic?

5.16. Name two reasons that AWGs and digitizers are used in mixed signal testing rather than CW sources and RMS voltmeters.

5.17. What is the purpose of the low-pass filter in the AWG illustrated in Figure 5.14?

5.18. Why is a programmable gain amplifier needed in the front end of the waveform digitizer illustrated in Figure 5.15?

5.19. What is the purpose of distributed digital signal processing hardware?

CHAPTER 6

Sampling Theory

6.1 ANALOG MEASUREMENTS USING DSP

6.1.1 Traditional versus DSP-Based Testing of AC Parameters

AC measurements such as gain and frequency response can be measured with relatively simple analog instrumentation, as mentioned in Section 5.4.1. To measure gain, an AC continuous sine wave generator can be programmed to source a single tone at a desired voltage level, V_{in}, and at a desired frequency. A true RMS voltmeter can then measure the output response from the DUT, V_{out}. Then gain can be calculated using a simple formula: $gain = V_{out}/V_{in}$.

The pure analog approach to AC testing suffers from a few problems, though. First, it is relatively slow when AC parameters must be tested at multiple frequencies. For example, each frequency in a frequency response test must be measured separately, leading to a lengthy testing process. Second, traditional analog instrumentation is unable to measure distortion in the presence of the fundamental tone. Thus the fundamental tone must be removed with a notch filter, adding to test hardware complexity. Third, analog testing measures RMS noise along with RMS signal, making results unrepeatable unless we apply averaging or band-pass filtering.

In the early 1980s, a new approach to production testing of AC parameters was widely adopted in the ATE industry. The new approach became known as *DSP-based testing*.[1] Digital signal processing (DSP) is a powerful methodology that allows faster, more accurate, more repeatable measurements than traditional AC measurements using an RMS voltmeter. A mixed-signal test engineer will never be fully competent without a strong background in signal processing theory. Unfortunately, a full treatment of sampling theory and DSP is well beyond the scope of this book. Other texts have covered the subject of signal processing in much more detail.[2-4]

The reader is assumed to already have a strong theoretical background in DSP, although this book will undoubtedly fall into the hands of the DSP novice as well. We will review the basics of sampling theory and DSP as they apply to mixed-signal testing, without giving the subject an in-depth treatment. Hopefully, this introductory coverage will both refresh the experienced reader's memory of DSP and allow the novice to understand the fundamentals of DSP-based testing.

Before we can discuss DSP-based testing, we must first understand sampling theory for both analog-to-digital converters and digital-to-analog converters. In this chapter, we will examine the basics of sampling theory before proceeding to a more detailed study of DSP-based testing in Chapter 7.

6.2 SAMPLING AND RECONSTRUCTION

6.2.1 Use of Sampling and Reconstruction in Mixed-Signal Testing

Sampling and reconstruction are the processes by which signals are converted from the continuous (i.e., analog) signal domain to the discrete (i.e., digital) signal domain and back again. Both sampling and reconstruction are used extensively in mixed-signal testing. The ATE tester samples and reconstructs signals to stimulate the DUT and measure its response. The DUT may also sample and reconstruct signals as part of its normal operation. Both mathematical and physical sampling and reconstruction occur as the DUT is tested. Figure 6.1 illustrates the various types of sampling and reconstruction that occur when the voice-band interface circuit of Figure 1.2 is tested.

In a purely mathematical world, a continuous waveform can be sampled and then reconstructed without loss of signal quality, as long as a few constraints are met. Unfortunately, a number of imperfections are introduced in the physical world that make the conversion between continuous time and discrete time fall short of the mathematical theory. Many of these imperfections will be discussed in this section.

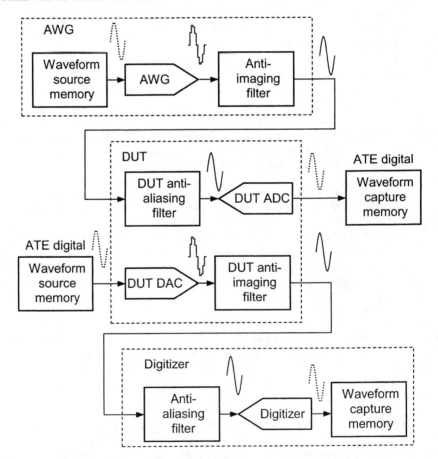

Figure 6.1. Various test signals associated with a voice-band interface circuit.

6.2.2 Sampling: Continuous-Time and Discrete-Time Representation

Many signals in the physical world around us are continuous (i.e., analog) in nature. Familiar examples of real-world analog signals include sound waves, light intensity, temperature, and pressure. Many modern electronic systems, such as the cellular telephone example in Chapter 1, must convert the continuous signals in the physical world into discrete digital representations compatible with digital storage, digital transmission, and mathematical processing. Continuous signals are often described by mathematical equations, such as

$$v(t) = A \sin(2\pi f_o t + \phi) \tag{6.1}$$

where $v(t)$ is a continuous function of time t, whose value in this particular case changes in a sinusoidal manner with amplitude A, frequency f_o, and phase shift ϕ.

Sampling is a process in which a continuous-time signal is converted into a sequence of discrete samples uniformly spaced at intervals of T_s seconds, often written as

$$v[n] = v(t)\big|_{t=nT_s} \tag{6.2}$$

where $v[n]$ defines the values of $v(t)$ at the sampling instants defined at $t=nT_s$. Such a process is depicted in Figure 6.2. We refer to T_s as the *sampling period* and its reciprocal $F_s=1/T_s$, as the *sampling frequency* or *sampling rate,* and n as an arbitrary integer. To simplify our notation, it is common practice to drop the T_s term in the argument of Eq. (6.2) as it is assumed to be constant for all time. The continuous waveform $v(t)$ is said to exist in continuous time, while the sampled waveform $v[n]$ is said to exist in discrete time. For example, substituting Eq. (6.1) into (6.2), we can write

$$v[n] = A \sin(2\pi f_o nT_s + \phi) = A \sin\left(2\pi \frac{f_o}{F_s} n + \phi\right) \tag{6.3}$$

For reasons that will become clear later in this chapter, we often impose the condition that the ratio f_o/F_s be a rational fraction, $f_o/F_s = M/N$, where M and N are integers, allowing one to write

$$v[n] = A \sin\left(2\pi \frac{M}{N} n + \phi\right) \tag{6.4}$$

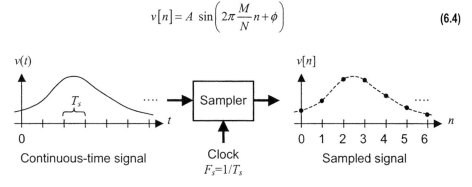

Figure 6.2. Continuous-time signal and its sampled equivalent.

Discrete signals such as this can then be stored in computer arrays and processed using DSP functions.

Up to this point we have defined a sampled waveform in the discrete-time domain as a sequence of numbers defined by $v[n]$. We can also define a sampled waveform as a continuous function of time. The use of this alternative notation is important in the next section where the samples are converted back into the original continuous-time signal. To enable such a description we must make use of the concept of impulse functions. Mathematically, an impulse function, denoted by $\delta(t)$, is defined as having zero amplitude everywhere except at $t=0$, where it is infinitely large in such a way that it contains unit area under its curve, as depicted by the following two rules

$$\delta(t) = 0, \ t \neq 0 \qquad (6.5)$$

and

$$\int_{-\infty}^{\infty} \delta(t)\, dt = 1 \qquad (6.6)$$

It is important to realize that no function in the ordinary sense can satisfy these two rules. However, we can imagine a sequence of pulselike functions that have progressively taller and thinner peaks, with the area under the curve remaining equal to unity as illustrated in Figure 6.3(a). If we take this argument to the limit, letting the pulse width go to zero while the pulse height goes to infinity, then we have what we refer to as an impulse function. It should be obvious from this description that we are going to encounter some difficulty in graphing the impulse function. Hence, an impulse is graphically represented by an arrow whose height is equal to the area (voltage × time) under the impulse, as shown in Figure 6.3(b).

An important property of impulse functions is the so-called *sifting property*, defined by

$$\int_{-\infty}^{\infty} v(t)\, \delta(t - t_0)\, dt = v(t_0) \qquad (6.7)$$

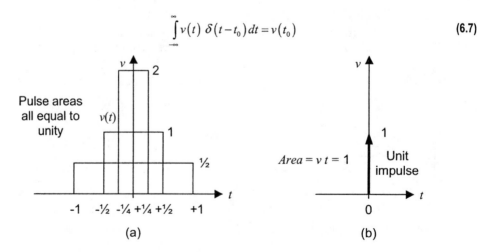

Figure 6.3. Impulse definition.

Here the impulse function selects or sifts out a particular value of the function $v(t)$, namely, the value at $t=t_0$, in the integration process. If $v_a(t)$ denotes a signal that has been uniformly sampled every T_s seconds, then we can make use of the sifting property and write the following mathematical representation for $v_a(t)$ in terms of a series of evenly spaced, equally sized impulse functions, commonly referred to as a *unit impulse train*

$$v_a(t) = \sum_{n=-\infty}^{\infty} v(t)\,\delta(t-nT_s) \tag{6.8}$$

Figure 6.4 illustrates the impulse representation of a sequence of samples from a continuous-time signal. Mathematically, the impulses are equal to the multiplication of the continuous-time signal times a unit impulse train.

Equivalently, through direct application of the sifting property of the impulse function, we can write Eq. (6.8) as

$$v_a(t) = \sum_{n=-\infty}^{\infty} v(nT_s)\,\delta(t-nT_s) \tag{6.9}$$

Note that $v_a(t)$ is not defined at the sampling instants because $\delta(t-nT_s)$ is not defined at $t=nT_s$. However, one must keep in mind that the values of $v_a(t)$ at the sampling instants are embedded in the area carried by each impulse function. It should now be clear that $v_a(t)$ and $v[n]$ are different but equivalent models of the sampling process in the continuous-time and discrete-time domains, respectively. In order to keep track of which domain we are working in (i.e., continuous or discrete) we shall make use of parentheses to encompass the argument of a continuous-time signal, $v(nT_s)$, and square brackets, $v[n]$, to denote a discrete-time signal.

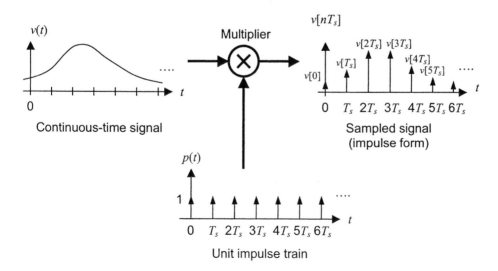

Figure 6.4. A continuous-time representation of a sampled signal as a series of impulses created by multiplying the original continuous-time signal by a unit impulse train.

Exercises:

6.1. Using MATLAB or an equivalent software program, plot 64 samples of a sine wave having unity amplitude, zero phase shift, and a period of 16 samples. We shall refer to this plotting range as the *observation interval*. (The *stem* command in MATLAB is an effective method for plotting discrete samples as a function of time.)

Ans. Setting $A=1$, $\phi=0$, $N=16$, and $M=1$, we get:

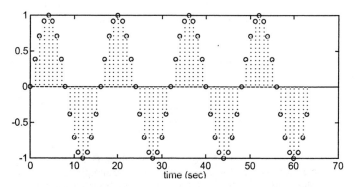

6.2. Using MATLAB or an equivalent software program, plot 16 samples of a sine wave having an amplitude of 2, $\pi/4$ phase shift, and a period of 16/3 samples.

Ans. Setting $A=2$, $\phi=\pi/4$, $N=16$, and $M=3$, we get:

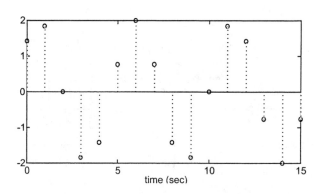

6.2.3 Reconstruction

The inverse operation of sampling is reconstruction. Reconstruction is a process in which a sampled waveform (impulse form) is converted into a continuous waveform by a circuit such as a digital-to-analog converter (DAC) and an anti-imaging filter. In effect, reconstruction is the operation that fills in the missing waveform that appears between samples. In essence, the combined effect of the DAC and filter can be modeled as a single reconstruction operation denoted with impulse response $p(t)$ as shown in Figure 6.5. Mathematically speaking, the

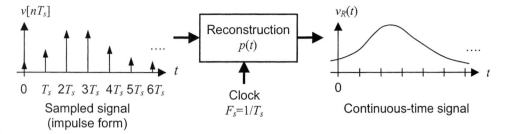

Figure 6.5. Reconstructing a continuous-time signal from a data sequence.

reconstruction operation performs interpolation between sampled values. A general formula that describes the operation of reconstruction is given by

$$v_R(t) = \sum_{n=-\infty}^{\infty} v(nT_s) p(t - nT_s) \tag{6.10}$$

The shape of the impulse response defines the shape of the waveform between adjacent samples. Thus $p(t)$ is commonly referred to as the *characteristic pulse shape* of the reconstruction operation. Eq. (6.10) states that each sample is multiplied by a delayed version of $p(t)$ and the resulting waveforms are added together to form $v_R(t)$. In other words, at each sample time $t=nT_s$, a pulse $p(t-nT_s)$ is generated with an amplitude proportional to the sample value $v(nT_s)$. Collectively, all the pulses are summed to form the output continuous signal $v_R(t)$. The general form of Eq. (6.10) appears often in the study of linear, time-invariant continuous-time systems. It is given a special name, *convolution*, and we say that the output is obtained by *convolving* the continuous-time equivalent signal of $v(nT_s)$ with $p(t)$. We shall have more to say about this in a moment.

Example 6.1

An input sequence $v[n]$ derived from a sinusoid has the following sampled values {0, 0.50, 0.87, 1.00, 0.87, 0.50, 0} corresponding to $n = 0, \ldots, 6$. Everywhere else the sequence is assumed to be zero. Using a triangular reconstruction pulse shape $p(t)$ defined as follows

$$p(t) = \begin{cases} 1 - |t - 1| & 0 \le t < 2 \\ 0 & \text{elsewhere} \end{cases} \tag{6.11}$$

plot the output waveform $v_R(t)$. Assume a sampling period, T_s, of 1 s.

Solution:

To begin, a plot of the characteristic pulse $p(t)$ is shown in Figure 6.6(a). As is evident, $p(t)$ is a triangular waveform with a pulse duration that lasts for 2 s and has a peak value of 1. Following Eq. (6.10), with the limits of summation changed from 0 to 6 (as all other sample values are assumed equal to zero), we can write the reconstructed waveform as

$$v_R(t) = \sum_{n=0}^{6} v(nT_s) p(t - nT_s)$$

or when expanded as

$$v_R(t) = v(0)p(t) + v(T_s)p(t-T_s) + v(2T_s)p(t-2T_s) + \cdots + v(6T_s)p(t-6T_s)$$

Now we can substitute an expression for each shifted $p(t)$ according to Eq. (6.11). However, it is more instructive to demonstrate this by superimposing all the pulses weighted by the sampled value on one time axis as shown in Figure 6.6(b). At any particular time point, we can add up the contribution from each pulse, and form a single point on the reconstructed waveform. This is shown in the figure for $t=2.6$ s. This same operation can be repeated for all the remaining time points. The result is a straight-line interpolation between adjacent sampled values.

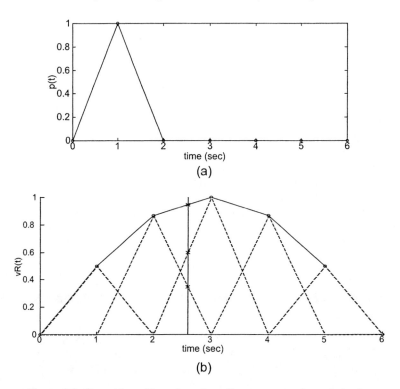

Figure 6.6. Convolving a triangular pulse with a sequence of sampled values.

Most DACs make use of a square characteristic pulse, as it is easiest to realize in practice. The sum of all shifted and scaled square pulses will result in a "staircase" continuous-time waveform, as shown in Figure 6.7. It is also evident that the staircase waveform is a rather poor approximation of the original waveform. A better approximation would certainly be obtained by increasing the number of steps per period used to reconstruct the waveform. However, the upper frequency range of the DAC limits this approach. It is also clear from Figure 6.7 that the reconstructed waveform $v_R(t)$ contains a large amount of undesirable high-frequency energy, as the reconstructed signal is made up of various sized pulses. To eliminate this high-frequency

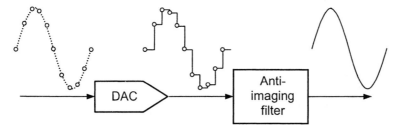

Figure 6.7. Illustrating the reconstruction operation with a DAC and an anti-imaging filter circuit.

energy, the DAC is usually followed by a postfiltering circuit, typically one with a low-pass characteristic having a cutoff frequency of at most one-half F_s. Such a filter is known under different names as a smoothing or anti-imaging filter. Collectively, the DAC and the anti-imaging filter are called a *reconstruction filter*.

Cascading a filter after the DAC effectively alters the characteristic pulse $p(t)$ of the reconstruction process and provides a much better approximation to the original waveform. In fact, perfect reconstruction can be obtained if the characteristic pulse of the overall reconstruction process has the following form

$$p(t) = \frac{\sin\left(\frac{\pi}{T_s}t\right)}{\frac{\pi}{T_s}t} \quad \text{for} \quad -\infty < t < \infty \tag{6.12}$$

This is a very long pulse, and its infinite length implies that to reconstruct a signal at time t exactly requires all the samples, not just those around that time. Substituting Eq. (6.12) into (6.10) allows us to write an exact interpolation formula for recovering the continuous-time information from the sampled values as

$$v_R(t) = \sum_{n=-\infty}^{\infty} v(nT_s) \frac{\sin\left[\frac{\pi}{T_s}(t-nT_s)\right]}{\frac{\pi}{T_s}(t-nT_s)} \tag{6.13}$$

It is interesting to note that $v_R(t)$ is equal to $v(nT_s)$ at all the sampling instants as the $\sin(x)/x$ term in Eq. (6.13) is equal to one.

In practice, a perfect reconstruction operation can only be approached, not actually realized. Consequently, some imperfections are introduced in the reconstruction process. There are two main sources of errors: (1) aperture effect due to the characteristic pulse shape, and (2) magnitude and phase errors related to the anti-imaging filter. Both types of errors lead to frequency-dependent magnitude and phase errors. If either error is an important parameter of a particular test, then they would need to be measured and corrected using a focused calibration procedure (see Chapter 10, "Focused Calibrations").

Convolution Using MATLAB

Convolution is a frequently used operation in the study of linear systems. For this reason, MATLAB has provided a built-in function called **CONV** to assist in such analysis. As with all computer operations, it works exclusively with discrete values. Hence, we can only approximate a continuous-time reconstructed waveform with a sequence of finely spaced sample values. Nonetheless, such an operation provides a useful aid to help us visualize the reconstructed waveform.

A set of N discrete values $v(nT_s)$ are obtained by sampling a waveform at a rate of T_{s1}. According to Eq. (6.10), the reconstructed waveform $v_R(t)$ is given by

$$v_R(t) = \sum_{n=-\infty}^{\infty} v(nT_{s1}) p(t - nT_{s1}) \tag{6.14}$$

This waveform is then sampled with a finer sampling period T_{s2} such that $T_{s1} = LT_{s2}$, where L is a positive integer. Substituting $t = kT_{s2}$ into Eq. (6.14) allows us to write the oversampled waveform in terms of the new sampling period T_{s2} as

$$\begin{aligned} v_R[k] &= v_R(t)\big|_{t=kT_{s2}} \\ &= \sum_{n=-\infty}^{\infty} v(nT_{s1}) p(kT_{s2} - nT_{s1}) \\ &= \sum_{n=-\infty}^{\infty} v(nLT_{s2}) p\big((k-nL)T_{s2}\big) \end{aligned} \tag{6.15}$$

Now, if we make the change of variable substitution, $m = nL$, and drop the time reference T_{s2}, we can rewrite Eq. (6.15) using discrete-time notation as

$$v_R[k] = \sum_{m=-\infty}^{\infty} v[m] p[k-m] \tag{6.16}$$

This equation is known as a *convolution summation* and is the basis of linear, time-invariant discrete-time systems. Here we speak about the sampled characteristic pulse, $p[k]$ convolving with the signal $v[k]$. To perform this operation using MATLAB we must first expand the time scale of the original sampled signal $v[k]$, as it has been resampled at its many zero locations.

If N samples are stored in a vector, say **V**, then we simply insert $L-1$ zeros between each sample value, increasing the size of the modified vector, say, **Vexpand,** to $(N-1) \times L$. Likewise, P samples of $p(k)$ are stored in a vector, **P**. Here the number of samples stored in **P** is determined as the ratio of the pulse duration, say, T_p, to the new sampling period, T_{s2}.

It is important to maintain the relative pulse duration with respect to the original sample values. Subsequently, we can then perform the convolution sum using the built-in function called **CONV** with vectors **Vexpand** and **P**, resulting in a new vector **Vr** of length $(N-1) \times L + P - 1$. This is executed in MATLAB using the following routine:

```
% Reconstruction Routine (V and P are input vectors)
Vexpand=zeros((N-1)*L,1); % expand the time scale
for m=1:N,
    Vexpand((m-1)*(L)+1)=V(m);
end
Vr = CONV(P, Vexpand); % convolve P with Vexpand
```

Example 6.2

Reconstruct and plot the sampled sinusoid given in Exercise 6.1 over one full period using a square characteristic pulse described as

$$p(t) = \begin{cases} 1 & 0 \leq t < 2 \\ 0 & \text{elsewhere} \end{cases}$$

Assume the sample rate is 1 s. Interpolate between sampled values using 16 sample points.

Solution:

From Exercise 6.1 we can write the sampled sequence vector as

V = [0, 0.3827, 0.7071, 0.9239, 1.0000, 0.9239, 0.7071, 0.3827, 0, -0.3827, -0.7071, -0.9239, -1.0000, -0.9239, -0.7071, -0.3827]

Likewise, using 16 points to define $p(t)$ over the 1-s pulse duration results in the vector:

P = [1, 1, 1, 1, 1, 1, 1, 1, 1, 1, 1, 1, 1, 1, 1, 1]

Executing the reconstruction MATLAB routine, we get the staircase plot shown in Figure 6.8.

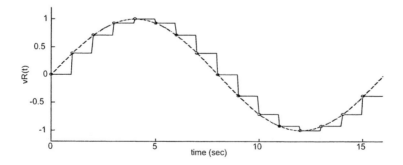

Figure 6.8. Unfiltered DAC output (MATLAB simulation).

Superimposed on the plot in Figure 6.8 is the original sinusoid. We clearly see that the reconstructed waveform is a poor approximation to the original sinusoid. Our next example will explore the same sample values but will use a more effective pulse shape.

Example 6.3

Repeat Example 6.2, but this time use a triangle characteristic pulse described as

$$p(t) = \begin{cases} 1 - |t-1| & 0 \le t < 2 \\ 0 & \text{elsewhere} \end{cases}$$

Assume the sample rate is 1 s and interpolate between sampled values using 16 sample points.

Solution:

From our previous example, we can write the sampled sequence vector as

V = [0, 0.3827, 0.7071, 0.9239, 1.0000, 0.9239, 0.7071, 0.3827, 0, -0.3827, -0.7071, -0.9239, -1.0000, -0.9239, -0.7071, -0.3827]

Likewise, using 32 points to define $p(t)$ (as it has a 2-s duration) results in the vector:

P = [0, 0.0625, 0.1250, 0.1875, 0.2500, 0.3125, 0.3750, 0.4375, 0.5000, 0.5625, 0.6250, 0.6875, 0.7500, 0.8125, 0.8750, 0.9375, 1.0000, 0.9375, 0.8750, 0.8125, 0.7500, 0.6875, 0.6250, 0.5625, 0.5000, 0.4375, 0.3750, 0.3125, 0.2500, 0.1875, 0.1250, 0.0625]

Executing the reconstruction MATLAB routine, we get the plot shown in Figure 6.9.

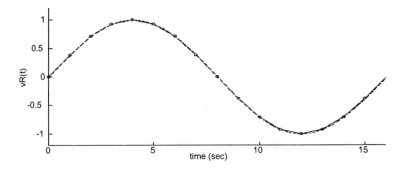

Figure 6.9. Low-pass filtered DAC output (MATLAB simulation).

Also superimposed in the plot is the original sinusoidal signal. As is evident, the reconstructed waveform is very similar to the original sinusoidal signal. In fact, little difference is visible. It is therefore fair to say that a reasonable approximation of the original continuous signal is obtained by joining adjacent sample value with straight lines (this assumes that there are at least 10 points per period).

Exercises:

6.3. Reconstruct the sampled signal displayed in Exercise 6.1 using the triangular pulse described by Eq. (6.11).

Ans. Interconnect sample points with straight lines

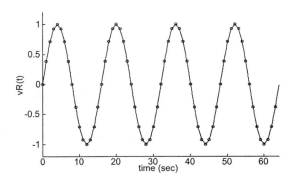

6.2.4 The Sampling Theorem and Aliasing

The sampling examples of the previous subsections are all performed in accordance with the *sampling theorem*. Shannon introduced the idea back in 1949 for application in communication systems. For this reason, it is sometimes referred to as the *Shannon sampling theorem*. However, interest and knowledge of the sampling theorem in engineering applications can be traced back to Nyquist in 1928, and as far back as 1915 in the literature of mathematicians. For a historical account of the sampling theorem, interested readers can refer to Jerri[6] for a detailed account. Specifically, the sampling theorem for band-limited signals can be stated in two separate but equivalent ways:

The Sampling Theorem

1. A continuous-time signal with frequencies no higher than F_{max} is completely described by specifying the values of the signal at instants of time separated by $1/(2F_{max})$ seconds.

2. A continuous-time signal with frequencies no higher than F_{max} may be completely recovered from a knowledge of its samples taken at the rate of $2F_{max}$ per second.

The sampling rate $2F_{max}$ is called the *Nyquist rate*, and its reciprocal is called the *Nyquist interval*. The Nyquist rate is the minimum sampling rate allowable by the sampling theorem. Although somewhat confusing at times, the *Nyquist frequency* refers to F_{max}.

The first part of the sampling theorem is exploited by ATE digitizers. Part 2 of the theorem is exploited by waveform generators. For example, a 10-kHz sine wave appearing at the output of

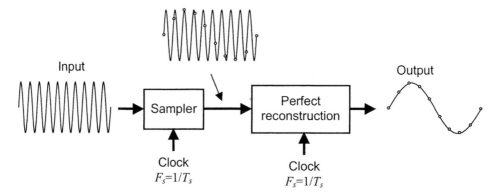

Figure 6.10. Undersampled sine wave and its reconstructed image.

a DUT can theoretically be sampled by the digitizer at 20.1 kHz with no loss of signal information. However, if it is sampled at a slightly lower frequency of 19.9 kHz, specific information about its characteristics are lost. To better understand this, consider the setup shown in Figure 6.10 consisting of a sampler driven by a sine wave, followed by a perfect reconstruction operation.

Ideally, if Shannon's theorem is satisfied, the output of this arrangement should correspond exactly with the input signal (i.e., have exactly the same amplitude, phase, and frequency). In the case shown here, less than two samples per period are taken from the input sine wave; hence it violates the sampling theorem. Such a waveform is said to be undersampled. Subsequently, the signal reconstructed from these samples, shown on the right-hand side in Figure 6.10, has the same amplitude as the input signal but has a much lower frequency (as an estimate of the reconstruction operation consider joining adjacent samples with straight lines). The sampling and reconstruction process has distorted the input signal. The phenomenon of a higher-frequency sinusoid acquiring the identity of a lower-frequency sinusoid after sampling is called *aliasing*.

To avoid aliasing in practice, it is important to limit the bandwidth of the signals that appear at the input to the digitizer to less than the one-half the Nyquist rate. In general, practical signals are not limited to a fixed range of frequencies, but have a frequency spectrum that decay to zero as the frequency approaches infinity. As a result, it is not always clear how to satisfy the sampling theorem. To avoid this ambiguity a low-pass *antialiasing* filter is placed before the digitizer to attenuate the high-frequency components in the signal so that their aliases become insignificant.

While aliasing is generally an effect that is to be avoided, the process of undersampling has been used to an advantage in several applications. As we will see in Section 9.2.5, undersampling is used to extend the measurement capabilities of an arbitrary waveform digitizer. Aliasing may also be advantageously utilized by a DUT as part of its normal operation. The cellular telephone base-band interface is one such example that might use undersampling to convert high-frequency inputs to lower-frequency signals to be digitized by a slow ADC.

Finally, we would like to address a commonly asked question: What happens if we sample a sinusoidal at exactly twice its frequency? The answer is that information may be lost. To understand this, consider that an arbitrary sinusoidal (i.e., one with arbitrary amplitude and

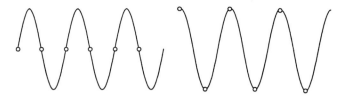

Figure 6.11. Sine and cosine waves sampled at twice the signal frequency.

phase) can always be represented as the sum of a sine and cosine signal operating at the same frequency

$$C\sin(2\pi f_o t + \phi) = A\cos(2\pi f_o t) + B\sin(2\pi f_o t) \tag{6.17}$$

Therefore, analyzing the effect of sampling a sine and cosine signal allows us to generalize the result for a signal having an arbitrary phase, ϕ. Figure 6.11 illustrates the samples derived from a sine and cosine signal sampled at twice their frequency. As is evident, all the samples from the sine wave are zero, whereas those from the cosine signal are not. Clearly, any information contained in the sine wave such as its amplitude would be lost and unobtainable from the samples. We can therefore conclude that one should not attempt to sample at exactly twice the Nyquist rate.

6.2.5 Quantization Effects

Mathematical sampling can be achieved with no loss of signal quality. A computer can come very close to mathematical perfection. For example, the following MATLAB routine can be used to create 64 samples of a sine wave with unity amplitude and zero phase shift:

```
pi=3.14159265359;
for k=1:64,
    v(k)=1*sin(2*pi/64*k);
end;
```

As the time index k is incremented in unit steps, the sampling period is by default equal to unity, resulting in a unity sampling frequency. Therefore, the frequency of the sampled sinusoid is 1/64 Hz, as $M=1$ and $N=64$. The quality of the sine wave is limited only by the tiny amounts of mathematical error in the computation process. This sampling process would result in a nearly perfect sampled representation of the sine wave. It would have almost no distortion and very little noise. The ADC included in a digitizer, on the other hand, will always introduce some amount of noise and distortion. The noise introduced by an ADC can be classified as: (1) quantization noise, and (2) circuit related noise such as thermal and shot noise. Distortion, on the other hand, is a result of nonlinear circuit behavior and component mismatches.

In a perfectly designed and manufactured ADC, the majority of the noise will be caused by the quantization error of the conversion process. Figure 6.12 shows a set of samples obtained from a sine wave that has been digitized by a 4-bit ADC. For example, the quantized waveform in Figure 6.12 could be stored in a computer memory as the sample set {7,11,14,14,11,7,4, 1,1,4,7,11,14,14,11,7,4,1,1,4,7,11,14,14,11,7,4,1,1,4}. Also shown in Figure 6.12 is the original analog waveform superimposed on a regular spaced set of grid lines, together with an expanded view of a single sample shown on the right-hand side.

The vertical grid lines correspond to the sampling instances, with time increasing from left to right. The horizontal grid lines correspond to the limited outputs available from the ADC. The distance between adjacent horizontal grid lines is known as the *least significant bit* (LSB). An LSB sets the largest distance that the ADC output will be from a sample obtained directly from the original waveform (see the expanded view on the right of Figure 6.12). In general, a D-bit ADC with a full-scale analog input range of FS has a corresponding LSB step size of

$$V_{LSB} = \frac{FS}{2^D - 1} \tag{6.18}$$

Exercises

6.4. To illustrate the effects of aliasing, compare 24 samples of a sinusoid with unity amplitude, zero phase shift, and a period of 12 s derived using a sampling rate of 1 Hz and a sampling rate of 1/8 Hz. Use MATLAB or an equivalent software program for your analysis.

Ans. Setting $A=1$, $\phi=0$, $N=12$, and $M=1$, we get:

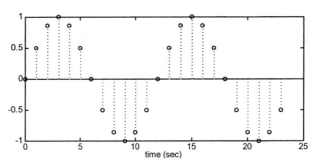

Setting $A=1$, $\phi=0$, $N=12$, and $M=8$, we get:

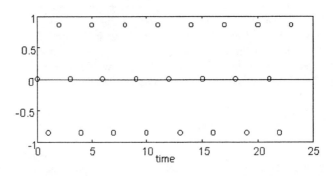

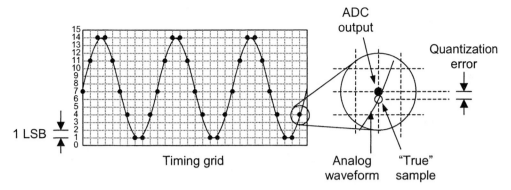

Figure 6.12. Quantized sine wave samples.

The quantization errors in Figure 6.12 do not look especially severe at first glance. However, if we were to reconstruct a continuous-time waveform from these samples, the analog waveform would contain a significant noise component as illustrated in Figure 6.13. If we separate the errors from the quantized samples, we can see how much noise has been introduced by the quantization process.

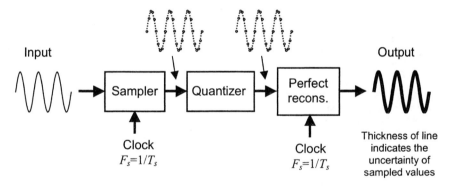

Figure 6.13. Illustrating the noise component that is associated with a quantization operation.

In Figure 6.14, the quantized waveform is equal to the sum of the ideal sampled waveform and an error waveform. The error waveform is the quantization noise added by ADC quantization process. Statistically speaking, the quantization errors of a random input signal exhibit a uniform probability distribution[5] from $-\frac{1}{2}$ LSB to $+\frac{1}{2}$ LSB, assuming a perfect ADC. Moreover, the ideal quantization error sequence v_q resembles a random sequence having an average and RMS value given by

$$v_{q-AVE} = 0 \quad \text{and} \quad v_{q-RMS} = \frac{V_{LSB}}{\sqrt{12}} \qquad (6.19)$$

Obviously, quantization error can be reduced using an ADC with more bits of resolution (consider combing Eqs. (6.18) and (6.19)). Higher resolution would provide more vertical

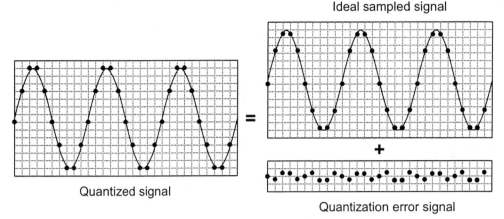

Figure 6.14. Representing the quantized waveform as a sum of the original sampled signal and a quantization error signal.

graticules on the plots in Figure 6.12, reducing the size of each LSB. Adding an extra bit of ADC resolution reduces the size of each LSB by one-half, thereby reducing the RMS value of the quantization noise by a factor of two, or 6 decibels (6 dB). A 16-bit ADC is theoretically capable of a 97.76 dB signal-to-noise ratio (SNR) with a full-scale sine wave input. A 15-bit ADC would therefore be capable of 91.76 dB SNR, etc. (See Chapter 8, "Analog Channel Testing" for an explanation of the decibel unit and SNR measurements.)

Example 6.4

Compute the quantization noise sequence that results from exciting a 3-bit ADC with a full-scale amplitude sinusoidal signal of unity amplitude, zero phase, $M=1$, and $N=64$. Also, compute the RMS value of the quantization noise and compare this result with its theoretical predicted value.

Solution:

To aid us in this investigation we shall make use of the following MATLAB routine for an ideal 3-bit quantizer performing a rounding operation typical of an ADC having a full-scale input range between -1 and +0.75:

```
%   3-Bit Quantizer (-1 <= X <= +0.75)
D=3;                        % # of bits of resolution
FS=1.75;                    % Full scale range
LSB=FS/(2^D-1);             % Least significant bit
Y = round(X/LSB)*LSB;       % rounds to nearest level
```

A quantizer is the element of the ADC that limits the continuous input signal, say X, to discrete values denoted by Y. In this case, values of -1, -0.75, -0.5, -0.25, 0, 0.25, 0.5, and 0.75. The ADC would then interpret these levels and provide an output digital representation, for example in a 2's complement form. The transfer characteristic, Y vs. X, for this quantizer is shown in Figure 6.15.

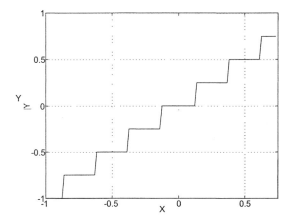

Figure 6.15. Ideal quantizer transfer characteristic.

Now passing a near full-scale sinusoid having the following parameters, $A=0.75$, $\phi=0$, $N=64$, and $M=1$ through the 3-bit quantizer, we get the error sequence in Figure 6.16.

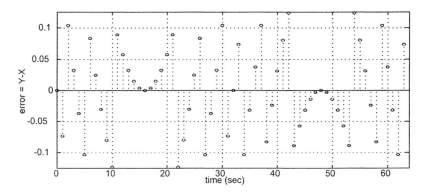

Figure 6.16. Quantization error sequence.

Here we see that the error sequence has symmetrical response bounded between ±0.125, and has a mean value of -1.0842e-18 or nearly zero. The RMS value of the error is computed to be 0.0670. According to the quantization theory presented earlier, the error sequence should have an average value of 0 and an RMS value of $0.25/\sqrt{12} = 0.0722$ based on an LSB of 0.25 V. For all intents and purposes, the results of this simulation agree reasonably well. The discrepancy is largely a result of the quantizer's low resolution of 3 bits. If we increased its resolution, we would discover a much closer correspondence between experiment and theory.

> **Exercises**
>
> **6.5.** What is the LSB of an ideal 8-bit ADC that has a full-scale input range of 0-1 V? What is the expected RMS value of the corresponding quantization noise?
>
> **Ans.** 3.9 mV, 1.13 mV.
>
> **6.6.** If an ideal 7-bit ADC has an RMS quantization noise component of 1.4 mV, what is the quantization noise for a 5-bit ADC having an identical full-scale input range?
>
> **Ans.** 5.74 mV.
>
> **6.7.** A 4-bit ADC with an analog input range from −1.5 to +1.5 V gives an output of code of 4 for a code range beginning at 0 and ending at 15. What are the minimum and maximum values of the input voltage corresponding to this output code?
>
> **Ans.** −0.7 V, −0.5 V.

6.2.6 Sampling Jitter

Another source of signal quality degradation is sampling jitter. Jitter is the error in the placement of each clock edge controlling the timing of each ADC or DAC sample. Figure 6.17 illustrates the effect of jitter on the sampling process of an ADC. Here we make use of the same regular spaced grid as that used in Section 6.2.5 except that this time we added an additional set of vertical dotted lines to indicate the actual clock edge subject to random clock jitter.

As is evident in this situation, the actual sample can differ quite significantly from the ideal sample and the size of this error is proportional to the magnitude of the jitter. Mathematically,

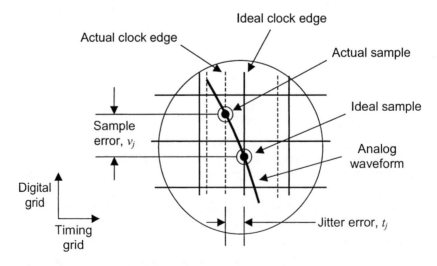

Figure 6.17. Illustrating the effect of clock jitter on the sampling process.

we can calculate the effects of jitter on the samples obtained by an ADC by associating jitter with a random timing variable, which we shall denote as t_j, and adding it to the sampling expression given in Eq. (6.2) according to:

$$v[n] = v(t)\big|_{t = nT_s + t_j} \tag{6.20}$$

Due to the nature of t_j, $v[n]$ is now a random variable as well. Calculating the effects of jitter can become mathematically complicated in all but the simplest examples. One example that allows us to draw some useful conclusions is the study of jitter on the sample points of a single sinusoid with peak amplitude A_o and frequency f_o. The phase shift is assumed equal to zero without loss of generality. Without jitter, the sample points are

$$v[n] = v(t)\big|_{t = nT_s} = A_o \sin(2\pi f_o nT_s) \tag{6.21}$$

With jitter present, according to Eq. (6.20), the samples become

$$v[n] = v(t)\big|_{t = nT_s + t_j} = A_o \sin(2\pi f_o (nT_s + t_j)) \tag{6.22}$$

We can separate this expression into two parts, one that includes the deterministic component and the other due to jitter. To see this, consider using the trigonometric identity $\sin(A+B)=\sin(A)\cos(B)+\cos(A)\sin(B)$ so that we can rewrite Eq. (6.22) as

$$v[n] = A_o \sin(2\pi f_o nT_s)\cos(2\pi f_o t_j) + A_o \cos(2\pi f_o nT_s)\sin(2\pi f_o t_j) \tag{6.23}$$

Since the magnitude of the jitter t_j is assumed to be small compared to the sampling period T_s, we can approximate Eq. (6.23) as

$$v[n] \approx A_o \sin(2\pi f_o nT_s) + A_o 2\pi f_o t_j \cos(2\pi f_o nT_s) \tag{6.24}$$

Here we made use of the fact that when x is small, $\cos(x) \approx 1$ and $\sin(x) \approx x$. Now we have the jitter term separated from the deterministic term, allowing us to claim that the error in the sample due to jitter, denoted as v_j, is

$$v_j[n] \approx A_o 2\pi f_o t_j \cos(2\pi f_o nT_s) \tag{6.25}$$

Recognizing that the derivative of a sine wave is a cosine wave further allows us to write the jitter-induced error in terms of the magnitude of the jitter and the slope of the signal at the sample point

$$v_j[n] \approx \left[\frac{dv(t)}{dt}\bigg|_{t = nT_s}\right] \cdot t_j \tag{6.26}$$

This result should be readily apparent from Figure 6.17. It suggests that a timing error will induce a larger sample error at the rapidly rising or falling points of a sine wave than at its peak or trough.

Assuming that the jitter t_j has an RMS value of $t_{j\text{-}RMS}$ and is independent of $v(t)$, then we can approximate the RMS value of the error sequence $v_j[n]$ as the product of the RMS value of t_j and the RMS value of the derivative of $v(t)$ at each sampling instant. For a sampled sinusoidal signal with peak amplitude A_o and frequency f_o, the RMS value of the jitter-induced error is

$$v_{j\text{-}RMS} \approx \frac{2\pi A_o f_o}{\sqrt{2}} t_{j\text{-}RMS} \tag{6.27}$$

At this point in our discussion we can use this result to set a limit on the maximum tolerable jitter allowable based on the ADC's speed and resolution. We first have to define the amount of jitter-induced noise that we are willing to tolerate. Let us define a 1-LSB upper limit on the tolerable amount of jitter-induced noise

$$\begin{aligned} v_{j\text{-}RMS} &< 1 \text{ LSB} \\ &< \frac{FS}{2^D - 1} \end{aligned} \tag{6.28}$$

Substituting Eq. (6.27) and rearranging allows us to bound the jitter according to

$$t_{j\text{-}RMS} < \frac{FS}{\sqrt{2}\pi A_o f_o \left(2^D - 1\right)} \tag{6.29}$$

Further, if we assume a full-scale input sinusoid, $FS = 2A_o$, then we can find a lower limit on the maximum allowable jitter given by

$$t_{j\text{-}RMS} < \frac{\sqrt{2}}{\pi f_o \left(2^D - 1\right)} \tag{6.30}$$

Conversely, for a D-bit ADC having an RMS sampling jitter $t_{j\text{-}RMS}$, the maximum sampling frequency that can be used (i.e., $F_{s\text{-}MAX} = 2 f_{o\text{-}MAX}$) is

$$F_{s\text{-}MAX} < \frac{2\sqrt{2}}{\pi t_{j\text{-}RMS} \left(2^D - 1\right)} \tag{6.31}$$

or we can conclude that the maximum conversion resolution (expressed in number of bits) available with a maximum sampling frequency $F_{s\text{-}MAX}$ and RMS sampling jitter $t_{j\text{-}RMS}$ is

$$D_{MAX} < \log_2\left(\frac{2\sqrt{2}}{\pi t_{j\text{-}RMS} \left(F_{s\text{-}MAX}\right)} + 1\right) \tag{6.32}$$

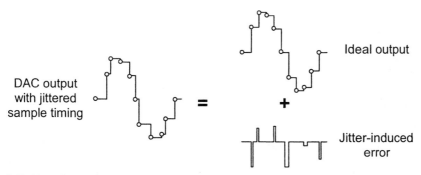

Figure 6.18. The effect of clock jitter on the actual DAC output can be separated into an ideal output and a jitter-induced error signal.

The effect of sampling jitter on the operation of a DAC can be described by similar mathematical expressions derived for the ADC. Consider that the effect of clock jitter on the output of a DAC can be separated from its ideal operation as shown in Figure 6.18. Here the actual output waveform is separated into an ideal waveform and one that contains the jitter-induced noise. Mathematically, the jitter-induced error can be described as

$$v_j(t) = \left[v(nT_s) - v((n-1)T_s)\right]\left[u(t - nT_s - t_j) - u(t - nT_s)\right] \quad (6.33)$$

where $u(t)$ is a unit step function. With an error pulse occurring on average once every clock period, we can consider that the effective energy contributed by each pulse at the sampling instant is

$$e_p[n] = (v[n] - v[n-1])^2 \frac{t_j}{T_s} \quad (6.34)$$

Further, we can relate this energy back to the original sample value by dividing Eq. (6.34) by T_s; that is, the pulse energy is distributed over a full clock period, and taking the square-root value, then we can write the jitter-induced error as

$$v_j[n] = (v[n] - v[n-1])\sqrt{\frac{t_j}{T_s}} \quad (6.35)$$

Recognizing that the difference operation normalized by T_s is a discrete-time representation of differentiation allows us to approximate the jitter-induced error (for high oversampling ratios) as

$$v_j[n] \approx \left[\frac{dv(t)}{dt}\bigg|_{t=nT_s}\right]\sqrt{t_j T_s} \quad (6.36)$$

This expression is similar to that given for the jitter-induced error of the ADC, except that t_j is replaced by $\sqrt{t_j T_s}$. Hence we can make use of Eqs. (6.27)–(6.32) with the appropriate change of variable. For example, the maximum conversion resolution available with a maximum sampling frequency $F_{s\text{-}MAX}$ and RMS sampling jitter $t_{j\text{-}RMS}$ is

$$D_{MAX} < \log_2\left(\frac{2}{\pi}\sqrt{\frac{2}{t_{j-RMS}F_{s-MAX}}}+1\right) \qquad (6.37)$$

In either the DAC or ADC case, according to Eq. (6.27) doubling the timing jitter doubles the noise level. Also, doubling the signal amplitude or signal frequency doubles the jitter-induced noise. Testers often have particular sampling frequencies or other conditions that produce minimum sampling jitter. For instance, a particular tester may produce minimum jitter if the digital pattern is exercised at the tester's master clock frequency divided by 2^N, where N is any integer. As another example, a particular digitizer may operate with minimum jitter when its phase-locked loop phase discriminator input is near 16 kHz. If extremely low noise measurements are to be performed, the test engineer should understand which sampling rates provide the least jitter in each of the tester's instruments and subsystems.

Exercises

6.8. What is the RMS value of the error induced by an ADC having an RMS sampling jitter of 100 ps while measuring a 1-V amplitude sinusoid with a frequency of 100 kHz?

Ans. 44.4 µV.

6.9. What is the maximum sampling jitter that a 6-bit ADC can tolerate when it has a full-scale input range of 0-3 V and is converting a 100-kHz, 1-V peak sinusoid?

Ans. 0.107 µs.

6.10. What is the maximum sampling jitter that a 5-bit DAC can tolerate when it has a maximum sampling rate of 10 MHz?

Ans. 84.3 ps.

6.11. If a 6-bit DAC has a sampling jitter of 500 ps RMS, what is its maximum sampling rate?

Ans. 408.5 kHz.

6.12. If an ADC is controlled by a clock circuit with a minimum clock period of 1 µs and RMS jitter of 2.5 ns, what is the maximum conversion resolution possible with the ADC?

Ans. 8.5 bits.

6.3 REPETITIVE SAMPLE SETS

6.3.1 Finite and Infinite Sample Sets

In many mixed-signal systems such as a cellular telephone, the waveforms sampled by the system's ADC sub-blocks are nonrepetitive. In the cellular telephone example, the caller's voice is a random signal that seldom, if ever, repeats. The cellular telephone digitizes the caller's

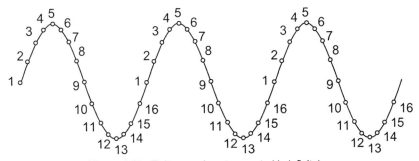

Figure 6.19. Finite sample set, repeated indefinitely.

voice and processes the samples in real time in a continuous process. For all intents and purposes, we can consider the cellular telephone sample sets to be infinite in length.

In the DSP-based testing environment, on the other hand, signals are often created and measured using a finite sample set of a few hundred or a few thousand samples. If desired, the finite sample sets in mixed-signal testers can be repeated endlessly, allowing easier debugging with spectrum analyzers and oscilloscopes. During production testing, however, the sample sets are only allowed to repeat long enough to collect the necessary measurement information. The use of repetitive, finite sample sets drives a number of ATE-specific limitations which the test engineer must understand. For example, Figure 6.19 shows a short sequence of 16 samples that repeats endlessly. Notice how sample 16 feeds smoothly into sample 1 at the end of each sequence. This smooth wraparound results from a property known as *coherence*. Coherence is one of the most important enabling factors for fast and accurate DSP-based testing.

6.3.2 Coherent Signals and Noncoherent Signals

In the example waveform of Figure 6.19, the last sample of the first iteration wraps smoothly into the first sample of the second iteration because there is exactly one sine wave cycle represented by the 16 samples. If we reconstruct this sample set at a sampling frequency F_s, then the sine wave would have a frequency of $F_s/16$. This frequency is known as the *fundamental frequency* or *primitive frequency*, F_f. In general, the fundamental frequency F_f of N samples collected at a sampling rate of F_s is

$$F_f = \frac{F_s}{N} \tag{6.38}$$

The period of the fundamental frequency is called the *primitive period* or *unit test period* (UTP)

$$UTP = \frac{1}{F_f} \tag{6.39}$$

The amount of time required to collect a set of N samples at a rate of F_s is also equal to one UTP

$$UTP = \frac{N}{F_s} \tag{6.40}$$

In practice, it usually takes an extra fraction of a UTP to allow the DUT and ATE hardware to settle to a stable state before a sample set is collected.

The fundamental frequency is often called the *frequency resolution*. The reason for this alternate terminology is that the only coherent frequencies that can be produced with a repeating sample set are those frequencies that are integer multiples of the fundamental frequency. Hence, in terms of N and the sampling frequency, the coherent frequencies F_c are

$$F_c = M \frac{F_s}{N} \tag{6.41}$$

where M is an integer 0, 1, 2, ..., N. The astute reader will recognize that we first made use of coherent frequencies in Section 6.2.1 in the development of Eq. (6.4), where $f_o = F_c$.

As an example, if we source the samples in Figure 6.19 at a rate of 16 kHz, then the fundamental frequency would be 16 kHz/16 = 1 kHz. The sine wave in Figure 6.19 would appear at 1 kHz. The next-highest frequency we could produce with 16 samples at this sampling rate is 2 kHz. If we wanted to produce a 1.5 kHz sine wave, then we would have a noncoherent sample set as shown in Figure 6.20.

If we wanted to produce a 1.5-kHz sine wave using a coherent sample set, then we would have to choose a sampling system with a fundamental frequency equal to 1.5 kHz/N, where N is any integer. We might choose $F_f = 500$ Hz, for example, and then use the third multiple of the fundamental frequency to produce the 1.5-kHz sine wave. A fundamental frequency of 500 Hz could be achieved using 32 samples instead of 16 (16 kHz/32 = 500 Hz). We would then calculate a sine wave with three cycles in 32 samples according to

```
pi=3.14159265359;
for k=1:32,
    sinewave(k) = sin(2*pi*3/32*(k-1));
end
```

Since the fundamental frequency determines the frequency resolution of a measurement, it might seem that minimizing the fundamental frequency would be a great idea. In the absence of test time constraints, a fundamental frequency of 1 Hz would provide good flexibility in test frequency choice. Remember, though, that the UTP drives the test time. Since one UTP is equal

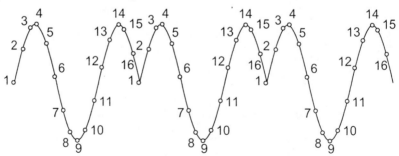

Figure 6.20. Noncoherent sample sets cannot be looped properly.

to $1/F_f$, a 1-Hz frequency resolution would require 1 s of data collection time. For most production tests, this would be unacceptable.

Many test situations call for the application of a coherent multitone signal to excite a device. Such a signal is created by simply adding together a set of *I* unique sine waves (i.e., having different coherent frequencies) according to the following formula

$$v[n] = \sum_{i=1}^{I} A_i \sin\left(2\pi \frac{M_i}{N} n + \phi_i\right) \qquad (6.42)$$

Here each sine wave is assigned a unique amplitude A_i, phase shift ϕ_i, and frequency designated by $(M_i/N)F_s$. The integers represented by M_i are commonly referred to as the *Fourier spectral bins*.

Any signal made up of a sum of coherent signals is also coherent. If one or more of the frequency components are noncoherent, though, the entire waveform will be noncoherent. Although noncoherent sample sets cannot be used to generate continuous signals through a looping process, they can be analyzed with DSP operations using a preprocessing operation called *windowing*. However, windowing is an inferior production measurement technique compared to coherent, nonwindowed testing. Windowing will be discussed in Chapter 7, "DSP-Based Testing."

Returning to the 16-kHz sampling example, we could create a multitone signal with frequencies at 1.5, 2.5, and 3.5 kHz using an expanded calculation given by the following MATLAB routine

```
pi=3.14159265359;
phase1=0, phase2=0, phase3=0;
for k=1:32,
    multitone(k) = sin(2*pi*3/32*(k-1) + phase1*pi/180) ...
        + sin(2*pi*5/32*(k-1) + phase2*pi/180) ...
        + sin(2*pi*7/32*(k-1) + phase3*pi/180;
end
```

The endpoints of this waveform would wrap smoothly from end to beginning because the waveform is coherent. The multitone signal calculated would be described as a three-tone multitone waveform with equal amplitudes at the third, fifth, and seventh spectral bins.

6.3.3 Peak-to-RMS Control in Coherent Multitones

Notice that in the multitone example in Section 6.3.2, all the frequency components are created at the same phase (0 degrees). The problem with this type of waveform is that it may have an extremely large peak-to-RMS ratio, especially as the number of tones increases. Consider the 7-tone multitone signal in Figure 6.21. The first waveform consists entirely of sine waves, while the second waveform consists entirely of cosine waves. These waveforms exhibit a spiked shape that is unacceptable for most testing purposes since it tends to cause signal clipping in the DUT's circuits.

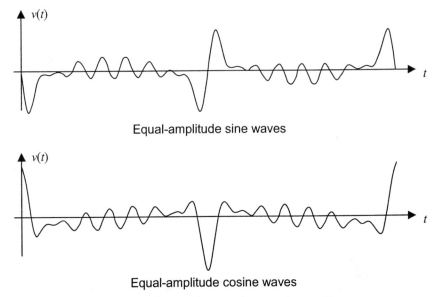

Figure 6.21. Pure sine and pure cosine seven-tone multitones.

The peak-to-RMS ratio of a multitone can be adjusted by shifting the phase of each tone to a randomly chosen value. The waveform in Figure 6.22 shows how randomly selected phases for the seven tones of Figure 6.21 produces a much less "spikey" waveform. Unfortunately, there is no equation to calculate phases to give a desired peak-to-RMS ratio. In many test programs, phases are chosen using a pseudorandom number generator with a uniform probability distribution between 0 and 2π radians. If the desired peak-to-RMS ratio is not achieved with one set of pseudorandom phases, then the program tries again until the desired ratio is found. These phase values can be generated each time the program loads, or they can be hard-coded into the test program once they have been determined through trial and error. This second approach prevents a pseudorandom algorithm from choosing one set of phases on a given day and another set of phases at a later date. For example, this might happen when an upgraded tester operating system includes a change to the pseudorandom algorithm. Theoretically, a different set of phases should not cause any shift in measurement results, but the use of hard-coded phases removes one more unknown factor from the measurement correlation effort.

What is the ideal peak-to-RMS ratio for a multitone signal? At first it might seem that it would be best to let the pseudorandom process search for a minimum peak-to-RMS ratio. This would provide the largest RMS voltage for a given peak-to-peak operating range. Larger RMS signals provide better noise immunity and improved repeatability. But this kind of signal is

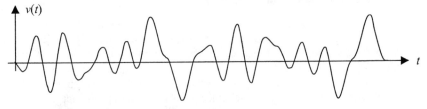

Figure 6.22. Seven-tone multitone created with random phases.

susceptible to large shifts in peak-to-RMS ratio if any of the filters in the ATE tester or DUT cause frequency-dependent phase shifts. A change in peak-to-RMS ratio could lead to a clipped signal, which would ruin the measurement accuracy.

In many end applications, the DUT will usually see a peak-to-RMS ratio of about 10–11 decibels (a ratio of about 3.35:1). Although the 10–11 dB range appears in many data sheets without explanation, it is based on the approximate peak-to-RMS levels encountered in typical analog signals. This range is roughly equal to the peak-to-RMS ratio of broadband signals having near-Gaussian-distributed amplitudes and random phases. As it happens, this ratio also tends to produce a multitone whose peak-to-RMS ratio is least sensitive to phase shifts from filters. For this reason, the pseudorandom phase selection process should be set to search for a peak-to-RMS ratio of between 10 and 11 decibels. A multitone signal must contain at least six tones to hit a peak-to-RMS ratio of 3.35:1. For signals having fewer tones, the target ratio is not terribly important, but it is still a good idea to use pseudorandom phase shifts for the tones rather than adding pure sine or cosine waveforms.

6.3.4 Spectral Bin Selection

One of the common mistakes a novice test engineer makes is to choose spectral bins by simply calculating the nearest integer multiple of the fundamental frequency. For example, if the test engineer wanted a 2-kHz sine wave using a 16-kHz sampling rate and 32 samples, then according to Eq. (6.41) the nearest Fourier spectral bin is

$$M = \frac{2 \text{ kHz}}{(16 \text{ kHz}/32)} = 4$$

which corresponds exactly with spectral bin 4. The problem with bin 4 is that it is not mutually prime with the number of samples, 32. (Mutually prime numbers are ones containing no common factors). The number $4 = 2^2$ is not mutually prime with the number $32 = 2^5$, so this choice of bin, sampling frequency, and number of points is a poor one.

One of the problems with non-mutually-prime spectral bins is that they may cause the quantization noise of a coherent signal to contain periodic errors instead of errors that are randomly distributed over the UTP. Consider the sine example with 4 cycles in 32 samples. If we look at a quantized version of this signal from a 4-bit ADC in Figure 6.23, we see that the quantization errors repeat four times in the sample set. The same problem occurs with DAC converters as well. Furthermore, a nonprime spectral bin hits fewer code levels on the DAC and ADC; so we are just testing the same points repeatedly. Repetitively exercising the same code levels results in less robust fault coverage in the DAC and ADC circuits.

Another problem with non-mutually-prime bins is that they tend to lead to overlaps between test tones, harmonic distortion components, and intermodulation distortion components. The use of mutually prime bins does not necessarily prevent intermodulation distortion overlaps, but it makes them less likely. Whether mutually prime bins are chosen or not, one should verify that all distortion components fall into spectral bins that do not coincide with bins containing important signal information.

Consider the example of a three-tone multitone at 1, 2, and 3 kHz. The problem with this multitone is that there is a great deal of distortion overlap. The second and third harmonic

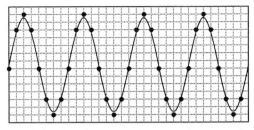
Non-mutually-prime spectral bin

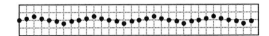
Periodic quantization errors
appear as gain error and
harmonic distortion

Figure 6.23. Non-mutually-prime spectral bin selection leads to periodic errors.

distortion of the 1-kHz tone falls on top of the 2- and 3-kHz test tones, respectively. Also, the second order intermodulation distortion between the 2-kHz tone and the 3-kHz tone appears at 1 and 4 kHz, corrupting the 1-kHz test tone. All these overlaps would cause errors in any measurement involving the 1-, 2-, or 3 kHz tones (gain, frequency response, distortion, etc.). A better approach is to use test frequencies close to the desired frequencies, but located at spectral bins that do not cause any intermodulation or harmonic distortion overlaps.

To avoid overlaps between harmonic distortion components and signal components, we should guarantee that the tones are not only mutually prime with the number of samples, but that they are also mutually prime with one another. For example, bins 3 and 9 are both mutually prime with 512 samples, but they are not mutually prime with one another. Consequently, the third harmonic of spectral bin 3 coincides with the tone at bin 9, resulting in an overlap.

Example 6.5

Select the spectral bins for a three-tone signal at 1, 2, and 3 kHz with no more than ±50 Hz error in the signal frequencies. The signal should take no more than 50 ms to repeat. Use a 16-kHz sampling rate.

Solution:

With a maximum UTP of 50 ms and a sampling rate of 16 kHz, the number of sample points is found from Eq. (6.40) to be

$$N \leq 50 \text{ ms} \times 16 \text{ kHz} = 800$$

An important constraint on the number of sample points used in most test systems is that N must be a power of two (i.e., 2^P, where P is an integer). The reason for this will be explained in more detail in Chapter 7, but it is because we will ultimately use the Fast Fourier Transform (FFT) algorithm to measure the response of the DUT to the three-tone signal. Therefore, we shall select N equal to 512. We want the highest possible N in order to achieve the greatest frequency resolution or the smallest fundamental frequency. Working with 512 samples, the fundamental frequency becomes

$$F_f = \frac{16 \text{ kHz}}{512} = 31.25 \text{ Hz}$$

Subsequently, the closest spectral bin numbers that correspond to the 1-, 2-, and 3-kHz signals are found using Eq. (6.41), together with Eq. (6.38), to be

$$M_1 = \frac{1 \text{ kHz}}{31.25 \text{ Hz}} = 32 \text{ (non-mutually-prime, shift to 31)}$$

$$M_2 = \frac{2 \text{ kHz}}{31.25 \text{ Hz}} = 64 \text{ (non-mutually-prime, shift to 63)}$$

$$M_3 = \frac{3 \text{ kHz}}{31.25 \text{ Hz}} = 96 \text{ (non-mutually-prime, shift to 97)}$$

In all three cases, the computed spectral bin values were all even numbers sharing a factor of 2 with the number of samples, 512. Shifting the result by one in either a positive or negative direction eliminates their dependence on the common factor of 2. The resulting test frequencies, f_1, f_2, and f_3 are then

$$f_1 = 31 \times 31.25 \text{ Hz} = 968.75 \text{ Hz}$$
$$f_2 = 63 \times 31.25 \text{ Hz} = 1968.75 \text{ Hz}$$
$$f_3 = 97 \times 31.25 \text{ Hz} = 3031.25 \text{ Hz}$$

In all three cases, the chosen test frequencies are within the desired ±50 Hz error margin and are therefore acceptable.

We now have to verify that there are no distortion overlaps using spectral bins 31, 63, and 97. First we list the harmonics of the three test tone bins (stopping at the Nyquist bin, which is located at 8 kHz, or bin 256). Harmonics are defined as all the frequencies at an integer multiple of the test tone and computed according to the following table:

Harmonic/Test Tone	M_A	$2M_A$	$3M_A$	$4M_A$	$5M_A$	$6M_A$	$7M_A$	$8M_A$
M_1	31	62	93	124	155	186	217	248
M_2	63	126	189	256	-	-	-	-
M_3	97	194	-	-	-	-	-	-

None of these harmonics overlaps with the other harmonics or with the test tones; so the harmonic distortion overlap criterion is met. Next we look for intermodulation components. Intermodulation components appear at the sum and difference of any two tones, that is, $2F_1$-F_2,

F_1+3F_2, etc. There are so many of these that it is usually sufficient to just limit the list to second- and third-order distortions. (Second-order distortions are those in which the magnitude of the integers in front of F_1 and F_2 add up to 2; third-order distortions are those in which the magnitude of the integers add up to 3; etc.). The following is a table listing the intermodulation interaction between the three test tones:

Intermod./ Test Tone	M_A	$|M_A-M_B|$	$|M_A+M_B|$	$|2M_A-M_B|$	$|2M_A+M_B|$	$|M_A-2M_B|$	$|M_A+2M_B|$
M_1	31	32	94	1	125	95	157
M_2	63	34	160	29	223	131	257
M_3	97	66	128	163	225	35	159

Here the letter A and B represent the current and next row spectral bin. When we are at the bottom of the table, the next spectral bin refers to the data at the top of the table. As is evident from this table, none of the intermodulation components falls on top of a test tone.

The lack of overlap between harmonic distortion components and test tones in this example is guaranteed by a choice of mutually prime bins. In addition, none of the harmonics interferes with any of the intermodulation components. The choice of mutually prime bins does not guarantee a lack of overlap between intermodulation components and test tones or distortion components, but it does reduce the likelihood of such overlaps. Since there are no overlaps in this example, we can measure gain, frequency response, harmonic distortion, intermodulation distortion, and signal-to-noise ratio with the same set of collected samples. The ability to measure multiple parameters using a single data collection cycle is an advantage of multitone testing. This technique saves a tremendous amount of test time compared with single-tone testing approaches.

As we have seen, multitone DSP-based testing only provides accurate measurements if the test engineer is careful with the selection of test tones. Careless selection of spectral bins will lead to answers that may be slightly incorrect. If we had chosen bin 62 for the 2-kHz tone, for example, then the second harmonic distortion from the 1-kHz tone would have affected the measured level of the 2-kHz tone by a small but significant amount.

In most cases, we choose a sample set consisting of an even number of samples. Thus the mutally prime rule prevents us from using even-numbered spectral bins. In the previous example, we chose bin 63 instead of 62 because it was mutually prime with the number of samples. There is a second reason that we chose 63 and not 62. Combinations of odd harmonics and even harmonics in a multitone signal result in a signal with asymmetrical positive and negative peaks relative to the DC offset of the signal. The DC offset of such an asymmetrical multitone is not centered between the maximum and minimum voltages. This gives poor fault coverage for the circuit under test because it exercises one side of the signal range more than the other.

Figure 6.24 shows the difference between an all-odd multitone and a mixed odd-even multitone. Of course a single-tone signal with an even harmonic does not have the asymmetry problem, but it may lead to the kind of quantization noise modulation illustrated in Figure 6.23.

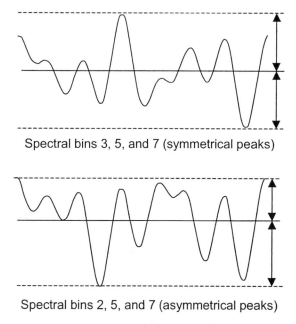

Spectral bins 3, 5, and 7 (symmetrical peaks)

Spectral bins 2, 5, and 7 (asymmetrical peaks)

Figure 6.24. Even spectral bins lead to asymmetrical peaks relative to the signal's DC offset.

The bottom line is that odd-numbered, mutually prime spectral bins should always be used whenever possible. If the situation is truly desperate, non-mutually-prime or even-numbered bins can be used as a last resort.

6.4 SYNCHRONIZATION OF SAMPLING SYSTEMS

6.4.1 Simultaneous Testing of Multiple Sampling Systems

Many DUTs contain both ADC and DAC channels, as in the case of the voice-band interface of Figure 6.1. These channels are often tested simultaneously in an ATE test program to minimize test time. Simultaneous testing requires a digital pattern loop containing the appropriate samples to excite the DAC channel. At the same time, an AWG converts another set of samples into analog form to excite the ADC channel. The response of the ADC is collected directly into ATE memory for later processing. The DAC response must be digitized before being stored into ATE memory. The response of each channel would then be analyzed through a postprocessing frequency-domain operation and judged suitable or not. For example, we might test the gain of the ADC channel and the DAC channel simultaneously using this approach.

Unfortunately, crosstalk between the ADC and DAC channels can lead to small gain errors if we use the same test tones in both channels. Often, the gain errors are small enough that we can live with them. However, if the DAC and ADC have channel-to-channel crosstalk specifications, we can save some test time by measuring the crosstalk during the gain test. All we have to do is select slightly different test tones on the DAC side from those used on the ADC side. Then the feedthrough from DAC to ADC will show up in different bins from the ADC signals and vice versa. This is made possible by operating the various components of the ATE

Exercises

6.13. What is the fundamental frequency of 512 samples collected at a rate of 1 MHz? What is the corresponding UTP?

Ans. 1/512 MHz, 512 μs.

6.14. How many cycles of a 2.1375-kHz sine wave are completed in a 7.953216-ms UTP?

Ans. 17 cycles.

6.15. What is the nearest coherent frequency to 20 kHz when 512 samples are collected at a rate of 44 kHz? How many cycles are completed in one UTP?

Ans. 20.023 kHz, 233 cycles.

6.16. Using a hand analysis, compute the peak-to-rms ratio of a two-tone multitone described by $A\sin(\omega_1 t) + B\sin(\omega_2 t)$.

Ans. $\sqrt{2}\dfrac{A+B}{\sqrt{A^2+B^2}}$

6.17. Select the spectral bins of a two-tone signal at 15 and 30 kHz such that minimum distortion overlap occurs. Assume that the sampling rate is 44.8 kHz and that the UTP must be less than 100 ms.

Ans. 1501, 2999.

and DUT at different test frequencies but ensuring that they have the same UTP. In turn, this also implies that the fundamental frequency F_f is the same for all components and will guarantee coherent sampling sets in both signal paths. Let us consider the following example.

Example 6.6

A DUT's DAC and ADC both operate at a 32-kHz sampling rate. Find a sampling system that tests the gain of a DAC channel and an ADC channel simultaneously, as shown in Figure 6.1. Use three tones at 1, 2, and 3 kHz, with a maximum test tone error of 100 Hz.

Solution:

Let us start by setting the number of samples in the waveform exciting the DUT's DAC and ADC at $N=1024$. Subsequently, the fundamental frequency F_f will be

$$F_f = \frac{32 \text{ kHz}}{1024} = 31.25 \text{ Hz}$$

The desired three tones will then be located in spectral bins of 32, 64 and 96, resulting in the desired test frequencies of 1, 2 and 3 kHz.

Beginning with the ADC channel test, we shall select the sampling rate of the AWG to be 16 kHz and impose the constraint that it has a fundamental frequency of 31.25 Hz. This in turn requires that the sample set consist of 512 samples. Using the sampling rate and spectral bins from the prior example, we will create an AWG waveform with 512 samples, having test tones at bins 31, 63, and 97. We will source this signal from the AWG at a 16-kHz sampling rate. To achieve the same fundamental frequency as the AWG signal, the ADC must collect 1024 samples (32 kHz sampling rate/1024 points = 16 kHz sampling rate/512 points). Using this sampling system we know that the ADC samples will form a coherent sample set.

Next let us consider the DAC channel test. By feeding 1024 samples to the DAC at 32 kHz, we would create a sampling system with the same fundamental frequency as the ADC. If we then set the digitizer sampling rate to 16 kHz and collect 512 samples from the DAC output, we would again achieve a fundamental frequency of 31.25 Hz, guaranteeing coherence. To allow simultaneous testing of the ADC and DAC gain, we need to select different spectral bins than those used to test the ADC. We can choose bins 33, 67, and 95 to meet all our testing criteria.

The preceding example demonstrates one of the reasons we prefer to use the same fundamental frequency for both ADC and DAC. It allows us to make coherent crosstalk measurements between two supposedly isolated signal paths. The other reason is that the UTP for the ADC and DAC is identical by definition, assuming we drive the ADC and DAC from the same digital pattern loop. For instance, if it takes 30 ms to collect the DAC channel samples, then it also takes 30 ms to collect the ADC channel samples. Identical UTPs drive identical fundamental frequencies, since the UTP is the inverse of the fundamental frequency. Sometimes, though, the ADC and DAC are not designed to sample at the same frequency. Fortunately, the sampling frequencies are often related by a simple integer multiple (i.e., 16 and 32 kHz). In these cases, we can simply collect more samples on one channel than on the other to achieve identical fundamental frequencies.

Matching all the fundamental frequencies in a particular test would be easy if we could simply request any arbitrary sampling rate from the ATE instruments. Unfortunately, many ATE testers have a limited choice of sampling frequencies. Henry Ford once said that you could purchase a Model T automobile in any color you wanted, as long as you wanted black. Sometimes a particular tester architecture gives us a similar choice of sampling rates. For example, we might have a choice of any sampling frequency we want as long we want a multiple of 4 Hz. In the remaining sections we shall examine some of the ATE clocking architectures that the test engineer might encounter.

6.4.2 ATE Clock Sources

Mixed-signal testers use a variety of different approaches to clock generation. The most common clock generation schemes involve phase-locked loops, frequency synthesizers, or flying adders. Each of these has strong points and weak points that the test engineer will have to deal with. Ultimately, though, all the clocks in a mixed-signal tester should be referenced to a single master clock so that all instrumentation can be synchronized to achieve coherent sampling systems during each test.

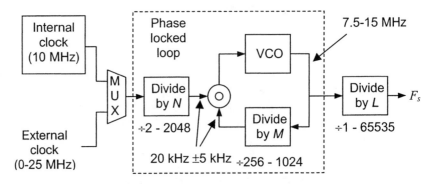

Figure 6.25. ATE PLL-based digitizer clocking source.

The phase-locked loop (PLL) frequency generator is a circuit that produces an output clock equal to a reference clock times M over N, where M and N are integers. An example ATE PLL-based clocking architecture is shown in Figure 6.25. It consists of several counter stages and a voltage-controlled oscillator (VCO). This PLL is used to generate the sampling clock for a digitizer. It can use either a fixed 10-MHz internal frequency reference or an externally supplied reference frequency.

The external reference is required if a DAC output is to be digitized, since a reference clock would have to come from the same digital pattern generator feeding the DAC its samples, frame syncs, and other digital signals. The PLL shown in Figure 6.25 operates by first dividing the reference frequency F_{REF} by N, then by multiplying the result by M through the divide-by-M counter in the negative feedback loop around the VCO. Finally, the frequency of the VCO output can be further divided by another counter stage, which divides the output by integer L resulting in the output sampling frequency F_s given by

$$F_s = \frac{M}{NL} F_{REF} \qquad (6.43)$$

This particular example imposes a number of restrictions on the test engineer. First, the externally supplied reference clock must be between 0 Hz and 20 MHz. Next, the value of N must be between 2 and 2048. The output of the divide-by-N stage should be as close to 20 kHz as possible for maximum stability of the PLL. Other frequencies will work, but will introduce additional jitter into the clock. The VCO output must be between 5 and 10 MHz. The value of M must be between 256 and 1024. Finally, the value of L must be between 1 and 65535. Every time the PLL is reconfigured it must be allowed to settle to a stable state, adding a bit of wait time between tests. Clearly, this clocking architecture is very inflexible and puts a large burden on the test engineer.

More modern testers allow the test engineer to select a wider range of frequencies using a frequency synthesizer. Frequency synthesizers work by taking a reference clock (10 MHz, for example) and passing it through a series of dividers and frequency mixers to produce a very stable output frequency with very little jitter. These synthesizers also take significant time to stabilize (25–50 ms), but that is the price paid for low jitter. Synthesizers are not entirely

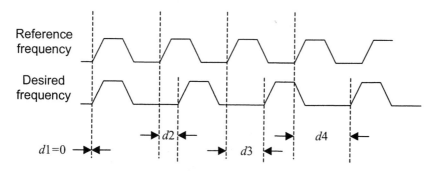

Figure 6.26. Clock generation using flying adder delays.

flexible either. For instance, a particular synthesizer may only be able to produce integer multiples of 4 Hz.

Flying adders can allow an even more flexible clocking source with little settling time, but they may introduce a little more jitter than a frequency synthesizer. A flying adder works by using a high-frequency reference clock and calculating the difference between the desired clock edges and the clock edges produced by the reference clock. Each desired clock edge is generated by delaying each reference clock edge by a carefully calculated amount of time as shown in Figure 6.26.

A new delay time has to be calculated for each delayed clock edge. The calculation is performed on the fly by an adder circuit, thus the terminology "flying adder." Because the edges are generated by programmable delay circuits, flying adders are sometimes more prone to jitter than frequency synthesizers. However, the frequency stabilization time for a flying adder clock circuit is nearly instantaneous.

6.4.3 The Challenge of Synchronization

The clocks generated by an ATE tester are sometimes modified further by the instruments or subsystems that use them. For example, the master clock operating a digital pattern generator may be divided by an integer before it is applied to the pattern generator's memory address counter. The clock divider may have restrictions similar to the ones mentioned in the PLL example. Digitizers and AWGs often perform even more convoluted operations on the clock before the final sampling clock is produced.

The following example shows how one particular tester's clocking architecture is arranged. The master clock for the tester must be set between 160 and 200 MHz. It must be a multiple of 4 Hz. This clock is divided by an integer A to produce the clock for the digital subsystem's pattern generator. However, the pattern generator can only run at frequencies less than 25 MHz. The master clock is also divided by another integer to produce a reference clock for a sigma-delta-based digitizer.

The digitizer clock input must fall within certain ranges, based on the desired cutoff frequency of the digitizer's antialiasing filter. The choice of filter cutoff frequency also places restrictions on two other integer dividers, the oversample divider (OSD) and the decimation factor (DF).

These integers must be chosen from a limited set of values listed in a digital filter table. The final sampling rate of the digitizer is then calculated by the following formula

$$F_s = \frac{F_{MCLK}}{A \times OSD \times DF} \tag{6.44}$$

Clearly, setting up the sampling rates in a mixed-signal tester can be a complicated process. With all the restrictions placed on a test engineer by the ATE clocking and divider architectures, it is sometimes impossible to find an acceptable sampling system for a given test. It can be a maddening process to calculate a coherent sampling system without violating any of the tester's clocking rules. The digital pattern must run at a particular frequency, specified in the DUT's data sheet. The DUT's DAC and ADC must run at particular frequencies. The tester's clocks must fall within certain ranges at each stage in the frequency divider chains. Sample sizes should be powers of two for use of efficient FFT routines. Constant reprogramming of the master clock may add extra test time because of frequency synthesizers. Finally, all the sampling rates and number of points must result in the same fundamental frequency for all the DACs and ADCs in the DUT and tester.

Since each tester presents a totally different set of clocking restrictions, this book will not attempt to teach a general approach to working through the clock and clock divider calculations. The basic rules of coherent sampling theory taught in this chapter apply to all testers, regardless of clocking peculiarities. The clock calculation task can be handled by software tools or by long-hand calculations on paper. Either way, the test engineer will have to spend time learning about a tester's clocking architecture and its restrictions so that the inevitable tradeoffs in test time and performance can be made in an informed manner.

Exercises

6.18. An ATE PLL-based clock source such as the one in Figure 6.25 is set with the divide-by-N counter equal to 1024, the divide-by-M equal to 512, and the divide-by-L equal to 64. With a reference frequency of 25 MHz, what is the output sampling frequency? Are all frequency constraints met in this configuration?

Ans. 195.3125 kHz. Yes, intermediate frequencies are 24.414 kHz and 12.5 MHz.

6.5 SUMMARY

In this chapter, we have presented an introduction to sampling theory and coherent sampling systems as they are applied in mixed-signal ATE testing. While the treatment of this material is not as thorough as one might encounter in a signal processing textbook, this level of coverage should be adequate for beginning mixed-signal test engineers. Thus far, we have only seen how coherent multitone sample sets are created, sourced, and captured. In the next chapter, we will explore the use of digital signal processing algorithms, such as the fast Fourier transform (FFT), in the analysis of the samples collected during a coherent mixed-signal test. As we will see, digital signal processing allows a combination of low test time and high accuracy not possible with conventional, purely analog instrumentation.

Problems

6.1. For the following parameters of a sampled sine wave (A, ϕ, N, and M), calculate the frequency of the resulting sine wave assuming a sampling rate of 1 Hz. What are the UTP and the fundamental frequency related to this collection of samples?

(a) $A=2$, $\phi=0$, $N=32$, and $M=1$

(b) $A=1$, $\phi=\pi/2$, $N=64$, and $M=13$

(c) $A=2$, $\phi=\pi/8$, $N=64$, and $M=5$

(d) $A=1$, $\phi=0$, $N=64$, and $M=33$

(e) $A=1$, $\phi=0$, $N=128$, and $M=65$

(f) $A=0.5$, $\phi=5\pi/8$, $N=32$, and $M=2.5$

6.2. Using MATLAB or an equivalent software program, plot the samples of the signals described in Problem 6.1 over its UTP.

6.3. Repeat Problems 6.1 and 6.2 with a sampling rate of 8 kHz.

6.4. Using the square characteristic pulse defined in Example 6.2, reconstruct the samples obtained in Problem 6.1 and determine the frequency of the resulting sine wave. Identify any situation where aliasing occurs.

6.5. Using the triangular characteristic pulse defined in Example 6.1, reconstruct the samples obtained in Problem 6.3 and determine the frequency of the resulting sine wave. Identify any situation where aliasing occurs.

6.6. Using a second-order hold function where the interpolation is done by passing a quadratic function through the points at $(n-2)T_s$, $(n-1)T_s$, and nT_s, reconstruct the samples obtained in Problem 6.1. How does the reconstructed waveform compare to a zero-order hold function (square pulse) and a first-order hold function (triangular pulse)?

6.7. If a digital sinusoidal signal described by $A=1$, $\phi=0$, $N=32$, and $M=5$ is played through a DAC and speaker arrangement whose sampling rate is 8 kHz, what analog frequency will be heard? Repeat for $A=1$, $\phi=0$, $N=32$, and $M=25$. *Hint:* Reconstruct the discrete signal using MATLAB and observe the frequency of the reconstructed signal.

6.8. What is the LSB of an ideal 12-bit ADC that has a full-scale input range of 0-3 V? What is the expected RMS value of the corresponding quantization noise?

6.9. If an ideal 8-bit ADC has a 400 µV RMS quantization noise component, what would be the noise component for a 5-bit ADC having the same input range?

6.10. A 6-bit ADC with an analog input range from -1.5 to $+1.5$ V gives an output of code of 37 for a code range beginning at 0 and ending at 63. What are the minimum and maximum values of the input voltage corresponding to this output code?

6.11. What is the RMS value of the error induced by an ADC having an RMS sampling jitter of 250 ps while measuring a 1-V amplitude sinusoid with a frequency of 20 kHz?

6.12. In each of the following questions, assume a 1 LSB maximum allowable voltage error. What is the maximum allowable sampling jitter that a 10-bit ADC can tolerate when it has a full-scale input range of 3 V and converting a 1-V amplitude sinusoid with a frequency of 20 kHz?

6.13. What is the maximum allowable sampling jitter that an 8-bit DAC can tolerate when it has a maximum sampling rate of 10 MHz?

6.14. If a 6-bit DAC has an RMS sampling jitter of 100 ps, what is the maximum sampling rate?

6.15. If an ADC or DAC is controlled by a clock circuit with a minimum clock period of 10 ns and RMS jitter of 250 ps, what is the maximum conversion resolution possible with either the ADC or DAC?

6.16. What is the RMS value of the uncertainty associated with a signal sampled by a 10-bit ADC having an RMS clock jitter of 100 ps. The ADC has a full-scale range of 5 V and Nyquist frequency of 1 MHz. *Hint:* Independent errors add in a mean-squared sense, i.e., $v^2_{total-RMS} = v^2_{q-RMS} + v^2_{j-RMS}$.

6.17. What is the RMS value of the uncertainty associated with a signal sampled by a 6-bit ADC having an RMS clock jitter of 1 ns. The ADC has a full-scale range of 5 V and Nyquist frequency of 1 MHz.

6.18. Plot the quantization error sequence that results after exciting a 6-bit ADC with a full-scale amplitude sine wave. Use the MATLAB routine given in Example 6.4 for the quantizer. Compute the mean and RMS value of the quantization noise sequence. Repeat for an 8-bit quantizer. How does the mean and RMS value compare with theory in the two cases?

6.19. What is the fundamental frequency of 1024 samples collected at a rate of 20 kHz? What is the corresponding UTP?

6.20. How many cycles of a 20-kHz sine wave are completed in a 0.8-ms UTP? How many cycles of a 20-kHz sine wave are completed in a 0.79-ms UTP? If the signals are repeated indefinitely, which ones are coherent?

6.21. What are the coherent frequencies associated with 16 samples collected at a rate of 1 kHz?

6.22. Using MATLAB or an equivalent software program, plot a multitone signal consisting of three unity amplitude sine waves with frequencies of 1, 5, and 11 kHz over a UTP of 1 ms. Assume that the phase shifts are all zero.

6.23. For the multitone signal described in Problem 6.22, using the random selection method described in Section 6.3.3 search for the phases of three tones such that the peak-to-RMS value is in a ratio of approximately 3.35:1. Provide a plot to illustrate your result. Investigate the sensitivity of the peak-to-RMS value by changing the phase of each tone by ±1% and computing the change in the peak-to-RMS value.

6.24. The Newman phase selection criterion[7] relies on selecting the phase of the kth tone of a multitone signal according the quadratic expression given by $\phi_k = (\pi/N)(k-1)^2$. Using this equation, determine the phases of the 3-tone multitone signal of Problem 6.22 and compare its peak-to-RMS value to the ideal value of 3.35. Investigate the sensitivity of the peak-to-RMS value by changing the phase of each tone by ±1% and computing the change in the peak-to-RMS value.

6.25. Using MATLAB or an equivalent software program, plot a multitone signal consisting of 100 unity-amplitude sine waves distributed across a frequency range of 100 kHz over a UTP of 1 ms. Determine the phases of each tone such that the peak-to-RMS value approaches the ideal value of 3.35.

6.26. Select the spectral bins of a four tone signal at 1, 2, 3, and 4 kHz such that minimum distortion overlap occurs. Assume that the sampling rate is 44.8 kHz and that the UTP must be less than 100 ms. The accuracy of the test frequencies should be less than ±100 Hz. Justify your answer by providing the appropriate distortion tables.

6.27. An ATE PLL-based clock source is set with the divide-by-N counter equal to 4096, the divide-by-M equal to 512, and the divide-by-L equal to 128. With a reference frequency of 200 MHz, what is the output sampling frequency?

6.28. An ATE PLL-based clock source has a divide-by-N counter with a range from 1 to 1024, a divide-by-M counter with a range from 1 to 256 and a divide-by-L with a range of 1 to 65535. With a reference frequency of 200 MHz, what is the range of the output sampling frequency?

References

1. Matthew Mahoney, *Tutorial DSP-Based Testing of Analog and Mixed-Signal Circuits*, The Computer Society of the IEEE, 1730 Massachusetts Avenue N.W., Washington D.C. 20036-1903, 1987, ISBN: 0818607858

2. Alan V. Oppenheim, Ronald W. Schafer, *Discrete-Time Signal Processing*, Prentice Hall, Englewood Cliffs, NJ, March 1989, ISBN: 013216292X

3. Alan V. Oppenheim et al., *Signals and Systems*, Prentice Hall, Englewood Cliffs, NJ, August 1997, ISBN: 0138147574

4. William McC. Siebert, *Circuits, Signals, and Systems*, The MIT Press, Cambridge, MA, September 1985, ISBN: 0262192292

5. Paul G. Hoel, Sidney C. Port, Charles J. Stone, *Introduction to Probability Theory*, Houghton Mifflin Company, Boston, MA, 1971, ISBN: 039504636X, p 118

6. A. J. Jerri, *The Shannon Sampling Theorem – Its Various Extensions and Applications: A Tutorial Review*, Proc. IEEE, Vol. 65, pp. 1565-96, 1997

7. E. V. Ouderaa, J. Schoukens, J. Renneboog, *Peak Factor Minimization of the Input and Output Signals of Linear Systems*, IEEE Transactions on Instrumentation and Measurement, Vol. 37, No. 2, June 1988

CHAPTER 7

DSP-Based Testing

7.1 Advantages of DSP-Based Testing

7.1.1 Reduced Test Time

In Chapter 6, we briefly touched on the advantages of DSP-based testing before beginning our review of sampling theory. In this chapter, we will take a more detailed look at digital signal processing and the power it gives us in testing mixed-signal devices. Although a full study of DSP is beyond the scope of this book, many good texts have been written on the subject.[1-3] In this chapter, we will review the basics of DSP, limiting our discussion to discrete (i.e., sampled) waveforms of finite length.

Coherent DSP-based testing gives us several advantages over traditional measurement techniques. The first advantage of DSP-based testing is reduced test time. Since we can create and measure signals with multiple frequencies at the same time, we can perform many parametric measurements in parallel. If we need to test the frequency response and phase response of a filter, for example, we can perform a series of gain and phase measurements at a dozen or so frequencies simultaneously.

DSP-based testing allows us to send all the filter test frequencies through the device under test (DUT) at the same time. Once we have collected the DUT's output response using a digitizer or capture memory, DSP allows us to separate each test tone in the output waveform from all the other test tones. We can then calculate a separate gain and phase measurement at each frequency without running many separate tests. We can also measure noise and distortion at the same time that we measure gain and phase shift, further reducing test time.

7.1.2 Separation of Signal Components

Separation of signal components from one another gives us a second huge advantage over non-DSP-based measurements. We can isolate noise and distortion components from one another and from the test tones. This allows for much more accurate and repeatable measurements of the primary test tones and their distortion components.

Using coherent test tones, we are always guaranteed that all the distortion components will fall neatly into separate Fourier spectral bins rather than being smeared across many bins (as is the case with noncoherent signal components). DSP-based testing is a major advantage in the elimination of errors and poor repeatability.

7.1.3 Advanced Signal Manipulations

In this chapter we will see how DSP-based testing allows us to manipulate digitized output waveforms to achieve a variety of results. We can perform interpolations between the samples to achieve better time resolution. We can apply mathematical filters to emphasize or diminish certain frequency components. We can remove noise from signals to achieve better accuracy. All these techniques are made possible by the application of digital signal processing to sampled DUT outputs.

7.2 DIGITAL SIGNAL PROCESSING

7.2.1 DSP and Array Processing

Before we embark on a review of digital signal processing, let us take some time to define exactly what a digital signal is. Then we shall look at the different ways we can process digital signals. There is a slight semantic difference between digital signal processing and array processing. An array, or vector, is a set of similar numbers, such as a record of all the heights of the students in a class. An example of array processing would be the calculation of the average height of the students. A digital signal is also a set of numbers (i.e., voltage samples), but the samples are ordered in time. Digital signal processing is thus a subset of array processing, since it is limited to mathematical operations on *time-ordered* samples. However, since most arrays processed on an ATE tester are in fact digital signals, most automated test equipment (ATE) languages categorize all array processing operations under the umbrella of DSP. So much for semantics!

ATE digital signal processing is often accomplished on a specialized computer called an *array processor*. However, tester computers have become faster over the years, making a separate array processor unnecessary in some of the newer testers. Depending on the sophistication of the tester's operating system, the presence or absence of a separate array processor may not even be apparent to the user.

There are many array processing functions that prove useful in mixed-signal test engineering. One simple example is the averaging function. The average of a series of samples is equivalent to the DC offset of the signal. If we want to measure the RMS noise of a digitized waveform, we must first remove the DC offset. Otherwise the DC offset would add to the RMS calculation, resulting in an erroneous noise measurement. We can compute the DC offset of the digitized waveform using an averaging function. Subtracting the offset from each sample in the waveform eliminates the DC offset. In MATLAB, we might accomplish the DC removal using the following simple routine:

```
% DC Removal Routine
    % Calculate the DC offset (average) of a waveform, x
        average=sum(x)/length(x);
    % Subtract offset from each waveform sample
        x=x-average;
```

This DC removal routine can be considered a digital signal processing operation, although a dedicated array processor is not needed to perform the calculations. Fortunately, we are able to take advantage of many built-in array processing operations in mixed-signal testers rather than writing them from scratch. These built-in operations are streamlined by the ATE vendor to allow the fastest possible processing time on the available computation hardware. For example, some of the computations may take place in parallel using multiple array processors, thus saving test time. Although tester languages vary widely, the following pseudocode is representative of typical ATE array processing operations:

```
float offset, waveform[256];
offset = average( waveform[ 1 to 256 ] );
waveform = waveform – offset;
```

Not only are the built-in operations simpler to read, they often execute faster than user-defined routines written in a language such as C. The following are examples of array processing functions that are typically included in an ATE array processor. The term "vector" refers to an array of values, while the term "scalar" refers to a single value.

vector average: average value of an array
vector RMS: root mean square of the values in an array
max/min: locate and report the maximum and minimum values in an array
vector add: add two arrays
add scalar to vector: add a constant to each element in an array
subtract scalar from vector: subtract a constant from each element in an array
vector multiply: multiply two arrays
multiply scalar by vector: multiply a constant by each element in an array
divide vector by scalar: divide each element in an array by a constant

Digital signal processing operations are somewhat more complicated than the simple array processing functions listed. Operations such as the discrete Fourier transform (DFT), fast Fourier transform (FFT), inverse FFT, and filtering operations will require a little more explanation.

7.2.2 Fourier Analysis of Periodic Signals

The tremendous advantage of DSP-based testing is the ability to measure many different frequency components simultaneously, minimizing test time. For example, we can apply a seven-tone multitone signal such as the one in Figure 6.22 to a low-pass filter. The filter will amplify or attenuate each frequency component by a different amount according to the filter's transfer function. It may also shift the phase of each frequency component.

It is easy to see how we can apply a multitone signal to the input of a filter. We simply compute the sample set in Figure 6.22 and apply it through an arbitrary waveform generator (AWG) to the input of the filter. A digitizer can then be used to collect samples from the output of the filter. But how do we then extract the amplitude of the individual frequency components from the composite signal at the filter's output?

The answer is a Fourier analysis. It is a mathematical method that gives us the power to split a composite signal into its individual frequency components. It is based on the fact that we can describe any signal in either the time domain or the frequency domain. Fourier analysis allows us to convert time-domain signal information into a description of a signal as a function of frequency. Fourier analysis allows us to convert a signal from the time domain to the frequency domain and back again without losing any information about the signal in either domain. This powerful capability is at the heart of mixed-signal testing.

Jean Baptiste Joseph Fourier was a clever French mathematician who developed Fourier analysis for the study of heat transfer in solid bodies. His technique was published in 1822. 175 years later, the importance of Fourier's work in today's networked world is astounding. Applications of his method extend to cellular telephones, disk drives, speech recognition systems, radar systems, and mixed-signal testing to name just a few.

7.2.3 The Trigonometric Fourier Series

Fourier's initial work showed how a mathematical series of sine and cosine terms could be used to analyze heat conduction problems. This became known as the *trigonometric Fourier series* and was probably the first systematic application of a trigonometric series to a problem. At the time of his death in 1830, he had extended his methods to include the Fourier integral leading to the concept of a *Fourier transform*. The Fourier transform is largely applied to the analysis of aperiodic signals. Mixed-signal test engineering is primarily concerned with the discrete form of the Fourier series and, specifically, coherent sample sets. Therefore, we shall limit our discussion mainly to the Fourier series.

Let $x(t)$ denote a periodic signal with period T such that it satisfies

$$x(t) = x(t+T) \tag{7.1}$$

for all values of $-\infty < t < \infty$. Using a trigonometric Fourier series expansion of this signal, we are able to resolve the signal into an infinite sum of cosine and sine terms according to

$$x(t) = a_0 + \sum_{k=1}^{\infty} \left[a_k \cos(k \times 2\pi f_o t) + b_k \sin(k \times 2\pi f_o t) \right] \tag{7.2}$$

The first term in the series a_0 represents the DC or average component of $x(t)$. The coefficients a_k and b_k represents the amplitudes of the cosine and sine terms, respectively. They are commonly referred to as the spectral or Fourier coefficients and are determined from the following integral equations:

$$a_0 = \frac{1}{T} \int_0^T x(t)\, dt$$

$$a_k = \frac{2}{T} \int_0^T x(t) \cos\left(k \frac{2\pi}{T} t\right) dt \tag{7.3}$$

$$b_k = \frac{2}{T} \int_0^T x(t) \sin\left(k \frac{2\pi}{T} t\right) dt$$

The frequency of each cosine and sine term is an integer multiple of the fundamental frequency $f_o = 1/T$, referred to as a harmonic. For instance, the quantity kf_o represents the kth harmonic of the fundamental frequency.

A more compact form of the trigonometric Fourier series is

$$x(t) = \sum_{k=0}^{\infty} c_k \cos(k \times 2\pi f_o t - \phi_k) \qquad (7.4)$$

where

$$c_k = \sqrt{a_k^2 + b_k^2} \quad \text{and} \quad \phi_k = \begin{cases} \tan^{-1}\left(\dfrac{b_k}{a_k}\right) & \text{if } a_k \geq 0 \\ \pi + \tan^{-1}\left(\dfrac{b_k}{a_k}\right) & \text{if } a_k < 0 \end{cases} \qquad (7.5)$$

This result follows from the trigonometric identity: $\cos(A+B) = \cos(A)\cos(B) - \sin(A)\sin(B)$.

We prefer this representation as it lends itself more directly to graphical form. Specifically, vertical lines can be drawn at discrete frequency points corresponding to $0, f_o, 2f_o, 3f_o$, and so on, with their heights proportional to the amplitudes of the corresponding frequency components, that is, c_k versus $k f_o$. Such a plot conveys the *magnitude spectrum* of $x(t)$. Likewise, a *phase spectrum* can be drawn in the exact same manner except that the heights of the vertical lines are proportional to the phases ϕ_k of the corresponding frequency components. The following example illustrates these two plots for a 5-V, 10-kHz clock signal.

Example 7.1

Determine the Fourier series representation of the 5-V, 10-kHz clock signal shown in Figure 7.1 and plot the corresponding magnitude and phase spectrum.

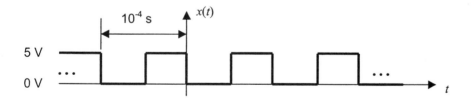

Figure 7.1. 10-kHz clock signal.

Solution:

The spectral coefficients are determined according to Eq. (7.3) as follows

$$a_0 = \frac{1}{10^{-4}} \left(\int_0^{0.5 \times 10^{-4}} 0 \, dt + \int_{0.5 \times 10^{-4}}^{10^{-4}} 5 \, dt \right) = \frac{1}{10^{-4}} \, 5 \left(10^{-4} - 0.5 \times 10^{-4} \right) = 2.5$$

$$a_k = \frac{2}{10^{-4}} \left(\int_0^{0.5\times10^{-4}} 0 \cos(k\, 10^4\, 2\pi t)\, dt + \int_{0.5\times10^{-4}}^{10^{-4}} 5 \cos(k\, 10^4\, 2\pi t)\, dt \right)$$

$$= \frac{5}{k\pi} \sin(k\, 10^{-4}\, 2\pi t)\Big|_{0.5\times10^{-4}}^{10^{-4}} = 0$$

$$b_k = \frac{2}{10^{-4}} \left(\int_0^{0.5\times10^{-4}} 0 \sin(k\, 10^4\, 2\pi t)\, dt + \int_{0.5\times10^{-4}}^{10^{-4}} 5 \sin(k\, 10^4\, 2\pi t)\, dt \right)$$

$$= -\frac{5}{k\pi} \cos(k\, 10^{-4}\, 2\pi t)\Big|_{0.5\times10^{-4}}^{10^{-4}} = \frac{5}{k\pi} \left[\cos(k\pi) - \cos(k\, 2\pi) \right]$$

$$= \frac{5}{k\pi}\left[(-1)^k - 1\right]$$

For even values of k, the term $[(-1)^k - 1] = 0$, hence the clock signal consists of only odd harmonics. Therefore,

$$b_k = \begin{cases} 0 & k \text{ even} \\ -\dfrac{10}{k\pi} & k \text{ odd} \end{cases}$$

The Fourier series representation then becomes

$$x(t) = 2.5 - \sum_{k=1,\, k\text{ odd}}^{\infty} \frac{10}{k\pi} \sin(k\, 2\pi\, 10^4\, t)$$

or, when written in the form of Eq. (7.4), becomes

$$x(t) = 2.5 + \sum_{k=1,\, k\text{ odd}}^{\infty} \frac{10}{k\pi} \cos\left(k\, 2\pi\, 10^4\, t + \frac{\pi}{2}\right)$$

as $\cos(x + \pi/2) = -\sin(x)$ and $\cos(x - \pi/2) = \sin(x)$. The corresponding magnitude and phase spectrum plots are then as shown in Figure 7.2.

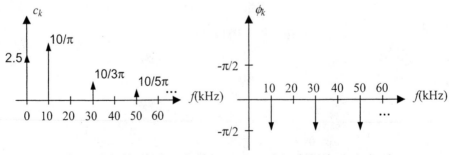

Figure 7.2. Magnitude and phase spectrum plots of 10-kHz clock signal.

It is instructive to view the behavior of the Fourier series representation for the clock signal of the previous example consisting of 10 and 50 terms. This we show in Figure 7.3 superimposed on one period of the clock signal. Clearly, as the number of terms in the series is increased, the Fourier series representation more closely resembles the clock signal. Some large amplitude oscillatory behavior occurs at the jump discontinuity. This is known as *Gibb's phenomenon* and is a result of truncating terms in the Fourier series representation.

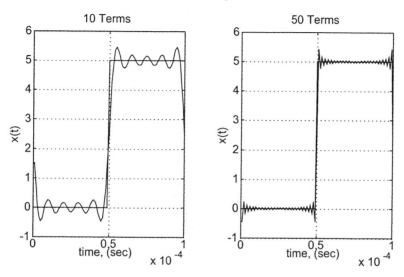

Figure 7.3. Observing the Fourier series representation of a clock signal for 10 and 50 terms. Superimposed on the plot is the original clock signal.

7.2.4 The Discrete-Time Fourier Series

As mentioned previously, sampling is an important step in mixed-signal testing. To understand the effects of sampling on an arbitrary periodic signal, consider sampling the general form of the Fourier series representation given in Eq. (7.2). Assuming that the sampling period is T_s, we can write

$$x(t)|_{t=nT_s} = a_0 + \sum_{k=1}^{\infty} \left[a_k \cos(k\; 2\pi f_o\; nT_s) + b_k \sin(k\; 2\pi f_o\; nT_s) \right] \quad (7.6)$$

As $F_s = 1/T_s$, we can rewrite Eq. (7.6) as

$$x(t)|_{t=nT_s} = a_0 + \sum_{k=1}^{\infty} \left\{ a_k \cos\left[k\left(\frac{2\pi f_o}{F_s}\right) n \right] + b_k \sin\left[k\left(\frac{2\pi f_o}{F_s}\right) n \right] \right\} \quad (7.7)$$

In this form we see that the frequency of the fundamental is a fraction f_o/F_s of 2π. Further, this term no longer has units of radians per second but rather just radians. To distinguish this from our regular notion of frequency, it is commonly referred to as a *normalized frequency*, as it has

Exercises

7.1. Find the trigonometric Fourier series representation of a square-wave $x(t)$ having a period of 2 s and whose behavior is described by:

$$x(t) = \begin{cases} +1 & \text{if } 0 < t < 1 \\ -1 & \text{if } 1 < t < 2 \end{cases}$$

Ans. $x(t) = \dfrac{4}{\pi}\left(\sin \pi t + \dfrac{1}{3}\sin 3\pi t + \dfrac{1}{5}\sin 5\pi t + \cdots\right)$

7.2. Express the Fourier series in Exercise 7.1 using cosine terms only.

Ans. $x(t) = \dfrac{4}{\pi}\left[\cos\left(\pi t - \dfrac{\pi}{2}\right) + \dfrac{1}{3}\cos\left(3\pi t - \dfrac{\pi}{2}\right) + \dfrac{1}{5}\cos\left(5\pi t - \dfrac{\pi}{2}\right) + \cdots\right]$

7.3. Find the Fourier series representation of a square-wave $x(t)$ having a period of 4 s and whose behavior is described by:

$$x(t) = \begin{cases} 1 & \text{if } -1 < t < 1 \\ 0 & \text{if } 1 < t < 3 \end{cases}$$

Ans. $x(t) = \dfrac{1}{2} - \dfrac{2}{\pi}\left(\cos\dfrac{\pi}{2}t - \dfrac{1}{3}\cos\dfrac{3\pi}{2}t + \dfrac{1}{5}\cos\dfrac{5\pi}{2}t - + \cdots\right)$

---

been normalized by the sampling frequency F_s. Except for the time reference T_s on the left-hand side of the equation, the information about the time scale is lost. This is further complicated by the fact that one usually uses the shorthand notation

$$x[n] = x(t)\big|_{t=nT_s} \qquad (7.8)$$

and eliminates the time reference altogether. The discrete-time signal $x[n]$ is simply a sequence of numbers with no reference to the underlying time scale. Hence, the original samples cannot be reconstructed without knowledge of the original sampling frequency. Therefore, a sampling period or frequency must always be associated with a discrete-time signal, $x[n]$.

For much of the work in this textbook, we are concerned with coherent sampling sets, that is, $UTP = NT_s$, or, equivalently, $f_o = 1/UTP = F_f = F_s/N$. Equation (7.7) can then be reduced to

$$x[n] = a_0 + \sum_{k=1}^{\infty}\left\{a_k \cos\left[k\left(\dfrac{2\pi}{N}\right)n\right] + b_k \sin\left[k\left(\dfrac{2\pi}{N}\right)n\right]\right\} \qquad (7.9)$$

where the frequency of the fundamental reduces to $1/N$, albeit normalized by F_s.

As the original continuous-time signal $x(t)$ is periodic and sampled coherently, $x[n]$ will be periodic with respect to the sample index n, according to

$$x[n] = x[n+N] \tag{7.10}$$

As the roles of n and k are interchangeable in the arguments of the sine and cosine terms of Eq. (7.9), it suggests that $x[n]$ will also repeat with index k over the period N. Through a detailed trigonometric development outlined in the appendix at the end of this chapter, we can rewrite the infinite series given by Eq. (7.9) as the sum of $N/2$ trigonometric terms as follows (here it is assumed N is even, as is often the case in testing applications)

$$x[n] = \tilde{a}_0 + \sum_{k=1}^{N/2-1} \left\{ \tilde{a}_k \cos\left[k\left(\frac{2\pi}{N}\right)n\right] + \tilde{b}_k \sin\left[k\left(\frac{2\pi}{N}\right)n\right] \right\} + \tilde{a}_{N/2} \cos(\pi n) \tag{7.11}$$

where

$$\tilde{a}_0 = \sum_{m=0}^{\infty} a_{0+mN} \quad \tilde{a}_k = \sum_{m=0}^{\infty} (a_{k+mN} + a_{N-k+mN}) \quad \tilde{a}_{N/2} = \sum_{m=0}^{\infty} a_{N/2+mN}$$

$$\tilde{b}_k = \sum_{m=0}^{\infty} (b_{k+mN} - b_{N-k+mN}) \tag{7.12}$$

Here $\tilde{a}_k$ represents the amplitude of the cosine component located at the kth Fourier spectral bin. Likewise, $\tilde{b}_k$ is the amplitude of the sine component located at the kth Fourier spectral bin. Of course, $\tilde{a}_0$ represents the DC or average value of the sample set. The equations of Eq. (7.12) represent the sum of all aliases terms that arise during the sampling process. Since we are assuming that the original continuous-time signal is frequency band-limited, the sums in Eq. (7.12) will converge to finite values.

As before, Eq. (7.11) can be written in a more compact form using magnitude and phase notation as

$$x[n] = \sum_{k=0}^{N/2} \tilde{c}_k \cos\left[k\left(\frac{2\pi}{N}\right)n - \tilde{\phi}_k\right] \tag{7.13}$$

where

$$\tilde{c}_k = \sqrt{\tilde{a}_k^2 + \tilde{b}_k^2} \quad \text{and} \quad \tilde{\phi}_k = \begin{cases} \tan^{-1}\left(\dfrac{\tilde{b}_k}{\tilde{a}_k}\right) & \text{if } \tilde{a}_k \geq 0 \\ \pi + \tan^{-1}\left(\dfrac{\tilde{b}_k}{\tilde{a}_k}\right) & \text{if } \tilde{a}_k < 0 \end{cases} \tag{7.14}$$

Equation (7.13) can then be graphically displayed using a magnitude and phase spectrum plot.

The importance of the above expressions cannot be understated, as it relates the spectrum of a sampled signal to the original continuous-time signal. Further, it suggests that a discrete time signal has a spectrum that consists of, at most, $N/2$ unique frequencies. Moreover, these frequencies are all harmonically related to the primitive or fundamental frequency of $1/N$ radians. This representation is known as a *discrete-time Fourier series (DTFS)* representation of $x[n]$ written in trigonometric form. It will form the basis for all the computer analysis in this text.

Example 7.2

Calculate the DTFS representation of the 10-kHz clock signal of Example 7.1 when sampled at a 100-kHz sampling rate.

Solution:

With a 100-kHz sampling rate, 10 points per period will be collected in one period of the 10-kHz clock signal (i.e., $N=10$). Using the equations for the spectral coefficients in (7.12), together with the Fourier series result of Example 7.1, we find the $\tilde{a}_k$ coefficients are as follows

$$\tilde{a}_0 = \sum_{m=0}^{\infty} a_{0+10m} = a_0 = 2.5$$

$$\tilde{a}_k = \sum_{m=0}^{\infty} (a_{k+10m} + a_{10-k+10m}) = 0 \quad k \in \{1,2,3,4\}$$

$$\tilde{a}_5 = \sum_{m=0}^{\infty} a_{5+10m} = 0$$

Subsequently, the $\tilde{b}_k$ coefficients for $k = 1, 2, ..., 4$ (by definition, $\tilde{b}_0 = 0$ and $\tilde{b}_5 = 0$) are found as follows

$$\tilde{b}_k = \sum_{m=0}^{\infty} (b_{k+10m} - b_{10-k+10m}) = \begin{cases} 0 & k \text{ even} \\ -\frac{10}{\pi} \sum_{m=0}^{\infty} \left(\frac{1}{(k+10m)} - \frac{1}{(10-k+10m)} \right) & k \text{ odd} \end{cases}$$

Here the summation involves the difference between two harmonic progressions where no closed-form summation formulas are known to exist. Subsequently, a numerical routine was written that summed the first 100 terms of this series. The result is

$$\tilde{b}_1 = -3.07516; \quad \tilde{b}_2 = 0; \quad \tilde{b}_3 = -0.72528; \quad \tilde{b}_4 = 0$$

According to Eq. (7.11), the discrete-time Fourier series representation for the clock signal then becomes

$$x[n] = 2.5 - 3.07516 \sin\left[1\left(\frac{2\pi}{10}\right)n\right] - 0.72528 \sin\left[3\left(\frac{2\pi}{10}\right)n\right]$$

or, using magnitude and phase notation, we write

$$x[n] = 2.5 + 3.07516 \cos\left[1\left(\frac{2\pi}{10}\right)n + \frac{\pi}{2}\right] + 0.72528 \cos\left[3\left(\frac{2\pi}{10}\right)n + \frac{\pi}{2}\right]$$

It is important for the reader to verify the samples produced by the above discrete-time Fourier series. Evaluating $x[n]$ for one complete period (i.e., $n=\{0, 1, 2, \ldots 9\}$), we obtain $x=\{2.5, 0.0027, 0.0017, 0.0017, 0.0027, 2.5, 4.9973, 4.9983, 4.9983, 4.9973\}$. As expected, all the samples correspond quite closely with samples from the original signal, as shown in Figure 7.4.

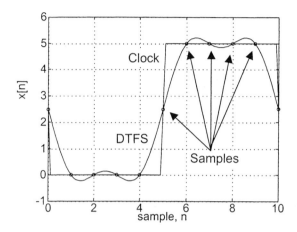

Figure 7.4. Comparing the samples of a DTFS representation and the original clock signal. Also shown is the DTFS as a continuous function of n.

The small difference can be contributed to the error that results from including only 100 terms in the summation of the $\tilde{b}_k$ coefficients. Of particular interest is the value that the discrete-time Fourier series assigns to the waveform at the jump discontinuity. In general, the sample value at a jump discontinuity is ambiguous and undefined. Fourier analysis resolves this problem by assigning the sample value of the discontinuity as the midway point of the jump. In this particular case, the midpoint of each jump discontinuity is $(5-0)/2 = 2.5$.

Also shown in the plot of Figure 7.4 is a graph of the DTFS representation as a continuous function of sample index, n. It is interesting to note that the samples are the intersection of this continuous-time function with the original clock signal.

When analyzing a waveform collected from a DUT we want to know the spectral coefficients $\tilde{a}_k$ and $\tilde{b}_k$ or $\tilde{c}_k$ and $\tilde{\phi}_k$ so that we can see how much signal power is present at each frequency and deduce the DUT's overall performance. Thus far, we have only considered how to compute the spectral coefficients from a Fourier series expansion of a continuous-time waveform. As we shall see, there is a more direct way to compute the spectral coefficients of a discrete-time Fourier series from N samples of the continuous-time waveform. In fact, we shall outline two methods: one that highlights the nature of the problem in algebraic terms and the other involving a closed-form expression for the coefficients in terms of the sampled values.

To begin, let us consider that we have N samples, denoted $x[n]$ for $n=0, 1, 2,...,N-1$, Consider that each one of these samples must satisfy the discrete-time Fourier series expansion of Eq. (7.11). This is rather unlike the Fourier series expansion for a continuous-time signal that consists of an infinite number of trigonometric terms.

In practice, the summation must be limited to a finite number of terms, resulting in an approximation error. The discrete-time Fourier series, on the other hand, consists of only N trigonometric terms and, hence, there is no error in its representation. Therefore, we can write directly from Eq. (7.11) at sampling instant $n=0$

$$x[0] = \tilde{a}_0 + \sum_{k=1}^{N/2-1} \{\tilde{a}_k \cos[0] + \tilde{b}_k \sin[0]\} + \tilde{a}_{N/2} \cos[0]$$
$$= \tilde{a}_0 + \tilde{a}_1 + \cdots + \tilde{a}_{N/2-1} + \tilde{a}_{N/2}$$
(7.15)

Next, at $n=1$, we write

$$x[1] = \tilde{a}_0 + \sum_{k=1}^{N/2-1} \left\{\tilde{a}_k \cos\left[k\left(\frac{2\pi}{N}\right)\right] + \tilde{b}_k \sin\left[k\left(\frac{2\pi}{N}\right)\right]\right\} + \tilde{a}_{N/2} \cos[\pi]$$
$$= \tilde{a}_0 + \tilde{a}_1 \cos\left[\left(\frac{2\pi}{N}\right)\right] + \tilde{b}_1 \sin\left[\left(\frac{2\pi}{N}\right)\right] + \tilde{a}_2 \cos\left[2\left(\frac{2\pi}{N}\right)\right] + \tilde{b}_2 \sin\left[2\left(\frac{2\pi}{N}\right)\right] + \cdots$$
$$+ \tilde{a}_{N/2-1} \cos\left[\left(\frac{N}{2}-1\right)\left(\frac{2\pi}{N}\right)\right] + \tilde{b}_{N/2-1} \sin\left[\left(\frac{N}{2}-1\right)\left(\frac{2\pi}{N}\right)\right] + \tilde{a}_{N/2} \cos[\pi]$$
(7.16)

Similarly, for $n=2$, we write

$$x[2] = \tilde{a}_0 + \sum_{k=1}^{N/2-1} \left\{\tilde{a}_k \cos\left[2k\left(\frac{2\pi}{N}\right)\right] + \tilde{b}_k \sin\left[2k\left(\frac{2\pi}{N}\right)\right]\right\} + \tilde{a}_{N/2} \cos(2\pi)$$
$$= \tilde{a}_0 + \tilde{a}_1 \cos\left[2\left(\frac{2\pi}{N}\right)\right] + \tilde{b}_1 \sin\left[2\left(\frac{2\pi}{N}\right)\right] + \tilde{a}_2 \cos\left[4\left(\frac{2\pi}{N}\right)\right] + \tilde{b}_2 \sin\left[4\left(\frac{2\pi}{N}\right)\right] + \cdots$$
$$+ \tilde{a}_{N/2-1} \cos\left[2\left(\frac{N}{2}-1\right)\left(\frac{2\pi}{N}\right)\right] + \tilde{b}_{N/2-1} \sin\left[2\left(\frac{N}{2}-1\right)\left(\frac{2\pi}{N}\right)\right] + \tilde{a}_{N/2} \cos(2\pi)$$
(7.17)

Continuing for all remaining sampling instants, up to $n=N-1$, we can write the Nth equation as

$$x[N-1] = \tilde{a}_0 + \sum_{k=1}^{N/2-1}\left\{\tilde{a}_k \cos\left[(N-1)k\left(\frac{2\pi}{N}\right)\right] + \tilde{b}_k \sin\left[(N-1)k\left(\frac{2\pi}{N}\right)\right]\right\} + \tilde{a}_{N/2}\cos\left[(N-1)\pi\right]$$

$$= \tilde{a}_0 + \tilde{a}_1 \cos\left[(N-1)\left(\frac{2\pi}{N}\right)\right] + \tilde{b}_1 \sin\left[(N-1)\left(\frac{2\pi}{N}\right)\right] + \tilde{a}_2 \cos\left[2(N-1)\left(\frac{2\pi}{N}\right)\right]$$

$$+ \tilde{b}_2 \sin\left[2(N-1)\left(\frac{2\pi}{N}\right)\right] + \cdots + \tilde{a}_{N/2-1}\cos\left[\left(\frac{N}{2}-1\right)(N-1)\left(\frac{2\pi}{N}\right)\right] \quad (7.18)$$

$$+ \tilde{b}_{N/2-1}\sin\left[\left(\frac{N}{2}-1\right)(N-1)\left(\frac{2\pi}{N}\right)\right] + \tilde{a}_{N/2}\cos\left[(N-1)\pi\right]$$

Finally, on observing the behavior of each one of these equations, we find that all the trigonometric terms have numerical values and the only unknown terms are the spectral coefficients. In essence, we have a system of N linear equations in N unknowns. Straightforward linear algebra can then be used to compute the spectral coefficients. For instant, if we define vectors

$$C = \begin{bmatrix} a_0 & a_1 & b_1 & a_2 & b_2 & \cdots & a_{N-1} \end{bmatrix}$$

and

$$X = \begin{bmatrix} x[0] & x[1] & x[2] & \cdots & x[N-1] \end{bmatrix}$$

together with matrix

$$W = \begin{bmatrix} 1 & \cos\left[(1)(0)\left(\frac{2\pi}{N}\right)\right] & \sin\left[(1)(0)\left(\frac{2\pi}{N}\right)\right] & \cdots & \sin\left[\left(\frac{N}{2}-1\right)(0)\left(\frac{2\pi}{N}\right)\right] & \cos\left[\left(\frac{N}{2}\right)(0)\left(\frac{2\pi}{N}\right)\right] \\ 1 & \cos\left[(1)(1)\left(\frac{2\pi}{N}\right)\right] & \sin\left[(1)(1)\left(\frac{2\pi}{N}\right)\right] & \cdots & \sin\left[\left(\frac{N}{2}-1\right)(1)\left(\frac{2\pi}{N}\right)\right] & \cos\left[\left(\frac{N}{2}\right)(1)\left(\frac{2\pi}{N}\right)\right] \\ 1 & \cos\left[(1)(2)\left(\frac{2\pi}{N}\right)\right] & \sin\left[(1)(2)\left(\frac{2\pi}{N}\right)\right] & \cdots & \sin\left[\left(\frac{N}{2}-1\right)(2)\left(\frac{2\pi}{N}\right)\right] & \cos\left[\left(\frac{N}{2}\right)(2)\left(\frac{2\pi}{N}\right)\right] \\ \vdots & \vdots & \vdots & & \vdots & \vdots \\ 1 & \cos\left[(1)(N-1)\left(\frac{2\pi}{N}\right)\right] & \sin\left[(1)(N-1)\left(\frac{2\pi}{N}\right)\right] & \cdots & \sin\left[\left(\frac{N}{2}-1\right)(N-1)\left(\frac{2\pi}{N}\right)\right] & \cos\left[\left(\frac{N}{2}\right)(N-1)\left(\frac{2\pi}{N}\right)\right] \end{bmatrix}$$

we can write

$$X = WC \quad (7.19)$$

Multiplying both sides by W^{-1}, the spectral coefficients are found as follows

$$C = W^{-1}X \quad (7.20)$$

The next example will illustrate this method on the clock signal samples from Example 7.2.

Example 7.3

Consider from Example 7.2 that the clock signal samples are $x=\{2.5, 0.0, 0.0, 0.0, 0.0, 2.5, 5.0, 5.0, 5.0, 5.0\}$. Compute the spectral coefficients of the DTFS using linear algebra.

Solution:

Using the procedure described, we can write the following system of linear equations in matrix form:

$$\begin{bmatrix} 2.5 \\ 0 \\ 0 \\ 0 \\ 0 \\ 2.5 \\ 5 \\ 5 \\ 5 \\ 5 \end{bmatrix} = \begin{bmatrix} 1 & 1 & 0 & 1 & 0 & 1 & 0 & 1 & 0 & 1 \\ 1 & 0.809 & 0.588 & 0.309 & 0.951 & -0.309 & 0.951 & -0.809 & 0.588 & -1 \\ 1 & 0.309 & 0.951 & -0.809 & 0.588 & -0.809 & -0.588 & 0.309 & -0.951 & 1 \\ 1 & -0.309 & 0.951 & -0.809 & -0.588 & 0.809 & -0.588 & 0.309 & 0.951 & -1 \\ 1 & -0.809 & 0.588 & 0.309 & -0.951 & 0.309 & 0.951 & -0.809 & -0.588 & 1 \\ 1 & -1.000 & 0.000 & 1.000 & 0.000 & -1.000 & 0.000 & 1.000 & 0.000 & -1 \\ 1 & -0.809 & -0.588 & 0.309 & 0.951 & 0.309 & -0.951 & -0.809 & 0.588 & 1 \\ 1 & -0.309 & -0.951 & -0.809 & 0.588 & 0.809 & 0.588 & 0.309 & -0.951 & -1 \\ 1 & 0.309 & -0.951 & -0.809 & -0.588 & -0.809 & 0.588 & 0.309 & 0.951 & 1 \\ 1 & 0.809 & -0.588 & 0.309 & -0.951 & -0.309 & -0.951 & -0.809 & -0.588 & -1 \end{bmatrix} \begin{bmatrix} \tilde{a}_0 \\ \tilde{a}_1 \\ \tilde{b}_1 \\ \tilde{a}_2 \\ \tilde{b}_2 \\ \tilde{a}_3 \\ \tilde{b}_3 \\ \tilde{a}_4 \\ \tilde{b}_4 \\ \tilde{a}_5 \end{bmatrix}$$

Using the matrix routines available in MATLAB, we obtain the following spectral coefficients:

$$\begin{bmatrix} \tilde{a}_0 \\ \tilde{a}_1 \\ \tilde{b}_1 \\ \tilde{a}_2 \\ \tilde{b}_2 \\ \tilde{a}_3 \\ \tilde{b}_3 \\ \tilde{a}_4 \\ \tilde{b}_4 \\ \tilde{a}_5 \end{bmatrix} = \begin{bmatrix} 2.500 \\ 0 \\ -3.078 \\ 0 \\ 0 \\ 0 \\ -0.7265 \\ 0 \\ 0 \\ 0 \end{bmatrix} \Rightarrow \begin{bmatrix} \tilde{c}_0 \\ \tilde{\phi}_0 \\ \tilde{c}_1 \\ \tilde{\phi}_1 \\ \tilde{c}_2 \\ \tilde{\phi}_2 \\ \tilde{c}_3 \\ \tilde{\phi}_3 \\ \tilde{c}_4 \\ \tilde{\phi}_4 \\ \tilde{c}_5 \\ \tilde{\phi}_5 \end{bmatrix} = \begin{bmatrix} 2.500 \\ 0 \\ 3.078 \\ -\pi/2 \\ 0 \\ 0 \\ 0.7265 \\ -\pi/2 \\ 0 \\ 0 \\ 0 \\ 0 \end{bmatrix}$$

On comparison with the results in Example 7.2, we see that they agree reasonably well (the small difference is attributed to the series truncation as explained in the last example). The latter set of magnitude and phase coefficients was derived using (7.14).

The preceding example highlights the fact that the spectral coefficients of the discrete-time Fourier series are determined by straightforward linear algebraic methods. Another method exists for finding the spectral coefficients and one that is much more insightful as it provides a closed-form solution for each spectral coefficient. To arrive at this solution, we need to consider the orthogonal property of cosines and sines. Specifically, consider the following set of orthogonal basis functions

$$\sum_{n=0}^{N-1} \cos\left[p\left(\frac{2\pi}{N}\right)n\right]\cos\left[k\left(\frac{2\pi}{N}\right)n\right] = \begin{cases} 0; & \text{for } p \neq k \\ N/2; & \text{for } p = k \neq 0 \\ N; & \text{for } p = k = 0 \end{cases}$$

$$\sum_{n=0}^{N-1} \sin\left[p\left(\frac{2\pi}{N}\right)n\right]\sin\left[k\left(\frac{2\pi}{N}\right)n\right] = \begin{cases} 0; & \text{for } p \neq k \\ N/2; & \text{for } p = k \neq 0 \end{cases} \quad (7.21)$$

$$\sum_{n=0}^{N-1} \sin\left[p\left(\frac{2\pi}{N}\right)n\right]\cos\left[k\left(\frac{2\pi}{N}\right)n\right] = 0 \quad \text{for all } p$$

Armed with these identities, we multiply Eq. (7.11) with $\cos\left[k\left(\frac{2\pi}{N}\right)n\right]$ and sum n from 0 to N-1 on both sides to obtain[*]

$$\tilde{a}_k = \begin{cases} \dfrac{1}{N}\sum_{n=0}^{N-1} x[n]\cos\left[k\left(\frac{2\pi}{N}\right)n\right], & k = 0, N/2 \\ \dfrac{2}{N}\sum_{n=0}^{N-1} x[n]\cos\left[k\left(\frac{2\pi}{N}\right)n\right], & k = 1, 2, \ldots, N/2 - 1 \end{cases} \quad (7.22)$$

Likewise, we multiply Eq. (7.11) with $\sin\left[k\left(\frac{2\pi}{N}\right)n\right]$ and sum n from 0 to N-1 on both sides to obtain

$$\tilde{b}_k = \frac{2}{N}\sum_{n=0}^{N-1} x[n]\sin\left[k\left(\frac{2\pi}{N}\right)n\right], \quad k = 1, 2, \ldots, N/2 - 1 \quad (7.23)$$

---

[*] In many textbooks, a single expression for a_k is usually written as was done for b_k. This is achieved by writing the discrete-time Fourier series as

$$x[n] = \frac{\tilde{a}_0}{2} + \sum_{k=1}^{N/2-1}\left\{\tilde{a}_k \cos\left[k\left(\frac{2\pi}{N}\right)n\right] + \tilde{b}_k \sin\left[k\left(\frac{2\pi}{N}\right)n\right]\right\} + \frac{\tilde{a}_{N/2}}{2}\cos(\pi n)$$

where the end terms in the series are scaled by a factor of 2.

Example 7.4

Repeat Example 7.3 but compute the spectral coefficients of the DTFS using the orthogonal basis method.

Solution:

With $x=\{2.5, 0.0, 0.0, 0.0, 0.0, 2.5, 5.0, 5.0, 5.0, 5.0\}$, we can compute from Eq. (7.22) the following coefficients

$$\tilde{a}_0 = \frac{1}{10}[2.5+0+0+0+0+2.5+5+5+5+5] = 2.5$$

$$\tilde{a}_1 = \frac{2}{10}\left\{2.5\cos\left[(1)\left(\frac{2\pi}{10}\right)(0)\right] + 2.5\cos\left[(1)\left(\frac{2\pi}{10}\right)(5)\right] + 5\cos\left[(1)\left(\frac{2\pi}{10}\right)(6)\right]\right.$$
$$\left. + 5\cos\left[(1)\left(\frac{2\pi}{10}\right)(7)\right] + 5\cos\left[(1)\left(\frac{2\pi}{10}\right)(8)\right] + 5\cos\left[(1)\left(\frac{2\pi}{10}\right)(9)\right]\right\}$$
$$= 0$$

Continuing, we find $\tilde{a}_2 = 0$, $\tilde{a}_3 = 0$, $\tilde{a}_4 = 0$, and $\tilde{a}_5 = 0$. Likewise, from Eq. (7.23)

$$\tilde{b}_1 = \frac{2}{10}\left\{2.5\sin\left[(1)\left(\frac{2\pi}{10}\right)(0)\right] + 2.5\sin\left[(1)\left(\frac{2\pi}{10}\right)(5)\right] + 5\sin\left[(1)\left(\frac{2\pi}{10}\right)(6)\right]\right.$$
$$\left. + 5\sin\left[(1)\left(\frac{2\pi}{10}\right)(7)\right] + 5\sin\left[(1)\left(\frac{2\pi}{10}\right)(8)\right] + 5\sin\left[(1)\left(\frac{2\pi}{10}\right)(9)\right]\right\} = -3.077$$

$$\tilde{b}_2 = \frac{2}{10}\left\{2.5\sin\left[(2)\left(\frac{2\pi}{10}\right)(0)\right] + 2.5\sin\left[(2)\left(\frac{2\pi}{10}\right)(5)\right] + 5\sin\left[(2)\left(\frac{2\pi}{10}\right)(6)\right]\right.$$
$$\left. + 5\sin\left[(2)\left(\frac{2\pi}{10}\right)(7)\right] + 5\sin\left[(2)\left(\frac{2\pi}{10}\right)(8)\right] + 5\sin\left[(2)\left(\frac{2\pi}{10}\right)(9)\right]\right\} = 0$$

$$\tilde{b}_3 = \frac{2}{10}\left\{2.5\sin\left[(3)\left(\frac{2\pi}{10}\right)(0)\right] + 2.5\sin\left[(3)\left(\frac{2\pi}{10}\right)(5)\right] + 5\sin\left[(3)\left(\frac{2\pi}{10}\right)(6)\right]\right.$$
$$\left. + 5\sin\left[(3)\left(\frac{2\pi}{10}\right)(7)\right] + 5\sin\left[(3)\left(\frac{2\pi}{10}\right)(8)\right] + 5\sin\left[(3)\left(\frac{2\pi}{10}\right)(9)\right]\right\} = -0.7265$$

and $\tilde{b}_4 = 0$ and $\tilde{b}_5 = 0$. Not surprising, the results are identical to those found in Example 7.3.

Exercises

7.4. Using the summation formulae in Eq. (7.16), determine the DTFS representation of the Fourier series representation of $x(t)$ given in Exercise 7.1. Use 10 points per period and limit the series to 100 terms. Also, express the result in magnitude and phase form.

Ans.
$$1.2301 \sin\left[\left(\frac{2\pi}{10}\right)n\right] + 0.2901 \sin\left[3\left(\frac{2\pi}{10}\right)n\right]$$
$$1.2301 \cos\left[\left(\frac{2\pi}{10}\right)n - \frac{\pi}{2}\right] + 0.2901 \cos\left[3\left(\frac{2\pi}{10}\right)n - \frac{\pi}{2}\right]$$

7.5. A sampled signal consists of the following 4 samples: {0.7071, 0.7071, -0.7071, -0.7071}. Using linear algebra, determine the DTFS representation for these samples. Also, express the result in magnitude and phase form.

Ans.
$$0.7071 \cos\left[\left(\frac{2\pi}{4}\right)n\right] + 0.7071 \sin\left[\left(\frac{2\pi}{4}\right)n\right]$$
$$1.0 \cos\left[\left(\frac{2\pi}{4}\right)n - \frac{\pi}{4}\right]$$

7.6. A sampled signal consists of the following set of samples {0, 1, 2, 3, 2, 1}. Determine the DTFS representation for this sample set using the orthogonal basis method. Also, express the result in magnitude and phase form.

Ans.
$$1.5 - 1.333 \cos\left[\left(\frac{2\pi}{6}\right)n\right] + 0.1667 \sin\left[3\left(\frac{2\pi}{6}\right)n\right]$$
$$1.5 + 1.333 \cos\left[\left(\frac{2\pi}{6}\right)n - \pi\right] + 0.1667 \cos\left[3\left(\frac{2\pi}{6}\right)n - \frac{\pi}{2}\right]$$

7.2.5 Complete Frequency Spectrum

One of the most important insights that can be obtained from the closed-form expression for the spectral coefficients is how they behave beyond the range $k = 0,\ldots,N/2$. To begin, consider evaluating Eq. (7.22) over the range $k = N/2,\ldots,N$. On doing so, we write

$$\tilde{a}_k = \begin{cases} \dfrac{1}{N}\sum_{n=0}^{N-1} x[n]\cos\left[k\left(\dfrac{2\pi}{N}\right)n\right], & k = N/2, N \\[2mm] \dfrac{2}{N}\sum_{n=0}^{N-1} x[n]\cos\left[k\left(\dfrac{2\pi}{N}\right)n\right], & k = N/2+1,\ldots,N-1 \end{cases} \quad (7.24)$$

Here, only the cosine terms are affected by the index k. Next, if we consider the change of variable substitution, $k \rightarrow N - k$, Eq. (7.24) can be re-written as

$$\tilde{a}_{N-k} = \begin{cases} \dfrac{1}{N} \sum_{n=0}^{N-1} x[n] \cos\left[(N-k)\left(\dfrac{2\pi}{N}\right)n\right], & k = 0, N/2 \\ \dfrac{2}{N} \sum_{n=0}^{N-1} x[n] \cos\left[(N-k)\left(\dfrac{2\pi}{N}\right)n\right], & k = 1, \ldots, N/2 - 1 \end{cases} \quad (7.25)$$

However, $\cos\left[(N-k)(2\pi/N)n\right] = \cos\left[k(2\pi/N)n\right]$, allowing us to write Eq. (7.25) as

$$\tilde{a}_{N-k} = \begin{cases} \dfrac{1}{N} \sum_{n=0}^{N-1} x[n] \cos\left[k\left(\dfrac{2\pi}{N}\right)n\right], & k = 0, N/2 \\ \dfrac{2}{N} \sum_{n=0}^{N-1} x[n] \cos\left[k\left(\dfrac{2\pi}{N}\right)n\right], & k = 1, \ldots, N/2 - 1 \end{cases} \quad (7.26)$$

Recognizing that Eq. (7.25) is equivalent to Eq. (7.26) allows us to conclude over the range of $k = 1, \ldots, N-1$ that

$$\tilde{a}_{N-k} = \tilde{a}_k \quad (7.27)$$

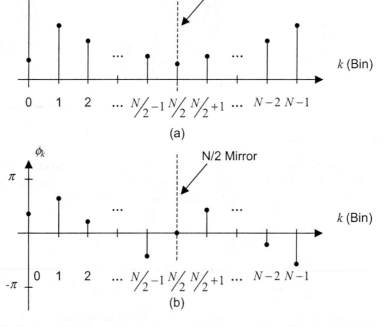

Figure 7.5. Illustrating the spectral symmetry about $N/2$ for $k=0,1,\ldots,N-1$: (a) magnitude spectrum and (b) phase spectrum.

Following the same reasoning as above in Eqs. (7.24)–(7.27), together with the trigonometric identity $\sin\left[(N-k)(2\pi/N)n\right] = -\sin\left[k(2\pi/N)n\right]$, one can write

$$\tilde{b}_{N-k} = -\tilde{b}_k \quad (7.28)$$

Converting this result into magnitude and phase form, we find using Eq. (7.14)

$$\begin{aligned}\tilde{c}_{N-k} &= \tilde{c}_k \\ \tilde{\phi}_{N-k} &= -\tilde{\phi}_k\end{aligned} \quad \text{for all } k \quad (7.29)$$

Here we see that the magnitude spectrum excluding the DC bin has even symmetry about bin $N/2$. Similarly, the phase spectrum exhibits an odd symmetry about $N/2$. These two situations are highlighted in Figure 7.5 for an arbitrary discrete-time periodic signal.

Example 7.5

Plot the magnitude and phase spectrum for the clock signal of Example 7.3 over 10 frequency bins.

Solution:

The magnitude spectrum for the clock signal of Example 7.2 is shown in Figure 7.6(a) and the corresponding phase spectrum appears in Figure 7.6(b).

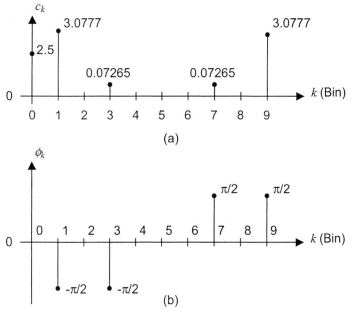

Figure 7.6. (a) Magnitude and (b) phase spectrum for the 10-kHz clock signal over 10 frequency bins.

Next, let us consider the periodicity of the spectrum. Consider replacing k in Eq. (7.22) by $k+N$ so that we write

$$\tilde{a}_{k+N} = \begin{cases} \dfrac{1}{N}\sum_{n=0}^{N-1} x[n]\cos\left[(k+N)\left(\dfrac{2\pi}{N}\right)n\right], & k=0, N/2 \\ \dfrac{2}{N}\sum_{n=0}^{N-1} x[n]\cos\left[(k+N)\left(\dfrac{2\pi}{N}\right)n\right], & k=1,\ldots,N/2-1 \end{cases} \qquad (7.30)$$

Due to the periodicity of the cosine function, Eq. (7.30) simplifies directly to

$$\tilde{a}_{k+N} = \begin{cases} \dfrac{1}{N}\sum_{n=0}^{N-1} x[n]\cos\left[k\left(\dfrac{2\pi}{N}\right)n\right], & k=0, N/2 \\ \dfrac{2}{N}\sum_{n=0}^{N-1} x[n]\cos\left[k\left(\dfrac{2\pi}{N}\right)n\right], & k=1,\ldots,N/2-1 \end{cases} \qquad (7.31)$$

which is equal to $\tilde{a}_k$. Hence, we write

$$\tilde{a}_{k+N} = \tilde{a}_k \quad \text{for all } k \qquad (7.32)$$

Exercises

7.7. Plot the magnitude and phase spectrum of the DTFS representation of a sampled signal described by

$$x[n] = 0.6150 \sin\left[\left(\dfrac{2\pi}{8}\right)n\right] - 0.2749 \sin\left[2\left(\dfrac{2\pi}{8}\right)n\right]$$
$$+ 0.1451 \sin\left[3\left(\dfrac{2\pi}{8}\right)n\right] + 0.0649 \cos\left[4\left(\dfrac{2\pi}{8}\right)n\right]$$

Ans.

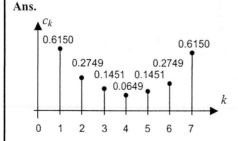 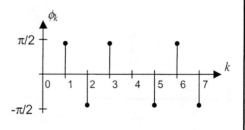

Following a similar line of reasoning, one can write

$$\tilde{b}_{k+N} = \tilde{b}_k \quad \text{for all } k \tag{7.33}$$

Through the direct application of Eq. (7.14), we can write

$$\begin{aligned}\tilde{c}_{k+N} &= \tilde{c}_k \\ \tilde{\phi}_{k+N} &= \tilde{\phi}_k\end{aligned} \quad \text{for all } k \tag{7.34}$$

We can therefore conclude from above that the spectrum of a periodic signal $x[n]$ with period N is also periodic with period N. Therefore, combining spectral symmetry, together with its periodicity, the frequency spectrum of a discrete-time periodic signal is defined for all frequencies. Figure 7.7 illustrates the full frequency spectrum of an arbitrary signal. To aid the reader, adjacent periods of the spectrum are indicated with dashed boxes.

At this point in our discussion, we should point out that most test vendors only provide spectral information corresponding to the Nyquist interval, $k=0,1,\ldots,N/2$. Although less important today, twenty years ago when DSP-based ATE started to appear on the market memory was expensive. Attempts to minimize memory usage were paramount. This led to the elimination of redundant spectral information.

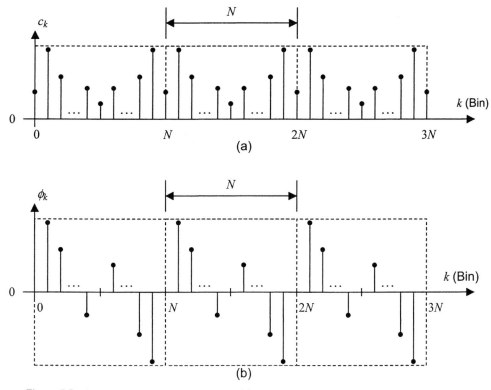

Figure 7.7. Illustrating the spectral periodicity: (a) magnitude spectrum and (b) phase spectrum.

7.2.6 Time and Frequency Denormalization

The time and frequency scale associated with a data sequence $x[n]$ is described in terms of normalized time and frequency, according to the sample indexes, n and k, respectively. To obtain the actual time and frequency scales associated with the original samples, one must perform the operation of time and frequency *denormalization*. To achieve this, knowledge of the sampling period T_s or sampling frequency F_s is required.

To reconstruct the original time scale, one simply multiplies the sample index, n by T_s, according to the translation

$$n \rightarrow nT_s \tag{7.35}$$

Conversely, the frequency scale is restored when one multiplies the sample index k by F_s/N, according to the translation

$$k \rightarrow k\frac{F_s}{N} \tag{7.36}$$

Figure 7.8 illustrates the frequency denormalization procedure for the magnitude of the spectrum for the clock signal described in Example 7.5. For this particular case, F_s=10 kHz and N=10, resulting in a frequency denormalization scale factor of 1 kHz.

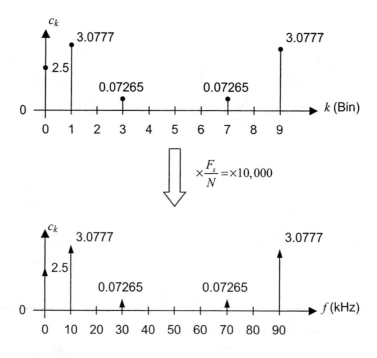

Figure 7.8. Illustrating the procedure of denormalizing a frequency axis.

Exercises

7.8. Plot the frequency-denormalized magnitude spectrum of the following DTFS representation of a sampled signal described by:

$$x[n] = 0.6150\sin\left[\left(\frac{2\pi}{8}\right)n\right] - 0.2749\sin\left[2\left(\frac{2\pi}{8}\right)n\right] + 0.1451\sin\left[3\left(\frac{2\pi}{8}\right)n\right] + 0.0649\cos\left[4\left(\frac{2\pi}{8}\right)n\right]$$

assuming as sampling rate of 16kHz.

Ans.

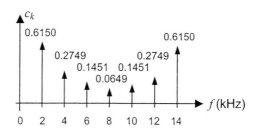

7.2.7 Complex Form of the DTFS

In most DSP textbooks, the DTFS is expressed in complex form using Euler's equation

$$e^{j\varphi} = \cos(\varphi) + j\sin(\varphi) \tag{7.37}$$

where j is a complex number equal to $\sqrt{-1}$. The main reason for this choice lies with the ease in which the exponential function can be algebraically manipulated when compared with trigonometric formulas. To convert the DTFS representation in Eq. (7.11) into complex form, consider that the cosine and sine functions can be written as

$$\cos(\varphi) = \frac{e^{j\varphi} + e^{-j\varphi}}{2}; \quad \sin(\varphi) = \frac{e^{j\varphi} - e^{-j\varphi}}{2j} \tag{7.38}$$

When the preceding two formulae are substituted into Eq.(7.11), we get

$$x[n] = \tilde{a}_0 + \sum_{k=1}^{N/2-1}\left(\frac{\tilde{a}_k - j\tilde{b}_k}{2}\right)e^{jk\left(\frac{2\pi}{N}\right)n} + \sum_{k=1}^{N/2-1}\left(\frac{\tilde{a}_k + j\tilde{b}_k}{2}\right)e^{-jk\left(\frac{2\pi}{N}\right)n} \\ + \left(\frac{\tilde{a}_{N/2}}{2}\right)e^{j\pi n} + \left(\frac{\tilde{a}_{N/2}}{2}\right)e^{-j\pi n} \tag{7.39}$$

Collecting the end terms inside each sum, we write

$$x[n] = \tilde{a}_0 + \sum_{k=1}^{N/2} \left(\frac{\tilde{a}_k - j\tilde{b}_k}{2} \right) e^{jk\left(\frac{2\pi}{N}\right)n} + \sum_{k=1}^{N/2} \left(\frac{\tilde{a}_k + j\tilde{b}_k}{2} \right) e^{-jk\left(\frac{2\pi}{N}\right)n} \quad (7.40)$$

Next, through the change of variable $k \rightarrow N-k$ in the rightmost summation term, we can write

$$\sum_{k=1}^{N/2} \left(\frac{\tilde{a}_k + j\tilde{b}_k}{2} \right) e^{-jk\left(\frac{2\pi}{N}\right)n} = \sum_{k=N/2}^{N-1} \left(\frac{\tilde{a}_{N-k} + j\tilde{b}_{N-k}}{2} \right) e^{jk\left(\frac{2\pi}{N}\right)n} \quad (7.41)$$

Substituting Eq. (7.41) into (7.40) leads to

$$x[n] = \tilde{a}_0 + \sum_{k=1}^{N/2} \left(\frac{\tilde{a}_k - j\tilde{b}_k}{2} \right) e^{jk\left(\frac{2\pi}{N}\right)n} + \sum_{k=N/2}^{N-1} \left(\frac{\tilde{a}_{N-k} + j\tilde{b}_{N-k}}{2} \right) e^{jk\left(\frac{2\pi}{N}\right)n} \quad (7.42)$$

The DTFS can then be written in complex form as

$$x[n] = \sum_{k=1}^{N-1} X(k) e^{jk\left(\frac{2\pi}{N}\right)n} \quad (7.43)$$

where

$$X(k) = \begin{cases} \tilde{a}_0 & k = 0 \\ \dfrac{\tilde{a}_k - j\tilde{b}_k}{2} & k = 1, 2, \ldots, N/2 - 1 \\ \tilde{a}_{N/2} & k = N/2 \\ \dfrac{\tilde{a}_{N-k} + j\tilde{b}_{N-k}}{2} & k = N/2 + 1, \ldots, N-1 \end{cases} \quad (7.44)$$

As is evident, the coefficients $X(k)$ in front of each exponential term are, in general, complex numbers. For the most part, the real component of each term relates to one-half the cosine coefficient of the trigonometric series. Conversely, the imaginary part of each term is related to one-half the sine coefficients. The exceptions are the $X(0)$ and $X(N/2)$ terms. These two terms are directly related to the cosine terms with a scale factor of one. The fact that the scale factor is not evenly distributed among each term can be a source of confusion for some.

Alternative forms of Eq. (7.44) can also be written. For instance, we can rewrite Eq. (7.44) in polar form, together with substitutions from Eq. (7.14), as

$$X(k) = \begin{cases} \tilde{c}_0 e^{j0} & k = 0 \\ \dfrac{1}{2}\tilde{c}_k e^{-j\tilde{\phi}_k} & k = 1, 2, \ldots, N/2 - 1 \\ \tilde{c}_{N/2} e^{j0} & k = N/2 \\ \dfrac{1}{2}\tilde{c}_{N-k} e^{+j\tilde{\phi}_k} & k = N/2 + 1, \ldots, N - 1 \end{cases} \qquad (7.45)$$

Exercises

7.9. A DTFS representation for a sequence of data is given by

$$x[n] = 0.25 + 1.0 \, \cos\left[\left(\frac{2\pi}{10}\right)n\right] + 0.5 \, \sin\left[\left(\frac{2\pi}{10}\right)n\right]$$
$$+ 0.2 \, \cos\left[3\left(\frac{2\pi}{10}\right)n\right] - 0.2 \, \sin\left[3\left(\frac{2\pi}{10}\right)n\right] + 0.2 \, \cos\left[5\left(\frac{2\pi}{10}\right)n\right]$$

Express $x[n]$ in complex form.

Ans.

$$x[n] = 0.25 + (0.5 - j0.25)e^{j\left(\frac{2\pi}{10}\right)n} + (0.1 + j0.1)e^{j3\left(\frac{2\pi}{10}\right)n}$$
$$+ 0.2 e^{j5\left(\frac{2\pi}{10}\right)n} + (0.1 - j0.1)e^{j7\left(\frac{2\pi}{10}\right)n} + (0.5 + j0.25)e^{j9\left(\frac{2\pi}{10}\right)n}$$

7.3 DISCRETE-TIME TRANSFORMS

7.3.1 The Discrete Fourier Transform

In the previous section it was shown how a sequence of N samples repeated indefinitely can be represented *exactly* with a set of $N/2$ harmonically related sinusoidal pairs and a DC component, or in terms of N harmonically-related complex exponential functions. In this section, we shall demonstrate that a similar set of harmonically related exponential functions can also be used to represent a sequence of N samples of finite duration. Such signals are known as *discrete-time aperiodic* signals.

Consider an arbitrary sequence $y[n]$ that is of finite duration over the time interval $n=0$ to $N-1$. Next, consider that $y[n]=0$ outside of this range. A signal of this type is shown in Figure 7.9(a). From this aperiodic signal, we can construct a periodic sequence $x[n]$ for which $y[n]$ is one

> **Exercises**
>
> **7.10.** A DTFS representation expressed in complex form is given by
>
> $$x[n] = 2 + (1+j1)e^{j\left(\frac{2\pi}{8}\right)n} + (1-j1)e^{j3\left(\frac{2\pi}{8}\right)n} + (1+j1)e^{j5\left(\frac{2\pi}{8}\right)n} + (1-j1)e^{j7\left(\frac{2\pi}{8}\right)n}$$
>
> Express $x[n]$ in trigonometric form.
>
> **Ans.**
>
> $$x[n] = 2 + 2.0\cos\left[\left(\frac{2\pi}{8}\right)n\right] - 2.0\sin\left[\left(\frac{2\pi}{8}\right)n\right] + 2.0\cos\left[3\left(\frac{2\pi}{8}\right)n\right] + 2.0\sin\left[3\left(\frac{2\pi}{8}\right)n\right]$$
>
> **7.11.** The following vector describes the spectral coefficients of a DTFS expressed in complex rectangular form
>
> $$X = [1 \quad 0.25 + j0.25 \quad 4 - j1 \quad 0 \quad 0.3 \quad 0 \quad 4 + j1 \quad 0.25 - j0.25]$$
>
> Write the DTFS in trigonometric form using a magnitude and phase notation.
>
> **Ans.**
>
> $$x[n] = 1 + 0.7071\cos\left[\left(\frac{2\pi}{8}\right)n + \frac{\pi}{4}\right] + 8.2462\cos\left[2\left(\frac{2\pi}{8}\right)n + 0.0780\pi\right] + 0.3\cos\left[4\left(\frac{2\pi}{8}\right)n\right]$$

period, as illustrated in Figure 7.9(b). This is known as the *periodic extension* of $y[n]$. In mathematical terms, we can describe $y[n]$ as

$$y[n] = \begin{cases} x[n] & n = 0, 1, \ldots, N-1 \\ 0 & \text{otherwise} \end{cases} \tag{7.46}$$

If we consider the complex form of the DTFS representation for $x[n]$, we can write $y[n]$ as

$$y[n] = \begin{cases} \sum_{k=1}^{N-1} X(k) e^{jk\left(\frac{2\pi}{N}\right)n} & n = 0, 1, \ldots, N-1 \\ 0 & \text{otherwise} \end{cases} \tag{7.47}$$

As we choose the period N to be larger, $y[n]$ matches $x[n]$ over a longer interval, and as $N \rightarrow \infty$, $y[n] = x[n]$ for any finite value of n. Thus, for very large N, the spectrum of $y[n]$ is identical to $x[n]$. However, we note that as $N \rightarrow \infty$, the form of the mathematics in Eq. (7.47) changes. The spectral makeup of the discrete-time signal no longer consists of harmonically related discrete frequencies, but rather becomes a continuous function of frequency. Under such

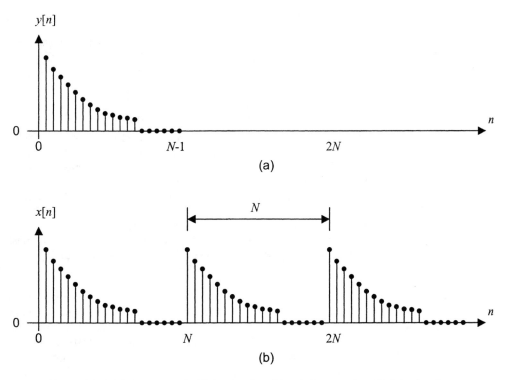

Figure 7.9. (a) Arbitrary signal of finite duration; (b) periodic extension of infinite duration.

conditions, the representation in Eq. (7.47) in the limit becomes known as a *Fourier transform* and is written with an integral operation as follows

$$y[n] = \frac{1}{2\pi} \int_{-\pi}^{\pi} Y(e^{j\omega}) e^{j\omega n} d\omega \qquad (7.48)$$

where

$$Y(e^{j\omega}) = \sum_{n=-\infty}^{\infty} y[n] e^{-j\omega n} \qquad (7.49)$$

As in all computer applications, of which mixed-signal testing is just one example, we must limit our discussion to finite values of N, and preferably (as test time is always a major concern), to small values of N. This implies that we really have no other choice but to work directly with Eq. (7.47). It has been shown that the spectral coefficients $X(k)$ associated with Eq. (7.47) are directly related to samples of $Y(e^{j\omega})$ uniformly spaced according to

$$X(k) = \left. \frac{Y(e^{j\omega})}{N} \right|_{\omega = \frac{2\pi}{N} k} \qquad (7.50)$$

Substituting Eq. (7.49) into (7.50), and limiting the summation to the maximum possible nonzero values of $y[n]$, $n=0,1,...,N-1$, we can write

$$X(k) = \frac{1}{N} \sum_{n=0}^{N-1} y[n] \, e^{-jk\left(\frac{2\pi}{N}\right)n} \tag{7.51}$$

Due to the importance of the interplay between the spectral coefficients of the DTFS representing the periodic extension of the aperiodic signal and its Fourier transform, the set of coefficients $\{X(0), X(1), ..., X(N-1)\}$ in Eq. (7.51) is referred to as the *discrete Fourier transform* (DFT) of $y[n]$. As the DFT is essentially a special interpretation of a DTFS, the algorithms in Section 7.2.5 can also be used to produce the spectral coefficients of the DFT. However, as explained in the next section, a more efficient algorithm is available to compute the DFT of a discrete-time aperiodic signal or, for that matter, the spectral coefficients of a DTFS. This algorithm is known as the *fast Fourier transform* (FFT) and represents one of the most significant developments in digital signal processing. However, before we move on to this topic, we shall first consider an important degenerate case of the DFT.

Consider the situation where an aperiodic signal has infinite duration or exists with nonzero values over a much longer time than the observation interval of N samples. This is illustrated in Figure 7.10(a) for an exponentially decaying waveform. Under such conditions it is impossible to represent this signal exactly with a periodic signal having a finite period. Instead, one can only approximate the waveform over the observation interval as shown in Figure 7.10(b) using a periodic extension of the finite duration signal shown in Figure 7.10(c). The error is a form of *time-domain aliasing* and is directly related to the jump discontinuity that occurs at the wraparound point of the periodically extended waveform. The spectral coefficients determined by the DFT would then correspond to samples of the Fourier transform of the signal shown in Figure 7.10(c), not Figure 7.10(a).

In practice it is important to keep the jump discontinuity to a minimum if the DFT is to reveal the spectral properties of the original waveform. The most common method is to extend the observation interval to large values of N so that the combined energy of the aliased components is made insignificant relative to the energy in the signal. Further, some reduction in the overall observation time can be achieved if the method of windowing is used. Windowing is a mathematical process that alters the shape of the signal over the observation interval and gradually forces it to decay to zero at both ends. This eliminates the discontinuities in the periodically extended waveform at the expense of decreased frequency resolution. The net result is a concentration of the aliasing energy or *frequency leakage* into a few spectral bins instead of having it spread across many different bins.

As test time is always paramount in mixed-signal test, one should avoid discontinuities in the periodically extended waveform. In the words of Chapter 6, one should restrict all signals to be coherent rather than noncoherent. We shall delay the introduction of our examples until we first describe the principles of the fast Fourier transform.

7.3.2 The Fast Fourier Transform

In the early 1960s J. Tukey invented a new algorithm for performing the DFT computations in a much more efficient manner. J. W. Cooley, a programmer at IBM, translated Tukey's algorithm

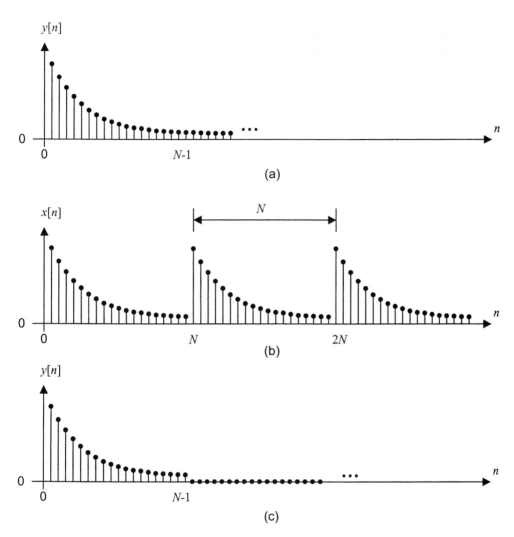

Figure 7.10. (a) Arbitrary signal of infinite duration; (b) periodic extension of infinite duration of short portion of signal; (c) signal approximation of finite duration.

into computer code and the Cooley-Tukey fast Fourier transform (FFT) was born.[4] It is now known that this algorithm actually dates back at least a century. The great German mathematician C. F. Gauss is known to have developed the same algorithm.

To understand the significance of the DFT algorithm, consider in Eq. (7.51) that N complex multiplications and N-1 additions are required to compute each spectral coefficient $X(k)$. For N spectral coefficients, another N multiplications and additions are necessary. Therefore, in total, $(N-1)N$ complex multiplications and additions will be required. As an example, to perform a DFT on 1024 samples, a computer has to perform over one million multiplications. To minimize test time, we would prefer that the DFT computation time be as small as possible; so obviously we need a more efficient way to perform the multiplications in a DFT. This is where the FFT comes in.

The FFT works by partitioning each of the multiplications and additions in the DFT in such a way that there are many redundant calculations. The redundancy is removed by "folding" the redundant calculations on top of one another and performing each calculation only once. The folding operation forms a so-called *butterfly network* because of the butterfly shapes in the calculation flowchart.[5] There are several different ways to split the calculations and fold the redundancies into one another. The butterfly network can be laid out in a decimation-in-frequency configuration or a decimation-in-time configuration. Fortunately, the details of the FFT algorithm itself are largely unimportant to the test engineer, since the FFT operation is built into the operating system of most mixed-signal testers.

Since many redundant calculations are eliminated by the FFT, it only requires $N \log_2(N)$ complex multiplications. For a 1024-point FFT, only 1024×10 or 10240 complex multiplications are required. Compared to the one million complex multiplications required by the complex version of the DFT, this represents a huge reduction in computation time. The difference between the FFT and DFT becomes more extreme with larger sample sets, since the DFT produces an exponential increase in computations as the sample size increases.

Although the FFT produces the same output as an equivalent DFT, the more common FFTs can only operate on a sample size that is equal to 2^n, where n is an integer. For instance, it is not possible to perform a standard Cooley-Tukey FFT on 380 samples, although a DFT would have no problem doing so. The limited choice of sample sizes is the major difference between the DFT and the FFT, other than the difference in computation time. Nevertheless, the savings in test time is so huge that test engineers usually have no choice but to use the FFT with its limited sample size flexibility. It is quite possible that improvements in computation speeds will eventually make the FFT obsolete in mixed-signal testers, allowing DFTs instead. Until then, the mixed-signal test engineer should be prepared to work with 2^n samples for most tests.

7.3.3 Interpreting the FFT Output

The output format of a mixed-signal tester's FFT depends somewhat on the vendor's operating system. In older testers, the format of the FFT output was arranged as an N-point array with the DC and Nyquist levels in the first two array elements followed by the cosine/sine pairs for each spectral bin. Today, most testers incorporate commercial DSP chips sets that compute the FFT using complex arithmetic and store the complex numbers in an array beginning with the DC bin followed by successive harmonic bins up to the Nyquist bin. The same format is also used for most numerical software packages such as MATLAB.

One has to be careful, though, when interpreting the FFT output. Many ATE versions of the FFT do not produce peak voltage outputs. Some produce voltage squared (power) outputs, some produce voltage outputs multiplied by the number of samples over 2 (i.e., a 1-V input with 1024 points produces an FFT output of 512 units), etc. This suggests that the test engineer must become familiar with the FFT routine that they intend to use and determine all the necessary scale factors. In addition, it has been the authors' preference to adjust the scale factors so that the FFT produces RMS levels instead of peak levels. For reasons that will become clear in the next chapter, many test metrics call for the combination of the power of several spectral components. Working with RMS values simplifies this approach.

To better understand the steps involved, let us consider the manner in which MATLAB performs the FFT and the corresponding scale factors needed to produce RMS spectral levels. First, the FFT that is performed in MATLAB is given by the equation

$$Y(k) = \sum_{n=0}^{N-1} y[n]\, e^{-jk\left(\frac{2\pi}{N}\right)n} \tag{7.52}$$

In turn, the complex spectral coefficients $X(k)$ of the DTFS representation of the periodic extension of $y[n]$ are given by

$$X(k) = \frac{Y(k)}{N} \tag{7.53}$$

and the corresponding cosine/sine coefficients are found according to Eq. (7.44) to be

$$\tilde{a}_k = \begin{cases} \operatorname{Re}\{X(k)\} & k = 0, N/2 \\ 2\operatorname{Re}\{X(k)\} & k = 1, 2, \ldots, N/2 - 1 \end{cases} \tag{7.54}$$

and

$$\tilde{b}_k = \begin{cases} 0 & k = 0, N/2 \\ -2\operatorname{Im}\{X(k)\} & k = 1, 2, \ldots, N/2 - 1 \end{cases} \tag{7.55}$$

where $\operatorname{Re}\{\ \}$ and $\operatorname{Im}\{\ \}$ denote the real and imaginary parts of a complex number, respectively. Again, we must alert the reader to the different scale factors in front of these terms.

The magnitude and phase representation is determined using Eq. (7.14) to be

$$\tilde{c}_k = \begin{cases} |X(k)| & k = 0, N/2 \\ 2|X(k)| & k = 1, \ldots, N/2 - 1 \end{cases} \tag{7.56}$$

where $|X(k)| = \sqrt{\operatorname{Re}\{X(k)\}^2 + \operatorname{Im}\{X(k)\}^2}$ and the phase is computed using

$$\tilde{\phi}_k = \begin{cases} \tan^{-1}\left(\dfrac{-\operatorname{Im}\{X(k)\}}{\operatorname{Re}\{X(k)\}}\right) & \text{if } \operatorname{Re}\{X(k)\} \geq 0 \\ \pi + \tan^{-1}\left(\dfrac{-\operatorname{Im}\{X(k)\}}{\operatorname{Re}\{X(k)\}}\right) & \text{if } \operatorname{Re}\{X(k)\} < 0 \end{cases} \tag{7.57}$$

In many situations we shall find it more convenient to report the spectral coefficients in terms of their RMS values. To do so, we divide the spectral coefficients $\tilde{c}_k$ (except the DC term described $k=0$) by $\sqrt{2}$ to obtain

$$\tilde{c}_{k\text{-RMS}} = \begin{cases} \tilde{c}_k & k = 0 \\ \dfrac{\tilde{c}_k}{\sqrt{2}} & k = 1, \ldots, N/2 \end{cases} \tag{7.58}$$

On substituting Eq. (7.56) into (7.58), we obtain

$$\tilde{c}_{k-RMS} = \begin{cases} |X(k)| & k=0 \\ \sqrt{2}|X(k)| & k=1,\ldots,N/2-1 \\ \dfrac{1}{\sqrt{2}}|X(k)| & k=N/2 \end{cases} \quad (7.59)$$

The following example will further illustrate this procedure.

Example 7.6

Using the FFT routine in MATLAB, compute the spectral coefficients $\{a_k\}$ and $\{b_k\}$ of a multitone signal having the following 8 samples, {0.1414, 1.0, −0.1414, −0.8, −0.1414, 1.0, 0.1414, −1.2}. These samples were derived from a signal with the following DTFS representation

$$x[n] = \cos\left[2\left(\frac{2\pi}{8}\right)n - \frac{\pi}{2}\right] + 0.2\cos\left[3\left(\frac{2\pi}{8}\right)n - \frac{\pi}{4}\right]$$

Also report the magnitude (RMS) and the phase of each spectral coefficient.

Solution:

With the samples of the multitone signal described as

$$x = [0.1414,\ 1.0,\ -0.1414,\ -0.8,\ -0.1414,\ 1.0,\ 0.1414,\ -1.2],$$

the FFT routine in MATLAB produces the following output, together with the scaled result

$$Y = FFT(x) = \begin{bmatrix} 0 \\ 0 \\ 0 - j4.0000 \\ 0.5657 - j0.5657 \\ 0 \\ 0.5657 + j0.5657 \\ 0 + j4.0000 \\ 0 \end{bmatrix} \Rightarrow X = \frac{Y}{8} = \begin{bmatrix} 0 \\ 0 \\ 0 - j0.5000 \\ 0.0707 - j0.0707 \\ 0 \\ 0.0707 + j0.0707 \\ 0 + j0.5000 \\ 0 \end{bmatrix}$$

Subsequently, the cosine/sine spectral coefficients are determined from Eqs. (7.54) and (7.55) to be

$$\begin{bmatrix} \tilde{a}_0 \\ \tilde{a}_1 \\ \tilde{a}_2 \\ \tilde{a}_3 \\ \tilde{a}_4 \end{bmatrix} = \begin{bmatrix} 0 \\ 0 \\ 0 \\ 0.1414 \\ 0 \end{bmatrix} \quad \text{and} \quad \begin{bmatrix} \tilde{b}_0 \\ \tilde{b}_1 \\ \tilde{b}_2 \\ \tilde{b}_3 \\ \tilde{b}_4 \end{bmatrix} = \begin{bmatrix} 0 \\ 0 \\ 1.0 \\ 0.1414 \\ 0 \end{bmatrix}$$

Finally, the corresponding magnitude (in RMS) and phase terms (in radians) are as follows

$$\begin{bmatrix} \tilde{c}_{0-RMS} \\ \tilde{c}_{1-RMS} \\ \tilde{c}_{2-RMS} \\ \tilde{c}_{3-RMS} \\ \tilde{c}_{4-RMS} \end{bmatrix} = \begin{bmatrix} 0 \\ 0 \\ 0.7071 \\ 0.1414 \\ 0 \end{bmatrix} \quad \text{and} \quad \begin{bmatrix} \tilde{\phi}_0 \\ \tilde{\phi}_1 \\ \tilde{\phi}_2 \\ \tilde{\phi}_3 \\ \tilde{\phi}_4 \end{bmatrix} = \begin{bmatrix} 0 \\ 0 \\ \pi/2 \\ \pi/4 \\ 0 \end{bmatrix}$$

At this point in our discussion of the FFT, it would be instructive to consider the spectral properties of a coherent sinusoidal signal and a noncoherent sinusoidal signal having equal amplitudes.

Example 7.7

Using MATLAB's FFT routine, compute the spectral coefficients of a coherent and noncoherent sinusoidal signal with parameters $A=1$, $\phi=0$, $M=3$, $N=64$, and $A=1$, $\phi=0$, $M=\pi(3.14156)$, $N=64$. For each case, plot the RMS magnitude of the spectrum in dB relative to a 1-V RMS reference level.

Exercises

7.12. Evaluate $x[n]=1.0\sin\left[3\left(2\pi/10\right)n+\pi/8\right]$ for $n=0,1,\ldots,9$. Using MATLAB, compute the FFT of the 10 samples of $x[n]$ and determine the corresponding $\{a_k\}$ and $\{b_k\}$ spectral coefficients.

Ans. $\{a_3+jb_3\}=\{0.3826+j0.9238\}$; all others are zero.

7.13. Evaluate $x[n]=(n-3)^2$ for $n=0,1,\ldots,9$. Using MATLAB, compute the FFT of the 10 samples of $x[n]$ and determine the corresponding $\{c_k\}$ and $\{\phi_k\}$ spectral coefficients. Express the magnitude coefficients in RMS form.

Ans. $\{c_k\}=\{10.50, 13.90, 5.615, 3.815, 3.172, 1.500\}$; $\{c_{k-RMS}\}=\{10.50, 9.834, 3.970, 2.697, 2.243, 1.060\}$; $\{\phi_k\}=\{0, -1.0868, 1.3726, 0.8659, 0.4220, 0.0000\}$.

Solution:

The following MATLAB routine was written to produce 64 samples of the coherent and noncoherent waveforms and to perform the corresponding FFT analysis:

```
% coherent signal definition - x -
N=64; M=3; A=1; P=0;  % signal definition
    for n=1:N,
        x(n)=A*sin(2*pi*M/N*(n-1)+P);
    end;
% noncoherent signal definition - z -
N=64; M=pi; A=1; P=0;
    for n=1:N,
        z(n)=A*sin(2*pi*M/N*(n-1)+P);
    end;
% perform Fourier analysis
    X=fft(x)/length(x);
    % magnitude of spectrum X – units of dBV
        % AC Terms
            magdBV_X = 20*log10(sqrt(2)*abs(X));
        % DC & Nyquist Terms
            magdBV_X(1) = 20*log10(abs(X(1)));
            magdBV_X(N/2+1) = 20*log10(1/sqrt(2)*abs(X(N/2+1)));
    Z=fft(z)/length(z);
    % magnitude of spectrum Z – units of dBV
        % AC Terms
            magdBV_Z = 20*log10(sqrt(2)*abs(Z));
        % DC & Nyquist Terms
            magdBV_Z(1) = 20*log10(abs(Z(1)));
            magdBV_Z(N/2+1) = 20*log10(1/sqrt(2)*abs(Z(N/2+1)));
% plot routine
    figure(1);
        subplot(1,2,1), stem(0:N-1, x, ':');
        subplot(1,2,2), plot(0:N/2, magdBV_X(1:N/2+1));
    figure(2);
        subplot(1,2,1), stem(0:N-1, z, ':');
        subplot(1,2,2), plot(0:N/2, magdBV_Z(1:N/2+1))
% end
```

The results of the analysis for the coherent waveform are shown in Figure 7.11. The time-domain waveform is shown on the left, while the corresponding magnitude of the spectrum is shown on the right. The spectrum is expressed in dB relative to a 1-V RMS reference level.

When this definition is used, the decibel units are referred to as dBV. Mathematically, it is written as

$$c_{k-RMS}(dBV) = 20 \log_{10}\left(\frac{c_{k-RMS}}{1-V\ RMS}\right) \quad (7.60)$$

Also, it is customary to plot the frequency-domain data as a continuous curve by interpolating between frequency samples instead of using a line spectrum. In some cases, one uses a zero-order interpolation operation to produce a step or bar graph of the spectrum, or as we shall use in the next few examples, a first-order interpolation operation.

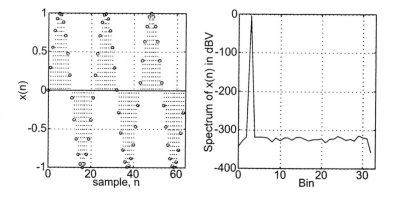

Figure 7.11. Coherent waveform time-domain plot and spectrum.

The results for the noncoherent waveform were also found and are shown in Figure 7.12.

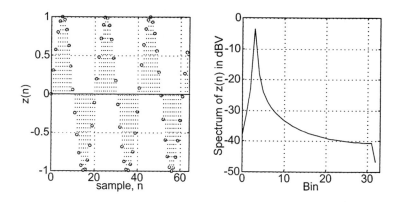

Figure 7.12. Noncoherent waveform time-domain plot and spectrum.

On comparing the magnitude of the two spectra, we clearly see a significant difference. In the case of the coherent sinusoidal waveform, a single spike occurs in bin 3 with over 300 dB of separation distance from all other spectral coefficients. A closer look at the numbers indicates that the tone has an RMS value of 0.707106 (-3.0103 dBV in the plot) or an amplitude of 1. This is exactly the value specified in the code that is used to generate the coherent sinusoid. In the case of the noncoherent sinusoidal waveform, no single spike occurs. Rather, the single-tone waveform appears to consist of many frequency components. If other signal components were present, then they would be corrupted by the power in these leaked components. What is worse, it is extremely difficult to determine the amplitude of the noncoherent sinusoidal signal with its power smeared across many frequency locations.

The most straightforward method to improve the measurement accuracy of a noncoherent waveform is to increase its observation interval. Generally speaking, this approach is used by most benchtop instruments found in one's laboratory, such as spectrum analyzers, multimeters, and digitizing oscilloscopes. Samples taken from the input signal are unrelated to the sampling rate of the instrument, and are therefore noncoherent. Instruments of this type are usually not expected to generate a result in a very short time, such as 25 ms. Rather, they are only required to produce a result every 1 or 2 s, which is usually more than adequate. Consequently, one can construct a less complex, noncoherent measurement system. Our next example will illustrate the effect of a longer observation interval on the spectrum of a noncoherent sinusoidal waveform.

Example 7.8

Extend the observation interval of the noncoherent waveform of Example 7.7 by collecting 8192 samples instead of 64. Plot the corresponding magnitude of the resulting spectrum. Determine the amplitude of the input signal from its spectrum.

Solution:

The MATLAB code for the noncoherent signal from Example 7.7 was modified as follows:

```
% noncoherent signal definition - z -
    NOI=8192;                   % observation interval
    N=64; M=pi; A=1; P=0;       % signal definition
    for n=1:NOI,
        z(n)=A*sin(2*pi*M/N*(n-1)+P);
    end;
```

Here we distinguish between N, the number of samples in one UTP, and N_{OI}, the number of samples collected over the entire observation interval. In other words, N_{OI}/N represents the number of UTPs that the signal will complete in the observation interval.

The Fourier analysis was then repeated and the corresponding spectrum was found as shown in Figure 7.13. On the left is the plot of the time-domain waveform over the last 64 samples of the full 8192 samples, as any more samples would fill the graph and mask all detail. On the right is the magnitude of its spectrum.

When we refer back to the spectrum derived in Example 7.7 and compare it to the one derived here, we see that the general shape of the magnitude of the spectrum is much more concentrated around a single frequency and that the frequency leakage components are much smaller. The astute reader may be wondering about the scale of the x axis. In this case we are plotting the index or bin of each frequency components from 0 to 8191, whereas in the previous case we plot from 0 to 63. It is important to realize that each bin is equivalent to $Bin(F_s/N_{OI})$ Hz. In other words, the frequency range is identical in each case; only the frequency granularity is different.

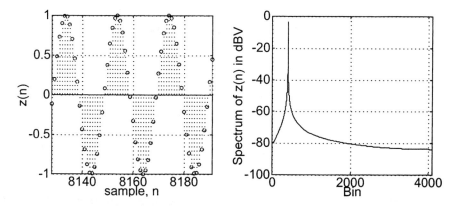

Figure 7.13. Noncoherent waveform time-domain plot and spectrum – expanded observation interval.

In order to estimate the amplitude of the input waveform, one cannot rely on the peak value of the spectrum as was done in the coherent waveform case. Rather, we must use several frequency components centered around the peak spectral concentration to estimate the waveform amplitude. To see this more clearly, we provide in Figure 7.14 an expanded view of the spectrum around the spectral peak.

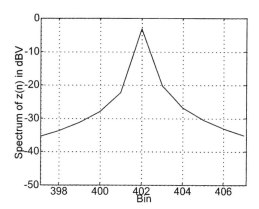

Figure 7.14. Expanded view of the noncoherent tone around the spectral peak.

Here we see that a peak spectral value of -3.2318 dBV occurs in bin 402 [bin $(M/N)N_{OI}$]. Ideally, the spectral peak value should be -3.0103 dBV [$= 20\log_{10}(0.707 \text{ V RMS}/1 \text{ V RMS})$]. To improve the estimate, we must take into consideration the power associated with the side tones. As the magnitude of these side tones drop off fairly quickly, let us consider that the power associated with the input signal is mainly associated with the power of the five tones before and after the spectral peak. On doing so, the amplitude estimate becomes −3.0355 dBV. Including more side tones into this calculation will only help to improve the estimate. Generally speaking, side tones less than −60 dB below the spectral peak value will improve the accuracy to within 0.1%.

We could also improve the accuracy of the estimate by further increasing the observation interval. For example, increasing the observation interval to 131,072 samples will improve the amplitude estimate to -3.0109 dB.

Extending the observation interval in the previous example certainly helped to decrease the amount of frequency leakage, which, in turn, helped to improve the measurement accuracy of the noncoherent waveform. However, as in most production test situations, one is always searching for a faster solution. In Section 7.3.1, the method of windowing was suggested as a possibility. The next example will investigate this further.

Example 7.9

Through the application of a Hanning window, compute the magnitude of the spectra of the noncoherent waveforms described in Examples 7.7 and 7.8 consisting of 64 and 8192 samples, respectively. Compare the spectra with the results from a rectangular window (i.e., the nonwindowed results obtained previously).

Solution:

Let us begin by investigating the spectral properties of the noncoherent waveform of Example 7.7, consisting of 64 samples (Figure 7.12). A MATLAB script was written to perform the windowing operation. The code is:

```
% noncoherent signal definition - z -
    NOI=64;                          % observation interval
    N=64; M=pi; A=1; P=0;            % signal definition
    for n=1:NOI,
        z(n)=A*sin(2*pi*M/N*(n-1)+P);
    end;
% windowing operation
    w= hanning(NOI)';
    epsilon=sqrt(sum(w.*w)/NOI);
    u= 1/epsilon * z .* w;
```

MATLAB provides a built-in function called *hanning* that generates the Hanning window. This window is shown in Figure 7.15(b) for 64 samples. Next, each sample of the waveform is multiplied by a corresponding sample from this windowed function, as denoted by the ".*" operation. In addition, the windowed data are scaled by the window shape factor[6] denoted by epsilon, ε. This serves to equalize the power in the windowed data with that in the original waveform. If the window time samples are denoted $w(n)$, then the window shape factor is simply given by

$$\varepsilon = \sqrt{\frac{1}{N}\sum_{n=0}^{N-1} w^2(n)} \qquad (7.61)$$

In this particular case, with $N=64$ the Hanning window has $\varepsilon=0.612$. This leads us to the new waveform shown in Figure 7.15(c). The window effectively squeezes the endpoints of the noncoherent waveform toward zero, forcing the endpoints of the waveform to meet smoothly. An FFT was then performed on the modified waveform from which the magnitude of the spectrum was derived. The windowed spectrum is shown in Figure 7.15(d). Superimposed on the plot is the spectrum of the original noncoherent waveform derived in Example 7.6 without windowing. Although no specific window operation was explicitly performed, the data are said to have been viewed through a rectangular window. As is evident, the Hanning windowed spectrum is much more concentrated about a single frequency than the rectangular windowed spectrum.

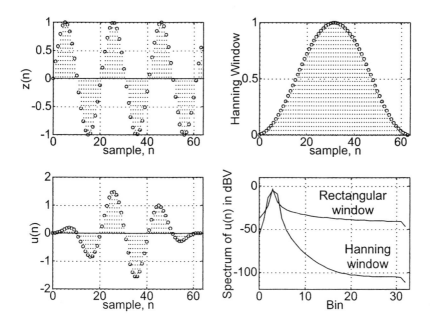

Figure 7.15. Windowing results (a) original noncoherent waveform, (b) Hanning window, (c) windowed data, and (d) spectrum magnitude.

In order to estimate the amplitude of the signal present in the windowed data, an expanded view of the spectrum about its peak is shown in Figure 7.16. As side tones are present about the peak value of the spectrum, we must consider these tones in the estimate of the signal amplitude. Performing a square-root-of-sum-of-squares calculation involving the RMS value of the signals present in bins 1 to 7, we obtain a combined RMS value of 0.7070674. This in turn implies an amplitude estimate of 0.999944. For all intents and purposes, this is unity and was found with only 64 samples. Unless improved frequency resolution is necessary, preprocessing with the Hanning window is just as effective as the coherent measurement in this example. However, repeatability of measurments is degraded in windowed systems in the presense of random noise.

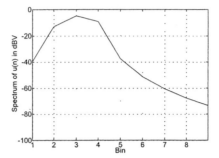

Figure 7.16. Expanded view of the windowed, noncoherent tone around the spectral peak.

If we increase the number of samples to 8192 and repeat the windowing operation, we find the spectrum much more closely concentrated about a single frequency. The results are shown in Figure 7.17, together with the spectrum resulting from a rectangular window.

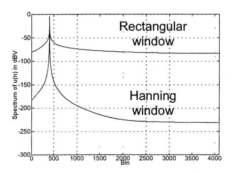

Figure 7.17. Spectral comparison of rectangular and Hanning windows.

Clearly, the Hanning window has a very narrow spectrum, very much like the coherent case. It would be suitable for making measurements in situations where more than one tone is present in the input signal. The drawback, of course, is that a longer observation interval is necessary and that side tones have to be dealt with.

In the previous example, windowing was used to improve the measurement of a noncoherent sinusoidal signal. In fact, an accurate estimate was obtained without extending the observation interval over and above that of a coherent sinusoidal signal. This example may incorrectly give the reader the impression that windowing can resolve the frequency leakage problem associated with noncoherent signaling with no added time expense. As we shall see in this next example consisting of a noncoherent multitone signal, this is indeed not the case.

Example 7.10

Determine the magnitude of the spectrum of a multitone signal consisting of three noncoherent tones with parameters, $A=1$, $\phi=0$, $M=\pi$, $N=64$; $A=1$, $\phi=0$, $M=\pi+1$, $N=64$; and $A=1$, $\phi=0$, $M=\pi+2$, $N=64$.

Solution:

The following MATLAB routine was written to generate the 3-tone multitone signal with a Hanning window over a 64-sample observation interval:

```
% noncoherent 3-tone multitone signal definition
    NOI=64;                % observation interval
    N=64;                  % signal definition
    M1=pi; A1=1; P1=0;
    M2=pi+1; A2=1; P2=0;
    M3=pi+2; A3=1; P3=0;
    for n=1:NOI,
        z(n)=A1*sin(2*pi*M1/N*(n-1)+P1) +
        A2*sin(2*pi*M2/N*(n-1)+P2) +
        A3*sin(2*pi*M3/N*(n-1)+P3);
    end;
% windowing operation
    w= hanning(NOI)';
    epsilon=sqrt(sum(w.*w)/NOI);
    u= 1/epsilon * z .* w;
```

The routine was executed and the corresponding magnitude of the frequency spectrum was derived using the frequency-domain conversion routine of Example 7.7. The result is shown in Figure 7.18 on the left-hand side. The plot extends over the Nyquist frequency range.

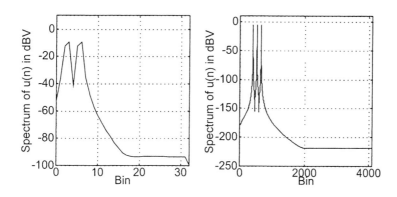

Figure 7.18. Windowed three-tone multitone spectra with $N=64$ and 8192.

Surprisingly, the spectrum appears to have only 2 spectral peaks. It is as if only 2 tones were present in the input signal. This is a direct result of frequency leakage. If the observation interval is increased, say, to 8192 samples, then better frequency resolution is obtained and a clear separation of each frequency component is evident. This is shown on the right-hand side of Figure 7.18. With the side tones around each spectral peak below 100 dB, each tone can be accurately measured with a relative error of no more than 0.001% using the method of the previous example.

Exercises

7.14. A signal has a 0.5-V amplitude. Express its amplitude in dBV units.

Ans. −9.03 dBV.

7.15. A signal has a period of 1 ms and is sampled at a rate of 128 kHz. If 128 samples are collected, what is the frequency resolution of the resulting FFT? If the number of samples collected increases to 8192 samples, what is the frequency resolution of the resulting FFT?

Ans. 1 kHz, 15.625 Hz.

7.16. The results of an FFT analysis of a noncoherent sinusoid indicates the following significant dBV values around the spectral peak: -38.5067, -36.3826, -33.6514, -29.7796, -22.9061, -7.7562, -25.3151. Estimate the amplitude of the corresponding tone.

Ans. 0.596.

7.4 THE INVERSE FFT

7.4.1 Equivalence of Time- and Frequency-Domain Information

A discrete Fourier transform produces a frequency-domain representation of a discrete-time waveform. This was shown to be equivalent to a discrete-time Fourier series representation of a periodically extended waveform. The transformation is lossless, meaning that all information about the original signal is maintained in the transformation. Since no information is lost in the transformation from the time domain to the frequency domain, it seems logical that we should be able to take a frequency-domain signal back into the time domain to reconstruct the original signal. Indeed, this is possible and can be seen directly from Eqs. (7.19) and (7.20). If we substitute the expression for the frequency-domain coefficients given by Eq. (7.20) back into (7.19), we clearly see that we obtain our original information

$$X = W\left[W^{-1}X\right] = WW^{-1}X = X \tag{7.62}$$

In practice, the form of the mathematics used to perform the frequency-to-time operation is very similar to that used to perform the time-to-frequency operation. In fact, the same FFT algorithm can be used, except for some possible array rearrangements and some predictable scale factor changes. When the FFT is used to perform the frequency-to-time transformation, it is referred to as an *inverse FFT*. The calculations are so similar that some testers perform an inverse FFT using the same syntax as the forward FFT. A flag is set to determine whether the FFT is forward or inverse.

It is worth noting that the magnitude of a spectrum alone cannot be converted back into the time domain. Phase information at each test frequency must be combined with the magnitude information in the form of a complex number using either rectangular or polar notation. The specific format will depend on the vendor's operating system. Our next example will illustrate this procedure using MATLAB.

Example 7.11

A discrete-time signal is described by its DTFS representation as

$$x[n] = 1 + 2 \cos\left[2\left(\frac{2\pi}{8}\right)n + \frac{\pi}{4}\right] + 0.5 \cos\left[3\left(\frac{2\pi}{8}\right)n\right]$$

Determine the time-domain samples using MATLAB's inverse FFT routine.

Solution:

The inverse FFT is performed in MATLAB using a vector **Y** that contains the samples of the Fourier transform $Y(k)$ in complex form. As described in Section 7.3, these are related to the coefficients of the DTFS according to

$$Y(k) = N\, X(k)$$

To obtain $X(k)$, we first note from $x[n]$ that the spectral coefficients of the DTFS written in trigonometric form are immediately obvious as

$$\begin{bmatrix} \tilde{c}_0 \\ \tilde{c}_1 \\ \tilde{c}_2 \\ \tilde{c}_3 \\ \tilde{c}_4 \end{bmatrix} = \begin{bmatrix} 1 \\ 0 \\ 2 \\ 0.5 \\ 0 \end{bmatrix} \quad \text{and} \quad \begin{bmatrix} \tilde{\phi}_0 \\ \tilde{\phi}_1 \\ \tilde{\phi}_2 \\ \tilde{\phi}_3 \\ \tilde{\phi}_4 \end{bmatrix} = \begin{bmatrix} 0 \\ 0 \\ -\pi/4 \\ 0 \\ 0 \end{bmatrix}$$

Subsequently, **X**=[$X(0)\ X(1)\ \ldots\ X(N-1)$] and **Y**=[$Y(0)\ Y(1)\ \ldots\ Y(N-1)$] with $N=8$ are determined from Eq. (7.45), and from their inter-relationship described, to be

$$X = \begin{bmatrix} 1 \\ 0 \\ 0.7071 + j0.7071 \\ 0.25 \\ 0 \\ 0.25 \\ 0.7071 - j0.7071 \\ 0 \end{bmatrix} \Rightarrow Y = 8X = \begin{bmatrix} 8 \\ 0 \\ 5.6568 + j5.6568 \\ 2 \\ 0 \\ 2 \\ 5.6568 - j5.6568 \\ 0 \end{bmatrix}$$

Submitting the **Y** vector to MATLAB's inverse FFT routine, we obtain the following time-domain samples

$$x = IFFT\{Y\} = \begin{bmatrix} 2.9142 \\ -0.7678 \\ -0.4142 \\ 2.7678 \\ 1.9141 \\ -0.0606 \\ -0.4142 \\ 2.0606 \end{bmatrix}$$

Of course, these time domain samples agree with those obtain by evaluating $x[n]$ directly at the sampling instances.

7.4.2 Parseval's Theorem

The RMS value X_{RMS} of a discrete time periodic signal $x[n]$ is defined as the square root of the sum of the individual samples squared, normalized by the number of samples N, according to

$$X_{RMS} = \sqrt{\frac{1}{N} \sum_{n=0}^{N-1} x^2[n]} \qquad (7.63)$$

Parseval's theorem for discrete-time periodic signals states that the RMS value X_{RMS} of a periodic signal $x[n]$ described by a DTFS written in trigonometric terms is given by

$$X_{RMS} = \sqrt{c_0^2 + \frac{1}{2} \sum_{k=1}^{N/2} c_k^2} \qquad (7.64)$$

When the magnitude of each spectral coefficient is expressed as an RMS value, Eq. (7.64) can be rewritten as a square-root-of-sum-of-squares calculation given by

$$X_{RMS} = \sqrt{\sum_{k=0}^{N/2} c_{k-RMS}^2} \qquad (7.65)$$

In this text, we shall make greatest use of this form of Parseval's theorem. It is the easiest form to remember, as there are no extra scale factors to keep track of.

The importance of Parseval's theorem is that it allows the computation of the RMS value associated with all aspects of a signal such as distortion, noise, etc., to be made directly from the DTFS description. For example, the RMS noise level associated with a signal is the square-root-of-sum-of-squares of all bins (excluding DC) that do not contain signal-related power given by

$$X_n = \sqrt{\sum_{\substack{k=1 \\ k \neq signal}}^{N/2} c_{k-RMS}^2} \qquad (7.66)$$

7.4.3 Applications of the Inverse FFT

One might ask what useful purpose is served by performing an FFT and then undoing it with an inverse FFT. Where would we use this type of double transform in mixed-signal testing? There are several applications for the inverse FFT. One is the removal of noise in a time-domain signal to produce a smoothed waveform. Another useful application of the inverse FFT is interpolation of a limited sample set into a more detailed sample set.

Consider the sampled waveform in Figure 7.19(a), which is a digitized 100-MHz digital clock. The digitized samples are corrupted with a large amount of noise, which we would like to remove. By performing an FFT on this clock signal, we can produce a frequency spectrum of the clock shown in Figure 7.19(b). Noise components in the time-domain signal are scattered all over the frequency-domain FFT spectrum.

The clock signal, by contrast, is located in a set of predictable FFT spectral bins. Since we have captured 2 cycles of the square wave, the fundamental energy is located in spectral bin 2 with harmonics at integer multiples of 2 (i.e., bins 2, 4, 6, 8, etc.). We can remove half of the noise power by simply setting all the nonsignal FFT elements to zero, which accounts for half the number of bins. This is a direct result of Parseval's theorem described in the previous subsection. The corresponding spectrum magnitude is shown in Figure 7.19(c).

An inverse FFT then restores the original clock signal with half of the noise removed [Figure 7.19(d)]. After cleaning up the square wave in this manner, we can measure rise and fall times, settling times, and other time-domain characteristics with twice the repeatability than if we had used the original noisy signal. When the signal bandwidth is lower (i.e., occupies fewer spectral bins), a greater improvement can be obtained because more spectral bins can be set to zero.

We can also interpolate points between the samples of a digitized waveform to allow more x axis resolution for time-domain measurements. Interpolation provides a higher degree of time resolution when measuring parameters such as rise and fall time. Whenever we perform an FFT or inverse FFT on N samples, we always get a result with N samples. By appending N_z additional zero-value bins to the FFT, immediately following the Nyquist frequency at bin $N/2$, we can effectively change the FFT frequency resolution from F_s/N to $F_s/(N+N_z)$.

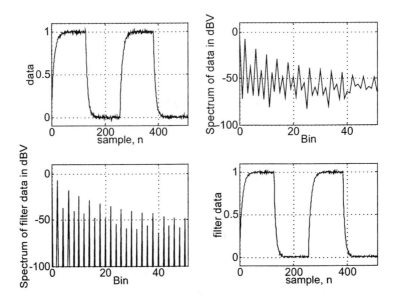

Figure 7.19. Time-domain smoothing using the inverse FFT: (a) original noisy clock signal, (b) magnitude of clock signal spectrum, (c) partially zeroed spectrum, (d) restored time-domain signal.

The parameter N_z should be a power of two in order to maintain compatibility with the inverse FFT algorithm. The inverse FFT, in turn, recreates a time-domain signal with N_z/N interpolated samples interspersed between the original sample set. In other words, the time resolution improves by the factor $(N+N_z)/N$.

Figure 7.20(a) illustrates a sampled sine wave with 64 samples, and Figure 7.20(b) shows the magnitude of the spectrum of this waveform from bin 0 to bin N-1. In Figure 7.20(c) the spectrum from Figure 7.20(b) has been expanded to include 4 times as many spectral bins. The expansion is accomplished by padding extra zeros in the middle of the original FFT data. For numerical reasons, we did not use zeros here; rather we used the smallest numbers that can could be represented on our computer. In this way, we avoided minus infinity on the logarithim plot of the spectrum. The inverse FFT of this expanded data results in a conversion back to the time domain with four times as many samples as the original waveform, as shown in Figure 7.20(d).

7.4.4 Frequency-Domain Filtering

Another useful feature of DSP-based testing is the ability to apply arbitrary filter functions to the collected samples, simulating electronic filters in traditional analog instrumentation. Filtering can be applied either in the time domain or in the frequency domain. Time-domain filtering is accomplished by convolving the sampled waveform by the impulse response of the desired filter. Frequency-domain filtering is performed by multiplying the signal spectrum by the desired filter's frequency response. Frequency-domain filtering is faster to implement than time-domain filtering, as the spectrum of a signal is already available in the computer. As such, time-domain filtering is rarely used in mixed-signal testing applications.

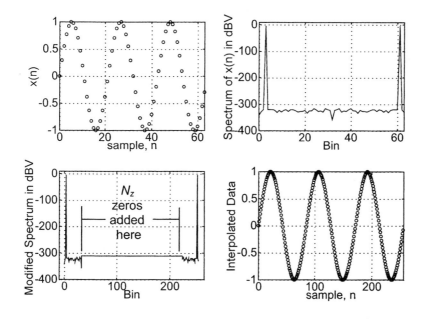

Figure 7.20. Time-domain interpolation using the inverse FFT.

Filter functions are selected to provide a predescribed frequency response such as a low-pass or band-pass behavior. For instance, a low-pass filter is used to suppress the high-frequency content of a signal while allowing the low-frequency energy to pass relatively unchanged. In another application, filters are used to alter the phase of a signal to improve its transient behavior (i.e., reduce ringing). In general, the z-domain transfer function $H(z)$ of an arbitrary discrete-time filter is described by

$$H(z) = \frac{a_0 + a_1 z^1 + \cdots + a_N z^N}{1 + b_1 z^1 + \cdots + b_N z^N} \tag{7.67}$$

The response of the filter to physical frequencies is determined by substituting $z = e^{j\omega T_s}$ into Eq. (7.67), where T_S is the sampling rate, to obtain

$$H(z)\big|_{z=e^{j\omega T_s}} = H\left(e^{j\omega T_s}\right) = \frac{a_0 + a_1 e^{j\omega T_s} + \cdots + a_N e^{j\omega N T_s}}{1 + b_1 e^{j\omega T_s} + \cdots + b_N e^{j\omega N T_s}} \tag{7.68}$$

For any single frequency ω_o, we see from Eq. (7.68) that the behavior of the filter can be collectively understood as a complex number represented in terms of real and imaginary parts as

$$H\left(e^{j\omega_o T_s}\right) = \text{Re}\left\{H\left(e^{j\omega_o T_s}\right)\right\} + j\,\text{Im}\left\{H\left(e^{j\omega_o T_s}\right)\right\} \tag{7.69}$$

Exercises

7.17. Using MATLAB, compute the IFFT of the following sequence of complex numbers: {1.875, 0.75-j0.375, 0.625, 0.75+j0.375}.

Ans. {1.0, 0.5, 0.25, 0.125}.

7.18. For the following signal $x[n]$, compute the RMS value of each frequency component (beginning with DC):

$$x[n] = 1 + 2\cos\left[2\left(\frac{2\pi}{8}\right)n + \frac{\pi}{4}\right] + 0.5\cos\left[3\left(\frac{2\pi}{8}\right)n\right]$$

Using Parseval's theorem, compute the RMS value of $x[n]$.

Ans. 1, 1.414, 0.3536; 1.7676.

7.19. A coherent signal is sampled at a rate of 1 MHz over a 1024-µs time interval. To increase the time resolution of the sampled waveform to 0.25 µs, how many zero-padded bins should be added to the spectral data before performing the inverse FFT?

Ans. 3072.

or in terms of magnitude and phase as

$$H\left(e^{j\omega_o T_s}\right) = \left|H\left(e^{j\omega_o T_s}\right)\right| e^{j\phi_H\left(e^{j\omega_o T_s}\right)} \tag{7.70}$$

From the convolution property of discrete-time Fourier transforms, the transform $Y_{out}\left(e^{j\omega T_s}\right)$ of the filter output is related to the transform $Y_{in}\left(e^{j\omega T_s}\right)$ of the input according to

$$Y_{out}\left(e^{j\omega T_s}\right) = H\left(e^{j\omega T_s}\right) Y_{in}\left(e^{j\omega T_s}\right) \tag{7.71}$$

If we limit the input to discrete frequencies

$$\omega = \frac{2\pi}{N}k, \quad k = 0, 1, \ldots, N-1 \tag{7.72}$$

then Eq. (7.71) can be rewritten as

$$Y_{out}\left(e^{j\frac{2\pi}{N}kT_s}\right) = H\left(e^{j\frac{2\pi}{N}kT_s}\right) Y_{in}\left(e^{j\frac{2\pi}{N}kT_s}\right) \tag{7.73}$$

or with the shorthand notation of Section 7.4 for $k=0, 1, \ldots, N-1$ as

$$Y_{out}(k) = H(k)Y_{in}(k) \tag{7.74}$$

Dividing both sides by N

$$\frac{Y_{out}(k)}{N} = H(k)\frac{Y_{in}(k)}{N} \tag{7.75}$$

leads us to the relationship between the input and output complex DTFS representation after filtering as

$$X_{out}(k) = H(k)X_{in}(k) \tag{7.76}$$

Returning to the trigonometric form of the DTFS representation written in polar form, we can write

$$\tilde{c}_{k-out}e^{-j\tilde{\phi}_{k-out}} = H(k)\tilde{c}_{k-in}e^{-j\tilde{\phi}_{k-in}} \tag{7.77}$$

or, with the magnitude and phase form for $H(k) = |H(k)|e^{j\phi_{k-H}}$ substituted, we write

$$\tilde{c}_{k-out}e^{-j\tilde{\phi}_{k-out}} = |H(k)|e^{j\phi_{k-H}}\tilde{c}_{k-in}e^{-j\tilde{\phi}_{k-in}} \tag{7.78}$$

Separating the magnitude and phase components leads to the result

$$\begin{aligned}\tilde{c}_{k-out} &= |H(k)|\tilde{c}_{k-in} \\ \tilde{\phi}_{k-out} &= \tilde{\phi}_{k-in} - \phi_{k-H}\end{aligned} \tag{7.79}$$

Following a similar line of reasoning but using the rectangular form of the DTFS representation, we write

$$\begin{aligned}\tilde{a}_{k-out} - j\tilde{b}_{k-out} &= \left[\text{Re}\{H(k)\} + j\,\text{Im}\{H(k)\}\right]\left(\tilde{a}_{k-in} - j\tilde{b}_{k-in}\right) \\ &= \left[\tilde{a}_{k-in}\,\text{Re}\{H(k)\} + \tilde{b}_{k-in}\,\text{Im}\{H(k)\}\right] \\ &\quad - j\left[-\tilde{a}_{k-in}\,\text{Im}\{H(k)\} + \tilde{b}_{k-in}\,\text{Re}\{H(k)\}\right]\end{aligned} \tag{7.80}$$

Separating into real and imaginary parts, we get

$$\begin{aligned}\tilde{a}_{k-out} &= \tilde{a}_{k-in}\,\text{Re}\{H(k)\} + \tilde{b}_{k-in}\,\text{Im}\{H(k)\} \\ \tilde{b}_{k-out} &= -\tilde{a}_{k-in}\,\text{Im}\{H(k)\} + \tilde{b}_{k-in}\,\text{Re}\{H(k)\}\end{aligned} \tag{7.81}$$

The procedure to filter a signal in the frequency domain is now clear. We first perform an FFT operation on the input signal to bring it into the frequency domain. From the FFT result one creates a DTFS representation in either rectangular or polar form. Next, each frequency component of the input is scaled by the corresponding filter response $H(k)$ to produce the filtered frequency-domain output. Finally, an inverse FFT can be performed on the filtered output to produce the filtered time-domain signal. Often, this last step can be eliminated because we can extract all desired information from the filtered spectrum in the frequency domain.

Example 7.12

Using the time-domain samples from the noisy clock signal of Section 7.5.2, apply a second-order Butterworth filter having the following z-domain transfer function

$$H(z) = \frac{0.0005 + 0.0011z^1 + 0.0005z^2}{1 - 1.9334z^1 + 0.9355z^2}$$

Plot the corresponding magnitude spectra and time-domain signals before and after filtering. Use MATLAB's built-in FFT and inverse FFT routines.

Solution:

Using the FFT data from Section 7.5.2 on the noisy clock signal, its spectrum was multiplied by the Butterworth filter transfer function and the resulting spectrum was converted back to the time domain via the inverse FFT. The results are summarized in Figure 7.21.

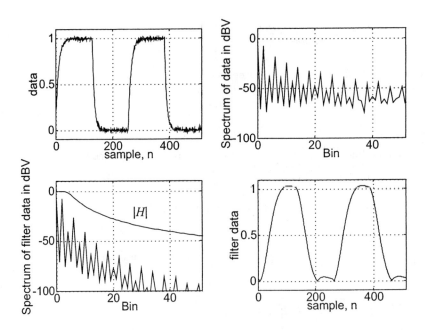

Figure 7.21. Transient and frequency spectra before and after filtering: (a) noisy clock signal, (b) magnitude of spectrum, (c) magnitude of filter response and magnitude of filtered signal spectrum, (d) filter transient signal.

7.4.5 Noise Weighting

Noise weighting is one common example of DSP-based filtering in mixed-signal test programs. Weighting filters are called out in many telecom and audio specifications because the human ear is more sensitive to noise in some frequency bands than others. The magnitude of the frequency response of the A-weighting filter is shown in Figure 7.22. It is designed to approximate the frequency response of the average human ear. For matters related to hearing, phase variations have little effect on the listener and are therefore ignored in noise-related tests. By weighting the noise from a telephone or audio circuit before measuring its RMS level, we can get a more accurate idea of how good or bad the telephone or audio equipment will sound to the consumer.

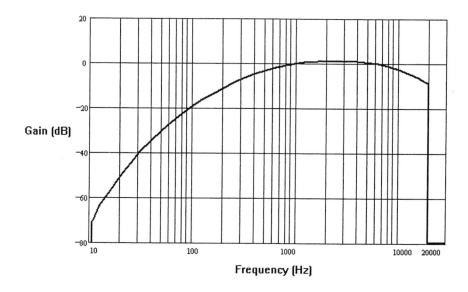

Figure 7.22. A-weighting filter magnitude response.

In traditional bench instruments, the weighting filter is applied to the analog signal before it is passed to an RMS voltmeter. In DSP-based testing, we can perform the same filtering function mathematically, after the unweighted DUT signal has been sampled. Application of a mathematical filter to a sampled waveform means the ATE tester does not have to include a physical A-weighting filter in its measurement instruments. The resulting reduction in tester complexity reduces tester cost and improves reliability. Application of the A-weighting filter is a simple matter of multiplying its magnitude by the magnitude of the FFT of the signal under test.

A very simple form of noise filtering can be used to measure the noise over a particular bandwidth. For example, if a specification calls for a noise level of 10 μV RMS over a band of 100 Hz to 1 kHz, then we can simply apply a brick wall band-pass filter to the FFT results, eliminating all noise components that do not fall within this frequency range. The remaining noise can then be measured by performing an inverse FFT followed by an RMS calculation. The same results can be also achieved by adding up the signal power in all spectral bins from 100 Hz to 1 kHz. To do this we simply square the RMS value of each frequency component from

100 Hz to 1 kHz, add them all together, and then take the square root of the total to obtain the RMS value of the noise according to

$$V_{N-rms} = \sqrt{\sum_{k=B_L}^{B_U} c_{k-RMS}^2} \qquad (7.82)$$

Here B_L and B_U are the spectral bins corresponding to the lower and upper frequencies of the brick wall filter (excluding any DC component). In this particular case, B_L and B_U correspond to the 100 Hz and 1 kHz frequencies, respectively.

Exercises

7.20. The gain and phase of a particular system at 1 kHz is 0.8 and $-\pi/4$, respectively. Determine the spectral coefficient of the DTFS that corresponds to the output signal at 1 kHz when excited by a signal with a spectral coefficient described by $0.25-j0.35$. Express the result in rectangular and polar form.

Ans. $-0.0566-j0.3394$; $0.3441e^{-j1.7341}$.

7.21. For the Butterworth transfer function described in Example 7.12, determine the gain and phase of the filter at the following three normalized frequencies: 0; $2(2\pi/8)$; and $3(2\pi/8)$.

Ans. 1.0000, $5.5537 \times 10^{-4} e^{-j3.1083}$, $9.5287 \times 10^{-5} e^{-j3.1278}$.

7.22. A signal with a DTFS representation given by

$$x[n] = 1 + 2\cos\left[2\left(\frac{2\pi}{8}\right)n + \frac{\pi}{4}\right] + 0.5\cos\left[3\left(\frac{2\pi}{8}\right)n\right]$$

passes through a system with a transfer function described by that in Example 7.12. What is the DTFS representation of the output signal?

Ans.

$$x[n] = 1 + 1.1107 \times 10^{-3} \cos\left[2\left(\frac{2\pi}{8}\right)n - 2.3229\right] + 4.7643 \times 10^{-3} \cos\left[3\left(\frac{2\pi}{8}\right)n - 3.1278\right]$$

7.5 SUMMARY

Coherent DSP-based testing allows the mixed-signal test engineer to perform AC measurements in a few tens of milliseconds. These same measurements might otherwise take hundreds or thousands of milliseconds using traditional analog bench instruments. The AWG, digitizer, source memory, and capture memory of a mixed-signal ATE tester allow us to translate signals between continuous time and sampled time. Digital signal processing operations such as the FFT and inverse FFT allow us to perform operations that are unavailable using traditional non-

DSP measurement methodologies. Time-domain interpolations, frequency-domain filtering, and noise reduction functions are just a few of the powerful operations DSP-based testing makes available to the accomplished mixed-signal test engineer.

We are fortunate in mixed-signal ATE to be able to use coherent sampling systems to bypass mathematical windowing. Bench instruments such as spectrum analyzers and digitizing oscilloscopes must use windowing extensively. The signals entering a spectrum analyzer or oscilloscope are generally noncoherent, since they are not synchronized to the instrument's sampling rate. However, a spectrum analyzer is not usually expected to produce an accurate reading in only 25 ms; so it can overcome the repeatability problem inherent in windowing by simply averaging results or collecting additional samples. ATE equipment must be fast as well as accurate; so windowing is normally avoided whenever possible. Fortunately, mixed-signal testers give us control of both the signal source and sampling processes during most tests. Synchronization of input waveforms and sampling processes affords us the tremendous accuracy/cost advantage of coherent DSP-based testing.

Despite its many advantages, DSP-based testing also places a burden of knowledge upon the mixed-signal test engineer. Matthew Mahoney, author of "DSP-Based Testing of Analog and Mixed-Signal Circuits,"[7] once told an amusing story about a frustrated student in one of his DSP-based testing seminars. Exasperated by the complexity of digital signal processing, the distressed student suddenly exclaimed "But this means we must know something!". Indeed, compared to the push-the-button/read-the-answer simplicity of bench equipment, DSP-based testing requires us to know a whole lot of "something."

The next two chapters will explore the use of DSP-based measurements in testing analog and sampled channels. In Chapter 8, "Analog Channel Testing," we will explore the various types of DSP-based tests that are commonly performed on nonsampled channels such as amplifiers and analog filters. Chapter 9, "Sampled Channel Testing" will then extend these DSP-based testing concepts to sampled channels such as ADCs, DACs, and switched capacitor filters.

APPENDIX A.7.1

Fourier Series Representation of a Coherent Sampled Signal

Sampling a continuous-time periodic signal using coherent sampling principles leads to a discrete-time periodic signal $x[n]$ with Fourier series representation given by

$$x[n] = a_0 + \sum_{k=1}^{\infty} \left\{ a_k \cos\left[k\left(\frac{2\pi}{N}\right)n\right] + b_k \sin\left[k\left(\frac{2\pi}{N}\right)n\right] \right\} \quad (7.83)$$

As this signal is periodic with period N over the index n, Eq. (7.83) suggests that it is also periodic in k as well, as the role of n and k are interchangeable. The insight provided by this observation can be used to re-arrange Eq. (7.83) in groups of N terms in the following manner

$$x[n] = \sum_{m=0}^{\infty} \left(a_{0+mN} + \sum_{k=1}^{N-1} \left\{ a_{k+mN} \cos\left[k\left(\frac{2\pi}{N}\right)n\right] + b_{k+mN} \sin\left[k\left(\frac{2\pi}{N}\right)n\right] \right\} \right) \quad (7.84)$$

Further, due to the symmetry of the cosine and sine functions, additional simplifications can be made. Consider regrouping the terms in Eq. (7.84), assuming N is even, as follows

$$x[n] = \sum_{m=0}^{\infty} a_{0+mN} + \sum_{m=0}^{\infty} \sum_{k=1}^{N/2-1} \left\{ a_{k+mN} \cos\left[k\left(\frac{2\pi}{N}\right)n\right] + b_{k+mN} \sin\left[k\left(\frac{2\pi}{N}\right)n\right] \right\}$$
$$+ \sum_{m=0}^{\infty} \sum_{k=1}^{N/2-1} \left\{ a_{N-k+mN} \cos\left[(N-k)\left(\frac{2\pi}{N}\right)n\right] + b_{N-k+mN} \sin\left[(N-k)\left(\frac{2\pi}{N}\right)n\right] \right\} \quad (7.85)$$
$$+ \sum_{m=0}^{\infty} \left\{ a_{N/2+mN} \cos\left[\left(\frac{N}{2}\right)\left(\frac{2\pi}{N}\right)n\right] + b_{N/2+mN} \sin\left[\left(\frac{N}{2}\right)\left(\frac{2\pi}{N}\right)n\right] \right\}$$

Recognizing that

$$\cos\left[(N-k)\left(\frac{2\pi}{N}\right)n\right] = \cos\left[k\left(\frac{2\pi}{N}\right)n\right]$$
$$\sin\left[(N-k)\left(\frac{2\pi}{N}\right)n\right] = -\sin\left[k\left(\frac{2\pi}{N}\right)n\right] \quad (7.86)$$
$$\sin\left[\left(\frac{N}{2}\right)\left(\frac{2\pi}{N}\right)n\right] = 0$$

we can write Eq. (7.85) as

$$x[n] = \sum_{m=0}^{\infty} a_{0+mN} + \sum_{m=0}^{\infty} \sum_{k=1}^{N/2-1} \left\{ [a_{k+mN} + a_{N-k+mN}] \cos\left[k\left(\frac{2\pi}{N}\right)n\right] \right.$$
$$\left. + [b_{k+mN} - b_{N-k+mN}] \sin\left[k\left(\frac{2\pi}{N}\right)n\right] \right\} + \sum_{m=0}^{\infty} \left\{ a_{N/2+mN} \cos\left[\left(\frac{N}{2}\right)\left(\frac{2\pi}{N}\right)n\right] \right\} \quad (7.87)$$

If N is odd, then the upper limit in the summation should be replaced by $(N-1)/2$. In this work, N will always be assumed to be even. In terms of the sampling theorem, $N/2$ represents the sample set's normalized Nyquist frequency. Subsequently, Eq. (7.87) can be rearranged such that the outer summation in the second and third term on the right-hand side is associated with each term inside the braces, so that we write

$$x[n] = \sum_{m=0}^{\infty} a_{0+mN} + \sum_{k=1}^{N/2-1} \left\{ \left(\sum_{m=0}^{\infty} (a_{k+mN} + a_{N-k+mN}) \right) \cos\left[k\left(\frac{2\pi}{N}\right)n\right] \right.$$
$$\left. + \left(\sum_{m=0}^{\infty} b_{k+mN} - b_{N-k+mN} \right) \sin\left[k\left(\frac{2\pi}{N}\right)n\right] \right\} + \left(\sum_{m=0}^{\infty} a_{N/2+mN} \right) \cos\left[\left(\frac{N}{2}\right)\left(\frac{2\pi}{N}\right)n\right] \quad (7.88)$$

Finally, by defining

$$\tilde{a}_0 = \sum_{m=0}^{\infty} a_{0+mN} \quad \tilde{a}_k = \sum_{m=0}^{\infty}\left(a_{k+mN}+a_{N-k+mN}\right) \quad \tilde{a}_{N/2} = \sum_{m=0}^{\infty} a_{N/2+mN}$$

$$\tilde{b}_k = \sum_{m=0}^{\infty}\left(b_{k+mN}+b_{N-k+mN}\right)$$

(7.89)

Equation (7.88) can be written in a less complicated manner as

$$x[n] = \tilde{a}_0 + \sum_{k=1}^{N/2-1}\left\{\tilde{a}_k \cos\left[k\left(\frac{2\pi}{N}\right)n\right] + \tilde{b}_k \sin\left[k\left(\frac{2\pi}{N}\right)n\right]\right\} + \tilde{a}_{N/2}\cos[\pi n] \quad (7.90)$$

This representation is known as a *discrete-time Fourier series (DTFS)* representation of $x[n]$ written in trigonometric form.

Problems

7.1. Find the trigonometric Fourier series representation of the functions displayed in (a)–(d). Assume that the period in all cases is 1 ms.

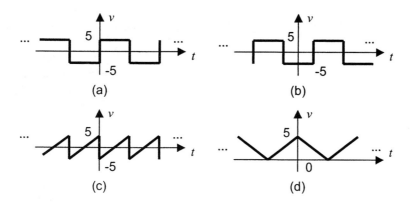

7.2. Find the trigonometric Fourier series representation of a sawtooth waveform $x(t)$ having a period of 2 s and whose behavior is described by: $x(t) = t$ if $-1 < t < 1$.

Using MATLAB, numerically compare your FS representation with $x(t)$.

7.3. Express the Fourier series in Problem 7.2 using cosine terms only.

7.4. Using the summation formulae in Eq. (7.12), determine the DTFS representation of the Fourier series representation of $x(t)$ given in Exercise 7.1. Use 10 points per period and limit the series to 100 terms. Also, express the result in magnitude and phase form.

7.5. A sampled signal consists of the following 5 samples: {0.1, 0.1, 1, -3, 4, 0}. Using linear algebra, determine the DTFS representation for these samples. Also, express the result in magnitude and phase form. Using MATLAB, numerically compare the samples from your DTFS representation with the actual samples given here.

7.6. A sampled signal consists of the following set of samples {0, -1, 2.4, 4, -0.125, 3.4}. Determine the DTFS representation for this sample set using the orthogonal basis method. Also, express the result in magnitude and phase form. Using MATLAB, numerically compare the samples from your DTFS representation with the actual samples given here.

7.7. Derive Eqs. (7.22) and (7.23) from first principles. Begin by multiplying the DTFS representation given in Eq.(7.11) by $\cos[k(2\pi/N)n]$, then sum n on both sides from 0 to N-1. Reduce the expression by using the set of trigonometric identities given in Eq. (7.21). Repeat using $\sin[k(2\pi/N)n]$.

7.8. Plot the magnitude and phase spectrum of the following FS representations of $x(t)$:

(a) $x(t) = \frac{2}{\pi}\left(\sin \pi t - \frac{1}{2}\sin 2\pi t + \frac{1}{3}\sin 3\pi t - \frac{1}{4}\sin 4\pi t + \cdots\right)$

(b) $x(t) = \frac{2}{\pi}\left[\cos\left(\pi t - \frac{\pi}{2}\right) + \frac{1}{2}\cos\left(2\pi t + \frac{\pi}{2}\right) \right.$
$\left. + \frac{1}{3}\cos\left(3\pi t - \frac{\pi}{2}\right) + \frac{1}{4}\cos\left(4\pi t + \frac{\pi}{2}\right) + \cdots\right]$

(c) $x(t) = -1.5 + \sum_{k=1,\,k\text{ odd}}^{\infty} \frac{1}{k\pi}\cos\left(k\,2\pi\times 10^3\, t + \pi/2\right)$

7.9. Plot the magnitude and phase spectrum of the following DTFS representations of $x[n]$:

(a) $x[n] = 0.6150\,\sin\left[\left(\frac{2\pi}{10}\right)n\right] - 0.2749\,\sin\left[2\left(\frac{2\pi}{10}\right)n\right]$
$+ 0.1451\,\sin\left[3\left(\frac{2\pi}{10}\right)n\right] - 0.0649\,\sin\left[4\left(\frac{2\pi}{10}\right)n\right]$

(b) $x[n] = 1 + \cos\left[\left(\frac{2\pi}{8}\right)n\right] + \sin\left[\left(\frac{2\pi}{8}\right)n\right] + \cos\left[2\left(\frac{2\pi}{8}\right)n\right] + \sin\left[2\left(\frac{2\pi}{8}\right)n\right]$
$- \cos\left[3\left(\frac{2\pi}{8}\right)n\right] - \sin\left[3\left(\frac{2\pi}{8}\right)n\right]$

7.10. A DTFS representation expressed in complex form is given by

$x[n] = 1 + 0.2e^{j\left[\left(\frac{2\pi}{8}\right)n - \frac{\pi}{4}\right]} + 0.3e^{j\left[3\left(\frac{2\pi}{8}\right)n + \frac{\pi}{3}\right]} + 0.3e^{j\left[5\left(\frac{2\pi}{8}\right)n - \frac{\pi}{3}\right]} + 0.2e^{j\left[7\left(\frac{2\pi}{8}\right)n + \frac{\pi}{4}\right]}$

Express $x[n]$ in trigonometric form. Plot the magnitude and phase spectra.

7.11. A DTFS representation expressed in complex form is given by

$$x[n] = 1 + (1.8 - j1.9)e^{j\left[2\left(\frac{2\pi}{8}\right)n\right]} + (0.75 + j0.25)e^{j\left[3\left(\frac{2\pi}{8}\right)n\right]}$$
$$+ (0.75 - j0.25)e^{j\left[5\left(\frac{2\pi}{8}\right)n\right]} + (1.8 + j1.9)e^{j\left[6\left(\frac{2\pi}{8}\right)n\right]}$$

Express $x[n]$ in trigonometric form and plot the corresponding magnitude and phase spectra.

7.12. Derive the polar form of Eq. (7.45) from (7.44).

7.13. Sketch the periodic extension for the following sequence of points:

(a) [0, 0.7071, 1.0, 0.7071, 0.0, -0.7071, -1.0000, -0.7071]

(b) [0, 0.7071, 1.0, 0.7071, 0.0, -0.7071, -1.0, -0.7071, 0.0, 0.7071]

7.14. Using the FFT algorithm in MATLAB, verify your answers to Problems 7.4, 7.5, and 7.6.

7.15. Given

$$x[n] = 0.25 + 0.5 \cos\left[\left(\frac{2\pi}{8}\right)n\right] + 0.1 \sin\left[\left(\frac{2\pi}{8}\right)n\right] + 2.1 \cos\left[2\left(\frac{2\pi}{8}\right)n\right]$$
$$- 0.9 \cos\left[3\left(\frac{2\pi}{8}\right)n\right] - 0.1 \sin\left[3\left(\frac{2\pi}{8}\right)n\right]$$

(a) Express $x[n]$ in complex form.

(b) Using MATLAB, write a script that samples $x[n]$ for $n = 0,1,\ldots,7$.

(c) Compute the FFT of the samples found in part (b) and write the corresponding DTFS representation in complex form. How does it compare with $x[n]$ found in part (a)?

7.16. Given

$$x[n] = (0.2 - j0.4)e^{j\left[\left(\frac{2\pi}{10}\right)n\right]} + (0.25 + j0.25)e^{j\left[3\left(\frac{2\pi}{10}\right)n\right]}$$
$$+ (0.25 - j0.25)e^{j\left[7\left(\frac{2\pi}{10}\right)n\right]} + (0.2 + j0.4)e^{j\left[9\left(\frac{2\pi}{10}\right)n\right]}$$

(a) Express $x[n]$ in trigonometric form.

(b) Using MATLAB, write a script that samples $x[n]$ for $n=0,1,\ldots,9$.

(c) Compute the FFT of the samples found in part (b) and write the corresponding DTFS representation in trigonometric form. How does it compare with $x[n]$ found in part (a)?

7.17. Evaluate $x[n] = n^3 - 2n^2 - 2n + 1$ for $n=0,1,\ldots,9$. Using MATLAB, compute the FFT of the 10 samples of $x[n]$ and determine the corresponding $\{c_k\}$ and $\{\phi_k\}$ spectral coefficients. Express the magnitude coefficients in RMS form.

7.18. Investigate the effects of increasing the observation window on the spectrum of a noncoherent sinusoidal signal. Consider generating a sinusoidal signal using parameters $A=1$, $\phi=0$, $M=9.9$, and $N=64$. Next, compare the magnitude spectrum of this signal when the following samples are collected: (a) 64 samples, (b) 512 samples, (c) 1024 samples, (d) 8192 samples. In all cases, estimate the amplitude of the sinusoidal signal.

7.19. An observation window consists of 128 points, plot the behavior of (a) rectangular window, (b) Blackman window, and (c) Kaiser window with $\beta=10$ all on the same graph.

7.20. Using the built-in window functions found in MATLAB, compute the window shape factors for the rectangular, Blackman, and Kaiser ($\beta=10$) windows.

7.21. Repeat Problem 7.18 but view the data first through a Blackman window. Estimate the amplitude of the sinusoidal signal.

7.22. Repeat Problem 7.18 but view the data first through a Kasier ($\beta=10$) window. Estimate the amplitude of the sinusoidal signal.

7.23. A signal has a period of 128 µs and is sampled at a rate of 1 MHz. If 128 samples are collected, what is the frequency resolution of the resulting FFT? If the number of samples collected increases to 8192 samples, what is the frequency resolution of the FFT?

7.24. Repeat Example 7.9 but this time use a Blackman window. By what factor does the accuracy of the calculation improve over the rectangular window?

7.25. Repeat Example 7.10 but this time use a Blackman window.

7.26. Using the FFT and IFFT routines found in MATLAB, together with the samples described in Exercise 7.12, verify that IFFT(FFT(x))=x.

7.27. Using the trigonometric identities described in Eq. (7.21), derive the trigonometric form of Parseval's theorem given in Eq. (7.64).

7.28. Using the trigonometric form of Parseval's theorem in Eq. (7.64) as a starting point, derive the corresponding complex form of the theorem given in Eq. (7.64).

7.29. The complex coefficients of a spectrum of a sampled signal are:

$\{X_k\}=\{0.5,\ 0.2\text{-}j0.4,\ 0,\ 0.25\text{+}j0.25,\ 0,\ 0.25\text{-}j0.25,\ 0,\ 0.2\text{+}j0.4\}$.

What is the RMS value of this signal?

7.30. The magnitude coefficients of a spectrum of a sampled signal are:

$\{c_k\}=\{0.1,\ 0.3,\ 0,\ 0.05,\ 0,\ 0.001\}$.

What is the RMS value of this signal?

7.31. Find the coherent sample set of $x[n]$ using MATLAB's IFFT routine assuming its spectrum is described by the following:

(a) $x[n] = 0.25 + 0.5\ \cos\left[\left(\frac{2\pi}{8}\right)n\right] + 0.1\ \sin\left[\left(\frac{2\pi}{8}\right)n\right] + 2.1\ \cos\left[2\left(\frac{2\pi}{8}\right)n\right]$

$-0.9\ \cos\left[3\left(\frac{2\pi}{8}\right)n\right] - 0.1\ \sin\left[3\left(\frac{2\pi}{8}\right)n\right]$

(b) $x[n] = 1 + 0.2e^{j\left[\left(\frac{2\pi}{8}\right)n - \frac{\pi}{4}\right]} + 0.3e^{j\left[3\left(\frac{2\pi}{8}\right)n + \frac{\pi}{3}\right]} + 0.3e^{j\left[5\left(\frac{2\pi}{8}\right)n - \frac{\pi}{3}\right]} + 0.2e^{j\left[7\left(\frac{2\pi}{8}\right)n + \frac{\pi}{4}\right]}$

(c) $$x[n] = (0.2 - j0.4)e^{j\left[\left(\frac{2\pi}{10}\right)n\right]} + (0.25 + j0.25)e^{j\left[3\left(\frac{2\pi}{10}\right)n\right]}$$
$$+ (0.25 - j0.25)e^{j\left[7\left(\frac{2\pi}{10}\right)n\right]} + (0.2 + j0.4)e^{j\left[9\left(\frac{2\pi}{10}\right)n\right]}$$

7.32. Verify the samples in Problem 7.31 by evaluating the function at each sampling instant.

7.33. A coherent signal is sampled with a frequency of 1 MHz over a 1024-μs time interval. If the spectrum of this signal is padded with 5120 zeros, and then converted back into the time domain, what is the effective time resolution of this signal?

7.34. A signal with a DTFS representation given by

$$x[n] = 0.1 + 2\cos\left[\left(\frac{2\pi}{16}\right)n - \frac{\pi}{5}\right] + 0.5\cos\left[5\left(\frac{2\pi}{16}\right)n + \frac{\pi}{5}\right]$$

passes through a system with the following transfer function

$$H(z) = \frac{0.0020 + 0.00402z^1 + 0.0020z^2}{1 - 1.5610z^1 + 0.64135z^2}$$

What is the DTFS representation of the output signal?

7.35. A signal with noise is described by the following DTFS representation,

$$x[n] = 0.25 + 2\cos\left[\left(\frac{2\pi}{8}\right)n\right] + 10^{-5}\cos\left[2\left(\frac{2\pi}{8}\right)n\right]$$
$$- 10^{-7}\cos\left[3\left(\frac{2\pi}{8}\right)n\right] - 10^{-5}\sin\left[3\left(\frac{2\pi}{8}\right)n\right] + 10^{-5}\cos\left[4\left(\frac{2\pi}{8}\right)n\right]$$

What is the RMS value of the noise signal that appears between bins 2 and 4?

References

1. Alan V. Oppenheim, Ronald W. Schafer, *Discrete-Time Signal Processing*, Prentice Hall, Englewood Cliffs, NJ, March 1989, ISBN: 013216292X

2. Alan V. Oppenheim et al., *Signals and Systems*, Prentice Hall, Englewood Cliffs, NJ, August 1997, ISBN: 0138147574

3. William McC. Siebert, *Circuits, Signals, and Systems*, The MIT Press, Cambridge, MA, September 1985, ISBN: 0262192292

4. Robert W. Ramirez, *The FFT Fundamentals and Concepts*, Prentice Hall, Englewood Cliffs, NJ 07632, January 1985, ISBN: 0133143864

5. John G. Proakis and Dimitris G. Manolakis, *Digital Signal Processing: Principles, Algorithms, and Applications* (Third Edition), Prentice Hall, Englewood Cliffs, NJ, 1996, ISBN: 0133737624

6. F. J. Harris., *On the Use of Windows for Harmonic Analysis with the Discrete Fourier Transform*, Proceedings of the IEEE, Vol. 66, No. 1, pp. 51-83, Jan. 1978.
7. Matthew Mahoney, *Tutorial DSP-Based Testing of Analog and Mixed-Signal Circuits*, The Computer Society of the IEEE, 1730 Massachusetts Avenue N.W., Washington D.C. 20036-1903, 1987, ISBN: 0818607858

CHAPTER 8

Analog Channel Testing

8.1 OVERVIEW

8.1.1 Types of Analog Channels

Analog channels include any nonsampled circuit with analog inputs and analog outputs. Examples of analog channels include continuous-time filters, amplifiers, analog buffers, programmable gain amplifiers (PGAs), single-ended to differential converters, differential to single-ended converters, and cascaded combinations of these circuits. Channels including ADCs, DACs, switched capacitor filters, and other sampling circuits will be discussed in Chapter 9, "Sampled Channel Testing." However, we will see in Chapter 14, "Design for Test (DfT)," that a sampled channel may be broken into subsections using DfT test modes, as illustrated in Figure 8.1.

DfT allows the filter and PGA in Figure 8.1 to be isolated from the rest of the sampled channel for more thorough testing. Since DfT allows portions of sampled channels to be reduced

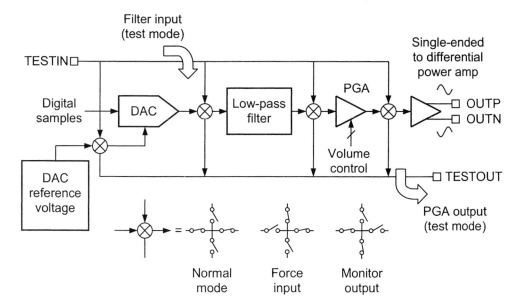

Figure 8.1. Analog bus DfT for a DAC mixed-signal channel.

to analog channel subsections, analog channel testing is extremely common in mixed-signal device testing.

It is critically important for the mixed-signal test engineer to gain a solid understanding of analog channel testing. The analog tests described in this chapter will represent at least half of many mixed-signal test programs. Although many analog channel tests can be measured without DSP-based techniques, this chapter will concentrate only on DSP-based methods of channel testing.

As previously explained in Chapter 7, DSP-based testing is the primary technique used in high-volume production testing of mixed-signal devices. A principal advantage of DSP-based testing is that many of the parameters described in this chapter can be measured simultaneously. Simultaneous measurements save test time and thereby reduce production costs.

8.1.2 Types of AC Parametric Tests

Analog and sampled channels share many AC parametric test specifications. Most of these specifications fall into a few general categories, including gain, phase, distortion, signal rejection, and noise. Each of these categories will be discussed in the sections that follow. We will examine the definition of each type of test, common test conditions, the causes and effects of parametric failures, common test techniques, and common measurement units (volts, decibels, etc.) for each test. Finally, an example of each test will be presented to clarify test definitions and techniques.

The test definitions in this chapter are all based on the assumption that the signals to be sourced or measured are voltages. Some circuits operate with currents rather than voltages. ATE testers are typically unable to measure or source AC currents directly. Some form of voltage-to-current or current-to-voltage circuit will be needed in cases where the DUT produces AC current outputs or requires AC current inputs.

In this chapter, we will assume that the tester's digitizer and AWG are perfect, having no gain errors at any frequency. This is a naive assumption at best, but one that will suffice for now. To make accurate AC measurements with general-purpose digitizers and arbitrary waveform generators (AWGs), proper use of software calibration is often required. Focused calibrations can be used to compensate for the various measurement errors inherent in the general purpose AWGs, digitizers, and other instruments in an ATE tester. In Chapter 10 "Focused Calibrations," we will examine the focused software calibration techniques for a variety of DC and AC measurements.

8.1.3 Review of Logarithmic Operations

Since many of the parameters in this chapter will be expressed in decibels, it is worth reviewing the basic characteristics of logarithms. Most logarithms in the test and measurement field are based on $\log_{10}$ calculations. The following is a quick review of logarithmic properties.

Conversion from voltage ratio to gain in decibels

$$G(\text{dB}) = 20 \, \log_{10} \left| \frac{V_1}{V_2} \right| = 20 \, \log_{10} \left| G(\text{V/V}) \right| \tag{8.1}$$

Conversion from gain in decibels to voltage ratio

$$G(\text{V/V}) = \left|\frac{V_1}{V_2}\right| = 10^{\frac{G(\text{dB})}{20}} \qquad (8.2)$$

Conversion from a ratio of squared voltages to gain in decibels, we use $10 \log_{10}(x^2) = 20 \log_{10}(x)$ to obtain

$$G(\text{dB}) = 10 \log_{10} \frac{|V_1|^2}{|V_2|^2} = 10 \log_{10} \frac{V_1^2}{V_2^2} \qquad (8.3)$$

Converting from power ratio to gain in decibels, we first make use of the fact that power is given by $P = V^2/R$, leading to

$$G(\text{dB}) = 10 \log_{10} \left(\frac{V_1^2/R}{V_2^2/R}\right) = 10 \log_{10} \frac{P_1}{P_2} \qquad (8.4)$$

Common voltage ratios and their dB equivalents (It is worth memorizing these ratios and their decibel equivalents, since they are used frequently in test engineering. Power ratios are half these numbers, i.e. a power ratio of 2 is equal to 3.01 dB.)

$1 = 0$ dB, $\sqrt{2} = 3.01$ dB, $2 = 6.02$ dB, $4 = 12.04$ dB, $8 = 18.06$ dB, etc.

$1/\sqrt{2} = -3.01$ dB, $1/2 = -6.02$ dB, $1/4 = -12.04$ dB, $1/8 = -18.06$ dB etc.

$10 = 20$ dB, $100 = 40$ dB, $1000 = 60$ dB, etc.

$1/10 = -20$ dB, $1/100 = -40$ dB, $1/1000 = -60$ dB, etc.

Multiplying ratios is equivalent to adding decibels:

2.0 x 4.0 = 8.0 resulting in 20×log(8.0) = 18.06 dB, or equivalently,

6.02 dB + 12.04 dB = 18.06 dB.

Dividing ratios is equivalent to subtracting decibels:

2.0/4.0 = 0.5 resulting in 20×log(0.5) = -6.02 dB, or equivalently,

6.02 dB − 12.04 dB = -6.02 dB.

8.2 GAIN AND LEVEL TESTS

8.2.1 Absolute Voltage Levels

Absolute voltage levels are perhaps the simplest AC parameters to understand, but they can be among the most difficult parameters to measure accurately using a general-purpose digitizer. Most electrical engineers are familiar with the use of bench equipment such as an oscilloscope or an AC voltmeter. The absolute voltage of a test tone is simply the RMS voltage of the signal

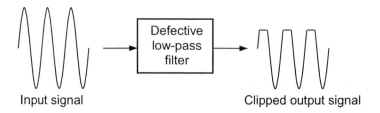

Figure 8.2. Absolute voltage level test detects gross circuit defects quickly.

under test, evaluated at the test tone's frequency. Energy at other frequencies is eliminated from the measurement. DSP-based measurement techniques allow noise, distortion, and other test tones to be easily eliminated from the RMS measurement. RMS voltmeters and oscilloscopes, by contrast, measure the total signal RMS, including noise, distortion, and other test tones. Spectrum analyzers offer a more frequency-selective voltage measurement capability, but they are not always as accurate as RMS voltmeters in measuring the absolute voltage level of a pure sinusoidal signal.

Absolute level specifications can be applied to any single-tone or multitone signal. The purpose of an absolute level test is to detect first order defects in a circuit, such as resistor or capacitor mismatch, DC reference voltage errors, and grotesque clipping or other distortion. For example, if the DC voltage reference for a DAC exhibits a 5% error, then the DAC's AC output amplitude will likely show a 5% absolute voltage level error as well. As a second example, consider a low-pass filter having a defective op amp (Figure 8.2). A clean sine wave input may become clipped very badly by the defective filter, resulting in an absolute voltage level at the filter output that is totally wrong. Obviously, it will also exhibit very high harmonic distortion as well. Absolute voltage level tests are a good way to find grossly defective circuits very quickly.

Loading conditions can be very important to many AC and DC parametric measurements, including absolute voltage level tests. If a buffer amplifier is designed to drive 32 Ω in parallel with 500 pF, then it makes no sense to test its absolute level in an unloaded condition. Generally, the test engineer must determine the worst-case loading conditions for a given output and test AC parameters using that loading condition. Device data sheets usually list a specific loading condition or a worst-case loading condition, which the output must drive during an absolute level test. The test engineer must design this load into the device interface board (DIB). In most cases, the load must be removable so that tests like continuity and leakage can also be performed on the DUT output. Electromechanical relays are often added to the DIB board to facilitate the removal of loads from DUT outputs.

Units of measure for absolute level tests vary somewhat, but they usually fall into a few common categories. Absolute levels may be specified in RMS volts, peak volts, peak-to-peak volts, dBV (decibels relative to 1.0 volt RMS), and dBm (decibels relative to 1.0 mW at a specified load impedance). When dealing with differential inputs or outputs, each of these measurement units can be defined from either a single-ended or differential perspective (Figure 8.3). It is critical for a test engineer to be able to communicate these units of measure without ambiguity. It is aggravating to ask an engineer to measure the voltage at the output of a differential circuit, only to get the reply "1.2 V." Does the answer refer to a single-ended or differential measurement? Is it peak, peak-to-peak, or RMS?

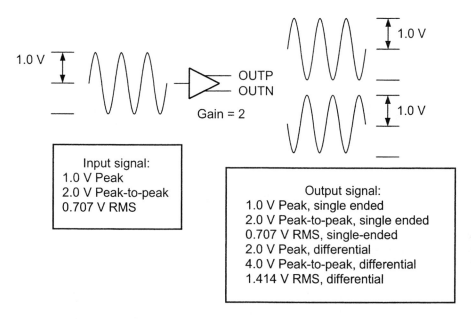

Figure 8.3. Equivalent voltage measurements for single-ended and differential signals.

When referring to a single-ended signal, it is acceptable to drop the single-ended / differential notation since there is no ambiguity. But when referring to a differential signal, it is imperative to specify the measurement type. Many times, correlation errors between the bench and ATE tester turn out to be simple misunderstandings about signal level definitions. Design errors can even be introduced if a customer specifies a differential signal level in the data sheet and the design engineer misinterprets it as a single-ended specification. Absolute voltage levels must be specified using a clear level definition, such as 1 V RMS, differential or 1 V peak, single-ended.

Decibel units can be abused as well. A signal level of +3.7 dB is meaningless, because it has no point of reference. What voltage level corresponds to 0 dB? The decibel unit represents a ratio of values, and as such it is inappropriate to refer to an absolute voltage level using decibels. Decibel units, when used to specify absolute levels, must include a definition of the 0 dB reference level. A common point of reference is 1 V RMS. When this definition of 0 dB is used, the measurement is expressed in dBV, or decibels relative to 1 V RMS. Differential reference levels are always used when measuring differential signals, and single-ended reference levels are always used when measuring single-ended signals. Therefore, dB units do not require a single-ended/differential notation. Nevertheless, we shall specify the measurement type explicitly in this text to avoid confusion.

Example 8.1

The positive side of a differential sine wave has a peak amplitude of 500 mV, as shown in Figure 8.4. Assuming the negative side is perfectly matched in amplitude and is 180 degrees out of phase, calculate the signal amplitude in dBV, differential.

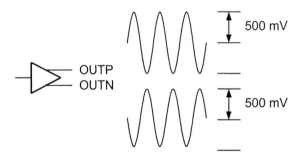

Figure 8.4. Differential sine wave at 500 mV peak, single ended.

Solution:

Since we need to compare this signal to 1.0 V RMS, differential, we start by converting the single-ended peak signal into differential RMS units. We rely on the fact that a sine wave always has a peak-to-RMS ratio of 1.414 (square root of two)

$$500 \text{ mV peak, single-ended} = 1.0 \text{ V peak, differential}$$
$$= 0.707 \text{ V RMS, differential}$$

Next, we convert RMS volts to dBV using the equation

$$\text{signal level (dBV)} = 20 \, \log_{10} \left| \frac{\text{RMS signal level}}{1.0 \text{ V RMS}} \right|$$

resulting in

$$\text{signal level (dBV)} = 20 \, \log_{10} \left| \frac{0.707 \text{ V RMS}}{1.0 \text{ V RMS}} \right| = -3.01 \text{ dBV}$$

Another commonly used absolute voltage unit is the dBm (decibels relative to 1.0 mW). 0 dBm is the voltage corresponding to 1 mW of power dissipation at a particular load impedance. A specified load impedance must always be linked to the dBm unit, since different load impedances will dissipate different amounts of power at a given voltage level. Sometimes the load impedance is specified in a note in the data sheet, but other times it is tied directly to the dBm unit (i.e., –30 dBm at 50 Ω).

Example 8.2

Convert a 250-mV single-ended RMS measurement into dBm units at 600 Ω.

Solution:

It is first necessary to convert the voltage level to a power level using the equation

$$power = \frac{V^2}{R}$$

Then the signal power is compared to the 1-mW reference using the equation

$$10 \log_{10}\left(\frac{power}{1 \text{ mW}}\right)$$

The total equation is therefore

$$\text{signal level (dBm)} = 10 \log_{10}\left(\frac{V^2/R}{1 \text{ mW}}\right)$$

Converting the 250 mV signal into dBm at 600 Ω

$$\text{signal level (dBm)} = 10 \log_{10}\left(\frac{(250 \text{ mV})^2/600}{1 \text{ mW}}\right) = -9.823 \text{ dBm at 600 } \Omega$$

Exercises

8.1. Convert a 1-V peak, single-ended signal into dBV units.

Ans. −3.01 dBV.

8.2. Convert a 1-V peak-to-peak, differential signal into dBV units.

Ans. −15.05 dBV.

8.3. Convert a 1-V RMS, differential signal into dBm units at 50 Ω.

Ans. +6.99 dBm

8.4. An FFT analysis reveals that a particular tone has a spectral coefficient of -0.2866-j0.133643 V. What is the amplitude of this tone? What is its RMS value? Express this value in dBV units.

Ans. 0.3162 V; 0.2236 V; -13.01 dBV.

On a mixed-signal ATE tester, absolute voltage levels are usually measured using a general-purpose digitizer in conjunction with Fourier analysis (i.e., DSP-based testing). Digitizers are often capable of measuring either single-ended signals or differential signals. To measure differential signals, the digitizer uses an instrumentation amplifier at its front end. The instrumentation amplifier converts a differential input signal into a single-ended signal before it is measured. If a digitizer lacks the capability to measure differential signals, then the test engineer must capture each side of the signal separately, using either two digitizers or using the same digitizer twice. The two signals must then be combined mathematically by subtracting the negative signal from the positive signal.

8.2.2 Absolute Gain and Gain Error

In analog channels, absolute gain is simply a ratio of output AC signal level divided by input signal level at a specified frequency

$$G = \frac{V_{out}}{V_{in}} \tag{8.5}$$

Absolute gain is frequently converted from V/V units to decibel units using the formula

$$G(\mathrm{dB}) = 20 \ \log_{10} |G(\mathrm{V/V})| \tag{8.6}$$

Often a channel's gain is specified using a minimum and maximum absolute gain. Sometimes, though, a channel's gain is specified in terms of its error relative to the ideal absolute gain. This parameter is called *gain error*.

Gain error ΔG is defined as the actual (measured) gain of a channel, in volts/volt, divided by its ideal (expected) gain. When working with decibels, gain error is defined as the actual gain in dB minus the ideal gain G_{IDEAL} in dB (since subtracting logarithms is equivalent to dividing ratios).

$$\Delta G(\mathrm{dB}) = G(\mathrm{dB}) - G_{IDEAL}(\mathrm{dB}) \tag{8.7}$$

For example, a channel may have an absolute gain of 12.35 dB (4.145 V/V), but its ideal gain should be 12.04 dB (4.0 V/V). Its gain error is therefore 12.35 dB - 12.04 dB = 0.31 dB (or equivalently, 4.145 V/V / 4.0 V/V = 1.036 V/V). While absolute gain and gain error can be specified using either V/V or decibels, gain error is usually specified in decibels.

Gain errors are frequently the result of component mismatch in the DUT. For example, mismatched resistors in an op amp gain circuit lead to gain errors. Excessive gain errors can lead to a number of system-level problems. In audio circuits, gain error can result in volumes that are too loud or too soft. Extreme gain errors can also lead to clipped (distorted) analog signals. In data transmission channels, the distortion caused by gain errors can lead to corrupted data bits. Absolute gain and gain error tests are well suited to the detection of gross functionality errors like dead transistors or incomplete signal paths.

Gain tests are commonly performed at a signal level below the maximum allowed signal level in a channel. The reason a full-scale signal is not used is that it might cause clipping (harmonic distortion). Distortion can be introduced if the gain error or offset of the signal causes the circuit under test to clip either the top or bottom of the test signal. Distortion in turn causes an error in the absolute gain reading, since energy from the test signal is transferred into distortion components. Since we want to be able to distinguish between gain errors and distortion errors, gain measurements are often performed a few decibels below full scale (typically 1 to 3 decibels below full scale). A separate distortion test can be performed at full scale to determine the extent of clipping near the full-scale signal range.

Gain and gain error tests are often tested with a single test tone, rather than a multitone signal. In audio circuits, a 1-kHz test tone is very commonly specified in the data sheet, since the average human ear is maximally sensitive near 1 kHz.

Example 8.3

A tester's AWG sources a sine wave to a low-pass filter with a single-ended input and differential output. A digitizer sampling at 8 kHz captures 256 samples of the sine wave at the input to the filter (Figure 8.5). The FFT of the captured waveform shows a signal amplitude of 1.25 V RMS at the thirty seventh FFT spectral bin. The digitizer is then connected to the output of the filter using electromechanical relays (Figure 8.6). The digitizer captures 1024 samples of the output of the filter (differentially) using a 16-kHz sampling rate. The output FFT shows a signal amplitude of 1.025 V RMS in one of the spectral bins. What is the frequency of the test signal? Which spectral bin in the FFT of the output signal most likely showed the 1.025 V RMS signal level? Assuming the digitizer is perfectly accurate in both sampling configurations, what is the gain (in decibels) of the low-pass filter at this frequency? The ideal filter gain at this frequency is –1.50 dB. What is the gain error of the filter at this frequency? Is the filter output too high or too low?

Solution:

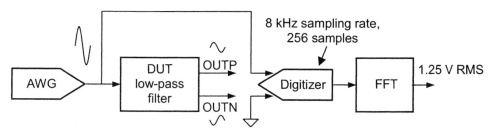

Figure 8.5. Sampling system configuration during LPF input measurement.

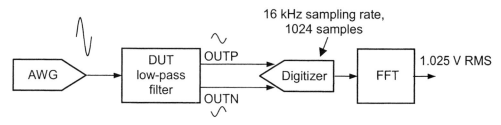

Figure 8.6. Sampling system configuration during LPF output measurement.

First we have to understand our sampling systems. When sampling the input signal, the digitizer samples at 8 kHz and captures 256 samples. This results in a fundamental frequency of 8 kHz / 256 = 31.25 Hz. The test frequency is therefore 37×31.25 Hz = 1156.25 Hz. Since the output of the filter must occur at the same frequency as the input, the output signal should also appear at 1156.25 Hz. When sampling the output signal, the digitizer samples at 16 kHz and captures 1024 samples. This sampling system results in a fundamental frequency of 16 kHz / 1024 = 15.625 Hz. The FFT spectral bin containing this signal energy must therefore be located at 1156.25 Hz/15.625 Hz = spectral bin 74. Any signal energy falling in any other FFT spectral bin is therefore either noise or distortion.

The gain is calculated using a logarithmic calculation as follows

$$G(\text{dB}) = 20 \, \log_{10} \left| \frac{V_{out}}{V_{in}} \right| = 20 \, \log_{10} \left| \frac{1.025 \text{ V}}{1.25 \text{ V}} \right| = -1.724 \text{ dB}$$

The gain error is therefore equal to -1.724 dB – (-1.50 dB) = -0.224 dB. This means that the filter output is lower than it should be.

8.2.3 Gain Tracking Error

Gain tracking, or gain tracking error, is defined as the variation in the gain G (expressed in dB) of a channel with respect to a reference gain G_{REF} (also expressed in dB) as the signal level V_{in} changes.

$$\Delta G(\text{dB}) = G(\text{dB}) - G_{REF}(\text{dB}) \quad \text{vs.} \quad V_{in} \tag{8.8}$$

Ideally, a channel should have a constant gain, regardless of the signal level (unless of course the signal is high enough to cause clipping). A perfectly linear analog channel has no gain tracking error. But small amounts of nonlinearity and other subtle circuit defects can lead to slight differences in gain at different signal levels. Gain tracking error is also introduced by the quantization errors in a DAC or ADC channel. As the signal level in a DAC or ADC quantized channel falls, the quantization errors become a larger percentage of the signal. Thus gain tracking errors in a quantized channel are most severe at low signal levels.

Gain tracking is calculated by measuring the gain at a reference level, usually the 0-dB level of the channel, and then measuring the gain at other signal levels (+3 dB, -6 dB, -12 dB, etc.). Gain tracking error at each level is calculated by subtracting the reference gain G_{REF} (dB) from the measured gain G (dB) corresponding to that level.

Gain tracking is often measured in 6-dB steps (factors of 2) for characterization. The number of steps is usually reduced to three or four levels after characterization of the device identifies which levels are most problematic. The reduction of levels saves considerable test time. Gain tracking error is almost always specified in decibels, although V/V would also be an acceptable unit of measure.

Example 8.4

The data sheet for a single-ended analog voltage follower defines a 0 dB reference level of 500 mV at the voltage follower input. The gain tracking specification for this device calls for a gain tracking error of ±0.25 dB at a +3.0 dB signal level and ±0.05 dB gain tracking error from 0.0 dB to –54.0 dB signal level. In this data sheet, gain tracking error is referenced to the gain at the 0-dB signal level. The gain error of the voltage follower is measured at each of the signal levels in Table 8.1, resulting in a series of absolute gain values. Calculate the gain tracking errors at each level and determine whether or not this device passes the gain tracking test. Calculate the input signal level at –54 dB.

Table 8.1. Absolute Gains for Gain Tracking Measurement

Signal Level	Absolute Gain	Signal Level	Absolute Gain
0 dB (500 mV)	0.02 dB (reference gain)		
+3 dB	-0.18 dB (slight clipping causes test tone attenuation)	-30 dB	-0.02 dB
-6 dB	0.02 dB	-36 dB	-0.03 dB
-12 dB	0.01 dB	-42 dB	-0.04 dB
-18 dB	0.00 dB	-48 dB	-0.05 dB
-24 dB	-0.01 dB	-54 dB	-0.07 dB

Solution:

We convert each absolute gain measurement into a gain tracking error measurement using the absolute gain at 0 dB as the reference level. To convert each absolute gain into gain tracking error, we subtract the reference gain (0.02 dB) from the absolute gain at each level. The gain tracking errors at each level are listed in Table 8.2.

Table 8.2. Gain Tracking Errors for Voltage Follower

Signal Level	Gain Error ΔG	Signal Level	Gain Error ΔG
0 dB (500 mV)	0.00 dB (by definition)		
+3 dB	-0.20 dB	-30 dB	-0.04 dB
-6 dB	0.00 dB	-36 dB	-0.05 dB
-12 dB	-0.01 dB	-42 dB	-0.06 dB
-18 dB	-0.02 dB	-48 dB	-0.07 dB
-24 dB	-0.03 dB	-54 dB	-0.09 dB

This voltage follower fails the specification limits of ±0.05 dB at all signal levels lower than −30 dB.

The signal level at −54 dB is calculated using the formula

$$-54 \text{ dB signal level} = 0 \text{ dB level} \times 10^{-54 \text{ dB}/20 \text{ dB}} = 500 \text{ mV RMS} \times 10^{-54 \text{ dB}/20}$$
$$= 0.998 \text{ mV RMS}$$

Exercises

8.5. The gain of an amplifier is measured with a 1-kHz tone test. An FFT analysis reveals that the input and output signals have spectral coefficients of -0.2866-j0.133643 and 0.313150-j0.044010 V, respectively. What is the gain of this amplifier?

Ans. 1 V/V.

8.6. The gain of an analog channel is assumed to be described by the following equation

$$G = 0.9 + 0.1\, V_{in} - 0.01\, V_{in}^2$$

What is the gain error of the channel at an input level of 3.0 V RMS if the ideal gain is 1 V/V? What is the gain error when the input is increased to 5.0 V RMS?

Ans. 1.11 V/V; 1.15 V/V.

8.7. Using the gain expression in Exercise 8.6, determine the gain tracking errors at input levels of 3 dB, 0 dB, -3 dB, -6 dB, and −12 dB, when the 0-dB reference level corresponds to a 100-mV RMS input level?

Ans. 0.0383 dB, 0 dB, -0.0274 dB, -0.0470 dB, -0.0709 dB.

8.2.4 PGA Gain Tests

A programmable gain amplifier (PGA) can be set to multiple gain settings using a digital control signal. PGAs are commonly used as volume control circuits in cellular telephones, televisions, etc. The absolute gain at each step in a PGA's gain curve is often less important than the difference in gain between adjacent steps. PGAs are often specified with an absolute gain at the first setting, a total gain difference between the highest and lowest setting, and the gain step size from each gain setting to the next.

For example, consider a 32-step PGA that has an ideal step size of 1.5 dB from each gain step to the next (Figure 8.7). If its lowest gain setting is 0 dB, then it should ideally be programmable to any of 32 different gains: 0 dB, 1.5 dB, 3.0 dB, 4.5 dB, ..., 46.5 dB. The gain range is defined as the highest gain (dB) minus the lowest gain (dB). Ideally, the gain range in this example should be 46.5 dB. PGA gain specifications, like other gain specifications, are usually tested at

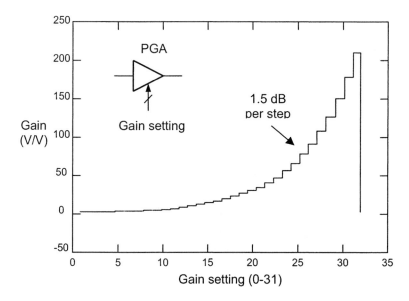

Figure 8.7. Programmable gain amplifier gain curve.

a particular frequency, such as 1 kHz. The absolute gain, gain step size, and gain range can all be measured by setting the PGA into each of its 32 settings and measuring the output voltage level divided by the input voltage level.

The absolute gain of each step can be measured by leaving the input signal unchanged and observing the change in output voltage. This eliminates the need for a focused calibration process, since the change in output level is equal to the gain step size regardless of any gain errors in the digitizer or AWG. Although it is possible to measure gain steps in this manner, it is probably best to adjust the input level at each step to produce a fairly constant output level at least 3 dB lower than the full-scale output level. This second technique avoids clipping while producing a strong output signal level at all gain settings. Remember that strong signals are less susceptible to noise, yielding better repeatability in the measurement.

It is worth noting that the 32 gain measurements could potentially be reduced to only 6 measurements, assuming the PGA is well designed. If the PGA is designed with 32 individual resistors in the op amp gain circuit, then each step must be measured individually (since any one resistor may be defective). However, if the PGA's gain is controlled by a sum of five binary-weighted resistors, each controlled by one of the PGA's control bits, then only the op amp and the five resistor paths need to be tested. This corresponds to the gains at 00000, 00001, 00010, 00100, 01000, and 10000.

The complete gain curve can be calculated by adding the gains in a binary-weighted fashion. This partial testing approach requires that superposition is proven to be a valid assumption through characterization of the PGA. For example, assume superposition is shown to be valid in a 3-bit PGA (8 gain stages) with 1.5-dB gain steps. Measuring the gain at each major transition yields the gain measurements in Table 8.3.

Table 8.3. Measured PGA Gains

Gain Setting	Measured Gain
0-dB setting (000)	0.01 dB
1.5-dB setting (001)	1.48 dB
3.0-dB setting (010)	3.25 dB
6.0-dB setting (100)	6.01 dB

The remaining gains can be calculated using superposition of the measured gain steps (Table 8.4).

Table 8.4. Calculated PGA Gains

Gain Setting	Calculated Gain
4.5-dB setting (011)	Gain = 0.01 dB + 1.48 dB + 3.25 dB = 4.74 dB
7.5-dB setting (101)	Gain = 0.01 dB + 1.48 dB + 6.01 dB = 7.5 dB
9.0-dB setting (110)	Gain = 0.01 dB + 3.25 dB + 6.01 dB = 9.27 dB
10.5-dB setting (111)	Gain = 0.01 dB + 1.48 dB + 3.25 dB + 6.01 dB = 10.75 dB

The binary-weighted PGA is an example of a subtle form of design for test (DfT). By choosing a design architecture that is based on a weighted structure with excellent superposition characteristics, the IC design engineer can reduce the test requirements from eight measurements to four. The remaining four measurements can be calculated in a fraction of the time it would take to measure them explicitly. The test engineer should get involved early in the design process to suggest such architectural features, or at least make the test impact of such design choices known to the design engineers.

Example 8.5

A single-ended 3-bit PGA is stimulated with a constant 100-mV RMS sine wave input at 1 kHz. The output response at each of the 8 gain settings is listed in Table 8.5. Calculate the absolute gain at each gain setting. Calculate the gain step size at each transition. Calculate the gain range. Is superposition a valid assumption for this PGA? Is it safe to modify the test program to measure only four gains and calculate the other four using a binary-weighted mathematical approach?

Solution:

Table 8.6 lists the absolute gain at each PGA step, calculated using the formula

$$G(\text{dB}) = 20 \ \log_{10} \left| \frac{V_{out}}{V_{in}} \right| = 20 \ \log_{10} \left| \frac{V_{out}}{100 \ \text{mV RMS}} \right|$$

Table 8.5. PGA Output Voltages with 100-mV Input

Gain Setting	Output Voltage	Gain Setting	Output Voltage
0 dB	100.00 mV RMS	6.0 dB	200.45 mV RMS
1.5 dB	119.26 mV RMS	7.5 dB	239.06 mV RMS
3.0 dB	140.44 mV RMS	9.0 dB	280.87 mV RMS
4.5dB	167.50 mV RMS	10.5dB	334.96 mV RMS

Table 8.6. PGA Measured Gains

Gain Setting	Measured Gain	Gain Setting	Measured Gain
0 dB	0.0 dB	6.0 dB	6.04 dB
1.5 dB	1.53 dB	7.5 dB	7.57 dB
3.0 dB	2.95 dB	9.0 dB	8.97 dB
4.5 dB	4.48 dB	10.5 dB	10.50 dB

Table 8.7 lists the gain step sizes, calculated by taking the difference in gain between adjacent steps.

Table 8.7. PGA Gain Step Sizes

PGA Transition	Gain Step Size
0 dB to 1.5 dB	1.53 dB − 0.00 dB = 1.53 dB
1.5 dB to 3.0 dB	2.95 dB − 1.53 dB = 1.42 dB
3.0 dB to 4.5 dB	4.48 dB − 2.95 dB = 1.53 dB
4.5 dB to 6.0 dB	6.04 dB − 4.48 dB = 1.56 dB
6.0 dB to 7.5 dB	7.57 dB − 6.04 dB = 1.53 dB
7.5 dB to 9.0 dB	8.97 dB − 7.57 dB = 1.44 dB
9.0 dB to 10.5 dB	10.50 dB − 8.97 dB = 1.53 dB

Gain range is calculated by subtracting the 0-dB gain measurement from the 10.5-dB gain measurement.

$$\text{gain range} = 10.52 \text{ dB} - 0.00 \text{ dB} = 10.52 \text{ dB}$$

If we calculate the gains at 4.5, 7.5, 9.0, and 10.5 dB using superposition instead of actual measurements, as we did in Table 8.4, we get the results in Table 8.8.

Table 8.8. Comparison of Measured Gains with Calculated Gains.

Gain Setting	Actual Gain	Calculated Gain using Superposition	Error
4.5 dB	4.48 dB	0.00 dB + 1.53 dB + 2.95 dB = 4.48 dB	0.00 dB
7.5 dB	7.57 dB	0.00 dB + 1.53 dB + 6.04 dB = 7.57 dB	0.00 dB
9.0 dB	8.97 dB	0.00 dB + 2.95 dB + 6.04 dB = 8.99 dB	0.02 dB
10.5 dB	10.50 dB	0.00 dB + 1.53 dB + 2.95 dB + 6.04 dB = 10.52 dB	0.02 dB

The question of whether or not superposition holds and whether or not we can change to a superposition calculation instead of a full measurement process is a trick question for two reasons. First, we have not specified the test limits. Do we have test limits that are tight or loose relative to the behavior of the typical DUT? If the limits are loose, we can probably tolerate the 0.02-dB errors in the superposition calculations. We can account for the errors by tightening our guardbands (tightening the normal test limits by 0.02 dB, for example). If the average device performance is very close to the test limits, however, then we cannot tolerate as much error and the superposition technique may not be acceptable.

Another reason this is a trick question is that we are looking at the results from a *single* DUT. A good engineer should *never* draw broad conclusions about device characteristics from a single DUT or even a limited sample of DUTs. This is one of the most common mistakes a novice test engineer makes. We would have to see superposition hold over at least three production lots (thousands of devices) before we would have confidence that we could change the test program to the faster superposition methodology. Also, we would need to confirm with the design engineer that the PGA is designed in such a way that superposition should be a valid assumption. This gives us confidence that we are making a sensible decision about the expected characteristics of the DUT.

Exercises

8.8. An analog channel with a differential input and differential output has an ideal gain of 6.02 dB. What is the gain error of the channel if a 202.43-mV RMS signal appears at the output when a 100-mV RMS sine wave is applied to its input?

Ans. +0.1055 dB.

8.9. An analog channel is excited by a 100-mV single-ended sinusoidal signal. A digitizer captures 512 samples using a 16-kHz sampling rate. An FFT analysis reveals a single peak value of 0.043 V RMS in the one hundred eleventh spectral bin. What is the frequency of the test signal? What is the absolute gain of the channel in V/V?

Ans. 3.46875 kHz; 0.43 V/V.

8.2.5 Frequency Response

Frequency response is similar to gain tracking in the sense that it is a measurement of gain under varying signal conditions, relative to a reference gain. Frequency response is most commonly used to measure the transfer function of a filter. Sometimes a frequency response test is used to measure the bandwidth of a circuit such as an op amp gain circuit to verify that its gain/bandwidth product is acceptable.

If a filter's transfer function is not within specifications, the system-level consequences depend on the filter's purpose. For example, a low-pass antialias filter removes high-frequency components from a digital audio recorder's ADC input. If the antialias filter transfer characteristics are not correct, the result can be unpleasant alias tones in the audio signal once it is reconstructed with a digital audio playback DAC.

Unlike gain tracking tests, in which the signal amplitude is varied, a frequency response test measures the variation in the gain of a circuit as the signal frequency is varied. One signal frequency is chosen as the reference frequency and the gain of the circuit at that frequency is the reference gain. All other gains are measured relative to the reference gain

$$\Delta G(\text{dB}) = G(\text{dB}) - G_{REF}(\text{dB}) \quad \text{vs. frequency} \tag{8.9}$$

For this reason, the gains computed according to Eq. (8.9) are called *relative gains*. Sometimes, the reference frequency is 0 Hz, meaning that all gains are measured relative to the circuit's DC gain. When DC is used as the reference frequency, it is often possible to measure the gain at a very low frequency (say, 100 Hz) rather than making a separate DC gain test as described in Chapter 3. This approach allows a single-pass DSP-based test, which saves test time. Sometimes the reference gain is defined as the midpoint between the highest and lowest gain in the frequency response curve. For example, if a filter's absolute gain varies from +0.25 dB to −0.31 dB across its in-band frequency range, then the reference gain G_{REF} is the average of these maximum and minimum gains

$$G_{REF} = \frac{(0.25 - 0.31)}{2} = -0.03 \text{ dB}$$

Frequency response is usually measured using a coherent multitone signal so that all signal frequencies can be measured simultaneously, saving test time. Sometimes the test must be broken into two parts, an in-band test and an out-of-band test. The reason it must sometimes be split is that the out-of-band components at the output of a filter may be extremely low in amplitude compared to the in-band components.

The out-of-band components must sometimes be amplified by the front-end circuitry of the ATE digitizer or by a local gain circuit on the DIB board before they can be measured accurately. This amplification technique is especially useful if the nonamplified out-of-band components would otherwise fall below the digitizer's quantization noise floor. Applying the same amplification to the in-band components might result in clipped signals. If so, the in-band and out-of-band components must be measured separately, at two different amplification settings. Ideally, though, the in-band and out-of-band components should be measured simultaneously to save test time.

Frequency response is usually measured with equal-level multitone signals in which the RMS amplitudes of the test tones are set equal to one another. To achieve a desired signal level, the amplitude of each tone is set to the desired total signal RMS amplitude divided by the square root of the number of test tones. For example, to produce a four-tone multitone signal with 100-mV RMS signal level, each tone must be set to $100 \text{ mV RMS}/\sqrt{4}$, or 50 mV RMS.

The frequencies of the tones should be chosen so that they do not produce harmonic or intermodulation distortion overlaps. The phase of each tone is randomly selected to produce a signal with an acceptable peak-to-RMS ratio. (For a review of these and other considerations in making a multitone measurement, refer to Chapter 6, "Sampling Theory" and Chapter 7, "DSP-Based Testing.")

As in all AC measurements, the DUT and tester must be allowed to settle to a stable state before valid data can be collected. Therefore, the tester's AWG must begin sending the input signal to the DUT for several milliseconds before the digitizer begins collecting samples. This precollection time is referred to as *settling time*. The settling time of an AC measurement is related to the signal frequency of interest and the filter characteristics of the DUT, AWG, and digitizer. In general, the lower the frequency being tested, the longer it takes to settle.

For example, a 10-Hz high-pass filter is difficult to test in production because it takes many tens or hundreds of milliseconds to settle to a steady state. Also, the higher the order of the filter, the longer it takes to settle. A band-pass filter with a quality factor (Q) of 10 takes much longer to settle than one with a Q of 1. For this reason, it is a good idea to get the DUT and AWG settling process started as early as possible in each test to reduce the test program's settling time overhead.

Once the filter has settled and its output has been digitized, frequency response is simple to calculate. Frequency response is calculated by first performing Fourier analysis (i.e., a DFT or FFT) on the waveforms collected at the DUT input and output. The gain at each frequency is then calculated by dividing the FFT's response to the DUT output signal by the DUT's input signal. The FFT allows a separate gain calculation at each frequency of interest. The reference gain (in dB) is then subtracted from each of the other gains to normalize them to the gain at the reference frequency. The absolute gain of the filter at the reference frequency is usually tested as a separate specification to guarantee that the filter's overall absolute gain is within specifications.

Example 8.6

A band-pass filter is formed by cascading a 60-Hz high-pass filter with a 3.4-kHz low-pass filter. The filter's frequency response specification is shown in Table 8.9. Specifications at intermediate points are determined by linear interpolation between the specified points, on a log/log scale. These upper and lower gain limits form the filter's gain mask (Figure 8.8).

The data sheet defines 1 kHz as the reference frequency. The data sheet also specifies that the -3.01-dB gain points must occur between 170 and 190 Hz (high-pass cutoff) and between 3550 and 3650 Hz (low-pass cutoff). Measure the frequency response of the filter, using a multitone signal. Calculate the signal level of each input tone that will result in a combined signal level of 1.0 V RMS.

Table 8.9. Filter Gain Mask Inflection Points

Freq.	Lower Limit	Upper Limit	Freq.	Lower Limit	Upper Limit
<10 Hz	None	-30 dB	300 Hz	-0.5 dB	+0.5 dB
50 Hz	None	-25 dB	3000 Hz	-0.5 dB	+0.5 dB
60 Hz	None	-23 dB	3400 Hz	-1.35 dB	0.0 dB
200 Hz	-1.28 dB	0.0 dB	4000 Hz	None	-14.0 dB
			>4600 Hz	None	-32.0 dB

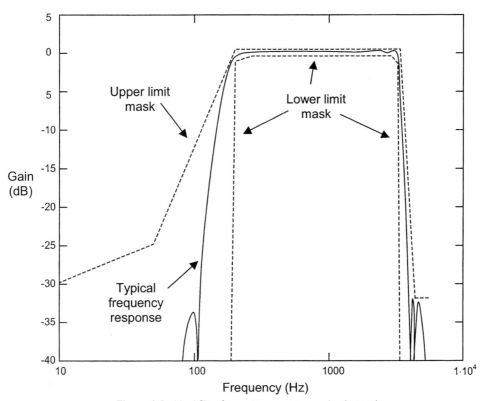

Figure 8.8. Ideal filter frequency response and gain mask.

Solution:

This specification points out a number of problems with frequency response testing. First, the in-band response has to be guaranteed over a range of frequencies, yet the data sheet does not specify which subset of frequencies should be measured in production. Obviously, we could sweep the frequency from 0 to 8 kHz in 1-Hz steps, but that would be unacceptably time-consuming for production testing. A limited number of tones must be chosen. Second, the 3-dB cutoff frequencies are specified. We do not have time to search for the exact frequency that results in a relative gain of -3.01 dB; so we have to find a compromise.

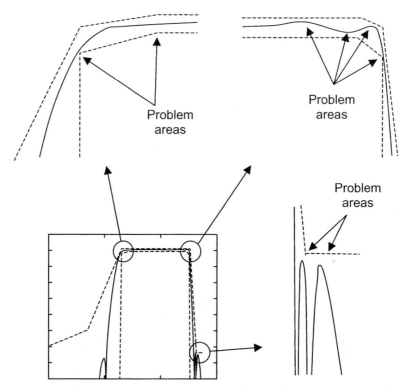

Figure 8.9. Problem areas in the filter frequency response.

The solution to the first problem is to determine the frequencies that are most likely to cause the filter to fail the upper and lower frequency-response limits of the filter through a detailed Monte Carlo or sensitivity analysis. This is usually a procedure that the design engineer has performed prior to releasing the design into production. We can see the potential problem areas by looking at a magnified view of the ideal frequency response plotted against the gain mask (Figure 8.9). We begin by choosing the problematic frequencies in the passband, corresponding to the peaks and valleys of the ideal frequency-response curve. In this example, the peaks and valleys of the in-band ripple are expected at 850, 1600, 2310, 2860, and 3150 Hz. Next we select frequencies at the center of the out-of-band side lobes, located at 100, 4430, and 4860 Hz. The gains at these frequencies are the most likely to cross the upper gain mask.

One solution to the −3-dB frequency problem is to place two frequencies at the upper and lower limits of the −3-dB points (170, 190, 3460, and 3560 Hz). We can then interpolate between the gains at these frequencies (using a log/log interpolation) to find the approximate location of the −3-dB frequency. (Log/log interpolations are explained in Section 8.7, "Weighting Filters.") Alternatively, we can simply require that one gain must be greater than −3.01 dB and the other must be less than −3.01 dB. The interpolation technique is more accurate, but the pass/fail technique is a bit faster.

Next we choose frequencies at the inflection points of the upper and lower masks. These correspond to the frequencies in Table 8.9. Since a 10-Hz tone would require a minimum of

100 ms of data collection time to capture just one cycle, and since the ideal characteristics are so far from the specified mask, we will eliminate the 10- and 50-Hz tests from our frequency list. This is another example where a robust design allows us to avoid costly testing. Of course, we should be prepared to verify that the average device does have this characteristic, but we can perform that characterization in a separate test and later eliminate it from the production test list.

We now have the following list of frequencies corresponding to the in-band and out-of-band regions of the filter: 60, 100, 170, 190, 200, 300, 850, 1000 (reference frequency), 1600, 2310, 2860, 3000, 3150, 3400, 3460, 3560, 4000, 4430, 4600, and 4860 Hz. Clearly the choice of a limited set of frequencies is a complicated one. The test engineer needs to work with the design team to predict which tones are most likely to cause filter response failures. These "problem areas" are the logical choice for production testing.

Of course, we have to adjust the frequencies of the actual test tones slightly to accommodate coherent DSP-based testing, as explained in Chapters 6 and 7. When correlating the ATE tester results to measurements made with bench equipment, the test engineer must communicate the exact frequencies used on the tester. Otherwise, differences in test conditions may introduce correlation errors between the tester and bench.

We can either test the filter with this 20-tone multitone signal all at once or we can split it into two tests. The decision is based on the repeatability we can achieve with a single-pass test. For this example a single-pass test is probably acceptable, since the lowest out-of-band gain specification is –32 dB. If the lowest gain were instead –80 dB, then we might worry about the measurement of such a small signal in the presence of the much larger in-band signals.

To calculate the desired signal level of each tone, we divide the total signal RMS by the square root of the number of tones. Each tone should therefore be set to 1.0 V RMS$/\sqrt{20}$ or 223.6 mV RMS to achieve a total signal amplitude of 1.0 V RMS at the filter's input.

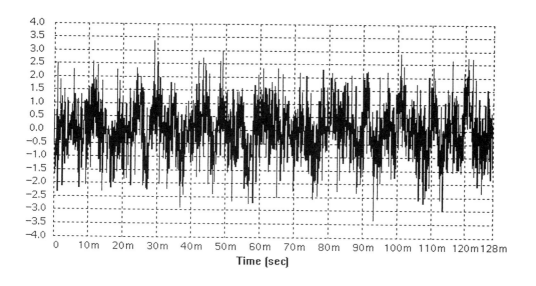

Figure 8.10. Filter input voltage versus time.

Figure 8.10 shows the 20-tone input signal, digitized at a sampling rate of 16 kHz and 2048 captured samples. Figure 8.11 shows the magnitude of the spectrum of the input signal. Figure 8.12 shows the magnitude of the spectrum of the digitized output. Both plots have been converted to a logarithmic scale (dBV) so that we can see the low amplitude signal components (including noise) as well as the high ones.

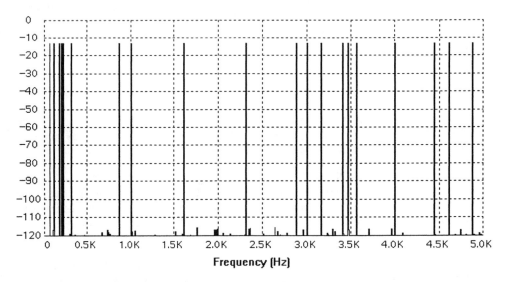

Figure 8.11. Filter input voltage (dBV) versus frequency.

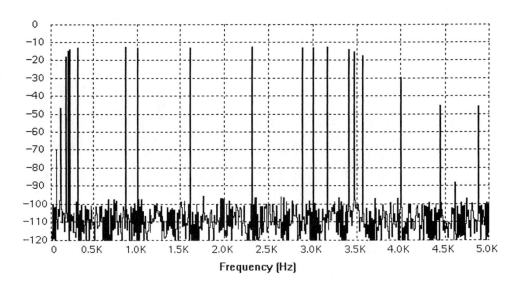

Figure 8.12. Filter output voltage (dBV) versus frequency.

The absolute gains are computed using the formula

$$G(\text{dB}) = 20 \, \log_{10} \left| \frac{V_{out}}{V_{in}} \right| \qquad (8.10)$$

The computed absolute gains and the frequency response (gains relative to the gain at 1 kHz) are listed in Table 8.10. Note that this example filter failed in two places, corresponding to the lower mask problem areas in Figure 8.9. We can see from the enlarged view of the problem areas in the filter mask that the gains at these frequencies were going to be very close to failure.

Table 8.10. Band-Pass Filter Absolute Gains and Relative Gain Measurements

Frequency	Absolute Gain	Relative Gain (Freq. Resp.)	Lower Limit	Upper Limit
1000 Hz (ref.)	0.168 dB	0.00 dB	-0.50 dB	+0.50 dB
60 Hz	-57.44 dB	-57.27 dB	NA	-23 dB
100 Hz	-33.62 dB	-33.79 dB	NA	-13.24 dB (interpolated)
170 Hz	-4.87 dB	-5.04 dB	NA	-3.01 dB
190 Hz	-1.89 dB	-2.06 dB	-3.01 dB	NA
200 Hz	-1.128 dB	-1.296 dB **Fail**	-1.280 dB	0.0 dB
300 Hz	0.013 dB	-0.155 dB	-0.50 dB	+0.50 dB
850 Hz	0.181 dB	0.013 dB	-0.50 dB	+0.50 dB
1600 Hz	0.017 dB	-0.151 dB	-0.50 dB	+0.50 dB
2310 Hz	0.266 dB	0.098 dB	-0.50 dB	+0.50 dB
2860 Hz	0.018 dB	-0.150 dB	-0.50 dB	+0.50 dB
3000 Hz	0.071 dB	-0.097 dB	-0.50 dB	+0.50 dB
3150 Hz	0.170 dB	0.002 dB	-0.853 dB	+0.292 dB
3400 Hz	-1.264 dB	-1.432 dB **Fail**	-1.35 dB	0.0 dB
3460 Hz	-2.29 dB	-2.46 dB	-3.01 dB	NA
3560 Hz	-4.55 dB	-4.72 dB	NA	-3.01 dB
4000 Hz	-16.82 dB	-16.99 dB	NA	-14.0 dB
4430 Hz	-32.21 dB	-32.38 dB	NA	-32.0 dB
4600 Hz	-75.23 dB	-75.40 dB	NA	-32.0 dB
4860 Hz	-32.52 dB	-32.69 dB	NA	-32.0 dB

Exercises

8.10. Using log/log interpolation, determine the –3-dB frequency of a filter corresponding to a measured gain of –1.56 dB at 3000 Hz and –4.32 dB at 3400 Hz.

Ans. 3202.4 Hz.

8.11. What is the total RMS value of a multitone signal consisting of twenty-five tones having an RMS value of 100 mV each?

Ans. 0.5 V RMS.

8.12. An eight-tone multitone signal of 250 mV RMS is required to perform a frequency response test. What is the peak amplitude of each tone if they are all equal in magnitude?

Ans. 125 mV peak.

8.13. The frequency response behavior of an amplifier is measured with a multitone signal consisting of four tones. An FFT analysis of the input and output samples reveal the following complex spectral coefficients:

Tone	Input	Output
1.1 kHz	-0.8402 + j0.5424	-0.1102 + j0.6636
2.1 kHz	0.6286 + j0.7778	0.4181 – j0.1002
3.1 kHz	-0.9180 – j0.3966	-0.2024 + j0.2308
4.1 kHz	-0.0134 + j0.9999	0.2294 + j0.0592

What is the relative gain (dB) of the amplifier at each frequency if the reference gain is based on the gain at 2.1 kHz?

Ans. 3.888 dB, 0 dB, -2.9252 dB, and -5.1747 dB.

In addition to the traditional multitone technique, frequency response can also be measured by applying a narrow impulse to the circuit under test and observing the filter's impulse response. The Fourier transform of the impulse response is the filter frequency response. The advantage of this approach is that it gives the full frequency response, at all frequencies in the FFT spectrum. The problem with the impulse response approach is that it is very difficult to measure the gain at any one frequency with any degree of accuracy, since the energy contained in a narrow impulse is very small. The small signal level makes the measurement very susceptible to noise. Also, an impulse contains energy at many frequencies, which interact with each other through distortion processes. Therefore, the gain at any one frequency may be corrupted by distortion components from other frequencies.

Another technique for frequency response testing is to assume a perfect filter with a known mathematical frequency response. If the mathematical frequency response has a limited number of independent variables that control its behavior, then the gain only needs to be measured at a few frequencies. Using N equations in N unknowns, the full filter transfer function can be estimated from a limited number of gain measurements. This technique has the advantage that only a few tones need to be measured to predict the complete transfer curve of the filter. This technique works well for simple filters with wide design margins, but may not work as well with more complex filters having very tight specifications. Although there are several different ways to approach frequency response testing, coherent DSP-based testing is still the most common methodology for measuring the frequency response characteristics of a filter or other circuit in high-volume production testing.

8.3 PHASE TESTS

8.3.1 Phase Response

The transfer function (frequency response) of a filter or other analog channel is defined not only by the gain variations over frequency (magnitude response) but also by the phase shift variations (phase response). In analog and mixed-signal testing, the frequency response test often measures only the magnitude response of a circuit. The phase information contained in the FFT results is frequently discarded because phase response is often an unspecified parameter. If phase response is specified in the data sheet, however, it can be calculated using the FFT results collected during the frequency response test.

Assuming that the tester's FFT routine returns the complex coefficients of the discrete-time Fourier series in the form of a vector X, then according to the development in Chapter 7, the amplitude of the kth tone is calculated according to

$$\tilde{c}_k = \begin{cases} \sqrt{\text{Re}\{X(k)\}^2 + \text{Im}\{X(k)\}^2} & k = 0, N/2 \\ 2\sqrt{\text{Re}\{X(k)\}^2 + \text{Im}\{X(k)\}^2} & k = 1, \ldots, N/2 - 1 \end{cases} \qquad (8.11)$$

where $\text{Re}\{X(k)\}$ and $\text{Im}\{X(k)\}$ denote the real and imaginary parts of the kth element of vector X. Correspondingly, the phase shift of the kth tone can be calculated using the formula

$$\tilde{\phi}_k = \begin{cases} \tan^{-1}\left(\dfrac{-\text{Im}\{X(k)\}}{\text{Re}\{X(k)\}}\right) & \text{if } \text{Re}\{X(k)\} \geq 0 \\ \pi + \tan^{-1}\left(\dfrac{-\text{Im}\{X(k)\}}{\text{Re}\{X(k)\}}\right) & \text{if } \text{Re}\{X(k)\} < 0 \end{cases} \qquad (8.12)$$

To aid the user, most testers include a DSP routine to convert the results of an FFT into polar notation (magnitude and phase). The polar conversion routines perform whatever corrections are necessary to compute a correct phase shift. Although the built-in polar conversion approach is easier than doing the conversion manually, it can be less efficient. Why should we calculate the magnitude and phases of all 511 complex spectral coefficients in a 1024-point FFT if only 10 of the phases are of interest? A full polar conversion or polar FFT is an inefficient process that can add unnecessary test time. The test engineer may find it more efficient to perform a manual polar conversion only on the tones of interest.

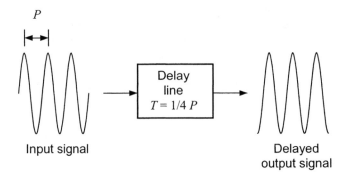

Figure 8.13. Analog delay line producing a -90 degree phase shift (phase lag).

The phase shift at each frequency in a multitone test signal is calculated by subtracting the input signal phase shift from the output signal phase shift, and then correcting for any 360 degree wraparound. A negative phase shift indicates a positive time delay (i.e., the output lags the input), while a positive phase shift indicates the opposite. Figure 8.13 shows an analog delay line with a time shift of $P/4$ producing a phase shift of -90 degrees at a frequency of $1/P$.

Using multitone DSP-based testing, we can calculate the real and imaginary components of the input and output signal of a circuit under test. The phase shifts can be calculated using a rectangular to polar conversion routine. All phase shifts can thus be measured with a single test, using the same data collected while measuring the magnitude portion of the frequency response.

Polar notation is limited to a phase shift of at most ±180 degrees. For example, a phase shift of -190 degrees translates into a shift of +170 degrees. The test engineer has to account for this "wrapping" effect by adding or subtracting integer multiples of 360 degrees to the polar conversion results. The idea is to eliminate jump discontinuities in the phase behavior of the device as these rarely, if ever, occur in practice. This process can get very confusing; so we shall look at an example to illustrate the phase measurement process.

Example 8.7

Measure the phase response of a 100-µs analog delay line in 1-kHz increments from 900 Hz to 9.9 kHz (10 tones). For simplicity, use a 1024-point sampling system with a 100-Hz fundamental frequency. Use odd harmonics so the time-domain signal will be symmetrical about the x axis (i.e., use spectral bins 9, 19, 29, 39, etc.). Use random phase shifts to produce a peak to RMS ratio of approximately 3.35:1.

Solution:

Figure 8.14 shows the digitized representation of the input signal to the delay line. The phases of the tones in this input signal were chosen to produce a peak to RMS ratio of 3.35:1. The real and imaginary spectral coefficients from an FFT analysis of the 10 digitized input tones are listed in Table 8.11, along with the phase calculated using Eq. (8.12). The digitized output signal is shown in Figure 8.15. The real and imaginary FFT results for the 10 digitized output tones are listed in Table 8.12, along with the calculated phases.

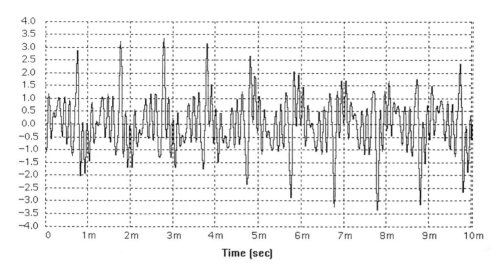

Figure 8.14. 100-μs delay line input voltage versus time.

Table 8.11. Calculated Signal Phases for Digitized Input

Frequency	Real Part	Imaginary Part	Phase (deg)
900 Hz	-0.199009	0.245755	231.0
1.9 kHz	-0.223607	0.223608	225.0
2.9 kHz	0.238660	0.207465	-41.0
3.9 kHz	0.286600	-0.133643	-25.0
4.9 kHz	-0.308123	-0.071136	167.0
5.9 kHz	-0.190311	0.252550	233.0
6.9 kHz	0.313150	0.044010	-8.0
7.9 kHz	-0.215668	-0.231275	133.0
8.9 kHz	-0.071136	-0.308123	103.0
9.9 kHz	-0.308123	-0.071136	167.0

Subtracting input phases from output phases should yield the phase shift at each frequency. Unfortunately, the polar conversion cannot distinguish between a shift of 360 degrees and a shift of 0 degrees. For this reason, we have to correct the output-input phase calculations in a two-step process. The raw output-minus-input phase shift calculations and the results of the correction steps are shown in Table 8.13.

In the first correction step (column three in Table 8.13), we have to account for the 360 degree ambiguity of the phase subtraction operation by first converting each phase shift to a value between -180 and +180 degrees. To do this, we either add 360 degrees or subtract 360 degrees from each phase number to limit our phase shift answers from −180 to +180 degrees. Then, we can see the discontinuity in the calculated phase shift between 4.9 and 5.9 kHz.

Observing the phase changes in the second column of Table 8.13, we see that there are several discontinuities in the calculated phase shift. In particular, between the frequencies of 1.9 and 2.9 kHz, 3.9 and 4.9 kHz, and 5.9 and 6.9 kHz. These discontinuities are caused by a mathematical wraparound effect, since the polar calculations cannot distinguish between a 360-degree shift and a 0-degree shift.

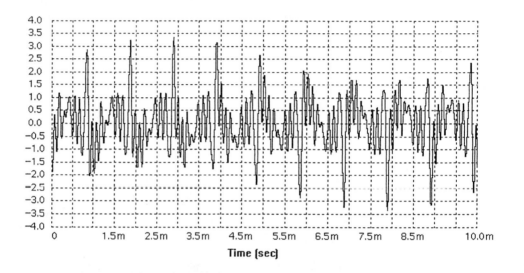

Figure 8.15. 100-µs delay line output voltage versus time.

Table 8.12. Calculated Signal Phases for Digitized Output

Frequency	Real Part	Imaginary Part	Phase (deg)
900 Hz	−0.298347	0.104827	−199.4
1.9 kHz	−0.293618	−0.117421	158.2
2.9 kHz	−0.252395	0.190517	217.0
3.9 kHz	−0.119020	0.292974	247.9
4.9 kHz	0.314893	0.033096	−6.0
5.9 kHz	0.285237	−0.133008	25.0
6.9 kHz	−0.105152	−0.298233	109.4
7.9 kHz	−0.258195	0.182579	215.3
8.9 kHz	−0.274162	−0.157590	150.1
9.9 kHz	−0.316175	−0.005777	179.0

To compensate for this 360-degree discontinuity, we either add or subtract 360 degrees from the wrapped result as needed to make the phase shift measurements continue in the correct direction. The results of this correction process are shown in the last column of Table 8.13. These numbers show a phase shift that starts at 0 degrees at 0 Hz (by definition - not listed) and continues in a negative direction as frequency increases.

Table 8.13. Phase Calculations and Corrections

Frequency	Phase Shift (Out – In)	±360 Phase Adj.	Actual Phase
900 Hz	199.4 – 231.0 = -31.6	0	-31.6
1.9 kHz	158.2 – 225.0 = -66.8	0	-66.8
2.9 kHz	217.0 – (-41) = 258.0	-360	-102.0
3.9 kHz	247.9 – 25 = 222.9	-360	-137.1
4.9 kHz	-6.0 – 167 = -173.0	0	-173.0
5.9 kHz	25.0 – 233 = -208.0	0	-208.0
6.9 kHz	109.4 – (-8) = 117.4	-360	-242.6
7.9 kHz	215.3 – 133 = 82.3	-360	-277.7
8.9 kHz	150.1 – 103 = 47.1	-360	-312.9
9.9 kHz	179.0 – 167 = 12.0	-360	-348.0

Sometimes phase calculations are specified not in degrees or radians, but in seconds or in fractions of a cycle. To convert phase shift ϕ in degrees or radians to phase shift in fractions of a cycle, we use the formula

$$\text{phase shift (fractions of cycle)} = \frac{\phi(\text{in degrees})}{360 \text{ degrees}} \tag{8.13}$$

or equivalently

$$\text{phase shift (fractions of cycle)} = \frac{\phi(\text{in radians})}{2\pi \text{ radians}} \tag{8.14}$$

To convert phase shift in degrees to phase shift in seconds, we use a similar calculation, taking the period ($1/f$) into account

$$\begin{aligned}\text{phase shift (sec)} &= \text{period (sec)} \times \frac{\phi(\text{in degrees})}{360 \text{ degrees}} \\ &= \frac{1}{f(\text{Hz})} \times \frac{\phi(\text{in degrees})}{360 \text{ degrees}}\end{aligned} \tag{8.15}$$

or equivalently

$$\text{phase shift (sec)} = \text{period (sec)} \times \frac{\phi(\text{in radians})}{2\pi \text{ radians}}$$
$$= \frac{1}{f(\text{Hz})} \times \frac{\phi(\text{in radians})}{2\pi \text{ radians}} \quad (8.16)$$

8.3.2 Group Delay and Group Delay Distortion

Group delay is a measurement of time shift versus frequency. It is commonly denoted as $\tau(f)$. Group delay is defined as the change in phase shift (fractions of a cycle) divided by the change in frequency (Hz)

$$\tau(f) = \frac{\Delta \text{phase shift (fractions of cycle)}}{\Delta f} = \frac{1}{360} \frac{\Delta\phi(\text{deg})}{\Delta f} = \frac{1}{2\pi} \frac{\Delta\phi(\text{rad})}{\Delta f} \quad (8.17)$$

where $\Delta\phi$ is the change in phase shift expressed in either degrees or radians. Group delay is expressed in units of time. Strictly speaking, group delay is defined as the derivative of phase with respect to frequency, $\tau(f) = (1/360)(d\phi/df)$. In reality, it is extremely difficult to resolve tiny changes in phase called for by the derivative operation.

Instead, we have to measure the phase at two points that are sufficiently separated in frequency to allow an accurate measurement of phase change. Group delay is typically measured with tone pairs centered around each frequency of interest. Many tone pairs can be measured simultaneously using DSP-based testing.

In a simple delay line, the phase shift through a circuit is directly proportional to frequency. The group delay of a delay line is therefore equal to the negative of the time delay through the circuit. A constant group delay indicates a circuit that shifts each signal component by a constant amount of time. This leaves the relative time shifts of the various signal components unchanged and therefore results in a signal that is identical in shape, but shifted in time (either delayed or advanced).

If, on the other hand, group delay varies over frequency, then the circuit will shift the various signal components relative to one another. This results in a change in shape as well as a shift in time. The variation in group delay over frequency is called *group delay distortion*. It is defined as the group delay minus the midpoint between the maximum and minimum group delay in the frequency range of interest

$$\tau_{distortion}(f) = \tau(f) - \frac{\tau_{max} + \tau_{min}}{2} \quad (8.18)$$

Group delay distortion may or may not be a problem in the system-level application. Signal clipping is one potential problem with a circuit that exhibits poor group delay characteristics. If different frequencies are shifted relative to each other, then the peak to RMS ratio may change enough to cause extreme peaks, which clip against the power supply rails of the analog channel.

Another problem that can arise from group delay distortion is incorrect data transmission in data communication channels, such as those used in modems, hard disk drive read/write channels, and cellular telephones. Since phase carries important information in many data communication protocols, group delay errors can lead to corrupted data.

Example 8.8

Calculate the group delay and group delay distortion of the 100-µs delay line from the data gathered in the previous example.

Solution:

Group delay and group delay distortion are simple calculations once we have calculated the phase shift at each frequency. The group delay and group delay distortion for each change in frequency is listed in Table 8.14. Frequency change is a constant 1000 Hz.

Group delay is equal to the change in phase (deg) divided by 360×1000 deg/sec. Group delay distortion at each frequency is equal to the group delay minus the midpoint between the maximum and minimum group delay. The midpoint is calculated as the average between −99.72 and −96.11 µs, (-99.72-96.11)/2 = -97.915 µs.

Exercises

8.14. An FFT analysis reveals that a test tone has a spectral coefficient of −0.2866 − j0.133643. What is the phase of this tone in degrees? Limit the phase to a range of ±180 degrees.

Ans. -155 degrees.

8.15. The phase shift (in degrees) of an analog channel as a function of increasing frequency appears as follows from an FFT analysis: 0, -35, -125, -165, 157, 56, -20, -55. What is the unwrapped phase shift of the channel (in degrees)?

Ans. 0, -35, -125, -165, -203, -304, -380, -415.

8.16. The frequency response behavior of an amplifier was measured with a multitone signal consisting of four tones. Using the spectral data provided in Exercise 8.13, determine the phase response of this amplifier (in degrees).

Ans. -47.73, -64.54, -72.12, -76.29.

8.17. Using the spectral data provided in Exercise 8.13, calculate the group delay of the amplifier. Also, determine the group delay distortion.

Ans. Group delay: -46.70, -21.07, -11.59 µs; Group delay distortion: -17.55, 8.07, 17.55 µs.

Table 8.14. 100-μs Delay Line Group Delay and Group Delay Distortion Calculations

Test Tone Pairs	Phase Change (deg)	Group Delay	Group Delay Distortion
900 Hz and 1.9 kHz	-66.8-(-31.6) = -35.2	-97.78 μs	135 ns
1.9 kHz and 2.9 kHz	-102.0-(-66.8) = -35.2	-97.78 μs	135 ns
2.9 kHz and 3.9 kHz	-137.1-(-102.0) = -35.1	-97.50 μs	415 ns
3.9 kHz and 4.9 kHz	-173.0-(-137.1) = -35.9	-99.72 μs	-1.8 μs
4.9 kHz and 5.9 kHz	-208.0-(-173.0) = -35.0	-97.22 μs	694 ns
5.9 kHz and 6.9 kHz	-242.6-(-208.0) = -34.6	-96.11 μs	1.8 μs
6.9 kHz and 7.9 kHz	-277.7-(-242.6) = -35.1	-97.50 μs	415 ns
7.9 kHz and 8.9 kHz	-312.9-(-277.7) = -35.2	-97.78 μs	135 ns
8.9 kHz and 9.9 kHz	-348.0-(-312.9) = -35.1	-97.50 μs	415 ns

8.4 DISTORTION TESTS

8.4.1 Signal to Harmonic Distortion

Harmonic distortion arises when a signal passes through a nonlinear circuit. The spectrum of the output of a nonlinear circuit includes not only the frequency components that appeared at the input, but also integer multiples (harmonics) of the input frequency components. Harmonic distortion is often measured with a single tone test signal, that is, a sine wave at a particular frequency (specified in the data sheet). To save test time, distortion can be measured in parallel with absolute gain using the FFT results from the gain test.

When passing a single test tone through the circuit under test, the harmonic distortion components appear at integer multiples of the test tone's frequency, F_t. F_t is often referred to as the *fundamental tone* (not to be confused with the fundamental frequency of the sampling system, F_f). Distortion that is symmetrical about the x axis gives rise to only odd harmonics ($3F_t$, $5F_t$, $7F_t$, etc.). Asymmetrical distortion, such as clipping on only the upper or lower portion of the waveform, gives rise to both odd harmonics and even harmonics ($2F_t$, $4F_t$, $6F_t$, etc.).

Signal to total harmonic distortion is defined as the ratio of the RMS signal level of the test tone divided by the total RMS of the odd and even harmonic distortion components. Signal-to-distortion is often expressed in decibel units, similar to gain. Since there are an infinite number of possible harmonics, the data sheet often calls out only a signal to second harmonic distortion and a signal to third harmonic distortion test. Also, the data sheet may call out a signal to total noise plus total harmonic distortion specification, which covers all harmonic distortion components and all noise components simultaneously (any spectral component that is not signal is either noise or distortion). To enable complete characterization, the test engineer will often write a test program that reports all these permutations of signal to noise and distortion measurements.

The definitions of the various signal to noise and distortion parameters are listed in Table 8.15. The symbol S denotes the fundamental signal component expressed in RMS volts, H_2 the RMS value of the second harmonic, H_3 the RMS value of the third harmonic, etc., and N is the RMS value of all the nonharmonic bins combined (noise will be explained in a later section). Note that we add RMS signal levels using a square-root-of-sum-of-squares calculation. This is a direct result of Parseval's theorem described in Section 7.4.2. Also note that the numbers will typically be positive since the signal is always larger than the distortion (unless the DUT is completely defective).

Table 8.15. Various Signal to noise and distortion Formulae; Individual Spectral Components Are Expressed in Terms of RMS Value. Note That DC Offset Is Not Included in Any of These Formulae.

Distortion Metric	Expression	
	(V/V)	(dB)
Signal to 2nd Harmonic Distortion (S/2nd)	$\dfrac{S}{H_2}$	$20 \log_{10}\left(\dfrac{S}{H_2}\right)$
Signal to 3rd Harmonic Distortion (S/3rd)	$\dfrac{S}{H_3}$	$20 \log_{10}\left(\dfrac{S}{H_3}\right)$
Signal to Total Harmonic Distortion (S/THD)	$\dfrac{S}{\sqrt{H_2^2 + H_3^2 + H_4^2 + H_5^2 + \cdots}}$	$20 \log_{10}\left(\dfrac{S}{\sqrt{H_2^2 + H_3^2 + H_4^2 + H_5^2 + \cdots}}\right)$
Signal-to-noise (S/N)	$\dfrac{S}{N}$	$20 \log_{10}\left(\dfrac{S}{N}\right)$
Signal to Total Harmonic Distortion plus Noise (S/THD+N)	$\dfrac{S}{\sqrt{H_2^2 + H_3^2 + H_4^2 + H_5^2 + \cdots + N^2}}$ or $\dfrac{S}{\sqrt{(\text{total signal RMS})^2 - S^2}}$	$20 \log_{10}\left(\dfrac{S}{\sqrt{H_2^2 + H_3^2 + H_4^2 + H_5^2 + \cdots + N^2}}\right)$ or $20 \log_{10}\left(\dfrac{S}{\sqrt{(\text{total signal RMS})^2 - S^2}}\right)$

Sometimes the data sheet will call out specifications in negative decibels, which simply means the test engineer swaps the numerator and denominator in the log calculations, or equivalently, changes the sign of the dB number reported. In the final row of the table, the S/(THD+N) calculation is defined in two different ways. Both are equivalent, although, the second definition is sometimes faster to compute than pulling all the harmonics apart from one another only to recombine them later. Testers usually include a very efficient RMS routine that can calculate the total signal RMS quickly. It should also be noted that the abbreviation S/(THD+N) is often replaced by the equivalent expression, SINAD, which stands for signal to noise and distortion.

Repeatable measurements of low-level distortion components can be extremely difficult and time-consuming to perform. For instance, if the specified distortion level is –85 dB, then the distortion component of interest may be very close to the noise floor of the DUT and/or ATE measurement hardware. This will lead to unrepeatable measurements of distortion that may or may not be tolerable.

The only way to improve the repeatability is to average or collect more samples with the ATE digitizer. The end result in either case is that data collection time (i.e., the number of samples collected) must quadruple to drop the nonrepeatability in half. The extra collection and DSP processing time obviously adds test time and drives up the cost of testing. Therefore, very low levels of distortion are inherently very costly to test, especially when they are close to failing test limits.

Example 8.9

A 1-kHz sine wave passes through a voltage follower and a digitizer captures 512 samples of the output signal at a sampling rate of 10 kHz. The fundamental frequency is equal to the sampling rate divided by number of samples, or 10 kHz/512 = 19.531 Hz. To achieve coherent testing, the 1-kHz sine wave is actually generated by an AWG at 51 times the digitizer's fundamental frequency, or 996.094 Hz. An FFT of the output signal shows several distortion components, listed in Table 8.16. Calculate $S/2^{nd}$, $S/3^{rd}$, and S/THD.

Table 8.16. Voltage Follower Output Levels at Fundamental and Second through Fifth Harmonics

FFT Spectral Bin	Frequency	RMS Voltage
51	1 kHz (fundamental tone)	1.025 V
102 (2×51)	2 kHz (second harmonic)	1.23 mV
153 (3×51)	3 kHz (third harmonic)	2.54 mV
204 (4×51)	4 kHz (fourth harmonic)	0.78 mV
255 (5×51)	5 kHz (fifth harmonic)	0.32 mV

Solution:

Notice that in a coherent DSP-based measurement system, the harmonic distortion components always fall into single FFT spectral bins. These can easily be calculated by multiplying the FFT bin of the fundamental tone, 51, by 2, 3, 4, 5, etc. Therefore, we do not have to multiply 996.094 Hz times 2, 3, 4, and 5 and then try to figure out which spectral bins they will fall into.

Working with spectral bins instead of frequencies is therefore easier in some cases than working with frequencies. The signal-to-distortion results are listed in Table 8.17.

Table 8.17. Signal-to-Distortion Results

$S/2^{nd}$	$S/3^{rd}$	S/THD
$20 \log_{10}\left(\dfrac{1.025 \text{ V}}{1.23 \text{ mV}}\right)$ $= 58.4$ dB	$20 \log_{10}\left(\dfrac{1.025 \text{ V}}{2.54 \text{ mV}}\right)$ $= 52.1$ dB	$20 \log_{10}\left(\dfrac{1.025 \text{ V}}{\sqrt{(1.23 \text{ mV})^2 + (2.54 \text{ mV})^2 + (0.78 \text{ mV})^2 + (0.32 \text{ mV})^2}}\right)$ $= 50.8$ dB

8.4.2 Intermodulation Distortion

Intermodulation distortion is very similar to harmonic distortion except that two or more tones are supplied to the DUT at once. The details of intermodulation tests vary widely from one type of device to another. Telecommunications and audio products usually specify a two- or four-tone test, while digital subscriber line (DSL) testing may require hundreds of test tones. Distortion components may appear at any sum or difference of any multiple of the test tones. Given any two test tones in a multitone signal, F_1 and F_2, there may be distortion components at any intermodulation frequency $F = | p \times F_1 \pm q \times F_2 |$, where p and q may be any positive integers. Second-order intermodulation components are those for which $p + q = 2$. Third-order intermodulation components are those for which $p + q = 3$, etc. Second- and third-order intermodulation components for two frequencies, F_1 and F_2, are shown in Figure 8.16.

Intermodulation distortion is expressed as a ratio of the signal RMS of any one test tone, denoted by, say, S_1, to the signal RMS of the intermodulation component(s)

$$20 \log_{10}\left(\frac{S_1}{\sqrt{IMD_1^2 + IMD_2^2 + IMD_3^2 + IMD_4^2 + \cdots}}\right) \qquad (8.19)$$

where IMD_1, IMD_2, etc., corresponds to the appropriate intermodulation distortion components.

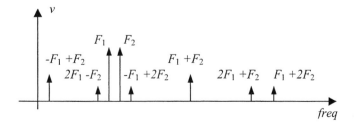

Figure 8.16. Second and third intermodulation frequencies for two-tone signal at F_1 and F_2.

The signal to intermodulation distortion calculations are usually specified with a limited number of distortion combinations, e.g., sum of all third-order intermodulation components. Alternatively, a signal to total noise plus distortion specification may be listed in the data sheet. When calculating specific combinations, the test engineer adds or subtracts test tone FFT bin numbers and their multiples to determine which FFT bins are likely to contain intermodulation components.

Example 8.10

A multitone test signal consists of a sum of two 1.0-V RMS sine waves, one at 1 kHz and the other at 1.1 kHz. Calculate the frequencies of the second-, third-, and fourth-order intermodulation components. The signal RMS at 100 Hz is 193 µV and the signal RMS at 2.1 kHz is 232 µV. Calculate the signal to second-order intermodulation distortion ratio, in dB.

Solution:

The second-order intermodulation components occur at

$$|1.0 \text{ kHz} \pm 1.1 \text{ kHz}| = 100 \text{ Hz and } 2.1 \text{ kHz}$$

The third-order intermodulation components occur at

$$|2 \times 1.0 \text{ kHz} \pm 1.1 \text{ kHz}| = 900 \text{ Hz and } 3.1 \text{ kHz}$$
$$|1.0 \text{ kHz} \pm 2 \times 1.1 \text{ kHz}| = 1.2 \text{ kHz and } 3.2 \text{ kHz}$$

The fourth-order intermodulation components occur at

$$|2 \times 1.0 \text{ kHz} \pm 2 \times 1.1 \text{ kHz}| = 200 \text{ Hz and } 4.2 \text{ kHz}$$
$$|3 \times 1.0 \text{ kHz} \pm 1.1 \text{ kHz}| = 1.9 \text{ kHz and } 4.1 \text{ kHz}$$
$$|1.0 \text{ kHz} \pm 3 \times 1.1 \text{ kHz}| = 2.3 \text{ kHz and } 4.3 \text{ kHz}$$

The signal to second-order intermodulation ratio is given by the equation

$$20 \log_{10}\left(\frac{1.0 \text{ V RMS}}{\sqrt{(193 \text{ }\mu\text{V})^2 + (232 \text{ }\mu\text{V})^2}}\right) = 70.4 \text{ dB}$$

8.5 SIGNAL REJECTION TESTS

8.5.1 Common-Mode Rejection Ratio

A number of signal rejection specifications are common to analog and sampled channel testing. Signal rejection tests are those which measure a channel's ability to prevent an undesired signal from propagating to the channel's output. The undesired signal may originate in the power supply, in another supposedly separate circuit, or in the channel itself.

> **Exercises**
>
> **8.18.** An FFT analysis of the output of an amplifier contains the following spectral amplitudes:
>
> | FFT Spectral Bin | RMS Voltage |
> |---|---|
> | 31 | 0.9560 V |
> | 62 (2×31) | 0.05 mV |
> | 93 (3×31) | 1.64 mV |
> | 124 (4×31) | 0.04 mV |
> | 155 (5×31) | 1.04 mV |
>
> In addition, the total RMS value of the output signal is 0.95601 V. Calculate S/2nd, S/3rd, S/THD and S/THD+N.
>
> **Ans.** 85.63, 55.31, 53.84, 46.79 dB.
>
> **8.19.** A multitone test signal consists of a sum of three 1.0-V RMS sine waves, one at 0.9 kHz, another at 2.1 kHz, and the third at 5.3 kHz. Determine the frequencies of all third-order intermodulation frequencies.
>
> **Ans.** 5.1, 3.3, 11.5, 9.7, 3.9, 0.3, 7.1, 3.5 kHz.

One such signal rejection test, common-mode rejection ratio (CMRR), is a measurement of how well a channel with a differential input can reject a common-mode signal. Ideally, a differential input circuit produces an output equal to GV_{diff}, where $V_{diff} = IN_P - IN_N$ is the differential input voltage and G is the gain of the input circuit. Provided that IN_P and IN_N are exactly equal (i.e., if the input signal is purely common-mode with no differential component), then the circuit should produce zero output. However, due to mismatched components in the input circuit, a small amount of common-mode signal usually feeds through to the output.

In Chapter 3, we studied DC CMRR testing for differential circuits. AC CMRR for analog channels is defined in a similar manner to DC CMRR. AC CMRR is defined as the AC gain of the channel with a common-mode input divided by the gain of the channel with a normal, differential input

$$CMRR(f) \equiv \left| \frac{G_{cm}(f)}{G_{diff}(f)} \right| = \left| \frac{\left(\frac{V_{out}}{V_{cm}} \right|_{V_{diff}=0} \right)}{\left(\frac{V_{out}}{V_{diff}} \right|_{V_{cm}\,constant} \right)} \right| \tag{8.20}$$

AC CMRR is a frequency-dependent parameter and is therefore denoted with a frequency argument.

The common-mode gain is measured by connecting the two inputs together and applying a single-ended signal to them. The common-mode gain is thus measured as if the channel had a single-ended input. Obviously, this gain should be very low, since IN_P and IN_N are equal and the output GV_{diff} should be zero. This very low gain is then divided by the differential AC gain of the channel to arrive at the CMRR value for the specified frequency. Often, the differential gain of the channel has already been measured at several frequencies during an absolute gain or frequency response test. In such a case, the differential gain results can be reused instead of repeating the same differential gain measurements again.

If high accuracy is not required, it is often acceptable to simply divide the measured common-mode gain by the ideal differential gain of the circuit, rather than measuring the differential gain at each frequency of interest. The measurement of CMRR is often performed at several frequencies, since a channel may have different characteristics at different frequencies. Multitone testing can be used to measure CMRR at several frequencies at once, saving test time.

Like many other AC parameters, CMRR is often expressed in decibel units rather than V/V. Since the gain of the circuit with a common-mode input is much smaller than the differential gain, CMRR should produce a negative decibel result. The calculation of CMRR for an analog channel is thus

$$CMRR(\text{dB}) \equiv 20 \ \log_{10}\left(\frac{|G_{cm}(f)|}{|G_{diff}(f)|}\right) \qquad (8.21)$$

Example 8.11

A differential gain circuit has a differential input, a differential output and an ideal gain of 6.02 dB (2.0 V/V). A differential multitone signal is applied to the input of the circuit at 300, 1020, and 3400 Hz. The level of each of the tones is 250 mV RMS, differential. The output of the circuit is digitized differentially, resulting in the following differential RMS output levels: 300 Hz: 510 mV RMS, 1020 Hz: 500 mV RMS, 3400 Hz: 480 mV RMS. Then the inputs are shorted together and a single-ended multitone is applied to the two inputs simultaneously.

The input signal level is 250 mV RMS, single-ended. The output of the channel is again digitized, resulting in the following differential RMS output levels: 300 Hz: 0.7 mV RMS, 1020 Hz: 0.8 mV RMS, 3400 Hz: 1.5 mV RMS. Calculate the CMRR at each frequency. If we observe the positive input during the differential gain measurement and then during the common-mode measurement, would the signal level change?

Solution:

The differential gain at each frequency is calculated as follows

$$\textbf{300 Hz:} \ Gain = \frac{510 \ \text{mV}}{250 \ \text{mV}} = 2.04 \ \text{V/V}$$

$$\textbf{1020 Hz:} \ Gain = \frac{500 \ \text{mV}}{250 \ \text{mV}} = 2.00 \ \text{V/V}$$

$$\textbf{3400 Hz:} \ Gain = \frac{480 \ \text{mV}}{250 \ \text{mV}} = 1.92 \ \text{V/V}$$

The common-mode gain at each frequency is calculated as follows

300 Hz: $Gain = \dfrac{0.7 \text{ mV}}{250 \text{ mV}} = 0.0028 \text{ V/V}$

1020 Hz: $Gain = \dfrac{0.8 \text{ mV}}{250 \text{ mV}} = 0.0032 \text{ V/V}$

3400 Hz: $Gain = \dfrac{1.5 \text{ mV}}{250 \text{ mV}} = 0.006 \text{ V/V}$

The CMRR at each frequency is thus

300 Hz: $CMRR = 20 \log_{10}\left(\dfrac{0.0028 \text{ V/V}}{2.04 \text{ V/V}}\right) = -57.2 \text{ dB}$

1020 Hz: $CMRR = 20 \log_{10}\left(\dfrac{0.0032 \text{ V/V}}{2.00 \text{ V/V}}\right) = -55.9 \text{ dB}$

3400 Hz: $CMRR = 20 \log_{10}\left(\dfrac{0.006 \text{ V/V}}{1.92 \text{ V/V}}\right) = -50.1 \text{ dB}$

Since the signal level during the differential test is 250 mV RMS differential (125 mV RMS, single-ended), we would see the signal level at the positive input increase by a factor of two during the common-mode test (250 mV RMS, single-ended).

8.5.2 Power Supply Rejection and Power Supply Rejection Ratio

Power supply rejection ratio is similar in nature to CMRR, except that the interference signal is applied through the power supply rather than through the normal inputs. In real-world applications, the power supply voltage is never perfect. It consists of a DC level plus AC variations caused by circuits pulling time-varying currents from the power supply circuits or from a battery.

PSRR is a measurement of a circuit's ability to reject a ripple signal added to its power supply voltage. PSRR is usually measured using a single tone or multitone signal, even though the device will seldom see a sinusoidal ripple on the power supply. PSRR is often specified for both the analog power supply and the digital power supply in mixed-signal devices. Usually, PSRR is much worse on the analog supply than on the digital supply, but not always.

Power supply rejection (PSR) is defined as the "gain" of a circuit with its input grounded or otherwise nonstimulated while an AC input signal is injected at a power supply pin. For instance, a test setup for a single-ended amplifier is shown in Figure 8.17. Here the PSR would be computed from the following formula

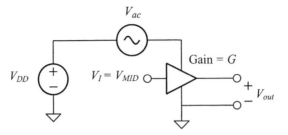

Figure 8.17. PSR and PSRR test setup.

$$PSR(f) \equiv \left| \frac{V_{out}}{V_{ac}} \bigg|_{V_{in}=V_{mid}} \right| \quad (8.22)$$

As with all AC test metrics, they are frequency dependent and thus should be expressed as a function of frequency. Of course, the AC input signal at the power supply pin must include a DC offset that corresponds to the normal power supply voltage so that the circuit remains powered up.

Power supply rejection ratio is defined as the PSR gain divided by the normal gain of the circuit according to

$$PSRR(f) \equiv \left| \frac{PSR(f)}{G(f)} \right| = \left| \frac{\left(\frac{V_{out}}{V_{ac}} \bigg|_{V_{in}=V_{mid}} \right)}{\left(\frac{V_{out}}{V_{in}} \bigg|_{V_{ac}=0} \right)} \right| \quad (8.23)$$

Both PSR and PSRR are usually specified in decibels, similar to CMRR. Like CMRR, the decibel results from a good DUT should be negative.

Example 8.12

A differential multitone signal at 300, 1020, and 3400 Hz is applied to the input of the gain circuit in Example 8.11. Each of the tones is set to a level of 250 mV RMS, differential. The output of the circuit is digitized differentially, resulting in the following differential RMS output levels: 300 Hz: 510 mV RMS, 1020 Hz: 500 mV RMS, 3400 Hz: 480 mV RMS. Then the inputs are shorted to a DC midsupply voltage and a single-ended multitone is added to the power supply. The input signal level is 100 mV RMS, single ended. The output of the channel is again digitized, resulting in the following differential RMS output levels: 300 Hz: 0.12 mV RMS, 1020 Hz: 0.15 mV RMS, 3400 Hz: 0.20 mV RMS. Calculate the PSR and PSRR at each frequency.

Solution:

The differential gain at each frequency is the same as in the CMRR test

300 Hz: $Gain = \dfrac{510 \text{ mV}}{250 \text{ mV}} = 2.04 \text{ V/V}$

1020 Hz: $Gain = \dfrac{500 \text{ mV}}{250 \text{ mV}} = 2.00 \text{ V/V}$

3400 Hz: $Gain = \dfrac{480 \text{ mV}}{250 \text{ mV}} = 1.92 \text{ V/V}$

The PSR at each frequency is calculated as follows

300 Hz: $PSR = 20\log_{10}\left(\dfrac{0.12 \text{ mV}}{100 \text{ mV}}\right) = -58.4 \text{ dB} \ (\text{or } 0.0012 \text{ V/V})$

1020 Hz: $PSR = 20\log_{10}\left(\dfrac{0.15 \text{ mV}}{100 \text{ mV}}\right) = -56.4 \text{ dB} \ (\text{or } 0.0015 \text{ V/V})$

3400 Hz: $PSR = 20\log_{10}\left(\dfrac{0.20 \text{ mV}}{100 \text{ mV}}\right) = -53.9 \text{ dB} \ (\text{or } 0.0020 \text{ V/V})$

The PSRR at each frequency is thus

300 Hz: $PSRR = 20\log_{10}\left(\dfrac{0.0012 \text{ V/V}}{2.04 \text{ V/V}}\right) = -64.6 \text{ dB}$

1020 Hz: $PSRR = 20\log_{10}\left(\dfrac{0.0015 \text{ V/V}}{2.00 \text{ V/V}}\right) = -62.5 \text{ dB}$

3400 Hz: $PSRR = 20\log_{10}\left(\dfrac{0.0020 \text{ V/V}}{1.92 \text{ V/V}}\right) = -59.6 \text{ dB}$

8.5.3 Channel-to-Channel Crosstalk

Crosstalk is another common measurement in analog channels, though its exact definition can be very DUT specific. Unlike CMRR or PSRR, crosstalk is a measurement with no exact definition. In general, crosstalk is the gain from one channel to a second supposedly independent channel. Ideally, of course, the channels should be perfectly isolated from one another so that there is no crosstalk.

In analog channels, crosstalk is often defined as the gain from one channel's input to another channel's output, divided by the gain of the second channel. Crosstalk is usually expressed in decibels, like CMRR and PSRR. Crosstalk may also be expressed as the gain from a channel's output to another channel's output divided by the second channel's gain. Other times, the crosstalk is not divided by the gain of the second channel at all. The test engineer has to clarify the definition of crosstalk in each case. For now, we will assume that crosstalk is defined as originally stated: the gain from one channel's input to another channel's output, divided by the gain of the second channel.

To aid the reader, a model of the channel-to-channel crosstalk is provided in Figure 8.18 where the crosstalk terms X_{rl} and X_{lr} are defined as

$$X_{rl}(f) \equiv \left|\frac{G_{rl}(f)}{G_r(f)}\right| = \frac{\left|\left(\frac{V_{out-r}}{V_l}\right)\Big|_{V_r=0}\right|}{\left|\left(\frac{V_{out-r}}{V_r}\right)\Big|_{V_l=0}\right|} \qquad X_{lr}(f) \equiv \left|\frac{G_{lr}(f)}{G_l(f)}\right| = \frac{\left|\left(\frac{V_{out-l}}{V_r}\right)\Big|_{V_l=0}\right|}{\left|\left(\frac{V_{out-l}}{V_l}\right)\Big|_{V_r=0}\right|} \qquad (8.24)$$

Crosstalk is often measured at several frequencies at once, using DSP-based multitone testing. The second channel's input is typically grounded (or connected to a midsupply voltage in single-supply circuits) during a crosstalk measurement. If crosstalk is specified from channel 1 to channel 2 and also from channel 2 to channel 1, however, then crosstalk can be measured from each channel to the other channel simultaneously to save test time.

To do this, the two channels have to be stimulated with slightly different frequencies so that each channel's crosstalk response can be isolated from its primary signal. The frequencies also have to be chosen so that the crosstalk components will not occur at the same frequencies as the harmonic and intermodulation distortion components of the primary signals. Otherwise, distortion will be misinterpreted as crosstalk.

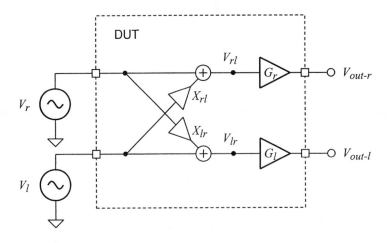

Figure 8.18. Modeling channel-to-channel crosstalk.

Example 8.13

A stereo audio channel consists of two identical analog signal paths (left and right). Each path is single ended (Figure 8.19). Using two digitizers and two AWGs, define a simultaneous sampling system that produces a multitone signal at approximately 300, 1020, and 3400 Hz on both the left and right channels. Use a 16-kHz sampling rate and 512 samples for the AWGs and digitizers. The actual test frequencies must be equal to the desired frequencies within a tolerance of plus or minus 10%.

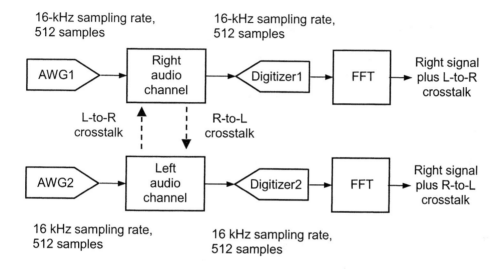

Figure 8.19. Simultaneous crosstalk measurements for stereo audio channels.

Apply the three-tone multitone with a total signal RMS of 500 mV. For simplicity, assume we have already measured a gain of exactly 6.02 dB through each channel. Using the following digitized signal levels, calculate crosstalk from L to R and from R to L. The frequencies below are approximate, since the exact frequencies should be slightly different for the left and right channels.

R Output:

300 Hz: 0.07 mV RMS, 1020 Hz: 0.08 mV RMS, 3400 Hz: 0.22 mV RMS

L Output:

300 Hz: 0.08 mV RMS, 1020 Hz: 0.09 mV RMS, 3400 Hz: 0.20 mV RMS

Solution:

First, we design the sampling system. We have two AWGs and two digitizers so we can perform two multitone crosstalk measurements simultaneously. The fundamental frequency for the 16-kHz sampling rate/512-sample system is 31.25 Hz. For the right channel, we can use spectral bins 11, 31, and 109, which gives frequencies of 343, 969, and 3406 Hz. For the left channel we can use spectral bins 9, 35, and 107, which gives frequencies of 281, 1103, and 3344 Hz. Note that we cannot use spectral bin 33 in the second multitone, even though it would give a frequency closer to 1020 Hz. If we tried to use bin 33, we would not be able to distinguish between third harmonic distortion from spectral bin 11 and crosstalk at spectral bin 33. These frequencies are within 10% of the desired frequencies. AWG #1 is set to produce the first three-tone multitone signal for the right channel, while AWG #2 is set to produce the second three-tone multitone for the left channel.

The two digitizers are set to capture the waveforms at the left and right channel outputs. Using the FFT from digitizer #1's waveform, we ignore the spectral components at spectral bins 11, 31, and 109 (the right channel's signal). Instead, we look for crosstalk components at 9, 33, and 107, which might result from the signal generated by AWG #2 (the left channel's signal). Similarly, we ignore spectral components at bins 9, 33, and 107 in the left channel's output, instead measuring the components at bins 11, 31, and 109, which might result from the right channel's signal.

Crosstalk is calculated as follows. The signal level of each input tone is equal to the total signal level divided by the square root of the number of tones (in this case, 3). Therefore, each tone's amplitude is 500 mV RMS/$\sqrt{3}$ or 288.68 mV RMS. Left-to-right crosstalk is defined as 20 $\log_{10}$(gain from L to R/gain in right channel) and right to left crosstalk is defined as 20 $\log_{10}$(gain from R to L/gain in left channel). The gain for both channels is exactly 6.02 dB, which is a gain of 2 V/V.

R-to-L crosstalk

300 Hz: $X_{rl} = 20 \; \log_{10}\left(\dfrac{0.08 \text{ mV}/288.68 \text{ mV}}{2}\right) = -77.2 \text{ dB}$

1020 Hz: $X_{rl} = 20 \; \log_{10}\left(\dfrac{0.09 \text{ mV}/288.68 \text{ mV}}{2}\right) = -76.1 \text{ dB}$

3400 Hz: $X_{rl} = 20 \; \log_{10}\left(\dfrac{0.20 \text{ mV}/288.68 \text{ mV}}{2}\right) = -63.2 \text{ dB}$

L-to-R crosstalk

300 Hz: $X_{lr} = 20 \; \log_{10}\left(\dfrac{0.07 \text{ mV}/288.68 \text{ mV}}{2}\right) = -78.3 \text{ dB}$

1020 Hz: $X_{lr} = 20 \; \log_{10}\left(\dfrac{0.08 \text{ mV}/288.68 \text{ mV}}{2}\right) = -77.2 \text{ dB}$

3400 Hz: $X_{lr} = 20 \; \log_{10}\left(\dfrac{0.22 \text{ mV}/288.68 \text{ mV}}{2}\right) = -68.4 \text{ dB}$

8.5.4 Clock and Data Feedthrough

The definition of clock and data feedthrough is even less standardized than crosstalk. Clock and data feedthrough is measured by digitizing the output of a channel and then applying one of several calculations to the resulting waveform. Digital feedthrough is usually very "spiky" in appearance. Clock feedthrough often has a signal bandwidth well into the megahertz range; so a high-bandwidth digitizer is typically used to measure clock feedthrough.

If the clock and data feedthrough is coherent with the digitizer, then an FFT can be performed on the output signal. In this case, the feedthrough may be specified in terms of a maximum spurious tone, relative to the level of a carrier tone. When specified in this manner, the spurious tone is specified in dBc (dB relative to the carrier level). Spurious tones are often a major concern in communication devices, such as cellular telephones. Because the energy in a spurious tone is concentrated around a single frequency, it can cause electromagnetic compatibility problems. A cellular telephone that generates a spurious tone in its transmit channel might interfere with other cellular telephones operating at the same frequency as the spurious tone. Such a telephone would fail the Federal Communications Commission (FCC) compliance tests.

Clock and data feedthrough may instead be specified in terms of total RMS voltage, excluding DC offset. The removal of DC offset can be accomplished by applying a DC blocking capacitor or high-pass filter to the signal before measuring its RMS level. More often, the DC component is removed mathematically by subtracting the average of the digitized signal from each point in the captured waveform.

The test engineer will have to make sure the exact definition of clock feedthrough or glitch energy is unambiguously defined. Usually the systems engineers or the customers will be the only ones who can define their intentions. Unfortunately, specifications are often lifted from a competitor's data sheet, which does not clearly specify the test conditions or test definition (what data pattern is being sent to or from the DUT, etc.). Again, the systems engineers or customers will have to help clarify the test requirements.

Often, clock and data feedthrough is not specified as a separate parameter. It is simply considered part of the noise in a signal-to-noise test. Noise testing is another major category of analog and sampled channel testing. Noise tests can be among the most difficult, time-consuming measurements in a mixed-signal test program.

8.6 NOISE TESTS

8.6.1 Noise

Random noise is generated by every real-world circuit. It can be generated by thermal noise in the case of resistors, $1/f$ noise in the case of CMOS transistors, or quantization noise in the case of DACs and ADCs. Noise can also be injected into a circuit by external forces, such as light falling on a bare die or electromagnetic interference coupling into a circuit under test. Excessive noise can result in a hissing noise in audio circuits, corrupted data in a modem or cellular telephone, and many other system-level failure mechanisms. Noise is generally, but not always, an undesirable property of a circuit under test. Noise is one of the leading causes of long test time, since averaging or added measurements are needed to remove the nonrepeatability caused by random noise.

The spectral density of noise energy is often described using color analogies. White noise, like white light, contains energy that is evenly distributed across the frequency spectrum. White noise is noise whose RMS voltage is constant in any band of frequencies from F to $F+\Delta F$, regardless of the value of F. Pink noise, by contrast, is noise that is weighted more heavily at low frequencies.

Often the level of noise is assumed to exhibit a Guassian (normal) statistical distribution. This is largely a result of the central limit theorem of large numbers. It is important to recognize, however, that the spectral properties of noise and its statistical distribution are separate concepts. They are combined for mathematical convenience.

Sometimes noise is defined as any signal component other than the primary test signal. (The 0 Hz, or DC component, is also excluded from the calculation of noise.) This definition of noise includes random noise as well as harmonic distortion, intermodulation distortion, clock feedthrough, sigma-delta converter self-tones, etc. Since test engineering is frequently concerned with characterization and diagnosis of failure mechanisms, a good test program should isolate all the known failure mechanisms into separate measurements. It is therefore preferable to measure distortion components separately from clock feedthrough, separately from random noise, etc. The signal to total noise, distortion, interference, etc. can also be calculated separately for additional characterization information. Often a data sheet will call out such a signal to total noise plus distortion test as an overall measure of quality.

There are several different ways to measure noise. Idle channel noise is the RMS voltage variation with no input signal (grounded inputs or inputs connected to a DC midpoint voltage). Signal-to-noise and signal-to-noise plus distortion are other figures of merit. There are also other definitions of noise performance such as spurious free dynamic range. Each of these tests looks for a different noise-based weakness in the circuit under test.

8.6.2 Idle Channel Noise

Idle channel noise (ICN) is a measurement of noise generated by the circuit itself, plus noise injected from external circuits or signal sources through a variety of coupling mechanisms. During an idle channel noise test, the input to the circuit under test is either shorted to ground or otherwise disabled into a quiet state. Ideally, the output should also fall into a noise-free state, but of course all outputs exhibit some amount of noise.

Using DSP-based testing methodologies, this output noise is measured by digitizing the output of the circuit and performing a noise calculation on the captured samples. Idle channel noise can be expressed in many different units of measure. The most straightforward idle channel noise measurement is to simply calculate the RMS voltage level from the captured samples.

It is important to realize that the bandwidth of the digitizer during this measurement is extremely important. A digitizer with a wide bandwidth will see more RMS noise than a digitizer with a narrow bandwidth. Since white noise is spread evenly across the frequency spectrum, a wider bandwidth will allow more noise components to be added into the final calculation. For this reason, it is critical to express the noise in terms of RMS voltage over a specified bandwidth. For example, a data sheet may specify idle channel noise as < 100 µV RMS from 100 Hz to 10 kHz.

The measurement of noise can be normalized by the bandwidth of the measurement using a unit of noise called the *root spectral density*, denoted S_n. This type of measurement is expressed in units of volts per root-Hz ($V/\sqrt{Hz}$) and it can be used to estimate RMS noise over other frequency bands. To convert a plain RMS voltage measurement into a $V/\sqrt{Hz}$ measurement, the RMS voltage is divided by the square root of the frequency span of the bandwidth B of the digitizer or RMS volt meter

$$S_n\left(\frac{V}{\sqrt{Hz}}\right) = \frac{\text{noise RMS}}{\sqrt{B}} \quad (8.25)$$

For example, if an RMS volt meter or digitizer allows only signal energy from 9 to 11 kHz to pass, and the RMS noise measurement is 100 µV RMS, then the noise can be expressed as $100\ \mu V/\sqrt{11\ kHz - 9\ kHz}$ or $2.236\ \mu V/\sqrt{Hz}$ from 9 to 11 kHz.

To estimate the RMS noise that would occur in the frequency span between 8 and 12 kHz, for example, one simply needs to multiply S_n by the square root of the frequency span (12 kHz − 8 kHz = 4 kHz). Using the previous noise result, we would then estimate noise from 8 to 12 kHz to be $2.236\ \mu V/\sqrt{Hz} \times \sqrt{4\ kHz} = 141.41\ \mu V\ RMS$. However, since noise is sometimes unevenly distributed, it might or might not be appropriate to estimate the noise from 100 to 200 kHz using the root spectral density measured between 9 and 11 kHz. A separate measurement of root spectral density could be performed near 100-200 kHz for this range of frequencies.

Noise measurements can be converted to decibel units as follows

$$\text{noise}(dB) = 20\ \log_{10}\left(\frac{\text{noise RMS}}{\text{reference RMS}}\right) \quad (8.26)$$

Thus expressing noise in decibel units is fairly straightforward, except for the determination of the reference voltage. There are a variety of references that the test engineer will encounter. One reference is simply 1.0 V RMS. When using this reference the noise is expressed in dBV (decibels relative to 1 V RMS). Noise can also be specified in dBm, referencing it to a 1.0 mW signal level at a particular load impedance.

Idle channel noise may also be specified relative to a full-scale sine wave (full scale must be defined in the data sheet). For example, if a circuit generates 1 mV RMS idle channel noise and its full scale range is ±500 mV, then the noise is measured relative to a sine wave at 500 mV peak. This corresponds to $500\ mV/\sqrt{2} = 354\ mV\ RMS$. The idle channel noise would then be calculated as

$$\text{noise}(dB) = 20\ \log_{10}\left(\frac{1\ mV\ RMS}{354\ mV\ RMS}\right) = -51\ dB,\ \text{relative to FS}$$

Idle channel noise can also be referenced to a 0-dB level, defined in the data sheet. For example, many central office telephone channels use a reference level that is about 3 dB lower than the full scale range of the channel. A sine wave at 0 dB produces a 0-dBm signal level at a specified point in the central office. Such a sine wave is said to be at a level of 0 dBm0. Idle channel noise measurements can also be referenced to this 0-dB level, resulting in the unit dBm0.

The dBm0 unit of measurement can be further refined by referencing the noise to a commonly accepted reference level of −90 dBm. This unit of measurement is called the dBrn (decibel referenced to noise). Referencing the measurement to the central office level of 0 dBm0 further gives us the dBrn0 unit. A dBrn0 measurement is therefore 90 dB higher than the equivalent dBm0 measurement. Finally, the noise can be weighted with a C-message filter before calculation of the RMS noise voltage, leading to the unit dBrnC0 (commonly pronounced duh-brink'-o). Weighting filters were discussed briefly in Chapter 7 and will be explained in more detail later in this chapter.

Example 8.14

A CODEC (coder decoder) is a device that is used by the telephone company's central office to digitize and reconstruct analog voice signals during a telephone call. The digitized voice information is transmitted between two central offices as the two customers speak to one another. A CODEC DAC channel is sent a data sequence of all zeros (idle signal).

The DAC output is digitized with a bandwidth of 20 kHz. The resulting signal is filtered with a software C-message weighting filter that limits the bandwidth of the noise to 0 Hz to 3.4 kHz. After filtering, the noise level is calculated as 100 µV RMS. The 0-dBm0 reference level is 1.4 V RMS for the DAC channel. Calculate the idle channel noise, in dBrnC0 units.

Solution:

First we take the 100 µV RMS signal and calculate its level in dBmC0 units. This gives us

$$ICN(\text{dBmC0}) = 20\ \log_{10}\left(\frac{100\ \mu\text{V RMS}}{1.4\ \text{V RMS}}\right) = -82.92\ \text{dBmC0}$$

Conversion to dBrnC0 is accomplished by adding 90 dB to this result

$$ICN(dBrnC0) = -82.92\ \text{dBmC0} + 90\ \text{dB} = +7.08\ \text{dBrnC0}$$

8.6.3 Signal to Noise, Signal to Noise and Distortion

Signal-to-noise ratio is another parameter that measures the noise performance of an analog or mixed-signal channel. It is different from idle channel noise in that it measures noise in the presence of a normal signal, usually a sine wave. When working with a purely analog channel, the presence of a signal should not change the amount of noise generated by the channel. However, in a DAC or ADC channel, quantization noise will not be present unless a signal is present. For this reason, it is necessary to measure not only idle channel noise, but also signal-to-noise ratio in mixed-signal channels. Signal-to-noise ratio is often measured using the same data collected during the gain and signal-to-distortion tests.

Signal-to-noise ratio can be defined as the ratio of the primary signal divided by all nonsignal AC components. However, this definition includes distortion components and other signal degradation factors that should be separated for characterization purposes. Therefore, signal-to-noise is more commonly measured by excluding harmonic distortion components, or excluding selected harmonic distortion components. For example, it is common to measure signal to second distortion, signal to third distortion, and signal to total harmonic distortion plus noise (S/THD+N). Table 8.15 summarizes these typical noise and distortion measurements. Notice from this table that individual signal components are added using a square-root-of-sum-of-squares calculation. *Note: The DC offset component is always excluded from these calculations.*

To calculate the total noise in nonharmonic bins, the test engineer can set the spectral coefficients to zero for all signal and harmonic distortion bins that are to be excluded. Then the test engineer can either perform an inverse FFT and calculate RMS of the time-domain signal, or better yet, simply calculate the square root of the sum of squares of all the remaining FFT bins. These two approaches are mathematically equivalent. However, the avoidance of an inverse FFT can save quite a bit of test time.

Poor repeatability is one of the biggest problems with noise measurements. Random noise by its very nature is nonrepeatable. Also, any signal component that is near the noise level will be unrepeatable as well. For example, a distortion component at 100 μV RMS may result in a very unrepeatable measurement if the noise level is also 100 μV RMS. The only way to get a repeatable noise or low-level distortion measurement is to collect hundreds or thousands of individual samples for the FFT calculation. This can lead to extremely long test times, depending on the level of repeatability needed. This is one reason why devices that are designed very close to the specification limits are expensive to test. A design with 6 dB of margin between the typical device performance and the specified limit can allow a much less expensive test than a device with 1 dB of margin, since 6 dB of margin requires less accuracy and repeatability.

Example 8.15

256 samples of a 1-kHz sine wave are captured with a digitizer at a sampling rate of 16 kHz. An FFT analysis of the output reveals the magnitude of the spectrum shown in Figure 8.20. The magnitude of the spectrum shows the signal and 5 significant distortion components. The data sheet for this device defines noise as anything other than the fundamental signal, second-, and third harmonic distortion components.

The fundamental test tone, second-, and third- harmonic distortion components are set to zero, leaving the spectrum in Figure 8.21. Taking the square root of sum of squares of the modified spectrum gives an RMS noise level of 100 μV RMS. Calculate the signal-to-noise ratio for this signal.

Solution:

The fundamental tone was at 1 V RMS; so the signal-to-noise ratio is equal to

$$SNR(\text{dB}) = 20 \log_{10}\left(\frac{1 \text{ V RMS}}{100 \text{ }\mu\text{V RMS}}\right) = 80 \text{ dB}$$

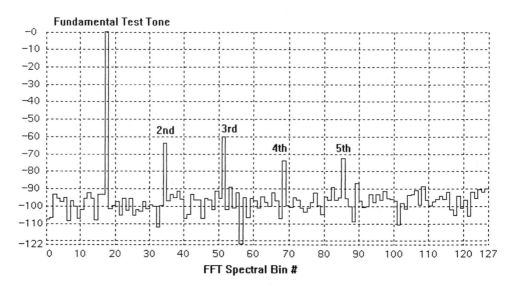

Figure 8.20. Magnitude spectrum (dBV) for 1-kHz test tone with noise and harmonic distortion.

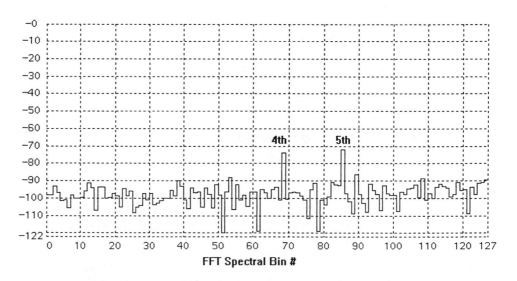

Figure 8.21. Magnitude spectrum with fundamental test tone, second and third distortion removed.

8.6.4 Spurious Free Dynamic Range

Spurious free dynamic range is a specification that is critical to audio circuits as well as telecommunication circuits that must pass FCC (or EC) certifications. A spur is defined as any nonsignal component that is confined to a single frequency. Spurs can be caused by harmonic and intermodulation distortion, clock feedthrough, sigma-delta converter self-tones, stray oscillations, or any of dozens of other undesirable processes.

> **Exercises**
>
> **8.20.** Determine the root spectral density of a noise signal that has an RMS value of 250 µV over a frequency range of 1 to 4 kHz.
>
> **Ans.** 4.564 $\mu V/\sqrt{Hz}$.
>
> **8.21.** A digitizer sampling at 8 kHz captures 16 samples of a noise signal. An FFT analysis reveals the following spectral coefficients:
>
> | FFT Spectral Bin | RMS Voltage |
> |---|---|
> | 0 | 0.0150 mV |
> | 1 | 0.2620 mV |
> | 2 | 0.4092 mV |
> | 3 | 0.5559 mV |
> | 4 | 0.1681 mV |
> | 5 | 0.7270 mV |
> | 6 | 0.4941 mV |
> | 7 | 0.2550 mV |
> | 8 | 0.2539 mV |
>
> Calculate the RMS level of the noise signal between 1 and 2.5 kHz. What is the corresponding root spectral density of the noise over this frequency interval? Repeat for the frequency range between 1 to 3 kHz.
>
> **Ans.** 1 mV RMS, 26.24 $\mu V/\sqrt{Hz}$; 1.1 mV RMS, 25.27 $\mu V/\sqrt{Hz}$.

Spurs are much more noticeable to the human ear than other types of noise. For this reason, sigma-delta converters sometimes include a randomizing circuit to inject random noise that spreads the inherent self-tones of the converter into many frequency bins. This degrades the signal-to-noise performance, but improves the spurious free dynamic range. The end result is less objectionable to the listener. In cellular telephone applications, spurs can be mixed into the transmitted signal, resulting in unwanted side lobes that might interfere with calls on other cellular telephones. For this reason, FCC compliance specifications limit unwanted spurs in transmitted signals.

A spur shows up in a magnitude FFT or on a spectrum analyzer display as a spike in the frequency domain. Spurious free dynamic range is often defined as the difference in decibels between the 0-dB signal level (the carrier level) and the maximum spur in the frequency domain. A spurious free dynamic range of 60 dBc would imply that no individual tone in the frequency domain is larger than 60 dB below the 0-dB carrier signal level as defined in the data sheet. For example, if a device has a 0-dB carrier signal level specification of 3.0 V RMS and the largest spur in the frequency domain is 1.5 mV RMS, then the spurious free dynamic range is $20 \log_{10}(3 \text{ V RMS}/1.5 \text{ mV RMS}) = 66$ dBc.

8.6.5 Weighting Filters

Weighting filters can be applied to any FFT output before calculations of gain, distortion, noise, etc. are performed. A weighting filter is usually designed to mimic the response of a human sense, such as hearing.[1] The most common weighting filters used in mixed-signal testing are designed to match the frequency response of the human ear.

Three filters are commonly used in telecommunication and audio applications: the ANSI A-weighting filter, the psophometric filter, and the C-message weighting filter (not to be confused with the ANSI C-weighting filter). The frequency response of each filter is shown in Figures 8.22–8.24. The corresponding filter gain specifications are listed in Tables 8.18–8.20.

To apply a particular filter to the FFT of a test signal, the test engineer must first calculate the gain at each FFT spectral bin. The spectral coefficients are then multiplied by the filter gain at that frequency to produce a weighted FFT. Since the weighting filters are only defined at a few frequencies, we have to use interpolation to find the gain at a particular frequency. The weighting filters specify a linear interpolation between points on a log/log plot. Since the gains are expressed in decibels, the gain is already in log format. The frequencies, on the other hand, must be converted into a logarithmic format before interpolation can be calculated.

Example 8.16

Calculate the gain of an A-weighting filter at the thirty seventh FFT spectral bin of a signal that was digitized at 16k Hz using 512 samples.

Solution:

The thirty seventh FFT spectral bin corresponds to $37 \times (16 \text{ kHz}/512) = 1156.25$ Hz. The nearest two points on the A-weighting curve are located at $F_1 = 1$ kHz ($G_1 = 0$ dB) and $F_2 = 1.25$ kHz ($G_2 = 0.6$ dB). First we have to convert these frequencies to logarithmic values

$$\log_{10}(F_{test}) = \log_{10}(1156.25) = 3.06305$$
$$\log_{10}(F_1) = \log_{10}(1000) = 3.0$$
$$\log_{10}(F_2) = \log_{10}(1250) = 3.09691$$

A standard linear interpolation is then performed to calculate the gain of the weighting filter at 1156.25 Hz as follows

$$\text{gain at } 1156.26 \text{ Hz} = G_1 + (G_2 - G_1)\left(\frac{\log_{10}(F_{test}) - \log_{10}(F_1)}{\log_{10}(F_2) - \log_{10}(F_1)}\right)$$
$$= 0.0 \text{ dB} + (0.6 \text{ dB} - 0.0 \text{ dB})\left(\frac{3.06305 - 3}{3.09691 - 3}\right)$$
$$= 0.39 \text{ dB}$$

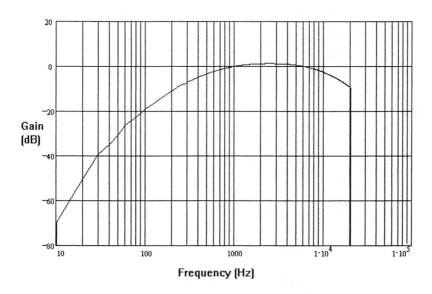

Figure 8.22. ANSI A-Message Weighting Filter

Table 8.18. ANSI A-Message Weighting Filter Gains

Freq (Hz)	Gain (dB)	Freq (Hz)	Gain (dB)	Freq (Hz)	Gain (dB)	Freq (Hz)	Gain (dB)
10	-70.4	100	-19.1	1000	0.0	10000	-2.5
12.5	-63.4	125	-16.1	1250	0.6	12500	-4.3
16	-56.7	160	-13.4	1600	1.0	16000	-6.6
20	-50.5	200	-10.9	2000	1.2	20000	-9.3
25	-44.7	250	-8.6	2500	1.3	>20000	Undefined
31.5	-39.4	315	-6.6	3150	1.2		
40	-34.6	400	-4.8	4000	1.0		
50	-30.2	500	-3.2	5000	0.5		
63	-26.2	630	-1.9	6300	-0.1		
80	-22.5	800	-0.8	8000	-1.1		

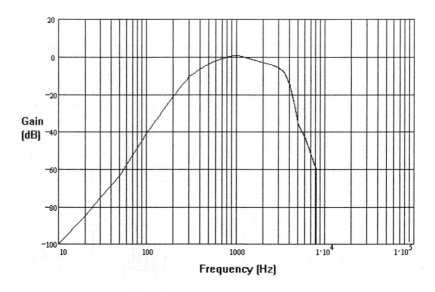

Figure 8.23. Psophometric weighting filter.

Table 8.19. Psophometric Weighting Filter Gains

Freq (Hz)	Gain (dB)	Freq (Hz)	Gain (dB)	Freq (Hz)	Gain (dB)
16.66	-70.4	1000	1.0	5000	-36.0
50	-63.0	1200	0.0	6000	-43.0
100	-41.0	1400	-0.9	8000	-60.0
200	-21.0	1600	-1.7	>8000	Undefined
300	-10.6	1800	-2.4		
400	-6.3	2000	-3.0		
500	-3.6	2500	-4.2		
600	-2.0	3000	-5.6		
700	-0.9	3500	-8.5		
800	0.0	4000	-15.0		
900	0.6	4500	-25.0		

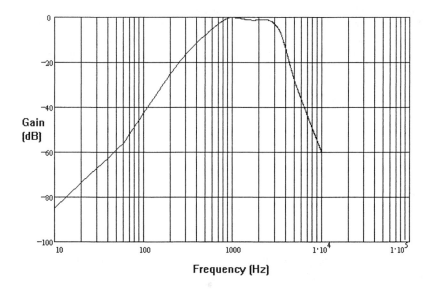

Figure 8.24. C-message weighting filter.

Table 8.20. C-Message Weighting Filter Gains

Freq (Hz)	Gain (dB)	Freq (Hz)	Gain (dB)	Freq (Hz)	Gain (dB)
60	-70.4	1000	1.0	3500	-36.0
100	-63.0	1200	0.0	4000	-43.0
200	-41.0	1300	-0.9	4500	-60.0
300	-21.0	1500	-1.7	5000	Undefined
400	-10.6	1800	-2.4	>5000	See Note
500	-6.3	2000	-3.0		
600	-3.6	2500	-4.2		
700	-2.0	2800	-5.6		
800	-0.9	3000	-8.5		
900	0.0	3300	-15.0		

Note: Gain above 5kHz shall decrease by at least 12 dB per octave until –60 dB.

For each different combination of sampling rate and sample size in a test program, a different set of filter gains must be calculated using this interpolation method. To save test time, weighting filter gain values should be calculated only once, during the first execution of the test program. Once all the gains for each weighting filter have been calculated, they are stored in an array for future use. On subsequent test program executions, the appropriate filter gains can be applied to the results of the FFTs to make weighted measurements. Application of the filter gains is a simple matter of multiplying the FFT results at each spectral bin by the filter gain at that bin.

Exercises

8.22. Calculate the gain of a Psophometric weighting filter at FFT spectral bin 413 of a signal that was digitized at 16 kHz using 2048 samples.

Ans. −6.97 dB.

8.23. Calculate the higher of the two frequencies of a C-message weighting filter that correspond to a gain of −0.55 dB.

Ans. 1260.20 Hz.

8.7 SIMULATION OF ANALOG CHANNEL TESTS

8.7.1 MATLAB Model of an Analog Channel

Mathematically modeling the behavior of an analog channel using MATLAB or some equivalent software is an important step for understanding the test methods described in this chapter. Through the appropriate software model, we can apply a coherent test signal and analyze the response of the channel using Fourier analysis as if one collected the data directly from a tester.

To begin we must first model the large-signal behavior of an analog channel in the presence of noise. This requires that we divide the analog channel into three components as shown in Figure 8.25. The first block models the presence of noise in the channel, the second block models the nonlinear input-output transfer characteristic, and the third block models the frequency response behavior of the channel.

To model the noise of the channel we make use of a normally (Gaussian) distributed random number generator available in MATLAB called *randn*. This number generator will produce a sequence of independent (uncorrelated) numbers whose statistics are normally distributed with zero mean and unity standard deviation. To create a noise signal with a DC value of V_{DC} and RMS value of V_{rms} we make use of the following linear transformation

$$Noise = V_{DC} + V_{rms}\xi \tag{8.27}$$

where ξ is a random number generated by *randn*. In order to run the same program multiple times and obtain a different number sequence each time but with identical statistics, the *seed*

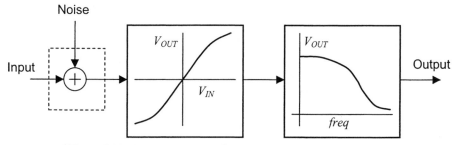

Figure 8.25. A nonlinear model of an analog channel with additive noise.

number that is used to initialize the random number generator should be changed. In MATLAB, this is achieved by passing the seed number s to $randn$ via the statement $randn(\text{'seed'}, s)$.

The nonlinear input-output transfer characteristic of the analog channel is simply modeled by expressing the output signal in terms of the input signal through a power-series representation described by

$$v_{out} = c_0 + c_1 v_{IN} + c_2 v_{IN}^2 + c_3 v_{IN}^3 + \cdots \qquad (8.28)$$

Finally, the frequency response behavior of the channel can be modeled using a digital filter described by an nth-order difference equation. While the topic of digital filters is beyond the scope of this text, we shall limit our discussion to a first-order digital filter described by the following difference equation:

$$v_{OUT}(n) = a_0 v_{IN}(n) + a_1 v_{IN}(n-1) - b_1 v_{OUT}(n-1) \qquad (8.29)$$

Through the application of z transforms, the normalized frequency response behavior of this filter is found to be

$$\left|\frac{V_{out}}{V_{in}}(f)\right| = \sqrt{\frac{[a_0 + a_1 \cos(2\pi f)]^2 + [a_1 \sin(2\pi f)]^2}{[1 + b_1 \cos(2\pi f)]^2 + [b_1 \sin(2\pi f)]^2}}$$

Parameters a_0, a_1, and b_1 are selected to control the shape of the filter response, e.g., low-pass, high-pass, etc. For example, with $a_0=0.5$, $a_1=0.5$ and $b_1=1$, we obtain a low-pass Butterworth response with unity DC gain. Now to put it all together, let us consider the following example.

Example 8.17

Consider a low-pass amplifier described by parameters $a_0=0.5$, $a_1=0.5$, $b_1=1$, $c_0=0$, $c_1=1$, $c_2=0.01$, and $c_3=0.001$. Simulate the behavior of the amplifier using MATLAB in the presence of a noise signal with zero mean and an RMS value of 1 mV. Compute the signal to third harmonic distortion ratio of the amplifier at approximately 1.4 kHz with a 1-V RMS input signal. Repeat the simulation 10 times using a different seed each time. Calculate the mean and standard deviation of the signal to third harmonic distortion ratio. The digitizer is set to collect 256 samples at a rate of 16 kHz.

Solution:

To begin, let us design the parameters of the 1.4-kHz test tone. With $N=256$ and $F_S=16$ kHz, the fundamental frequency is 16 kHz / 256 or 62.5 Hz. Using $M=23$, we will obtain a test frequency of 1.4375 kHz, which is quite close to the desired 1.4 kHz. The MATLAB code that describes the amplifier and its input test signal is listed on the following pages. Also listed is the code for the Fourier analysis of the output signal.

```
%
% Coherent signal definition - x -
%
        N=256; M=23; A=sqrt(2); P=0;
        for n=1:N,
                x(n)=A*sin(2*pi*M/N*(n-1)+P);
        end;
%
% Amplifier model: input signal: x,  output signal: z
%
% noise model
        s=1; randn('seed',s); % initialize the RN generator
        VDC=0; VRMS=1e-3;
        for n=1:N,
                NOISE=VDC+VRMS*randn;
                x(n)=x(n)+NOISE;
        end;
% input-output transfer characteristic
        c0=0; c1=1; c2=0.01; c3=0.001;
        or n=1:N,
                y(n)=c0+c1*x(n)+c2*x(n)^2+c3*x(n)^3;
        end;
% frequency response behavior
        a0=0.5; a1=0.5; b1=1;           % filter coefficients
        z(1)= a0*y(1);                   % y(0)=0, z(0)=0 initial conditions
        for n=2:N,
                z(n)=a0*y(n)+a1*y(n-1)-b1*z(n-1);
        end;
%
% FFT analysis on the output signal
%
        Z=fft(z)/length(z);
        % magnitude of spectrum Z - units of dBV
                % AC Terms
                        magdBV_Z = 20*log10(sqrt(2)*abs(Z));
                % DC & Nyquist Terms
                        magdBV_Z(1) = 20*log10(abs(Z(1)));
                        magdBV_Z(N/2+1) = 20*log10(1/sqrt(2)*abs(Z(N/2+1)));
%
```

```
% plot spectrum
        plot(0:N/2, magdBV_Z(1:N/2+1))
%
% compute distortion metrics
        S=10^(magdBV_Z(1*M+1)/20);
        H3=10^(magdBV_Z(3*M+1)/20);
        SN3rd=20*log10(S/H3);
%
% end
```

The spectral data that result from the first MATLAB simulation are shown in Figure 8.26. Here we find that the RMS value of the fundamental is 0.50071 V, and the RMS value of the third harmonic is 0.28759 mV. Hence, the signal to third harmonic is

$$S/3^{rd} \, (\text{dB}) = 20 \, \log_{10}\left(\frac{S}{H_3}\right) = 20 \, \log_{10}\left(\frac{0.50071 \text{ V RMS}}{0.28759 \text{ mV RMS}}\right) = 64.816 \text{ dB}$$

Next, we ran the simulation nine more times with a different seed each time. The results are tabulated in Table 8.21.

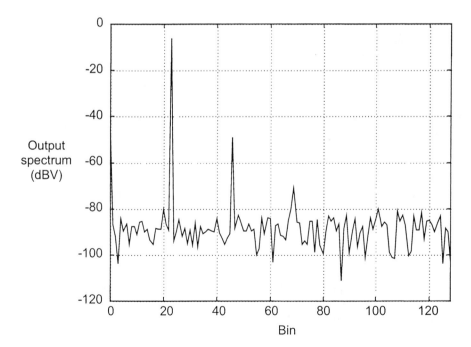

Figure 8.26. MATLAB analog channel model output spectrum (dBV).

Table 8.21. MATLAB Analog Channel Model Test Results

RN Seed	Signal (V RMS)	Third Harmonic (mV RMS)	Signal to Third Harmonic Ratio (dB)
1	0.50071	0.28759	64.816
34	0.50076	0.22247	65.720
65	0.50073	0.22915	66.037
1111	0.50077	0.25223	66.017
653	0.50073	0.27865	65.805
897	0.50080	0.22068	65.987
7775	0.50079	0.24979	65.995
5554	0.50079	0.31756	65.648
88898	0.50077	0.25935	65.655
23157	0.50073	0.28159	65.582
		Mean Value:	65.726
		Standard Deviation:	0.36339

As is evident from Table 8.21, the average value of the signal to third harmonic is 65.726 dB with a standard deviation of 0.36339 dB. If the data sheet called for a device with a signal to third harmonic ratio greater than, say, 65 dB, then to guard against the variation that occurs with noise the test limit should be raised by an amount related to the standard deviation of the underlying noise distribution (see Section 4.3.3). For instance, if the effect of noise is assumed to be normal with a mean value of 0 dB and a standard deviation of 0.36339 dB (same as that in the table), then the limit should be raised by, say, three standard deviations to 65 dB + 3 × 0.36339 dB, or 66.0902 dB. In this way, only 13 out of 10,000 different devices would escape the test and be labeled as good devices. Unfortunately, as this example demonstrates, a good device can be incorrectly rejected.

8.8 SUMMARY

This chapter provides a basic foundation for DSP-based mixed-signal testing, even though all the tests we have discussed so far are purely analog in nature. These same tests are performed on sampled channels, which may include DACs and ADCs. These channels require some slight differences in measurement definition. For example, we cannot calculate the gain of a DAC in volts per volt, because a DAC does not have a voltage input. Nevertheless, many of the measurement techniques are almost identical in nature to purely analog tests. In Chapter 9, "Sampled Channel Testing," we will continue developing DSP-based testing techniques as they apply to channels containing DACs, ADCs, switched capacitor filters, and other sampling circuits.

Problems

8.1. Perform the following signal conversions:

(a) 1.414 V peak, single-ended signal into dBV, differential.

(b) 0.5 V peak-to-peak, single-ended signal into dBm units at 600 Ω.

(c) 100 mV RMS, differential signal into dBV units.

(d) 700 mV RMS, differential signal into peak, single-ended signal.

8.2. An FFT analysis reveals that a signal is present in the eleventh bin with a spectral coefficient described by a_{11}=0.707 and b_{11}=0.100 V. What is the amplitude of this signal? What is its phase? What is its RMS value? Express the signal amplitude in dBV units.

8.3. An FFT analysis reveals that a particular tone has a spectral coefficient of -0.5-j0.5 V. What is the amplitude of this tone? What is its phase? What is its RMS value? Express the signal amplitude in dBV units.

8.4. The small-signal gain of a channel is defined as the derivative of the output voltage with respect to the input voltage. Determine the gain of the following channels:

(a) $V_{out} = 0.1 + 0.99\, V_{in}$

(b) $V_{out} = 0.1 + 2V_{in} + 0.1\, V_{in}^2 + 0.01\, V_{in}^3$

(c) $V_{out} = a_0 + a_1 V_{in} + a_2 V_{in}^2 + a_3 V_{in}^3 + \cdots + a_N V_{in}^N$

(d) $V_{out} = 4 \tan^{-1}(V_{in})$

8.5. The gain of an analog channel as a function of the RMS signal level at its input is assumed to be described by the following equation,

$$G = 2.1 - 0.1\, V_{in-RMS} + 0.01\, V_{in-RMS}^2$$

(a) What is the gain error of the channel at an input level of 2.5 V RMS if the ideal gain is 2.0 V/V? What is the gain error when the input is increased to 4.0 V RMS?

(b) What are the gain tracking errors at input levels of 3, 0, -3, -6, and –12 dB, when the 0-dB reference level corresponds to a 3.0 V RMS input level? Plot the gain tracking error as a function of input level in dB.

8.6. An analog channel is excited by a 1-V RMS single-ended sinusoidal signal from an arbitrary waveform generator. The output of the channel is digitized at a rate of 32 kHz and 1024 samples are collected and stored in memory. An FFT analysis reveals a signal amplitude of 2.1 V RMS in spectral bin 301. What is the frequency of the test signal? What is the absolute gain of the channel in V/V? In dB?

8.7. A digitizer captures 2048 samples of a signal with a sampling rate of 32 kHz. Signals are present in the following FFT bins: 31, 53, 54, 527, 544, 749, 1011. What are the frequencies of the corresponding signals?

8.8. What is the total RMS value of a multitone signal consisting of sixteen tones having an RMS value of 25 mV each and twenty-four tones having an RMS value of 10 mV each?

8.9. A fifteen-tone multitone signal of 1 V RMS is required to perform a frequency response test. What is the amplitude of each tone if they are all equal in magnitude?

8.10. The frequency response behavior of an amplifier is measured with a 1-V RMS multitone signal consisting of eight equal-amplitude tones. An FFT analysis of the output samples reveal the following output signal amplitudes (RMS), in increasing frequency order: 353, 335, 314, 331, 349, 257, 158, 81 mV. What is the absolute gain of the channel at each frequency, in V/V? What is the relative gain (dB) of the amplifier at each frequency if the reference gain is based on the gain of the first tone?

8.11. A data sheet calls for the −1 dB gain point of a low-pass filter to occur between 19,500 and 20,000 Hz. A frequency response measurement was made and the gain at 19,486.3 Hz was found to be −0.01 dB and the gain at 20,001.97 Hz was −2.4 dB. Using log/log interpolation, determine the frequency at which the gain is −1 dB.

8.12. The frequency response behavior of a frequency selective analog channel is measured with a multitone signal consisting of eight tones. An FFT analysis of the input and output samples reveal the following complex spectral coefficients (in V):

FFT Bin	Input	Output
0	$0.9609 - j0.2768$	$0.9609 - j0.2768$
1	$-0.0107 - j0.9999$	$-0.4896 - j0.8431$
2	$-0.9418 - j0.3363$	$-0.7132 + j0.6186$
3	$0.7735 - j0.6338$	$-0.8270 - j0.5488$
4	$-0.7078 - j0.7064$	$0.8208 - j0.4663$
5	$0.5466 - j0.8374$	$0.0004 + j0.0063$
6	$0.5371 + j0.8435$	$0.0309 - j0.0019$
7	$0.2319 + j0.9727$	$0.0228 - j0.00006$
8	$-0.7535 + j0.6575$	$0.0000 - j0.0000$

(a) What is the RMS value of the input and output signals as a function of frequency?

(b) What is the phase of the input and output signals as a function of frequency? Limit the range of the phase of each tone to ±180 degrees.

(c) What is absolute gain of this amplifier as a function of frequency?

(d) What is the relative gain of this channel as a function of frequency if the reference gain is based on the gain of the tone corresponding to the fourth FFT bin?

(e) What is the phase shift (in degrees) of the analog channel as a function of frequency? Provide an unwrapped description of the phase.

(f) What is the group delay of this channel? Also, determine the group delay distortion of this channel.

8.13. An FFT analysis of the output of an amplifier contains the following harmonically related spectral amplitudes:

FFT Spectral Bin	RMS Voltage
101	0.555 V
202	10 mV
303	1 mV
404	0.1 mV
505	0.01 mV

(a) What is the signal to second harmonic distortion ratio?

(b) What is the signal to third harmonic distortion ratio?

(c) What is the signal to total harmonic distortion ratio?

(d) If the RMS value of the noise component is 125 mV, calculate the signal to total harmonic distortion plus noise ratio.

8.14. An amplifier's input-output behavior can be described mathematically by the following third-order polynomial, $V_{out} = a_0 + a_1 V_{in} + a_2 V_{in}^2 + a_3 V_{in}^3$. What is the signal to third harmonic distortion ratio of this amplifier if the input sinusoidal signal is described by $V_{in}(t) = A \sin(2\pi ft)$.

8.15. A multitone test signal consists of a sum of four 1.0 V RMS sine waves, one at 1.3 kHz, another at 2.1 kHz, another at 3.2 kHz, and the fourth at 5.3 kHz. Determine the frequencies of second-order and third-order intermodulation frequencies.

8.16. An amplifier's input-output behavior can be described mathematically by the following third-order polynomial, $V_{out} = a_0 + a_1 V_{in} + a_2 V_{in}^2 + a_3 V_{in}^3$. What are the third-order intermodulation products (amplitude and frequency) produced by this amplifier if the input is described by $V_{in}(t) = A_1 \sin(2\pi f_1 t) + A_2 \sin(2\pi f_2 t)$.

8.17. A digitizer sampling at 4 kHz captures 16 samples of a 1-kHz sinusoidal signal corrupted by noise. An FFT analysis reveals the following spectral amplitudes:

FFT Spectral Bin	RMS Voltage
0	2300 μV
1	32 μV
2	14 μV
3	12 μV
4	0.707 V
5	31 μV
6	42 μV
7	11 μV
8	4 μV

(a) What is the total RMS level of the noise?

(b) Determine the signal-to-noise ratio of the channel.

(c) What is the root spectral density of the noise over the 2-kHz Nyquist interval.

8.18. Determine the root spectral density of a noise signal that has an RMS value of 543 µV evenly distributed over a frequency range of 19 to 23 kHz.

8.19. What is the RMS value of a noise signal measured over a 12-kHz bandwidth if its root spectral density is $10 \ \mu V/\sqrt{Hz}$? Express this result in dBm units at 50 Ω.

8.20. Calculate the gain of a C-message weighting filter at the FFT spectral bin 73 of a signal that was digitized at 8 kHz using 512 samples.

8.21. Using the MATLAB model of the analog channel described in Section 8.7 with parameters $a_0=0.5$, $a_1=0.5$, $b_1=1$, $c_0=0$, $c_1=1$, $c_2=0.01$, and $c_3=0.001$, simulate the behavior of the channel subject to a 1-V RMS, sixteen-tone multitone signal in the presence of a noise signal with an RMS value of 1 mV. Assume that 512 samples are collected at a rate of 8 kHz. The frequencies of the sixteen tones should be uniformly distributed between 500 and 3500 Hz and include the reference frequency of 1 kHz. Subsequently, compute:

(a) The absolute gain of this amplifier as a function of frequency.

(b) The relative gain of this channel as a function of frequency.

(c) The phase shift (in degrees) of the analog channel as a function of frequency. Provide an unwrapped description of the phase.

(d) The group delay and group delay distortion of this channel.

8.22. Using the MATLAB model of the analog channel described in Section 8.7 with parameters $a_0=0.5$, $a_1=0.5$, $b_1=1$, $c_0=0.5$, $c_1=1$, $c_2=0.02$, and $c_3=0.004$, simulate the behavior of the channel subject to a 1-V RMS, 1.4-kHz sinusoidal signal in the presence of a noise signal with an RMS value of 200 µV. Subsequently, compute the following parameters associated with the output signal assuming that the digitizer collects 512 samples at a rate of 8 kHz:

(a) The signal to second harmonic distortion ratio.

(b) The signal to third harmonic distortion ratio.

(c) The signal to total harmonic distortion ratio.

(d) The total RMS level of the noise that appears at the output.

(e) The signal-to-noise ratio.

(f) The signal to total harmonic distortion plus noise.

8.23. Using the MATLAB model of the analog channel described in Section 8.7 with parameters $a_0=0.5$, $a_1=0.5$, $b_1=1$, $c_0=0$, $c_1=1$, $c_2=0.02$ and $c_3=0.004$, simulate the behavior of the amplifier subject to a 1 V RMS, two-tone multitone signal in the presence of a noise signal with an RMS value of 200 µV. Assume that 512 samples are collected at a rate of 8 kHz. Set one tone to 1140 Hz and the other to 1328 Hz. Subsequently, compute:

(a) The signal to second-order intermodulation distortion ratio.

(b) The signal to third-order intermodulation distortion ratio.

8.24. Using the MATLAB model of the analog channel described in Section 8.7 with parameters $a_0=0.5$, $a_1=0.5$, $b_1=1$, $c_0=0.5$, $c_1=1$, $c_2=0.02$, and $c_3=0.004$, simulate the behavior of the channel subject to a 2.8-kHz sinusoidal signal in the presence of a noise signal with an RMS value of 500 µV. Subsequently, compute the signal to harmonic distortion plus noise ratio as a function of the input signal level. Perform the simulation over an input range of −80 to 0 dB in 5-dB increments. Assume that the digitizer collects 512 samples at a rate of 8 kHz.

References

1. Bob Metzler, *Audio Measurement Handbook*, Audio Precision, Inc., Beaverton, OR, 97075, August, 1993, p. 147

CHAPTER 9

Sampled Channel Testing

9.1 OVERVIEW

9.1.1 What Are Sampled Channels?

Sampled channels are similar to analog channels in many ways. Both channel types consist of a signal transmission path from one or more inputs to one or more outputs. Unlike analog channels, though, sampled channels operate on discrete waveforms rather than continuous ones. Discrete waveforms consist of a sequence of instantaneous voltage samples that are either represented as digital values or as sampled-and-held analog voltages.

Examples of sampled channels include digital-to-analog converters (DACs), analog-to-digital converters (ADCs), switched capacitor filters (SCFs), sample-and-hold (S/H) amplifiers, and cascaded combinations of these and other circuits. The test requirements for sampled channels are very similar to those described in Chapter 8, "Analog Channel Testing." However, the sampled nature of the signals transmitted by sampled channels forces some additional considerations in their test requirements. Because sampled circuits may be subject to quantization errors, aliased interference tones, and reconstruction image tones, they require additional test considerations that are not applicable to continuous analog channels.

Sampled channels are often tested for gain error, distortion, signal-to-noise ratio, PSRR, CMRR, and all the other tests common to analog channels. The similarities and differences between analog channel testing and sampled channel testing will be discussed later in this chapter. First, let us look at examples of sampled channels and how they are applied in system-level applications like cellular telephones and disk drive read channels.

9.1.2 Examples of Sampled Channels

Sampled channels form the interface between the physical world around us and the mathematical world of computers and digital signal processors. A digital cellular telephone (Figure 9.1) contains at least six sampled channels – three for the transmit channel and three for the receive channel. The transmit (XMIT) channel is the signal path that transmits the near-end speaker's voice, while the receive (RECV) channel is the signal path that receives the far-end speaker's voice. The transmit channel of a digital cellular telephone includes a number of ADCs, DACs, filters, and signal processing circuits that convert the speaker's voice into a modulated stream of digital data. The data stream is mixed with a high-frequency carrier signal so that it can be transmitted via radio waves to the cellular base station.

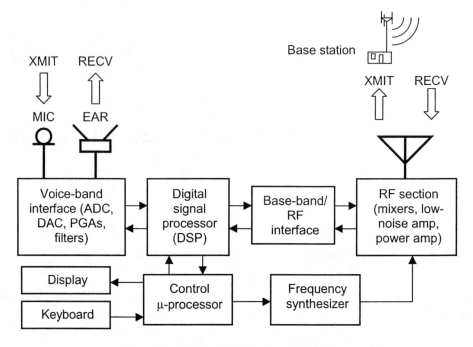

Figure 9.1. Digital cellular telephone block diagram.

The first step in voice transmission is to digitize the speaker's voice using an ADC connected to the cellular telephone's microphone. The voice-band interface circuit is a sampled channel that contains a number of circuits that amplify, filter, and digitize the speaker's voice (Figure 9.2). The digitized voice signal is then routed to either a digital signal processor or a specialized modulator logic block, which converts the ones and zeros of the digitized voice samples into an amplitude/phase modulation protocol similar to that used in computer modems.

Unlike modems, which transmit data over telephone lines at audio frequencies, cellular telephones must transmit the data into an antenna using a radio frequency (RF) power amplifier. The RF transmission frequency is 900 MHz or higher, depending on the type of cellular telephone. Since it would be impossible to directly generate this 900-MHz modulated signal by feeding samples into a DAC (at a rate of 1.8 GHz or more), it is necessary to use an RF mixer circuit to upconvert an intermediate frequency (IF) signal to the RF frequency range. Since the IF signal is much lower in frequency than the RF signal, it can be generated using a DAC channel operating at a lower sampling frequency (Figure 9.3).

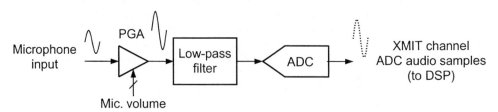

Figure 9.2. Voice-band XMIT (ADC) channel.

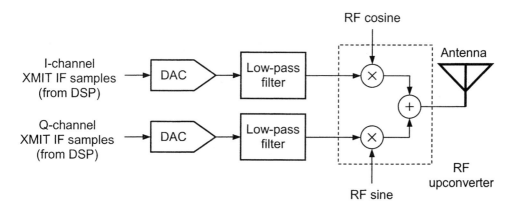

Figure 9.3. XMIT I-channel and Q-channel.

The digital signal processor converts each amplitude/phase pair of the modulation protocol into a sine amplitude sample and a cosine amplitude sample. The sine and cosine amplitude samples are then converted into analog waveforms using two separate mixed-signal channels: the in-phase channel (I-channel) and the quadrature channel (Q-channel). The I-channel waveform is sent to an RF mixer circuit, controlling the amplitude of an RF cosine waveform. Similarly, the Q-channel waveform controls the amplitude of an RF sine waveform. By adding the RF cosine and sine waveforms together, the RF section of the cellular telephone reconstructs the amplitude/phase-modulated data waveform. This composite RF waveform is suitable for transmission through the cellular telephone's antenna.

The receive channel works in a very similar manner, except that the direction of the signal is reversed and the DACs are replaced by ADCs (Figures 9.4 and 9.5). The received data bits are downconverted to IF I and Q channel signals by the RF section, digitized by an IF ADC channel, demodulated into voice samples by the digital signal processor, and converted into audio voice waveforms by the voice-band interface circuit. Thus there are six separate sampled channels in this cellular telephone example: the voice-band interface transmit channel, the transmit I-channel, the transmit Q-channel, the receive I-channel, the receive Q-channel, and the voice-band interface receive channel.

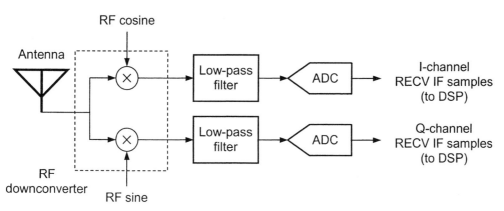

Figure 9.4. RECV I-channel and Q-channel.

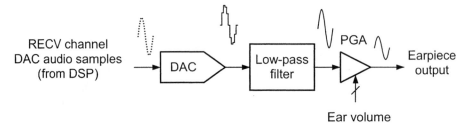

Figure 9.5. Voice-band RECV (DAC) channel.

Other examples of devices containing sampled channels are disk drive read channels, digital audio record and playback devices, and digital telephone answering devices (DTADs). A disk drive read channel is a sampled channel that is used to recover 1/0 data from the read coil of a disk drive read head. The read channel must digitize the high-frequency signal generated by the magnetic variations of the data stored on the spinning disk drive platter.

The read coil signal is typically noisy and distorted by the physical properties of the disk's magnetic storage media. The read channel must clean up and correct the signal using a variety of analog and/or digital filters before data can be recovered. One of the major challenges in read channel testing is that the signals operate at extremely high frequencies. ATE testers are more adept at measuring and sourcing low-frequency signals with a high degree of accuracy. At higher frequencies, the signal source and measurement quality degrades, making the high-frequency read channel difficult to test.

Digital audio channels are similar to the voice-band interface transmit and receive channels of a cellular telephone. However, the sampled channels of a digital audio circuit must record and play back much higher-quality audio than that required for a telephone conversation. One of the major challenges in digital audio testing is that the signals are low in frequency but they have very tight signal-to-distortion and signal-to-noise specifications. Digital telephone answering devices, on the other hand, are more similar to telephones in that they do not need to record and play back especially high-quality audio signals.

Clearly, the uses for sampled channels are very diverse, and a wide variety of testing requirements are needed, depending on the end application of the channel. Despite the wide differences in functionality and quality requirements, though, most sampled channels have a series of common test specifications, including gain error, signal-to-noise, PSRR, etc. Before we look at these common tests in detail, let us look at the different kinds of circuits that can be classified as sampled channels.

9.1.3 Types of Sampled Channels

Sampled channels fall into four basic categories: digital in/analog out, analog in/digital out, digital in/digital out, and analog in/analog out. For convenience, we will introduce the notation DIAO, AIDO, DIDO, and AIAO to refer to these four types of sampled channels. Let us look at some examples of each of the four categories of mixed-signal channels to see how they operate on continuous and sampled waveforms.

Falling into the digital in/analog out (DIAO) category are DACs and cascaded combinations of DACs and other circuits. DIAO channels are characterized by one or more digital signal

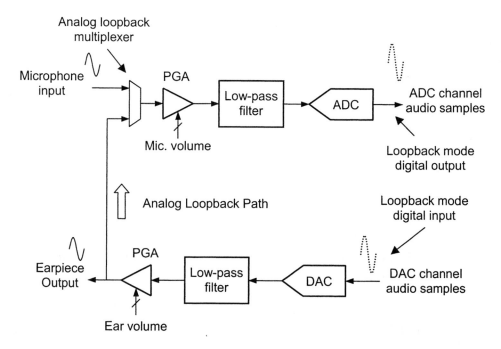

Figure 9.6. Digital input / digital output (DIDO) sampled channel (loopback test mode).

inputs and one or more analog signal outputs. Note that a programmable gain amplifier does not fall into this category, even though it does have a digital input. A PGA's transmission channel consists of an analog input and an analog output. The digital control lines feeding a PGA are not generally used to transmit signal information; so we cannot really consider the PGA to have a digital *signal* input. Furthermore, the PGA does not sample or reconstruct its signals in any way; so it is considered an analog channel rather than a sampled channel. A very high-speed digital line driver might also be considered a simple DIAO channel, as far as testing is concerned, since it converts a digital input into an output with analog characteristics (rise time, overshoot, undershoot, etc.).

ADCs and cascaded combinations of ADCs and other circuits fall into the second category, analog in/digital out (AIDO). AIDO circuits are characterized by one or more analog signal inputs and one or more digital signal outputs. Comparators and slicers fall into this category since they act as one-bit ADCs.

The third category is digital in/digital out (DIDO). While it may not be obvious that a DIDO circuit would have anything to do with a mixed-signal sampled channel, DIDO circuits are actually quite common in mixed-signal testing. Consider the analog loopback mode illustrated in Figure 9.6. In this test mode, a DIAO DAC channel's analog output is connected through a special test multiplexer to the analog input of an AIDO ADC channel. The cascaded circuit has a digital input and a digital output, yet it is a mixed-signal sampled channel. Another example of a DIDO sampled channel is a digital filter, which accepts digital samples at its input, filters the samples using a mathematical algorithm, and produces digital samples at its output. While a digital filter can be tested as a sampled channel by measuring its frequency response, it should be noted that it can be tested much more efficiently and thoroughly using traditional digital methodologies, since it does not contain any analog circuit elements.

An analog in/analog out (AIAO) sampled channel can be formed by placing a device in digital loopback mode, in which the digital output of an AIDO channel is looped back into a DIAO input. The resulting circuit has an analog input and output, but it may exhibit quantization errors, imaging, and aliasing, just like any other sampled circuit. Another example of AIAO sampled channels is the switched capacitor filter (SCF). Switched capacitor filters operate on sampled-and-held or continuous analog input signals, sampling them and applying a high-pass, low-pass, or other filter characteristic to the analog samples. The output of a switched capacitor filter is a sampled-and-held version of the filtered waveform. Since the switched capacitor output is a "stepped" waveform, it is considered a sampled channel, even though it never converts the analog signal into the digital domain. A third example of an AIAO sampled channel is the simple sample-and-hold (S/H) amplifier. This sampled channel converts a continuous analog input waveform into a sampled-and-held analog waveform. In theory, S/H amplifiers and switched capacitor filters do not introduce quantization errors, since they hold nonquantized voltages using capacitors. But since they introduce a sample-and-hold operation, they suffer from all the same imaging, aliasing, and $\sin(x)/x$ rolloff characteristics associated with ADC and DAC channels.

In general, sampled channels are more difficult to test than analog channels. Difficulties arises from a number of factors, such as quantization noise, image and alias tones, sampling rate synchronization headaches, and extra complexity in the form of coherent sampling loops in the digital pattern. One of the main differences between testing analog channels and testing sampled channels is that the sampling rate of the ATE tester's digital patterns, AWG sampling rates, and digitizer sampling rates must mesh with the sampling rate of the DUT. Otherwise, coherent DSP-based testing is not possible. Achieving a coherent DSP-based sampling system that meets the requirements of both the DUT's various sampling circuits and the ATE tester's various instruments can be quite challenging. In the next section we shall look at some of the sampling considerations we face as we design cost-effective tests for sampled channels. We will also look at the structure of the digital patterns that source and capture digital signal samples on the digital side of mixed-signal sampled channels.

9.2 SAMPLING CONSIDERATIONS

9.2.1 DUT Sampling Rate Constraints

When making a coherent DSP-based analog channel measurement, we only need to insure that the fundamental frequency of the AWG is related to the fundamental frequency of the digitizer by an integer ratio (usually a ratio of 1/1), and that the various Nyquist frequencies are above the maximum frequency of interest. Other than these constraints, we are fairly free to choose whatever sampling frequencies we want. Once we begin testing sampled channels, however, we are often burdened with very specific sampling rate constraints placed upon us by the DUT specifications.

The data sheet for a mixed-signal DUT often requires a specific sampling frequency or list of sampling frequencies for each of the DUT's sampled channels. For example, the transmit (ADC) and receive (DAC) channels in a codec device may be specified at a sampling rate of exactly 8 kHz. The tester's sampling systems must mesh with this sampling rate so the waveforms generated and digitized by the DUT's sampled channels are coherent with the tester's AWGs, digitizers, and digital patterns. The most cost-effective testing for this type of device is simultaneous testing of both the transmit and receive channel at the same time. This type of test requires that all sampling rates must be coherent, including the transmit channel, receive channel, digital pattern frame syncs, digital source data rate, digital capture data rate, AWG, and digitizer.

In some cases, it may be acceptable to force-fit the sampling rate of a DUT into a sampling system that is more agreeable to the ATE tester's instruments. For example, we may find that a device whose master clock is specified at 38.88 MHz may actually fit the tester's constraints better if we shift it slightly to 38.879962 MHz. However, it is seldom acceptable to shift a clock by more than a fraction of a percent. If we shifted the 38.88 MHz clock to 40 MHz, for instance, we would run the risk of generating correlation errors between the tester and stand-alone bench equipment operating at 38.88 MHz. Correlation errors in mixed-signal tests are often the result of minor discrepancies in test conditions, such as sampling rates and output loading conditions. It is much safer to use the exact specified master clock and sampling rate combinations outlined in the device data sheet.

It would seem that a tester costing one million dollars or more would be able to produce any sampling rate the test engineer desires. Unfortunately, most testers have constraints that limit the clock frequencies that can be generated or utilized by each instrument. Even when a tester's clocks are highly programmable, certain frequencies may result in a better quality of test than others. For example, a digital pattern rate that is divided from the tester's master clock by a factor of 2^N may generate less sampling jitter than one generated using other divide ratios. The lower jitter may result in superior signal-to-noise ratio measurements. Of course, these constraints are highly tester dependent. A frequency that is not ideal on one type of tester may cause no problem on another. This is one of the reasons that it is so difficult to convert mixed-signal test programs from one tester platform to another.

Yet another headache in selecting sampling rates is the long settling time of low-jitter frequency generators and/or phase-locked loops (PLLs). These often take a long time (25-50 ms) to settle to a stable frequency after their frequency setting is changed. It is often better to select a single master clock frequency and let the tester's low-jitter master clock generator stabilize to this frequency. Ideally, all sampling rates in the test program can be derived from this single master clock frequency using digital clock divider circuits. Since digital divider circuits require no additional setting time, test time can be minimized using this approach. Unfortunately, this may or may not be possible. Sometimes it is necessary to switch the master clock back and forth between various frequencies, adding undesirable test time as the clock source stabilizes. Again, this constraint is highly tester dependent and may not be an issue on some types of testers.

9.2.2 Digital Signal Source and Capture

When testing mixed-signal devices, the tester must apply digital signal samples to the DUT's inputs and collect digital signal samples from its outputs. The DUT usually requires these samples to be applied and captured at a particular sampling rate. A repeating digital pattern, called a *sampling frame*, is often required by the DUT to control the timing of the digital signal samples. Figure 9.7 shows an example of a digital pattern consisting of a repeating frame, digital signal inputs to a DAC channel, and digital signal outputs from an ADC channel. The pattern also includes a feature not found in purely digital patterns: source and capture shift and load commands. These pattern commands, called *microcode instructions*, determine when the data for source memory will be substituted in place of the normal pattern data. They also determine when the digital samples from the DUT will be stored into capture memory. If the DUT has a serial input or output, the microcode SHIFT commands determine the time at which of each bit of the data word is shifted into or out of the DUT. Other microcode instructions initialize loop counters, determine loop endpoints, stop the pattern, etc.

```
                                                    R R P
                                                    F F C
                                                    I I M
                                                    R W D
                                                    D R C
                            pin     pin             X X K
                            RFID11  RFID11 RFIA RFID
LABEL     COMMAND           dig_src dig_cap (H)  (H)  S S S  COMMENT
          SET_LOOP 512      START   TRIG   .d0  .−   1 1 1  Prepare to collect 512 samples
ADC_DAC:                                   .−   .d0  1 1 0  Send Frame Sync (PCMDCK)
                            SEND           .d8  .W   − 0 1  Source one DAC sample
                                           .−   .X   − 1 −
                                    STORE  .d9  .−   0 − −  Capture one ADC sample
          REPEAT 7                         .−   .−   1 − −  Finish the 12-cycle frame
          END_LOOP ADC_DAC                 .−   .−   − − −  Loop back to ADC_DAC
          HALT                             .−   .−   − − −  until 512 samples
                                                            are collected
```

Figure 9.7. Mixed-signal digital pattern example.

The waveforms stored in source memory are computed by the tester computer during an initialization run. Signals such as sine waves, multitones, and ramps can be stored into various locations in the source memory, ready to be inserted one sample at a time into the looping frame pattern. In Figure 9.7, a multibit parallel data word is written into the RFID pin group using a W character linked with the SEND command on the third vector. Likewise, the samples collected during the looping frame pattern are stored one sample at a time into capture memory using the STORE command on the fifth vector. The captured samples can then be moved into an array processor or tester computer for DSP analysis (DFTs, FFTs, average value, RMS value, etc). Naturally, the example pattern in Figure 9.7 is specific to a particular ATE tester. Different testers will use a variety of notations to specify the STORE and SEND operations, but all true mixed-signal testers provide a looped source and capture capability.

The sampling frame usually consists of one or more high-frequency clocks and one or more frame synchronization signals that determine the timing of the input and output sample data stream. For example, a digital audio device data sheet may specify that DAC channel samples are to be applied in parallel on DUT pins DAC7-DAC0 on the second rising edge of the master clock (MCLK) after the frame synchronization signal (FSYNC) goes low. Likewise, the data sheet may specify that the ADC channel samples will be available at ADC7-ADC0 on the third rising edge of the master clock after FSYNC goes low. Finally, the FSYNC frequency may be specified to run at a rate of MCLK divided by 16. Figure 9.8 shows these timing relationships.

In addition to the timing of the digital signal data inputs and outputs, the master clock and frame sync are usually required to operate at very specific frequencies. Like AWGs and digitizers, the digital source and capture instruments have their own fundamental frequencies, defined as the sampling rate divided by the number of samples sourced or captured. Coherent DSP-based testing requires that the fundamental frequency of the digital source, digital capture, AWG, and digitizer must all be equal (or at least related by an integer ratio).

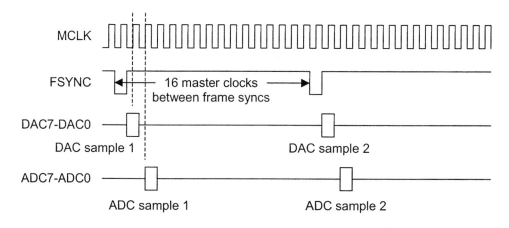

Figure 9.8. Sampling frame timing diagram.

Unlike digital circuits, mixed-signal channels are usually specified at a particular frequency, rather than a particular period. When making coherent DSP-based measurements, there is a huge difference between a 25-ns master clock period and a 38.88-MHz (25.72016460905... ns) period. The frequency of these clocks (relative to the sampling frequencies of the tester's AWGs and digitizers) must be quite accurate to synchronize the fundamental frequencies of all the various sampling systems, including the DUT's sampling rates. A frequency accuracy of one part in a million or better is required to achieve acceptable signal-to-noise performance in coherent DSP-based tests. For this reason it is usually not acceptable to round off the clock periods to the nearest nanosecond or even the nearest picosecond. Digital circuits, by contrast, are often tested at frequencies higher or lower than the normal operating frequency. Their timing can often be rounded off to the nearest nanosecond to simplify the automated test pattern generation (ATPG) process.

Depending on the tester's architecture, it may or may not be possible to generate certain sampling rates. For example, a device with a master clock of 41.327 MHz may produce a fundamental frequency that cannot be easily matched by the tester's high-performance digitizer. It may be necessary to switch to a non-power-of-2 sample size or it may be necessary to shift the master clock slightly to allow a coherent measurement. When shifting the master clock to accommodate the tester's instruments, the test engineer should take care. The performance of the DUT may shift slightly as well. For example, the cutoff frequency of a switched capacitor filter changes with its master clock. Shifting the master clock by 1% will shift the 3-dB point of the filter by 1% as well. There are many more subtle problems that occur when the test conditions are shifted in this manner; so it is best to test the DUT at exactly the specified frequencies to avoid correlation errors.

An interesting question arises when we try to test mixed-signal devices on digital testers: Do we really need source and capture memory to test mixed-signal devices? Let us look at the alternative. A mixed-signal DAC pattern could be written as a "flattened" series of samples with a repeating digital sample frame, as shown in Figure 9.9. Notice that the frame sync pattern is identical from one DUT sample to the next. If a frame consists of 1024 digital samples, each of which requires a 100 vector frame, then the pattern would require 102400 vectors. Mixed-signal test programs may consist of dozens of these patterns. Testing mixed-signal devices using patterns such as the one shown in Figure 9.9 would require many megabytes of vector memory, an obvious waste of digital pattern memory. Also, flattened patterns such as the one in Figure 9.9 pose a debugging problem. What if we needed to quickly characterize a DAC's performance using a frequency of 2 kHz instead of 1 kHz? Likewise, what if we suddenly needed to try a frame of 102 vectors rather than a frame of 100 vectors? We would have to manually insert the new vectors or digital samples into the huge flattened vector set or use a cumbersome digital pattern compiler to recreate a totally new digital pattern. Fortunately, mixed-signal tester architectures provide source and capture memory to allow compact, easily modified digital patterns for mixed-signal tests.

A true mixed-signal tester uses source and capture memory to implement a type of vector compression that is ideally suited to mixed-signal sampling frames. Notice that the frames in Figure 9.9 consist of the same basic pattern of ones and zeros for each sample. Only the value of the DAC input data and ADC output data changes from one frame to the next. A mixed-signal tester allows digital samples to be inserted into or extracted from a repeating frame loop, as previously shown in Figure 9.7. The samples are specified with a digital signal sample notation, such as W or X instead of 1 or 0. The Ws and Xs are place holders for digital signal samples, which are either read from source memory or written to capture memory. Figure 9.10 shows how source memory and vector memory are combined to weave together the digital frame data with the digital samples. Similarly, Figure 9.11 shows how DUT output data is captured within a repeating capture frame.

Digital samples can be sourced to and captured from the DUT in either a parallel format or a serial format. In parallel format, the data for each sample are loaded into the device with a single clock cycle, as shown in the previous examples. Serial format, by contrast, involves a serial shifting operation that takes multiple data clock cycles. Versatile mixed-signal testers include built-in hardware features that ease the conversion from parallel to serial and serial to parallel data formats.

```
                                                      R R P
                                                      F F C
                                                      I I M
                                                      R W D
                                                      D R C
                       pin      pin                   X X K
                       RFID11   RFID11   RFIA   RFID
LABEL     COMMAND      dig_src  dig_cap  (H)    (H)   S S S  COMMENT

          START        TRIG              .d0    .-    - - -  Prepare to collect 512 samples
ADC_DAC                                  .-     .d0   1 1 0
          SEND                  .d8      .d94   1 0 1  Source 1st DAC sample
                                         .-     .X    1 1 1
                       STORE             .d9    .-    0 1 1  Capture 1st ADC sample
          REPEAT 8                       .-     .-    1 1 1  Finish the 12-cycle frame
                                         .-     .d0   1 1 0
          SEND                  .d8      .dE89  1 0 1  Source 2nd one DAC sample
                                         .-     .X    1 1 1
                       STORE             .d9    .-    0 1 1  Capture 2nd ADC sample
          REPEAT 8                       .-     .-    1 1 1  Finish the 12-cycle frame
                                         .-     .d0   1 1 0
          SEND                  .d8      .d299  1 0 1  Source 3rd DAC sample
                                         .-     .X    1 1 1
                       STORE             .d9    .-    0 1 1  Capture 3rd ADC sample
          REPEAT 8                       .-     .-    1 1 1  Finish the 12-cycle frame

...Repeat this pattern for 512 samples
```

Figure 9.9. Flattened mixed-signal DAC frames.

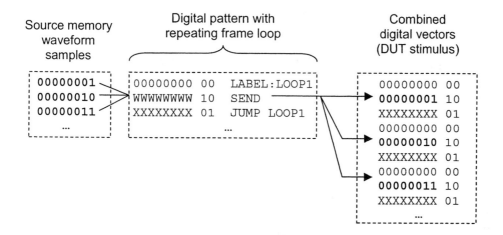

Figure 9.10. Source memory sample insertion.

When supplying serial data to a DUT's digital signal input, data are loaded into source memory in a parallel format. The digital subsystem hardware performs a parallel to serial shift operation controlled by digital pattern microcode instructions such as SEND and SHIFT. Likewise, when capturing serial data from a DUT's digital signal output, data can be translated from serial format to parallel format before they are stored into capture memory. Using the hardware serial/parallel conversion features, the test engineer does not have to spend additional time writing software translation routines.

In addition to the parallel/serial data conversion, a mixed-signal tester should be capable of shifting the digital signal data into or out of the DUT with the most significant bit (MSB) first or the least significant bit (LSB) first. Mixed-signal testers accommodate the MSB-first / LSB-first translation using built-in digital logic in the digital subsystem.

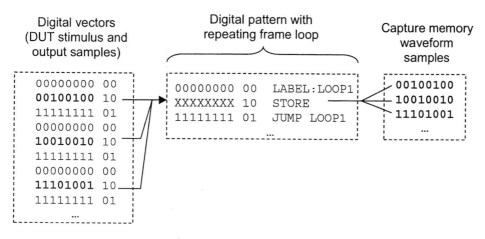

Figure 9.11. Capture memory sample extraction.

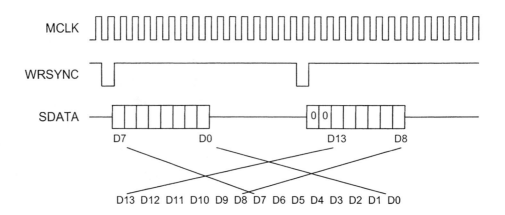

Figure 9.12. Scrambled bit order in a two-byte interface.

Unfortunately, mixed-signal ATE testers cannot predict every data format. Occasionally the test engineer will run across an "oddball" data format that requires an explicit software bit scrambling operation. For example, a device may have a 14-bit digital signal format that must be shifted into the device serially, with the least significant 8 bits shifted into the DUT MSB first, followed by the most significant 6 bits shifted into the DUT MSB first. Figure 9.12 shows why this is an odd format. Notice that the order of the bits is scrambled. Before we want to use the tester's parallel to serial shift hardware, we have to produce data samples that have the same scrambled bit order. This scrambling operation must be performed in software before the samples are loaded into the source memory. This is not a catastrophic problem; it just adds some overhead to the test development process and makes the code a little harder to follow.

9.2.3 Simultaneous DAC and ADC Channel Testing

When a DUT contains two or more channels that can be tested simultaneously, the test engineer will often test both channels at once to save test time. For example, a central office codec may have a transmit (ADC) channel and a receive (DAC) channel that both operate at a sampling rate of 8 kHz. The various parameters of these two channels can be tested in parallel, saving test time. For example, the absolute gain, distortion, and signal-to-noise of the DAC channel can be tested while the same tests are being performed on the ADC channel. The AC measurement system for simultaneous ADC and DAC testing is shown in Figure 9.13. In addition to the digital source and capture memories shown in Figure 9.13, the digital subsystem must also provide any necessary reset functions, initialization patterns, master clocks, frame syncs, etc.

In Figure 9.13, we have shown four different sampling systems operating simultaneously. The AWG is one sampling system and the digitizer is another. The third sampling system is formed by the source memory and the DAC channel. The fourth sampling system consists of the ADC and the capture memory. In order to ensure that each sampling system forms a coherent DSP-based measurement, the sampling rate and the number of samples collected must be chosen in consideration with the other sampling systems.

Consider the situation involving the AWG and ADC, together with the capture memory. The AWG produces a test tone with frequency F_t that stimulates the ADC channel of the DUT.

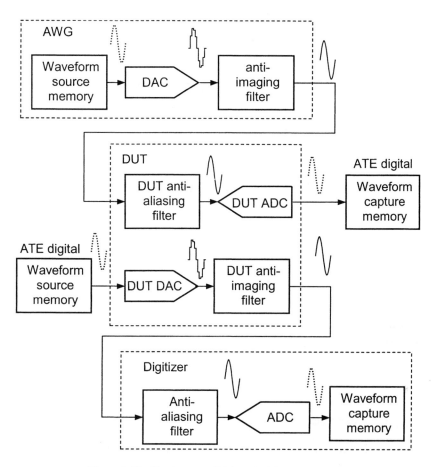

Figure 9.13. Simultaneous DAC and ADC channel testing.

Assuming that the AWG is operating at a sampling rate of $F_{s\text{-}AWG}$ and cycles through N_{AWG} samples, we can express the test tone frequency as

$$F_t = M_{AWG} \frac{F_{s\text{-}AWG}}{N_{AWG}} \tag{9.1}$$

where M_{AWG} is a spectral bin number for the AWG. We can also express Eq. (9.1) as

$$F_t = M_{AWG} F_{f\text{-}AWG} \tag{9.2}$$

where the fundamental frequency for the AWG is given by

$$F_{f\text{-}AWG} = \frac{F_{s\text{-}AWG}}{N_{AWG}} \tag{9.3}$$

As the ADC together with the capture memory forms another coherent sampling system with sampling rate $F_{S\text{-}ADC}$ and N_{ADC} samples, we can also write the test tone frequency as

$$F_t = M_{ADC} \frac{F_{s\text{-}ADC}}{N_{ADC}} \tag{9.4}$$

or

$$F_t = M_{ADC} F_{f\text{-}ADC} \tag{9.5}$$

where M_{ADC} is a spectral bin number for the ADC and $F_{f\text{-}ADC}$ is the fundamental frequency for the ADC given by

$$F_{f\text{-}ADC} = \frac{F_{s\text{-}ADC}}{N_{ADC}} \tag{9.6}$$

The number of samples collected in the capture memory N_{ADC} should be made a power of two in order for the sample set to be compatible with the FFT analysis that will follow.

Recognizing that Eqs. (9.2) and (9.5) are equal in a coherent sampling system allows us to write the following expression

$$M_{AWG} F_{f\text{-}AWG} = M_{ADC} F_{f\text{-}ADC} \tag{9.7}$$

Any two sampling systems that satisfy this equation will be coherent with one another.

Following a similar development, we can state that the requirement for coherence between the DAC and digitizer is

$$M_{DAC} F_{f\text{-}DAC} = M_{DIGITIZER} F_{f\text{-}DIGITIZER} \tag{9.8}$$

Since the ADC and DAC samples are usually collected using a single digital pattern loop, their sampling rates and number of samples are not independent. The simplest sampling combination that meets all coherence requirements is one in which all four sampling systems have the same fundamental frequency. This requirement is met by any combination of sampling rates and number of samples in which

$$F_{f\text{-}AWG} = F_{f\text{-}ADC} = F_{f\text{-}DAC} = F_{f\text{-}DIGITIZER} \tag{9.9}$$

Of course, this also implies that the test frequency is equal in both the ADC and DAC channels as

$$M_{AWG} = M_{ADC} = M_{DAC} = M_{DIGITIZER} \tag{9.10}$$

In the 8-kHz codec example, the simplest sampling system is formed by choosing an 8-kHz sampling rate on all four systems. For example, if we want to collect 512 samples from the ADC channel while sending 512 samples to the DAC channel, we could use the following sampling system

$$F_{s\text{-}ADC} = F_{s\text{-}AWG} = 8 \text{ kHz}$$
$$N_{ADC} = N_{AWG} = 512$$
$$F_{s\text{-}DAC} = F_{s\text{-}DIGITIZER} = 8 \text{ kHz}$$
$$N_{DAC} = N_{DIGITIZER} = 512$$

In this case, each of the four sampling systems has a fundamental frequency of 8 kHz/512 = 15.625 Hz. Often test engineers will double or quadruple the sampling rate of the digitizer and/or AWG to give a higher Nyquist frequency. A higher Nyquist frequency allows a digitizer to detect more frequency components in the output spectrum of a DAC. This is useful for testing the DAC's anti-imaging low-pass filter. Similarly, a higher Nyquist frequency allows an AWG to produce a cleaner test stimulus, free of images. To maintain fundamental frequency compatibility with these higher sampling rates, we have to double or quadruple the number of samples we collect with the digitizer or source with the AWG. For example, we can use the following sampling system to double the fundamental frequency of the tester's digitizer while quadrupling the fundamental frequency of the AWG:

ADC:

$F_{s\text{-}ADC} = 8 \text{ kHz}$

$N_{ADC} = 512$

$F_{f\text{-}ADC} = 8 \text{ kHz} / 512 = 15.625 \text{ Hz}$

AWG:

$F_{s\text{-}AWG} = 8 \text{ kHz} \times 4 = 32 \text{ kHz}$

$N_{AWG} = 512 \times 4 = 2048$

$F_{f\text{-}AWG} = 32 \text{ kHz} / 2048 = 15.625 \text{ Hz}$ (matches ADC)

DAC:

$F_{s\text{-}DAC} = 8 \text{ kHz}$

$N_{DAC} = 512$

$F_{f\text{-}DAC} = 8 \text{ kHz} / 512 = 15.625 \text{ Hz}$

Digitizer:

$F_{s\text{-}DIGITIZER} = 8 \text{ kHz} \times 2 = 16 \text{ kHz}$

$N_{DIGITIZER} = 512 \times 2 = 1024$

$F_{f\text{-}DIGITIZED} = 16 \text{ kHz} / 1024 = 15.625 \text{ Hz}$ (matches DAC)

9.2.4 Mismatched Fundamental Frequencies

The example of the previous subsection dealt with the situation where the fundamental frequencies were made equal. In this subsection, we shall investigate the fact that they need only be related by a ratio of two integers, as can be seen from Eq. (9.8) when rearranged

$$\frac{F_{f-DIGITIZER}}{F_{f-DAC}} = \frac{M_{DAC}}{M_{DIGITIZER}} \qquad (9.11)$$

For instance, in the previous example we could have used the following sampling system:

DAC:

$F_{s-DAC} = 8$ kHz

$N_{DAC} = 512$

$F_{f-DAC} = 8$ kHz $/ 512 = 15.625$ Hz

Digitizer:

$F_{s-DIGITIZER} = 8$ kHz $\times (3/2) = 12$ kHz

$N_{DIGITIZER} = 512$

$F_{f-DIGITIZER} = 8$ kHz $\times (3/2) / 512 = 15.625$ Hz $\times (3/2) = 23.4375$ Hz

Clearly, the digitizer fundamental frequency is 3/2 times that of the fundamental frequency of the DAC. Therefore, the spectral bins in the DAC channel are also related to the spectral bins in the digitizer channel in much the same way, i.e. $M_{DAC} = (3/2) M_{DIGITIZER}$. For example, if we use bin 9 in the DAC channel, it will produce a frequency of 9×15.625 Hz, or 140.625 Hz. Since the digitizer's fundamental frequency is 15.625 Hz $\times (3/2)$, this same frequency will appear at spectral bin 6 in the FFT of the digitizer's samples.

Note that this integer-ratio sampling approach will often force us into even-numbered spectral bins; so it is sometimes inferior to the more straightforward approach using equal fundamental frequencies. However, the test engineer will occasionally find that the use of an integer-ratio sampling approach is the best way to achieve a coherent sampling set given the constraints of the DUT and ATE tester.

Example 9.1

A DAC must be tested with a sampling rate of 5 MHz and a test tone frequency of approximately 5 kHz. The output of the DAC is sampled by a digitizer with a maximum sampling rate of 20 kHz. The ATE tester's source memory will be needed for other DAC tests; so we need to use the minimum number of DAC samples possible (no more than 1024 samples). Due to sampling constraints inherent to the tester, both the DAC sampling rate and the digitizer sampling rate must be integer multiples of 1 Hz. Find a coherent sampling system compatible with these constraints.

Solution:

First, let us first consider the DAC. The DAC operates at a sampling rate of approximately 5 MHz and is exercised by a tone at approximately 5 kHz. Hence we can write

$$\sim 5 \text{ kHz} = M_{DAC} \frac{\sim 5 \text{ MHz}}{N_{DAC}} \quad (9.12)$$

In order to determine values for the two unknowns (N_{DAC} and M_{DAC}), we shall consider that the test and sampling frequencies are exactly 5 kHz and 5 MHz, respectively, as we are attempting to approach these frequencies as closely as possible. In addition, we also know that N_{DAC} has to be less than 1024, due to source memory limitations. This can only be satisfied if $M_{DAC} = 1$, resulting in $N_{DAC} = 1000$. Thus, by storing 1000 samples of a single cycle of a sine wave in the source memory we can produce a 5-kHz test tone to exercise the DAC.

Next, let us consider the digitizer. According to the information supplied, we can relate the sampling rate, number of samples and test frequency according to

$$\sim 5 \text{ kHz} = M_{DIGITIZER} \frac{\sim 20 \text{ kHz}}{N_{DIGITIZER}} \quad (9.13)$$

Our first attempt at solving for the two unknowns ($N_{DIGITIZZER}$ and $M_{DIGITIZER}$) is to consider setting the fundamental frequency of the digitizer equal to the fundamental frequency of the DAC, 5 MHz/1000 or 5 kHz. However, on doing so, we create a problem. A 20-kHz sampling rate would only allow us to collect only 4 samples with a fundamental frequency of 5 kHz. This is clearly not enough samples. One possible solution is to collect 256 samples at 20 kHz, and look for power in spectral bin 64 (64 × 20 kHz / 256 = 5 kHz). Unfortunately, this bin would result in the same samples over and over, as $N_{DIGITIZER}$ and $M_{DIGITIZER}$ are not mutually prime. So instead, we shall select $M_{DIGITIZER} = 67$, the nearest prime number, and $N_{DIGITIZZER} = 256$. In turn, we must make an adjustment to the sampling rates F_{DAC} and $F_{DIGITIZER}$ in order for the DAC and digitizer to be coherent. This we obtain through the development in Section 9.4.3, where we found

$$M_{DAC} \frac{F_{s-DAC}}{N_{DAC}} = M_{DIGITIZER} \frac{F_{s-DIGITIZER}}{N_{DIGITIZER}} \quad (9.14)$$

Substituting known values, we can write

$$1 \times \frac{F_{s-DAC}}{1000} = 67 \times \frac{F_{s-DIGITIZER}}{256} \quad (9.15)$$

or

$$F_{s-DIGITIZER} = \frac{256}{67 \times 1000} F_{s-DAC} \quad (9.16)$$

With the DAC sampling rate set to 5 MHz, the digitizer's sampling rate would have to be 19,104.47 Hz to satisfy coherence.

Unfortunately, we are not done yet. The ATE tester imposes a further constraint where by all sampling rates have to be integer multiples of 1 Hz. In other words, all sample rates must be integer numbers. In order to satisfy this constraint, the sample rates of the DAC and digitizer will have to be altered such that they are both integer numbers. To achieve this, we must first express the rational fraction in Eq. (9.16) as a product of prime numbers and eliminate any common terms. That is,

$$F_{s-DIGITIZER} = \frac{2^8}{67 \times 5^3 \times 2^3} F_{s-DAC} = \frac{2^5}{67 \times 5^3} F_{s-DAC} \qquad (9.17)$$

Next, select F_{s-DAC} as a multiple of 67×5^3 which is nearest the desired frequency of 5 MHz. That is,

$$F_{s-DAC} = 67 \times 5^3 \times n \qquad (9.18)$$

where

$$n = \left[\frac{5 \text{ MHz}}{67 \times 5^3}\right] = [597.015] = 597 \qquad (9.19)$$

and [ ] indicates a rounding to the nearest integer operation. Therefore, F_{s-DAC} = 4,999,875 Hz. Subsequently, from Eq. (9.17) we find $F_{s-DIGITIZER}$ = 19,104 Hz.

Summarizing, the final solution is:

	DAC	Digitizer
Sampling rate	4,999,875 Hz	19,104 Hz
Number of samples	1000	256
Spectral bin	1	67
Fundamental frequency	4,999.875 Hz	74.625 Hz
Test tone frequency	4,999.875 Hz (approx. 5 kHz)	4,999.875 Hz (matches DAC)

In conclusion, we used a different fundamental frequency for the DAC and digitizer, and a different Fourier spectral bin for each to achieve a coherent sampling system. As there are numerous steps to follow to obtain coherence, test engineers will often devise software tools to aid in the selection of sampling rates, sample sizes, etc. for a given set of DUT and tester constraints. This is particularly important when more that two sampling systems are required to be made coherent.

Exercises

9.1. Two sampling systems described in the following table are operating with a test tone of approximately 2,000 Hz:

	System #1	System #2
Sampling rate	32,000 Hz	23,000 Hz
Number of samples	2,000	1,024
Spectral bin	125	89

Slightly adjust the sampling rate of each system so that the two sampling systems will become coherent. What are the sampling frequencies?

Ans. Coherence requires $F_{s1} = 1.390625 \times F_{s2}$. If F_{s1}=32 kHz, then F_{s2}= 23.011236 kHz.

9.2. Repeat Exercise 9.1 but this time limit the sampling frequencies to integer multiples of 1 Hz.

Ans. One possible solution: F_{s1}=31,951 Hz and F_{s2}= 22,976 Hz.

9.2.5 Undersampling

Undersampling is a technique that allows a digitizer or ADC to measure signals beyond the Nyquist frequency. A digitizer sampling at a frequency of F_s has a Nyquist frequency equal to $F_s/2$. Any input signal frequency, F_t, which is above the Nyquist frequency will appear as an alias component somewhere between 0 Hz and the Nyquist frequency. We normally try to filter the input signal so that it has no components above the Nyquist frequency. However, we may remove the filter if we want to allow our digitizer or DUT to collect samples from a signal that includes components above the Nyquist frequency. This technique is called *undersampling*. Undersampling can also be used to measure the out-of-band rejection of a low-pass antialiasing filter before an ADC as shown in Figure 9.14. Frequencies that extend beyond the Nyquist rate of the ADC may not be fully attenuated by the filter, and as a result they may appear as low-amplitude in-band alias tones. The amplitude of these alias components can be used to compute the filter cutoff performance at frequencies beyond the Nyquist frequency.

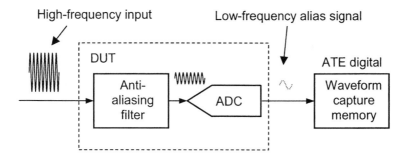

Figure 9.14. Low-pass antialiasing filter and ADC.

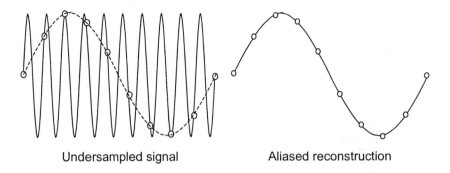

Figure 9.15. Undersampling a high-frequency sine wave.

Undersampling is often used when we have a digitizer or a DUT ADC with a limited sampling rate, but we want to capture a signal with components beyond the Nyquist frequency. Provided that the bandwidth of the digitizer's front end is adequate, we can use undersampling to collect samples from the high frequency signal (Figure 9.15).

There are several things to consider when using undersampling. First, all the noise components from 0 Hz to the digitizer or ADC input bandwidth will be additively folded back into the range from 0 Hz to $F_s/2$. This means the signal-to-noise ratio of the aliased signal will probably be degraded compared to a fully sampled measurement.

Second, the digitizer's front end may be less linear or may have a gain error at the frequency of the test signal. These problems can usually be corrected using focused software calibration techniques. Finally, the aliased image of two or more tones in a multitone signal may fold back to the same frequency. Care must be taken when selecting frequencies so they each fall into a unique in-band FFT bin.

To calculate the expected alias frequency of a test tone denoted F_{ta}, perform the following steps:

1. Repetitively subtract F_s from the test frequency F_t until the result is between 0 and F_s. Call this result, F'_{ta}. Mathematically, this is equivalent to

$$F'_{ta} = F_t - nF_s \quad \text{where} \quad n = \left\lfloor \frac{F_t}{F_s} \right\rfloor \tag{9.20}$$

where [ ] indicates a rounding down to the nearest integer operation.

2. Next, check whether F'_{ta} is above or below the Nyquist frequency $F_s/2$. If it is below the Nyquist frequency then F'_{ta} is the frequency of the aliased tone. Otherwise, it is an image and the aliased tone will appear at frequency $F_s - F'_{ta}$. Mathematically, we can express these two conditions as

$$F_{ta} = \begin{cases} F'_{ta} & \text{if } 0 < F'_{ta} \leq F_s/2 \\ F_s - F'_{ta} & \text{if } F_s/2 < F'_{ta} \leq F_s \end{cases} \tag{9.21}$$

Example 9.2

A sine wave with a frequency of 65 kHz is sampled by an ADC at a sampling rate F_s of 20 kHz. An FFT is performed on the samples collected by the ADC. At what frequency do we expect the 65-kHz tone to appear? Assuming N samples were collected, at what spectral bin will this frequency appear? Repeat the example with an input tone of 75 kHz.

Solution:

We subtract 20 kHz from 65 kHz until we get an answer less than 20 kHz

$$65 \text{ kHz} - 20 \text{ kHz} - 20 \text{ kHz} - 20 \text{ kHz} = 5 \text{ kHz}.$$

This result is less than the Nyquist frequency of 10 kHz; so we would expect to see the aliased tone at 5 kHz. The spectral bin is calculated using the equation Eq. (9.4)

$$5 \text{ kHz} = M \frac{20 \text{ kHz}}{N}$$

or rewriting

$$M = \frac{N \times 5 \text{ kHz}}{20 \text{ kHz}} = \frac{N}{4}$$

Repeating the example with a 75-kHz input tone, we keep subtracting 20 kHz until we get a number less than F_s: F'_{ta}=15 kHz. This frequency is above the Nyquist; so we expect to see an aliased sine wave at 20 kHz − 15 kHz = 5 kHz. Again, this tone shows up in spectral bin $N/4$. Notice that both input frequencies (65 and 75 kHz) fold back into the same FFT spectral bin. This is the nature of aliasing − it is a many-to-one mapping process. Care must be taken when undersampling multitone signals to avoid overlaps between different frequencies. Each undersampled tone must be selected so that it falls into a unique spectral bin. It would be a mistake to undersample a multitone signal with both 65- and 75-kHz components at a sampling rate of 20 kHz, since the aliased versions of the two tones would overlap at 5 kHz.

9.2.6 Reconstruction Effects in AWGs, DACs, and Other Sampled-Data Circuits

Reconstruction is performed by first converting the discrete samples into a stepped or staircase-like waveform using some form of sampled-and-held process, such as a DAC, followed by a filtering operation to smooth the signal and remove the frequency images. Recall from Chapter 6 that perfect reconstruction cannot be realized in practice. It can only be approached. As a result, some signal artifacts are introduced into the reconstructed waveform. These, in turn, limit the quality of the signal produced by an AWG, DAC, or other sampled-data circuit. In particular, we shall investigate the effects of images and $\sin(x)/x$ rolloff.

Ideally, a perfect smoothing operation has a brick-wall frequency response, that is, one that rejects all signal frequencies completely except those in some frequency region that are allowed to pass unattenuated. Such frequency response behavior is impossible to realize with any real circuit. Instead, one must tolerate some of the image energy appearing in the reconstructed signal. For a single tone with frequency F_t, the image tones, denoted by F_{image}, will appear in the reconstructed waveform at the following frequencies

$$F_{image} = nF_s \pm F_t \quad \text{where} \quad n = 1, 2, 3, \ldots \tag{9.22}$$

The imaging process follows the same mapping rules as the aliasing process, only in reverse. While aliasing is a many-to-one mapping process, imaging is a one-to-many mapping process. Using the previous aliasing example, a 5-kHz sine wave reconstructed at a 20-kHz sampling rate would produce images at 65 and 75 kHz, as well as many other frequencies such as 15, 25, 35, 45 kHz, ..., etc.

A common test performed on a DAC followed by a low-pass anti-image filter is to test the circuit with a test tone set very close to the Nyquist frequency of the channel. This is typically the worst-case condition, as the anti-imaging filter must pass the test tone at F_t undisturbed, but reject most of the image tone that appears very close to it at F_s-F_t. In other words, this single test acts to verify that the filter's pass-band and stop-band regions meet the maximum and minimum attenuation requirements, respectively.

In addition to the images generated by the reconstruction process, we also have to take into account the frequency-domain effects introduced by the DAC through the sampled-and-hold operation. Assuming that the each sample is held for the full duration of the sampling period, that is, $T_s=1/F_s$, then one can easily show that the frequency response behavior $G(f)$ of the sampled-and-held operation introduced by the DAC is given by

$$G_{\sin(x)/x}(f) = \frac{\sin\left(\pi \frac{f}{F_s}\right)}{\left(\pi \frac{f}{F_s}\right)} \tag{9.23}$$

A plot of the magnitude of $G_{\sin(x)/x}(f)$ is shown in Figure 9.16. From this plot we see that reconstructed tones near the Nyquist frequency are attenuated much more than tones located near DC. Further, the magnitude of an image decreases monotonically toward zero as the frequency of the image increases. With $x = \pi f/F_s$, Eq. (9.23) takes on the form $\sin(x)/x$, leading one to describe this filter behavior as a sin-x-over-x rolloff.

The effects of the sample-and-hold operation can be corrected for by multiplying the spectrum of the discrete-time signal by the inverse of $G_{\sin(x)/x}(f)$. For example, the effects of the $\sin(x)/x$ rolloff on a single tone at frequency F_t can be corrected in software by boosting its amplitude by a correction factor given by

$$\text{correction factor} = \frac{1}{\left|G_{\sin(x)/x}(F_t)\right|} = \left|\frac{\left(\pi \frac{F_t}{F_s}\right)}{\sin\left(\pi \frac{F_t}{F_s}\right)}\right| \tag{9.24}$$

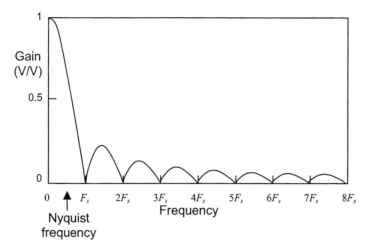

Figure 9.16. Sin(x)/x gain versus frequency.

Compensation can also be performed in hardware. In fact, some DAC channels include a sin(x)/x correction factor in the low-pass anti-imaging filter that automatically corrects for most of the rolloff. Therefore, one must know ahead of time whether sin(x)/x correction factors need to be included in the software description of the test samples.

Example 9.3

Calculate the sample set for a 3.0-V peak, 8-kHz sine wave that is to be reconstructed at 20 kHz using a conventional DAC. Boost the signal level to correct for sin(x)/x rolloff.

Solution:

First we calculate the correction factor

$$\text{correction factor} = \left| \frac{\left(\pi \frac{8\text{ kHz}}{20\text{ kHz}}\right)}{\sin\left(\pi \frac{8\text{ kHz}}{20\text{ kHz}}\right)} \right| = 1.321$$

Next we multiply the peak value we want by the correction factor to get a sample set that will be attenuated to 3.0 V peak by sin(x)/x rolloff. An example MATLAB procedure is:

```
% Correcting for DAC sin(x)/x effect
%
pi =3.14159265359;
N=2000; M=800; A=3.0; P=0.0;
x = pi*8e3/20e3;
correction_factor = x/sin(x);
```

```
for n=1:N,
    sinewave(n) = correction_factor * A * sin(2*pi*M/N*(n-1)+P);
end
```

These samples can be reconstructed using a sample-and-hold process (AWG or DAC) followed by a 10-kHz low-pass filter to remove the sampling images. The sample-and-hold process will attenuate the 8-kHz sine wave by 1.321, resulting in a 3.0-V peak waveform at 8 kHz.

Exercises

9.3. A 1 V RMS sine wave with a frequency of 65 kHz is sampled by an ADC at a sampling rate of 32 kHz. Sketch an RMS magnitude spectrum that includes the five lowest frequencies that could alias into the same FFT spectral bin as the 65-kHz tone.

Ans.

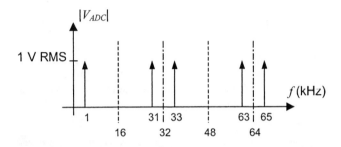

9.4. Samples from a 1-V RMS sine wave with a frequency of 1 kHz are reconstructed with a DAC at a frequency of 32 kHz. Sketch the RMS magnitude of the reconstructed waveform spectrum that includes the test tone plus the four lowest image tones.

Ans.

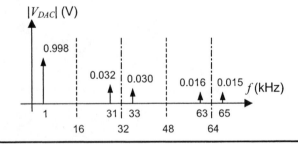

9.3 ENCODING AND DECODING

9.3.1 Signal Creation and Analysis

In the previous example, we produced a sample set whose values were expressed in Volts. When we create a sample set for a DAC or when we analyze captured samples from an ADC, we have to work with code steps (also called *quanta* or LSBs). The term LSB is commonly used to refer to a single step in a DAC or ADC transfer curve, but this terminology can be a bit ambiguous. If we tell a design engineer that his ADC generates 3 LSBs of noise, do we mean 3 steps or do we mean that the 3 least significant bits of the ADC digital output are toggling randomly? If the three least significant bits (D2-D0) were toggling, that would correspond to 8 steps of noise. Unfortunately, the term "LSB" is much more common in data sheet specifications than the less ambiguous "quantum"; so we shall use the term LSB when referring to a single division in a DAC or ADC transfer curve.

To convert a series of desired voltages into a series of DAC codes, we have to know the DAC's ideal gain in bits per volt as well as its encoding format. This information is used to encode a series of floating point voltage samples into integer DAC samples. Similarly, before we can analyze the output of an ADC, we need to know its ideal gain and format. This information is used to convert the ADC output from integer values into floating point voltage samples. The encoding and decoding process depends on the encoding scheme of the DAC or ADC. Let us look at some common encoding schemes and see how we would create and analyze encoded and decoded waveforms.

9.3.2 Data Formats

There are several different encoding formats for ADCs and DACs including unsigned binary, sign/magnitude, two's complement, one's complement, mu-law, and A-law. One common omission in device data sheets is DAC or ADC data format. The test engineer should always make sure the data format has been clearly defined in the data sheet before writing test code.

The most straightforward data format is unsigned binary. Unsigned binary format places the lowest voltage at code 0 and the highest voltage at the code with all 1's. For example, an 8-bit DAC with a full-scale voltage range of 1.0 to 3.0 V would have the code-to-voltage relationship shown in Table 9.1.

Table 9.1. Unsigned Binary Format for an 8-Bit DAC

Code	Voltage
00000000 (decimal 0)	1.0 V
00000001 (decimal 1)	1.0 V + 1 LSB = 1.007843 V
...	
01111111 (decimal 127)	1.0 V + 127 LSBs = 1.996078 V
10000000 (decimal 128)	1.0 V + 128 LSBs = 2.003922 V
...	
11111111 (decimal 255)	1.0 V + 255 LSBs = 3.0 V

One LSB is equal to the full-scale voltage range, $V_{FS+} - V_{FS-}$ divided by the number of DAC codes minus one

$$V_{LSB} = \frac{V_{FS+} - V_{FS-}}{\# \text{ DAC codes} - 1} \quad (9.25)$$

In this example, the voltage corresponding to one LSB is equal to (3.0 V − 1.0 V)/255 = 7.843 mV. Sometimes the full-scale voltage is defined with one an additional imaginary code above the maximum code (i.e., code 256 in our 8-bit example). If so, then the LSB size would be (3.0 V − 1.0 V)/256 = 7.8125 mV. This source of ambiguity should be clarified in the data sheet.

The C-code for encoding and decoding floating point numbers to unsigned binary format might appear as follows:

```
int encode_sample(vs, mfs, lsb_size)
float vs, mfs, lsb_size; {
    /* vs = voltage sample, mfs = voltage at code 0, lsb-size in volts */
    return ((int)(0.5 + (vs-mfs)/(lsb_size)));
}
float decode_sample(code,mfs,lsb_size)
int code;
float mfs, lsb_size; {
    /* code = unsigned binary code, mfs = voltage at code 0, lsb_size in volts */
    return (mfs + lsb_size*code);
}
```

Here we choose to present the conversion software routine using C code instead of the usual MATLAB code representation seen previously in other chapters. This decision is based on the fact that the above procedures are required to be resident in the tester and are usually written with a C code syntax. Converting the above procedure into a MATLAB routine is a relatively straightforward exercise.

Another common data format is two's complement. It can be used to express both positive and negative values. Positive numbers are encoded the same as unsigned binary in two's complement, except that the most significant bit must always be zero. When the most significant bit is one, the number is negative. To multiply a two's complement number by −1, all bits are inverted and one is added to the result. The two's complement encoding scheme for an 8-bit DAC is shown in Table 9.2. As is evident from the table, all outputs are made relative to the DAC's midscale value of 2.0 V. This level corresponds to input digital code 0. Also evident from this table is the LSB is equal to 5 mV. The midscale (MS) value is computed from either of the following two expressions using knowledge of the lower and upper limits of the DAC's full-scale range, denoted V_{FS-} and V_{FS+}, respectively, together with the LSB voltage obtained from Eq. (9.25)

$$V_{MS} = V_{FS-} + \frac{\text{\# DAC codes}}{2} V_{LSB} \tag{9.26}$$

or

$$V_{MS} = V_{FS+} - \left(\frac{\text{\# DAC codes}}{2} - 1\right) V_{LSB} \tag{9.27}$$

Note that the two's complement encoding scheme is slightly asymmetrical since there are more negative codes than positive ones.

Table 9.2. Two's Complement Format for an 8-Bit DAC.

Code	Voltage
10000000 (decimal −128)	2.0 V − 128 LSBs = 1.360 V
10000001 (decimal −127)	2.0 V − 127 LSBs = 1.365 V
...	
11111111 (decimal −1)	2.0 V-1 LSB = 1.995 V
00000000 (decimal 0)	2.0 V (midscale voltage)
00000001 (decimal 1)	2.0 V + 1 LSB = 2.005 V
...	
01111110 (decimal 126)	2.0 V + 126 LSBs = 2.630 V
01111111 (decimal 127)	2.0 V + 127 LSBs = 2.635 V

The C-code for encoding and decoding floating point numbers to two's complement format might appear as follows:

```
int encode_sample(vs, lsb_size)
float vs, lsb_size; {
        /* vs = voltage sample, lsb-size in volts */
        return ((int)(0.5 + vs/lsb_size);
}

float decode_sample(code,lsb_size)
int code;
float lsb_size; {
        /* code = unsigned binary code, lsb_size in volts */
        return (lsb_size*code);
}
```

One's complement format is similar to two's complement, except that it eliminates the asymmetry by defining 11111111 as minus zero instead of minus one, thereby making 11111111 a redundant code. One's complement format is not commonly used in data converters because it is not quite as compatible with mathematical computations as two's complement or unsigned binary formats.

Sign/magnitude format is occasionally used in data converters. In sign/magnitude format, the most significant bit is zero for positive values and one for negative values. Like one's complement, sign/magnitude format also has a redundant negative zero value. Table 9.3 shows sign/magnitude format for the 8-bit DAC example. The midscale level corresponding to input code 0 for this type of converter is

$$V_{MS} = V_{FS-} + \left(\frac{\#\text{ DAC codes}}{2} - 1\right)V_{LSB} = V_{FS+} - \left(\frac{\#\text{ DAC codes}}{2} - 1\right)V_{LSB} \quad (9.28)$$

where V_{LSB} is given by

$$V_{LSB} = \frac{V_{FS+} - V_{FS-}}{\#\text{ DAC codes} - 2} \quad (9.29)$$

Table 9.3. Sign/Magnitude Format for an 8-Bit DAC

Code	Voltage
11111111 (decimal −127)	2.0 V − 127 LSBs = 1.365 V
11111110 (decimal −126)	2.0 V − 126 LSBs = 1.370 V
...	
10000001 (decimal −1)	2.0 V-1 LSB = 1.995 V
10000000 (decimal -0)	2.0 V
00000000 (decimal 0)	2.0 V
00000001 (decimal 1)	2.0 V + 1 LSB = 2.005 V
...	
01111110 (decimal 126)	2.0 V + 126 LSBs = 2.630 V
01111111 (decimal 127)	2.0 V + 127 LSBs = 2.635 V

The C-code for encoding and decoding floating point numbers to sign/magnitude format might appear as follows:

```
int encode_sample(vs, lsb_size)
float vs, lsb_size; {
    /* vs = voltage sample, lsb-size in volts */
    if(vs>=0) return ((int)(0.5 + vs/lsb_size));
    else return (MSB | (int)(0.5 + -1*vs/lsb_size));
    /*Exampe: MSB is 10000000 for 8-bit converter */
}

float decode_sample(code,mfs,lsb_size)
int code;
float mfs, lsb_size; {
    /* code = unsigned binary code, mfs = voltage at code 0, lsb_size in volts */
    if(MSB & code) return (lsb_size*code);
    else return (-1*(MSB^code)*lsb_size);
}
```

Two other data formats, mu-law and A-law, were developed in the early days of digital telephone equipment. Mu-law is used in North American and related telephone systems, while A-law is used in European telephone systems. Today the mu-law and A-law data formats are sometimes found not only in telecommunications equipment but also in digital audio applications, such as PC sound cards. These two data formats are examples of companded encoding schemes.

Companding is the process of compressing and expanding a signal as it is digitized and reconstructed. The idea behind companding is to digitize or reconstruct large amplitude signals with coarse converter resolution while digitizing or reconstructing small amplitude signals with finer resolution. The companding process results in a signal with a fairly constant signal to quantization noise ratio, regardless of the signal strength.

Compared with a traditional linear converter having the same number of bits, a companding converter has worse signal-to-noise ratio when signal levels are near full scale, but better signal-to-noise ratios when signal levels are small. This tradeoff is desirable for telephone conversations, since it limits the number of bits required for transmission of digitized voice. Companding is therefore a simple form of lossy data compression.

Figure 9.17 shows the transfer curve of a simple 4-bit companded ADC followed by a 4-bit DAC. In a true logarithmic companding process such as the one in Figure 9.17, the analog signal is passed through a linear-to-logarithmic conversion before it is digitized. The logarithmic process compresses the signal so that small signals and large signals appear closer in magnitude. Then the compressed signal may be digitized and reconstructed using an ADC and DAC. The reconstructed signal is then passed through a logarithmic-to-linear conversion to recover a companded version of the original signal.

The mu-law and A-law encoding and decoding rules are a sign/magnitude format with a piecewise linear approximation of a true logarithmic encoding scheme. They define a varying

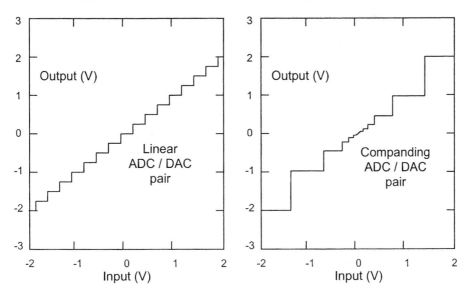

Figure 9.17. Comparison of linear and companding 4-bit ADC-to-DAC transfer curves.

LSB size that is small near 0 and larger as the voltage approaches plus or minus full scale. Each of the piecewise linear sections is called a *chord*. The steps in each chord are of a constant size. The piecewise approximation was much easier to implement in the early days of digital telecommunications than a true logarithmic companding scheme, since the piecewise linear sections could be implemented with traditional binary weighted ADCs and DACs. Today, the A-law and mu-law encoding and decoding process is often performed using lookup tables combined with linear sigma-delta ADCs and DACs having at least 13 bits of resolution.

The mu-law and A-law ADC decision levels are shown in Tables 9.4 and 9.5. The DAC levels are midway between the ADC levels. The logarithmic nature of these curves is apparent in Figures 9.18 and 9.19. The negative decision levels are mirror images of the positive levels.

Notice that the decision levels are listed in integer units called quanta, which must be converted to volts during a separate scaling process. In mu-law encoding, a sine wave with a peak level of 8159 quanta corresponds to a power level at the central office of +3.17 dBm0. Therefore, a 0 dBm0 mu-law signal level corresponds to a peak quanta level of $8159 \times 10^{-3.17/20}$ or, equivalently, 5664.2. Likewise, a peak level of 4096 quanta in A-law corresponds to a central office power level of +3.14 dBm0. Thus, a 0 dBm0 A-law signal level corresponds to a peak level of $4096 \times 10^{-3.14/20}$ quanta equal to 2853.4 quanta.

A more complete discussion of A-law and mu-law codec testing can be found in Matthew Mahoney's book, "DSP-based Testing of Analog and Mixed-Signal Circuits."[1]

9.3.3 Intrinsic Errors

Whenever a sample set is encoded and then decoded, quantization errors are added to the signal. In high-resolution encoding and decoding processes, these errors may be less than the errors generated by noise in the signal. But in low-resolution converters, or in signals that are very small relative to the full-scale range of the converter, the quantization errors can make a sine wave appear to be larger or smaller than it would otherwise be in a higher-resolution system.

These intrinsic errors can be compensated for in the final measured result by knowing ahead of time the gain error of a perfect ADC/DAC process as it encodes and decodes the signal under test. This is achieved by modeling the perfect DAC/ADC in software using, say, MATLAB. This process is made somewhat difficult, as these errors are dependent on the exact input signal characteristics, including signal level, frequency, offset, phase shift, and number of samples. All these parameters must be modeled correctly, otherwise the results will be incorrect. In the case of a DAC, the procedure is relatively straightforward. Unfortunately, the same cannot be said for an ADC.

Let us first consider the AWG encoding process of a single sinusoidal signal. One begins by writing a numerical routine that generates the samples using a floating point number representation. Next, the samples are encoded into the corresponding format of the DAC found within the AWG. In the process of encoding the samples, they are scaled by a factor of $1/V_{LSB}$, followed by a quantization or rounding operation, as only a finite number of bits can be used to represent each sample. Subsequently, the samples are stored in the waveform source memory and passed to the DAC to be decoded (i.e., produce the output analog waveform). The DAC restores the samples to their original level, as the DAC has an ideal gain of V_{LSB} volts per bit. We can model the encoding/decoding process with the block diagram shown in Figure 9.20(a).

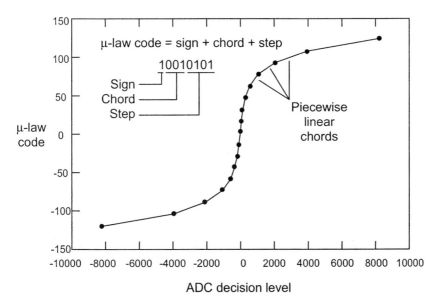

Figure 9.18. µ-law ADC (encoder) decision levels.

Table 9.4. µ-law Encoder Decision Levels

Step	Chord							
	0	1	2	3	4	5	6	7
0	1	35	103	239	511	1055	2143	4319
1	3	39	111	255	543	1119	2271	4579
2	5	43	119	271	575	1183	2399	4831
3	7	47	127	287	607	1247	2527	5087
4	9	51	135	303	639	1311	2655	5343
5	11	55	143	319	671	1375	2783	5599
6	13	59	151	335	703	1439	2911	5855
7	15	63	159	351	735	1503	3039	6111
8	17	67	167	367	767	1567	3167	6367
9	19	71	175	383	799	1631	3295	6623
10	21	75	183	399	831	1695	3423	6879
11	23	79	191	415	863	1759	3551	7135
12	25	83	199	431	895	1823	3679	7391
13	27	87	207	447	927	1887	3807	7647
14	29	91	215	463	959	1951	3935	7903
15	31	95	223	479	991	2015	4063	8159

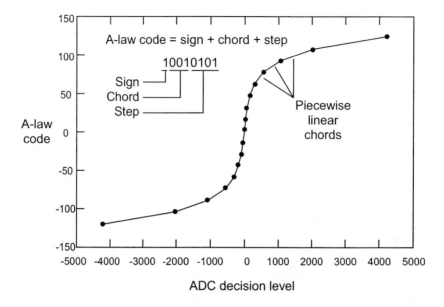

Figure 9.19. A-law ADC (encoder) decision levels.

Table 9.5. A-law Encoder Decision Levels

Step	Chord							
	0	1	2	3	4	5	6	7
0	2	34	68	136	272	544	1088	2176
1	4	36	72	144	288	576	1152	2304
2	6	38	76	152	304	608	1216	2432
3	8	40	80	160	320	640	1280	2560
4	10	42	84	168	336	672	1344	2688
5	12	44	88	176	352	704	1408	2816
6	14	46	92	184	368	736	1472	2944
7	16	48	96	192	384	768	1536	3072
8	18	50	100	200	400	800	1600	3200
9	20	52	104	208	416	832	1664	3328
10	22	54	108	216	432	864	1728	3456
11	24	56	112	224	448	896	1792	3584
12	26	58	116	232	464	928	1856	3712
13	28	60	120	240	480	960	1920	3840
14	30	62	124	248	496	992	1984	3968
15	32	64	128	256	512	1024	2048	4096

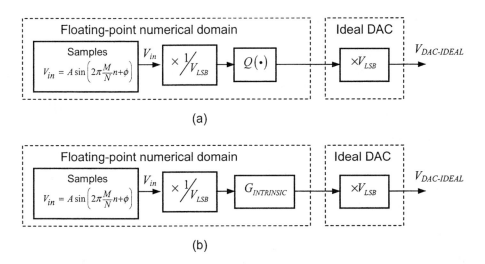

Figure 9.20. Modeling the AWG waveform encoding/decoding process: (a) nonlinear model; (b) equivalent linear model for a particular sample set.

Of particular interest is the quantization operation, denoted $Q(\cdot)$ in the figure. This block is the source of signal dependent errors mentioned above.

To quantify these errors, we first define the gain of the entire encoding process as the ratio of the output $V_{DAC-IDEAL}$ over the input V_{in}. This gain is referred to as the intrinsic gain of the encoding process for a particular sample set and is defined as

$$G_{intrinsic} = \frac{V_{DAC-IDEAL}}{V_{in}} \text{ V/V} \tag{9.30}$$

Ideally, the entire encoding/decoding procedure should have a gain of unity (i.e. output equals input). Rather, the output contains intrinsic errors causing the output to differ from the input. Subsequently, the intrinsic gain error $\Delta G_{intrinsic}$ for a particular input becomes

$$\Delta G_{intrinsic} \equiv G_{intrinsic} - 1 = \frac{V_{DAC-IDEAL} - V_{in}}{V_{in}} \text{ V/V} \tag{9.31}$$

The fact that the output signal $V_{DAC-IDEAL}$ is related to the input V_{in} by a gain constant $G_{intrinsic}$ suggests that we can model the nonlinear quantizer operation as a linear operation with gain $G_{intrinsic}$ as shown in Figure 9.20(b). This representation is, of course, valid only for a particular sample set. If the sample set is changed, then a new $G_{intrinsic}$ gain constant must be derived.

For example, let us say that we want to generate a low level sine wave at the first Fourier spectral bin with no offset and 0 radians of phase shift using an 8-bit two's complement DAC. If we want to generate a very low level sine wave with an RMS amplitude of 8 LSBs, we could calculate the quantized/reconstructed sample set (normalized by an LSB) shown in Figure 9.21 using the following MATLAB code:

```
% Coherent signal definition -x-
%
N=64; M=1; A=sqrt(2)*8; P=0;
for n=1:N,
        x(n) = A*sin(2*pi*M/N*(n-1) + P);
end
%
% Quantize input
%
LSB=1; % normalize all results in terms of LSBs
for n=1:N,
        q(n) = LSB*round(x(n)/LSB);
end
%
```

If we perform an FFT on the output signal, we see that we get an RMS level of 7.861 LSBs, corresponding to an intrinsic error of (7.861-8.0) = -0.139 LSBs. Subsequently, the intrinsic gain and intrinsic gain error becomes 0.9826 and –0.0173 V/V, respectively. If we shift the phase of this sine wave by $\pi/3$ radians as shown in Figure 9.22, we get a different signal level of 8.026 LSBs, corresponding to an intrinsic error of +0.026 LSBs. The intrinsic gain and gain error will then be 1.00325 and 0.00325 V/V, respectively.

Attempting to apply this same approach to uncover the intrinsic errors associated with an ADC is much more difficult. A block diagram illustrating the decoding/encoding process for the digitizer is shown in Figure 9.23. Unlike the situation with the AWG, the input signal to the digitizer changes with each test and, hence; so does the quantization error. Thus, knowing the intrinsic error corresponding to a particular sample set provides no additional insight into the quantization errors that occur during a particular test.

Intrinsic error is the result of consistent quantization errors. In general, intrinsic error is less of a problem with higher-resolution converters and/or larger sample sizes. The intrinsic error of a DAC or ADC quickly approaches the noise floor of the measurement as the number of samples increases, as long as we use Fourier spectral bins that are mutually prime with respect to the sample size. Spectral bins that are not mutually prime will produce the same samples repeatedly, which in turn produces the same quantization errors over and over. This is another reason to use mutually prime spectral bins, since it tends to minimize intrinsic errors.

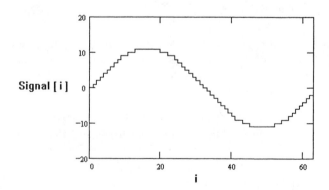

Figure 9.21. Quantized sine wave: intrinsic error = -0.139 LSBs.

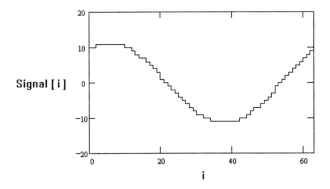

Figure 9.22. Quantized sine wave with phase shift: intrinsic error = +0.026 LSBs.

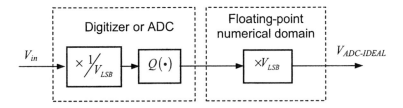

Figure 9.23. Modeling the digitizer/ADC waveform decoding/encoding process.

Exercises

9.5. What is the ideal gain of a 10-bit DAC with a full-scale voltage range of 0 to 5 V?

Ans. 4.888 mV/bit.

9.6. A 7-bit DAC has a full-scale voltage range of 0.5 to 2.5 V. The input is formatted using a 2's complement number representation. What is the midscale voltage level? What is the expected output voltage level if the input digital code is 1001101?

Ans. 1.5079 V; 0.7047 V.

9.7. The digital decimal representation of a µ-law converted signal is 10110011. In what decimal range is the input signal?

Ans. −303 to -287.

9.8. A set of samples derived from a 0.707 V RMS sinusoidal signal is found to have an intrinsic gain error of 0.05 V/V. What is the RMS amplitude of the captured test signal?

Ans. 0.742 V.

9.4 SAMPLED CHANNEL TESTS

9.4.1 Similarity to Analog Channel Tests

Sampled channel tests are very similar to the analog channel tests described in Chapter 8. In DSP-based testing, the inputs and outputs of an analog channel are actually stimulated and measured using sampled channels. Let us look at the similarities and differences between the DSP-based analog channel gain test shown in Figure 9.24 and a DAC-to-ADC sampled channel test shown in Figure 9.25.

Figure 9.24 shows the full stimulus-to-analysis path for an analog channel test. A continuous mathematical sine wave is, in effect, "sampled" by calculating 512 evenly spaced values using a C code loop such as:

```
int i;
float sample[512], pi=3.14159265359, A = 1.414 V;
for (i=0;i<512;i++)
            sample[i] = A * sin(2*pi*i/512);
load_waveform_into_AWG(sample,512);
```

Whether the routine "load_waveform_into_AWG( )" is supplied by the ATE vendor as part of a library or whether the routine is written by the test engineer, it must perform a very important first step. Since most AWGs produce their waveforms by applying integer values to a DAC, the load_waveform_into_AWG( ) routine must first convert the continuous floating-point waveform, sample[ ], into a quantized integer waveform. It must also calculate the necessary AWG attenuation settings and mathematical scaling factors that make this waveform come out of the AWG at the proper voltage level. This scaled and quantized integer waveform is then compatible with the AWG's waveform memory.

To create a continuous analog stimulus waveform, the quantized waveform is passed through the AWG's DAC, then through a reconstruction filter and various signal conditioning circuits (programmable attenuators, etc.). The resulting continuous waveform passes through the DUT's analog channel and into the input stage and low-pass anti-imaging filter of a digitizer. The digitizer's integer samples are stored into a bank of memory. These samples are then available for DSP operations, such as the FFT.

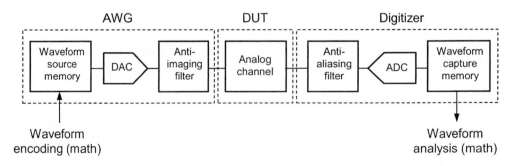

Figure 9.24. DSP-based analog channel test.

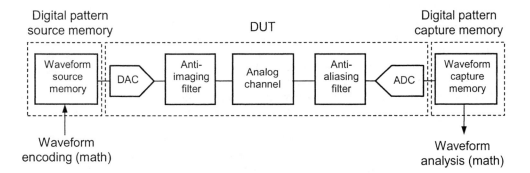

Figure 9.25. DAC-to-ADC sampled channel test.

Now consider the sampled channel test in Figure 9.25. The same process occurs in this case, only the position of the components is shifted from the tester to the DUT. A continuous mathematical waveform is sampled by a mathematical process as before, but this time the integers are stored in the source memory of the digital pattern subsystem rather than being stored in the AWG waveform memory. The digital samples from the source memory are then shifted into the sampled channel's DAC input.

The DAC output is low-pass filtered, producing a continuous waveform. This waveform is then passed through an analog channel, such as a diagnostic loopback path inside the DUT. The continuous signal is then filtered and resampled by the DUT's ADC channel, producing digital output samples. These are captured and stored into the digital subsystem's capture memory, where they are available for DSP operations.

The sampled channel gain test is therefore almost identical to DSP-based analog channel testing. We could also show how DAC channels, ADC channels, switched capacitor filters, and any other sampled channel could be reduced to a similar measurement system. The only difference is that the location of DACs, ADCs, filters, and other signal conditioning circuits may move from the ATE tester to the DUT or vice versa. Unfortunately, this means that we have to apply more rigorous testing to sampled channels, since all the effects of sampling (aliasing, imaging, quantization errors, etc.) vary from one DUT to the next. These sampling effects are often a major failure mode for sampled channels. Let us look at each of the analog channel tests described in Chapter 8, and see how they must be modified as we apply them to sampled channels.

9.4.2 Absolute Level, Absolute Gain, Gain Error, and Gain Tracking

The process for measuring absolute level in DACs and other analog output sampled channels is identical to that for analog channels. The only difference is the possible compensation for intrinsic DAC errors. In this way, a measurement is made independent of the sample set used. Moreover, when compared to bench measurements made with noncoherent test equipment, better correlation is made possible. Otherwise, absolute voltage level measurements are performed the same way as any other AC output measurement. ADC absolute level is equally easy to measure. The difference is that we express the output measurement in terms of RMS LSBs (or RMS quanta, RMS bits, RMS codes, or whatever terminology is preferred) rather than RMS volts.

In some sampled channels, such as switched capacitor filters and sample-and-hold amplifiers, absolute gain is measured using the same voltage-in/voltage-out process as in analog channels. By contrast, measurement of absolute gain in mixed-signal channels is complicated by the fact that the input and output quantities are dissimilar. Gain in mixed-signal channels is defined not in volts per volt, but in bits per volt or volts per bit, where the term "bit" refers to the LSB step size. For example, if we have an 8-bit two's complement DAC with a full-scale range of –1 to +1 V, its ideal gain is 2.0/255 or 7.84 mV/bit. Notice that there are only 255 steps between codes in an 8-bit converter, not 256. Therefore one LSB is equal to 2.0 V/255 steps or 7.84 mV.

However, we sometimes see a data sheet that defines the upper voltage of a DAC as 1 LSB above the maximum valid DAC code. For example the data sheet in this 8-bit DAC example might define –1.0 V as code –128 and 1.0 V as code 128. Code 128 does not actually exist in an 8-bit two's complement DAC; so it is an imaginary point on the DAC curve. In this case, we would calculate the ideal gain as 2.0 / 256 = 7.81 mV/bit. If the data sheet is not clear on this definition, the test engineer should request clarification. Similar issues exist on an ADC's gain specification. Data sheets do not always define the gain of a converter in bits per volt or volts per bit. Often they define the ideal LSB step size instead, which is equal to the gain in the case of a DAC or inverse of the gain in the case of an ADC.

The absolute gain of a DAC is expressed as the ratio of its output signal V_{DAC} divided by the input signal V'_{in} (expressed in LSBs)

$$G_{DAC} = \frac{V_{DAC}}{V'_{in}} \text{ V/V} \tag{9.32}$$

As $V'_{in} = 1/V_{LSB}(G_{intrinsic})V_{in}$, we can write the DAC's absolute gain as

$$G_{DAC} = \frac{V_{LSB}}{G_{intrinsic}} \frac{V_{DAC}}{V_{in}} \text{ V/bit} \tag{9.33}$$

It is interesting to note that G_{DAC} also includes the sin(x)/x frequency rolloff of the sampled-and-hold action of the DAC. Thus G_{DAC} is a frequency-dependent parameter.

Similarly, the absolute gain of an ADC without intrinsic error compensation is given by

$$G_{ADC} = \frac{D_{ADC}}{V_{in}} \text{ bits/V} = \frac{1}{V_{LSB}} \times \frac{V_{ADC}}{V_{in}} \text{ bits/V} \tag{9.34}$$

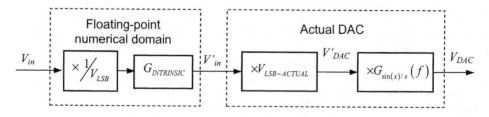

Figure 9.26. Modeling the test setup for a DAC.

where D_{ADC} is the ADC digital output expressed in LSBs, V_{ADC} is the corresponding output signal level in volts and V_{in} is the ADC input voltage signal.

Converter gain cannot be specified in decibels, because it is a ratio of dissimilar quantities (e.g., volts per bit). Converter gain error, however, can be expressed in decibels. Gain error ΔG is equal to the actual gain G_{ACTUAL} divided by the ideal gain G_{IDEAL} (which includes any sampled-and-hold effects)

$$\Delta G = \frac{G_{ACUTAL}}{G_{IDEAL}} \quad \text{V/V} \tag{9.35}$$

Either result can be converted into decibel units using the standard conversion expression

$$\Delta G = 20 \log_{10}\left(\frac{G_{ACTUAL}}{G_{IDEAL}}\right) \text{ dB} \tag{9.36}$$

For example, the ideal gain of a DAC, which includes the sampled-and-hold effect, is

$$G_{DAC-IDEAL}(f) = V_{LSB} \times G_{\sin(x)/x}(f) \quad \text{V/bit} \tag{9.37}$$

Substituting the above result into Eq. (9.35), together with the measured result in Eq. (9.33), gives

$$\Delta G_{DAC}(f) = \frac{1}{G_{intrinsic}} \frac{1}{G_{\sin(x)/x}(f)} \frac{V_{DAC}}{V_{in}} \quad \text{V/V} \tag{9.38}$$

As $V_{DAC\text{-}IDEAL} = G_{intrinsic} V_{in}$, we can write

$$\Delta G_{DAC}(f) = \frac{1}{G_{\sin(x)/x}(f)} \frac{V_{DAC}}{V_{DAC-IDEAL}} \quad \text{V/V} \tag{9.39}$$

In the case of an ADC, the ideal gain is assumed to be $1/V_{LSB}$ bits per volt, as we have no accurate means of computing its intrinsic gain. Subsequently, the ADC gain error becomes

$$\Delta G_{ADC} = V_{LSB} \frac{D_{ADC}}{V_{in}} \quad \text{V/V} = \frac{V_{ADC}}{V_{in}} \quad \text{V/V} \tag{9.40}$$

Example 9.4

An 8-bit two's complement DAC with a single-ended output has an ideal LSB size of $3.0 \text{ V} / 2^8$ = 11.719 mV. The ideal output range is 1.0 to 4.0 V − 1 LSB, that is, the 4.0-V level corresponds to imaginary code +128. A sample set corresponding to a 1-kHz sine wave at 0.8 V RMS is desired. Assuming a perfect DAC, write a MATLAB routine that produces a 512-point sample set that will generate a 1-kHz sine wave at 800 mV RMS, at a DAC sampling rate of

16 kHz. If the DAC output is digitized and the actual RMS voltage is determined to be 780 mV RMS instead of 800 mV, what is the gain and gain error of the DAC at 1 kHz? Include $\sin(x)/x$ rolloff and intrinsic error in the gain and gain error calculations.

Solution:

First we need to calculate the sample set for the DAC. The peak amplitude of the tone is set at $\sqrt{2} \times 0.8$ V. We also have to compute the Fourier spectral bin for a 1-kHz tone with a 16-kHz sampling rate and 512 samples. The Fourier spectral bin is found by dividing 1 kHz by the fundamental frequency of the sample set: $M = 1$ kHz $/ (16$ kHz $/ 512) = 32$. Of course, 32 is a poor choice since it is not a mutually prime number and will generate excessive intrinsic error. We shift the spectral bin to 31 to achieve a prime bin. The resulting MATLAB code is therefore:

```
% DAC Encoding/Decoding Procedure
%
        D=8;       ----------% 8-bit DAC
        LSB=3.0/2^D;  -% least-significant bit using an imaginary bit at 4.0 V
%
% Coherent signal definition -x-
%
        N=512; M=31; A=sqrt(2)*0.8; P=0;
        for n=1:N,
                x(n) = A*sin(2*pi*M/N*(n-1) + P);
        end
%
% Quantize result and perform DAC operation
%
        for n=1:N,
                q(n) = LSB*round(x(n)/LSB);
        end
% end
```

We next need to calculate the absolute gain of a perfect DAC so we can compare our DAC's gain to the ideal gain. An FFT is performed on the scaled sample set. Using the FFT output, we calculate the voltage level at bin 31. It should be 800 mV RMS, but because of intrinsic quantization error, the sample set produces an RMS output of 800.127 mV.

Thus, according to Eq. (9.30), the sample set has an intrinsic gain of

$$G_{intrinsic} = 1.00016 \quad \text{V/V}$$

Next, we need to consider the sampled-and-hold effect of the DAC on this sample set. As the frequency of the tone is $F_t = 16$ kHz $/ 512 \times 31$ or 968.75 Hz, and the sampling frequency is 16 kHz, we know from Section 9.2.6 that the gain of the DAC at this frequency due to the sampled-and-hold operation is

$$G_{\sin(x)/x}(968.75 \text{ Hz}) = \left| \frac{\sin\left(\pi \frac{968.75 \text{ Hz}}{16 \text{ kHz}}\right)}{\left(\pi \frac{968.75 \text{ Hz}}{16 \text{ kHz}}\right)} \right| = 0.994 \text{ V/V}$$

The expected RMS amplitude from the ideal DAC will then be 0.994 × 800.127 mV = 795.33 mV.

Finally, we load the sample set into digital source memory and start a digital pattern that sends the samples to the DAC at 16 kHz. A digitizer collects the output, and an FFT shows a signal level of 780 mV RMS. A perfect DAC would produce a sine wave at a signal level of 795.33 mV RMS. Since our DAC produced 780 mV RMS, the gain error according to Eq. (9.39) would be

$$\Delta G_{DAC} = \frac{780 \text{ mV}}{795.33 \text{ mV}} = 0.98072 \text{ V/V} = -0.1691 \text{ dB}$$

The absolute gain of a DAC is determined from Eq. (9.33) with intrinsic error compensation. Substituting the appropriate values, we obtain

$$G_{DAC} \text{ at 1 kHz} = \frac{3 \text{ V}}{256 \text{ bits}} \times \frac{780 \text{ mV}}{800 \text{ mV}} \times \frac{1}{1.00016 \text{ V/V}} = 11.42 \text{ mV/bit}$$

Fortunately, intrinsic error is usually very small; so it is often ignored in gain and gain error calculations. However, intrinsic error may become much larger if a low-resolution converter is tested, if a low-level signal is to be tested, or if the number of samples is small. The test engineer should at least verify that intrinsic errors will be negligible before dropping intrinsic error correction from the gain calculations.

Intrinsic error does not apply to S/H amplifiers and switched capacitor filters, since these do not quantize the signal as they sample it. They are, however, subject to $\sin(x)/x$ rolloff since they produce a stepped version of the waveform samples. ADCs, on the other hand, quantize signals and are therefore subject to intrinsic error. However, ADCs collect instantaneous voltage values and therefore do not see $\sin(x)/x$ rolloff. Since many of the tests in this chapter are subject to intrinsic error and/or $\sin(x)/x$ effects, the test engineer should always keep these potential error sources in mind as the performance of sampled systems are measured. Often, the correlation errors between bench equipment and tester can be traced to a bench measurement that does not take such effects into account.

Sampled channel gain tracking error is measured in a similar manner to analog channel gain tracking error. Intrinsic error is especially important in gain tracking measurements. It is important to realize that some of the variation in gain at different levels is caused by differences in quantization error. The intrinsic errors in each sample set should be extracted from the measurement of each level before calculating gain tracking error.

Intrinsic error and $\sin(x)/x$ corrections are usually not performed on any of the remaining tests in this chapter. These corrections are usually applied only to gain and absolute signal level measurements. In the remaining tests, sin(x)/x rolloff and intrinsic errors are usually considered part of the specified measurement. As usual, if there is any doubt about this issue, the test engineer should ask for clarification.

9.4.3 Frequency Response

Frequency response measurements of sampled channels differ from analog channel measurements mainly because of imaging and aliasing considerations. Since sampled channels often include an anti-imaging filter, the quality of this filter determines how much image energy is allowed to pass to the output of the channel. Frequency response tests in channels containing DACs, switched capacitor filters, and S/H amplifiers should be tested for out-of-band images that appear past the Nyquist frequency. In coherent DSP-based measurements, these images will appear at specific Fourier spectral bins, as explained in Section 9.2.6.

Notice that the digitizer used to measure these frequencies must sample at a high enough frequency to allow measurements past the Nyquist rate of the sampled channel. Also notice that each sampling process in a sampled channel has its own Nyquist frequency. An 8-kHz DAC followed by a 16-kHz switched capacitor filter has two Nyquist frequencies, one at 4 kHz and the other at 8 kHz. The images from the DAC must first be calculated. These images may themselves be imaged by the 16-kHz switched capacitor filter. Each of the primary test tones and the potential images should be measured.

The specification for a low-pass anti-imaging filter in a sampled channel may be stated in terms of the frequency response of the filter itself or it may be stated in terms of image attenuation of the total sampled channel. Since a sample-and-hold process introduces sin(x)/x rolloff, the images should appear even lower than the filter's gain curve would indicate. If image attenuation is specified rather than filter frequency response, then the test is a simple matter of comparing the amplitude of each in-band test tone with the amplitude of its image or images.

If the filter's frequency response is specified, then it can be measured in one of two ways. The best way is to provide design for test (DfT) access to the input and output of the filter and measure its frequency response using standard analog channel testing. The alternate test approach is to measure the attenuation of each test tone's first image compared to the ideal attenuation expected from sin(x)/x rolloff. The additional attenuation is due to the filter.

Example 9.5

The 16-kHz DAC in the previous example is followed by a low-pass filter with a cutoff frequency of 8 kHz. The 1-kHz test tone is passed through the DAC and filter. Calculate the frequency of the first image of the 1-kHz test tone. A digitizer samples the filter output and sees a signal level of 780 mV at the 1-kHz frequency and a signal level of 5 mV at the out-of-band image frequency. Calculate the gain of the filter at the image frequency relative to the gain of the filter at 1 kHz.

Solution:

The image of the 1-kHz tone is located at F_i = 16 kHz - 968.75 Hz = 15031.25 Hz. Note that we will have to use a digitizer with a Nyquist frequency greater than 15031.25 Hz to see this

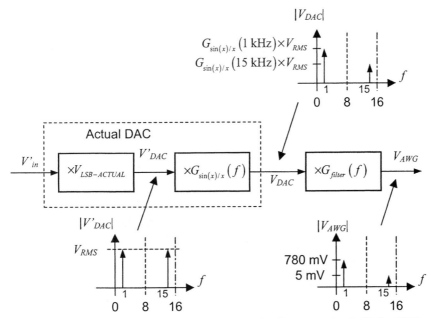

Figure 9.27. Modeling the frequency domain effects associated with the AWG.

frequency. If we used a digitizer with a 16-kHz sampling rate, then the digitizer would alias the 15031.25-Hz image back onto the test tone at 968.75 Hz, making the image measurement impossible. A logical choice for the digitizer sampling rate would be 32 or 64 kHz, since these are integer multiples of the DAC sampling rate.

Next, let us plot the spectral information that we are given in the problem, together with a block diagram that identifies each significant component of the AWG. This we do in Figure 9.27. Of particular interest is the spectrum corresponding to the output of the DAC, denoted V'_{DAC}. This particular spectrum is periodic; thus the test tone and all of its images will have equal amplitude, say, V_{RMS}. To keep the presentation relatively straightforward, we show only the test tone and its first image. Subsequently, the sample-and hold operation will modify the magnitude of the test tone and its image according to

$$V_{DAC}(968.75\ \text{Hz}) = G_{\sin(x)/x}(968.75\ \text{Hz})V_{RMS}$$

and

$$V_{DAC}(15031.25\ \text{Hz}) = G_{\sin(x)/x}(15031.25\ \text{Hz})V_{RMS}$$

where the two gain factors are computed as follows

$$G_{\sin(x)/x}(968.75\ \text{Hz}) = \left|\frac{\sin\left(\pi \frac{968.75\ \text{Hz}}{16\ \text{kHz}}\right)}{\pi \frac{968.75\ \text{Hz}}{16\ \text{kHz}}}\right| = 0.993981\ \text{V/V} = -0.052\ \text{dB}$$

and

$$G_{\sin(x)/x}(15031.25 \text{ Hz}) = \left| \frac{\sin\left(\pi \frac{15031.25 \text{ Hz}}{16 \text{ kHz}}\right)}{\left(\pi \frac{15031.25 \text{ Hz}}{16 \text{ kHz}}\right)} \right| = 0.064061 \text{ V/V} = -23.868 \text{ dB}$$

Similarly, the low-pass filter will alter the magnitude of the test tone according to

$$V_{AWG}(968.75 \text{ Hz}) = G_{filter}(968.75 \text{ Hz})V_{DAC}$$

or

$$V_{AWG}(968.75 \text{ Hz}) = G_{filter}(968.75 \text{ Hz})G_{\sin(x)/x}(968.75 \text{ Hz})V_{RMS}$$

Likewise, for its image, we write

$$V_{AWG}(15031.25 \text{ Hz}) = G_{filter}(15031.25 \text{ Hz})V_{DAC}$$

or

$$V_{AWG}(15031.25 \text{ Hz}) = G_{filter}(15031.25 \text{ Hz})G_{\sin(x)/x}(15031.25 \text{ Hz})V_{RMS}$$

Therefore, we can expect the relative gain of the first image amplitude to the test tone amplitude to be

$$\frac{V_{AWG}(15031.25 \text{ Hz})}{V_{AWG}(968.75 \text{ Hz})} = \frac{G_{filter}(15031.25 \text{ Hz})G_{\sin(x)/x}(15031.25 \text{ Hz})}{G_{filter}(968.75 \text{ Hz})G_{\sin(x)/x}(968.75 \text{ Hz})}$$

Subsequently, the gain of the filter at the image frequency relative to the gain of the filter at 1 kHz is derived from this equation to be

$$\frac{G_{filter}(15031.25 \text{ Hz})}{G_{filter}(968.75 \text{ Hz})} = \frac{V_{AWG}(15031.25 \text{ Hz})G_{\sin(x)/x}(968.75 \text{ Hz})}{V_{AWG}(968.75 \text{ Hz})G_{\sin(x)/x}(15031.25 \text{ Hz})}$$

Substituting known parameter values gives

$$\frac{G_{filter}(15031.25 \text{ Hz})}{G_{filter}(968.75 \text{ Hz})} = \frac{V_{AWG}(15031.25 \text{ Hz})G_{\sin(x)/x}(968.75 \text{ Hz})}{V_{AWG}(968.75 \text{ Hz})G_{\sin(x)/x}(15031.25 \text{ Hz})}$$

$$= \frac{5 \text{ mV}}{780 \text{ mV}} \frac{0.993981}{0.064041}$$

$$= 0.00994625 \text{ V/V}$$

$$= -20.05 \text{ dB}$$

Unfortunately, this answer may or may not correlate perfectly with a frequency response measurement of the continuous low-pass filter at these same frequencies. The potential error source is the shape of the DAC steps. If they are very sharp (i.e., very fast settling time), then the idealized $\sin(x)/x$ correction should be valid. However, if the steps of the DAC waveform are not sharp, there will be an additional low-pass filtering effect. The filter gain may not be accurately measurable using the $\sin(x)/x$ correction method in the previous example. This is the reason we usually prefer to measure both the absolute level of images relative to the reference tone, as well as measuring the true filter characteristics using a DfT test mode and analog channel test methodologies (see Chapter 14, "Design for Test"). This approach allows us to verify the filter characteristics separate from the DAC characteristics. The additional information gives us a more thorough characterization of the DAC/filter combination.

Unlike DAC channels, ADC channels do not suffer from $\sin(x)/x$ errors. Therefore, they can be measured without any additional compensation. However, ADC channels must be tested for alias components rather than images. These alias components are likely to appear if the low-pass anitaliasing filter of the ADC channel is inadequate. Again, we prefer to measure both the filter in isolation (using a DfT test mode) and the alias components of the composite ADC/filter channel. The location of alias components in the ADC output spectrum is determined using the technique outlined in Section 9.1.1.

Exercises

9.9. An 8-bit unsigned binary formatted DAC has a full-scale range of 0.5 to 3.5 V. A sample set corresponding to 5-kHz sine wave at 0.75 V RMS is desired, assuming a DAC sampling rate of 32 kHz. An ideal analysis reveals an intrinsic gain error of –0.07 V/V. If the DAC output is digitized and the actual RMS output is found to be 0.81 V, what are the gain and gain error of the DAC at 5 kHz?

Ans. 13.7 mV/bit; 1.21 V/V (1.65 dB).

9.10. A 9-bit two's complement formatted ADC operating at an 8-kHz sampling rate has a full-scale range of 1 to 4 V. With a 1-V RMS sinusoidal signal at 3 kHz applied to its input, an analysis of the output codes indicates an RMS output value of 167.27 LSBs. Determine the gain and gain error of the ADC at 5 kHz.

Ans. 167.27 bits/V; 0.982 V/V (-0.16 dB).

9.4.4 Phase Response (Absolute Phase Shift)

This is one of the more difficult parameters to measure in a mixed-signal channel (AIDO or DIAO). The problem with this measurement is that it is difficult to determine the exact phase relationship between analog signals and digital signals in most mixed-signal testers. The phase relationships are often not guaranteed to any acceptable level of accuracy. Also, the phase shifts through the analog reconstruction and anti-imaging filters of the AWGs and digitizers are not guaranteed by most ATE vendors. The solution to this problem is a complicated focused calibration process that is beyond the scope of this book. These problems are pointed out only as a warning to the new test engineer who might think that analog waveforms coming from an AWG or analog waveforms captured by a digitizer are exactly lined up with the digital samples coming out of a DUT or going into a DUT. Fortunately, phase response of mixed-signal channels is not a common specification. Group delay and group delay distortion specifications are much more common; so we will look more closely at these measurements.

9.4.5 Group Delay and Group Delay Distortion

These tests are much easier to measure than absolute phase shift, since they are based on a change-in-phase over change-in-frequency calculation. We can measure the phase shifts in a mixed-signal channel in the same way we measured them in the analog channel. The only difference between analog channel group delay measurements and mixed-signal channel measurements is a slight difference in the focused calibration process for this measurement. The modified calibration process removes the group delay distortion of the AWG or digitizer. The calibration processes for analog and mixed-signal channels will be discussed in more detail in Chapter 10, "Focused Calibrations."

9.4.6 Signal to Harmonic Distortion, Intermodulation Distortion

These tests are also nearly identical to the analog channel tests, except for the obvious requirement to work with digital waveforms rather than voltage waveforms. $\sin(x)/x$ attenuation is usually considered part of the measurement in distortion tests. In other words, if our third harmonic is down by an extra 2 dB because of $\sin(x)/x$ rolloff, then we consider the extra 2 dB to be part of the performance of the channel.

In ADC channels, we have to realize that some of the distortion components may fold back according to the rules of aliasing. We have to test these components just like any other distortion components. The following example shows how this is done.

Example 9.6

512 samples are collected from an ADC channel sampling at 8 kHz. A 3-kHz test tone (spectral bin 193) is applied to the input of the ADC. Calculate the frequency and spectral bin numbers of the second and third harmonics of this tone.

Solution:

The second and third harmonics would normally appear near 6 and 9 kHz. But since the Nyquist frequency of this channel is 4 kHz, we know these tones will fold back in band due to aliasing. Note that we may see distortion at these frequencies *even if the low-pass antialiasing filter is set to cut off everything above 4 kHz*. The reason for this is that the filter itself may be the source of the distortion, producing 6- and 9-kHz energy at its output which then gets aliased back into the 0-4 kHz band.

We could calculate alias frequencies using the traditional approach outlined in Section 9.2.5, but let us use a slightly different approach. Instead of converting all the tones from bin numbers into frequencies, let us work with bin numbers instead. We know our test tone is actually at a frequency slightly different from 3 kHz because we chose a prime bin number (bin 193). The

Nyquist frequency exists at bin 256 (one half the sample size). The second and third harmonics would appear at bins $193 \times 2 = 386$ and $193 \times 3 = 579$, if these bins existed in a 512 point FFT (which they do not). We use a rule similar to the one in Section 9.2.5: keep subtracting the number of samples from the bin number until the result is less than the number of samples. If the result is less than the Nyquist bin (number of samples divided by 2), the result is the alias bin number. If the result is larger than the Nyquist bin, subtract the result from the number of samples to get the alias bin number.

The second harmonic is at (nonexistent) bin 386, which is already less than 512 but greater than the Nyquist bin. Therefore, we subtract 386 from 512 to arrive at the alias bin of the second harmonic, bin 126. If there is any second harmonic distortion at the input to the ADC, or if the ADC itself generates second harmonic distortion, it will appear at bin 126, which corresponds to 1968.75 Hz. (If we were working with a multitone signal, we would need to make sure we did not have any other test tones at this frequency.)

We have to subtract 512 from the third harmonic once, to get an answer less than 512. The result is $579 - 512 = 67$, which is less than the Nyquist bin; so we are done. The third harmonic should appear at bin 67. Bin 67 corresponds to 1046.875 Hz.

When measuring DAC harmonic distortion frequencies, we do not have to worry about calculating alias frequencies as we did in the preceding ADC example. This is because the distortion appears at the expected frequencies, rather than appearing at aliased frequencies. However, we do need to check to make sure we have no overlap between distortion components of one tone and images of other test tones caused by reconstruction.

9.4.7 Crosstalk

Crosstalk measurements in sampled systems are virtually identical to those in analog channels. The difference is that we have to worry about the exact definition of signal levels. If we have two identical DAC channels or two ADC channels, then we can say the crosstalk from one to the other is defined as the ratio of the output of the inactive channel divided by the output of the active channel. But what if the channels are dissimilar? If we have one DAC channel that has a differential output and it generates crosstalk into an ADC channel with a single-ended input, then what is the definition of crosstalk? Is it the single-ended level of the DAC divided by the ADC digital output, converted into equivalent input volts? Generally, crosstalk is defined by a ratio of voltages (converted to decibels). When working with digital samples, they are usually converted into volts using either the ideal gain of the converter or the actual gain of the converter. For example, the ratio of an ADC output, converted to RMS volts, relative to a DAC output in RMS volts, would be a good guess for the definition of DAC to ADC crosstalk. But this is not a solid rule to follow. The point is that the test engineer has to make sure the data sheet clearly spells out the definition of crosstalk when dissimilar channels are involved.

One difference between analog channel crosstalk measurements and quantized channel crosstalk measurements relates to ADC quantization. A very low-level crosstalk signal may not be large enough by itself to toggle one LSB of a low-resolution ADC in a quiet state. For this reason, an ADC with an inactive DC input can mask low-level crosstalk. An AC dithering signal is sometimes applied to the ADC input rather than a DC signal. A low level sine wave added to the input to an ADC allows the very small crosstalk signal to appear as part of a multitone signal when it otherwise might be invisible by itself. However, a dithering signal is sometimes unnecessary. The noise in the DUT is often high enough to act as a dithering signal, overcoming the quantization masking effect. If the output of the ADC toggles by several LSBs because of noise, or if the crosstalk is well above 1 LSB in amplitude, then dithering is unnecessary. However, if the ADC output is only toggling between one or two LSBs, dithering may improve accuracy and repeatability of a crosstalk measurement.

DAC quantization, unlike ADC quantization, does not act as a hiding place for crosstalk signals. When measuring crosstalk into a DAC output, the DAC input is set to zero or midscale, and the crosstalk appears as a nonquantized continuous-time signal.

9.4.8 CMRR

DACs do not have differential inputs; so there is no such thing as DAC CMRR. ADC channels with differential inputs, on the other hand, often have CMRR specifications. ADC CMRR is tested the same way as analog channel CMRR, except that the outputs are measured in RMS LSBs and gains are measured in bits per volt. Otherwise the calculations are identical. Like crosstalk, ADC CMRR tests may be affected by quantization masking effects. A dithering source can be used at the input to the ADC to uncover CMRR components below the 1 LSB level.

9.4.9 PSR and PSRR

Unlike analog channels, DAC and ADC channels do not have both PSR and PSRR specifications. A DAC has no analog input, and therefore no V/V gain. For this reason, it has PSR, but no PSRR. For similar reasons, ADCs have PSRR but no PSR. Unfortunately, data sheets usually list DAC channel PSRR, meaning PSR, but that is a minor semantic issue. Here is the definition of the ADC supply rejection

$$PSRR_{ADC} = 20\ \log_{10}\left(\frac{D_{ADC}/V_{ripple}}{G_{ADC}}\right)\ \text{dB} \qquad (9.41)$$

where

V_{ripple} = RMS voltage added to power supply (typically around 50-100 mV RMS)

D_{ADC} = ADC digital output expressed in RMS LSBs with V_{ripple} added to power supply

ADC gain = gain of ADC from normal input to output, in bits per volt

The definition of DAC PSR is even simpler, and is given by

$$PSR_{DAC} = 20\ \log_{10}\left(\frac{V_{out}}{V_{ripple}}\right)\ \text{dB} \qquad (9.42)$$

ADC PSRR is typically measured with the input grounded or otherwise set to a midscale DC level. However, like crosstalk, the ripple from a power supply may not be large enough to appear at the output of a low-resolution ADC with an inactive DC input. A dithering signal can be added to the input of the ADC to allow an accurate measurement of PSRR. DAC PSR is often measured with the DAC set to a static midscale value. However, it is important to realize that DACs may be more sensitive to supply ripple near one end of their scale, usually the most positive setting. PSR specs apply to worst-case conditions, which means the DAC should be set to the DC level that produces the worst results. This level can be determined by characterization, combined with knowledge of the DAC architecture. ADCs may also suffer from worst results near one end of the scale or the other; so they should be characterized as well.

9.4.10 Signal-to-Noise Ratio and ENOB

In sampled channels, signal-to-noise ratio (SNR) is again tested in a manner almost identical to that in analog channels. The output of the converter is captured using a digitizer or capture memory. The resulting waveform is analyzed using an FFT and the signal-to-noise ratio is calculated as in an analog channel. In this case, we do not care whether we are working with volts or LSBs, because SNR is a ratio of similar values.

Excessive noise in an ADC or DAC can make it appear to have fewer bits of resolution than it actually has. For example, a 23-bit ADC that has only 98 dB of signal-to-noise ratio with a full-scale sine-wave input might as well have only 16 bits of resolution. This is because a perfect 16-bit converter has a SNR of about 98 dB. The apparent resolution of a converter based on its signal-to-noise ratio is specified by a calculation called the *equivalent number of bits,* or *effective number of bits,* (ENOB). The ENOB is related to the SNR by the equation

$$ENOB = \frac{SNR(\text{dB}) - 1.761 \text{ dB}}{6.02 \text{ dB}} \tag{9.43}$$

9.4.11 Idle Channel Noise

Idle channel noise (ICN) in DAC channels is measured the same way as in analog channels, except the DAC is set to midscale, positive full scale, or negative full scale, whichever produces the worst results. Usually there is not much difference in ICN results at different settings; so the DAC is simply set to midscale. Like analog channel ICN, DAC channel ICN is usually measured in RMS volts over a specified bandwidth. Often DAC ICN is specified with a specific weighting filter, as discussed in Chapter 8.

ICN testing of ADCs again involves quantization effects. Unfortunately, a dithering source in this case would introduce additional quantization noise, destroying the ICN measurement. If we instead apply a DC level, the ADC will produce different amounts of noise depending on the exact DC level and the ADC's own DC offset. If the ADC input is midway between two decision levels, we may get a fixed DC code out of the ADC and our ICN measurement will be zero LSBs RMS. If the input is equal to one of the ADC's decision levels, then we will get a random dithering between two levels, resulting in an unweighted ICN measurement of 1/2 LSBs RMS. So the exact DC offset will make the ICN measurement vary wildly. Despite this seeming flaw in test definition, this is how ADC ICN is measured.

Correlation can be a nightmare in ADC ICN tests. Extreme care must be taken to provide the exact DC input voltage specified in the data sheet during an ICN measurement. Otherwise the test results will be completely wrong. Because of the extreme sensitivity of an ADC ICN measurement to DC offset, some ADC channels include an auto-zero or squelch function to reduce the ICN of a DC input to zero regardless of the input offset. ICN in these devices is zero by design, as long as the auto-zero or squelch function is operational.

> **Exercises**
>
> **9.11.** An ADC channel is sampled at 16 kHz and 256 samples are collected. A 7.68-kHz test tone (spectral bin 123) is applied to the input of the ADC. Calculate the frequency and spectral bin numbers of the second and third harmonics of this tone.
>
> **Ans.** 0.625 kHz, bin 10; 7.0625 kHz, bin 113.
>
> **9.12.** A 2's complement formatted ADC has a nominal gain of 73.14 bits/V. With its input shorted to a DC midsupply voltage and a 100-mV RMS sinusoidal signal added to the power supply, the RMS digital output of the channel is found to be 11.5 LSBs. Determine the PSRR of the ADC.
>
> **Ans.** 78.5 dB.
>
> **9.13.** A sampled channel has 9.3 equivalent number of bits of resolution. What is the corresponding SNR of the channel?
>
> **Ans.** 57.74 dB.

9.5 SUMMARY

DSP-based measurements of sampled channels are very similar to the equivalent tests in analog channels. The most striking differences relate to bit/volt gains and scaling factors, quantization effects, aliasing, and imaging. We also have to deal with a new set of sampling constraints, since the DUT must now be synchronized with the ATE tester's sampling system. Coherent testing requires that we interweave the DUT's various sampling rates with the sampling rates of the ATE tester instruments. Often this represents one of the biggest challenges in setting up an efficient mixed-signal test program.

Another difference between analog channel tests and sampled channel tests is in the focused calibration process, which we have only mentioned briefly. The next chapter, "Focused Calibrations," will explore this subject in more detail. Focused calibrations provide the additional accuracy that the ATE vendor may not include with an off-the-shelf, general-purpose ATE tester. While focused calibrations are sometimes unnecessary in measurements requiring limited accuracy, a good command of focused calibration techniques is a must for the professional test engineer.

Problems

9.1. A codec is operating at a 32-kHz sampling rate. An AWG has a maximum operating frequency of 100 kHz and an allocated source memory capacity of 4096 samples. The digitizer has a maximum operating frequency of 200 kHz and an allocated captured memory capacity of 2048 samples. For all intents and purposes, the digital source and capture memory is assumed unlimited. Select the appropriate test parameters so that the

AWG and the ADC are coherent. Likewise, determine the test parameters of the DAC and the digitizer.

9.2. A DAC and digitizer are arranged according to the values listed in the following table to work with a test tone of approximately 6.15 kHz

	DAC	Digitizer
Sampling rate	24,000 Hz	44,000 Hz
Number of samples	2,000	1,024
Spectral bin	513	143

Slightly adjust the sampling rate of each system so that the two sampling systems are coherent. Due to sampling constraints inherent to the tester, both the DAC and digitizer sampling rates must be integer multiples of 1 Hz. Determine the new sampling rates of the DAC and the digitizer.

9.3. An AWG and ADC are arranged according to the values listed in the following table to work with a test tone of approximately 15 kHz

	AWG	ADC
Sampling rate	128,000 Hz	44,000 Hz
Number of samples	1500	1,024
Spectral bin	175	349

Slightly adjust the sampling rate of each system so that the two sampling systems are coherent. Due to sampling constraints inherent to the tester, both the AWG and ADC sampling rates must be integer multiples of 3 Hz. Determine the new sampling rates of the AWG and the ADC.

9.4. For the codec test setup shown in Figure 9.13, the ADC must be tested with a sampling rate of 16 kHz and a sine wave of approximately 4.2 kHz. The AWG has a memory of only 1024 samples. What should the AWG sample rate be to establish coherence with the ADC if the sample rate must be a multiple of 1 Hz? How many samples should be collected by the waveform capture memory?

9.5. A 1-V RMS sine wave with a frequency of 55 kHz is sampled by an ADC at a sampling rate of 24 kHz. Sketch the magnitude of the sampled spectrum that includes the six lowest frequencies related to this test tone.

9.6. A 1-V RMS sine wave with a frequency of 63 kHz is sampled by an ADC at a sampling rate of 24 kHz. Sketch the magnitude of the sampled spectrum that includes the six lowest frequencies related to this test tone.

9.7. A 1-V RMS two-tone multitone signal with frequencies of 55 and 63 kHz is sampled by an ADC at a sampling rate of 24 kHz. Sketch the RMS magnitude of the spectrum that includes the six lowest frequencies that could alias into the same spectral bin as the test tones. Do these two test frequencies overlap below the Nyquist frequency? Do the test tones overlap if the test tones are shifted to 55 and 65 kHz?

9.8. A 1-V RMS tone of 6.5 kHz is reconstructed with a DAC at a sampling rate of 16 kHz. Determine the RMS amplitude of the in-band test tone. What gain factor should be used to correct for the sampled-and-hold operation?

9.9. Samples from a 0.3-V RMS sine wave with a frequency of 5 kHz is reconstructed with a DAC operating a sampling rate of 12 kHz. What is the RMS amplitude of the in-band test tone and the RMS amplitude of the three lowest-frequency images?

9.10. Samples from a 2.5-V RMS sine wave with a frequency of 1 kHz are reconstructed with a DAC operating a sampling rate of 32 kHz, followed by a first-order low-pass filter having the following frequency response,

$$G(f) = \frac{1}{\sqrt{\left(\frac{f}{1000}\right)^2 + 1}}$$

What is the RMS amplitude of the in-band test tone and the RMS amplitude of the lowest-frequency image?

9.11. Plot the transfer characteristic of a 3-bit DAC formatted according to the following:

(a) unsigned binary format

(b) 2's-complement format

(c) sign/magnitude format

Pay particular attention to the behavior around input code 000. Label the key points of interest such as lower and upper limits, midscale level and the LSB.

9.12. What is the ideal gain of a 10-bit DAC with a full-scale voltage range of 1.5 V?

9.13. A 6-bit DAC has a full-scale voltage range of 2.0 to 4.0 V. The input is formatted using a 2's-complement number representation. What is the midscale voltage level? What is the expected output voltage level if the input digital code is 1001001? Repeat for a sign/magnitude format.

9.14. Plot the first two positive chords of a mu-law ADC and DAC on separate graphs. Subsequently, plot the combined ADC-DAC transfer characteristic corresponding to these two chords.

9.15. Using MATLAB determine the intrinsic error of a quantized low-level sine wave with an RMS amplitude of 4 LSBs. Assume the sine wave is described by parameters $M=1$, $N=32$, $A=4\sqrt{2}$, and $P=0$. Repeat for $M=8$.

9.16. A set of samples derived from a 0.6-V RMS sinusoidal signal is found to have an intrinsic gain error of -0.045 V/V. What is the actual amplitude of the test signal?

9.17. A sample set corresponding to an approximately 3-kHz sine wave at 0.5 V RMS is desired. Assuming a perfect DAC operating at a sampling rate of 16 kHz, write a MATLAB routine, or some other equivalent software routine, that produces a 1024-point sample set that will generate this desired signal. What is the intrinsic gain of this sampling set? If the samples are held for the duration of the sampling period by the DAC, what is the effect on the amplitude of the desired signal?

9.18. A sample set corresponding to an approximately 4-kHz sine wave at 0.5 V RMS is desired. Assuming a perfect DAC, write a MATLAB routine, or some other equivalent software routine, that produces a 512-point sample set that will generate a 4-kHz sine wave at 500 mV RMS, at a DAC sampling rate of 16 kHz. What is the intrinsic gain error of this sampling set?

9.19. An 8-bit sign/magnitude formatted DAC has a full-scale range of 1.0 to 4.0 V. A sample set corresponding to a 12-kHz sine wave at 0.65 V RMS is desired, assuming a DAC sampling rate of 32 kHz. An ideal analysis reveals an intrinsic gain error of –0.045 V/V. If the DAC output is digitized and the actual RMS output is found to be 0.71 V, what are the gain and gain error (in dB) of the DAC at 12 kHz?

9.20. A 10-bit 2's-complement formatted ADC has a full-scale range of 0 to 5 V. A signal level of 1 VRMS is applied to the ADC input at a frequency of 4.2 kHz. The ADC output is measured and the actual RMS output is found to be 203.5 LSBs. What are the gain and gain error of the ADC at 4.2 kHz?

9.21. A 64-kHz DAC is followed by a low-pass filter with a cutoff frequency of 32 kHz. A 15-kHz test tone is passed through the DAC and filter. Calculate the frequency of the second image of the 15-kHz test tone. A digitizer samples the filter output and sees a signal level of 250 mV at the 15-kHz frequency and a signal level of 0.1 mV at the second image frequency. Calculate the gain of the filter at the second image frequency relative to the gain of the filter at 15 kHz.

9.22. A 12-kHz DAC is followed by a low-pass filter with a cutoff frequency of 6 kHz. A 15-kHz spurious signal appears at the input of the DAC. At what in-band frequency does the spurious tone appear?

9.23. An ADC channel is sampled at 16 kHz and 1024 samples are collected. A 2.5-kHz test tone (spectral bin 161) is applied to the input of the ADC. Calculate the frequency and spectral bin numbers of the fifteenth and twenty-third harmonics of this tone.

9.24. A two-tone multitone signal consisting of frequencies 3 kHz (spectral bin 191) and 3.2 kHz (spectral bin 205) is applied to the input of an ADC. The ADC channel is sampled at 8 kHz and 512 samples are collected. Calculate the frequency and spectral bin numbers of the second- and third-order intermodulation distortion components. Is their any frequency overlap between these distortion components and the images created by reconstruction?

9.25. A 1-V RMS sinusoidal signal at 1 kHz is applied to the input of an ADC. The output of the ADC is analyzed, resulting in an RMS digital output of 52.33 LSBs. Then the input is shorted to a DC midsupply voltage and a 100-mV RMS sinusoidal signal is added to the power supply. The output of the channel is again analyzed, resulting in an RMS digital output of 11.25 LSBs. Determine the gain of the ADC and its corresponding PSRR.

9.26. A digitizer sampling at 4 kHz captures 16 samples of a 0.5-kHz sinusoidal signal corrupted by noise from a DAC channel. An FFT analysis reveals the spectral amplitudes in the following table. Calculate the total RMS level of the noise associated with this channel measurement. Determine the signal-to-noise ratio of the channel. Express this result in terms of its equivalent number of bits (ENOB).

FFT Spectral Bin	RMS Voltage
0	2300 µV
1	32 µV
2	0.707 V
3	12 µV
4	15 µV
5	31 µV
6	12 µV
7	11 µV
8	4 µV

References

1. Matthew Mahoney, *DSP-Based Testing of Analog and Mixed-Signal Circuits*, The Computer Society of the IEEE, 1730 Massachusetts Avenue N.W., Washington, D.C. 20036-1903, 1987, ISBN: 0818607858, pp. 179-199

CHAPTER 10

Focused Calibrations

10.1 OVERVIEW

10.1.1 Traceability to National Standards

Before discussing focused calibrations, we should review the purpose and process of instrument calibration in general. Calibration is the process of transferring accuracy standards from one source or measurement instrument to another. In many countries, a central standards agency has been established so that all measurements can be referenced through calibration processes to a common set of standards. In the United States, that agency is the National Institute of Standards and Technology (NIST), formerly the National Bureau of Standards (NBS). From the NIST, the accuracy standards are transferred through a number of calibration processes to achieve a high accuracy ATE test measurement. Figure 10.1 shows a typical chain of standards transferal from the NIST standards to an ATE measurement, including a final focused calibration stage.

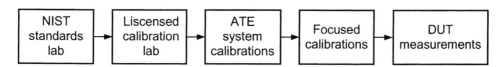

Figure 10.1. Transferal of accuracy standards from the NIST to ATE measurements.

In the United States, the NIST is in charge of maintaining standards defining the volt, ampere, second, meter, inch, pound, etc. It is also chartered with the task of licensing calibration laboratories, which in turn maintain "copies" of the calibration standards to be used to maintain the accuracy of bench instruments such as ATE calibration reference sources or high-accuracy voltmeters. ATE calibration reference sources are used by the tester as the "golden" standard for the volt, ohm, ampere, second, etc. These highly accurate instruments are removed from the tester on a periodic basis (typically once every six months) and sent to one of the NIST-licensed laboratories for recalibration. Once it has been recalibrated, a calibration reference source can be reinstalled into the ATE tester. Using the freshly calibrated reference source, the tester can perform daily or weekly system calibrations to maintain accuracy on a day-to-day basis.

There are three ways to calibrate a measurement instrument. The first calibration technique is to remove errors through adjustment of physical controls such as potentiometers and variable capacitors. This process, called *hardware calibration*, is one of the techniques used by the NIST-licensed laboratories to bring measurement instruments and calibration reference sources into compliance with accuracy specifications. Hardware calibrations are not commonly used in ATE testers since this process cannot be automated easily.

A second calibration technique allows hardware errors to be corrected using digitally controlled adjustment circuits such as programmable gain amplifiers and timing verniers. This second technique is basically a software-controlled version of hardware calibrations. Since the programmable circuits allow an automated calibration process, this technique is commonly used in ATE testers to adjust electrical parameters such as gain and propagation delay. Using this process, the tester measures its own error and then eliminates the error using the digitally controlled adjustment circuit.

The third and most powerful technique used in ATE testers is to leave hardware errors as they are and simply compensate for them using software algorithms. The second and third techniques are known collectively as *software calibrations*, since either process can be implemented with automatic software algorithms without the need for a calibration technician to twist knobs.

Full ATE system calibrations can be initiated manually if needed, but they are also executed automatically by the tester's operating system. System calibrations occur at periodic intervals, such as once per week, or when a specific event occurs. One such event is the loading of a test program, which may initiate a series of system calibrations on the appropriate tester instruments. Another event that can initiate a system calibration is a temperature change that exceeds a certain number of degrees. Since accuracy drift is most commonly caused by a change in temperature, temperature-based recalibration is an important feature. If the tester's instrumentation experiences a temperature shift, production testing must automatically stop until the instruments have been recalibrated using the high-accuracy calibration reference sources. The level of automation in system recalibration is dependent upon the tester model, but most modern testers can calibrate themselves automatically. Less advanced testers may require a calibration process using calibration-specific hardware called *load boards*. Load boards must be physically attached to the tester during the calibration process.

Although the periodic ATE system calibrations improve the accuracy of the ATE tester to meet its guaranteed specifications, the test engineer may need extra accuracy above and beyond the normal specifications of the ATE instruments. The extra accuracy is achieved using focused calibrations. Therefore, many measurements are calibrated in a two-step process. The first step involves the standard system calibrations, which occur automatically. The second step involves the test engineer's own focused calibrations, which must be explicitly written into the test program. Since the system calibration processes and the focused calibration processes are implemented using software corrections rather than manually adjusted knobs, the term "software calibration" is commonly used to describe both the system and focused calibration processes.

10.1.2 Why Are Focused Calibrations Needed?

Focused calibrations, or focused cals as they are often called, are measurements made by a test program on its initial execution, after automatic system calibrations are complete but before DUT testing begins. There are at least three reasons a test engineer may choose to perform focused calibrations. The first use of focused calibrations is to transfer accuracy standards from one ATE instrument (such as a high-accuracy voltmeter) to another, less accurate instrument. Focused calibrations can also be used to transfer ATE accuracy standards to an uncalibrated circuit, such as a DIB buffer amplifier or an on-chip measurement circuit. Finally, focused calibrations can be used to store values that must be measured at least once, but do not need to be measured for each DUT. By measuring these values only once, and then reusing the measured value during subsequent test executions, we can substantially reduce test time. An example would be a one-time measurement of the actual signal level of a sine wave to be applied at the input to an op amp gain circuit. All three types of focused calibration are commonly used in mixed-signal test programs. In this chapter, we will look at examples of each of these focused calibration techniques.

It is easy to understand why we need to transfer accuracy standards to DIB circuits and why we need to reduce test time by measuring unchanging values only once. But why should we have to write extra calibration routines to improve the accuracy of the standard tester instruments? ATE testers can cost two million dollars or more, depending on their configuration. For a two million dollar price tag we would naturally expect a premium level of accuracy without any extra effort on our part. The answer to this question relates to calibration time.

ATE vendors could certainly calibrate the tester in every possible measurement setup. Unfortunately, we would be unhappy with the time it would take to calibrate a general-purpose mixed-signal tester for all possible conditions. The problem is that a general purpose mixed-signal tester can be configured to perform any of a seemingly infinite number of measurements. Each measurement setup has its own peculiarities, leading to a unique set of measurement errors. For example, ATE instruments such as digitizers and AWGs may have a slightly different offset, gain, and phase shift characteristic for each and every gain setting, input mode setting, sampling rate, filter setting, test tone frequency, etc. If the tester's operating system tried to calibrate these instruments for absolutely any possible hardware configuration and any possible signal type, the calibration time would be unreasonably long. The tester might waste hours calibrating itself for millions of measurement setups that would never even get used by a particular test program.

Instead, the tester's operating system performs a subset of the possible calibrations to achieve a fairly high level of accuracy across all the possible setups. The resulting "unfocused" system calibration process results in an acceptable tradeoff between accuracy and calibration time. For example, a digitizer's gain error might be measured at an 8-kHz sampling rate with a 1-kHz test tone. That gain error is then automatically removed from any subsequent measurements, even if the sampling rate and test tone is different, say, a 13-kHz sampling rate with a 2-kHz test tone. The differences in the digitizer's performance at a 13-kHz sampling rate and a 2-kHz test tone will introduce small gain errors in the final measurement. Advanced testers may utilize a calibration processes that measures errors at several test conditions, applying an interpolation process to estimate errors at intermediate conditions. Even with this advanced calibration technique, though, small errors may still remain.

When accuracy requirements are not particularly demanding, the standard system calibration process may suffice without additional focused calibrations. When test limits are tight relative to the device characteristics, however, the standard system calibrations may leave an unacceptable amount of residual error in the measurements. We have to "focus" on the exact measurement conditions that have to be calibrated with a high level of accuracy; thus the term "focused calibrations." Most of the residual errors remaining after a standard system calibration can be removed through the focused calibration process, which narrows the billions of possible measurement setups down to a short list of test conditions. The short list consists of only those measurement setups used in a particular test program, such as the 13 kHz/2 kHz example. Since only the test engineer knows what that list should be, the test engineer has to write the focused calibration routines for a particular test program.

10.1.3 Types of Focused Calibrations

In Section 4.2, we took a brief look at focused calibrations, mainly to introduce the concept that the tester's results are not always accurate enough to meet our needs in mixed-signal testing. In this chapter, we will take a much more detailed look at calibration processes and common techniques for performing focused calibrations. Included in this discussion are focused calibration approaches for DC measurements, AC amplitude measurements, phase measurements, distortion measurements, and noise measurements. This is by no means a complete list of all types of focused calibrations, but the techniques discussed will form a good base of knowledge from which the test engineer can draw. Many focused calibration techniques must be developed as needed for a particular situation. Development of new focused calibration techniques is an important skill that mixed-signal test engineers should master.

10.1.4 Mechanics of Focused Calibration

Analog measurements are never perfectly accurate. Some measurement errors are tolerable, but others are not. It is up to the test engineer to understand which errors are unimportant and which errors must be reduced. The purpose of focused calibrations is to reduce the errors inherent in the source and measurement signal paths for each test. A source signal path includes all the mathematical computations, tester instruments, and electrical circuits between the idealized mathematical signal representation in the ATE computer and the input node under test. As we will see in Chapter 14, "Design for Test," the input node under test may actually be an internal node of the DUT, accessed through special on-chip test circuits. In this case, the on-chip circuits are *part of the measurement path* and must be calibrated on a DUT-by-DUT basis as if they were part of the tester. A measurement signal path includes all the circuits, computations, and instruments between the signal under test and the final measurement result. Again, this path may include DUT circuits such as buffer amplifiers and test access signal paths. Figure 10.2 shows a typical source path and a typical measurement path for an AC gain test.

Each combination of instrument configuration and interconnection path represents a unique signal path. Each unique signal path must be calibrated separately. For example, a digitizer set to an input range of ±1 V represents a different signal path than a digitizer set to a range of ±2 V. This is because the digitizer's input range adjustment circuit is implemented as a programmable gain amplifier (PGA). A PGA has entirely different electrical characteristics (offset, gain, frequency response, etc.) at each gain setting. Therefore, we should not use a calibration process from the 1-V range setting to correct errors introduced by the 2-V setting.

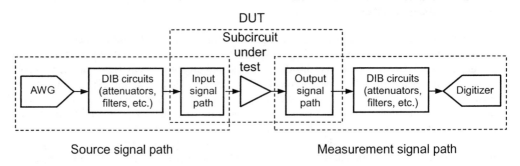

Figure 10.2. Source and measurement signal paths for AC gain test.

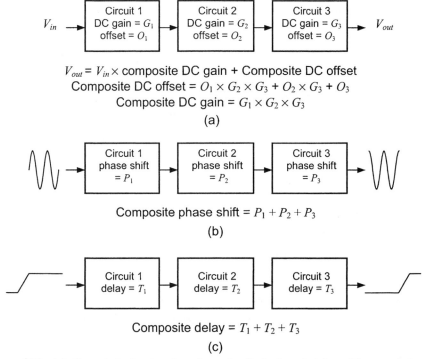

Figure 10.3. (a) Cascaded signal path gain and offset characteristics, (b) cascaded phase-shift characteristics, (c) cascaded delay characteristics.

The focused calibration process varies with the type of measurement (gain, offset, timing, etc.). However, the basic concept of calibration is the same in all cases. The object of calibration is to measure the nonideal characteristics of the source path or measurement path, and then extract these characteristics from the measured result using a software adjustment. For example, if a voltmeter has a DC offset of +2 mV when set to a 2-V range, then we know that we must subtract 2 mV from any DC measurement we make with this voltmeter when it is set to the 2-V range. Similarly, if a DIB voltage follower has a gain of 1.01 V/V at 1 kHz, then we know that at a frequency of 1 kHz we must divide the RMS voltage level measured at its output by 1.01 to calculate the actual RMS voltage present at its input.

When measuring DC offsets of a cascaded signal path, we can generally assume that composite path produces an offset that is the sum of the characteristics of each circuit element in the signal path. However, since the offset from each stage is amplified by the gain of the following stage, the total offset is more than a simple summation of all offsets in the path [e.g., Figure 10.3(a)]. Gains are easier to cascade, since they are multiplicative in nature. The gain through a cascaded signal path is equal to the product of the gains of each element in the path. Since gains often vary with frequency, a separate measurement must be made at each frequency of interest. Phase shifts, unlike DC offsets and gains, are additive. The phase shift through a signal path at a particular frequency is equal to the sum of the phase shifts through each of the individual elements in the signal path [e.g., Figure 10.3(b)]. Likewise, timing parameters such as delay time are additive in cascaded circuits [e.g., Figure 10.3(c)]. Characteristics such as distortion and noise are difficult to extract from a signal path, since they are neither additive nor multiplicative.

Having said all this, it is important to note that these simple cascading rules may not be valid in all cases. For example, the output stage of Circuit 1 in Figure 10.3(b) will be loaded down by the input impedance of Circuit 2. If the loading is severe enough, the phase shift through Circuit 1 may be affected by the circuit-to-circuit interactions. Likewise, the output slew rate of Circuit 1 in Figure 10.3(c) may affect the propagation delay through Circuit 2. Furthermore, the output slew rate of Circuit 1 may be altered by the input capacitance of Circuit 2. The test engineer should be prepared to verify the basic cascading rules for a given signal path before performing a cascaded focused calibration.

We have a choice of generating a composite calibration factor for an entire signal path or generating a series of cascaded calibration factors for the individual blocks of the signal path. When working with a cascaded calibration, we can measure the characteristics of each block in the signal path, maintaining individual calibration factors for each block. The advantage of measuring each block individually and then cascading the calibration factors is that we can arrange the blocks in many combinations without needing a separate calibration factor for each configuration. In fact, this is the way a tester's system calibrations are often performed, since the tester's calibration software cannot predict how the tester will be configured for a particular test. By cascading the appropriate calibration factors for a particular test setup, the tester can handle many permutations of hardware configuration with a limited number of calibration factors.

Alternatively, we can choose to work with a single calibration for each unique signal path. For example, we can calibrate the combination of a DC voltmeter and a DUT voltage follower as if the voltage follower were part of the voltmeter's front end. The advantage of measuring the whole signal path as a single instrument during focused calibrations is that it simplifies the calibration process. Instead of keeping track of several calibration factors that must be combined to calculate the total gain and offset errors of the cascaded signal path, we can keep a single gain and offset calibration factor for the composite path. Composite calibration factors are also superior in that they do not make assumptions about the interactions between circuit blocks. For example, two circuits may have a gain of 2 and 3 V/V when measured separately. But when cascaded, they may have a gain of 5.999 V/V, which is almost equal to the ideal 6 V/V. The reason for this type of nonideal cascaded behavior is often related to interactions between the output of one circuit and the input to the next, as previously discussed. If we calibrate the cascaded circuit as a single element, though, we will get the correct gain of 5.999 V/V. When cascaded calibration factors and composite calibration factors produce equally accurate results, the choice of calibration technique is a matter of personal taste. The test engineer should be ready to encounter either type of calibration approach when reading test code from other engineers.

So far, we have discussed calibration as a method to eliminate measurement or source errors. Sometimes measurement path errors are irrelevant, since they cancel each other out in the final calculation. For example, an AC gain measurement of an analog channel does not actually require accurate voltage measurements. Even though the gain of an analog circuit is defined as output voltage divided by input voltage, the absolute voltages are irrelevant. We simply need to measure the output signal amplitude and the input signal amplitude with the same instrument, dividing the measured output level by the measured input level. The input and output levels could be expressed in volts or digitizer LSBs. Whatever gain or scaling errors are inherent in the measurement instrument will cancel in the ratio of output over input. Since the errors cancel out in the final gain computations, a focused calibration to achieve highly accurate absolute voltage measurements is unnecessary. However, the resulting code can be difficult for the novice test engineer to work with, since signals are never converted into fully accurate, familiar units such as RMS volts.

As previously mentioned, the input signal can be measured once and then reused to save test time. Rather than measuring the unchanging input signal each time a new DUT is tested, we can measure the input voltage level once, during the focused calibration process, and store it as a calibration factor. On subsequent program runs, we can measure gain by simply dividing the measured output voltage by the calibration factor (measured input voltage). Strictly speaking, this technique is not really a calibration process, since it does not transfer accuracy standards from one instrument or circuit to another. Premeasurement of input signals is a test time reduction process rather than a calibration process. Nevertheless, the stored value is still called a calibration factor for lack of a better term. The use of calibration factors for premeasurement of inputs is one of the simplest, most powerful ways to achieve fast, highly accurate results for analog channel testing.

It is important to note that we have to use the same instrument with the exact same settings for both input and output measurements if we want the error cancellation approach to work. If we change the voltage range, sampling rate, or any other measurement setting of the measurement instrument, we destroy the error cancellation effect. Consequently, we would have to work with absolute volts or otherwise calibrate the gain variation from one instrument setting to the other.

10.1.5 Program Structure

The focused calibration process is an integral part of most test programs. The test program typically performs the focused calibrations for all the tests during a first-pass program execution. Calibration factors are stored as global program variables whose values are retained from one program execution to the next. After the calibration factors are generated and stored, subsequent executions of the test program retrieve the calibration factors and use them to make the necessary DUT measurements.

Focused calibration factors should be regenerated on a periodic basis to account for potential drift in the tester instruments and DIB circuits. Focused calibrations should also be regenerated when test conditions change, as frequently happens during the test debugging process. The details of recalibration vary significantly from one type of tester to another, but we can make some general comments. For example, the test program should allow the test engineer to force a recalibration at any time, to aid in the debugging process. If a test tone has to be modified from 1 to 2 kHz or if the digitizer's input range has to be altered, then the test engineer has to modify the focused calibration portion of the test program as well as the normal DUT testing portion of the program. A forced recalibration is then required to compensate for the new digitizer errors inherent in the modified test configuration.

Another common rule of recalibration is that we need to regenerate focused calibrations any time the tester performs automatic calibrations of its own. For example, a tester may calibrate all its instruments every four hours and also any time the temperature changes by more than three degrees Celsius. After the tester performs its automated system calibrations, the test program should regenerate the focused calibrations. For example, a digitizer's gain error is compensated by the tester's operating system using the automatic system calibrations. If the test engineer

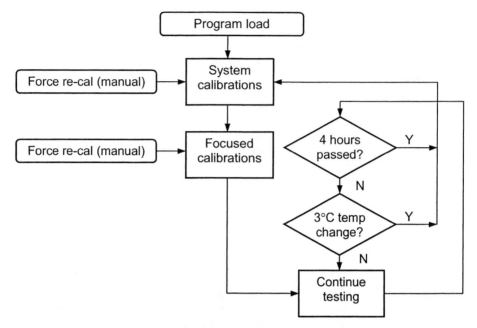

Figure 10.4. Example flowchart of system calibrations and focused calibrations.

refines the accuracy of the digitizer by performing a focused calibration, then the digitizer results actually pass through two calibration processes, one general system calibration and one focused calibration. If the characteristics of the digitizer change because of a periodic system recalibration, then the focused calibration is no longer valid. Therefore, focused calibrations involving a particular ATE instrument should be considered invalid after system calibrations are performed on that instrument. Figure 10.4 shows an example flowchart for the system calibrations, the focused calibrations, and the DUT testing portion of the test program.

10.2 DC CALIBRATIONS

10.2.1 DC Offset Calibration

The offset of an instrument or circuit can be measured by setting its input to midscale and observing the offset from the ideal level at its output. DC offset calibrations can be performed on DC voltage sources, DC voltmeters, AWGs, digitizers, and various types of DIB circuits such as buffer amplifiers and filters. The definition of midscale varies, depending on the type of circuit or instrument to be calibrated. In an op amp buffer circuit having a bipolar power supply, for example, the definition of midscale is usually ground (0 V). In op amp circuits having a single power supply, however, the definition of midscale is generally halfway between ground and the power supply voltage. For example, an op amp having a single +5-V power supply (V_{DD}) has a midscale voltage defined as $V_{DD}/2 = 2.5$ V. Likewise, the definition of the ideal output is usually defined as either 0 V or $V_{DD}/2$ depending on the power supply configuration.

The DC offset of tester instruments can also be measured to generate offset calibration factors. For example, the DC offset of a voltmeter can be measured by grounding its input, as detailed in Section 4.2.3. Similarly, the DC offset of a voltage source can be measured by setting it to force 0 V and then measuring its actual output as described in Section 4.2.5. Using similar techniques, the DC offset of AWGs and digitizers can be calibrated by measuring their offsets and then compensating for their offset during subsequent measurements.

Example 10.1

An AWG needs to produce a single-ended 1.0-V peak sine wave with 2.5-V DC offset. The AWG has an offset specification of ±10 mV, but we need an input offset accuracy of ±1 mV for this measurement. The tester has a high accuracy DC voltmeter with a total error (gain plus offset) of ±100 μV when set to its 5.0-V range. Determine a calibration process necessary to achieve the ±1 mV DC offset accuracy from the AWG.

Solution:

There are many ways to approach this problem. The simplest approach is to take advantage of superposition. We can assume that the AWG has the same DC offset, whether there is a sine wave at its output or not. Using this assumption, we can set the AWG to a DC level of 2.5 V by simply creating and loading a short waveform containing the value 2.5 in each sample:

```
calibrate_AWG_offset( ) /* Run this routine only once, before testing DUT's */
{
    int i;
    float waveform[32];
    for(i=0;i<32;i++)
        waveform[i]=2.5;
    configure_AWG_to_state_XYZ( );
    load_and_start_AWG_waveform(waveform,32);
    set meter input = AWG;
    OffsetCal = read_meter( ) – 2.5V
    /* i.e. actual offset – ideal offset */
}
```

In this pseudocode example, the global variable OffsetCal is a calibration factor whose contents remain unchanged between one execution of the program and the next. To produce a calibrated waveform from the AWG, we subtract this offset from the desired signal when we calculate the actual DUT signal we want (sine wave with a 2.5-V offset):

```
calculate_and_load_calibrated_waveform( )
{
int i;
float waveform[256];
for(i=0;i<256;i++)
    waveform[i]=(2.5V-OffsetCal) + sin(2.0*PI*i/256);
```

```
    configure_AWG_to_state_XYZ( );
    load_and_start_AWG_waveform(waveform,256);
    /* ... Now measure DUT response */
}
```

The resulting waveform should have an offset very near 2.5 V. Of course the accuracy of this calibration relies on the fact that we set the AWG to the exact same conditions during the calibration measurement as we plan to use during the actual DUT test. If we need a 10-kHz sampling rate during the DUT test, for example, then we should measure the AWG's offset using a 10-kHz sampling rate as well. In theory, sampling rate should not affect offset, but the theory may or may not hold true in practice. It is usually best to assume that any change in hardware configuration will give different performance. Other AWG settings, such as low-pass reconstruction filter cutoff frequency, input gain (range setting), etc., must also match the DUT test for maximum accuracy. The accuracy of this calibration also relies on the assumption that if we ask for a 10-mV increase in DC offset, then we will get approximately 10 mV of increased offset. Unless the AWG is out of specification (i.e., failing system calibrations), this is usually a safe assumption.

The preceding example demonstrates how the accuracy standard of one instrument could be transferred to another, less accurate instrument. The DC voltmeter in an ATE tester is generally more accurate when measuring DC offsets than the AWG or digitizer. AWGs and digitizers may generate different DC offsets at different sampling rates, filter settings, etc.. By using the DC voltmeter to measure the actual performance of the AWG under a specific set of conditions, we have transferred the accuracy of the meter to the AWG without having to calibrate the AWG in every possible configuration.

10.2.2 DC Gain and Offset Calibrations

The DC gain together with the offset of a circuit or instrument is very easy to measure. Assuming that the input-output DC behavior of a circuit is described by the first-order linear equation

$$V_{OUT} = G_{DC}\, V_{IN} + \mathit{offset} \tag{10.1}$$

we can deduce the two unknowns G_{DC} and *offset* by applying two different input DC voltage levels to the circuit (V_{IN1}, V_{IN2}) and measure the corresponding output levels (V_{OUT1}, V_{OUT2}). Subsequently, we can compute the gain and offset using the following equations

$$G_{DC} = \frac{\Delta V_{OUT}}{\Delta V_{IN}} = \frac{V_{OUT2} - V_{OUT1}}{V_{IN2} - V_{IN1}} \tag{10.2}$$

and

$$\mathit{offset} = V_{OUT1} - G_{DC}\, V_{IN1} \tag{10.3}$$

or, alternatively

$$\text{offset} = V_{OUT2} - G_{DC} V_{IN2} \tag{10.4}$$

The nature of the input and output variables need not be expressed in volts. Rather, they can also be expressed in terms of LSBs. For example, a DIB circuit may be an ADC whose output is in LSBs, not volts. Hence, the gain will be expressed in units of bits per volt and the offset in terms of bits.

The gain and offset errors of the DIB circuit or measurement instrument can then be corrected using the inverse equation

$$V_{IN} = \frac{V_{OUT} - \text{offset}}{G_{DC}} \tag{10.5}$$

In most situations, we seldom know exactly the functional behavior between the output and input signals of any circuit or instrument. So, as explained in Section 4.2.3, our corrected level is really only an estimate of the true level. In order to make this distinction clear, we introduce a new variable called the calibrated signal, $V_{CALIBRATED} \sim V_{IN}$ and write

$$V_{CALIBRATED} = \frac{V_{OUT} - \text{offset}}{G_{DC}} \tag{10.6}$$

We shall refer to this equation as the *calibration equation*; it relates the measured value to the corrected or calibrated value. The gain and offset values will then be stored as calibration factors in global program variables for later use with this equation.

Example 10.2

An analog voltage follower is placed on the DIB to buffer a weak DUT output before passing it to the ATE digitizer. Devise a calibration process that will correct the DC offset errors of the digitizer and voltage follower. Apply the calibration to the measurement of a single-ended DUT output signal with a 1.0-V RMS sine wave plus 2.5-V offset.

Solution:

The combined instrument (voltage follower plus digitizer) can be calibrated the same way as we would calibrate the digitizer by itself. Two accurate DC voltages are applied to the voltage follower, one voltage at a time. If necessary, these two voltages can be measured with a high-accuracy voltmeter, rather than simply trusting the DC source to produce the programmed voltage levels. This depends on whether the meter is more or less accurate than the DC voltage source. Each DC signal is digitized separately and the DC gain and offset is calculated using Eqs. (10.2) and (10.3), or (10.4). As the digitizer output will consist of a series of samples rather than just one value, they should be averaged to eliminate any variation in sample value that is caused by noise. Subsequently, the calibrated signal is determined from Eq. (10.5).

Note that the DUT signal is supposed to be the sum of a sine wave at 1.0 V RMS and a DC offset near 2.5 V. Since the average value of a coherent sine wave without offset is always zero, we can still measure the DC offset of the composite signal by computing the average of all the samples. The sine wave component averages to zero, and we are left with only the DC offset of the waveform. However, there is a better way to calculate DC offset of the waveform. Since the DC + AC signal is probably part of an AC test, we will probably need to perform an FFT on the digitized signal at some point. If we are already performing an FFT on the signal anyway, then we can simply read the first spectral bin of the FFT, since the first spectral bin corresponds to the DC offset of the signal. This saves the extra computation time of a separate average calculation.

10.2.3 Cascading DC Offset and Gain Calibrations

A series of individually calibrated circuits or instruments can be calibrated collectively by combining individual calibration factors. For example, a voltage buffer can be modeled such that its input and output behavior is described by

$$V_{BUF} = G_{BUF}\, V_{IN-BUF} + \mathit{offset}_{BUF} \tag{10.7}$$

Similarly, a digitizer can also be modeled by a first-order equation given by

$$V_{DIG} = G_{DIG}\, V_{IN-DIG} + \mathit{offset}_{DIG} \tag{10.8}$$

Now, if these two stages are cascaded, $V_{IN\text{-}DIG} = V_{BUF}$, then the composite behavior becomes

$$V_{DIG} = G_{DIG}\left(G_{BUF}\, V_{IN-BUF} + \mathit{offset}_{BUF}\right) + \mathit{offset}_{DIG} \tag{10.9}$$

or

$$V_{DIG} = G_{COMP}\, V_{IN-BUF} + \mathit{offset}_{COMP} \tag{10.10}$$

where

$$G_{COMP} = G_{DIG}\, G_{BUF} \tag{10.11}$$

and

$$\mathit{offset}_{COMP} = G_{DIG}\, \mathit{offset}_{BUF} + \mathit{offset}_{DIG} \tag{10.12}$$

We can therefore consider this composite circuit to have a gain of $G_{DIG}\,G_{BUF}$ and an offset of $G_{DIG}\,\text{offset}_{BUF} + \text{offset}_{DIG}$. Subsequently, the calibration equation becomes

$$V_{CALIBRATED} = \frac{V_{DIG} - \text{offset}_{COMP}}{G_{COMP}} \qquad (10.13)$$

Extending this approach to include additional stages in cascade is relatively straightforward. For instance, the composite gain of three stages in cascade with gains, G_1, G_2, and G_3, and offsets O_1, O_2, and O_3, from input to output, respectively, is simply

$$G_{COMP} = G_1 G_2 G_3 \qquad (10.14)$$

$$\text{offset}_{COMP} = G_2 G_3 O_1 + G_3 O_2 + O_3 \qquad (10.15)$$

The derivation details are left as an exercise in the problem set at the end of the chapter.

Example 10.3

A signal path consists of two cascaded DIB circuits and a medium-accuracy ATE voltmeter as shown in Figure 10.5. The DIB circuits have a DC gain of $G_1 = 1.002$ and $G_2 = 2.102$ and an offset of $O_1 = 10$ mV and $O_2 = 20$ mV, respectively. The voltmeter has an offset of $O_3 = -1$ mV and a DC gain of $G_3 = 0.997$. (The gain of the voltmeter is measured by forcing two input voltages from a DC source and measuring them with the voltmeter. Gain is defined as output measurement change divided by actual input voltage change as measured by a more accurate meter.) These offsets and gains are measured and stored as calibration factors on the first execution of the test program. A DUT output is applied to the input to the first DIB circuit and the voltmeter measures the output of the second DIB circuit. The voltmeter produces a reading of 2.523 V. What is the actual output voltage of the DUT?

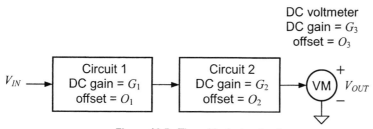

Figure 10.5. Three-block signal path.

Solution:

We start by calculating the gain and offset of the composite signal path, using Eqs. (10.14) and (10.15).

$$G_{COMP} = G_1 G_2 G_3 = 1.002 \times 2.102 \times 0.997 = 2.09989 \text{ V/V}$$

$$\text{offset}_{COMP} = G_2 G_3 O_1 + G_3 O_2 + O_3 = 39.9 \text{ mV}$$

Next we apply Eq. (10.10) to write the input-output DC behavior of this signal path

$$2.523 \text{ V} = V_{IN} \times 2.09989 + 39.9 \text{ mV}$$

or, rewriting, we get

$$V_{IN} = \frac{(2.523 \text{ V} - 39.9 \text{ mV})}{2.09989} = 1.1825 \text{ V}$$

Thus the DUT output is equal to 1.1825 V.

Exercises

10.1. A ×10 inverting amplifier is used to boost a signal before it is applied to a digitizer. A calibration sequence determined that the amplifier has a gain of -10.9 V/V and an offset of 25 mV. In addition, the digtizier was found to have a gain of 1.13 V/V with an offset of −5.4 mV. What is the composite gain and offset for this cascade combination? Write the calibration equation for this test setup.

Ans. $G_{COMP} = -12.31 \text{V/V}$; $\text{offset}_{COMP} = 22.8$ mV. $V_{CALIBRATED} = \dfrac{V_{DIG} - 22.8 \text{ mV}}{-12.31 \text{ V/V}}$

10.3 AC AMPLITUDE CALIBRATIONS

10.3.1 Calibrating AWGs and Digitizers

If we want to perform highly accurate AC voltage measurements using multitone DSP-based testing techniques, we often have to make sure our digitizer and/or AWG is calibrated for absolute voltage accuracy. Sometimes the absolute voltages are unimportant, as in the DC gain example, but often we need to source or measure an accurate voltage at each frequency of interest. To guarantee the highest level of AC source and measurement accuracy, we need to transfer the accuracy standards from a more accurate tester instrument to the AWG and digitizer using a focused calibration process. The details of this focused calibration process depends entirely on the architecture of the ATE tester. For instance, a tester that has an accurate AC voltmeter allows a different type of AC amplitude calibration than a tester that lacks one. Nevertheless, there are several common techniques from which the test engineer can choose.

The first commonly used AC amplitude calibration technique is to calibrate the digitizer first, using a DC calibration step involving a highly accurate DC voltmeter. Next, the digitizer calibrates the AWG, followed by an antialiasing filter response calibration. Each step is

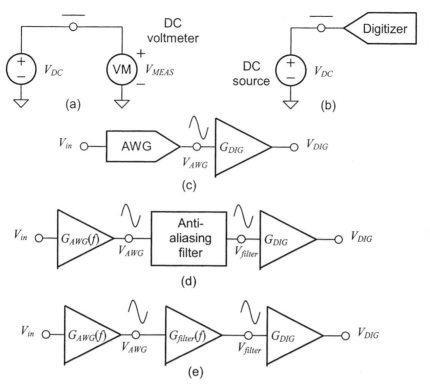

Figure 10.6. Highlighting the steps of the first calibration method: (a) transfer voltmeter accuracy to DC voltage source, (b) DC calibrate digitizer with DC voltage source using two different voltage levels, (c) measure AWG excitation using digitizer and determine frequency response model of AWG, (d) insert antialiasing filter, and repeat AC response measurement to determine filter's frequency response model, (e) linear model of complete test setup.

summarized in Figure 10.6. This technique was common on older testers where a highly accurate sine wave signal source or RMS voltmeter was not available. The second commonly used technique is to calibrate the AWG using a highly accurate RMS voltmeter and then use the AWG to calibrate the digitizer, as summarized in Figure 10.7. Another possibility is to use a highly accurate AC signal source to calibrate the digitizer, and then calibrate the AWG with the digitizer. This third possibility is not commonly used because ATE testers are much more likely to have a high-accuracy RMS voltmeter than a high-accuracy sine wave generator. Let us look at the first two techniques in detail.

The first method begins by measuring the DC gain of the digitizer as shown in Figure 10.6(a) and (b). With the digitizer's antialiasing filter bypassed, the digitizer's DC gain is determined using the two accurately defined DC voltage levels (V_{DC1} and V_{DC2}), according to

$$G_{DIG} = \frac{\Delta V_{DIG}}{\Delta V_{DC}} = \frac{V_{DIG2} - V_{DIG1}}{V_{DC2} - V_{DC1}} \qquad (10.16)$$

Next, the AC frequency response of the AWG is derived by generating a multitone signal that contains the frequencies of interest as shown in Figure 10.6(c). Subsequently, the frequency response behavior of the AWG is determined by comparing the digitizer's frequency domain data with the ideal input signal levels. Since $V_{DIG} = G_{DIG} V_{AWG}$, we can write

$$G_{AWG}(f) = \frac{V_{AWG}}{V_{in}} = \frac{1}{G_{DIG}} \frac{V_{DIG}}{V_{in}} \tag{10.17}$$

This first technique is predicated on the assumption that the digitizer's gain is flat across the frequency band of interest. In other words, $G_{DIG}(f) = G_{DIG}$ for $f < F_s$. This is not a perfectly safe assumption, but it is the best we can do on older testers that do not have high-accuracy AC instruments. It may also have to suffice for high-frequency measurements involving frequencies beyond the range of the tester's high accuracy RMS voltmeter. Next, the digitizer's antialiasing filter is enabled [Figure 10.6(d)] and the same signal is digitized again (denoted V'_{DIG}).

The antialiasing filter's frequency response is then determined from

$$G_{filter}(f) = \frac{V_{filter}}{V_{AWG}} = \frac{1}{G_{AWG}(f) G_{DIG}} \frac{V'_{DIG}}{V_{in}} \tag{10.18}$$

Furthermore, as $V_{DIG} = G_{AWG}(f) G_{DIG} V_{in}$, we can write Eq. (10.18) as

$$G_{filter}(f) = \frac{V'_{DIG}}{V_{DIG}} \tag{10.19}$$

Once the overall characteristics of the AWG, filter, and digitizer are known at each frequency, that is, the input-output model given by Figure 10.6(e), the AWG output can be adjusted to produce signal levels closer to the desired value. These levels can either be corrected by boosting or attenuating the requested signal level in math, or the errors can simply be recorded as calibration factors so that their effects can be removed from the final test result. Whether the AWG signal levels are corrected or whether their errors are simply recorded depends on how accurate the absolute voltage of each test tone must be. Even if the signal levels are corrected, though, they should still be measured a second time and any residual errors should be stored as calibration factors.

Example 10.4

A three-tone multitone signal at 1, 2, and 3 kHz is to be generated by an AWG and applied to a DUT input. The level of each tone should be 500 mV RMS. The DUT will amplify or attenuate each tone and produce an analog output. The output will be sampled by a digitizer and the level of each tone will be measured in RMS volts. Devise a calibration scheme based on DC calibration of the digitizer with a high-accuracy voltmeter and a DC source. Also, devise a calibration scheme for the AWG so that each tone is as close to 500 mV RMS as possible.

Solution:

First we apply two DC voltages to the digitizer (antialiasing filter bypassed) and measure their levels with a high-accuracy voltmeter and with the digitizer. This produces two accurately measured input voltages, $V_{DC1} = +1.010$ V and $V_{DC2} = -0.998$ V, and two digitizer DC measurements, $V_{DIG1} = +0.987$ V and $V_{DIG2} = -1.001$ V. The DC gain of the digitizer is then found to be equal to

$$G_{DIG} = \frac{V_{DIG2} - V_{DIG1}}{V_{DC2} - V_{DC1}} = 0.990 \text{ V/V}$$

Next we create a three-tone multitone from the AWG using a calculation routine such as:

```
for(i=0;i<512;i++)
waveform[i]=   1.414*500mV*sin(2*PI*31*i/512) /* 1kHz tone*/
             + 1.414*500mV*sin(2*PI*67*i/512) /* 2kHz tone */
             + 1.414*500mV*sin(2*PI*97*i/512); /* 3kHz tone */
```

We then load this waveform into the AWG, and source it to the digitizer, whose antialiasing filter is still bypassed. We collect samples with the digitizer and measure the signal level of each tone using an FFT. The signal levels are recorded in variables $V_1 = 0.498$ V, $V_2 = 0.480$ V, and $V_3 = 0.472$ V. The gain of the AWG at each frequency is then computed according to Eq. (10.17)

$$G_{AWG}(1 \text{ kHz}) = \frac{1}{G_{DIG}} \frac{V_1}{V_{in}} = \frac{1}{0.990 \text{ V/V}} \frac{0.498 \text{ V}}{0.500 \text{ V}} = 1.00606 \text{ V/V}$$

$$G_{AWG}(2 \text{ kHz}) = \frac{1}{G_{DIG}} \frac{V_2}{V_{in}} = \frac{1}{0.990 \text{ V/V}} \frac{0.480 \text{ V}}{0.500 \text{ V}} = 0.969696 \text{ V/V}$$

$$G_{AWG}(3 \text{ kHz}) = \frac{1}{G_{DIG}} \frac{V_3}{V_{in}} = \frac{1}{0.990 \text{ V/V}} \frac{0.472 \text{ V}}{0.500 \text{ V}} = 0.953535 \text{ V/V}$$

Next we enable the digitizer's antialiasing filter and measure the signal levels again. These signal levels are recorded in variables $V'_1 = 0.497$ V, $V'_2 = 0.475$ V, and $V'_3 = 0.460$ V. The gain of the antialiasing filter at each frequency is therefore given by Eq. (10.19) according to

$$G_{filter}(1 \text{ kHz}) = \frac{V'_1}{V_1} = \frac{0.497 \text{ V}}{0.498 \text{ V}} = 0.9980 \text{ V/V}$$

$$G_{filter}(2 \text{ kHz}) = \frac{V'_2}{V_2} = \frac{0.475 \text{ V}}{0.480 \text{ V}} = 0.9896 \text{ V/V}$$

$$G_{filter}(3 \text{ kHz}) = \frac{V'_3}{V_3} = \frac{0.460 \text{ V}}{0.472 \text{ V}} = 0.9746 \text{ V/V}$$

The total gain of the digitizer at each frequency is therefore calculated as

$$G_1 = G_{filter}(1 \text{ kHz})G_{DIG} = 0.9880 \text{ V/V}$$
$$G_2 = G_{filter}(2 \text{ kHz})G_{DIG} = 0.9798 \text{ V/V}$$
$$G_3 = G_{filter}(3 \text{ kHz})G_{DIG} = 0.9649 \text{ V/V}$$

Now whenever we make a measurement at 1, 2, or 3 kHz using this configuration, we can divide the uncalibrated result by the focused gain calibration factors G_1, G_2, and G_3 to correct for the digitizer's gain errors. These focused calibration factors can be used when we test the DUT output at each frequency.

Next we have to adjust the AWG signal levels so that each tone is equal to 500 mV RMS. We do this by dividing the desired RMS level of each tone by the AWG gain at each frequency:

```
G_AWG_1kHz=1.00606;    /* Note: These values would normally be calculated */
G_AWG_2kHz=0.969696;   /* and stored during the focussed calibration process */
G_AWG_3kHz=0.953535;   /* rather than being hard-coded as shown here. */
for(i=0;i<512;i++)
    waveform[i]= (1.414*500mV/G_AWG_1kHz)*sin(2*PI*31*i/512)
               + (1.414*500mV/G_AWG_2kHz)*sin(2*PI*67*i/512)
               + (1.414*500mV/G_AWG_3kHz)*sin(2*PI*97*i/512);
```

When this waveform is loaded into the AWG, we expect each signal level from the AWG to be very close to 500 mV RMS. There will still be small errors; so it is a good idea to measure the actual signal level at each frequency again, using the calibrated digitizer for maximum accuracy. These actual voltage levels should be saved as calibration factors for later use, anytime the program needs to know the exact signal levels produced by the AWG when sourcing this multitone signal. The digitizer gain factors G_1, G_2, and G_3 should also be saved as calibration factors for later use, when measuring the output of the DUT at these frequencies.

The second, more common method of AWG and digitizer calibration relies on a highly accurate RMS voltmeter, capable of measuring sine waves from the AWG. The calibration steps were highlighted in Figure 10.7. Using this technique, we first calibrate the gain of the AWG at each test tone as shown in Figure 10.7(a). The gain of the AWG is defined as the ratio of the actual sine wave output level divided by the desired level

$$G_{AWG}(f) = \frac{V_{AWG}}{V_{in}} \tag{10.20}$$

For instance, if we request a 1-kHz sine wave with a signal level of 1.0 V RMS and we get a 1-kHz sine wave with a signal level of 0.9 V RMS, then the gain of the AWG is 0.8 V/V at 1 kHz. We can correct this error by requesting a signal level of 1.0/0.8 = 1.25 V RMS the next time we want 1.0 V RMS. The actual signal level produced by the AWG can be measured using a highly accurate RMS voltmeter. We repeat this process for each frequency of interest to build up a frequency response calibration table for the AWG. Once we know the gain of the AWG at each frequency of interest, we can produce a composite multitone signal containing all the frequencies of interest at highly accurate signal levels.

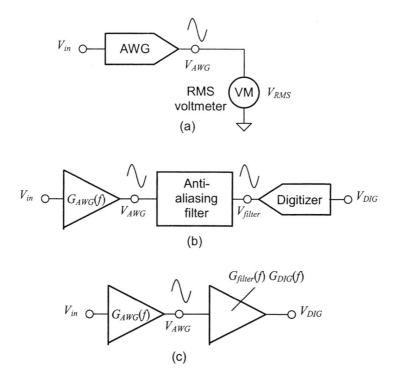

Figure 10.7. Second calibration method: (a) measure AWG AC response using highly accurate voltmeter and create linear gain model; (b) measure frequency response behavior of filter and digitizer combined using AWG and create linear gain model; (c) linear model of complete test setup.

While measuring the tones from the AWG with the high-accuracy AC voltmeter, we also measure them with the digitizer (antialiasing filter enabled) as shown in Figure 10.7(b). The digitizer gain at each frequency can be easily calculated by dividing the digitizer result (FFT output at each frequency of interest) by the RMS signal level as measured by the AC voltmeter

$$G_{DIG}(f) = \frac{V_{DIG}}{V_{AWG}} \qquad (10.21)$$

The only problem with this technique is that most AC voltmeters are less accurate at higher frequencies than at audio-band frequencies. The test engineer should realize that this will limit the accuracy of higher-frequency focused calibrations.

Example 10.5

Repeat the previous example using a high-accuracy AC voltmeter rather than a DC voltmeter as the accuracy standard.

Solution:

This technique is more straightforward than the previous one, and is often more accurate as well. Instead of producing a three-tone multitone signal, we produce single tones from the AWG at each tone of interest (1, 2, and 3 kHz). As the AWG sources each tone, we measure its output with the digitizer and also with a high-accuracy RMS voltmeter. The voltmeter gives three readings

$$V_{AWG}(1 \text{ kHz}) = 503.0 \text{ mV}$$
$$V_{AWG}(2 \text{ kHz}) = 484.8 \text{ mV}$$
$$V_{AWG}(3 \text{ kHz}) = 476.7 \text{ mV}$$

These results are assumed to be accurate. Therefore the AWG's gain is calculated using Eq. (10.20) as follows

$$G_{AWG}(1 \text{ kHz}) = \frac{0.5030 \text{ V}}{0.500 \text{ V}} = 1.006 \text{ V/V}$$
$$G_{AWG}(2 \text{ kHz}) = \frac{0.4848 \text{ V}}{0.500 \text{ V}} = 0.970 \text{ V/V}$$
$$G_{AWG}(3 \text{ kHz}) = \frac{0.4767 \text{ V}}{0.500 \text{ V}} = 0.953 \text{ V/V}$$

As in the previous example, the filtered digitizer measures the signal levels $V'_1 = 0.497$ V, $V'_2 = 0.475$ V, and $V'_3 = 0.460$ V. The combined filter and digitizer gain can be calculated simply by comparing its output at each frequency by the known input

$$G_1 = G_{filter}(1 \text{ kHz}) G_{DIG}(1 \text{ kHz}) = \frac{0.497 \text{ V}}{0.503 \text{ V}} = 0.9880 \text{ V/V}$$
$$G_2 = G_{filter}(2 \text{ kHz}) G_{DIG}(2 \text{ kHz}) = \frac{0.475 \text{ V}}{0.4848 \text{ V}} = 0.9798 \text{ V/V}$$
$$G_3 = G_{filter}(3 \text{ kHz}) G_{DIG}(3 \text{ kHz}) = \frac{0.460 \text{ V}}{0.4767 \text{ V}} = 0.9649 \text{ V/V}$$

The remaining focused calibration steps are the same as before. Notice that this calibration process is considerably easier to follow than the previous one.

So far, we have assumed that the output of the digitizer is in volts and the output of the FFT gives us RMS volts. This is not necessarily true in practice. Once a waveform has been captured by a digitizer, it often requires a mathematical scaling and correction process before it produces anything that even resembles absolute volts. A typical digitizer produces a nonscaled waveform, which may be represented by two's complement integers or by a floating-point waveform normalized to 1 unit peak, where the unit is undefined. For example, if the digitizer is set to a ±2-V range, and a 1-V peak input is applied, then its output may appear as a sine wave with a 0.5 unit peak amplitude. If we want to convert this unscaled output into something that approximates absolute voltage, we have to multiply by a scaling factor (in this example, the scaling factor would be 2).

Depending on the tester's operating system, the FFT of this signal may or may not produce RMS volts in each spectral bin. More often than not, an FFT routine introduces scaling factors of its own, usually equal to the square root of the number of samples or some similar factor. We have to remove any such scaling factor from the FFT output by dividing the FFT result by the scaling factor.

Notice that we can combine all the scaling factors and the focused calibration adjustments after the FFT operation to save test time. Using this post-FFT scaling technique, we only have to correct a few values, rather than scaling the whole time domain waveform. The post-FFT scaling technique makes the code a little harder to follow, since we never see a correctly scaled time domain waveform. However, the post-FFT process can save quite a bit of calculation time. The details of scaling digitizer and FFT outputs is highly dependent on the type of digitizer; so we will not cover these operations in any more detail. The scaling process should be part of the training offered by the ATE vendor.

It is worth noting that the digitizer tones and AWG tones may not be identical during a given test. For example, the simultaneous ADC and DAC crosstalk test described in Chapter 9 uses a different frequency for the DAC channel (digitizer) than the ADC channel (AWG). When using the AWG to calibrate the digitizer or vice versa, we often have to perform one calibration for the AWG signal and then perform a second calibration for the digitizer signal.

10.3.2 Low-Level AWG and Digitizer Amplitude Calibrations

The RMS voltmeter calibration approach works nicely for high-amplitude signals. However, low-amplitude signals cannot be accurately calibrated in such a simple manner. Low-amplitude sine waves and multitones are often sourced and/or measured during tests such as gain tracking, signal to distortion, crosstalk, CMRR, and PSRR.

Our objective in calibrating a low-amplitude test tone is to measure only the test tone, not the noise and distortion that inevitably accompany the test tone. Unfortunately, RMS voltmeters measure the total RMS voltage at their input (up to a certain bandwidth), including distortion and noise. Some of the noise even comes from the voltmeter itself. The distortion is not much of a problem because it is generally many orders of magnitude lower than the test tone amplitude, even when the test tone amplitude is small. However, the electrical noise inherent to all analog measurements is basically independent of the signal level (quantization noise notwithstanding). As a result, electrical noise can introduce significant RMS voltage measurement errors into low-amplitude sine wave measurements.

When measuring the RMS voltage of a high-amplitude sine wave corrupted by small amounts of noise, a true RMS voltmeter's reading is dominated by the RMS signal level of the test tone. Small amounts of noise introduce almost no amplitude error because of the way sine waves and random noise combine into a composite RMS signal level. The total RMS of a sine wave plus a random noise signal is given by

$$RMS_{total} = \sqrt{RMS_{signal}^2 + RMS_{noise}^2} \quad (10.22)$$

RMS_{total} is the RMS of the composite sine wave plus noise signal that we could expect to measure using an RMS voltmeter. RMS_{noise} is the amplitude of the noise corrupting the sine wave, and RMS_{signal} is the RMS amplitude of the sine wave itself (the amplitude we actually want to measure). If the RMS amplitude of the sine wave is much larger than RMS amplitude of the noise, then the RMS_{noise} component of Eq. (10.22) becomes insignificant due to the squaring operations. In this case, the amplitude of the composite signal is very close to the amplitude of the sine wave we wish to measure. Therefore, we can use the RMS voltmeter to make an accurate measurement of a high-level sine wave, even if it contains a small amount of noise. If, on the other hand, the sine wave amplitude is not large compared to the noise level, the noise portion of the equation above introduces a significant measurement error. DSP-based testing provides a solution to this problem.

We first generate a high-amplitude sine wave from the AWG to calibrate the gain of the digitizer at each test frequency of interest, using the voltmeter-based calibration technique of the previous example. After we have generated a set of gain calibration factors to correct the digitizer's amplitude error at each frequency of interest, we can then produce a low-level sine wave or multitone signal from the AWG that approximates the signal we need during the DUT test. Since we have just finished characterizing the digitizer's amplitude error at each frequency, we can apply the gain calibration factors to accurately measure the signal level of the lowered AWG test tones. The digitizer/FFT combination, unlike the RMS voltmeter, can differentiate between signal, noise, and distortion components, giving a much more accurate measurement of the low-level test tone components generated by the AWG. Each AWG test tone amplitude can then be adjusted to produce a highly accurate multitone signal.

If we need to calibrate the digitizer for measurements of very low-level signals, we can use this AWG calibration technique to produce a highly accurate single tone or multitone signal. After producing the low-level AWG signal, the digitizer can be reconfigured into a mode more suitable for low-level signal measurements. For instance, the digitizer's input range may be dropped from ±1.0 V to ±100 µV, and a high-pass DC blocking filter may be added to its signal path to prevent signal clipping on this lowered range. (Small DC offsets in the signal can cause digitizer clipping on low signal ranges because the digitizer's front end is set to a high gain.) Digitizing the multitone signal with this new digitizer configuration and comparing the outputs of an FFT with the known signal amplitudes, we can generate calibration factors for the digitizer that are tailored for this new input configuration. These calibration factors can later be used to correct measurements of low-level DUT signals measured using the digitizer in this particular input mode.

10.3.3 Amplitude Calibrations for ADC and DAC Tests

So far, we have mainly discussed calibrations for purely analog channels. When we measure analog channels, we can often ignore absolute voltage measurements since we are working with ratios of output voltage divided by input voltage. Voltage errors in these cases simply cancel out in the final result. When working with DACs and ADCs, though, we usually have to worry about absolute voltages. The absolute voltage calibrations discussed in the previous sections are adequate for measuring the output of DACs using a digitizer. However, signals generated by an AWG are never exactly correct and we need to compensate for ADC input signal level errors.

We have seen how an AWG waveform can be calibrated to produce a fairly accurate signal level at each frequency of interest, and how the residual errors can be measured and stored as calibration factors. But what would we do with these residual errors? The answer is that we fine-tune our final measurements by realizing that the input was not quite what we requested.

Example 10.6

We wish to measure the gain of an ADC at 1, 2, and 3 kHz. We use the technique of Example 10.4 to create a fairly accurate three-tone multitone signal with approximately 500 mV RMS at each frequency. The RMS voltage levels of each tone are measured and stored as calibration factors. The calibration factors are:

$$1-\text{kHz amplitude} = 501 \text{ mV RMS}$$
$$2-\text{kHz amplitude} = 500 \text{ mV RMS}$$
$$3-\text{kHz amplitude} = 499 \text{ mV RMS}$$

The calibrated AWG multitone is applied to the ADC and samples of the ADC output are collected by the tester's capture memory. An FFT is performed on the ADC resulting in the following amplitudes, in RMS LSBs:

$$1-\text{kHz output amplitude} = 127.52 \text{ LSBs RMS}$$
$$2-\text{kHz output amplitude} = 120.32 \text{ LSBs RMS}$$
$$3-\text{kHz output amplitude} = 118.88 \text{ LSBs RMS}$$

Calculate the ADC gain at each frequency, expressed in bits per volt.

Exercises

10.2. A digitizer produces readings of 1.19 and 4.44 V when two known DC voltage levels of 1.234 and 4.32 V are applied as input, respectively. Next, an AWG is set to produce a two-tone multitone signal at 2 and 3 kHz, each having a 1 V RMS level. The digitizer, with the antialiasing filter bypassed, indicates RMS readings of 0.93 V at 2 kHz and 1.21 V at 3 kHz. Subsequently, the antialiasing filter is connected in the measurement path and the digitizer RMS output now indicates 0.925 V at 2 kHz and 1.19 V at 3 kHz. Determine the individual gains of the digitizer, AWG, and filter at 2 and 3 kHz from the calibrated data given.

Ans. $G_{DIG} = G_{DIG}(2 \text{ kHz}) = G_{DIG}(3 \text{ kHz}) = 1.0531$ V/V; $G_{AWG}(2 \text{ kHz}) = 0.8831$ V/V, $G_{AWG}(3 \text{ kHz}) = 1.1486$ V/V; $G_{filter}(2 \text{ kHz}) = 0.9946$ V/V, $G_{filter}(3 \text{ kHz}) = 0.9835$ V/V.

10.3. A three-tone multitone signal at 1, 2, and 3 kHz is to be generated by an AWG and applied to a DUT input. The level of each tone should be 300 mV RMS. At each frequency, the AWG output is measured with a highly accurate RMS voltmeter. The measured values at 1, 2, and 3 kHz are 299, 310, and 267 mV, respectively. What RMS amplitude values should be used to program the AWG? If the corresponding digitized output at each frequency is 311, 326, and 294 mV, respectively, what are the RMS levels of a DUT whose values read by the digitizer at each frequency is 517, 520, and 450 mV, respectively.

Ans. AWG programmed amplitudes at 1, 2, and 3 kHz: 310, 290.3, and 337.1 mV. DUT RMS levels at 1, 2, and 3 kHz: 497.1, 494.5, and 408.7 mV.

Solution:

We are tempted to simply divide the ADC outputs by 500 mV to calculate the gain of the ADC at each frequency. Fortunately, we kept a record of the small errors in the AWG amplitudes; so we can make a more precise measurement of gain. The gain at each frequency is given by the output/input calculations

$$G_{ADC}(1\text{ kHz}) = \frac{127.52 \text{ LSBs}}{501 \text{ mV}} = 254.53 \text{ bits/V}$$

$$G_{ADC}(2\text{ kHz}) = \frac{120.32 \text{ LSBs}}{500 \text{ mV}} = 240.64 \text{ bits/V}$$

$$G_{ADC}(3\text{ kHz}) = \frac{118.88 \text{ LSBs}}{499 \text{ mV}} = 238.24 \text{ bits/V}$$

These answers are nearly the same as if we divided by the ideal voltage level of 500 mV RMS. However, we never know when a tester's AWG will produce more significant errors after calibration, leading to correlation problems at a later time. Consequently it is always a good idea to measure the actual input signal levels with as much accuracy as possible so that the final measurement results can be adjusted for any residual input errors.

10.4 OTHER AC CALIBRATIONS

10.4.1 Phase Shifts

In the same way that signal paths modify the amplitude of multitone signals, they also modify the phase shift of each test tone. Whereas cascaded gains are multiplicative in nature, phases are additive. The total phase shift through multiple cascaded circuits is equal to the sum of the phase shifts through each of the individual circuits. Therefore, phase errors from cascaded DIB circuits and measurement instruments are fairly easy to extract.

In fact, most phase measurements are implemented by simply digitizing the input and output signal of a circuit and then computing their frequency spectra. The input and output phase at each frequency is derived, from which the phase shift $\Delta\phi(f)$ at each frequency is simply calculated as

$$\Delta\phi(f) = \phi_{output}(f) - \phi_{input}(f) \tag{10.23}$$

Whatever phase shifts are introduced by the measurement path are present in both the input measurement and the output measurement. Therefore, the phase shift of the measurement path is irrelevant. It cancels out in the final calculation, just like the gain errors in an analog gain measurement. Like the gain measurement, the error cancellation depends on the entire measurement path remaining in the same configuration in both the output measurement and the input measurement. However, if the paths need to change for some reason, we can simply measure the difference in phase shift between the two configurations, and subtract the difference from the measured phase shift. The following example will illustrate this approach.

Example 10.7

A DUT includes an analog channel with an attenuation of 100 V/V. Since the attenuation is so high, we have to measure a very small signal at its output and measure a very high signal at its input. This large change in amplitude requires us to set our digitizer to a low input range for the DUT output measurement and a high input range for the DUT input measurement. Determine a calibration scheme that will allow an accurate phase shift measurement of this DUT circuit at a frequency of 1 kHz.

Solution:

The input to the DUT is set at a fairly high signal level of 1.0 V RMS, resulting in an output signal level of about 10 mV RMS. In order to keep track of the test data, let us sketch a diagram of the test setup. This we do in Figure 10.8(a), where we illustrate the digitization of the input and output nodes of the DUT using two separate instances of the digitizer; each representing a different range setting on the actual digitizer. (The dashed line is to indicate that there is really only one physical digitizer.) Now, consider the linear equivalent model of the measurement setup shown in Figure 10.8(b). Here the AWG, DUT and the digitizer on each range setting are modeled with gains G_{AWG}, G_{DUT}, G_{DIG1} and G_{DIG2}, respectively. In general, these gains are complex numbers, having a magnitude and phase shift component.

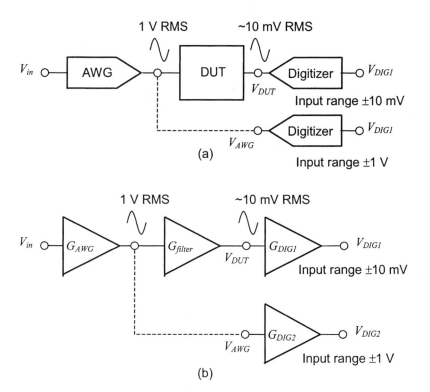

Figure 10.8. DUT phase measurement setup: (a) sampling input and output signals to DUT using a digitizer with two separate range settings; (b) equivalent linear model representation.

According to linear system theory, we can write

$$V_{DIG1} = G_{AWG}\, G_{DUT}\, G_{DIG1}\, V_{IN} \tag{10.24}$$

from which we can deduce the total phase difference from output of the first digitizer with respect to the input is

$$\phi_{DIG1} - \phi_{IN} = \Delta\phi_{AWG} + \Delta\phi_{DUT} + \Delta\phi_{DIG1} \tag{10.25}$$

Here we distinguish between the phase of the input and output signals, ϕ_{IN} and ϕ_{DIG1}, obtained from an FFT analysis and the phase shift caused by each stage of the test setup as $\Delta\phi_{AWG}$, $\Delta\phi_{DUT}$, and $\Delta\phi_{DIG1}$. Similarly, the phase shift difference between the output of the second digitizer and the input is

$$\phi_{DIG2} - \phi_{IN} = \Delta\phi_{AWG} + \Delta\phi_{DIG2} \tag{10.26}$$

For the phase measurement at hand, we are interested in the phase shift caused by the DUT, $\Delta\phi_{DUT}$. To obtain this, subtract Eq. (10.26) from (10.25), and rearrange to get

$$\Delta\phi_{DUT} = (\phi_{DIG1} - \phi_{DIG2}) - (\Delta\phi_{DIG1} - \Delta\phi_{DIG2}) \tag{10.27}$$

Through the FFT analysis of the digitizer's two outputs, we have information about the first two terms on the right side of Eq. (10.27). The latter two terms represent the phase shift mismatch of the digitizer on the two range settings. To obtain this phase mismatch, an additional set of measurements must be made that involve the digitizer alone. Consider removing the DUT from the test setup and have the AWG generate a signal suitable for the digitizer on the 10-mV and 1-V ranges, as shown in Figure 10.9(a). For the case described, we shall select a signal level of 10 mV. Next, the 10-mV signal is digitized on the two range settings. Subsequently, the phase mismatch can be found according to the linear representation shown in Figure 10.9(b) to be

$$\Delta\phi_{DIG1} - \Delta\phi_{DIG2} = \phi_{DIG3} - \phi_{DIG4} \tag{10.28}$$

Substituting Eq. (10.28) into (10.27), we write the final result in terms of the data collected from the spectral analysis of the digitizer outputs as

$$\Delta\phi_{DUT} = (\phi_{DIG1} - \phi_{DIG2}) - (\phi_{DIG3} - \phi_{DIG4}) \tag{10.29}$$

Note that only one of these measurements, ϕ_{DIG1}, has to be measured for each DUT. The other three values can be measured once during the focused calibration process and stored as calibration factors. In practice, we would probably combine the three premeasured calibration factors into a single calibration term $\phi_{CAL} = (\phi_{DIG2} + \phi_{DIG3} - \phi_{DIG4})$ to reduce the number of separate calibration factors.

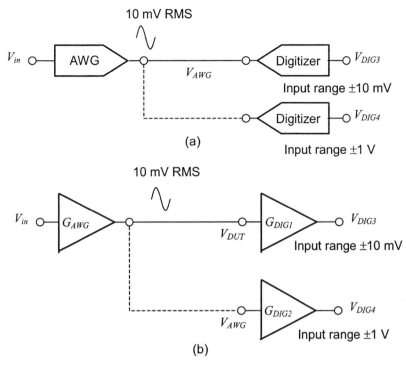

Figure 10.9. Measuring phase shift difference between two ranges of a digitizer: (a) digitizing a signal with two range settings on a digitizer; (b) equivalent linear model representation.

Exercises

10.4. The phase difference between the input to a DUT and its output was measured with a digitizer on separate ranges to be $-\pi/16$ radians. In addition, it was determined through a separate measurement that the phase shift difference between the two ranges is $+\pi/16$ radians. Both phase measurements were made in the exact same manner, that is, the same phase reference. What is the phase shift created by the DUT at this frequency?

Ans. 0 radians.

10.5. Write the calibration equation for a DIB circuit that is described by the following output-input equation, $V_{DIB} = \tan(V_{IN})$.

Ans. $V_{CALIBRATED} = \tan^{-1}(V_{DIB})$.

10.4.2 Digitizer and AWG Synchronization

Thus far, we have treated the AWG and digitizer as if the timing of their waveform creation and capture circuits were well controlled relative to one another and relative to the digital patterns. In other words, if we produce a 1-kHz sine wave with the AWG and repetitively digitize it, we might expect the phase shift of the digitized waveform to always come out the same. Unfortunately, the AWG and digitizer may very well start sourcing and digitizing samples at a different time each time we make a DSP-based measurement. If we do not take care to resynchronize the AWG sample timing with the digitizer sample timing, we may find that the phase of the digitized waveform is different every time we execute the same measurement.

During gain measurements, we do not usually worry about the relative timing of the AWG and digitizer, but during phase measurements the phase synchronization is critically important. Each tester has a different way of resynchronizing waveform instruments like AWGs, digitizers, source memory, and capture memory. Some testers may not require resynchronization at all. These details should be covered by the ATE vendor's training class.

10.4.3 DAC and ADC Phase Shifts

As mentioned in Chapter 9, "Sampled Channel Testing," the calibration of absolute phase shifts is beyond the scope of this book. To measure the phase shift through an ADC or DAC, we have to know exactly where the phase of each test tone is located relative to the digital samples sent to or captured from the converter circuit. Since mixed-signal testers do not typically align analog waveforms and digital signals precisely, we cannot use an output-minus-input calculation for phase shift through ADCs and DACs.

One possible solution to this problem is to produce a square wave or similar signal from the digital pattern generator at the frequency of interest and capture it using the digitizer. Since a square wave has known amplitude and phase characteristics, this technique can be used to determine the digitizer's phase shift relative to the digital pattern generator. This only works if the digitizer sampling times can be repeatedly synchronized to the digital pattern generator. Once the digitizer's phase shift at each frequency of interest is calibrated relative to the digital pattern, it can be used to calibrate the AWG.

10.4.4 Distortion Tests

It is difficult to extract distortion components from cascaded signal paths, since distortion is neither an additive nor a multiplicative process. However, the distortion of a tester instrument's DAC or ADC can be compensated to a large degree by thoroughly characterizing the input/output transfer characteristics of the converter. By building a software table of input voltage versus output code (in the case of ADCs) or input code versus output voltage (in the case of DACs) we can compensate for the nonlinearities of voltmeters, digitizers, AWGs, and DC sources. For example, if we know the transfer characteristics of a digitizer at a particular frequency, we can multiply the collected samples by the inverse transfer curve to extract much of the distortion caused by the digitizer. Unfortunately, distortion characteristics vary with frequency; so there are many subtleties involved in this process.

Fortunately, much of this type of software calibration for distortion removal is already performed by the tester's operating system (at least in advanced mixed-signal testers). Test engineers do not commonly involve themselves with this kind of advanced focused calibration process. However, ATE vendors and bench equipment manufacturers are quite familiar with these types of software calibration techniques, since they provide tester performance that might not otherwise be achievable.

10.4.5 Noise Tests

Noise tests, like distortion measurements, are not typically calibrated as thoroughly as AC amplitude and phase measurements. In theory, we should measure the frequency response of a digitizer across the full spectrum of the noise measurement, correcting each noise component in the FFT spectrum by the digitizer gain at that frequency. In practice we find that the in-band frequency characteristics of most digitizers are flat enough to produce a reasonably good measurement of noise without any additional focused calibrations. This assumption is verified by correlating the ATE tester's noise measurement to bench equipment as part of the usual tester-to-bench correlation effort.

Occasionally, a test engineer wants to subtract the noise generated by the measurement path from the total noise measured. This idea is fraught with problems, since noise is a random process and its exact time-varying nature is unknown until after the measurement is made. It is fair to assume, though, that noise from the tester will not cancel noise from the DUT. This is because uncorrelated random noise always adds constructively rather than destructively. Therefore, whatever noise is measured by the tester is guaranteed to be a worst-case measurement. The noise floor of the tester's instruments is often the limiting factor in the accuracy with which measurements can be made. This is a fact of life for mixed-signal test engineers.

10.5 ERROR CANCELLATION TECHNIQUES

10.5.1 Avoiding Absolute Calibration

At this point in the chapter, the reader is probably happy to see the words "avoiding" and "calibration" in the same phrase. The professional test engineer is equally happy to avoid generating hundreds of unnecessary calibration factors. We have already seen how gain measurements and phase shift measurements can sometimes be made without relying on absolute voltage and phase values, thus simplifying the focused calibration process. This is because the gain errors and phase errors sometimes cancel or "wash out" in the final calculation. We try to take advantage of error cancellation techniques whenever possible, since it leads to simplified test techniques providing very high accuracy. Let us look at a couple of techniques that allow us to avoid focused calibration altogether.

10.5.2 Gain and Phase Matching

Many DUT circuits, like stereo audio L and R channels and cellular telephone I and Q channels, require a high level of performance matching between two supposedly identical circuits. Gain matching is defined as the ratio between one channel's gain and the gain of the other channel. Phase matching is defined as the difference in phase shift from one channel to the other. One approach to measuring these parameters is to use two AWGs and two digitizers to measure the gain and phase of the two channels. If we do this, we have to calibrate the performance mismatch bewteen the two AWGs and digitizers so their mismatch can be subtracted from the

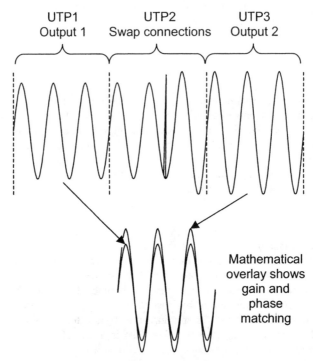

Figure 10.10. Three-UTP technique for AWG/digitizer alignment. Mathematical overlay shows gain and phase differences.

final answer. A simpler solution is to use a single AWG (or source memory pattern) to apply the same signal to both channels at once and then measure the two outputs using a single digitizer, one channel at a time. Whatever gain errors and phase shifts are introduced by the AWG and/or digitizer on one channel will be introduced on the other as well. In this case, calibration is unnecessary.

Sometimes the AWG and digitizer cannot be accurately synchronized for a phase-matching measurement, due to a limitation in the tester's clocking architecture. This makes phase-matching measurements very difficult. In such a case, a three-UTP approach can be used to keep the digitizer and AWG sample timing aligned between the two phase shift measurements. A UTP (see Chapter 6) is a unit test period. One UTP represents the time it takes to cycle through all the samples in a DSP-based measurement waveform. If we allow the AWG and digitizer to continue through three UTPs instread of stopping after the first UTP, then we would normally get three identical waveforms.

Now if we switch the digitizer from one channel to the other in the middle of the second UTP, we produce three sets of samples: one set for Channel 1, one set of Channel 2, and one set of garbage samples that contain an abrupt change from Channel 1 to Channel 2, as shown in Figure 10.10. The garbage samples are discarded. Because the AWG and digitizer never stopped between the first and third UTPs, their sample timing is perfectly aligned. Therefore, we can measure the difference in phase shift from one channel to the other very accurately by simply subtracting the phase shift of one set of samples from the phase shift of the other set. Focused calibration in this case is unnecessary, since the sample timing between the AWG and digitizer remains constant from one channel's sample set to the other. Any phase shifts caused by the AWG and digitizer therefore cancel out in the final calculation.

10.5.3 Differential Gain and Differential Phase

Focused calibrations can sometimes be eliminated using DIB circuits, as in the case of differential gain and phase measurements. Video circuits often include a differential gain and phase specification. NTSC color TV signals consist of a high frequency sine wave riding on a low frequency intensity signal. The phase and amplitude of the high frequency signal determine hue and saturation (color) while the lower frequency signal determines brightness.

It is important that a video channel give the same gain and phase shift at different DC offsets so that the hue and saturation signals are not affected by the slower variations in the brightness signal. What this means in circuit terms is that the AC amplitude and phase shift of the high frequency sine wave have to be unaffected by varying DC offset. Differential gain is defined as the change in AC amplitude with varying DC offset. Differential phase is defined as the change in phase with varying DC offset.

When measuring the differential gain and phase of a video circuit, it would seem obvious that we want to digitize a high-frequency sine wave from the circuit's output twice, once with one DC offset and once with another DC offset. In theory, differential gain and phase could be calculated directly from these two captured waveforms. In practice, our digitizer may have a nonideal differential gain and phase of its own. This means that we cannot distinguish between gain and phase shifts caused by the digitizer and those caused by the video circuit.

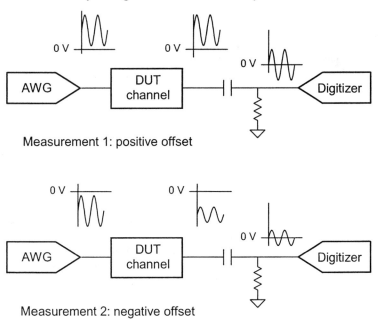

Figure 10.11. DC blocking capacitor avoids digitizer's differential gain and phase errors.

One solution to this problem is to apply a pair of equal-amplitude sine waves to the digitizer with DC offsets corresponding to the two outputs expected from the video circuit. The difference in digitizer gain and phase can be noted and stored as calibration factors for use during the video circuit output measurements. A simpler technique is to simply block the DC offset of the video circuit output using an *RC* high-pass filter as shown in Figure 10.11. Using this technique, the digitizer sees the same DC offset (zero volts) regardless of the video circuit's DC offset. This removes the digitizer's differential gain and differential phase characteristics from the measurement. We might also want to use an *RC* blocking circuit terminated to a DC source in the AWG signal path. The DC source could be set to the two DC offsets rather than adjusting the DC offset of the AWG signal. This would remove any differential gain and phase errors inherent in the AWG.

10.6 SUMMARY

While it might be reasonable to assume that a multimillion dollar ATE tester is perfectly accurate, it is certainly not wise to make that assumption. Sometimes the tester's accuracy is adequate for a given test, but often the test engineer must improve upon the basic accuracy of the tester using focused calibrations. Focused calibrations are the key to fast, accurate ATE measurements. They are also the source of much confusion. We have discussed some of the common calibration techniques found in mixed-signal test programs. These techniques represent a good starting point for the novice test engineer. Other focused calibration techniques will have to be learned or even invented as the test engineer's expertise develops.

Problems

10.1. A unity-gain amplifier is used to buffer a signal before it is applied to a digitizer. A calibration sequence determined that the amplifier has a gain of 1.09 V/V and an offset of 5.6 mV. In addition, the digitizer was found to have a gain of 0.98 V/V with an offset of −11.3 mV. What is the composite gain and offset for this cascade combination? Write the calibration equation for this test setup.

10.2. Three stages are connected in cascade. Each stage has a gain and an offset. From first principles, derive the composite gain and offset of this arrangement. What about when four stages are connected in cascade? Extend the formula for N stages connected in cascade?

10.3. A digitizer produces readings of 0.56 and 3.78 V when two known DC voltage levels of 0.54 and 3.65 V are applied as input, respectively. Next, an AWG is set to produce a three-tone multitone signal at 2, 3, and 4 kHz. Each tone has an RMS amplitude of 0.5, 0.707, and 1.0 V, respectively. The digitizer, with the antialiasing filter bypassed, indicates RMS readings of 0.486 V at 2 kHz, 0.721 V at 3 kHz, and 1.05 V at 4 kHz. Subsequently, the antialiasing filter is connected in the measurement path and the digitizer RMS output now indicates 0.471 V at 2 kHz, 0.714 V at 3 kHz, and 0.987 V at 4 kHz. Determine the individual gains of the digitizer, AWG and filter at 2, 3, and 4 kHz from the calibrated data given.

10.4. The calibration factors associated with an AWG, digitizer, and antialiasing filter at 1 kHz are 1.12, 1.06, and 0.998 V/V, respectively. A 2.5-V RMS sine wave at 1 kHz is to be generated by the AWG. What amplitude should be used to program the AWG? If the digitized value of a DUT is 1.65 V RMS, what is the actual RMS value of the DUT output? Assume that the antialiasing filter is connected in the signal path.

10.5. A three-tone multitone signal at 1, 2, and 3 kHz is to be generated by an AWG and applied to an ADC. The level of each tone should be 800 mV RMS. At each frequency, the AWG output is measured with a highly accurate RMS voltmeter. The measured values at 1, 2, and 3 kHz are 788, 821, and 799 mV, respectively. If the corresponding digitized output at each frequency is 41.37 LSBs RMS, 43.06 LSBs RMS, and 41.53 LSBs RMS, respectively, calculate the gain of the ADC at each frequency.

10.6. A three-tone multitone signal at 1, 2, and 3 kHz is to be source to a DAC via the source memory. Each tone is to have an amplitude of 1 V RMS. The digitizer was calibrated and has the following calibration scale factors: 1.13 V/V at 1 kHz, 0.998 V/V at 2 kHz, and 0.987 V/V at 3 kHz. If the digitizer output at each frequency is 1.02 V RMS, 0.998 V RMS, and 1.08 V RMS, what is the gain of the DAC at each frequency?

10.7. The input and output signals to a DUT was captured by a digitizer on one range setting and an FFT was performed on each signal. At a frequency of 1 kHz, the input signal has a phase value of 25 degrees and the output has a phase value of -45 degrees. What is the phase shift created by the DUT?

10.8. The input and output signals to a DUT was captured by a digitizer on range setting 1 and setting 2, respectively. Subsequently, an FFT was performed on each signal. At a frequency of 1 kHz, the input signal has a phase value of 25 degrees and the output has a phase value of -45 degrees. Further, the AWG sourced a 1 kHz signal directly to the digitizer while on range setting 1, where an FFT analysis revealed a phase value of 10 degrees. Next, the range of the digitizer was change to the second setting where an FFT analysis revealed a phase value of 6 degrees. What is the phase shift created by the DUT at 1 kHz?

10.9. A 1 V RMS, 1-kHz sinusoidal signal is passed through a DIB circuit whose input-output behavior is described by the equation,

$$V_{DIB} = \tan(V_{in})$$

A digitizer sampling at 16 kHz captures 64 samples of the output, V_{DIB}. Using MATLAB, or an equivalent software package, determine the signal-to-noise ratio of the output signal. Next, determine a calibration expression for the DIB and pass the collected samples through the calibration equation. Perform the same signal-to-noise ratio analysis and compare the result to that obtained previously. How does it compare if a 1-mV RMS random noise component is added to the collected samples?

CHAPTER 11

DAC Testing

11.1 BASICS OF CONVERTER TESTING

11.1.1 Intrinsic Parameters versus Transmission Parameters

In Chapter 8, "Analog Channel Testing," and Chapter 9, "Sampled Channel Testing," we discussed common channel parameters such as gain, gain tracking, signal-to-noise ratio, and signal to total harmonic distortion. These parameters are called *transmission parameters*, or *performance parameters*, since they describe the effect of the analog or sampled channel on the quality of transmitted signals such as voice or modulated data. In both analog and sampled channels, transmission parameters are determined by the quality of all the channel's subcircuits. For example, the signal-to-noise ratio of a DAC channel might be determined by the quality of a low-pass reconstruction filter, an output buffer amplifier, and of course the DAC itself.

In this chapter, we will focus on the so-called *intrinsic parameters* of DACs, such as absolute error, integral nonlinearity (INL) and differential nonlinearity (DNL). Intrinsic parameters are those parameters that are intrinsic to the circuit itself. They are not dependent on the nature of the test stimulus. For example, the difference between the actual DC voltage level measured at a DAC's output and the ideal voltage level is called *absolute voltage error*. Absolute voltage error can be measured at each digital code, resulting in a set of intrinsic voltage error values. Since these errors are determined purely by the quality of the DAC circuitry and not by the nature of the transmitted signal, absolute voltage error is considered to be an intrinsic parameter. Transmission parameters, by contrast, are dependent on the nature of the transmitted signal. For instance, the amplitude and frequency of the sine wave used in a signal-to-distortion test will often affect the measured result.

When testing a DAC or ADC, it is common to measure both intrinsic parameters and transmission parameters for characterization. However, it is often unnecessary to perform the full suite of transmission tests and intrinsic tests in production. The production testing strategy is often determined by the end use of the DAC or ADC. For example, if a DAC is to be used as a programmable DC voltage reference, then we probably do not care about its signal-to-distortion ratio at 1 kHz. We care more about its worst-case absolute voltage error. On the other hand, if that same DAC is used in a voice-band codec to reconstruct voice signals, then we have a different set of concerns. We do not care as much about the DAC's absolute errors as we care about their end effect on the transmission parameters of the composite audio channel, comprising the DAC, low-pass filter, output buffer amplifiers, etc.

This example highlights one of the differences between digital testing and specification-oriented mixed-signal testing. Unlike digital circuits which can be tested based on what they *are*

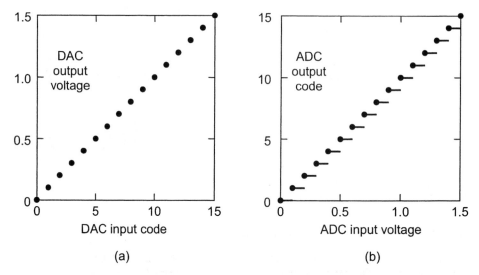

Figure 11.1. (a) DAC code-to-voltage transfer curve (b) ADC voltage-to-code transfer curve.

(NAND gate, flip-flop, counter, etc.), mixed-signal circuits are often tested based on what they *do* in the system-level application (precision voltage reference, audio signal reconstruction circuit, video signal generator, etc.). Therefore, a particular analog or mixed-signal subcircuit may be copied from one design to another without change, but it may require a totally different suite of tests depending on its intended functionality in the system-level application.

11.1.2 Comparison of DACs and ADCs

Although this chapter is devoted to DAC testing, many of the concepts presented are closely tied to ADC testing. For instance, the code-to-voltage transfer characteristics for a DAC are similar to the voltage-to-code characteristics of an ADC. However, it is very important to note that a DAC represents a one-to-one mapping function whereas an ADC represents a many-to-one mapping. This distinction is illustrated in Figure 11.1(a) and (b). For each digital input code, a DAC produces only one output voltage.

An ADC, by contrast, produces the same output code for many different input voltages. In fact, because an ADC's circuits generate random noise and because any input signal will include a certain amount of noise, the ADC decision levels represent *probable* locations of transitions from one code to the next. We will discuss the probabilistic nature of ADC decision levels and their effect on ADC testing in Chapter 12. While DACs also generate random noise, this noise can be removed through averaging to produce a single, unambiguous voltage level for each DAC code. Therefore, the DAC transfer characteristic is truly a one-to-one mapping of codes to voltages.

The difference between DAC and ADC transfer characteristics prevents us from using complementary testing techniques on DACs and ADCs. For example, a DAC is often tested by measuring the output voltage corresponding to each digital input code. The test engineer might be tempted to test an ADC using the complementary approach, applying the ideal voltage levels for each code and then comparing the actual output code against the expected code. Unfortunately, this approach is completely inappropriate in most ADC testing cases, since it does not characterize the location of each ADC decision level. Furthermore, this crude testing approach will often fail perfectly good ADCs simply because of gain and offset variations that are within acceptable limits.

Although there are many differences in the testing of DACs and ADCs, there are enough similarities that we have to treat the two topics as one. In Chapter 12 we will see how ADC testing is similar to DAC testing and also how it differs. In this chapter, however, we will concentrate mainly on DAC testing. Also, we will concentrate mostly on voltage output DACs. Current output DACs are tested using the same techniques, using either a current mode DC voltmeter or a calibrated current-to-voltage translation circuit on the device interface board (DIB).

11.1.3 DAC Failure Mechanisms

The novice test engineer may be inclined to think that all N-bit DACs are created equal and are therefore tested using the same techniques. As we will see, this is not the case. There are many different types of DACs, including binary-weighted architectures, resistive divider architectures, pulse-width-modulated (PWM) architectures, and pulse-density-modulated (PDM) architectures (commonly known as *sigma-delta* DACs). Furthermore, there are hybrids of these architectures, such as the multibit sigma-delta DAC and segmented resistive divider DACs. Each of these DAC architectures has a unique set of strengths and weaknesses. Each architecture's weaknesses determines its likely failure mechanisms, and these in turn drive the testing methodology. As previously noted, the requirements of the DAC's system-level application also determine the testing methodology.

Before we discuss testing methodologies for each type of DAC, we first need to outline the DC and dynamic tests commonly performed on DACs. The DC tests include the usual specifications like gain, offset, power supply sensitivity, etc. They also include converter-specific tests such as absolute error, monotonicity, integral nonlinearity (INL), and differential nonlinearity (DNL), which measure the overall quality of the DAC's code-to-voltage transfer curve. The dynamic tests are not always performed on DACs, especially those whose purpose is to provide DC or low-frequency signals. However, dynamic tests are common in applications such as video DACs, where fast settling times and other high-frequency characteristics are key specifications.

11.2 Basic DC Tests

11.2.1 Code-Specific Parameters

DAC specifications sometimes call for specific voltage levels corresponding to specific digital codes. For instance, an 8-bit two's complement DAC may specify a voltage level of 1.360 V ± 10 mV at digital code –128 and a voltage level of 2.635 V ± 10 mV at digital code +127. (See Section 9.3.2 for a description of converter data formats such as unsigned binary and two's complement.) Alternatively, DAC code errors can be specified as a percentage of the DAC's full-scale range rather than an absolute error. In this case, the DAC's full-scale range must first be measured to determine the appropriate test limits. Common code-specific parameters include the maximum full-scale (V_{FS+}) voltage, minimum full-scale (V_{FS-}) voltage, and midscale (V_{MS}) voltage. The midscale voltage typically corresponds to 0 V in bipolar DACs or a center voltage such as $V_{DD}/2$ in unipolar (single power supply) DACs. It is important to note that although the minimum full-scale voltage is often designated with the V_{FS-} notation, it is not necessarily a negative voltage.

11.2.2 Full-Scale Range

Full-scale range (V_{FSR}) is defined as the voltage difference between the maximum voltage and minimum voltage that can be produced by a DAC. This is typically measured by simply measuring the DAC's positive full-scale voltage, V_{FS+}, then measuring the DAC's negative full-scale voltage, V_{FS-}, and subtracting

$$V_{FSR} = V_{FS+} - V_{FS-} \tag{11.1}$$

11.2.3 DC Gain, Gain Error, Offset, and Offset Error

It is tempting to say that the DAC's offset is equal to the measured midscale voltage, V_{MS}. It is also tempting to define the gain of a DAC as the full-scale range divided by the number of spaces, or steps, between codes. These definitions of offset and gain are approximately correct, and in fact they are sometimes found in data sheets specified exactly this way. They are quite valid in a perfectly linear DAC. However, in an imperfect DAC, these definitions are inferior because they are very sensitive to variations in the V_{FS-}, V_{MS}, and V_{FS+} voltage outputs while being completely insensitive to variations in all other voltage outputs.

Figure 11.2 shows a simulated DAC transfer curve for a rather bad 4-bit DAC. Notice that code 0 does not produce 0 V, as it should. However, the overall curve has an offset near 0 V. Also, notice that the gain, if defined as the full-scale range divided by the number of spaces between codes, does not match the general slope of the curve. The problem is that the V_{FS+}, V_{FS-}, and V_{MS} voltages are not in line with the general shape of the transfer curve.

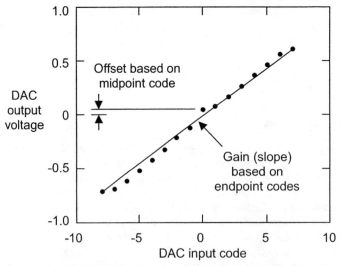

Figure 11.2. Endpoint/midpoint-referenced gain and offset for a 4-bit DAC.

A less ambiguous definition of gain and offset can be found by computing the best-fit line for these points and then computing the gain and offset of this line. For high-resolution DACs with reasonable linearity, the errors between these two techniques become very small. Nevertheless, the best-fit line approach is independent of DAC resolution; so it is the preferred technique.

A best-fit line is commonly defined as the line having the minimum squared errors between its ideal, evenly spaced samples and the actual DAC output samples. For a sample set $S(i)$, where i ranges from 0 to $N-1$ and N is the number of samples in the sample set, the best-fit line is defined by its slope (DAC gain) and offset using a standard linear equation having the form

$$Best\_fit\_line = gain \times i + offset \quad \text{for } i = 0, 1, \ldots, N-1 \quad (11.2)$$

The equations for slope and offset can be derived using various techniques. One technique minimizes the partial derivatives with respect to slope and offset of the squared errors between the sample set S and the best-fit line. Another technique is based on linear regression.[1] The equations derived from the partial derivative technique are

$$gain = \frac{N K_4 - K_1 K_2}{N K_3 - K_1^2} \qquad offset = \frac{K_2}{N} - gain \frac{K_1}{N} \quad (11.3)$$

where

$$K_1 = \sum_{i=0}^{N-1} i \quad K_2 = \sum_{i=0}^{N-1} S(i) \quad K_3 = \sum_{i=0}^{N-1} i^2 \quad K_4 = \sum_{i=0}^{N-1} i\, S(i)$$

The derivation details are left as an exercise in the problem set found at the end of this chapter. These equations translate very easily into a computer program, such as the following MATLAB routine:

```
% Store DAC output voltages in vector S %
%
% Initialize routine
k1=0; k2=0; k3=0; k4=0;
N=length(S);
% perform best-fit analysis
for i=0:N-1,
        k1 = k1 + i;
        k2 = k2 + S(i+1);
        k3 = k3 + i*i;
        k4 = k4 + i*S(i+1);
end
Gain = (N*k4 - k1*k2) / (N*k3 - k1*k1);
Offset = k2/N-Gain * (k1/N);
for i=0:N-1,
        Best_fit_line(i+1) = Gain*i + Offset;
end
```

The values in the array *Best_fit_line* represent samples falling on the least-squared-error line. The program variable *Gain* represents the gain of the DAC, in volts per bit. This gain value is the average gain across all DAC samples. Unlike the gain calculated from the full-scale range divided by the number of code transitions, the slope of the best-fit line represents the true gain of the DAC. It is based on all samples in the DAC transfer curve and therefore is not especially sensitive to any one code's location. Gain error, ΔG, expressed as a percent, is defined as

$$\Delta G = \left(\frac{G_{ACTUAL}}{G_{IDEAL}} - 1 \right) \times 100\% \qquad (11.4)$$

Likewise, the best-fit line's calculated offset is not dependent on a single code as it is in the midscale code method. Instead, the best-fit line offset represents the offset of the total sample set. The DAC's offset is defined as the voltage at which the best-fit line crosses the *y* axis. The DAC's offset error is equal to its offset minus the ideal voltage at this point in the DAC transfer curve. The *y* axis corresponds to DAC code 0.

In unsigned binary DACs, this voltage corresponds to *Best_fit_line(1)* in the MATLAB routine. However, in two's complement DACs, the value of *Best_fit_line(1)* corresponds to the DAC's V_{FS-} voltage, and therefore does not correspond to DAC code 0. In an 8-bit two's complement DAC, for example, the 0 code point is located at $i = 128$. Therefore, the value of the program variable *Offset* does not correspond to the DAC's offset. This discrepancy arises simply because we cannot use negative index values in MATLAB code arrays such as *Best_fit_line(-128)*. Therefore, to find the DAC's offset, one must determine which sample in vector *Best_fit_line* corresponds to the DAC's 0 code. The value at this array location is equal to the DAC's offset. The ideal voltage at the DAC 0 code can be subtracted from this value to calculate the DAC's offset error.

Example 11.1

A 4-bit two's complement DAC produces the following set of voltage levels, starting from code –8 and progressing through code +7:

-780 mV, -705 mV, -530 mV, -455 mV, -400 mV, -325 mV, -150 mV, -75 mV,

120 mV, 195 mV, 370 mV, 445 mV, 500 mV, 575 mV, 750 mV, 825 mV

These code levels are shown in Figure 11.3. The ideal DAC output at code 0 is 0 V. The ideal gain is equal to 100 mV/bit. Calculate the DAC's gain (volts per bit), gain error, offset, and offset error.

Solution:

We calculate gain and offset using the previous MATLAB routine, resulting in a gain value of 109.35 mV/bit and an offset value of –797.64 mV. The gain error is found from Eq. (11.4) to be

$$\Delta G = \left(\frac{109.35 \text{ mV}}{100 \text{ mV}} - 1 \right) \times 100\% = 9.35\%$$

Because this DAC uses a two's complement encoding scheme, this offset value is the offset of the best-fit line, not the offset of the DAC at code –8.

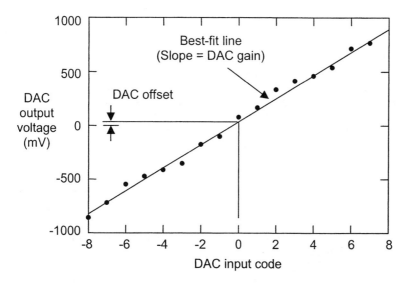

Figure 11.3. A 4-bit DAC transfer curve and best-fit line.

The DAC's offset is found by calculating the best-fit line's value at DAC code 0, which corresponds to $i = 8$

$$\text{DAC offset} = gain \times 8 + \textit{offset}$$
$$= 109.35 \text{ mV} \times 8 - 797.64 \text{ mV}$$
$$= 77.16 \text{ mV}$$

$$\text{DAC offset error} = \text{DAC offset} - \text{ideal offset}$$
$$= 77.16 \text{ mV} - 0 \text{ V} = 77.16 \text{ mV}$$

Clearly, when the ideal offset is 0 V, the DAC offset and offset error are identical. Many DACs have an ideal offset of $V_{DD}/2$ or some other nonzero value. These DACs are commonly used in applications requiring a single power supply. In such a case, the offset should be nonzero, but the offset error should always be zero.

11.2.4 LSB Step Size

The least significant bit (LSB) step size is defined as the average step size of the DAC transfer curve. It is equal to the gain of the DAC, in volts per bit. Although it is possible to measure the approximate LSB size by simply dividing the full-scale range by the number of code transitions, it is more accurate to measure the gain of the best-fit line to calculate the average LSB size. Using the results from the previous example, the 4-bit DAC's LSB step size is equal to 109.35 mV.

11.2.5 DC PSS

DAC DC power supply sensitivity (PSS) is easily measured by applying a fixed code to the DAC's input and measuring the DC gain from one of its power supply pins to its output. PSS for a DAC is therefore identical to the measurement of PSS in any other circuit, as described in Section 3.8.1. The only difference is that a DAC may have different PSS performance depending on the applied digital code. Usually, a DAC will exhibit the worst PSS performance at its full-scale and/or minus full-scale settings because these settings tie the DAC output directly to a voltage derived from the power supply. Worst-case conditions should be used once they have been determined through characterization of the DAC.

Exercises

11.1. A 4-bit unsigned binary DAC produces the following set of voltage levels, starting from code 0 and progressing through to code 15:

    1.0091, 1.2030, 1.3363, 1.5617, 1.6925, 1.9453, 2.0871, 2.3206,

    2.4522, 2.6529, 2.8491, 2.9965, 3.1453, 3.3357, 3.4834, 3.6218

The ideal DAC output at code 0 is 1 V and the ideal gain is equal to 200 mV/bit. The data sheet for this DAC specifies offset and offset using a best-fit line, evaluated at code 0. Gain is also specified using a best-fit line. Calculate the DAC's gain (volts per bit), gain error, offset, and offset error.

Ans. $G = 177.3$ mV/bit; $\Delta G = -11.3\%$; $\textit{offset} = 1.026$ V; $\textit{offset error} = 26.1$ mV.

11.2. Estimate the LSB step size of the DAC described in Exercise 11.1 using its measured full-scale range (i.e. using the endpoint method). What are the gain error and offset error?

Ans. LSB = 174.2 mV; $\Delta G = -12.9\%$; $\textit{offset error} = 9.1$ mV.

11.3 TRANSFER CURVE TESTS

11.3.1 Absolute Error

The ideal DAC transfer characteristic or transfer curve is one in which the step size between each output voltage and the next is exactly equal to the desired LSB step size. Also, the offset error of the transfer curve should be zero. Of course, physical DACs do not behave in an ideal manner; so we have to define figures of merit for their actual transfer curves.

One of the simplest, least ambiguous figures of merit is the DAC's maximum and minimum absolute error. An absolute error curve is calculated by subtracting the ideal DAC output curve from the actual measured DAC curve. The values on the absolute error curve can be converted to LSBs by dividing each voltage by the ideal LSB size, V_{LSB}. The conversion from volts to LSBs is a process called *normalization*.

Mathematically, if we denote the *i*th value on the ideal and actual transfer curves as $S_{IDEAL}(i)$ and $S(i)$, respectively, then we can write the normalized absolute error transfer curve $\Delta S(i)$ as

$$\Delta S(i) = \frac{S(i) - S_{IDEAL}(i)}{V_{LSB}} \qquad (11.5)$$

Example 11.2

Assuming an ideal gain of 100 mV per LSB and an ideal offset of 0 V at code 0, calculate the absolute error curve for the 4-bit DAC of the previous example. Express the results in terms of LSBs.

Solution:

The ideal DAC levels are −800, −700, ..., +700 mV. Subtracting these ideal values from the actual values, we can calculate the absolute voltage errors $\Delta S(i)$ as:

+20 mV, -5 mV, +70 mV, +45 mV, 0 mV, -25 mV, +50 mV, +25 mV,

+120 mV, +95 mV, +170 mV, +145 mV, +100 mV, +75 mV, +150 mV, +125 mV

The maximum absolute error is +170 mV and the minimum absolute error is −25 mV. Dividing each value by the ideal LSB size (100 mV), we get the normalized error curve shown in Figure 11.4. This curve shows that this DAC's maximum and minimum absolute errors are +1.7 and −0.25 LSBs, respectively. In a simple 4-bit DAC, this would be considered very bad performance, but this is an imaginary DAC designed for instructional purposes. In high-resolution DACs, on the other hand, absolute errors of several LSBs are common. The larger

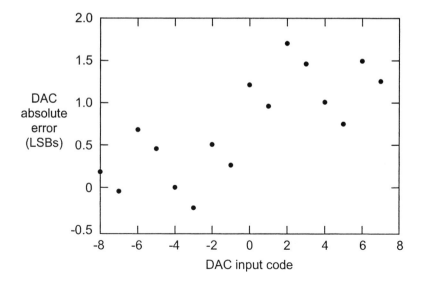

Figure 11.4. Normalized DAC error curve.

normalized absolute error in high-resolution DACs is a result of the smaller LSB size. Therefore, absolute error testing is often replaced by gain, offset, and linearity testing in high-resolution DACs.

11.3.2 Monotonicity

A monotonic DAC is one in which each voltage in the transfer curve is larger than the previous voltage, assuming a rising voltage ramp for increasing codes. (If the voltage ramp is expected to decrease with increasing code values, we simply have to make sure that each voltage is less than the previous one.) While the 4-bit DAC in the previous examples has a terrible set of absolute errors, it is nevertheless monotonic. Monotonicity testing requires that we take the discrete first derivative of the transfer curve, denoted here as $S'(i)$, according to

$$S'(i) = S(i+1) - S(i) \qquad (11.6)$$

If the derivatives are all positive for a rising ramp input or negative for a falling ramp input, then the DAC is said to be monotonic.

Example 11.3

Verify monotonicity in the previous DAC example.

Solution:

The first derivative of the DAC transfer curve is calculated, yielding the following values

75 mV, 175 mV, 75 mV, 55 mV, 75 mV, 175 mV, 75 mV, 195 mV,

75 mV, 175mV, 75 mV, 55 mV, 75 mV, 175 mV, 75 mV

Notice that there are only 15 first derivative values, even though there are 16 codes in a 4-bit DAC. This is the nature of the discrete derivative, since there are one fewer *changes* in voltage than there are voltages. Since each value in this example has the same sign (positive), the DAC is monotonic.

11.3.3 Differential Nonlinearity

Notice that in the monotonicity example the step sizes are not uniform. In a perfect DAC, each step would be exactly 100 mV corresponding to the ideal LSB step size. Differential nonlinearity (DNL) is a figure of merit that describes the uniformity of the LSB step sizes between DAC codes. DNL is also known as *differential linearity error* or DLE for short. The DNL curve represents the error in each step size, expressed in fractions of an LSB. DNL is

> **Exercises**
>
> **11.3.** Assuming an ideal gain of 200 mV/bit and an ideal offset of 1 V at code 0, calculate the absolute error transfer curve for the 4-bit DAC of Exercise 11.1. Normalize the result to 1 LSB.
>
> **Ans.** 0.0455, 0.0150, -0.3185, -0.1915, -0.5375, -0.2735, -0.5645, -0.3970,
>
> -0.7390 -0.7355 -0.7545 -1.0175 -1.2735 -1.3215 -1.5830 -1.8910
>
> **11.4.** Compute the discrete first derivative of the DAC transfer curve given in Exercise 11.1. Is the DAC output monotonic?
>
> **Ans.** 0.1939, 0.1333, 0.2254, 0.1308, 0.2528, 0.1418, 0.2335, 0.1316,
>
> 0.2007, 0.1962, 0.1474, 0.1488, 0.1904, 0.1477, 0.1384
>
> The DAC is monotonic since there are no negative values in the discrete derivative.

computed by calculating the discrete first derivative of the DACs transfer curve, subtracting one LSB (i.e., V_{LSB}) from the derivative result, then normalizing the result to one LSB

$$DNL(i) = \frac{S(i+1) - S(i) - V_{LSB}}{V_{LSB}} \text{ LSBs} \qquad (11.7)$$

As previously mentioned, we can define the average LSB size in one of three ways. We can define it as the actual full-scale range divided by the number of code transitions (number of codes minus 1) or we can define the LSB as the slope of the best-fit line. Alternatively, we can define the LSB size as the ideal DAC step size.

The choice of LSB calculations depends on what type of DNL calculation we want to perform. There are four basic types of DNL calculation method: best-fit, endpoint, absolute, and best-straight-line. Best-fit DNL uses the best-fit line's slope to calculate the average LSB size. This is probably the best technique, since it accommodates gain errors in the DAC without relying on the values of a few individual voltages. Endpoint DNL is calculated by dividing the full-scale range by the number of transitions. This technique depends on the actual values for the maximum full-scale (V_{FS+}) and minimum full-scale (V_{FS-}) levels. As such it is highly sensitive to errors in these two values, and is therefore less ideal than the best-fit technique. The absolute DNL technique uses the ideal LSB size derived from the ideal maximum and minimum full-scale values. This technique is less commonly used, since it assumes the DAC's gain is ideal.

The best-straight-line method is similar to the best-fit line method. The difference is that the best-straight-line method is based on the line that gives the best answer for integral nonlinearity (INL) rather than the line that gives the least squared errors. Integral nonlinearity will be discussed later in this chapter. Since the best-straight-line method is designed to yield the best possible answer, it is the most relaxed specification method of the four. It is used only in cases where the DAC or ADC linearity performance is not critical. Thus the order of methods from most relaxed to most demanding is best-straight line, best-fit, endpoint, and absolute.

The choice of technique is not terribly important in DNL calculations. Any of the three techniques will result in nearly identical results, as long as the DAC does not exhibit grotesque gain or linearity errors. DNL values of ±1/2 LSB are usually specified, with typical DAC performance of ±1/4 LSB for reasonably good DAC designs. A 1% error in the measurement of the LSB size would result in only a 0.01 LSB error in the DNL results, which is tolerable in most cases. The choice of technique is actually more important in the integral nonlinearity calculation, which we will discuss in the next section.

Example 11.4

Calculate the DNL curve for the 4-bit DAC of the previous examples. Use the best-fit line to define the average LSB size. Does this DAC pass a ±1/2 LSB specification for DNL? Use the endpoint method to calculate the average LSB size. Is this result significantly different from the best-fit calculation?

Solution:

The first derivative of the transfer curve was calculated in the previous monotonicity example. The first derivative values are

75 mV, 175 mV, 75 mV, 55 mV, 75 mV, 175 mV, 75 mV, 195 mV,

75 mV, 175 mV, 75 mV, 55 mV, 75 mV, 175 mV, 75 mV

The average LSB size, 109.35 mV, was calculated in Example 11.1 using the best-fit line calculation. Dividing each step size by the average LSB size yields the following normalized derivative values (in LSBs)

0.686, 1.6, 0.686, 0.503, 0.686, 1.6, 0.686, 1.783,

0.686, 1.6, 0.686, 0.503, 0.686, 1.6, 0.686

Subtracting one LSB from each of these values gives us the DNL values for each code transition of this DAC expressed as a fraction of an LSB

-0.314, 0.6, -0.314, -0.497, -0.314, 0.6, -0.314, 0.783,

-0.314, 0.6, -0.314, -0.497, -0.314, 0.6, -0.314

Note that there is one fewer DNL value than there are DAC codes.

Figure 11.5 shows the DNL curve for this DAC. The maximum DNL value is +0.783 LSB, while the minimum DNL value is −0.497. The minimum value is within the −1/2 LSB test limit, but the maximum DNL value exceeds the +1/2 LSB limit. Therefore, this DAC fails the DNL specification of ±1/2 LSB.

The average LSB size calculated using the endpoint method is given by

$$1 \text{ LSB} = \frac{V_{FS+} - V_{FS-}}{\text{number of codes} - 1}$$

$$= \frac{825 \text{ mV} - (-780 \text{ mV})}{16 - 1}$$

$$= 107 \text{ mV}$$

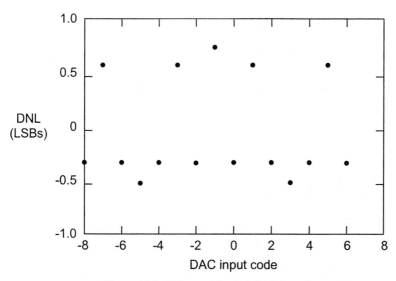

Figure 11.5. DNL curve for 4-bit DAC (best-fit method).

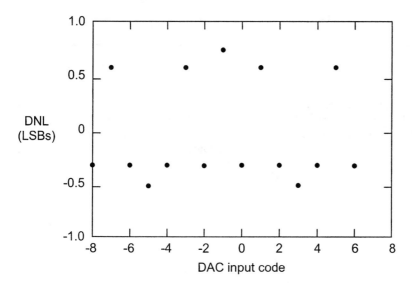

Figure 11.6. DNL curve for 4-bit DAC (endpoint method).

The DNL curve calculated using the endpoint method gives the following values, which have been normalized to an LSB size of 107 mV

-0.299, 0.636, -0.299, -0.486, -0.299, 0.636, -0.299, 0.822,

-0.299, 0.636, -0.299, -0.486, -0.299, 0.636, -0.299

The corresponding DNL curve is shown in Figure 11.6. Using the endpoint calculation, we get slightly different results. Instead of a maximum DNL result of +0.783 LSB and a minimum

DNL of −0.497 LSB, we get +0.822 and −0.486 LSB, respectively. This might be enough of a difference compared to the best-fit technique to warrant concern. Unless the endpoint method is explicitly called for in the data sheet, the best-fit method should be used since it is the least sensitive to abnormalities in any one DAC voltage.

Exercises

11.5. Calculate the DNL curve for the 4-bit DAC of Exercise 11.1. Use the best-fit line to define the average LSB size. Does this DAC pass a ±1/2 LSB specification for DNL?

Ans. 0.0937, -0.2481, 0.2714, -0.2622, 0.4259, -0.2002, 0.3170, -0.2577,

0.1320, 0.1067 -0.1686, -0.1607, 0.0739, -0.1669, -0.2194; pass

11.6. Calculate the DNL curve for the 4-bit DAC of Exercise 11.1. Use the endpoint method to calculate the average LSB size. Does this DAC pass a ±1/2 LSB specification for DNL?

Ans. 0.1132 -0.2347, 0.2941 -0.2491, 0.4514 -0.1859, 0.3406 -0.2445,

0.1523, 0.1264 -0.1537 -0.1457, 0.0931 -0.1520 -0.2054; pass

11.3.4 Integral Nonlinearity

The integral nonlinearity curve is a comparison between the actual DAC curve and one of three lines: the best-fit line, the endpoint line, or the ideal DAC line. The INL curve, like the DNL curve, is normalized to the LSB step size. As in the DNL case, the best-fit line is the preferred reference line, since it eliminates sensitivity to individual DAC values. The INL curve can be calculated by subtracting the reference DAC line (best-fit, endpoint, or ideal) from the actual DAC curve, dividing the results by the average LSB step size according to

$$INL(i) = \frac{S(i) - S_{REF}(i)}{V_{LSB}} \tag{11.8}$$

Note that using the ideal DAC line is equivalent to calculating the absolute error curve. Since a separate absolute error test is often specified, the ideal line is seldom used in INL testing. Instead, the endpoint or best-fit line is generally used. As in DNL testing, we are interested in the maximum and minimum value in the INL curve, which we compare against a test limit such as ±1/2 LSB.

Example 11.5

Calculate the INL curve for the 4-bit DAC in the previous examples. First use an endpoint calculation, then use a best-fit calculation. Does either result pass a specification of ±1/2 LSB? Do the two methods produce a significant difference in results?

Solution:

Using an endpoint calculation method, the INL curve for the 4-bit DAC of the previous examples is calculated by subtracting a straight line between the V_{FS-} voltage and the V_{FS+} voltage from the DAC output curve. The difference at each point in the DAC curve is divided by the average LSB size, which in this case is calculated using an endpoint method. As in the endpoint DNL example, the average LSB size is equal to 107 mV. The results of the INL calculations are (again, these values are expressed in LSBs)

0.0, -0.299, 0.336, 0.037, -0.449, -0.748, -0.112, -0.411,

0.411, 0.112, 0.748, 0.449, -0.037, -0.336, 0.299, 0.0

Figure 11.7 shows this endpoint INL curve. The maximum INL value is +0.748 LSB, and the minimum INL value is –0.748. This DAC does not pass an INL specification of ±1/2 LSB.

Using a best-fit calculation method, the INL curve for the 4-bit DAC of the previous examples is calculated by subtracting the best-fit line from the DAC output curve. Each point in the difference curve is divided by the average LSB size, which in this case is calculated using the best-fit line method. As in the best-fit DNL example, the average LSB size is equal to 109.35 mV. The results of the INL calculations are

0.161, -0.153, 0.448, 0.133, -0.364, -0.678, -0.077, -0.392,

0.392, 0.077, 0.678, 0.364, -0.133, -0.448, 0.153, -0.161

The maximum value is +0.678, and the minimum value is –0.678. These INL results are better than the endpoint INL values, but still do not pass a ±1/2 LSB test limit. The best-fit INL curve is shown in Figure 11.8 for comparison with the endpoint INL curve. The two INL curves are somewhat similar in shape, but the individual INL values are quite different. Remember that the DNL curves for endpoint and best-fit calculations were nearly identical. So, as previously stated, the choice of calculation technique is much more important for INL curves than for DNL curves. Notice also that while an endpoint INL curve always begins and ends at zero, the best-fit curve

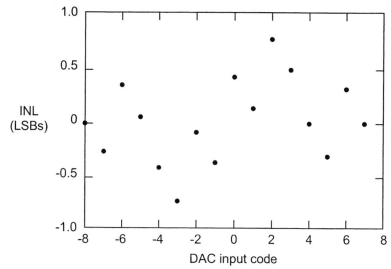

Figure 11.7. INL curve calculated using the endpoint linearity method.

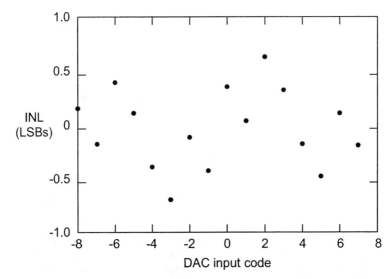

Figure 11.8. INL curve calculated using the best-fit linearity method.

does not necessarily behave this way. A best-fit curve will usually give better INL results than an endpoint INL calculation. This is especially true if the DAC curve exhibits a bowed shape in either the upward or downward direction. The improvement in the INL measurement is another strong argument for using a best-fit approach rather than an absolute or endpoint method, since the best-fit approach tends to increase yield.

The INL curve is the integral of the DNL curve, thus the term "integral nonlinearity"; DNL is a measurement of how consistent the step sizes are from one code to the next. INL is therefore a measure of accumulated errors in the step sizes. Thus, if the DNL values are consistently larger than zero for many codes in a row (step sizes are larger than 1 LSB), the INL curve will exhibit an upward bias. Likewise, if the DNL is less than zero for many codes in a row (step sizes are less than 1 LSB), the INL curve will have a downward bias. Ideally, the positive error in one code's DNL will be balanced by negative errors in surrounding codes and vice versa. If this is true, then the INL curve will tend to remain near zero. If not, the INL curve may exhibit large upward or downward bends, causing INL failures.

The INL integration can be implemented using a running sum of the elements of the DNL. The ith element of the INL curve is equal to the sum of the first i-1 elements of the DNL curve plus a constant of integration. When using the best-fit method, the constant of integration is equal to the difference between the first DAC output voltage and the corresponding point on the best-fit curve, all normalized to one LSB. When using the endpoint method, the constant of integration is equal to zero. When using the absolute method, the constant is set to the normalized difference between the first DAC output and the ideal output. In any running sum calculation it is important to use high-precision mathematical operations to avoid accumulated math error in the running sum. Mathematically, we can express this process as

$$INL(i) = \sum_{k=0}^{i-1} DNL(k) + C \qquad (11.9)$$

where

$$C = \begin{cases} \dfrac{S(0) - Best\_fit\_line(0)}{V_{LSB}} & \text{for best-fit linearity method} \\ 0 & \text{for endpoint linearity method} \\ \dfrac{S(0) - S_{IDEAL}(0)}{V_{LSB}} & \text{for absolute linearity method} \end{cases}$$

and $i=0$ indicates the DAC level corresponding to V_{FS-}.

Conversely, DNL can be calculated by taking the first derivative of the INL curve

$$DNL(i) = INL'(i) = INL(i+1) - INL(i) \qquad (11.10)$$

This is usually the easiest way to calculate DNL. The first derivative technique works well in DAC testing, but we will see in the next chapter that the DNL curve for an ADC is easier to capture than the INL curve. In ADC testing it is more common to calculate the DNL curve first, and then integrate it to calculate the INL curve. In either case, whether we integrate DNL to get INL or differentiate INL to get DNL, the results are mathematically identical.

Integral nonlinearity and differential nonlinearity are sometimes referred to by the names integral linearity error (ILE) and differential linearity error (DLE). However, the terms INL and DNL seem to be more prevalent in data sheets and other literature. We will use the terms INL and DNL throughout this text.

11.3.5 Partial Transfer Curves

A customer or systems engineer may specify that only a portion of a DAC or ADC transfer curve must meet certain specifications. For example, a DAC may be designed so that its V_{FS-} code corresponds to 0 V. However, due to analog circuit clipping as the DAC output signal

---

Exercises

11.7. Calculate the INL curve for a 4-bit unsigned binary DAC whose DNL curve is described by the following values (in LSBs)

0.0937, -0.2481, 0.2714, -0.2622, 0.4259, -0.2002, 0.3170, -0.2577,

0.1320, 0.1067, -0.1686, -0.1607, 0.0739, -0.1669, -0.2194

The DAC output for code 0 is 1.0091 V. Assume that the best-fit line has a gain of 177.3 mV/bit and an offset of 1.026 V.

Ans. -0.0959, -0.0022, -0.2503, 0.0210, -0.2412, 0.1847, -0.0155, 0.3016,

0.0438, 0.1759, 0.2825, 0.1139, -0.0467, 0.0272, -0.1397, -0.3591

approaches ground, the DAC may clip to a voltage of 100 mV. If the DAC is designed to perform a specific function that never requires voltages below 100 mV, then the customer may not care about this clipping. In such a case, the DAC codes below 100 mV are excluded from the offset, gain, INL, DNL, etc. specifications. The test engineer may then treat these codes as if they do not exist. This type of partial DAC and ADC testing is becoming more common as more DACs and ADCs are designed into custom applications with very specific requirements. General-purpose DACs are unlikely to be specified using partial curves, since the customer's application needs are unknown.

11.3.6 Major Carrier Testing

The techniques discussed thus far for measuring INL and DNL are based on a testing approach called *all-codes testing*. In all-codes testing, all valid codes in the transfer curve are measured to determine the INL and DNL values. Unfortunately, all-codes testing can be a very time-consuming process.

Depending on the architecture of the DAC, it may be possible to determine the location of each voltage in the transfer curve without measuring each one explicitly. We will refer to this as selected-code testing. Selected-code testing can result in significant test time savings, which of course represents substantial savings in test cost. There are several selected-code testing techniques, the simplest of which is called the *major carrier* method.

Many DACs are designed using an architecture in which a series of binary-weighted resistors or capacitors are used to convert the individual bits of the converter code into binary-weighted currents or voltages. These currents or voltages are summed together to produce the DAC output. For instance, a binary-weighted unsigned binary DAC's output can be described as a sum of binary-weighted voltage or current values, $W_0, W_1, ..., W_n$, multiplied by the individual bits of the DAC's input code, $D_0, D_1, ..., D_n$. The DAC's output value is therefore equal to

$$\text{DAC output} = D_0 W_0 + D_1 W_1 + \cdots + D_n W_n + \text{DC base} \tag{11.11}$$

where

DAC code bits $D_0 - D_n$ take on the value of 1 or 0

$W_1 = 2W_0$
$W_2 = 2W_1$
...
$W_n = 2W_{n-1}$

DC base is the DAC output value with a V_{FS-} input code

If this idealized model of the DAC is sufficiently accurate, then we only need to measure the values of $W_0, W_1, ..., W_n$ to predict every voltage in the DACs transfer curve. This DAC testing method is called the major carrier technique. The major carrier approach can be used for ADCs as well as DACs. The assumption of sufficient DAC or ADC model accuracy is only valid if the actual superposition errors of the DAC or ADC are low. This may or may not be the case. The superposition assumption can only be determined through characterization, comparing the all-codes DAC output levels with the ones generated by the major carrier method.

The most straightforward way to measure the value W_0 is to set DAC code bit D_0 to one and all other bits to zero. Likewise, the other major carrier values W_n can be measured by setting D_n to one and all other bits to zero. However, the resulting output levels are widely different in magnitude. This makes them difficult to measure accurately with a voltmeter, since the voltmeter's range must be adjusted for each measurement. A better approach that alleviates the accuracy problem is to measure the step size of the major carrier transitions in the DAC curve, which are all approximately 1 LSB in magnitude. A major carrier transition is defined as the voltage (or current) transition between the DAC codes 2^n-1 and 2^n. For example, the transition between binary 00111111 and 01000000 is a major carrier transition. Major carrier transitions can be measured using a voltmeter's sample-and-difference mode, giving highly accurate measurements of the major carrier transition step sizes.

Once the step sizes are known, we can use a series of inductive calculations to find the values of $W_0, W_1, ..., W_n$. We start by realizing that we have actually measured the following values:

DC base = measured DAC output with minus full-scale code

$$V_0 = W_0$$
$$V_1 = W_1 - W_0$$
$$V_2 = W_2 - (W_1 + W_0)$$
$$V_3 = W_3 - (W_2 + W_1 + W_0)$$
$$\dots$$
$$V_n = W_n - (W_{n-1} + W_{n-2} + W_{n-3} + \dots + W_0)$$

The value of the first major transition, V_0, is a direct measurement of the value of W_0 (the step size of the least significant bit). The value of W_1 can be calculated by rearranging the second equation: $W_1 = V_1 + W_0$. Once the values of W_0 and W_1 are known, the value of W_2 is calculated by rearranging the third equation: $W_2 = V_2 + W_1 + W_0$, and so forth. Once the values of $W_0 - W_n$ are known, the complete DAC curve can be reconstructed for each possible combination of input bits D_0-D_n using the original model of the DAC described by Eq. (11.11).

The major carrier technique can also be used on signed binary and two's complement converters, although the codes corresponding to the major carrier transitions must be chosen to match the converter encoding scheme. For example, the last major transition for our two's complement 4-bit DAC example happens between code 1111 (decimal –1) and 0000 (decimal 0). Aside from these minor modifications in code selection, the major carrier technique is the same as the simple unsigned binary approach.

Example 11.6

Using the major carrier technique on the 4-bit DAC example, we measure a DC base of –780 mV setting the DAC to V_{FS-} (binary 1000, or -8). Then we measure the step size between 1000 (-8) and 1001 (-7). The step size is found to be 75 mV. Next we measure the step size between 1001 (-7) and 1010 (-6). This step size is 175 mV. The step size between 1011 (-5) and 1100 (-4) is 55 mV and the step size between 1111 (-1) and 0000 (0) is 195 mV. Determine the values of W_0, W_1, W_2, and W_3. Reconstruct the voltages on the ramp from DAC code –8 to DAC code +7.

Solution:

Rearranging the set of equations $V_n = W_n - (W_{n-1} + W_{n-2} + W_{n-3} + \cdots + W_0)$ to solve for W_n, we obtain:

DC baseline = measured DAC output with V_{FS-} code = -780 mV

$W_0 = V_0 = 75$ mV

$W_1 = V_1 + W_0 = 175$ mV $+ 75$ mV $= 250$ mV

$W_2 = V_2 + W_1 + W_0 = 380$ mV

$W_3 = V_3 + W_2 + W_1 + W_0 = 900$ mV

For a two's complement DAC, we have to realize that the most significant bit is inverted in polarity compared to an unsigned binary DAC. Therefore, the DAC model for our 4-bit DAC is given by

$$\text{DAC output} = D_0 W_0 + D_1 W_1 + D_2 W_2 + \overline{D_3} W_3 + \text{DC base} \quad (11.12)$$

Using this two's complement version of the DAC model, the 16 voltage values of the DAC curve are reconstructed as shown in Table 11.1.

Table 11.1. DAC Transfer Curve Calculated Using the Major Carrier Technique

DAC Code	Calculation	Output Voltage
1000	DC Base	-780 mV
1001	W_0+DC Base	-705 mV
1010	W_1+DC Base	-530 mV
1011	W_1+W_0+DC Base	-455 mV
1100	W_2+DC Base	-400 mV
1101	W_2+W_0+DC Base	-325 mV
1110	W_2+W_1+DC Base	-150 mV
1111	W_2+W_1+W_0+DC Base	-75 mV
0000	W_3+DC Base	120 mV
0001	W_3+W_0+DC Base	195 mV
0010	W_3+W_1+DC Base	370 mV
0011	W_3+W_1+W_0+DC Base	445 mV
0100	W_3+W_2+DC Base	500 mV
0101	W_3+W_2+W_0+DC Base	575 mV
0110	W_3+W_2+W_1+DC Base	750 mV
0111	W_3+W_2+W_1+W_0+DC Base	825 mV

Notice that these values are exactly equal to the all-codes results in Figure 11.4. The example DAC was created using a binary-weighted model with perfect superposition; so it is no surprise the major carrier technique works for this imaginary DAC. Real DACs and ADCs often have superposition errors that make the major carrier technique unusable.

11.3.7 Other Selected-Code Techniques

Besides the major carrier method, other selected-code techniques have been developed to reduce the test time associated with all-codes testing. The simplest of these is the segmented method. This method only works for certain types of DAC and ADC architectures, such as the 12-bit segmented DAC shown in Figure 11.9. Although most segmented DACs are actually constructed using a different architecture than that in Figure 11.9, this simple architecture is representative of how segmented DACs can be tested.

The example DAC uses a simple unsigned binary encoding scheme with twelve data bits, D11-D0. It consists of two portions, a 6-bit coarse resolution DAC and a 6-bit fine resolution DAC. The LSB step size of the coarse DAC is equal to the full-scale range of the fine DAC plus one fine DAC LSB. In other words, if the combined 12-bit DAC has an LSB size of V_{LSB}, then the fine DAC also has a step size of V_{LSB}, while the coarse DAC has a step size of $2^6 \times V_{LSB}$. The output of these two 6-bit DACs can therefore be summed together to produce a 12-bit DAC

$$\text{DAC output} = \text{coarse DAC contribution} + \text{fine DAC contribution} \qquad (11.13)$$

Both the fine DAC and the coarse DAC are designed using a resistive divider architecture (see Section 11.5.1), rather than a binary-weighted architecture. Since major carrier testing can only be performed on binary-weighted architectures, an all-codes testing approach must be used to verify the performance of each of the two 6-bit resistive divider DACs. However, we would like to avoid testing each of the 2^{12}, or 4096 codes of the composite 12-bit DAC. Using superposition, we will test each of the two 6-bit DACs using an all-codes test. This requires only 2×2^6, or 128 measurements. We will then combine the results mathematically into a 4096-point all-codes curve using a linear model of the composite DAC.

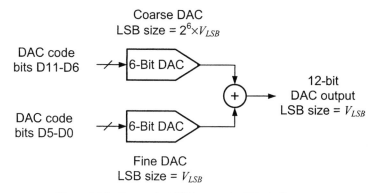

Figure 11.9. Segmented DAC conceptual block diagram.

Let us assume that through characterization, it has been determined that this example DAC has excellent superposition. In other words, the step sizes of each DAC are independent of the setting of the other DAC. Also, the summation circuit has been shown to be highly linear. In a case such as this, we can measure the all-codes output curve of the coarse DAC while the fine DAC is set to 0 (i.e., D5-D0 = 000000). We store these values into an array $V_{DAC\text{-}COARSE}(n)$, where n takes on the values 0 to 63, corresponding to data bits D11-D6. Then we can measure the all-codes output curve for the fine DAC while the coarse DAC is set to 0 (i.e., D11-D6 = 000000). These voltages are stored in the array $V_{DAC\text{-}FINE}(n)$, where n corresponds to data bits D5-D0.

Although we have only measured a total of 128 levels, superposition allows us to recreate the full 4096-point DAC output curve by a simple summation. Each DAC output value $V_{DAC}(i)$ is equal to the contribution of the coarse DAC plus the contribution of the fine DAC

$$V_{DAC}(i) = V_{DAC-FINE}(i \text{ AND } 000000111111) + V_{DAC-COARSE}\left(\frac{i \text{ AND } 111111000000}{64}\right) \quad (11.14)$$

where i ranges from 0 to 4095.

Thus a full 4096-point DAC curve can be mathematically reconstructed from only 128 measurements by evaluating this equation at each value of i from 0 to 4095. Of course, this technique is totally dependent on the architecture of the DAC. It would be inappropriate to use this technique on a nonsegmented DAC or a segmented DAC with large superposition errors.

A more advanced selected-codes technique was developed at the National Institute of Standards and Technology (NIST). This technique is useful for all types of DACs and ADCs. It does not make any assumptions about superposition errors or converter architecture. Instead, it uses linear algebra and data collected from production lots to create an empirical model of the DAC or ADC. The empirical model only requires a few selected codes to recreate the entire DAC or ADC transfer curve. Although the details of this technique are beyond the scope of this book, the original NIST paper is listed in the references at the end of this chapter.[2]

Another similar technique uses wavelet transforms to predict the overall performance of converters based on a limited number of measurements.[3] Again, this topic is beyond the scope of this book.

Before we end this section we would like to point out that an appendix to this chapter lists several MATLAB routines to enable the reader to automatically characterize the DAC's transfer curve according to the methods described in this section.

11.4 DYNAMIC DAC TESTS

11.4.1 Conversion Time (Settling Time)

So far we have discussed only low-frequency DAC performance. The DAC DC tests and transfer curve tests measure the DAC's static characteristics, requiring the DAC to stabilize to a stable voltage or current level before each output level measurement is performed. If the DAC's output stabilizes in a few microseconds, then we might step through each output state at a high frequency, but we are still performing static measurements for all intents and purposes.

A DAC's performance is also determined by its dynamic characteristics. One of the most common dynamic tests is settling time, commonly referred to as *conversion time*. Conversion time is defined as the amount of time it takes for a DAC to stabilize to its final static level *within a specified error band* after a DAC code has been applied. For instance, a DAC's settling time may be defined as 1 μs to ±1/2 LSB. This means that the DAC output must stabilize to its final value plus or minus a 1/2 LSB error band no more than 1 μs after the DAC code has been applied.

This test definition has one ambiguity. Which DAC codes do we choose to produce the initial and final output levels? The answer is that the DAC must settle from any output level to any other level within the specified time. Of course, to test every possibility, we might have to measure millions of transitions on a typical DAC. As with any other test, we have to determine what codes represent the worst-case transitions. Typically settling time will be measured as the DAC transitions from minus full-scale (V_{FS-}) to plus full-scale (V_{FS+}) and vice versa, since these two tests represent the largest voltage swing.

The 1/2 LSB example uses an error band specification that is referenced to the LSB size. Other commonly used definitions require the DAC output to settle within a certain percentage of the full-scale range, a percentage of the final voltage, or a fixed voltage range. So we might see any of the following specifications:

settling time = 1 μs to ± 1% of full-scale range

settling time = 1 μs to ± 1% of final value

settling time = 1 μs to ±1 mV

The test technique for all these error-band definitions is the same; we just have to convert the error-band limits to absolute voltage limits before calculating the settling time. The straightforward approach to testing settling time is to digitize the DAC's output as it transitions

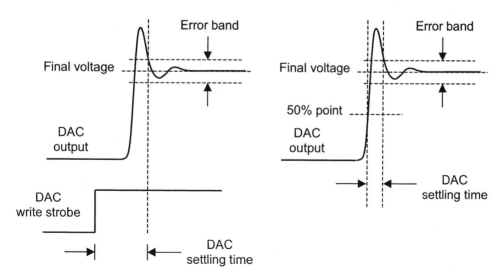

Figure 11.10. DAC settling time measurement (a) referenced to a digital signal; (b) referenced to the DAC output 50% point.

from one code to another and then use the known time period between digitizer samples to calculate the settling time. We measure the final settled voltage, calculate the settled voltage limits (i.e., ±1/2 LSB), and then calculate the time between the digital signal transition that initiates a DAC code change and the point at which the DAC first stays within the error band limits, as shown in Figure 11.10(a).

In extremely high frequency DACs it is common to define the settling time not from the DAC code change signal's transition but from the time the DAC passes the 50% point to the time it settles to the specified limits as shown in Figure 11.10(b). This is easier to calculate, since it only requires us to look at the DAC output, not at the DAC output relative to the digital code.

11.4.2 Overshoot and Undershoot

Overshoot and undershoot can also be calculated from the samples collected during the DAC settling time test. These are defined as a percentage of the voltage swing or as an absolute voltage. Figure 11.11 shows a DAC output with 10% overshoot and 2% undershoot on a V_{FS-} to V_{FS+} transition.

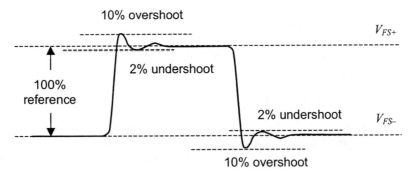

Figure 11.11. DAC overshoot and undershoot measurements.

11.4.3 Rise Time and Fall Time

Rise and fall time can also be measured from the digitized waveform collected during a settling time test. Rise and fall times are typically defined as the time between two markers, one of which is 10% of the way between the initial value and the final value and the other of which is 90% of the way between these values as depicted in Figure 11.12. Other common marker definitions are 20% to 80% and 30% to 70%.

11.4.4 DAC-to-DAC Skew

Some types of DACs are designed for use in matched groups. For example, a color palette RAM DAC is a device that is used to produce colors on video monitors. A RAM DAC uses a random access memory (RAM) lookup table to turn a single color value into a set of three DAC output values, representing the red, green, and blue intensity of each pixel. These DAC outputs must change almost simultaneously to produce a clean change from one pixel color to the next. The degree of timing mismatch between the three DAC outputs is called *DAC-to-DAC skew*. It is measured by digitizing each DAC output and comparing the timing of the 50% point of each

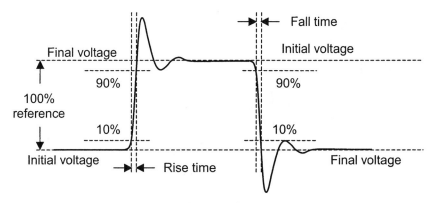

Figure 11.12. DAC rise and fall time measurements.

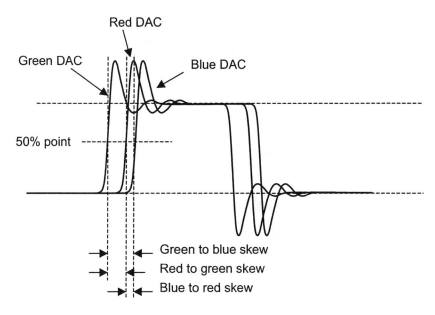

Figure 11.13. DAC-to-DAC skew measurements.

output to the 50% point of the other outputs. There are three skew values (R-G, G-B, and B-R), as illustrated in Figure 11.13. Skew is typically specified as an absolute time value, rather than a signed value.

11.4.5 Glitch Energy (Glitch Impulse)

Glitch energy, or glitch impulse, is another specification common to high-frequency DACs. It is defined as the total area under the voltage-time curve of the glitches in a DAC's output as it switches across the largest major transition (i.e., 01111111 to 10000000 in an 8-bit DAC) and back again. As shown in Figure 11.14, the glitch area is defined as the area that falls *outside* the

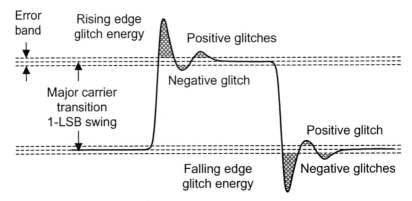

Figure 11.14. Glitch energy measurements.

rated error band. These glitches are caused by a combination of capacitive/inductive ringing in the DAC output and skew between the timing of the digital bits feeding the binary-weighted DAC circuits. The parameter is commonly expressed in picosecond-volts (ps-V) or equivalently, picovolt-seconds (pV-s). (These are not actually units of energy, despite the term *glitch energy*.) The area under the negative glitches is considered positive area, and should be added to the area under the positive glitches. Both the rising-edge glitch energy and the falling-edge glitch energy should be tested.

11.4.6 Clock and Data Feedthrough

Clock and data feedthrough is another common dynamic DAC specification. It measures the crosstalk from the various clocks and data lines in a mixed-signal circuit that couple into a DAC output. There are many ways to define this parameter; so it is difficult to list a specific test technique. However, clock and data feedthrough can be measured using a technique similar to all the other tests in this section. The output of the DAC is digitized with a high-bandwidth digitizer. Then the various types of digital signal feedthrough are analyzed to make sure they are below the defined test limits. The exact test conditions and definition of clock and data feedthrough should be provided in the data sheet. This measurement may require time-domain analysis, frequency-domain analysis, or both.

11.5 DAC ARCHITECTURES

11.5.1 Resistive Divider DACs

There are many different types of DAC architectures, each with its own set of strengths and weaknesses. In this section, we will look at some basic architectures as examples. This section is by no means an exhaustive list of every type of DAC, nor is it meant to present DAC architectures in any significant detail. The purpose of this section is to illustrate some basic DAC structures and their probable strengths and weaknesses.

Perhaps the simplest DAC architecture is the resistive divider DAC, illustrated in Figure 11.15. This type of DAC uses a series-connected string of resistors to produce a set of

Exercises

11.9. The step response of a DAC obtained from an oscilloscope is as follows

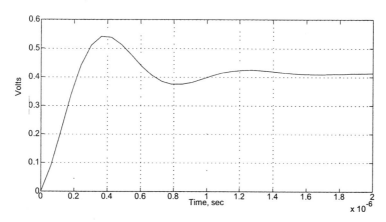

The data sheet states that the settling time is 1 µs (error band = ± 20 mV). Does this DAC settle fast enough to meet the settling time specification? Also, determine the overshoot of this signal and its rise time. Estimate the total glitch energy during the positive-going transition.

Ans. Actual settling time = 0.85 µs. (Yes); overshoot=30%; rise time=0.3 µs. Glitch energy=0.5(0.3)(0.13)+0.5(0.5)(-0.033)+0.5(0.6)(0.01)=14.25 ns-V (triangle appoximation).

voltages evenly spaced between V_{REF-} and V_{REF+}. The digital input of the DAC determines which of these voltages is connected through an analog switch to a buffer amplifier. Although the resistive divider architecture may be simple to understand, it quickly loses its appeal in high-resolution DACs. Each additional bit of DAC resolution requires twice as many resistors and analog switches. For example, a 12-bit resistive divider DAC would require 4095 resistors and 4096 switches.

The large silicon area consumed by a high-resolution resistive divider DAC may make this architecture impractical for DACs exceeding seven or eight bits of resolution. In addition to the excessive silicon area required to implement a high-resolution resistive divider DAC, the test time for traditional DAC transfer curve tests may also become unacceptably lengthy. This is because any one of the resistors or analog switches can be defective; so each and every DAC voltage level must be tested explicitly. For this reason, selected-code techniques cannot be used to save test time on resistive divider DACs. Also, transmission parameters such as gain and distortion are not sufficient in themselves to guarantee all the switches and resistors. This is because these tests typically do not exercise each of the possible DAC output levels.

For low-resolution converters, the resistive divider DAC may be more appealing than other architectures for one reason in particular. The resistive divider DAC is inherently monotonic and is usually very linear. Since the voltage levels produced by a string of resistors in a voltage divider network are always monotonic, the resistive divider DAC is monotonic by design. Also, since the value of each resistor in the divider chain can be fabricated with reasonably good

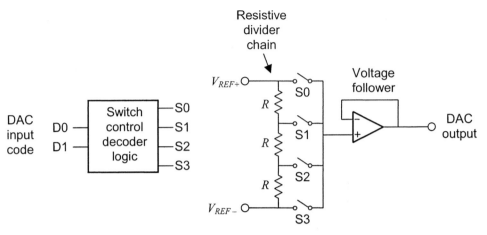

Figure 11.15. Resistive divider DAC architecture.

tolerance, the size of each DAC step is substantially equal to the other steps. Constant step size leads to good DNL characteristics. Although DNL and monotonicity are low failure modes by design, all codes must still be measured to detect defects in any of the resistors or switches. Therefore, the resistive divider architecture does not necessarily reduce test time, though it does typically lead to very high yields.

As a side note, this type of architecture might be implemented with capacitive divider networks rather than resistive dividers. In fact, many of the DAC architectures discussed in this section can be implemented with either resistors or capacitors. We will discuss only the resistive version of each converter type, with the understanding that the capacitive versions share many of the same general testing characteristics.

11.5.2 Binary-Weighted DACs

If the resolution of a DAC exceeds six or seven bits, a binary-weighted DAC often provides a more efficient use of silicon area than the resistive divider architecture. One common binary-weighted architecture is shown in Figure 11.16. This circuit is known as an *R/2R resistive ladder* DAC.

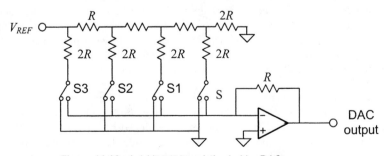

Figure 11.16. A 4-bit *R/2R* resistive ladder DAC.

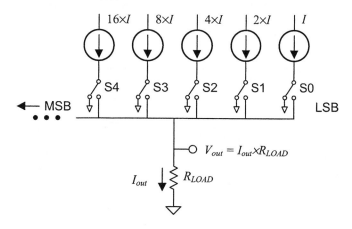

Figure 11.17. Current-steering DAC.

Another binary-weighted architecture is the current-steering DAC shown in Figure 11.17. The current-steering DAC produces a current output rather than a voltage output. The current can be converted to voltage using a load resistor, as shown.

Binary-weighted DACs are based on a summation of binary-weighted currents or voltages. For example, the currents in the current switching DAC are set to binary-weighted values, I, $2 \times I$, $4 \times I$, $8 \times I$, etc. The minimum current is equal to 0 and the maximum current is equal to $(2^N-1) \times I$ where N is the number of bits in the DAC's input code. The least significant bit (LSB) controls the smallest current source, enabling or disabling its output so that it contributes either zero current or current equal to I. The second to least significant bit controls the next largest current source, enabling or disabling the $2 \times I$ current value, and so on.

Binary-weighted architectures provide two main advantages. First, they are efficient in their use of silicon area. For instance, a 9-bit current steering DAC only requires one more current source and switch than an 8-bit current steering DAC. Also, a binary-weighted architecture allows major carrier testing, as described in Section 11.3.6, assuming the summation of the individual binary-weighted currents or voltages add without superposition error. The major carrier method reduces INL and DNL test time, compared to the brute force all-codes testing method.

11.5.3 PWM DACs

Pulse-width modulation (PWM) DACs are very simple DACs that are mostly digital in nature, using very little analog circuitry. Figure 11.18 shows a block diagram for a simple PWM DAC. PWM DACs adjust their output voltages using a high-frequency pulse train of varying duty cycle. The duty cycle controls the amount of time the 1-bit DAC spends at the V_{FS+} level and how much time it spends at the V_{FS-} level.

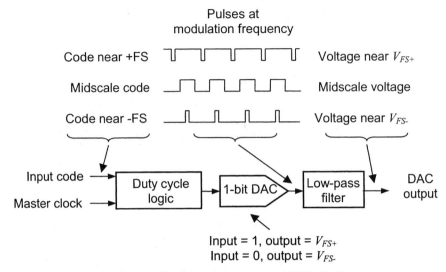

Figure 11.18. Pulse width modulation (PWM) DAC.

If the duty cycle approaches 50/50, the filtered output of the one-bit DAC settles to a voltage midway between V_{FS-} and V_{FS+}. A duty cycle of 100% high results in a voltage equal to V_{FS+}, while a duty cycle of 0% high results in a voltage equal to V_{FS-}. The DAC output is therefore proportional to the duty cycle of the digital input to the one-bit DAC. In many PWM DAC architectures, the duty cycle is produced by purely digital circuits driven by a high-frequency master clock. For this reason, the DNL and monotonicity of some PWM DACs are guaranteed by design, as long as the digital logic functions correctly. Other PWM architectures use analog circuits to generate the varying pulse widths. These may not be guaranteed to produce monotonic curves, depending on the exact implementation.

INL, on the other hand, is a potential weakness of all PWM DACs. Depending on the nature of the design architecture, PWM DACs can sometimes be tested by sampling a few evenly spaced points on the DAC transfer curve to verify good INL characteristics. This shortcut works for DACs in which the INL curve may be bowed or curved, but does not exhibit sudden code-to-code discontinuities. The pulse duty cycle circuits can then be verified using time measurement techniques and/or purely digital patterns to verify monotonicity and DNL.

PWM DACs are similar in nature to resistive divider DACs. To obtain a high-resolution DAC, a PWM DAC must be able to adjust the pulse edges by tiny amounts of time. This requires a very high-frequency clock to drive the duty cycle generator circuits, assuming a purely digital circuit is to be used. Otherwise, an analog pulse width generator must be used, which is much more likely to exhibit variations from DUT to DUT.

PWM DACs are typically used in low-cost, low-resolution applications where extreme quality is not a concern. Example applications of PWM DACs include toy speech products and talking greeting cards. Since extreme low cost is often a concern, expensive all-codes testing is usually not an option in testing PWM DACs. Often, the all-codes INL and DNL testing is replaced by more cost effective channel testing, such as signal-to-distortion ratio (S/D) tests and signal-to-noise ratio (SNR) tests using a small number of samples.

11.5.4 Sigma-Delta DACs

One thing in particular limits the resolution of PWM DACs that use purely digital circuits to control pulse widths. Very high-resolution DACs require very high-frequency master clocks to drive the digital counters controlling the width of the digital pulses (i.e., duty cycle). A 16-bit PWM DAC, for example, requires a pulse time resolution of $1/65536^{th}$ of the period of the pulse waveform. Since the pulses must be low-pass filtered to generate an analog output, the pulse frequency must be substantially higher (say, a factor of 100) than the highest frequency in the reconstructed analog signal. Therefore, a 16-bit PWM DAC for audio applications having a 20-kHz bandwidth would require a master clock frequency of 65536×100×20000, or 131 GHz! Clearly, present technology does not support such a design.

A common solution to this modulation ratio problem is provided by the sigma-delta DAC architecture, shown in Figure 11.19. Although the digital logic is more complicated than that of a simple PWM DAC, the sigma-delta architecture allows a much smaller ratio of master clock to audio bandwidth. A modulation ratio of only 100 to one (clock rate divided by audio bandwidth) allows 16-bit performance from a sigma-delta architecture, compared to a ratio of 6.5 million to one for a 16-bit PWM architecture.

The sigma-delta DAC accomplishes this reduction in master clock frequency using a noise-shaping algorithm that moves the quantization noise of the one-bit DAC to high frequencies. The noise-shaping algorithm reduces noise components in the low-frequency spectrum of the reconstructed signal as depicted in Figure 11.20. Because the noise-shaping algorithm uses a process called *pulse density modulation* (PDM), sigma-delta converters are also known as PDM converters. The noise-shaped signal can be cleaned up using a low-pass filter to separate the high-frequency quantization noise from the low-frequency signal. The operation of a sigma-delta DAC is another subject that falls outside the scope of this book. Other texts have explained this architecture in detail.[4,5] Fortunately, we can make some observations about the typical applications and test approaches for sigma-delta DACs without getting into their detailed operation.

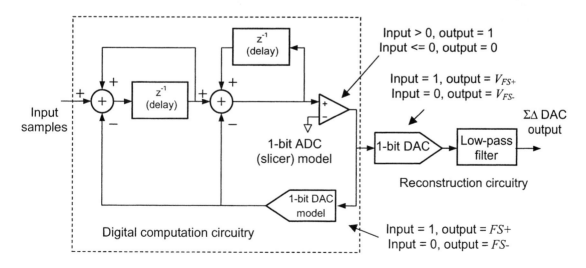

Figure 11.19. Second-order sigma-delta (pulse density modulation) DAC block diagram.

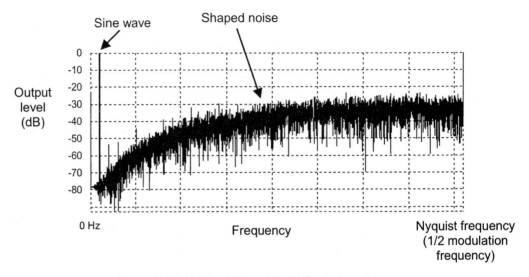

Figure 11.20. Unfiltered sigma-delta DAC output spectrum.

Sigma-delta DACs are well suited for applications requiring relatively low-frequency, high-quality, AC signal creation. Audio applications such as digital audio and cellular telephony are well suited to sigma-delta technology. Therefore, sigma-delta DACs are generally tested for AC parameters such as S/THD and SNR rather than the DC transfer curve tests like INL and DNL.

In fact, sigma-delta DACs are somewhat poorly suited to most DC applications because they generate interference signals called *self-tones*. Self-tones are low amplitude periodic waves that are generated by the sigma-delta noise-shaping algorithm itself when certain DC signal levels are applied to the DAC input. Most input codes will produce self-tones in the unfiltered DAC output. Fortunately, the low-pass reconstruction filter eliminates most self-tones, since a majority of them occur outside the filter's passband. AC signals, by contrast, change the DAC input codes often enough that self-tones do not have a chance to appear.

Only certain DC levels produce self-tones that are in the pass band of the low-pass filter. These DC levels and the corresponding self-tone frequencies are very predictable, since they are controlled by the DAC's sigma-delta algorithm. For this reason, DAC self-tones are fairly easy to measure. However, it is often unnecessary to measure DAC self-tones in AC applications, as long as these DC levels can be avoided. The test engineer should consult the design and systems engineers to determine whether or not self-tones should be measured in a particular application.

11.5.5 Companded DACs

Companded DACs, such as the codec discussed in Section 9.3.2, are seldom used in DC applications. They are more commonly used in applications such as low-cost voice compression and decompression for use in telephone central offices. INL and DNL are virtually meaningless in the case of companded DACs and ADCs. The usual test list for companded DACs includes mostly AC sampled channel tests such as SNR, S/THD, ICN, etc. with an emphasis on testing these parameters at various signal levels to detect flaws in the inherently nonlinear companding circuits.

11.5.6 Hybrid DAC Architectures

Many DACs do not fall into any of the categories listed in this chapter, but are instead hybrids of two or more of the basic architectures. For example, a sigma-delta DAC can be built using a resistive divider multibit DAC instead of a simple one-bit DAC. This gives lower quantization noise and therefore better performance, as long as the multibit DAC is very linear. The multibit DAC may be implemented using a PWM DAC rather than a resistive divider DAC, leading to another hybrid design. Yet another hybrid DAC example is the segmented DAC example of Section 11.3.7. It combines the characteristics of two resistive divider DACs into a single DAC characteristic.

Each of these hybrid designs requires unique testing methodologies. Regardless of the DAC architecture, the secret to effective testing of any DAC is to understand its weaknesses and design a suite of tests that specifically targets those weaknesses. Weaknesses in the system-level application must also be understood as well. The test engineer, systems engineer, and design engineer should work together closely to define the most efficient test approach for each DAC, taking its intended application into consideration.

11.6 Tests for Common DAC Applications

11.6.1 DC References

As previously mentioned, the test list for a given DAC often depends on its intended functionality in the system-level application. Many DACs are used as simple DC references. An example of this type of DAC usage is the power level control in a cellular telephone. As the cellular telephone user moves closer or farther away from a cellular base station (the radio antenna tower), the transmitted signal level from the cellular telephone must be adjusted. The transmitted level may be adjusted using a transmit level DAC so that the signal is just strong enough to be received by the base station without draining the cellular telephone's battery unnecessarily.

If a DAC is only used as a DC (or slow-moving) voltage or current reference, then its AC transmission parameters are probably unimportant. It would probably be unnecessary to measure the 1-kHz signal to total harmonic distortion ratio of a DAC whose purpose is to set the level of a cellular telephone's transmitted signal. However, the INL and DNL of this DAC would be extremely important, as would its absolute errors, monotonicity, full-scale range, and output drive capabilities (output impedance).

DACs used as DC references are usually measured using the intrinsic parameters listed in this chapter, rather than the transmission parameters outlined in Chapter 9. Notable exceptions are signal-to-noise ratio and idle channel noise (ICN). These may be of importance if the DC level must exhibit low noise. For example, the cellular telephone's transmitted signal might be corrupted by noise on the output of the transmit level control DAC, and therefore we might need to measure the DAC's noise level.

Dynamic tests are not typically performed on DC reference DACs, with the exception of settling time. The settling time of typical DACs is often many times faster than that required in DC reference applications; so even this parameter is frequently guaranteed by design rather than being tested in production.

11.6.2 Audio Reconstruction

Audio reconstruction DACs are those used to reproduce digitized sound. Examples include the voice-band DAC in a cellular telephone and the audio DAC in a PC sound card. These DACs are more likely to be tested using the transmission parameters of Chapter 9, since their purpose is to reproduce arbitrary audio signals with minimum noise and distortion.

The intrinsic parameters (i.e., INL and DNL) of audio reconstruction DACs are typically measured only during device characterization. Linearity tests can help track down any transmission parameter failures caused by the DAC. It is often possible to eliminate the intrinsic parameter tests once the device is in production, keeping only the transmission tests.

Dynamic tests are not typically specified or performed on audio DACs. Any failures in settling time, glitch energy, etc. will usually manifest themselves as failures in transmission parameters such as signal-to-noise, signal-to-distortion, and idle channel noise.

11.6.3 Data Modulation

Data modulation is another purpose to which DACs are often applied. The cellular telephone again provides an example of this type of DAC application. The IF section of a cellular telephone base-band modulator converts digital data into an analog signal suitable for transmission, similar to those used in modems (see Section 9.1.2). Like the audio reconstruction DACs, these DACs are typically tested using sine wave or multitone transmission parameter tests.

Again, the intrinsic tests like INL and DNL may be added to a characterization test program to help debug the design. However, the intrinsic tests are often removed after the transmission parameters have been verified. Dynamic tests such as settling time may or may not be necessary for data modulation applications.

Data modulation DACs also have very specific parameters such as error vector magnitude (EVM) or phase trajectory error (PTE). Parameters such as these are very application-specific. They are usually defined in standards documents published by the IEEE, NIST, or other government or industry organization. The data sheet should provide references to documents defining application-specific tests such as these. The test engineer is responsible for translating the measurement requirements into ATE-compatible tests that can be performed on a production tester. ATE vendors are often a good source of expertise and assistance in developing these application-specific tests.

11.6.4 Video Signal Generators

As discussed earlier, DACs can be used to control the intensity and color of pixels in video cathode ray tube (CRT) displays. However, the type of testing required for video DACs depends on the nature of their output. There are two basic types of video DAC application, RGB and NTSC. An RGB (red-green-blue) output is controlled by three separate DACs. Each DAC controls the intensity of an electron beam, which in turn controls the intensity of one of the three primary colors of each pixel as the beam is swept across the CRT. In this application, each DAC's output voltage or current directly control the intensity of the beam. RGB DACs are typically used in computer monitors.

The NTSC format is used in transmission of standard (i.e., non-HDTV) analog television signals. It requires only a single DAC, rather than a separate DAC for each color. The picture intensity, color, and saturation information is contained in the time-varying offset, amplitude, and phase of a 3.54-MHz sinusoidal waveform produced by the DAC. Clearly this is a totally different DAC application than the RGB DAC application. These two seemingly similar video applications require totally different testing approaches.

RGB DACs are tested using the standard intrinsic tests like INL and DNL, as well as the dynamic tests like settling time and DAC-to-DAC skew. These parameters are important because the DAC outputs directly control the rapidly changing beam intensities of the red, green, and blue electron beams as they sweep across the computer monitor. Any settling time, rise time, fall time, undershoot, or overshoot problems show up directly on the monitor as color or intensity distortions, vertical lines, ghost images, etc.

The quality of the NTSC video DAC, by contrast, is determined by its ability to produce accurate amplitude and phase shifts in a 3.54-MHz sine wave, while changing its offset. This type of DAC is tested with transmission parameters like gain, signal-to-noise, differential gain, and differential phase (see Section 10.5.3).

11.7 SUMMARY

DAC testing is far less straightforward than one might at first assume. Although DACs all perform the same basic function (digital-to-analog conversion), the architecture of the DAC and its intended application determine its testing requirements and methodologies. A large variety of standard tests have been defined for DACs, including transmission parameters, DC intrinsic parameters, and dynamic parameters. We have to select DAC test requirements carefully to guarantee the necessary quality of the DAC without wasting time with irrelevant or ineffective tests.

ADC testing is very closely related to DAC testing. Many of the DC and intrinsic tests defined in this chapter are very similar to those performed on ADCs. However, due to the many-to-one transfer characteristics of ADCs, the measurement of the ADC input level corresponding to each output code is much more difficult than the measurement of the DAC output level corresponding to each input code. Chapter 12, "ADC Testing," explains the various ways the ADC transfer curve can be measured, as well as the many types of ADC architectures and applications the test engineer will likely encounter.

APPENDIX A.11.1

MATLAB Routines for DAC Characterization

This appendix lists two MATLAB routines that can be used to characterize a DAC according to the metrics described in this chapter. The first routine calculates intrinsic parameters such as DC offset, gain, INL, and DNL from a given set of samples. The second routine calculates an estimated all-codes linearity curve based on the DC base and the major carrier transition levels, as described in Section 11.1.6.

```
%%%%%%%%%         DNL & INL DAC Characterization          %%%%%%%%
%                                                                %
% Given a set of DAC output levels in vector S_Actual, the following  %
% routine computes   the DAC metrics such as: (1) Absolute Gain and   %
% DC Error, (2) Transfer Curves such as absolute error, DNL an INL.   %
%                                                                %
%%%%%%%%%%%%%%%%%%%%%%%%%%%%%%%%%%%%%%%%%%%


% Initialization
    clear; % clear the workspace
    for k=1:3,
        figure(k), clg; % reset graphics
    end

% Main Routine
% Chapter 11 running example  - two's complement DAC
S_Actual= [ -780e-3, -705e-3, -530e-3, -455e-3, -400e-3, -325e-3, -150e-3, ...,
-75e-3, 120e-3, 195e-3, 370e-3, 445e-3, 500e-3, 575e-3, 750e-3, 825e-3];
    codeword=-8:7;  % code -8, code -7, ...., code 7
    D=4;   % 4-bit DAC
    FS_neg_ideal=-0.8; FS_pos_ideal=0.7; % 2 Volt full-scale range
    format=2; % 1=unsigned binary, 2=two's complement
% plot sample set versus codeword
    figure(1); plot(codeword, S_Actual,'ro'); hold on
        text(codeword(2),max(S_Actual),'S_Actual: red circles')
        xlabel('DAC Input Code');
        ylabel('DAC Output Level');

% Define Ideal DAC Charateritics
    LSB_Ideal=(FS_pos_ideal-FS_neg_ideal)/(2^D-1);
    Gain_Ideal=LSB_Ideal;

    if format==1,
        % unsigned binary
        for k=1:length(S_Actual),
            S_Ideal(k)=LSB_Ideal*codeword(k)+FS_neg_ideal;
            if codeword(k)==0,
                DAC_Offset_Ideal=[ k, S_Ideal(k)];
            end
        end
    elseif format==2,
        % two's complement
        for k=1:length(S_Actual),
            S_Ideal(k)=LSB_Ideal*codeword(k);
```

Chapter 11 • DAC Testing

```
                    if codeword(k)==0,
                          DAC_Offset_Ideal=[ k, S_Ideal(k)];
                    end
            end
      else
            disp('I do not recognize this DAC format')
      end

% plot sample set versus codeword
            figure(1); plot(codeword, S_Ideal,'r-'); hold on
            text(codeword(2),FS_pos_ideal-2*LSB_Ideal,'S_Ideal: red line')

%  Compute Best-fit line coefficients

      % Initialize routine
            k1=0; k2=0; k3=0; k4=0;
            N=length(S_Actual);
      % perform best-fit analysis
            for i=0:N-1,
                    k1 = k1 + i;
                    k2 = k2 + S_Actual(i+1);
                    k3 = k3 + i*i;
                    k4 = k4 + i*S_Actual(i+1);
            end
            Gain_Bestfit = (N*k4 - k1*k2) / (N*k3 - k1*k1);
            LSB_Bestfit=Gain_Bestfit;
            Offset_Bestfit = k2/N-Gain_Bestfit * (k1/N);
      % prepare for plotting & offset calculation
            for i=1:N,
                    Best_fit_line(i) = Gain_Bestfit*(i-1) + Offset_Bestfit;
                    if codeword(i)==0,
                          DAC_Offset_Bestfit=[ i, Best_fit_line(i)];
                    end
            end
            figure(1); plot(codeword, Best_fit_line,'b-'); hold on
            text(codeword(2),FS_pos_ideal-3*LSB_Ideal,'Best_fit_line: blue line')

%  Compute Endpoint line coefficients

      % Initialize routine
            S_Endpoint=[];
      % perform endpoint line analysis
            FS_neg_actual=S_Actual(1);
            FS_pos_actual=S_Actual(length(S_Actual));
            Gain_Endpoint=(FS_pos_actual-FS_neg_actual)/(2^D-1);
```

```
                LSB_Endpoint=Gain_Endpoint;
                for k=1:length(S_Actual),
                        S_Endpoint(k)=LSB_Endpoint*(k-1)+FS_neg_actual;
                        if codeword(k)==0,
                                DAC_Offset_Endpoint=[ k, S_Endpoint(k)];
                        end
                end

        % prepare for plotting & offset calculation
                figure(1); plot(codeword, S_Endpoint,'g-'); hold off
                text(codeword(2),FS_pos_ideal-4*LSB_Ideal,'S_Endpoint: green line')

% DC Gain, Gain Error, Offset and Offset Error

disp('***** DC Gain, Gain Error, Offset and Offset Error *****')
        disp('Ideal Line:')
                Gain_Ideal
        disp('Best-Fit Line:')
                disp('[ Gain_Bestfit Offset_Bestfit ]')
                [ Gain_Bestfit Offset_Bestfit ]
                Gain_Error_percent=100*( Gain_Bestfit/Gain_Ideal - 1)
                DAC_Offset_Bestfit
                Offset_Error = DAC_Offset_Bestfit(1,2) - DAC_Offset_Ideal(1,2)
        disp('End Point Line:')
                Gain_Endpoint
                Gain_Error_percent=100*( Gain_Endpoint/Gain_Ideal - 1)
                DAC_Offset_Endpoint
                Offset_Error = DAC_Offset_Endpoint(1,2) - DAC_Offset_Ideal(1,2)


        disp('========== Transfer Curve Tests ==========')
        disp(' ')
        disp(' ')
        disp('************ (1) Absolute Error Curve ***********')
        disp(' ')
                % Compute The Absolute Error Curve
                        Delta_S = (S_Actual-S_Ideal);
                        Delta_S_normalized = Delta_S/LSB_Ideal;
                        disp('[ S_Ideal S_Actual Delta_S Delta_S_normalized ]')
                                [ S_Ideal' S_Actual' Delta_S' Delta_S_normalized' ]

        disp('************ (2) Monotonicity Test ***********')
        disp(' ')
                % Compute The Discrete First Derivative of S_Actual
                        for k=1:length(S_Actual)-1,
                                S_derivative(k,1) = S_Actual(k+1) - S_Actual(k);
```

```
                    end
                disp('S_derivative')
                        S_derivative

            disp('***** (3) Differential Nonlinearity (DNL) Curve *****')
            disp(' ')
            % Best-Fit Line
                    for k=1:length(S_Actual)-1,
                            DNL_Bestfit(k,1) = S_derivative(k,1)/LSB_Bestfit - 1;
                            transitions(k)=k;
                    end
                    disp('Best-Fit Line: ')
                            DNL_Bestfit

            % Endpoint Line
                    for k=1:length(S_Actual)-1,
                            DNL_Endpoint(k,1) = S_derivative(k,1)/LSB_Endpoint - 1;
                            transitions(k)=k;
                    end
                    disp('Endpoint Line: ')
                            DNL_Endpoint
            % plot DNL versus Code Transition
                    figure(2); plot(transitions, DNL_Bestfit,'bo'); hold on
                    text(transitions(2),max(DNL_Bestfit),'DNL_Bestfit: blue circles')
                    figure(2); plot(transitions, DNL_Endpoint,'go'); hold off
            text(transitions(2),max(DNL_Bestfit)-0.1,'DNL_Endpoint: green circles')
                            xlabel('DAC Code Transition');
                            ylabel('DNL [LSBs]');

    disp('****** (4) Integral Nonlinearity (INL) Curve ******')
    disp(' ')
            % Best-Fit Line:
                    INL_Bestfit= (S_Actual'-Best_fit_line') / LSB_Bestfit
            % Endpoint Line:
                    INL_Endpoint = (S_Actual'-S_Endpoint') / LSB_Endpoint
            % plot DNL versus Code Transition
                    figure(3); plot(codeword, INL_Bestfit,'bo'); hold on
                    text(codeword(2),max(INL_Bestfit),'INL_Bestfit: blue circles')
                    figure(3); plot(codeword, INL_Endpoint,'go');
            text(codeword(2),max(INL_Bestfit)-0.1,'INL_Endpoint: green circles')
                            xlabel('DAC Input Code');
                            ylabel('INL [LSBs]');
% end
```

```
%%%%%%%%%%%%        Major Carrier Testing           %%%%%%%%
%                                                           %
% Given a vector V containing the major carrier levels from a DAC and the  %
% DC Base value, the following routine computes the complete DAC           %
% transfer curve.                                                          %
%                                                           %
%%%%%%%%%%%%%%%%%%%%%%%%%%%%%%%%%%%%%%%%%%%%%



% Initialization
    V=[75e-3 175e-3 55e-3 195e-3];
    DC_Base=-780e-3; % DC Base value
    D=length(V);  % number of bits in DAC



% Main Routine
    disp('************ DAC Characterization Using Major Carrier Testing ************')
    disp(' ')
        % Compute the DAC binary weights (W0, W1, ..., Wn)
            sum_W=0;
            for k=1:D,
                W(k)=V(k)+sum_W;
                sum_W=sum_W+W(k);
            end
        % Convert unsigned codeword in integer form to binary number
            binaryword=[ ];
            for k=1:2^D,
                quotient=(k-1);
                for n=1:D,
                    binaryword(k,n)=rem(quotient,2);
                    quotient=fix(quotient/2);
                end
            end
            for k=1:2^D,
                DAC_Output(k,1)= binaryword(k,:) * W' + DC_Base;
            end
            disp('Major Carrier Transitions (V(0), ..., V(D-1))')
                V
            disp('Binary-weighted Values (W(0), ..., W(D-1)) & DC Base')
                [ W DC_Base ]
            disp('Comparison: [ D0 D1 D2 D3 DAC_Output ]')
                [ binaryword DAC_Output ]
% end
```

Problems

11.1. Given a set of N points denoted by $S(i)$, derive the parameters of a straight line described by

$$Best\_fit\_line(i) = gain \times i + offset \quad \text{for } i = 0, 1, \ldots, N-1$$

that minimizes the following mean-square error criteria

$$\overline{e^2} = \sum_{i=0}^{N-1}\left[S(i) - Best\_fit\_line(i)\right]^2 = \sum_{i=0}^{N-1}\left[S(i) - gain \times i + offset\right]^2$$

Hint: Find partial derivatives $\delta\overline{e^2}/\delta gain$ and $\delta\overline{e^2}/\delta offset$, set them both to zero, and solve for the two unknowns, *gain* and *offset*, from the system of two equations.

11.2. The output levels of a 4-bit DAC produces the following set of voltage levels, starting from code 0 and progressing through to code 15:

0.0465, 0.3255, 0.7166, 1.0422, 1.5298, 1.8236, 2.1693, 2.5637,

2.8727, 3.3443, 3.6416, 4.0480, 4.3929, 4.7059, 5.0968, 5.5050

What is the full-scale range of this DAC?

11.3. A 4-bit DAC has a full-scale voltage range of 0 to 1.0 V. The input is formatted using an unsigned binary number representation. List all possible ideal output levels. What output level corresponds to the DAC input code 0?

11.4. A 5-bit DAC has a full-scale voltage range of 2.0 to 4.0 V. The input is formatted using a 2's complement number representation. List all possible ideal output levels. What output level corresponds to the DAC input code 0?

11.5. A 4-bit unsigned binary DAC produces the following set of voltage levels, starting from code 0 and progressing through to code 15

0.0465, 0.3255, 0.7166, 1.0422, 1.5298, 1.8236, 2.1693, 2.5637,

2.8727, 3.3443, 3.6416, 4.0480, 4.3929, 4.7059, 5.0968, 5.5050

The ideal DAC output at code 0 is 0 V and the ideal gain is equal to 400 mV/bit.

(a) Calculate the DAC's gain (volts per bit), gain error, offset and offset error.

(b) Estimate the LSB step size using its measured full-scale range. What is the gain error and offset error?

(c) Calculate the absolute error transfer curve for this DAC. Normalize the result to one LSB.

(d) Is the DAC output monotonic?

(e) Compute the DNL curve for this DAC. Use the best-fit line to define the average LSB size. Does this DAC pass a ±1/2 LSB specification for DNL?

(f) Repeat part (e) but this time use the endpoint method to calculate the average LSB size. Does this DAC pass a ±1/2 LSB specification for DNL?

11.6. Compute the INL curve for this DAC whose DNL curve is described by the following values. A 4-bit two's complement DAC produces the following set of voltage levels, starting from code -8 and progressing through to code +7

-0.9738, -0.8806, -0.6878, -0.6515, -0.3942, -0.3914, -0.2497, -0.1208,

-0.0576, 0.1512, 0.2290, 0.4460, 0.4335, 0.5999, 0.6743, 0.8102

The ideal DAC output at code 0 is 0 V and the ideal gain is equal to 133.3 mV/bit.

(a) Calculate the DAC's gain (volts per bit), gain error, offset and offset error.

(b) Estimate the LSB step size using its measured full-scale range. What is the gain error and offset error?

(c) Calculate the absolute error transfer curve for this DAC. Normalize the result to one LSB.

(d) Is the DAC output monotonic?

(e) Compute the DNL curve for this DAC. Use the best-fit line to define the average LSB size. Does this DAC pass a ±1/2 LSB specification for DNL?

(f) Repeat part (e) but this time use the endpoint method to calculate the average LSB size. Does this DAC pass a ±1/2 LSB specification for DNL?

(g) Compute the INL curve for this DAC whose DNL curve is described by the following values

11.7. Calculate the INL curve for a 4-bit unsigned binary DAC whose DNL curve is described by the following values

-0.0815, -0.1356, -0.1133, 0.0057, 0.0218, 0.1308, -0.0361, -0.0950,

0.1136, -0.1633, 0.2101, 0.0512, 0.0119, -0.0706, -0.0919

The DAC output for code 0 is -0.4919 V. Assume that the best-fit line has a gain of 63.1 mV/bit and an offset of -0.5045 V. Does this DAC pass a ±1/2 LSB specification for INL?

11.8. Calculate the DNL curve for a 4-bit DAC whose INL curve is described by the following values

0.1994, 0.1180, -0.0177, -0.1310, -0.1253, -0.1036, 0.0272, -0.0089,

-0.1039, 0.0096, -0.1537, 0.0565, 0.1077, 0.1196, 0.0490, -0.0429

Does this DAC pass a ±1/2 LSB specification for DNL?

11.9. Using the MATLAB routine listed in the appendix of this chapter, check your answers to Problems 11.6 – 11.8.

11.10. The step sizes between the major carries of a 5-bit unsigned binary DAC were measured to be as follows

code 0→1: 0.1939 V, code 1→2: 0.1333 V, code 3→4: 0.1308 V, code 7→8: 0.1316 V, code 15→16: 0.1345 V

Determine the values of W_0, W_1, W_2, W_3, and W_4. Reconstruct the voltages on the ramp from DAC code 0 to DAC code 31 if the DC base value is 100 mV.

11.11. The step sizes between the major carries of a 4-bit two's complement DAC were measured to be as follows:

code -8→-7: 0.1049 V, code -7→-6: 0.1033 V, code -5→-4: 0.0998 V, code -1→0: 0.1016 V

Determine the values of W_0, W_1, W_2, and W_3. Reconstruct the voltages on the ramp from DAC code -8 to DAC code +7 if the DC base value is 500 mV.

11.12. Using the MATLAB routine listed in the appendix of this chapter, check your answers to Problems 11.10 and 11.11.

11.13. Can a major carrier test technique be used to describe a 4-bit unsigned DAC if the output levels beginning with code 0 were found to be the following

0.0064, 0.0616, 0.1271, 0.1812, 0.2467, 0.3206, 0.3856, 0.4406,

0.5021, 0.5716, 0.6364, 0.6880, 0.7662, 0.8262, 0.8871, 0.9480

What if the DAC output levels were described by the following

0.0064, 0.0616, 0.1271, 0.1823, 0.2478, 0.3030, 0.3684, 0.4236,

0.4851, 0.5403, 0.6058, 0.6610, 0.7264, 0.7816, 0.8471, 0.9023

Explain your reasoning.

11.14. The step response of a DAC obtained from an oscilloscope is as follows

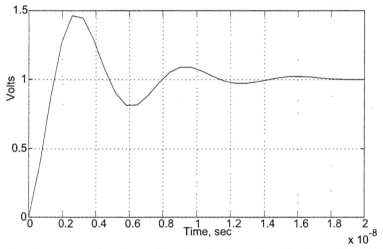

The data sheet states that the settling time is 10 ns (error band = ±50 mV). Does this DAC settle fast enough to meet this specification? Also, determine the overshoot of this signal and its rise time. Estimate the total glitch energy during the positive-going transition.

11.15. Using MATLAB or equivalent software, evaluate the following expression for the step response of a circuit using a time step of no larger than 1 ns

$$v(t) = 1 - \frac{e^{-\zeta \omega_n t}}{\sqrt{1-\zeta^2}} \sin\left(\omega_n t \sqrt{1-\zeta^2} - \cos^{-1} \zeta\right)$$

where

$$\omega_n = 2\pi \times 100 \text{ MHz} \quad \text{and} \quad \zeta = 0.3$$

Determine the time for circuit to settle to within 1% of its final value.

Determine the rise time.

11.16. Categorize the DAC designs described in Section 11.5 into low, medium, or high speed.

References

1. George W. Snedecor, William G. Cochran, *Statistical Methods*, Eighth Edition, Iowa State University Press, 1989, ISBN: 0813815614, pp. 149-176
2. G. N. Stenbakken, T. M. Souders, *Linear Error Modeling of Analog and Mixed-Signal Devices*, Proc. International Test Conference, 1991
3. T. Yamaguchi, M. Soma, *Dynamic Testing of ADCs Using Wavelet Transforms*, Proc. International Test Conference, 1997, pp. 379-88
4. James C. Candy, Gabor C. Temes, *Oversampling Delta-Sigma Data Converters: Theory, Design, and Simulation*, IEEE Press, New York, NY, January 1992, ISBN: 0879422858
5. Steven R. Norsworthy, Richard Schreier, Gabor C. Temes, *Delta-Sigma Data Converters: Theory, Design, and Simulation*, IEEE Press, New York, NY, November 1996, ISBN: 0780310454

CHAPTER **12**

ADC Testing

12.1 ADC Testing versus DAC Testing

12.1.1 Comparison of DACs and ADCs

As mentioned in Chapter 11, "DAC Testing," there are many similarities between DAC testing and ADC testing. There are also a few notable differences. The differences between ADC and DAC testing of transmission parameters such as gain and signal-to-noise ratio were discussed in Chapter 9, "Sampled Channel Testing." In this chapter, we will examine the differences as they relate to intrinsic parameters such as DC offset, INL, and DNL.

The primary difference between DAC and ADC testing relates to the fundamental difference in their transfer curves. As discussed in Chapter 11, the DAC transfer curve is a one-to-one mapping function, while the ADC transfer curve is a many-to-one mapping function (Figure 12.1). In this chapter, we will see that the ADC curve in Figure 12.1 is actually an idealized one. The output codes generated by a real-world ADC are affected by noise from the input circuits. As a result, an ADC curve is statistical in nature rather than deterministic. In other words, for a given input voltage, it may not be possible to predict exactly what output code will be produced. Before we can study testing methods for ADCs, we should first examine the statistical nature of a true ADC transfer curve.

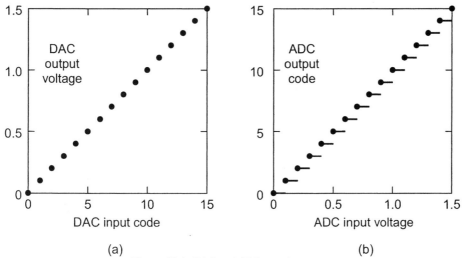

Figure 12.1. DAC and ADC transfer curves.

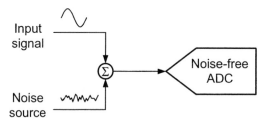

Figure 12.2. ADC model including input noise.

12.1.2 Statistical Behavior of ADCs

To understand the statistical nature of ADCs, we have to model the ADC as a combination of a perfect ADC and a noise source with no DC offset. The noise source represents the combination of the noise portion of the real-world input signal plus the self-generated noise of the ADCs input circuits. Figure 12.2 shows this noisy ADC model.

Applying a DC level to the noisy ADC, we can begin to understand the statistical nature of ADC decision levels. A noise-free ADC might be described by a simple output/input relationship such as

$$\text{output code} = Quantize(\text{input voltage}) \tag{12.1}$$

where the function $Quantize(\;)$ represents the noise-free ADC's quantization process. The noisy ADC can be described using a similar equation

$$\text{output code} = Quantize(\text{input voltage} + \text{noise voltage}) \tag{12.2}$$

Now consider the case of a noisy ADC with a DC input voltage. If the DC voltage is exactly between two ADC decision levels, and the noise voltage never exceeds ±½ LSB, then the ADC will always produce the same output code. The noise voltage never gets large enough to push the total voltage across either of the adjacent decision levels. The probability density function (pdf) plot depicted in Figure 12.3 illustrates this situation. This plot shows the probability that the total input signal V (DC plus noise) will fall within a particular range. It is assumed that the pdf of the noise component is a Gaussian-distributed random variable N with zero mean and a standard deviation of σ (i.e., the RMS noise voltage) described by

$$pdf(N) = \frac{1}{\sigma\sqrt{2\pi}} e^{\frac{-N^2}{2\sigma^2}} \tag{12.3}$$

When the noise is combined with the DC input signal, the total signal pdf can be described by

$$pdf(V) = \frac{1}{\sigma\sqrt{2\pi}} e^{\frac{-(V-DC)^2}{2\sigma^2}} \tag{12.4}$$

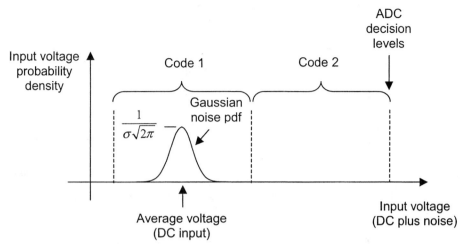

Figure 12.3. Probability density plot for DC input between two decision levels.

In essence, $pdf(V)$ has the same form as $pdf(N)$, but the mean value is different. The total area bounded by the curve described by Eq. (12.4) and the voltage axis is one. Hence, the area under the curve between two ordinates $V = a$ and $V = b$, where $a < b$, represents the probability that the total signal at the input of the ADC lies between a and b. This probability is denoted by $P(a < V < b)$. Figure 12.3 depicts a situation where all the area under the pdf is bounded between two ADC decision levels. This suggests that the probability the input signal will fall between the two ADC decision levels is equal to one, or 100%. (Actually, there is a tiny probability that the input signal will exceed one of the thresholds, since the Gaussian pdf extends to infinity in both directions. In practice, the probability is so low it can be considered zero.)

On the other hand, if the DC input voltage is exactly equal to a decision level, then even a tiny amount of noise voltage will cause the quantization process to randomly dither between the two codes on each side of the decision level. Assuming the statistical distribution of noise is symmetrical, as in the case of the Gaussian pdf, the ADC will produce an equal number of each code. This is shown by the pdf diagram in Figure 12.4. Since the area under the pdf is equally split between code 1 and code 2, we would expect 50% of the ADC conversions to produce code1 and 50% of the conversions to produce code 2.

For input voltages that are close but not equal to the decision levels, the process gets more complicated. Consider an input DC level that is ΔV volts below one of the ADC's decision levels, such as that shown in Figure 12.5. Any time the noise voltage exceeds ΔV, the ADC quantizer will trip to the next highest value. In effect, this is an erroneous conversion result caused by the noise. The probability that the noise voltage will *not* exceed ΔV and trip the quantizer into the next code is equal to the area underneath the portion of the noise pdf that is less than ΔV. This area, denoted $F(\Delta V)$, is commonly referred to as the cumulative distribution function, or cdf. In this particular case, $F(\Delta V)$ is equal to the integral of the probability density function of the noise signal from minus infinity to ΔV

$$F(\Delta V) = \int_{-\infty}^{\Delta V} pdf(N) dN = \frac{1}{\sigma\sqrt{2\pi}} \int_{-\infty}^{\Delta V} e^{\frac{-N^2}{2\sigma^2}} dN \qquad (12.5)$$

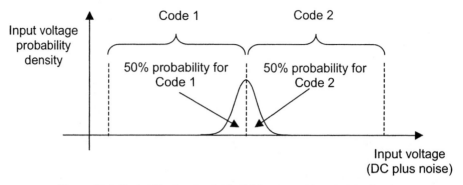

Figure 12.4. Probability density plot for DC input equal to a decision level.

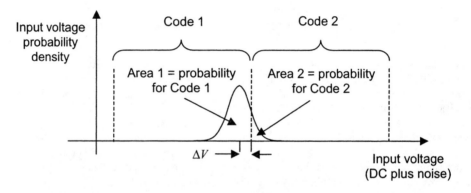

Figure 12.5. Probability density plot for DC input equal to a decision level minus ΔV.

Unfortunately, no closed-form solution exists for this definite integral. Moreover, the integration is dependent on the values of both ΔV and σ. This makes the integration rather specific to the problem at hand. However, if we make the change of variable $Z=N/\sigma$, Eq. (12.5) can be rewritten as

$$F(\Delta V) = \frac{1}{\sqrt{2\pi}} \int_{-\infty}^{\Delta V/\sigma} e^{\frac{-Z^2}{2}} dZ \tag{12.6}$$

where the two parameters collapse into one single variable in the upper limit of integration. By tabulating a single function, say

$$\Phi(x) = \frac{1}{\sqrt{2\pi}} \int_{-\infty}^{x} e^{\frac{-Z^2}{2}} dZ \tag{12.7}$$

we can relate the cdf behavior of a Gaussian random variable with zero mean and a standard deviation of unity to a random variable having an arbitrary standard deviation σ according to

$$F(\Delta V) = \Phi\left(\frac{\Delta V}{\sigma}\right) \tag{12.8}$$

In other words, to determine the value of a particular cdf involving a Gaussian random variable with a zero mean and a standard deviation of σ at a particular point, say, V_1, we simply evaluate Eq. (12.7) using a normalized value for V_1, that is, $x = V_1/\sigma$.

The function $\Phi(x)$ has been extensively tabulated to varying degrees of accuracy, and a short tabulation of it is given in Table 12.1. Rows of $\Phi(\Delta V/\sigma)$ are interleaved between rows of ΔV expressed as multiples of the standard deviation of the noise, σ. A plot of $\Phi(\Delta V/\sigma)$ versus ΔV is provided in Figure 12.6. From this cdf plot we see that the probability that ΔV will fall below 0 V is 0.5, or 50%. The probability that it will fall below $+1.0\sigma$ is equal to approximately 0.8413, or 84.13%, as obtained from Table 12.1. The probability that ΔV will fall above $+1.0\sigma$ (i.e., that it will *not* fall below $+1.0\sigma$) is equal to $1-0.8413 = 0.1587$, since the probability that something will *not* happen is always 1 minus the probability that it *will* happen.

Table 12.1. Gaussian Cumulative Distribution Function (cdf) Values

ΔV	-3.0σ	-2.9σ	-2.8σ	-2.7σ	-2.6σ	-2.5σ	-2.4σ	-2.3σ	-2.2σ	-2.1σ
$\Phi(\Delta V/\sigma)$	0.0013	0.0019	0.0026	0.0035	0.0047	0.0062	0.0082	0.0107	0.0139	0.0179
ΔV	-2.0σ	-1.9σ	-1.8σ	-1.7σ	-1.6σ	-1.5σ	-1.4σ	-1.3σ	-1.2σ	-1.1σ
$\Phi(\Delta V/\sigma)$	0.0228	0.0287	0.0359	0.0446	0.0548	0.0668	0.0808	0.0968	0.1151	0.1357
ΔV	-1.0σ	-0.9σ	-0.8σ	-0.7σ	-0.6σ	-0.5σ	-0.4σ	-0.3σ	-0.2σ	-0.1σ
$\Phi(\Delta V/\sigma)$	0.1587	0.1841	0.2119	0.2420	0.2743	0.3085	0.3446	0.3821	0.4207	0.4602
ΔV	0.0	0.1σ	0.2σ	0.3σ	0.4σ	0.5σ	0.6σ	0.7σ	0.8σ	0.9σ
$\Phi(\Delta V/\sigma)$	0.5000	0.5398	0.5793	0.6179	0.6554	0.6915	0.7257	0.7580	0.7881	0.8159
ΔV	1.0σ	1.1σ	1.2σ	1.3σ	1.4σ	1.5σ	1.6σ	1.7σ	1.8σ	1.9σ
$\Phi(\Delta V/\sigma)$	0.8413	0.8643	0.8849	0.9032	0.9192	0.9332	0.9452	0.9554	0.9641	0.9713
ΔV	2.0σ	2.1σ	2.2σ	2.3σ	2.4σ	2.5σ	2.6σ	2.7σ	2.8σ	2.9σ
$\Phi(\Delta V/\sigma)$	0.9772	0.9821	0.9861	0.9893	0.9918	0.9938	0.9953	0.9965	0.9974	0.9981

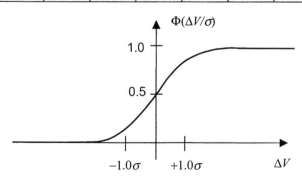

Figure 12.6. Cumulative distribution function of a Gaussian-distributed random variable (i.e., noise).

Example 12.1

An ADC input is set to 2.453 V DC. The noise of the ADC and DC signal source is characterized to be 10 mV RMS and is assumed to be perfectly Gaussian. The transition between code 134 and 135 occurs at 2.461 V DC for this particular ADC. Therefore, the value 134 is the expected output from the ADC. What is the probability that the ADC will produce code 135 instead of 134? If we collected 200 samples from the output of the ADC, how many would we expect to be 134 and how many would be 135? How might we determine that the transition between code 134 and 135 occurs at 2.461 V DC? How might we characterize the effective RMS input noise?

Solution:

With an input of 2.453 V DC, the ADC's input noise would have to exceed (2.461 V – 2.453 V) = +8 mV to cause the ADC to trip to code 135. This value is equal to $+0.8\sigma$, since $\sigma = 10$ mV. From Table 12.1 the Gaussian cdf of $+0.8\sigma$ is equal to 0.7881. Therefore, there is a 78.81% probability that the noise will *not* be sufficient to trip the ADC to code 135. Thus we can expect 78.81% of the conversions to produce code 134 and 21.19% of the conversions to produce code 135. If we collect 200 samples from the ADC, we would expect 78.81% of the 200 conversions (approximately 158 conversions) to produce code 134. We would expect the remaining 21.19% of the conversions (42 samples) to produce code 135.

To determine the transition voltage, we simply have to adjust the input voltage up or down until 50% of the samples are equal to 134 and 50% are equal to 135. To determine the value of σ, we can adjust the input voltage until we get 84.13% of the conversions to produce code 134. The difference between this voltage and the transition voltage is equal to 1.0σ, which is equal to the effective RMS input noise of the ADC.

Exercises

12.1. If V is normally distributed with zero mean and a standard deviation of 2 mV, find $P(V < 4$ mV$)$. Repeat for $P(V > -1$ mV$)$. Repeat for $P(-1$ mV $< V < 4$ mV$)$.

Ans. $P(V < 4$ mV$) = 0.9772$; $P(V > -1$ mV$) = 0.6915$; $P(-1$ mV $< V < 4$ mV$) = 0.6687$.

12.2. If V is normally distributed with zero mean and a standard deviation of 100 mV, what is the value of ΔV such that $P(V < \Delta V) = 0.9641$.

Ans. $\Delta V = 180$ mV.

12.3. An ADC input is set to 1.4 V DC. The noise of the ADC and DC signal source is characterized to be 15 mV RMS and is assumed to be perfectly Gaussian. The transition between code 90 and 91 occurs at 1.4255 V DC. If 500 samples of the ADC output are collected, how many do we expect to be code 90 and how many would be code 91?

Ans. # of code 90 = 95.54% or ~ 478 and # of code 91 = 4.46% (~22).

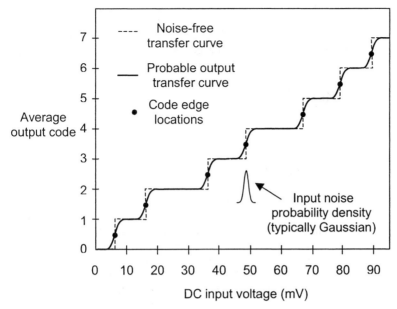

Figure 12.7. ADC probable output code transfer curve.

Because the circuits of an ADC generate random noise, the ADC decision levels represent *probable* locations of transitions from one code to the next. In the previous example, we saw that an input noise level of 10 mV would cause a 2.453 V DC input voltage to produce code 134 only 79% of the time and code 135 21% of the time. Therefore, with an input voltage of 2.453V, we will get an *average* output code of 134×079 + 135×0.21 = 134.21. Of course, the ADC cannot produce code 134.21. This value only represents the average output code we can expect if we collect many samples.

If we plot the average output code from a typical ADC versus DC input levels, we will see the true transfer characteristics of the ADC. Figure 12.7 shows a true ADC transfer curve compared to the idealized, noise-free transfer curve. The center of the transition from one code to the next (i.e., the decision level) is often called a *code edge*. The wider the distribution of the Gaussian input noise, the more rounded the transitions from one code to the next will be. In fact, the true ADC transfer characteristic is equal to the convolution of the Gaussian noise probability density function with the noise-free transfer curve.

Code edge measurement is one of the primary differences between ADC and DAC testing. DAC voltages can simply be measured one at a time using a DC voltmeter or digitizer. By contrast, ADC code edges can only be measured using an iterative process in which the input voltage is adjusted until the output samples dither equally between two codes. Because of the statistical nature of the ADC's transfer curve, each iteration of the search requires 100 or more conversions to achieve a repeatable average value. Since this brute-force approach would lead to very long test times in production, a number of faster methodologies have been developed to locate code edges. Unfortunately, these production techniques generally result in somewhat less exact measurements of code edge voltages.

In the next section, we will examine the various ways in which the code edges of an ADC can be measured, both for characterization and production. Once the code edges have been located, we can apply all the same tests to ADCs that we applied to DACs. Tests such as INL, DNL, DC gain, and DC offset are commonly performed using the code edge information.

12.2 ADC CODE EDGE MEASUREMENTS

12.2.1 Edge Code Testing versus Center Code Testing

To measure ADC intrinsic parameters such as INL and DNL, we first have to convert the many-to-one transfer curve of the ADC into a one-to-one mapping function similar to that of a DAC. Then we simply apply the same testing methods and criteria from Chapter 11 to the one-to-one transfer curve of the ADC. There are two ways to convert the many-to-one transfer curve of an ADC into a one-to-one curve. These two methods are known as *center code testing* and *edge code testing*. Figure 12.8 illustrates the difference between edge code testing and center code testing. Code centers are defined as the midpoint between the code edges. For example, consider a case in which the decision level between code 57 and code 58 corresponds to an input voltage of 100 mV and the decision level between codes 58 and 59 corresponds an input of 114 mV. In this example, the center of code 58 corresponds to the average of these two voltages, (114 mV + 100 mV)/2 = 107 mV.

Figure 12.8 highlights the problem with center code testing. Notice that the code centers fall very nearly on a straight line, while the code edges show much less linear behavior. The averaging process in the definition of code centers produces an artificially low DNL result compared to edge code testing. Because the code widths in Figure 12.8 alternate between wide and narrow codes, the averaging process effectively smoothes these variations out, leaving a

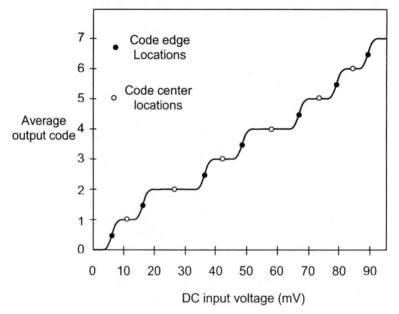

Figure 12.8. Code edges and code centers.

transfer characteristic that looks like it has fairly evenly spaced steps. Because center code testing produces an artificially low DNL value, this technique should be avoided. The edge code method is a more discerning test, and is therefore the preferred means of translating the transfer curve of an ADC to the one-to-one mapping needed for INL and DNL measurements.

We can search for code edges in one of several different ways. Three common techniques are the step search or binary search method, the hardware servo method, and the histogram method. In the next section, we will see how each of these techniques is applied, and we will examine the strengths and weaknesses of each method. Since all the various ADC edge measurement techniques are slower than simply measuring an output voltage, ADC testing is generally much slower than DAC testing.

12.2.2 Step Search and Binary Search Methods

The most obvious method to find the edge between two ADC codes is to simply adjust the input voltage of the ADC up or down until the output codes are evenly divided between the first code and the second. To achieve repeatable results, we need to collect about 50 to 100 samples from the ADC so that we have a statistically significant number of conversions. The input voltage adjustment could be performed using a simple step search, but a faster method is to use a binary search to quickly find the input voltage corresponding to the ADC code edge. (Step searches and binary searches were discussed in Chapter 3.)

Binary searches are an acceptable production test method for comparators and slicer circuits, which are effectively one-bit ADCs. However, if we try to apply a binary search technique to multibit ADCs in production, we run into a major problem. If we use a binary search with, say, five iterations, we have to collect 100 samples for each iteration. This would result in a total of 500 collected samples *per code edge*. An N-bit ADC has 2^N-1 code edges. Therefore, the test time for most ADCs would be far too high. For example, a 10-bit ADC operating at a sampling rate of 100 kHz would require a total data collection time of 500 codes times $2^{10}-1$ edges times the sample period (1/100 kHz). Thus the total collection time would be $500 \times 1023 \times 10$ μs = 5.115 s! Clearly, this is not a production-worthy solution.

12.2.3 Servo Method

A much better method for measuring code edges in production is the use of a servo circuit. Figure 12.9 shows a simplified block diagram of an ADC servo measurement setup. The output codes from the ADC are compared against a value programmed into the search value register. If the ADC output is greater than or equal to the expected value, the integrator ramps downward. If it is less than the expected value, the integrator ramps upward.

Eventually, the integrator finds the desired code edge and fluctuates back and forth across its transition level. The average voltage at the ADC input, $V_{CodeEdge}$, represents the lower edge of the code under test. This voltage can easily be measured using a DC voltmeter with a low-pass filtered input. The servo search process is repeated for each code edge in the ADC transfer curve.

The servo method is actually a fast hardware version of the step search. Unlike the step search or binary search methods, the servo method does not perform averaging before moving from one input voltage to the next. The continuous up/down adjustment of the servo integrator

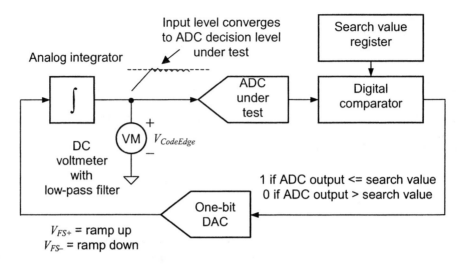

Figure 12.9. ADC servo test setup.

coupled with the averaging process of the filtered voltmeter act together to remove the effects of the ADC's input noise. Because of its speed, the servo technique is generally more production-worthy than the step search or binary search methods.

Although the servo method is faster than the binary search method, it is also fairly slow compared with a more common production testing technique, the histogram method. Histogram testing requires an input signal with a known voltage distribution. There are two commonly used histogram methods: the linear ramp method and the sinusoidal method.

12.2.4 Linear Ramp Histogram Method

The simplest way to perform a histogram test is to apply a rising or falling linear ramp to the input of the ADC and collect samples from the ADC at a constant sampling rate. The ADC samples are captured as the input ramp slowly moves from one end of the ADC conversion range to the other. The ramp is set to rise or fall slowly enough that each ADC code is "hit" several times, as shown in Figure 12.10. The number of occurrences of each code is directly proportional to the width of the code. In other words, wide codes are hit more often than narrow codes. For example, if the voltage spacing between the upper and lower decision levels for code 2 are twice as wide as the spacing for code 1, then we expect code 2 to occur twice as often as code 1. The reason for this is that it takes the linear ramp input signal twice as long to sweep through code 2 as it takes to sweep through code 1. Of course, this method assumes that the ramp is perfectly linear and that the ADC sampling rate is constant throughout the entire ramp. This condition is easily maintained in mixed-signal ATE testers.

The number of occurrences of each code is plotted as a histogram, as illustrated in Figure 12.11. Ideally, each code should be hit the same number of times, but this would only be true for a perfectly linear ADC. The histogram shows us which codes are hit more often, indicating that they are wider codes. For example, we can see from the histogram in Figure 12.11 that codes 2 and 4 are twice as wide as codes 1 and 6.

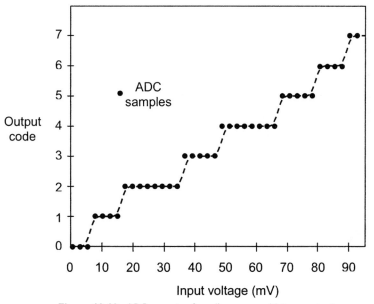

Figure 12.10. ADC samples from linear ramp histogram test.

Let us denote the number of hits that occur for the ith code word of an N-bit DAC as $H(i)$ for $i = 0, 1, \ldots, 2^N-1$. Next, let us define the average number of hits for each code word, excluding the number of hits included in the two end codes, as

$$H_{Average} = \frac{1}{2^N - 2} \sum_{i=1}^{2^N - 2} H(i) \qquad (12.9)$$

Dividing $H(i)$ by $H_{Average}$ we obtain the width of each code word in units of LSBs as

$$\text{code width}(i) = \frac{H(i)}{H_{Average}}, \quad i = 1, 2, \ldots, 2^N - 2 \qquad (12.10)$$

Excluding the highest and lowest code count is necessary, as these two codes do not have a defined code width. In effect, the end codes are infinitely wide. For example, code 0 in an unsigned binary ADC has no lower decision level, since there is no code corresponding to -1. In many practical situations, the input ramp signal extends beyond the upper and lower ranges of the ADC resulting in an increase code count for these two code words. These meaningless hits should be ignored in the linear ramp histogram analysis.

12.2.5 Conversion from Histograms to Code Edge Transfer Curves

To calculate absolute or best-fit INL and DNL curves, we have to determine the absolute voltage for each decision level. Unfortunately, an LSB code width plot such as the one in Figure 12.11 tells us the width of each code in LSBs rather than volts. To convert the code width plot into voltage units, we need to measure the average LSB size of the ADC, in volts. This can be done using a binary search or servo method to find the upper and lower code edge voltages, V_{UE} and

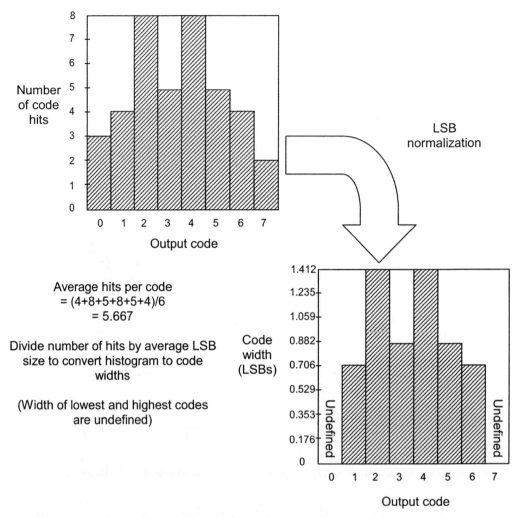

Figure 12.11. LSB normalization translates ADC code histogram into LSB code widths.

V_{LE}. In an N-bit ADC, there are 2^N-2 LSBs between these two code edges. Therefore, the average LSB size can be calculated as follows

$$V_{AveCodeWidth} = V_{LSB} = \frac{V_{UE} - V_{LE}}{2^N - 2} \qquad (12.11)$$

The code width plot can then be converted to volts by multiplying each value by the average code width, in volts

$$V_{CodeWidth}(i) = V_{LSB} \times LSB\ code\ width(i) \qquad (12.12)$$

The following example will illustrate this approach.

Example 12.2

A binary search method is used to find the transition between code 0 and code 1 of the ADC in Figure 12.10. The code edge is found to be 53 mV. A second binary search determines the code edge between codes 6 and 7 to be 2.77 V. What is the average LSB size for this 3-bit ADC? Based on the histogram in Figure 12.11, what is the width of each of the 8 codes, in volts?

Solution:

The average LSB size is equal to

$$V_{LSB} = \frac{2.77\text{ V} - 0.053\text{V}}{2^3 - 2} = 452.8\text{ mV}$$

Therefore, the code width for each code is:

Code 0: Undefined (infinite width)

Code 1: 0.706 LSBs × 452.8 mV = 319.68 mV

Code 2: 1.412 LSBs × 452.8 mV = 639.35 mV

Code 3: 0.882 LSBs × 452.8 mV = 399.37 mV

Code 4: 1.412 LSBs × 452.8 mV = 639.35 mV

Code 5: 0.882 LSBs × 452.8 mV = 399.37 mV

Code 6: 0.706 LSBs × 452.8 mV = 319.68 mV

Code 7: Undefined (infinite width)

If we wish to calculate the absolute voltage level of each code edge, we simply perform a running sum on the code widths expressed in volts, starting with the voltage V_{LE}, as follows

$$V_{CodeEdge}(i) = \begin{cases} V_{LE}, & i=0 \\ V_{UE} + \sum_{k=1}^{i} V_{CodeWidth}(k), & i=1,2,\cdots,2^N-2 \end{cases} \quad (12.13)$$

Alternatively, we can write a recursive equation for the code edges as follows

$$V_{CodeEdge}(i) = V_{CodeEdge}(i-1) + V_{LSB}V_{CodeWidth}(i), \quad i=1,2,\ldots,2^N-2 \quad (12.14)$$

where we begin with $V_{CodeEdge}(0) = V_{LE}$. The resulting code edge transfer curve is equivalent to a DAC output transfer curve, except that it will only have 2^N-1 values rather than 2^N values.

Example 12.3

Using the results of Example 12.2, reconstruct the 3-bit ADC transfer curve for each decision level.

Solution:

The transition from code 0 to code 1 was measured using a binary search. It was 53 mV. The other codes edges can be calculated using a running sum:

Code 0 to Code 1: 53 mV

Code 1 to Code 2: 53 mV + 319.68 mV = 372.68 mV

Code 2 to Code 3: 372.68 mV + 639.35 mV = 1011.9 mV

Code 3 to Code 4: 1011.9 mV + 399.37 mV = 1411.5 mV

Code 4 to Code 5: 1411.5 mV + 639.35 mV = 2050.8 mV

Code 5 to Code 6: 2050.8 mV + 399.37 mV = 2450.4 mV

Code 6 to Code 7: 2450.4 mV + 319.68 mV = 2770.0 mV

12.2.6 Accuracy Limitations of Histogram Testing

The accuracy of any code width or edge in units of LSBs is inversely proportional to the average number of hits per code, that is, accuracy=$1/H_{Average}$. For instance, if we measure an average of 5 hits per code, then the code width or code edge would, on average, have one-fifth of an LSB of resolution. If one LSB is equivalent to 452.8 mV, as in the last example, then the code width and edge would have a possible error of ±45.28 mV. To improve the accuracy of the histogram test, the average number of hits per code must be increased.

For characterization of the ADC, we would prefer to ramp the input very slowly; so that each code is hit hundreds of times instead of just 5 or 6 times. This would result in better measurement resolution and repeatability, since the input voltage steps would be spaced much closer together. Also, the random nature of the ADC decision levels would be averaged out by the large sample size.

In production testing, however, we can only afford to collect a relatively small number of samples from each code, typically 16 or 32. Otherwise the test time becomes excessive. Therefore, even a perfect ADC will not produce a flat histogram in production testing because the limited number of samples collected gives rise to a limited code width resolution and repeatability. We can see that the samples in Figure 12.10 are spread too far apart to resolve small fractions of an LSB.

In addition to the accuracy limitation caused by limited resolution, we also face a repeatability limitation. If we look carefully at Figure 12.10, we notice that several of the codes occur so close to a decision level that the ADC noise will cause the results to vary from one test execution

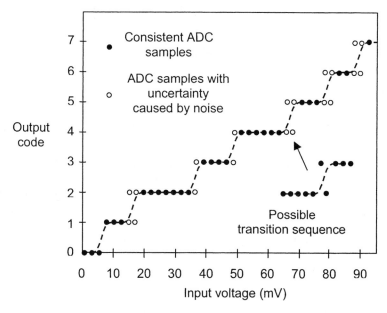

Figure 12.12. Uncertainty caused by random noise.

to the next. This variability will happen *even if our input signal is exactly the same during each test execution*. Figure 12.12 illustrates the uncertainty in output codes caused by noise in the ADC circuits.

In many cases, we find that the raw data sequence from the ADC may zigzag up and down as the output codes near a transition from one code to the next. In Figure 12.12, for instance, we see that it is possible to achieve an ADC output sequence 4, 4, 4, 4, 4, 5, 4, 5, 5, 5 rather than the ideal sequence 4, 4, 4, 4, 4, 4, 5, 5, 5, 5. Unfortunately, this is the nature of histogram testing of ADCs. The results will be variable and somewhat unrepeatable unless we collect many samples per code. In histogram testing, as in many other tests, there is an inherent tradeoff between good repeatability and low test time. It is the test engineer's responsibility to balance the need for low test time with the need for acceptable accuracy and repeatability.

12.2.7 Rising Ramps versus Falling Ramps

Most ADC architectures include one or more analog comparators in their design. Since comparators may be subject to hysteresis, we occasionally find a discrepancy between code edges measured using a rising ramp and code edges measured using a falling ramp. The most complete way to test an ADC is to test parameters such as INL and DNL using both a rising ramp and a falling ramp. Both methods must produce a passing result before the ADC is considered good. However, the extra test doubles the test time; so we prefer to use only one ramp. If characterization shows that we have a good match between the rising ramp and falling ramp, then we can drop back to a single test for production. Alternatively, if characterization shows that either the rising ramp or falling ramp always produces the worst-case results, then we can use only the worst-case test condition to save test time.

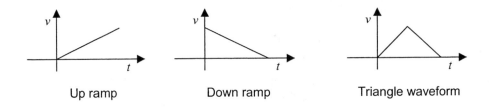

Figure 12.13. Types of linear histogram inputs.

A compromise solution is to ramp the signal up at twice the normal rate and then ramp it down again (Figure 12.13). This triangle waveform approach tests both the falling and rising edge locations, averaging their results. It takes no longer than a single ramp technique, but it cancels the effects of hysteresis. A separate test could then be performed to verify that the ADC's hysteresis errors are within acceptable limits. The hysteresis test could be performed at only a few codes, saving test time compared to the two-pass ramp solution.

Exercises

12.4. A linear histogram test was performed on an unsigned 4-bit ADC resulting in the following distribution of code hits beginning with code 0

4, 5, 5, 7, 8, 4, 2, 4, 4, 3, 6, 3, 4, 6, 5 , 9

A binary search was performed on the first transition between codes 0 and 1 and found the code edge to be at 125 mV. A second binary search was performed and found the code edge between codes 14 and 15 to be 3.542 V. What is the average LSB size for this 4-bit ADC? Determine the width of each code, in volts.

Ans. LSB=224.1 mV; code 0: undefined, code 1: 258.9 mV, code 2: 258.9 mV, code 3: 362.4 mV, code 4: 414.2 mV, code 5: 207.1 mV, code 6: 103.5 mV, code 7: 207.1 mV, code 8: 207.1 mV, code 9: 155.3 mV, code 10: 310.6 mV, code 11: 155.3 mV, code 12: 207.1 mV, code 13: 310.6 mV, code 14: 258.9 mV, code 15: undefined.

12.5. For the distribution of code hits obtained for the 4-bit ADC listed in Exercise 12.4, determine the location of the code edges.

Ans. Beginning with code 0-1 transistion: 0.1250 V, 0.3839 V, 0.6427 V, 1.0051 V, 1.4193 V, 1.6264 V, 1.7300 V, 1.9370 V, 2.1441 V, 2.2995 V, 2.6101 V,   2.7654 V, 2.9725 V, 3.2831 V, 3.5420 V.

12.2.8 Sinusoidal Histogram Method

Sinusoidal histogram tests were originally used to compensate for the relatively poor linearity of early AWG instruments. Since it is easier to produce a pure sinusoidal waveform than to produce a perfectly linear ramp, early testers often relied on sinusoidal histogram testing for high-resolution ADCs. A second, more common reason to use the sinusoidal histogram method

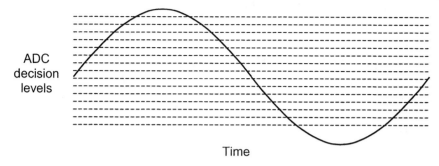

Figure 12.14. Sinusoidal input for 4-bit ADC.

is that it allows better characterization of the dynamic performance of the ADC. The linear histogram technique is basically a static performance test. Because the input voltage is ramped slowly, the input level only changes by a fraction of an LSB from one ADC sample to the next. Sometimes we need to test the ADC transition levels in a more dynamic, real-world situation. To do this, we can use a high-frequency sinusoidal input signal. Our goal is to make the ADC respond to the rapidly changing inputs of a sinusoid rather than the slowly varying voltages of a ramp. In theory, we could use a high-frequency triangle wave to achieve this result, but high-frequency linear triangles are much more difficult to produce than high-frequency sinusoids.

Ramp inputs have an even distribution of voltages over the entire ADC input range. Sinusoids, on the other hand, have an uneven distribution of voltages. A sine wave spends much more time near the upper and lower peak than at the center. As a result, we would expect to get more code hits at the upper and lower codes than at the center of the ADC's transfer curve, even when testing a perfect ADC. Fortunately, the distribution of voltage levels in a pure sinusoid is well defined; so we can compensate for the uneven distribution of voltages inherent to sinusoidal waveforms.

Figure 12.14 shows a sinusoidal waveform that is quantized by a 4-bit ADC. Notice that there are only 15 decision levels in a 4-bit ADC and that the sine wave is programmed to exceed the upper and lower decision levels by a fairly wide margin. The reason we program the sine wave to exceed the ADC's full-scale range is that we have to make sure that the sine wave passes through all the codes if we want to get a histogram of all code widths. If we expand the time scale to view a quarter period of the waveform, we can see how the distribution of output codes is nonuniform due to the sinusoidal distribution of voltages, as shown in Figure 12.15.

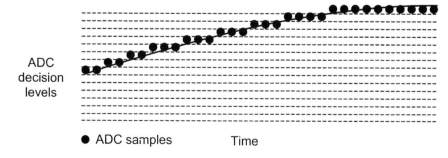

Figure 12.15. Closeup of 4-bit ADC sinusoidal input.

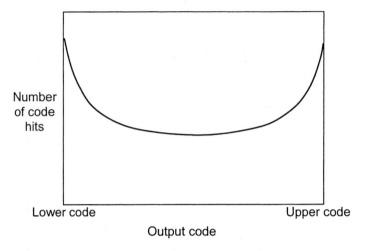

Figure 12.16. Sinusoidal histogram for an ideal ADC.

Clearly we get more code hits near the peaks of the sine wave than at the center, even for this simple example. Thus, the sinusoidal histogram of a perfect ADC exhibits a "bathtub" shape, as illustrated in Figure 12.16. If we try to use this histogram result the same way we use the linear ramp histogram results, the upper and lower codes would appear to be much wider than the middle codes. Clearly, we need to normalize our histogram to remove the effects of the sinusoidal waveform's nonuniform voltage distribution.

The normalization process is slightly complicated because we do not really know what the gain and offset of the ADC will be a priori. Additionally, we may not know the exact offset and amplitude of the sinusoidal input waveform. Fortunately, we have a piece of information at our disposal that tells us the level and offset of the signal as the ADC sees it.

The number of hits at the upper and lower codes in our histogram can be used to calculate the input signal's offset and amplitude. The mismatch between these two numbers tells us the offset, while the number of total hits tells us the amplitude. For example, in Figure 12.14, we can see that we will get more hits at the lower code than at the upper code. The lower codes will be hit more often because the sinusoid has a negative offset.

The equations that relate the number of upper and lower hits to the offset (relative to midscale) and peak amplitude of the sinusoid expressed in terms of LSBs are

$$\mathit{offset} = \left(\frac{C_2 - C_1}{C_2 + C_1}\right)\left(2^{N-1} - 1\right) \qquad (12.15)$$

and

$$\mathit{peak} = \frac{2^{N-1} - 1 - \mathit{offset}}{C_1} \qquad (12.16)$$

where

$$C_1 = \cos\left(\pi \frac{H(2^N-1)}{N_S}\right) \quad (12.17)$$

$$C_2 = \cos\left(\pi \frac{H(0)}{N_S}\right) \quad (12.18)$$

These equations are adapted from Mahoney's textbook "DSP-Based Testing of Analog and Mixed Signal Circuits."[2] As before, $H(2^N-1)$ is the number of times the upper code is hit, $H(0)$ is the number of times the lower code is hit, and N is the converter resolution, in bits. In addition, N_S is the total number of samples (including end code counts) and should be large enough that each ADC code is hit at least 16 times. The common rule of thumb is to collect at least 32 samples for each code in the ADC's transfer curve. For example, an 8-bit converter would require $2^8 \times 32 = 8192$ samples. Of course, some codes will be hit more often than 32 times and some will be hit less often than 32 times due to the curved nature of the sinusoidal input.

Once we know the values of *peak* and *offset*, we can calculate the ideal sine wave distribution of code hits, denoted $H_{sinewave}$, that we would expect from a perfectly linear ADC excited by a sinusoid. The equation for the ith code count, once again, excluding the upper and lower code counts, is

$$H_{sinewave}(i) = \frac{N_s}{\pi}\left[\sin^{-1}\left(\frac{i+1-2^{N-1}-offset}{peak}\right) - \sin^{-1}\left(\frac{i-2^{N-1}-offset}{peak}\right)\right], \quad (12.19)$$

$$i = 1, 2, \ldots, 2^N - 2$$

As Mahoney points out, $H_{sinewave}(i)$ represent probable numbers of hits per code, and are therefore not necessarily integers. The ideal hit counts for each ADC code should therefore be calculated using floating-point calculations.

We obtain the width of the ith code word in units of LSBs by dividing the actual ith code count by $H_{sinewave}(i)$

$$LSB\ code\ width(i) = \frac{H(i)}{H_{sinewave}(i)}, \quad i = 1, 2, \ldots, 2^N - 2 \quad (12.20)$$

Figure 12.17 illustrates the sinusoidal histogram normalization process for an idealized 4-bit ADC. Once we have calculated the normalized histogram, we are ready to convert the code widths into a code edge plot, using the same steps as we used for the linear ramp histogram method

This example is based on an ideal ADC with equal code widths. Even with this idealized simulation, the normalized histogram does not result in equal code width measurements. This simulated example was based on a sample size of 32 samples per ADC code (16 ADC codes × 32 samples per code = 512 collected samples). As we can see in Figure 12.17, many of the codes were hit fewer than 20 times in this simulation. Like the linear ramp histogram method, the measurement resolution of a sinusoidal histogram is limited by the number of hits per code. If we had collected hundreds of samples for each code in this 4-bit ADC example, the results would have been much closer to a flat histogram. Also, the repeatability of code width measurements will improve with a larger sample size. Unfortunately, a larger sample size requires a longer test time. Again, we are faced with a tradeoff between low test time and high accuracy.

Throughout this section, we have only dealt with unsigned binary ADC examples. The histogram technique works equally well with any other type of converter. For instance, if we want to test a two's complement ADC, we have a small problem. Histogram functions typically cannot deal with negative values. The get around this problem, we simply add 2^{N-1} to the collected samples to shift them into an unsigned binary format. Then we can treat the converter as if it were an unsigned binary ADC. Of course, we have to keep track of our array indices so that we know which code edge corresponds to which array element, but that is a minor bookkeeping task.

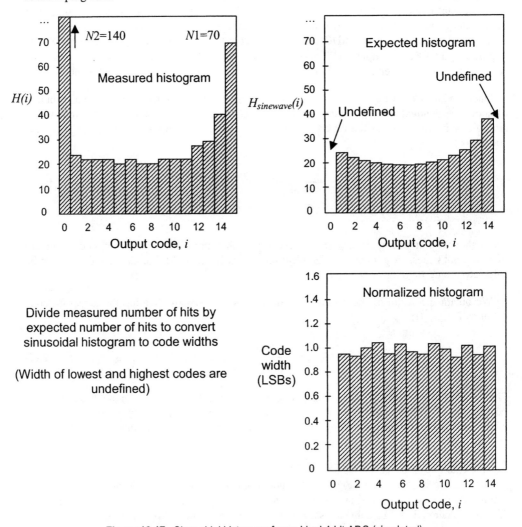

Figure 12.17. Sinusoidal histogram for an ideal 4-bit ADC (simulated).

> **Exercises**
>
> **12.6.** The distribution of code hits for an unsigned 4-bit ADC excited by a sinusoidal signal beginning with code 0 is as follows
>
> 76, 20, 19, 19, 19, 18, 19, 18, 18, 19, 19, 19, 24, 25, 32, 60
>
> What is the offset and amplitude of the input sinusoidal signal? What is expected or ideal sinusoidal distribution of code hits? Finally, what is the distribution of code widths (in LSBs) for this ADC?
>
> **Ans.** Offset = -0.2290 LSBs; peak = 8.0073 LSBs.
>
> Ideal sinusoidal distribution (code 1 to 14): 27.53, 22.57, 20.04, 18.55, 17.64, 17.13, 16.91, 16.97, 17.30, 17.95, 19.06, 20.89, 24.13, 31.25.
>
> Code widths (code 1 to 14): 0.7264, 0.8415, 0.9478, 1.0240, 1.0199, 1.1090, 1.0640, 1.0606, 1.0982, 1.0579, 0.9964, 1.1484, 1.0359, 1.0238.

12.3 DC TESTS AND TRANSFER CURVE TESTS

12.3.1 DC Gain and Offset

Once we have produced a code edge transfer curve for an ADC, we can test the ADC much as we would test a DAC. Since a code edge transfer curve is a one-to-one mapping function, we can apply all the same DC and transfer curve tests outlined in Chapter 11, "DAC Testing." There are a few minor differences to consider. For example, an N-bit ADC has one fewer code edge than an N-bit DAC has outputs. A more important difference is that the ideal ADC transfer curve may be ambiguously defined. The test engineer should realize that there are several ways to define the ideal performance of an ADC.

Figure 12.18 shows two alternate definitions of an 8-bit ADC's ideal performance. The first alternative is to define the ideal location of the first code edge, V_{LE}, at a voltage corresponding to +½ LSB above the V_{FS-} level. The second alternative is to define the ideal location of the first code edge at a voltage corresponding to +1 LSB above V_{FS-}.

It is very important when measuring DC offsets and other absolute voltage levels that we understand exactly what the ideal transfer curve is supposed to be. Otherwise, we may introduce errors of ±½ LSB. Unfortunately, there is little consistency from one ADC data sheet to the next as to the intended ideal performance. This is another issue that the test engineer must clarify before writing the test program.

Once the ideal curve has been established, DC gain and offset can be measured in a manner similar to DAC DC gain and offset. The gain and offset are measured by calculating the slope and offset of the best-fit line. If the converter is defined using the definition illustrated in Figure 12.18(a), we have to remember that the ideal line would have an offset of +½ LSB.

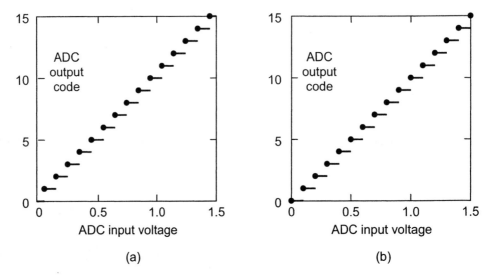

Figure 12.18. Alternate definitions of ADC transfer curves.

Unfortunately, there are many other ways to define gain and offset. In some data sheets, the offset is defined simply as the offset of the first code edge from its ideal position and the gain is defined as the ratio of the actual voltage range divided by the ideal voltage range from V_{FS-} to V_{FS+}. Other definitions abound; so the test engineer is responsible for determining the correct methodology for each ADC to be tested. Of course, ambiguities in the data sheet should be clarified to prevent correlation headaches caused by misunderstandings in data sheet definitions.

12.3.2 INL and DNL

Except for the fact that an ADC code edge transfer curve has one fewer values than an equivalent DAC curve, we can calculate ADC INL and DNL exactly the same way as DAC INL and DNL. If we use the histogram method, we can take a shortcut in measuring INL and DNL. Specifically, once the code widths are known, the endpoint DNL expressed in units of LSBs can be determined by subtracting one LSB from each code width as follows

$$DNL(i) = LSB\ code\ width(i) - 1, \quad i = 1, 2, \ldots, 2^N - 2 \qquad (12.21)$$

Subsequently, as described in Chapter 11, the DNL curve can then be integrated using a running sum to calculate the endpoint INL curve in units of LSBs according to the following

$$INL(i) = \sum_{k=1}^{i-1} DNL(k), \quad i = 1, 2, \ldots, 2^N - 2 \qquad (12.22)$$

Using this shortcut method, we never even have to compute the absolute voltage level for each code edge, unless we need that information for a separate test, such as gain or offset.

As with DAC INL and DNL testing, a best-fit approach is the preferred method for calculating ADC INL and DNL. As discussed in Chapter 11, "DAC Testing," best-fit INL and DNL testing results in a more meaningful, repeatable reference line than endpoint testing, since the best-fit reference line is less dependent on any individual code's edge location. We can convert an endpoint INL curve to a best-fit INL curve by first calculating the best-fit line for the endpoint INL curve. Subtracting the best-fit line from the endpoint INL curve yields the best-fit INL curve. Then the best-fit DNL curve is calculated by taking the discrete time first derivative of the best-fit INL curve.

Notice that the histogram method captures an endpoint DNL curve and then integrates the DNL curve to calculate endpoint INL. This is unlike the DAC methodology and the ADC servo/search methodologies, which start with a measurement of absolute voltage levels to measure INL and then calculate the DNL through discrete time first derivatives.

12.3.3 Monotonicity and Missing Codes

One final difference between ADC testing and DAC testing relates to differences in their weaknesses. For example, a DAC may be nonmonotonic, while an ADC will usually be monotonic *if it is tested statically*. For an ADC to be nonmonotonic, one or more of its code widths has to be negative. (One example of this is an ADC whose DC reference voltage is somehow drastically perturbed as the input voltage varies. However, this failure mechanism is quite rare.) Nevertheless, an ADC can *appear* to be nonmonotonic when its input is changing rapidly.[3]

For this reason, we do not typically test ADCs for monotonicity when we use slowly changing inputs (as in search or linear ramp INL and DNL tests). However, when testing ADCs with rapidly changing inputs, the ADC may behave as if it were nonmonotonic due to slew rate limitations in its comparator(s). These monotonicity errors show up as signal-to-noise ratio failures in some ADCs and as sparkling in others. (Sparkling is a dynamic failure mode discussed in Section 12.4.3.)

Exercises

12.7. A linear histogram test was performed on an unsigned binary 4-bit ADC resulting in the following distribution of code hits beginning with code 0

4, 5, 5, 7, 8, 7, 9, 5, 6, 3, 6, 7, 9, 6, 5, 9

Determine the endpoint DNL curve for this ADC.

Ans. DNL for code 1 to 14: -0.2045, -0.2045, 0.1136, 0.2727, 0.1136, 0.4318, -0.2045, -0.0455, -0.5227, -0.0455, 0.1136, 0.4318, -0.0455, -0.2045.

12.8. For the code distribution described in Exercise 12.7, determine the endpoint INL curve for this 4-bit ADC.

Ans. INL at code edge 1 to 15: 0, -0.2045, -0.4091, -0.2955, -0.0227, 0.0909, 0.5227, 0.3182, 0.2727, -0.2500, -0.2955, -0.1818, 0.2500, 0.2045, 0.

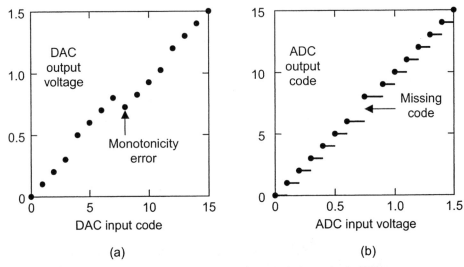

Figure 12.19. Monotonicity errors in DACs and missing codes in ADCs.

Unlike DACs, ADCs are often tested for missing codes. A missing code is one whose voltage width is zero. This means that the missing code can never be hit, regardless of the ADC's input voltage. A missing code appears as a missing step on an ADC transfer curve, as illustrated in Figure 12.19. Since DACs always produce a voltage for each input code, DACs cannot have missing codes. Although a true missing code is one that has zero width, missing codes are often defined as any code having a code width smaller than some specified value, such as 1/10 LSB. Technically, a code having a width of 1/10 LSB is not missing, but the chances of it being hit are low enough that it is considered to be missing from the ADC transfer curve.

12.4 Dynamic ADC Tests

12.4.1 Conversion Time, Recovery Time, and Sampling Frequency

DACs have many dynamic tests such as settling time, rise and fall time, overshoot and undershoot. ADCs do not exhibit these same features, since they do not have an analog output. Instead, an ADC may have any or all the following timing specifications: maximum sampling frequency, maximum conversion time, and minimum recovery time. There are many ways to design ADCs and ADC digital interfaces. Let us look at a few of the common interfacing strategies.

One common interface scheme is shown in Figure 12.20. The ADC begins a conversion cycle when the CONVERT signal is asserted high. After the conversion cycle is completed, the ADC asserts a DATA_READY signal that indicates the conversion is complete. Then the data are read from the ADC using a READ signal.

Maximum conversion time is the maximum amount of time it takes an ADC to produce a digital output after the CONVERT signal is asserted. The ADC is guaranteed to produce a valid output within the maximum conversion time. It is tempting to say that an ADC's maximum sampling frequency is simply the inverse of the maximum conversion time. In many cases this is

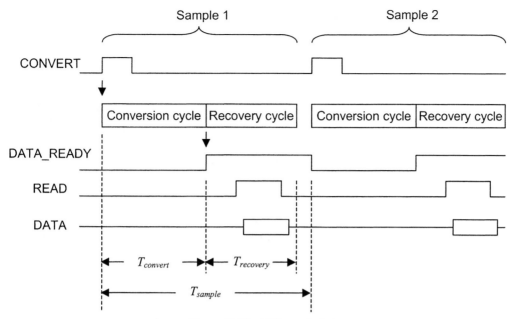

Figure 12.20. ADC sample timing.

true. Some ADCs require a minimum recovery time, which is the minimum amount of time the system must wait before asserting the next CONVERT signal. The maximum sampling frequency is therefore given by the equation

$$F_{max} = \frac{1}{T_{convert} + T_{recovery}} \quad (12.23)$$

We typically test $T_{convert}$ by measuring the period of time from the CONVERT signal's active edge to the DATA_READY signal's active edge. We have to verify that the $T_{convert}$ time is less than or equal to the maximum conversion time specification. For this measurement, we can use a time measurement system (TMS) instrument, or we can sometimes use the tester's digital pattern compare function if we can tolerate a less accurate pass/fail test. We can verify the F_{max} specification (and thus the $T_{recovery}$ specification) by simply operating the converter at its maximum sampling rate, F_{max}, and verifying that it passes all its dynamic performance specifications at this frequency.

In many ADC designs, the CONVERT signal is generated automatically after the ADC output data is read, as shown in Figure 12.21. This type of converter requires no externally supplied CONVERT signal. The first sample read from the ADC must therefore be discarded, since no conversion is performed until after the first READ pulse initiates the first conversion cycle.

Sometimes ADCs simply perform continuous conversions at a constant sampling rate. The CONVERT signal is generated at a fixed frequency derived from the device master clock. This architecture is very common in ADC channels such as those in a cellular telephone voice band interface or multimedia audio device. The continuous conversions can usually be disabled by a register bit or other control mechanism to minimize power consumption when conversions are not needed. These devices sometimes generate a DATA_READY signal that must be used to

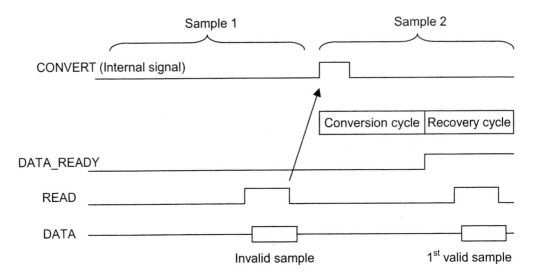

Figure 12.21. ADC conversion cycles with internally generated CONVERT signal.

synchronize the tester with an asynchronous data stream. DUT-defined timing can be a difficult situation to deal with, since ATE testers are not designed to operate in a slave mode with the DUT driving digital timing.

Clearly, there are many ways to design ADCs. The test engineer has to deal with many different permutations of interfacing possibilities, each with its own testing requirements.

12.4.2 Aperture Jitter

In Chapter 6, "Sampling Theory," we saw how sampling jitter can introduce noise in a digitized signal. Typically, aperture jitter is guaranteed by acceptable signal-to-noise ratio (SNR) performance. It may or may not be tested in production, depending on the required sampling rate of the ADC. Very high-frequency ADCs typically must be tested for aperture jitter in production.

12.4.3 Sparkling

Sparkling is a phenomenon that happens most often in high-speed flash converters, such as those described in Section 12.5.3, due to digital timing race conditions. It is the tendency for an ADC to occasionally produce a conversion that has a larger than expected offset from the expected value. We can think of a sparkle sample as one that is a statistical outlier from the Gaussian distribution in Figure 12.7. Sparkling shows up in a time-domain plot as sudden variations from the expected values. It got its name from early flash ADC applications, in which the sample outliers produced white sparkles on a video display. Sparkling is specified as a maximum acceptable deviation from the expected conversion result. For example, we might see a specification that states sparkling will be less than 2 LSBs, meaning that we will never see a sample that is more than 2 LSBs from the expected value (excluding gain and offset errors, of course). Sparkling should not be confused with noise-induced errors such as those illustrated in Figure 12.12.

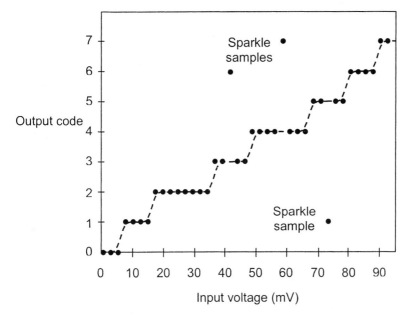

Figure 12.22. Sparkling in a linear ramp histogram sample set.

Test methodologies for sparkling vary, mainly in the choice of input signal. We might look for sparkling in our ramp histogram raw data, such as that shown in Figure 12.22. We might also apply a very high-frequency sine wave to the ADC and look for time-domain spikes in the collected samples.

Since it is a random digital failure process, sparkling often produces intermittent test results. Sparkling is generally caused by a weakness in the ADC design that must be eliminated through good design margin rather than being screened out by exhaustive testing. Nevertheless, ADC sparkling tests are often added to a test program as a quick sanity check, making use of samples collected for one of the required parametric tests.

12.5 ADC ARCHITECTURES

12.5.1 Successive Approximation Architectures

Many ADCs are designed using a successive approximation architecture, in which a DAC output is adjusted with a binary search algorithm until it is substantially equal to the ADC input voltage (Figure 12.23). The comparison between the input voltage and the DAC's binary search voltage is performed using an analog comparator. Successive approximation register (SAR) logic controls the binary search process, moving the DAC value up or down depending on the result of the comparison. Once the binary search process is complete, the SAR value (i.e., the DAC's input code) represents the ADC's conversion result. In many ADC designs, the analog input is "frozen" by a sample-and-hold amplifier so that it does not change while the successive approximation search is in progress. This allows the ADC to digitize AC signals as well as DC signals.

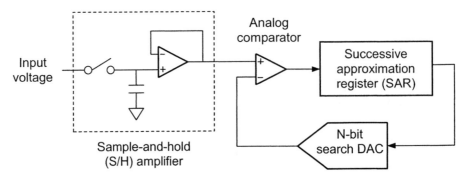

Figure 12.23. *N*-bit successive approximation ADC block diagram.

Successive approximation ADCs can be designed with virtually any type of DAC, including binary-weighted, resistive divider, pulse-width modulated, and hybrid architectures. Therefore, successive approximation ADCs may suffer from all the same nonideal performance problems that plague DACs. For instance, if the search DAC exhibits poor INL or DNL, then the ADC will have the same problem, since the successive approximation voltages are nonideal. In addition to the DAC's weaknesses, the S/H amplifier and the analog comparator may have poor linearity, hysteresis errors, poor power supply rejection ratio, etc. Also, the S/H amplifier may not slew from one voltage level to the next quickly enough, or it may exhibit voltage droop while the successive approximation process is underway. A successive approximation ADC's performance is limited by the aggregate of all these potential problems. Thus ADCs are typically more difficult to design and more difficult to test than their DAC counterparts.

12.5.2 Integrating ADCs (Dual-Slope and Single-Slope)

If a successive approximation ADC is analogous to a binary search, then a dual-slope ADC is analogous to a step search. A dual-slope ADC is much simpler but much slower than a successive approximation ADC in the same way that a step search is much slower than a binary search. Instead of a search DAC, it uses a simple integrator to ramp upward for a fixed amount of time, $T_{integration}$, starting from the time it crosses a fixed threshold voltage, as illustrated in Figure 12.24.

The slope of integration is directly proportional to the analog input voltage. Therefore, the larger the input voltage, the higher the integration voltage will be at the end of the fixed time period. Then the integrator is ramped downward at a fixed slope until it reaches the threshold voltage again. The time it takes to discharge is directly proportional to the integrator's peak voltage, which in turn is proportional to the ADC input voltage. The time period T_{count} is measured by a digital counter, whose output therefore represents the ADC conversion result. Because the integrator ramps up and then down, this type of converter is called a *dual-slope ADC*.

Single-slope ADCs work in a similar manner, but only count the time it takes the integrator output to ramp from an initial voltage to a threshold voltage. The integrator only ramps in one direction. Single-slope ADCs are simpler in nature than dual-slope ADCs, but they typically suffer from worse offset errors. Dual-slope ADCs are also more immune to linearity errors in the integrator because the linearity errors in the upward ramp cancel the linearity errors in the downward ramp.

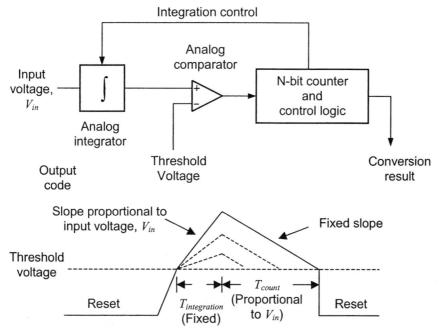

Figure 12.24. *N*-bit dual-slope ADC.

The conversion time of a single- or dual-slope ADC is typically quite long, perhaps 100 ms or more. Therefore, all-codes testing would be prohibitively expensive for production testing of most integrating ADCs. By their nature, integrating ADCs have excellent DNL characteristics, since each code width is dependent on a smoothly ramping analog integrator rather than a binary-weighted sum of components such as capacitors or resistors. However, integrating ADCs may be susceptible to INL errors. The INL curve is dominated by the linearity of the comparator and the linearity of the integrator's ramp, both of which tend to have a simple bend rather than a complex shape, as illustrated in Figure 12.25. Therefore, INL is generally guaranteed in production by simply measuring five points on the curve and comparing them to a straight reference line (best-fit endpoint, etc.).

Because of the long conversion times, integrating ADCs may not be a very good design choice from a testability standpoint. However, from a circuit cost and complexity standpoint, the dual-slope design is highly desirable. For this reason, integrating ADCs are often used in applications that do not require fast conversion time but do require low cost and minimal circuit area.

12.5.3 Flash ADCs

Flash ADCs are somewhat analogous to resistive divider DACs. A flash conversion is a brute force means of comparing the input signal against all possible decision levels, simultaneously. This requires $2^N - 1$ comparators for an *N*-bit ADC as depicted in Figure 12.26. Digital decode logic determines which of the comparators producing a logic one has the highest threshold voltage. The number of the comparator represents the ADC output code. Like the resistive

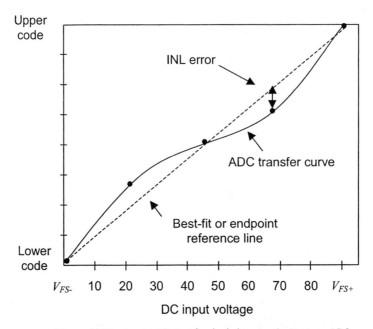

Figure 12.25. 5-point INL test for dual-slope or single-slope ADC.

divider DAC, the flash converter must be tested for all codes, since any resistor or any comparator may be defective.

The flash ADC is much faster than a successive approximation ADC because the decision levels are compared all at once. No S/H amplifier is required for a flash ADC because there is no need to hold the input constant. Since a separate comparator is required for each decision level, the flash ADC becomes prohibitively expensive as resolution increases beyond a few bits. The flash ADC architecture is mostly used in very high-frequency applications that can tolerate the high silicon area required by the many comparators. However, multiple flash ADCs can also be used to construct a multipass successive approximation architecture called a *semiflash ADC*.

12.5.4 Semiflash ADCs

Semiflash ADCs are somewhat analogous to segmented DACs. A semiflash converter is constructed from two or more flash converters to produce a higher-resolution analog-to-digital conversion. The semiflash converter provides a compromise between the high conversion rates possible with flash converters and the lower silicon area of successive approximation ADCs.

A two-stage semiflash converter is shown in Figure 12.27. The first flash ADC digitizes the input level with a limited resolution of N_1 bits. Its N_1-bit conversion result is then fed into an N_1-bit DAC. The difference (i.e. quantization error) between the input voltage and the DAC output is then amplified and digitized by a second flash ADC with N_2 bits of resolution. The most significant N_1 bits of the semiflash converter output are available from the coarse ADC, while the N_2 least significant bits are available from the fine ADC. The composite ADC resolution is therefore N_1+N_2 bits.

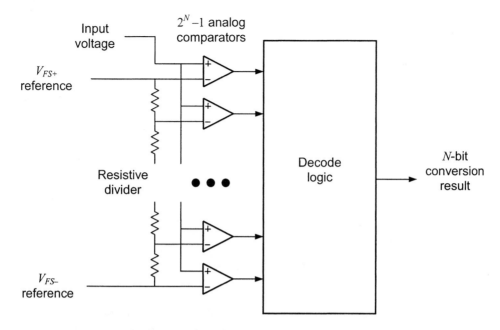

Figure 12.26. *N*-bit flash ADC block diagram.

A semiflash converter performs two very fast conversions with a slower difference-and-gain step in between. For this reason, semiflash converters are generally slower than flash converters but faster than successive approximation converters. Like flash converters, semiflash converters can suffer from sparkling.

12.5.5 PDM (Sigma-Delta) ADCs

Sigma-delta ADCs are similar to sigma-delta DACs in terms of their operating theory. Sigma-delta analog-to-digital converters, such as that shown in Figure 12.28, use a crude ADC (typically an analog comparator) combined with a noise-shaping process to produce an oversampled pulse density modulated (PDM) data stream. This data stream is then digitally filtered and decimated to produce high-resolution ADC samples.

The high-resolution and excellent linearity of sigma-delta ADCs make them ideal for audio and modulated data applications like modems, PC sound cards, and cellular telephones. Sigma-delta DACs and ADCs are well beyond the scope of this book, though many books have been written on the subject.[4,5]

Sigma-delta ADCs, like their DAC counterparts, are typically used to digitize continuous signals in sampled channels. Tests such as INL and DNL are not well suited for sigma-delta converters. Instead, channel tests like gain, offset, signal-to-noise ratio, idle channel noise, etc., are commonly specified.

Like sigma-delta DACs, sigma-delta ADCs may produce self-tones when their inputs are set to certain DC levels. Self-tones appear as spikes in the frequency spectrum of an ADC output.

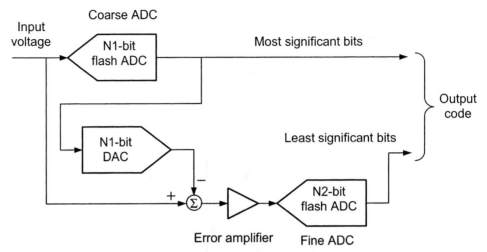

Figure 12.27. Semiflash ADC block diagram.

Unfortunately, the self-tones do not occur at predictable frequencies as they do in sigma-delta DACs. Instead, the DC offset of the ADC and other circuit variations will make self-tones appear at unpredictable frequencies, and only with certain input voltages. This makes worst-case self-tone testing very difficult. A thorough self-tone test would require a sweep through all possible DC inputs to find worst-case test conditions for each device. Therefore, maximum self-tone amplitudes, when they are tested in production at all, can only be tested at a limited set of DC input levels.

Because self-tones are so noticeable in audio applications, many sigma-delta ADC designs use a dithering source to eliminate self-tones altogether. The dithering source adds random noise to the ADC input to prevent the possibility of DC input levels. Of course, dithering degrades the signal-to-noise ratio and idle channel noise performance of the ADC.

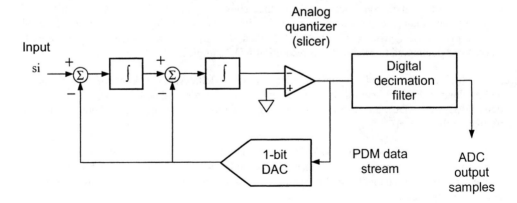

Figure 12.28. Second order sigma-delta (PDM) ADC.

12.6 Tests for Common ADC Applications

12.6.1 DC Measurements

Like DACs, ADCs can be used for a variety of purposes. The ADC's application often determines its required parameters. For example, an ADC may be used to measure absolute voltage levels, as in a DC voltmeter or battery monitor. In this type of application, we do not usually care about transmission parameters like signal-to-noise ratio. We will typically only need to know the DC gain, DC offset, INL, DNL, and worst-case absolute voltage errors in decision levels, relative to the ideal decision levels. Idle channel noise will sometimes be specified, to ensure that results obtained from the ADC are not unrepeatable due to excessive noise.

Successive approximation ADCs and integrating ADCs are the most common converter type used for DC measurements. Sigma-delta designs are seldom used due to their inherent tendency to produce self-tones with certain DC inputs.

12.6.2 Audio Digitization

Audio digitization is a very common application for ADCs, especially high-resolution ADCs. When the resolution exceeds 12 or 13 bits, it becomes very expensive to perform transfer curve tests such as INL and DNL because of the large number of code edges that must be measured. Fortunately, transmission parameters such as frequency response, signal to distortion ratio, idle channel noise, etc., are more meaningful measures of audio digitizer performance. These sampled channel tests are much less time-consuming to measure than INL and DNL, especially when testing ADCs with 16 or more bits of resolution. Sigma delta ADCs have become the most common architecture for audio digitization application.

As previously mentioned, self-tones are a potential source of trouble when sigma-delta ADCs are used in audio digitization applications. Because of the way the human mind processes sound, very low amplitude self-tones are much easier to hear than white noise at equivalent signal levels. It is impractical to test self-tones at every possible DC input level. Self-tones should at least be tested with the analog input tied to ground or $V_{DD}/2$ (or whatever voltage represents the converter's midscale input level). When characterization indicates that a particular ADC design is not prone to self-tone generation, then this test is often eliminated in production.

12.6.3 Data Transmission

Data transmission applications differ from audio applications mainly in terms of the sampling rates and the frequency range of the transmitted signals. Data transmission ADCs, such as those found in modems, hard disk drive read channels, and cellular telephone intermediate frequency (IF) sections, often digitize signals that are well above the audio band. These applications typically require lower-resolution ADCs, but may require much higher sampling rates. Aperture jitter is often a prime concern for these applications, especially if the signal frequency band extends past a few tens of megahertz. Excessive aperture jitter can introduce apparent noise in the digitized signal, ruining the performance of the ADC.

Signal-to-noise ratio, group delay distortion, and other transmission parameters are often specified in data transmission applications. Also, data transmission specifications such as error vector magnitude (EVM), phase trajectory error (PTE), and bit error rate (BER) may also need to be tested. These parameters are so numerous that we cannot possibly cover them in this book. The test engineer will have to learn about these and other application-specific testing requirements by studying the relevant standards documents. ATE vendors can also be a tremendous source of expertise when learning about new testing requirements and methodologies.

Most ADC architectures are well suited for low-frequency data transmission applications (with the exception of integrating converters). High-frequency applications may require fast successive approximation ADCs, semiflash ADCs, or even full-flash ADCs, depending on the required sampling rates.

12.6.4 Video Digitization

NTSC video signal digitization is another key application for high-speed ADCs. These applications require the faster ADC types (flash, semiflash, or pipelined successive approximation ADCs). The test list for these types of converters usually includes transmission parameters as well as differential gain and differential phase measurements. Like other high-speed applications; aperture jitter is a key performance specification for video digitization applications. Sparkling is particularly noticeable in video applications; so this potential weakness should be thoroughly characterized and/or tested in production.

12.7 SUMMARY

ADC testing is very closely related to DAC testing. Many of the DC and intrinsic tests defined in this chapter are very similar to those performed on DACs. The most important difference is that the ADC code edge transfer curve is harder and much more time consuming to measure than the DAC transfer curve. However, once the many-to-one statistical mapping of an ADC has been converted to a one-to-one code edge transfer curve, the DC and transfer curve tests are very similar in nature to those encountered in DAC testing. This chapter by no means represents an exhaustive list of all possible ADC types and testing methodologies. There are a seemingly endless variety of ADC architectures and methods for defining their performance. Hopefully, this chapter will provide a solid starting point for the beginning test engineer.

Problems

12.1. If V is normally distributed with zero mean and a standard deviation of 50 mV, find $P(V < 40$ mV$)$. Repeat for $P(V > 10$ mV$)$. Repeat for $P(-10$ mV $< V < 40$ mV$)$. Use Table 12.1 and use linear interpolation between values in the table.

12.2. If V is normally distributed with mean 10 mV and standard deviation 50 mV, find $P(V < 40$ mV$)$. Repeat for $P(V > 10$ mV$)$. Repeat for $P(-10$ mV $< V < 40$ mV$)$. Use Table 12.1 and use linear interpolation between values in the table.

12.3. If V is normally distributed with zero mean and standard deviation 200 mV, what is the value of ΔV such that $P(V < \Delta V)=0.6$. Use Table 12.1 and use linear interpolation between values in the table.

12.4. An ADC input is set to 3.340 V DC. The noise of the ADC and DC signal source is characterized to be 15 mV RMS and is assumed to be perfectly Gaussian. The transition between code 324 and 325 occurs at 3.350 V DC for this particular ADC, therefore the value 324 is the expected output from the ADC. What is the probability that the ADC will produce code 325 instead of 324? If we collected 400 samples from the output of the ADC, how many would we expect to be code 324 and how many would be code 325?

12.5. An ADC input is set to 1.000 V DC. The transition between code 65 and 66 occurs at 1.025 V DC for this particular ADC. If 200 samples of the ADC output are collected and 176 of them are code 65 and the remaining code 66, what is the RMS value of the noise at the input of this particular ADC?

12.6. An ADC input is set to 2.000 V DC. The noise of the ADC and DC signal source is characterized to be 10 mV RMS and is assumed to be perfectly Gaussian. The transition between code 115 and 116 occurs at 1.990 V DC and the transition between code 116 and 117 occurs at 2.005 V DC for this particular ADC. If 500 samples of the ADC output are collected, how many do we expect to be code 115, code 116 and code 117?

12.7. A linear histogram test was performed on an unsigned binary 3-bit ADC resulting in the following distribution of code hits beginning with code 0

5, 6, 4, 6, 7, 7, 5, 6

A binary search was performed on the first transition between codes 0 and 1 and found the code edge to be at 10 mV. A second binary search was performed and found the code edge between codes 6 and 7 to be 1.25 V. What is the average LSB size for this 3-bit ADC? Determine the width of each code, in volts. Also, determine the location of the code edges. Plot the transfer curve for this ADC.

12.8. A linear histogram test was performed on a two's complementary 4-bit ADC resulting in the following distribution of code hits beginning with code -8

12, 15, 13, 12, 10, 12, 12, 14, 14, 13, 15, 19, 16, 14, 20, 19

A binary search was performed on the first transition between codes -8 and -7 and found the code edge to be at 75 mV. A second binary search was performed and found the code edge between codes 6 and 7 to be 4.56 V. What is the average LSB size for this 4-bit ADC? Determine the width of each code, in volts. Also, determine the location of the code edges. Plot the transfer curve for this ADC.

12.9. A linear histogram test was performed on an unsigned binary 3-bit ADC resulting in the following distribution of code hits beginning with code 0

6, 6, 5, 6, 4, 6, 5, 6

A binary search was performed on the first transition between codes 0 and 1 and found the code edge to be at 32 mV. A second binary search was performed and found the code edge between codes 6 and 7 to be 3.125 V. What is the average LSB size for this 3-bit ADC? What is the measurement accuracy of this test, in volts?

12.10. A sinusoidal histogram test was performed on an unsigned binary 4-bit ADC resulting in the following distribution of code hits beginning with code 0

76, 20, 19, 19, 19, 18, 19, 18, 18, 19, 19, 19, 24, 25, 32, 62

A binary search was performed on the first transition between codes 0 and 1 and found the code edge to be at 14 mV. A second binary search was performed and found the code edge between codes 6 and 7 to be 1.725 V. What is the average LSB size for this 4-bit

ADC? Determine the width of each code, in volts. Also, determine the location of the code edges. Plot the transfer curve for this ADC.

12.11. A linear histogram test was performed on a two's complementary 4-bit ADC resulting in the following distribution of code hits beginning with code -8

20, 15, 14, 12, 11, 12, 12, 14, 14, 13, 15, 16, 16, 14, 20, 23

Determine the endpoint DNL and INL curves for this ADC. Compare these results to those obtained with a best-fit reference line.

12.12. Determine the endpoint DNL and INL curves for the histogram data provided in Problem 12.8. Compare these results to those obtained with a best-fit reference line.

12.13. Determine the endpoint DNL and INL curves for the histogram data provided in Problem 12.9. Compare these results to those obtained with a best-fit reference line.

12.14. Categorize the ADC designs described in Section 12.5 into low-, medium- or high-speed architectures.

12.15. Derive the sinewave histogram normalization equations, Eq. (12.15) – (12.19).

References

1. Mark J. Kiemele, Stephen R. Schmidt, Ronald J. Berdine, *Basic Statistics, Tools for Continuous Improvement*, Fourth Edition, Air Academy Press, 1155 Kelly Johnson Blvd., Suite 105, Colorado Springs, CO 80920, 1997, ISBN: 1880156067, pp. 9-71

2. Matthew Mahoney, *Tutorial DSP-Based Testing of Analog and Mixed-Signal Circuits*, The Computer Society of the IEEE, 1730 Massachusetts Avenue N.W., Washington, D.C. 20036-1903, 1987, ISBN: 0818607858, pp. 147-54

3. Reference 2, p.137.

4. James C. Candy, Gabor C. Temes, *Oversampling Delta-Sigma Data Converters: Theory, Design, and Simulation*, IEEE Press, New York, NY, January 1992, ISBN: 0879422858

5. Steven R. Norsworthy, Richard Schreier, Gabor C. Temes, *Delta-Sigma Data Converters: Theory, Design, and Simulation*, IEEE Press, New York, NY, November 1996, ISBN: 0780310454

CHAPTER 13

DIB Design

13.1 DIB BASICS

13.1.1 Purpose of a Device Interface Board

On any given day, a general-purpose ATE tester may be required to test a wide variety of device types. A mixed-signal tester may test video converters in the morning, modem chips in the afternoon, and standalone ADCs in the evening. Obviously, the electrical testing requirements of each type of device are unique to that device. Also, the mechanical requirements of each device are unique. The tester's various electrical resources must be connected to each of the DUT's pins, regardless of the mechanical configuration of the DUT package. For example, an 8-bit DAC might be available in several different packages such as the small outline IC (SOIC), quad flat pack (QFP), and leadless chip carrier (LCC). These package types are illustrated in Figure 13.1. Also, the tester needs to be connected to the bare die during wafer probing. Clearly, a general-purpose tester cannot be expected to provide all electrical resources and mechanical fixtures to test any arbitrary device type in any package.

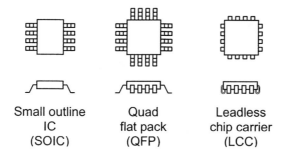

Figure 13.1. Common IC package types.

The device interface board (DIB) provides a means of customizing the general-purpose tester to specific DUTs and families of DUTs. The DIB serves two main purposes. First, it gives the test engineer a place to mount DUT-specific circuitry that is not available in the ATE tester. This circuitry can be placed near the DUT to enhance electrical performance during critical tests. Second, the DIB provides a temporary electrical interface to each DUT during electrical performance testing. When testing packaged devices, the temporary connection is achieved using a hand-test socket or a handler-specific mechanism called a *contactor assembly*. Thus a DIB is often intended for use with only one type of DUT mounted in a particular mechanical package.

When testing bare die on a wafer, the temporary DUT connection is made using the tiny probes of a probe card. A probe interface board (PIB) is usually required to interface the probe card to the tester's resources. Together, the PIB and probe card serve the same purpose as a DIB and contactor assembly. If the same device is offered in three different packages, then three different DIBs and a PIB may be required. Clearly, electromechanical hardware design represents a large portion of the test engineering task.

DUTs that are purely digital in nature typically require a very simple DIB that simply provides point-to-point connectivity between the DUT pins and the tester's power supplies and digital pin card electronics. Analog and mixed-signal DUTs usually require much more elaborate DIBs. A mixed-signal DIB often contains a variety of active and passive circuits that must be connected to or disconnected from various DUT pins as the test program progresses. For example, the harmonic distortion of an analog output may be specified with a 1-kΩ load connected between the analog output and ground. The same output may also have an off-state output leakage specification. The 1-kΩ load resistor must be disconnected during the leakage test to prevent current from leaking through the resistor to ground. Using electromechanical relays, the DIB can modify the DUTs electrical environment under test program control. The relays act as electrical switches that can be turned on and off by commands in the test program (Figure 13.2).

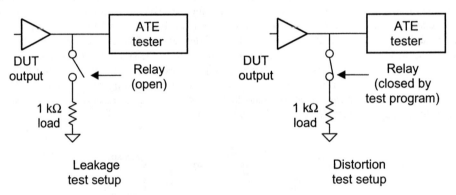

Figure 13.2. Electromechanical relays modify the DUT's electrical environment.

13.1.2 DIB Configurations

Thus far, we have talked about DIBs as if they are the same for each type of tester. In reality, the mechanical details of interface hardware vary widely from one tester type to another. Mechanical configurations may even vary within the same company, even when the various test development organizations all use the exact same tester. Figure 13.3 shows three possible interfacing schemes. The first is the simple DIB interfacing scheme. In this type of configuration, the DIB and contactor assembly form the entire interface between the tester and the DUT.

The second scheme shows a socket adapter/swap block stackup that is often used to test families of similar devices. In this configuration, the socket adapter (also called a *family board* or *mother board*) contains the support circuitry required to test a family of devices, such as video ADCs. The swap block (daughter board) provides the customization needed to test a particular

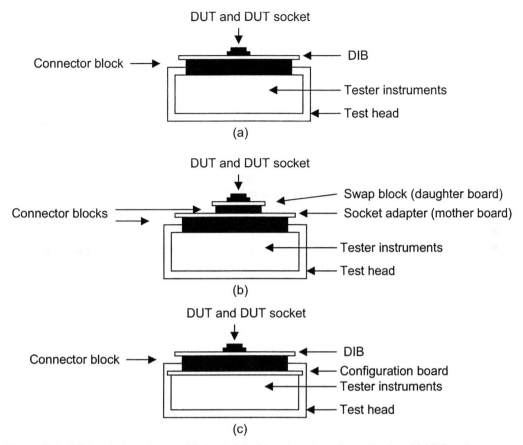

Figure 13.3. DIB interfacing schemes: (a) simple DIB, (b) socket adapter/swap block, and (c) DIB and configuration board.

video ADC in a particular type of package. For example, one swap block might be compatible with 64-pin LCCs while another is compatible with 64-pin QFPs. This scheme allows a relatively complex and expensive socket adapter to be reused for a family of similar DUTs. Of course, this scheme is not limited to only two layers of interfacing, but the attachment of a fully custom swap block to a semicustom socket adapter is the most common multilayer configuration.

A third possibility is the configuration board/DIB scheme. The configuration board is similar to the family board, except that it is located inside the test head. It is generally only intended to customize the tester for a particular organization's needs. Unlike the family board, which may be changed daily, the configuration board is usually left in place for all device types. All DUT-specific circuitry is located on the DIB.

The terminology used to describe each layer of interface hardware varies widely from one ATE vendor to another. The terminology even varies from one vendor's tester to another. Throughout this book, and for the remainder of this chapter, we will use the term DIB to refer to all the layers of custom and semicustom interfacing hardware, including mother boards, daughter boards, swap blocks, and DIBs. In other words, we will treat the subject of DIB design as if all testers used the configuration illustrated in Figure 13.3(a).

13.1.3 Importance of Good DIB Design

One of the major causes of long test program development time is poor mixed-signal DIB design and printed circuit board layout. A DIB schematic shows only an idealized view of the DIB. Resistors are shown as ideal resistances, capacitors as ideal capacitances, and traces as perfect connections with no parasitic resistance, inductance, or capacitance. In reality, the exact mechanical layout of the components and traces on the DIB may make the difference between failing test results and passing results.

The performance of analog and mixed-signal devices is highly dependent on the quality of the surrounding circuit design. It is important to be able to distinguish between legitimate DUT failures and failures caused by poor design of the DIB. Consequently, a DIB should represent the best-case environment for the DUT, rather than a worst-case environment. Unfortunately, it is difficult to provide the DUT with a perfect environment using a general-purpose tester with bulky electromechanical interconnections. For example, the pins of the DUT socket will typically add more inductance and capacitance to the DUT's environment than the DUT will encounter when it is soldered directly onto a printed circuit board in the end application. Nevertheless, the test engineer must try to design a DIB that does not present the DUT with unfair electrical handicaps.

There are so many performance considerations in mixed-signal DIB design that many people consider it a mystical black art. Actually, DIB design is more of a "light gray" art, since many of the major considerations are fairly well understood. In this chapter, we will examine some of the main considerations in mixed-signal DIB design such as power supply and grounding connections, shielding schemes, parasitic circuit elements, component selection, common DIB circuits, and common DIB mistakes.

First, let us look at one of the DIB's most important electrical components: the printed circuit board. Although the printed circuit board is often thought to be nothing more than a mechanical frame onto which the circuit components are fastened together, its physical construction is absolutely key to the performance of many mixed-signal DUTs.

13.2 PRINTED CIRCUIT BOARDS

13.2.1 Prototype DIBs versus PCB DIBs

One of the common debates in test engineering is the choice between hand-wired prototype DIBs versus printed circuit board (PCB) DIBs. Hand-wired DIBs can be quickly constructed from prefabricated blank prototype boards. The alternate approach is to produce a production-worthy custom PCB version of the DIB without first building a hand-wired prototype. Each approach has advantages and disadvantages.

The hand-wired approach results in rapid turn-around at relatively low production cost. However, the resulting board is typically not very production worthy, since the loose wires are easily broken. Also, hand-wired DIBs may not give the same high-quality electrical performance that can be achieved using PCB-based DIBs. When multiple DIBs are required, then the PCB approach is usually the superior solution. PCB DIBs are easily manufactured in quantity, they are mechanically robust during debug and production, they provide superior electrical performance, and they provide good consistency (i.e., correlation) from one board to another. Correlation between hand-wired DIBs can be very problematic, since each is electrically unique depending on the exact length and physical layout of the wires on each board. At very high frequencies, hand-wired boards are often useless, since they can produce incorrect readings due to their inferior electrical characteristics.

The downside to PCB-based DIBs is primarily longer cycle time and higher initial cost. It may take several weeks or even months to get a PCB DIB designed, laid out, and fabricated. Also, PCB DIBs are more expensive than hand-wired DIBs, at least in small quantities. However, assuming the test engineer is skilled enough to produce a useable PCB DIB design on the first pass, the PCB DIB is actually a less expensive approach. After all, a PCB DIB will eventually be required for a robust production solution anyway; so why should the company spend money to have the first few boards hand-wired?

Rapid turnaround is a problem that can be solved by good methodology. PCB-based DIBs can be designed, laid out, and fabricated in a matter of a week or two if the test engineer is skilled in the proper use of computer-aided design (CAD) tools. To achieve a rapid turn-around with minimal errors, a CAD-based design, layout, and fabrication approach must be established between the test organization and the PCB layout organization (which may either be an external vendor or an internal support group).

13.2.2 PCB CAD Tools

A streamlined PCB design and layout process requires the use of netlist-based CAD tools. A netlist is a database describing each interconnection in the circuit. For example, one line of a typical netlist file might tell the PCB layout tool that circuit node 55 interconnects resistor R1 pin 1, inductor L1 pin 2, and amplifier U37 pin 15. In addition to the point-to-point interconnection information, the netlist also includes such information as the footprint, or shape, of each component in the circuit. A footprint represents the mechanical specification of the component's package. Information such as (X,Y) pin locations, pad sizes, hole sizes, and package outline shapes to be printed on the finished PCB are included in the footprint description for each type of component.

Using a netlist-compatible schematic capture tool, the test engineer draws the circuit schematic on a computer workstation or PC. Then the schematic database (including the netlist) is transferred to the PCB designer for use in the DIB layout process. Once the netlist has been extracted from the database, the PCB designer begins laying out the DIB from a standard DIB template. The DIB template database represents a head start DIB design, which includes the shape of the DIB and its standard mechanical mounting holes as well as many preplaced standard components, such as tester connectors.

The netlist directs the PCB layout software to import all the required DIB components from a standard parts library. The PCB designer then places these components and connects them as shown in the schematic. The netlist prevents errors in point-to-point interconnections by refusing to let the layout designer place traces where they do not belong. The netlist also guarantees that none of the desired connections are mistakenly omitted. Once the DIB layout is completed, each layer of the design is plotted onto transparent film for use in PCB fabrication. These plots are commonly known as *Gerbers*, or *Gerber plots*, named after the company that pioneered some of the early plotting equipment (Gerber Scientific). Figure 13.4 illustrates the CAD-based DIB design, layout, and fabrication process.

488   *An Introduction to Mixed-Signal IC Test and Measurement*

13.2.3 Multilayer PCBs

Low-cost PCBs can be designed and fabricated using one or two layers of copper trace, as shown in Figure 13.5. Traces on opposite sides of a double-layer PCB can be connected using a copper plated through-hole called a *via*. Double-layer PCB fabrication starts with a blank PCB consisting of a sheet of insulator (e.g., fiberglass) plated with a thin layer of copper on both sides. The component lead holes and vias are drilled first. Then the holes are plated with copper to form the layer-to-layer interconnects. Finally, the traces are printed and etched using a photolithographic process similar to that used in IC fabrication.

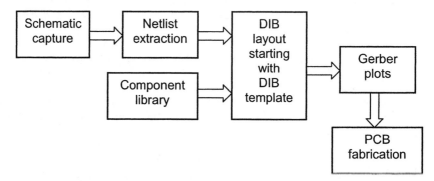

Figure 13.4. CAD-based DIB design and fabrication process.

Multilayer PCBs having four or more layers can be formed by stacking multiple two-layer boards together, as shown in Figure 13.6. The internal, or buried, layers are first printed and etched. Then the layers are all stacked and pressed together under heat to form a single board. Finally, the vias are drilled and plated and the outer layers are etched to form the finished PCB.

Most mixed-signal DIBs are formed using 6- to 10-layer PCBs. The arrangement of layers in a PCB is known as the *stackup*. The stackup of a DIB may vary from one type of DUT to another, but some general guidelines are commonly followed. The internal layers are typically

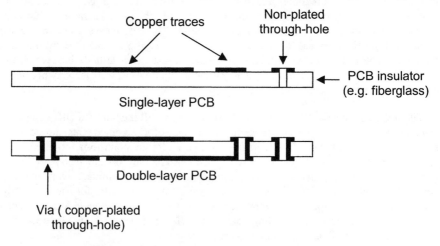

Figure 13.5. Single- and double-layer PCBs.

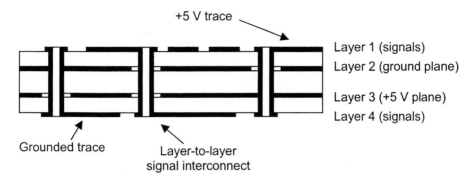

Figure 13.6. Multilayer PCBs.

used for ground and power distribution, as well as for various noncritical signal traces. The outer layers are usually reserved for critical signals or those signal traces that might need to be modified after the DIB has been fabricated. External traces are also easier to access for observation during the test program debug process. If desired, test point vias can be added to a DIB to access buried signals for debugging purposes.

In addition to the trace layers and insulator layers in a PCB, the outer layers are usually coated with a material called a *solder mask*. This thin, nonconductive layer keeps solder from flowing all over the traces when the DIB components are soldered onto the PCB. The soldermask helps to prevent unwanted solder shorts between adjacent traces.

A silkscreened pattern may also be printed on the outer layers of the PCB. The silkscreened patterns show the outline and reference numbers for all the DIB components, such as resistors, capacitors, relays, and connectors. The silkscreened patterns are quite useful during the DIB component assembly process, and they are equally useful during the test program debugging process.

13.2.4 PCB Materials

Printed circuit boards can be constructed using a variety of materials. The most common trace material is copper, due to its excellent electrical conductivity. The most common insulator material is FR4 (fire retardant, type 4) fiberglass. Fiberglass is an inexpensive material that exhibits good electrical properties up to several hundred megahertz. As frequencies approach 1 GHz, more exotic materials such as Teflon®[*] or cynate ester may be needed.

Teflon® exhibits excellent microwave characteristics, including low signal loss and a low dielectric constant. However, it suffers from poor mechanical stiffness. A DIB made exclusively of Teflon® insulator would be too weak to stand up to the force of DUT insertions by a handler. Cynate ester is a material with reasonably good high frequency properties and yet it is stiff enough to withstand the mechanical stress of production testing. A hybrid stackup consisting

---

[*] Teflon® is a registered trademark of DuPont

of sandwiched layers of Teflon® and cynate ester provides a compromise between the good electrical properties of Teflon® and the good mechanical properties of cynate ester.

13.3 DIB TRACES, SHIELDS, AND GUARDS

13.3.1 Trace Parasitics

One of the most important DIB components is the printed circuit board trace. It is easy to think that wires and traces are not components at all, but are instead represented by the connecting lines that appear in a schematic. However, PCB traces (and wires in general) are slightly resistive, slightly inductive, and slightly capacitive in nature.

The nonideal circuit characteristics are known as *parasitic elements*, though they are often simply referred to as *parasitics*. Often, trace parasitics can be ignored, especially when working with low frequencies and low to moderate current levels. Other times, the parasitics will have a significant effect on a circuit's behavior. The test engineer should always be aware of the potential problems that trace parasitics might pose.

Trace resistance on DIBs seldom exceeds a few ohms. Inductance can be anywhere from one or two nanohenrys to several microhenrys. Capacitance can range from one or two picofarads to tens of picofarads. Although these values are very approximate, they can be used as a thumbnail estimate to determine whether the parasitic elements might be large enough to affect the DUT's performance. To estimate trace parasitics with a little more accuracy, we need to review the equations for trace resistance, inductance, and capacitance.

13.3.2 Trace Resistance

The parasitic resistance of a PCB trace is directly proportional to the length of the trace, and inversely proportional to the height and width of the trace. The equation for resistance in a uniform conductive material with a rectangular cross section is

$$R = \frac{L_{TRACE}}{\sigma WT} \tag{13.1}$$

where R = trace resistance, L_{TRACE} = trace length, W = trace width, T = trace thickness, and σ is the conductivity of the trace material.

Most PCB traces are constructed using copper, which has a conductivity of about 5.7×10^7 $(\Omega \text{ m})^{-1}$. The trace thickness is usually about 1 mil, although PCBs can be fabricated with a copper sheet thickness of 3 mils or more if desired. When working with equations such as Eq. (13.1), we will consistently convert all units of length to meters, since electrical units such as resistance, current, and voltage are metric units. Since the mil is an English unit (1 mil = 1/1000 in.), we will convert it to meters before using any of our electrical equations. The conversion factor is

$$1 \text{ mil} = 1/39000 \text{ meter} = 2.56 \times 10^{-5} \text{ meter} \tag{13.2}$$

Example 13.1

Calculate the parasitic resistance of a PCB trace that is 15 in. long, 1 mil thick, and 20 mils wide.

Solution:

First we convert all units of length into meters

$L_{TRACE} = 15 \text{ in.} \times (1 \text{ m} / 39 \text{ in.}) = 0.385 \text{ m}$

$T = 1 \text{ mil} \times (1 \text{ m} / 39000 \text{ mils}) = 2.56 \times 10^{-5} \text{ m}$

$W = 20 \text{ mil} \times (1 \text{ m} / 39000 \text{ mils}) = 5.12 \times 10^{-4} \text{ m}$

Applying Eq. (13.1) to a copper trace with $\sigma = 5.7 \times 10^7$ $(\Omega \text{ m})^{-1}$, we get a total parasitic trace resistance of

$$R = \frac{0.385}{5.7 \times 10^7 \times 5.12 \times 10^{-4} \times 2.56 \times 10^{-5}} = 515 \text{ m}\Omega$$

13.3.3 Trace Inductance

The inductance of a DIB trace depends on the shape and size of the trace, as well as the geometry of the signal path through which the currents flow to and from the load impedance. Figure 13.7 shows a signal source feeding a load impedance through a pair of signal lines. In this example, the current is forced to return to the source through a dedicated current return line. The signal line and the current return line form a loop through which the load current flows. The larger the area of this loop, the higher the inductance of the signal path. This inductance can be modeled as a single inductor in series with the signal source, as shown in Figure 13.8.

A parasitic inductance such as the one in Figure 13.8 is generally an undesirable circuit component. We wish to minimize the effects of parasitic trace inductance on the DUT and DIB circuits. There are a number of ways to reduce this inductance. The first way is to minimize the area enclosed by the load current path. One easy way to do this is to lay a dedicated current

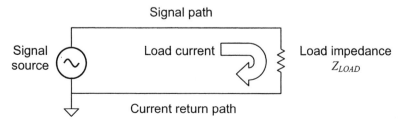

Figure 13.7. Signal source, load impedance, and current path.

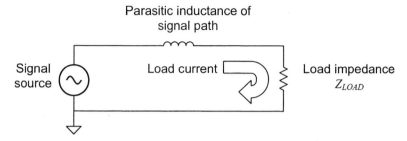

Figure 13.8. Schematic representation of signal path inductance.

return trace beside each signal trace. Of course, if we did this with every signal, we would have a very cluttered PCB layout. An easier way to obtain low inductance is to use one or more solid ground planes as current return paths.

By routing each signal trace over a solid ground plane, the load current can return underneath the trace along a path with very low cross-sectional area. The cross-sectional area can be minimized by placing the trace and ground plane very close together in the PCB layer stackup. Another way to reduce inductance is to make the trace as wide as is practical, since a wide trace over a ground plane has minimal inductance. A thicker trace will also have somewhat less inductance, though PCBs are normally fabricated with trace thicknesses of 1 to 3 mils. We have less control over trace thickness than we have over trace width and layer spacings.

The inductance of a trace over a ground plane (a configuration known as a *stripline*) is dominated by the ratio of the trace-to-ground spacing, D, divided by the trace width, W (Figure 13.9). The parasitic inductance of a wide trace routed over a ground or power plane can be estimated using the equation

$$L_\ell = \mu_o \mu_r \frac{D}{W} \tag{13.3}$$

where L_ℓ = Inductance per unit length (henrys per meter), μ_o = magnetic permeability of free space (400π nH per meter), μ_r = magnetic permeability of the PCB material divided by μ_o, W = trace width, and D = separation between trace and ground plane.

The value of μ_r is very nearly equal to 1.0 in all common PCB materials; so we can drop it from our calculations. The total inductance of the trace is directly proportional to the length of

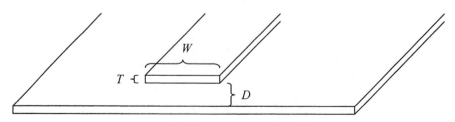

Figure 13.9. Cross section of a long trace over a ground plane (stripline).

the trace

$$L = L_{TRACE} L_\ell \qquad (13.4)$$

where L = total inductance and L_{TRACE} = trace length (meters).

Thus trace inductance increases as trace length increases and also increases as trace width decreases. Therefore, if we want to minimize parasitic inductance in PCB traces, we should make them as wide as possible, as short as possible, and as close to the ground or power plane as possible.

Unfortunately, Eq. (13.3) is only valid for traces in which $W \gg D$. In most PCB designs, the width of the trace is not much larger than the trace-to-ground spacing. In these cases, the magnetic fields between the trace and ground plane are not uniform, making Eq. (13.3) invalid. Figure 13.10 shows a more accurate relationship between the space-to-width ratio and the inductance per meter of a trace over a ground plane.[*] The dotted line represents the inductance

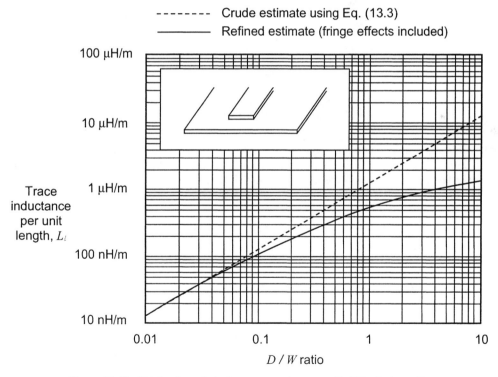

Figure 13.10. Stripline trace inductance per meter versus D / W ratio ($\mu_r = 1$).

---

[*] The graph in Figure 13.10 was derived using a mathematical approximation; so its values should not be taken as absolutely accurate. However, the approximations are adequate for estimating the effects of parasitic elements on DIB circuits. The derivation of this graph and others in this chapter are based on electromagnetic field theory, a subject that is beyond the scope of this book.

per meter as estimated using Eq. (13.3). As we can see, the more accurate estimation converges with the estimations using Eq. (13.3) as the value of W becomes much larger than D.

It should be noted that the inductance of a stripline is approximately the same as the inductance of a trace over a second trace of equal size and shape (Figure 13.11). This is because most of the higher-frequency current returning through a stripline's ground plane returns directly underneath the stripline trace (i.e., the path of least inductance). However, the two-conductor configuration in Figure 13.11 is seldom used in DIB design, since a ground plane permits a much easier means of achieving the same low inductance.

Example 13.2

Calculate the parasitic inductance of a trace having a 16-mil width, running over a ground plane for 6 in. The spacing between the trace and the plane is 8 mils. If this trace is connected in series with a 50-Ω resistor to ground (as in Figure 13.8), what is the 3-dB bandwidth of the low-pass filter formed by the trace inductance and the 50-Ω resistance? Can a 50-MHz sine wave be passed through the trace to the resistor without significant loss of amplitude? What is the phase shift caused by the inductance at 50 MHz?

Solution:

Combining Eqs. (13.3) and (13.4)

$$L = 6 \text{ in.} \times \frac{1 \text{ m}}{39 \text{ in.}} \times \pi \times 400 \frac{\text{nH}}{\text{m}} \times 1.0 \times \frac{8 \text{ mils}}{16 \text{ mils}} = 97 \text{ nH}$$

However, we can see from the graph in Figure 13.10 that the actual inductance per meter at a D/W ratio of 0.5 is lower than that predicted from Eq. (13.3). If we use the more accurate value of inductance per meter from this graph (about 350 nH per meter), then we get a more accurate prediction of the trace inductance. The refined estimation of inductance is

$$L = 6 \text{ in.} \times \frac{1 \text{ m}}{39 \text{ in.}} \times 350 \frac{\text{nH}}{\text{m}} = 54 \text{ nH}$$

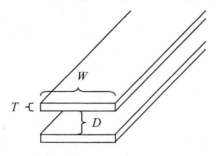

Figure 13.11. Parallel traces on adjacent PCB layers.

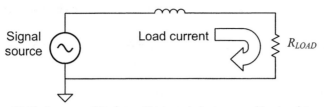

Figure 13.12. Low-pass filter formed by trace inductance and load resistance.

The trace inductance and load resistance form an *RL* low-pass filter, as shown in Figure 13.12. The 3-dB cutoff frequency, F_c, of this *RL* low-pass filter is given by

$$F_C = \frac{R_{LOAD}}{2\pi L} \qquad (13.5)$$

Thus the 3-dB bandwidth of the low-pass filter formed by the trace inductance and load resistor is equal to

$$F_C = \frac{50}{2\pi \times 54 \times 10^{-9}} = 147 \text{ MHz}$$

At a frequency *f*, the trace inductance and load resistance form a voltage divider having a transfer function equal to

$$H(f) = \frac{R_{LOAD}}{R_{LOAD} + Z_L(f)} \qquad (13.6)$$

where Z_L is the complex impedance of the trace inductance.

Substituting $Z_L(f) = L\,j2\pi f$, calculating the magnitude of *H(f)*, and combining the result with Eq. (13.5) gives us the gain of the *RL* low-pass filter at any frequency *f*

$$\text{gain}(f) = |H(f)| = \left[\frac{1}{\sqrt{1+\left(\frac{f}{F_C}\right)^2}}\right] \text{ V/V} \qquad (13.7)$$

At 50 MHz, the gain calculated using Eq. (13.7) is equal to = 0.947 V/V. Therefore, we get an attenuation due to the low-pass nature of the *RL* circuit that attenuates the 50-MHz sine wave by a factor of 0.947 V/V. This attenuation is probably unacceptable, unless we do not mind a 5% error in the amplitude of the signal at the load resistor.

The phase shift of the *RL* low-pass filter is given by

$$\phi(f) = \angle H(f) = -\frac{180}{\pi}\tan^{-1}\left(\frac{f}{F_C}\right) \text{ degrees} \qquad (13.8)$$

Thus, at 50 MHz, the phase shift produced by the trace inductance and resistance is equal to −18.7 degrees. This is a fairly serious phase error. If we wanted to measure phase mismatch or group delay at a frequency near 50 MHz, the parasitic inductance of this example would be completely unacceptable.

If we want to achieve less attenuation and phase shift due to the trace inductance in Example 13.2, we must either shorten the trace, widen the trace, or reduce the spacing between the trace and ground. The spacing between the trace and ground in this example is 8 mils, which is about as thin as we can reliably fabricate a PCB. Rather than trying to fabricate a board with even thinner layer spacing, it is much easier to simply widen the trace.

As we will see in the following sections, widening the trace or reducing the trace-to-ground spacing has the unfortunate side effect of increasing the parasitic capacitance of the trace to ground. The extra capacitance may be just as undesirable as having too much inductance. Therefore, the best solution is to keep traces as short as possible, since this reduces both the trace inductance and trace capacitance to ground.

13.3.4 Trace Capacitance

The capacitance between two parallel traces such as those in Figure 13.11 can be estimated using the standard parallel plate capacitance equation. The parasitic capacitance between two metal plates of area A is given by the equation

$$C = \varepsilon_r \varepsilon_o \frac{A}{D} \tag{13.9}$$

Exercises

13.1. Calculate the parasitic resistance of a 5.7-in. PCB trace having a width of 2.5 mm and a thickness of 2 mils. If this trace feeds a 2.5-V DC signal to a 1-Ω load resistance, what will be the error of the voltage at the load as a percentage of the source voltage? (Assume a zero-resistance current return path.) How much power is dissipated by the trace?

Ans. $R = 20$ mΩ; $V_{ERR} = 49$ mV $= 1.96\%$; $power = 120$ mW.

13.2. Using Eq. (13.3), calculate the parasitic inductance of a 23 cm PCB trace having a width of 12 mils, and a spacing of 15 mils to the current-return ground plane. If this trace feeds a 125 kHz, 1.25-V RMS sinusoidal signal to a 10-Ω load resistance, what will be the error of the RMS voltage at the load as a percentage of the source voltage? (Assume zero trace resistance.) Compare your answers with those obtained using the refined inductance estimate of Figure 13.10.

Ans. Using Eq. (13.3), $L = 360.1$ nH; $V_{ERR} = 503$ µV $= 0.04\%$; Using Figure 13.10, $L = 138$ nH; $V_{ERR} = 73.4$ µV $= 0.006\%$; Eq. (13.3) yields significant error.

where A = area of either plate ($L_{TRACE} \times W$ for rectangular traces), D = distance between the plates, ε_o = electrical permittivity of free space (8.8542×10^{-12} farads/m), and ε_r = relative permittivity of the dielectric material between the plates.

The value of ε_r depends on the PCB insulator material. Air, for example, has a relative permittivity very near 1.0, while FR4 fiberglass has a relative permittivity of about 4.5. Teflon®, by contrast, has a relative permittivity of about 2.7. Therefore, Teflon® PCBs exhibit less capacitance per unit area than FR4 PCBs. This is one reason that Teflon® is superior for extremely high frequency applications, since it leads to lower values of unwanted parasitic capacitance.

Equation (13.9) is only accurate for capacitor plates in which the length and width of the plates is much larger than the dielectric thickness, D. If W is about 10 times larger than D, then we can use Eq. (13.9) to estimate the capacitance per unit length of the trace

$$C_\ell = \frac{C}{L_{TRACE}} = \varepsilon_r \varepsilon_o \frac{A/L_{TRACE}}{D} = \varepsilon_r \varepsilon_o \frac{W}{D} \qquad (13.10)$$

To calculate the total capacitance between two traces, we multiply the capacitance per unit length by the trace length. This is true for the configuration in Figure 13.11 as well as for any other configuration illustrated in this chapter.

$$C = L_{TRACE} C_\ell \qquad (13.11)$$

When either the length or width is less than about 10 times the dielectric thickness, the so-called *fringe effects* in the electric field between the plates cause Eqn (13.10) to become inaccurate. Unfortunately, trace capacitance can seldom be accurately calculated using Eq. (13.10) since the width of the trace is often less than 10 times the trace to trace spacing. The graph in Figure 13.13 shows a more accurate estimation of the capacitance per meter between two parallel traces. The dotted line shows the capacitance per unit length as calculated by Eq. (13.10). Note that as the value of W becomes much larger than D, the refined estimation converges with the estimation from Eq. (13.10).

The chart in Figure 13.13 assumes a relative permittivity, ε_r, of 4.5. If our PCB material has a different relative permittivity, ε', then we simply multiply the capacitance per unit length obtained from Figure 13.13 by the ratio of $\varepsilon'/4.5$ to calculate the correct capacitance per unit length. Equivalently, we can multiply the total capacitance [calculated using Eq. (13.11)] by $\varepsilon'/4.5$ to achieve the same result.

Example 13.3

An insulator thickness of 10 mils separates a pair of 12 mil wide traces on adjacent layers in a multilayer FR4 DIB. One trace is 7 in. long, while the other is 5 in. long. The traces run directly over one another for a distance of 3 in., as shown in Figure 13.13, but do not cross each other at any other point. The upper trace carries a 10-MHz sine wave at 1.0 V RMS, while the lower trace is connected to an amplifier with an input impedance of 100 kΩ. What is the signal level of

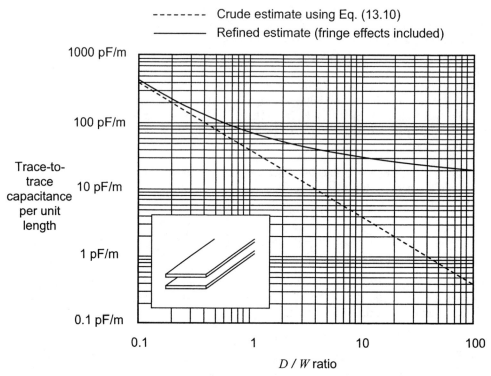

Figure 13.13. Trace capacitance per meter versus D/W ratio ($\varepsilon_r = 4.5$).

the crosstalk coupling from the 10-MHz signal into the input of the amplifier? Would a Teflon® PCB reduce the capacitance enough to give significantly better performance?

Solution:

First, we draw a model of the signal source and amplifier stage, including the parasitic trace-to-trace capacitance (Figure 13.14). The parasitic capacitance between the two traces will interact with the 100-kΩ input impedance of the amplifier to form a first-order high-pass filter.

To estimate the value of the capacitance between the two traces, we first need to calculate the D/W ratio of this parasitic capacitor. The value of D is 10 mils, while the value of W is 12 mils. Thus the D/W ratio is equal to 0.833. From the graph in Figure 13.13 we can estimate a trace-to-trace capacitance of about 85 pF per meter. Applying Eq. (13.11), we calculate the total capacitance:

$$C = 3 \text{ in.} \times \frac{1 \text{ m}}{39 \text{ in.}} \times 85 \text{ pF/m} = 6.5 \text{ pF}$$

The 3-dB cutoff frequency of an RC high-pass filter is given by

$$F_c = \frac{1}{2\pi RC} \quad (13.12)$$

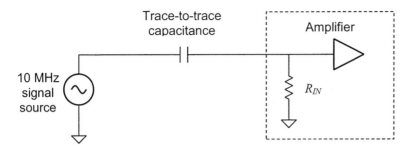

Figure 13.14. Parasitic capacitance between a signal source and a high impedance amplifier input.

Using Eq. (13.12), we calculate the cutoff frequency

$$F_c = \frac{1}{2\pi \times 100 \times 10^3 \times 6.5 \times 10^{-12}} = 245 \text{ kHz}$$

Since the 10-MHz signal is well into the passband of the high-pass filter, the 10-MHz sine wave will feed directly into the amplifier at nearly 1.0 V RMS! Clearly, this would be a very bad DIB design. Using a Teflon® PCB rather than an FR4 fiberglass PCB, we would multiply the 6.5 pF capacitance by a factor of 2.7/4.5 (2.7 = relative permittivity of Teflon®, 4.5 = relative permittivity of FR4 fiberglass). This would result in a capacitance of 3.9 pF. The value of F_c would change to 408 kHz, which would not significantly reduce the crosstalk. To solve the crosstalk problem in this example, the traces must be moved farther away from one another. Also, the sensitive 100-kΩ line should be shortened to a fraction of an inch to minimize capacitive coupling from the 10-MHz signal source as well as any other potential sources of crosstalk.

Example 13.3 shows how important it is to keep high-impedance nodes protected from potential sources of crosstalk. The best form of crosstalk prevention is to simply keep the sensitive trace as short as possible. Another method for reducing crosstalk is to place a ground plane underneath the critical signal traces, thus preventing layer-to-layer crosstalk such as that in Example 13.3. Each of the traces would then see a parasitic capacitance to ground, but the ground plane would block the trace-to-trace capacitance altogether. The effect of a ground plane on trace-to-trace capacitance is illustrated in Figure 13.15. The trace-to-trace capacitance is replaced by two parasitic capacitances to ground. This effectively shunts the offending source to ground so that it cannot inject its signal into the sensitive node.

The amount of capacitance between a trace and a ground plane (Figure 13.9) is tricky to calculate. If the length and width of the trace are much larger than the trace-to-ground separation, then we can simply use Eq. (13.10) to calculate the capacitance per unit length. For most practical situations, though, the width of the trace is not much larger than the trace-to-ground separation. We again have to resort to a more accurate estimation, as shown in Figure 13.16. The dotted line shows the capacitance per unit length as calculated using Eq. (13.10). The solid line represents a more accurate calculation that takes the fringing effects of the electric fields into account. As expected, the two lines converge as W becomes much larger than D.

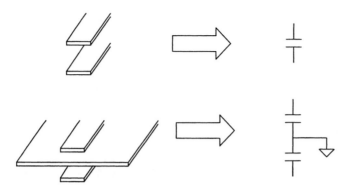

Figure 13.15. Ground planes prevent layer-to-layer crosstalk.

Next we consider the capacitance between two parallel traces on the same PCB layer (Figure 13.17). This configuration occurs very frequently in PCB designs, since many traces run parallel to each other for several inches on a typical DIB.

If the trace-to-trace spacing, S, is equal to or larger than the trace width, W, we can approximate this configuration as two circular wires having the same cross-sectional area as the

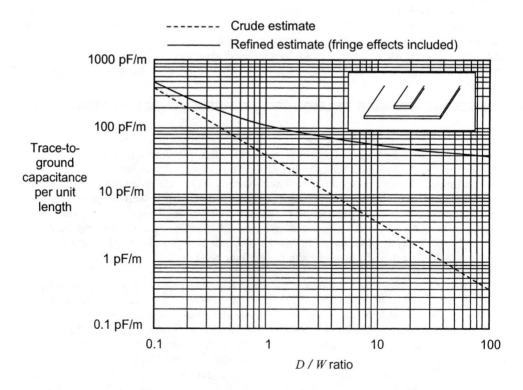

Figure 13.16. Trace capacitance per meter versus D / W ratio ($\varepsilon_r = 4.5$).

traces and having a center-to-center spacing of $S+W$. The equation for the capacitance per unit length of two circular conductors having this geometry is given by[1]

$$C_l = \left[\frac{12.1 \times 10^{-12} \cdot \varepsilon_r}{\log \left[\frac{\left(\frac{S+W}{2}\right)}{\sqrt{\frac{TW}{\pi}}} + \sqrt{\frac{\left(\frac{S+W}{2}\right)^2}{\frac{TW}{\pi}} - 1} \right]} \right] \quad (13.13)$$

where C_l = capacitance per unit length (farads per m), ε_o = electric permeability of free space, ε_r = relative permeability of the PCB material, W = width of the rectangular trace, and T = thickness of the rectangular trace.

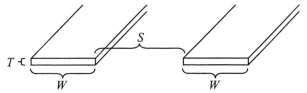

Figure 13.17. Coplanar PCB traces.

Comparing this crude circular conductor approximation to a more accurate estimation based on flattened rectangular traces, we can see how closely the two approximations agree with one another (Figure 13.18). The dotted line shows the trace-to-trace capacitance per unit length using

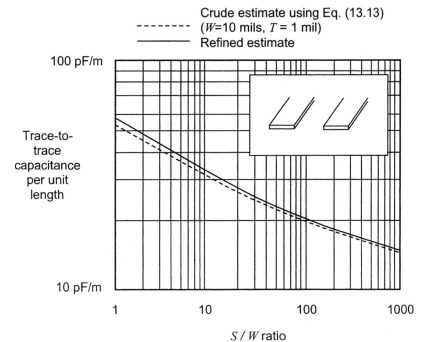

Figure 13.18. Capacitance between two coplanar traces of equal width (ε_r = 4.5).

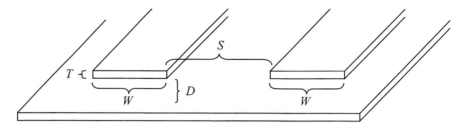

Figure 13.19. Coplanar traces over a ground plane.

Eq. (13.13), while the solid line shows a more accurate calculation based on a flat trace geometry. These estimates are close enough to each other that Eq. (13.13) can probably be used in many cases as a reasonably good approximation.

We can reduce the effects of trace-to-trace crosstalk between coplanar traces using a ground plane. Figure 13.19 shows a pair of coplanar traces with a width of W separated from one another by a distance S and spaced a distance D over a ground plane.

The ground plane forms two parasitic capacitances to ground that serve to shunt the interference signal to ground (Figure 13.20). While this may not eliminate the crosstalk, it reduces it by a significant amount. From Figures 13.16 and 13.18, we can see that the trace-to-ground capacitance will be several times larger than the trace-to-trace capacitance for values of $S > D$. Thus, if we lay out our traces so that the trace-to-trace spacing, S, is larger than our trace-to-ground spacing, D, we can make the shunt capacitance larger than the trace-to-trace coupling capacitor. This forms a capacitive voltage divider with good interference rejection.

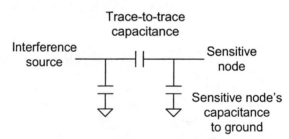

Figure 13.20. Equivalent circuit for coplanar traces over a ground plane.

13.3.5 Shielding

Electrostatic shields can also be used to reduce coplanar trace-to-trace crosstalk. A shield is any conductor that shunts electric fields to ground (or a similar low-impedance node) so that the fields do not couple into a sensitive trace, causing crosstalk. The electric fields can originate from external noise sources such as radio waves or 60-Hz power line radiation, or they can originate from other signals on the DIB. The ground plane in Figure 13.15 is one type of electrostatic shield. Ideally, a shield should completely enclose the sensitive node. A coaxial cable is one example of a fully shielded signal path. In most cases, it is impractical to completely shield every signal on a DIB using coaxial cables. However, we can achieve a close

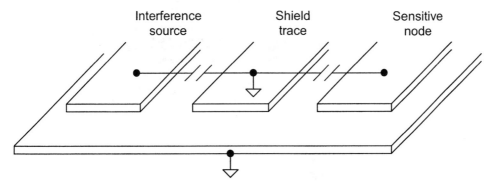

Figure 13.21. Electrostatic shielding reduces trace-to-trace crosstalk.

approximation of a fully shielded signal path by placing shield traces around sensitive signal traces. This configuration is called *coplanar shielding*.

Figure 13.21 illustrates how coplanar shielding can reduce crosstalk between a interference source and a sensitive DIB signal. The shield trace is connected to the ground plane to provide an extra level of protection for the sensitive node. Sometimes, a shield trace is routed all the way around a sensitive node, as illustrated in Figure 13.22. This type of shielding helps to reduce the coupling of electromagnetic interference from all directions.

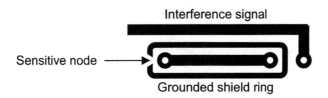

Figure 13.22. Shield trace routed around a sensitive node.

13.3.6 Driven Guards

Electrostatic shields suffer from one small drawback. The shield forms a parasitic load capacitance between the sensitive signal and ground. The parasitic capacitance is a both a blessing and a curse. It is a blessing because it shunts interference signals to ground, but a curse because it loads the sensitive node with undesirable capacitance. The capacitive loading problem can be largely eliminated using a driven guard instead of a shield. A driven guard is a shield that is driven to the same voltage as the sensitive signal. The guard is driven by a voltage follower connected to the sensitive node (Figure 13.23). The interference signal is shunted to the low-impedance output of the voltage follower, reducing its ability to couple into the sensitive signal node. A common mistake made by novice test engineers is to connect the tester's driven guards to analog ground. As seen in Figure 13.23, this is obviously a mistake.

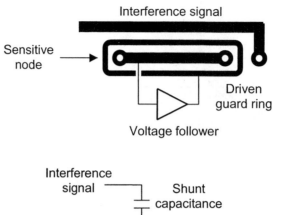

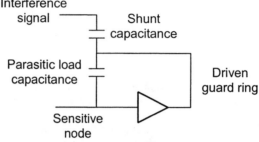

Figure 13.23. Driven guard – PCB layout and equivalent circuit.

The voltage follower drives the guard side of the parasitic load capacitance to the same voltage as the sensitive signal line. Since the parasitic load capacitance always sees a potential difference of 0 V, it never charges or discharges. Thus the loading effects of the parasitic capacitance on the signal trace are eliminated by the voltage follower.

Of course, all voltage followers exhibit a finite bandwidth. Therefore, the parasitic capacitance can only be eliminated at frequencies within the voltage follower's bandwidth. For this reason, driven guards are typically used on relatively low frequency applications that cannot tolerate any crosstalk (e.g., high-performance audio circuits).

13.4 TRANSMISSION LINES

13.4.1 Lumped- and Distributed-Element Models

In Example 13.2, we treated the PCB trace feeding the 50-Ω load resistor as if it had no capacitance to ground. In reality, the *RL* low-pass filter formed by the trace inductance and load resistance also includes a parasitic capacitance to ground, as shown in Figure 13.24. This simplistic model of a transmission line is known as a *lumped-element* model. The series inductance per unit length of the 6-in. trace in Example 13.2 was found to be 350 nH per meter. From the graph in Figure 13.16, we can determine that the capacitance per unit length of this trace is about 160 pF per meter, assuming FR4 PCB material. Therefore, the capacitance in Figure 13.24 is about 160 pF/m × [6 in. × (1 m /39 in.)] = 24.6 pF. At 50 MHz, this capacitance has an impedance of 130 Ω, which is not insignificant when placed in parallel with the 50 Ω load resistance. Therefore, we should consider both the trace inductance and capacitance when evaluating the effects of trace parasitics on circuit performance.

Unfortunately, even the refined lumped-element model of Figure 13.24 becomes deficient at higher frequencies. In reality, the parasitic trace inductance and capacitance can only be modeled as a lumped inductance and capacitance at relatively low frequencies. At higher frequencies, we have to realize that the inductance and capacitance are distributed along the length of the trace. The effect of this distributed inductance and capacitance causes the true model of the trace to look more like an infinite series of infinitesimally small inductors and capacitors, as shown in Figure 13.25. This model is known as a *distributed-element* model. If we let the number of inductors and capacitors approach infinity as their values approach zero, the PCB trace becomes a circuit element known as a *transmission line*. The transmission line exhibits unique electrical properties, which the test engineer needs to understand.

As the voltage at the input to a transmission line changes, it forces current through the first inductor into the first capacitor. In turn the rising voltage on the first capacitor forces current through the second inductor into the second capacitor and so on. The signal thus propagates from one *LC* pair to the next as a continuous flow of inductive currents and capacitive voltages. Notice that the transmission line is symmetrical in nature, meaning that signals can propagate in either direction through this same inductive/capacitive process.

Exercises

13.3. Using Eq. (13.10), calculate the parasitic capacitance of a 14-cm-long, 25-mil-wide stripline trace with a spacing of 8 mils to the ground plane, fabricated on an FR4 PCB. Compare your answer with that obtained using the refined capacitance estimate of Figure 13.16.

Ans. Using Eq. (13.10), $C = 17.4$ pF; using Figure 13.16, $C = 28$ pF; Eq. (13.10) yields significant error.

13.4. Using Eq. (13.13), calculate the parasitic capacitance between two 4-cm-long, 16-mil-wide coplanar traces separated by a spacing of 30 mils, fabricated on an FR4 PCB. Compare your answer with that obtained using the refined capacitance estimate of Figure 13.18.

Ans. Using Eq. (13.13), $C = 1.7$ pF; using Figure 13.18, $C = 2$ pF; Eq. (13.13) agrees reasonably well.

13.5. Using Eq. (13.10), calculate the parasitic capacitance of a 7.3-inch-long, 25-mil-wide stripline trace with a spacing of 10 mils to the ground plane, fabricated on a Teflon® PCB. This trace feeds a 50-kHz, 1.25-V RMS sinusoidal signal from a DUT output having a 100-kΩ output resistance to a buffer amplifier having an input capacitance of 2 pF. How much will the combined capacitance of the trace and buffer amplifier input capacitance attenuate the DUT signal? (Express your answer as a voltage gain in decibels.) Would this be an acceptable attenuation if the signal were the output of a gain test having ±0.5-dB limits? Compare your answer with that obtained using the refined capacitance estimate of Figure 13.16.

Ans. Using Eq. (13.10), $C = 11$ pF + 2 pF, $G(50$ kHz$) = -0.69$ dB, not acceptable; using Figure 13.16, $C = 20$ pF + 2 pF, $G(50$ kHz$) = -1.72$ dB, not acceptable.

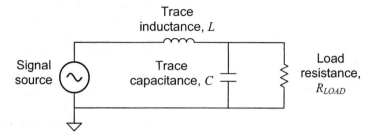

Figure 13.24. Parasitic trace inductance and capacitance (lumped-element model).

A transmission line can be formed by parallel trace pairs, as shown in Figure 13.25, a stripline (a single trace over a ground plane), or a coaxial cable. Each of these types of transmission lines behaves according to many of the same equations. For example, one of the key parameters of a transmission line is its *characteristic impedance*, defined as

$$Z_o = \sqrt{\frac{L_\ell}{C_\ell}} \; \Omega \qquad (13.14)$$

where Z_o = characteristic impedance of the transmission line, L_ℓ = trace inductance per unit length, and C_ℓ = trace capacitance per unit length.

Notice that the characteristic impedance of a trace or cable is not dependent on its length. It is dependent only on the inductance per unit length and the capacitance per unit length. Therefore, a 6-in. trace of a particular width and spacing to ground has the same characteristic impedance as one that is 10 ft long.

Signals injected into a transmission line travel down the line at a speed determined by the inductance and capacitance per unit length. The equation for the signal velocity is

$$v_{signal} = \sqrt{\frac{1}{L_\ell C_\ell}} \; \text{m/s} \qquad (13.15)$$

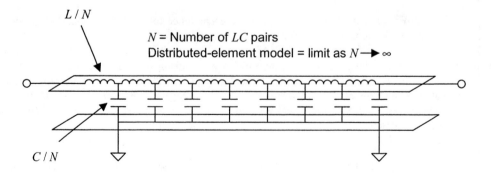

Figure 13.25. Distributed-element model of a transmission line.

The total time it takes a signal to travel down a transmission line is therefore equal to the length of the line divided by the signal velocity. This time is commonly called the transmission line's *propagation delay*

$$T_d = \frac{l_{line}}{v_{signal}} = l_{line}\sqrt{L_\ell C_\ell} \text{ s} \tag{13.16}$$

Combining Eqs. (13.11), (13.14), and (13.16), we can find the total distributed capacitance of a transmission line as a function of its propagation delay and characteristic impedance

$$C = \frac{T_d}{Z_o} \text{ farads} \tag{13.17}$$

The wavelength of a sine wave travelling along a transmission line is given by

$$\lambda_{signal} = \frac{v_{signal}}{f_{signal}} \text{ m/cycle} \tag{13.18}$$

If the wavelength of a signal's *highest frequency component of interest* is significantly larger than the length of the transmission line, then we can use a lumped-element model such as the one in Figure 13.24. Note that the highest frequency component of a digital signal such as a square wave is determined by its rise and fall time, not by its period! A common rule of thumb is that the wavelength of the highest frequency component must be 10 times the transmission line length before we can treat the line as a lumped-element model. Otherwise, we must treat the signal path as a transmission line with distributed rather than lumped parasitic elements. Another way to state this is that the period of the signal should be at least 10 times larger than the transmission line's propagation delay before we can treat the parasitic elements as lumped rather than distributed. Another practical rule of thumb is that the highest frequency component of interest in a digital signal is roughly equal to 1/3 the inverse of its rise or fall time.

Example 13.4

Determine the characteristic impedance of the PCB trace in Example 13.2. What is the velocity of a signal travelling along this transmission line? At what fraction of the speed of light does it travel? What is the propagation delay of this line? If we wish to transmit a 50-MHz sine wave along this trace, should we treat the parasitic capacitance and inductance as lumped elements, or should we treat the trace as a transmission line?

Solution:

The inductance per unit length was previously found to be 350 nH per meter. The capacitance per unit length was found to be 160 pF per meter. Using Eq. (13.14), the characteristic impedance of the trace is

$$Z_o = \sqrt{\frac{350 \frac{nH}{m}}{160 \frac{pF}{m}}} = 46.77 \; \Omega$$

The velocity of a signal traveling along this line is given by Eq. (13.15)

$$v_{signal} = \sqrt{\frac{1}{350\frac{nH}{m} \times 160\frac{pF}{m}}} = 133.63 \times 10^6 \text{ m/s}$$

The speed of light, c, is 300×10^6 m/s. Therefore, signals travel down this transmission line at a speed of 133.63/300 times the speed of light, or $0.445c$.

The propagation delay can be calculated using either of two methods. First we can divide the length of the transmission line by the signal velocity

$$T_d = \frac{l_{line}}{v_{signal}} = \frac{6 \text{ in} \times \frac{1 \text{ m}}{39 \text{ in}}}{133.63 \times 10^6 \text{ m}/_s} = 1.15 \text{ ns}$$

An alternative calculation uses Eq. (13.17)

$$T_d = C \cdot Z_o = 24.6 \text{ pF} \times 46.77 \text{ }\Omega = 1.15 \text{ ns}$$

The wavelength of a 50-MHz signal, as calculated using Eq. (13.18) is

$$\lambda_{signal} = \frac{133.63 \times 10^6 \text{ m/s}}{50 \text{ MHz}} = 2.67 \text{ m}$$

Since the length of the trace is only 6 in. and the wavelength of the 50-MHz signal is much larger (2.67 m × 39 in./m = 104 in.), we can safely treat this line as a lumped-element model.

13.4.2 Transmission Line Termination

Transmission lines can behave in a fairly complicated manner. Although their behavior is well defined, a full study of transmission line behavior is beyond the scope of this book. Fortunately, we can easily predict the basic behavior of a transmission line as long as we provide proper resistive termination at one or both of its ends.

To understand the purpose of transmission line termination, let us first examine the behavior of an unterminated line. An unterminated transmission line behaves as a sort of electronic echo chamber. If we transmit a stepped voltage down an unterminated transmission line, it will bounce back and forth between the ends of the line until its energy is dissipated and the echoes die out. The energy can be dissipated as electromagnetic radiation as well as heat in the source resistance and parasitic resistance in the line. The resulting reflections appear as undesirable ringing on the stepped signal. Properly chosen termination resistors placed at either the source side or the load side of a transmission line cause it to behave in a much simpler manner than it would behave without termination. The purpose of termination resistors is to dissipate the energy in the transmitted signal so that reflections do not occur.

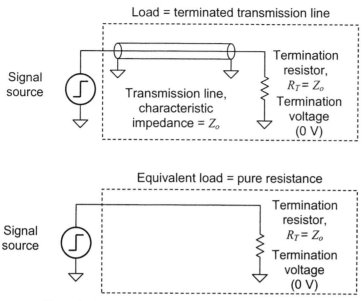

Figure 13.26. Terminated transmission line and resistive equivalent.

The simplest termination scheme to understand is the far-end termination scheme shown in Figure 13.26. As shown in this diagram, transmission lines are commonly drawn in circuit schematics as if they were coaxial cables, even if they are constructed using a PCB trace. This is because the basic behavior of a transmission line is dependent only on its characteristic impedance and propagation delay rather than its physical construction.

If the termination resistor R_T is equal to the characteristic impedance of the transmission line, then the transmitted signal will not reflect at all. The energy associated with the currents and voltages propagating along the transmission line is completely dissipated by the termination resistor. As far as the signal source is concerned, *a terminated transmission line looks just like a resistor whose value is equal to Z_o.* The distributed inductance and capacitance of the transmission line *completely disappear as far as the source is concerned.* The only difference between a purely resistive load and a terminated transmission line is that the signal reaching the termination resistor is delayed by the propagation delay of the transmission line. Also it is important to note that while the termination resistor is usually connected to ground (0 V), it can be set to any DC voltage and the transmission line will still be properly terminated. The terminated line then appears to the source as a pure resistance connected to the DC termination voltage.

The ability to treat a terminated transmission line as a purely resistive element is very useful. Many tester instruments are connected to the DUT through a 50-Ω transmission line, which is terminated with a 50-Ω resistor at the instrument's input (Figure 13.27). As far as the DUT is concerned, this instrument appears as a 50-Ω resistor attached between its output and ground. If the DUT output is unable to drive such a low impedance, then we can add a resistor, R_S, between the DUT output and the terminated transmission line. The DUT output then sees a purely resistive load equal to $R_S + Z_o$. Although the signal amplitude is reduced by a factor of $Z_o / (Z_o + R_S)$, we can compensate for this gain error using focused calibration (see Chapter 10).

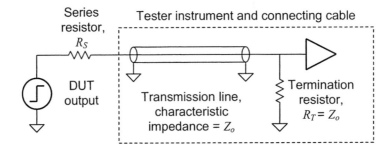

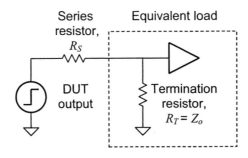

Figure 13.27. Tester instrument with terminated transmission line.

If we observe the signals at the DUT output, the input to the transmission line, and the input to the tester instrument, we can see the effects of the resistive divider and the propagation delay of the transmission line (Figure 13.28). The signal is attenuated by the series resistor and termination resistor, and it is also delayed by a time equal to T_d.

If we observe the voltage at a particular point along the transmission line, we will see a delayed version of the signal at the DUT output. For example, if we look at a point halfway down the transmission line, we will see a signal that has a rising transition that occurs halfway between the rising edge at the transmission line input and the rising edge of the transition at its output. Of course, if the signal had a falling edge instead of a rising edge, this same delay would

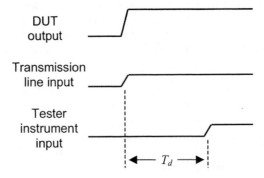

Figure 13.28. DUT output, transmission line input, and instrument input.

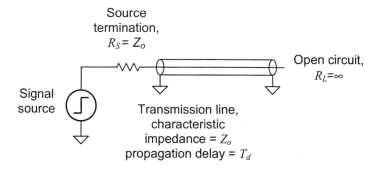

Figure 13.29. Transmission line with source termination.

occur. In fact *any* signal that is transmitted down the transmission line, whether it be a voltage step, sine wave, or complex signal, will exhibit the same attenuation and time delay described previously.

The next common method of transmission line termination is the source termination scheme shown in Figure 13.29. In this scheme, the transmitted signal is allowed to reflect off the unterminated far end of the transmission line. As the signal returns to the source end of the transmission line, the source resistance $R_S = Z_o$ absorbs all the energy in the currents and voltages of the transmitted signal. No further reflections occur. Source termination is used in the digital pin card driver electronics of most ATE testers to prevent ringing in the high-speed digital signals generated by the tester's digital subsystem.

While the signal propagates down the transmission line and back, the source cannot tell whether the far end is terminated or unterminated. *For a short period of time, the transmission line appears to the source as if it were a pure resistance of Z_o Ω.* Therefore, during the period of time that the signal travels down the transmission line and back, the voltage at the transmission line input will be ½ that at the source output (since $R_S = Z_o$). Once the reflected signal returns to the source, the source resistor absorbs all the reflected energy and the voltage at the transmission line input becomes equal to the voltage at the output of the source. Therefore, the source only sees a load of $2 \times Z_o$ for a period of $2 \times T_d$. Afterwards, the source sees an open circuit.

At the far end of the transmission line, the voltage remains at the termination voltage until the incident signal arrives. The incident signal arrives at ½ the amplitude of the source signal, but it immediately adds to the reflected signal whose amplitude is also ½ the amplitude of the source signal. Therefore, the far end sees the unattenuated source signal with a delay of T_d. Figure 13.30 shows a stepped voltage as it appears at the source output, transmission line input, and transmission line output.

If we observe the voltage at a particular point along the source terminated transmission line, we will see a signal that is similar in shape to the signal at the transmission line input. However, the spacing between the first edge and the second edge will be closer together. At a point halfway down the transmission line, for instance, the edges will be spaced by a time equal to T_d, as illustrated in Figure 13.30. As shown, the edges observed at an intermediate point on the transmission line are always centered around the time the incident signal reflects off the unterminated transmission line output.

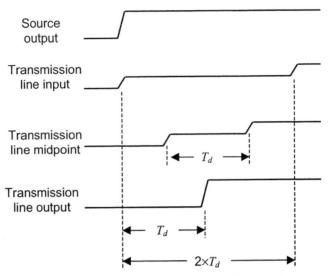

Figure 13.30. Source output, transmission line input, and transmission line output.

Transmission lines can consist of multiple controlled-impedance segments, each having the same characteristic impedance. For example, the digital channel drivers from a tester are routed to the DIB through a series of cascaded coaxial cables and controlled impedance PCB traces (Figure 13.31). To create a cascaded controlled impedance transmission line, the DIB's traces must also exhibit the same characteristic impedance as the tester's transmission lines. If any of the impedances of the transmission line segments are not matched, then the point where they connect will generate signal reflections. Therefore, we have to make sure we lay out our DIB with a characteristic impedance equal to the tester instrument's characteristic impedance to avoid unwanted signal reflections.

One of the common mistakes made by novice test engineers is to observe the output of a digital channel at the point where the DIB connects to the test head. Such an observation point represents an intermediate point along the cascaded transmission line. As a result, a rising edge will appear as a pair of transitions such as those shown in Figure 13.30 rather than a single transition. The novice test engineer often thinks the tester driver is defective, when in fact it is working perfectly well, in accordance with the laws of physics. The only way to see the expected DUT signal is to observe it at the DUT's input.

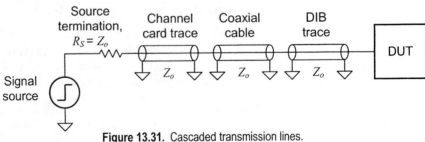

Figure 13.31. Cascaded transmission lines.

Notice that we can measure the propagation delay of a transmission line by measuring the time between the first and second step transitions at the source end of a source-terminated transmission line. This time is equal to $2 \times T_d$. We can divide the measured time by two to calculate the transmission line's propagation delay. This is how modern testers measure the propagation delays from the digital channel card drivers to the DUT's digital inputs. The tester can automatically compensate for the electrical delay in each transmission line, thereby removing timing skew from the digital signals. This measurement process is known as *time-domain reflectometry*, or TDR. On older testers, TDR deskew calibrations were not used. The timing edges of the digital channels were deskewed at a point inside the test head and any delays caused by the various transmission lines were not taken into account. The test engineer had to lay out DIB traces that were equal in length to avoid channel-to-channel skew. This is the reason that DIBs are often round rather than square. The round board allows a radial layout, like spokes on a bicycle wheel. The spokes can be laid out with equal lengths, leading to matched delay times.

Exercises

13.6. A 12-in. stripline trace is fabricated on an FR4 PCB with a width of 15 mils. It is separated from its ground plane by a spacing of 12 mils. Using the refined estimates of Figures 13.10 and 13.16, calculate the stripline's parasitic capacitance per meter and inductance per meter. What is the stripline's characteristic impedance? Repeat the exercise for the same stipline fabricated using a Teflon® PCB.

Ans. FR4: C_l =110 pF/m, L_l=500 nH /m, Z_o = 67.4 Ω; Teflon®; C_l = 66 pF/m, L_l=500 nH /m, Z_o = 87.0 Ω.

13.7. What is the velocity of a signal propagating along the stipline of Exercise 13.6? Express your answer in m/s and in a percentage of the speed of light. What is the stipline's propagation delay? Is the delay for the Teflon® PCB longer or shorter than the FR4 PCB?

Ans. FR4: v_{signal} = 1.35×10^8 m/s, 0.45 c, T_d = 2.28 ns; Teflon®; v_{signal} = 1.74×10^8 m/s, 0.58 c, T_d = 1.77 ns; Teflon® PCB propagation delay is shorter (signal velocity is higher) than FR4.

13.8. What is the wavelength, in inches, of a 500-MHz sine wave travelling along the FR4 stripline in Exercise 13.6? Can we treat the parasitic reactances of the stripline as lumped elements or do we have to treat the stripline as a transmission line at this frequency? Could we approximate the stripline using a lumped-element model if we used Teflon® instead?

Ans. FR4: λ = 10.5 in. – transmission line. Teflon®: λ = 13.6 in. – transmission line.

13.9. A 900-MHz sinusoidal signal is transmitted from a DUT output to tester digitizer along a 1.5-foot terminated 50-Ω coaxial cable having a signal velocity of 0.65c. What load resistance is presented to the DUT during the time the signal propagates? What resistance is presented to the DUT after the signal has settled? What is the distributed capacitance of this coaxial cable? What is the phase shift, in degrees, between the DUT output and the digitizer input? (Allow your answer to exceed 360 degrees.)

Ans. R_{LOAD} = 50 Ω, R_{LOAD} = 50 Ω, C = 47.3 pF, phase shift = 766.9 degrees.

13.4.3 Parasitic Lumped Elements

So far, we have only considered transmission lines that have been terminated with a pure resistance. Unfortunately, there are usually a few picofarads of parasitic capacitance or a few nanohenrys of parasitic inductance located at various points along a transmission line. These parasitic reactances arise from interconnections from one cascaded transmission line segment to another and from electromechanical relays located inside the tester instrument. There are also small amounts of inductance and capacitance at the ends of the transmission line. These are caused by the DUT's input or output parasitics, the DUT socket parasitics, or the pin card input and output parasitics. These parasitic elements are known as *lumped elements*, to distinguish them from the distributed inductance and capacitance of the transmission line itself. Lumped reactances can cause undesirable ringing, overshoot, and/or undershoot in our signals. Unfortunately, there often is not much we can do about the problem, other than to be aware of its existence.

Tester instruments are often specified using both lumped and distributed transmission line parameters. For example, a digital channel card is typically specified with a distributed capacitance and a lumped capacitance. The lumped capacitance is usually only a few picofarads, while the distributed capacitance [as defined by Eq. (13.17)] may be 30 pF or more. To accurately model the tester's load on a DUT output, we should use a model that includes both the transmission line and the lumped elements. We can use this type of model for simulations using software tools such as SPICE. Figure 13.32 shows a typical model for a tester's digital pin card electronics. The distributed capacitance listed in a tester's specifications may or may not include that associated with the DIB trace; so the test engineer should read the channel card specifications carefully. A number of good books have been written on the subject of transmission lines and high-speed digital design. The reader is encouraged to refer to Johnson[2] and Wadell.[3]

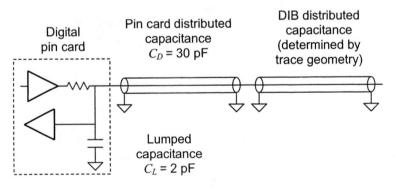

Figure 13.32. Typical digital channel card model.

13.5 GROUNDING AND POWER DISTRIBUTION

13.5.1 Grounding

The term "grounding" refers to the electrical interconnection and physical layout of the various ground nodes in an electronic system such as an ATE tester and DIB. In a circuit schematic, grounds are treated as perfect zero volt reference points exhibiting zero impedance. In a real

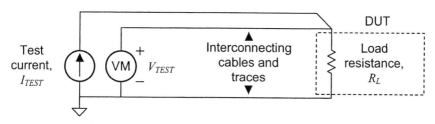

Figure 13.33. Resistor measurement test setup: idealized model.

system, there can be only one point that is defined as true ground. All other ground nodes are connected to true ground through resistive and inductive traces, wires, or ground planes. Often, these parasitic resistors and inductors play a significant role in the performance of the DUT and the ATE tester instruments.

Grounding is one of the more difficult DIB design subjects to master. Proper grounding is essential to a DUT's electrical performance. Poor grounding can lead to resistive voltage drops, inductive transients, capacitive crosstalk, radio frequency (RF) interference, and a host of other electrical ailments that result in poor measurement accuracy or noisy DUT performance. Grounding is not a subject taught in most college curricula. Instead, it is often learned through experience (i.e., bad experience!). Fortunately, a few books have been written on the subject of grounding and other related topics. Henry Ott's book,[4] *Noise Reduction Techniques in Electronic Systems*, provides an excellent introduction to the subjects of grounding, shielding, electromagnetic interference (EMI), and electromagnetic compatibility (EMC).

One way to achieve proper grounding is to pay close attention to the flow of currents through the traces, wires, and planes in the DIB and tester. The first thing we have to consider is DC measurement errors caused by resistive drops in ground connections. Figure 13.33 shows a simple test setup including a DC current source, a DC voltmeter, and a DUT (a simple load resistor in this case). We wish to measure the value of the resistor by forcing a current, I_{TEST}, and measuring the voltage drop across the resistor, V_{TEST}. The resistor's value is calculated by dividing V_{TEST} by I_{TEST}.

As we have seen in previous sections, the ground path from the DUT to the DC source will not be a zero-ohm path. If the ground path is several feet long, it may exhibit several ohms of series resistance, R_G. Therefore, the parasitic model illustrated in Figure 13.34 should be used to

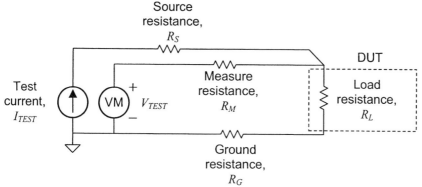

Figure 13.34. Resistor measurement test setup: parasitic model.

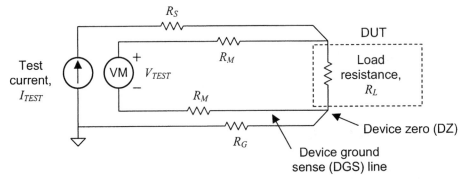

Figure 13.35. Resistor measurement test setup using device ground sense (DGS) line.

predict the measurement results obtained using the grounding scheme of Figure 13.33. Since the DC meter in this example is connected to the tester's internal ground, it measures the voltage across the series combination of the DUT resistance R_L and the ground interconnection resistance, R_G. Therefore, the measured resistance is equal to $R_L + R_G$, resulting in an error of several ohms. Notice that the value of R_S is unimportant, since the current source will force I_{TEST} through the DUT load resistance regardless of the value of R_S. Also, notice that the value of R_M is unimportant, since it carries little or no current (assuming the DC voltmeter's input impedance is sufficiently high).

Obviously, accurate mixed-signal testers can not be constructed using such a simple grounding scheme. Instead, they use a signal, which we will call *device ground sense*, or DGS, to carry the DUT's 0-V reference back to each tester instrument. Since the DGS signal is carried on a network of zero-current wires, the series resistances of these wires do not result in voltage measurement errors. Figure 13.35 shows a resistor measurement test setup using the DGS reference signal. Note that any number of tester instruments can use DGS as a zero-volt reference, provided that they do not pull current through the DGS line. Consequently, each tester instrument typically contains a high input impedance voltage follower to buffer the voltage on DGS.

The DGS line is often routed all the way to a point near the DUT, which serves as the true 0-V reference point of the entire test system. This single point is known by several names, including *star ground*, and *device zero*. We will use the term "device zero", or DZ, to refer to this point in the circuit. All measurement instruments should be referenced to the DIB's DZ voltage.

13.5.2 Power Distribution

As shown in Figure 13.36, the power supplies and sources in a mixed-signal tester are typically connected using a 4-wire Kelvin connection (see Section 5.2.2). Each Kelvin connection includes a separate wire for high-force (HF), low-force or ground current return (LF), high-sense (HS), and low-sense (LS). The low-sense line is equivalent to DGS. Therefore all supplies may use a single DGS line as their low-sense reference, resulting in only three DIB connections per source.

All currents from a power supply must return through its low-force line, in accordance with Kirchhoff's current law. Therefore, we can control the path of DUT power supply currents by

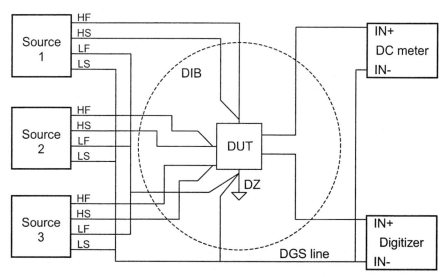

Figure 13.36. Power and ground distribution in a mixed-signal ATE tester.

forcing them through separate DIB traces back to the low-force line of the supply. Separate return of each supply's current prevents unexpected voltage drops across the various current paths in the ground network. Some testers do not provide separate ground return lines for each supply. In these testers, it is impossible to isolate the currents from each power supply and DC source. Instead, the return currents are lumped together and returned to the tester through a common current return path, as shown in Figure 13.36.

The PCB traces in a DZ grounded DIB should be laid out as shown in Figure 13.37(a). If the DGS line or the force and sense lines of a power supply are not connected properly, the full effectiveness of the zero-volt sense lines will not be realized.

13.5.3 Power and Ground Planes

At this point in the chapter, the reader should be catching on that ground planes are a good idea for many reasons. The ground plane provides a low inductance connection between all the grounds on a DIB. Similarly, power supplies can be routed using power planes to reduce the series inductance between all power supply nodes. Power and ground planes can be divided into sections, forming what are known as *split planes*. Each section of a split plane can carry a different signal, such as +12 V, +5 V, and –5 V. This provides the electrical superiority of copper planes without requiring a separate PCB layer for each supply. Typically, power is applied through split planes while grounds are connected to solid (nonsplit) planes, but even the ground planes can be split if desired.

There are usually at least two separate ground planes in a mixed-signal DIB. One plane forms the ground for the transmission line traces carrying digital signals. This plane is subject to rapidly changing current flows from the digital signals, and therefore exhibits fairly large voltage spikes caused by the interactions of the currents with its own inductance. This ground plane is often called DGND (digital ground) in the DIB schematics. The second plane, AGND (analog

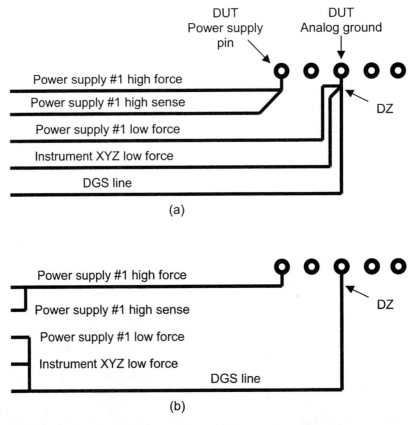

Figure 13.37. Physical layout of Kelvin connections and DZ grounding system: (a) proper connections; (b) improper connections.

ground), is for use by analog circuits. Ideally, this plane should carry only low-frequency, low-current signals that will not give rise to voltage spikes.

A third plane is occasionally used as a DIB-wide zero volt reference. This "quiet ground" plane (QGND) can be used by any analog circuits on the DIB that need a low noise ground reference. The QGND plane must be connected in such a way that it does not carry any currents exceeding a few milliamps. To guarantee this, the QGND plane should be tied only to the DZ node at a single point and to relatively high-impedance DUT pins and DIB circuit nodes. Often, the analog ground plane and the quiet ground plane are combined into a single plane, resulting in a DIB with only two ground planes (analog and digital). Figure 13.38 shows a cross section of a DIB with a three-plane grounding scheme.

13.5.4 Ground Loops

A star grounding scheme is formed by connecting the grounds of multiple circuits to a single ground point, rather than connecting them in a daisy chain. Star grounds prevent a common grounding error known as a *ground loop*. A ground loop is formed whenever the metallic traces and wires in a ground network are connected so that a loop is formed (Figure 13.39). In a world without magnetic fields, ground loops would not be problematic. Unfortunately, fluctuating

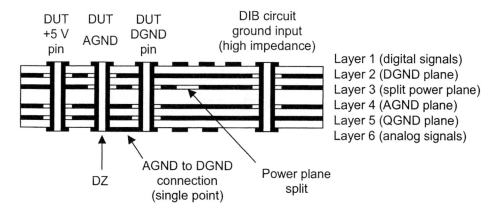

Figure 13.38. Three-plane grounding scheme.

magnetic fields surround us in the form of 60-Hz power fields and various electromagnetic radio signals. A fluctuating magnetic field passing through a loop of wire gives rise to a fluctuating electric current in the wire. The fluctuating current, in turn, gives rise to a fluctuating voltage in the wire due to the wire's parasitic resistance and inductance. Thus AC voltages can be induced into ground wires if we carelessly connect them in a loop.

Ground loops are most commonly formed when we connect instruments such as oscilloscopes and spectrum analyzers to our DIB. The tester housing and its electrical ground must be connected to earth ground for safety reasons to prevent electrical shock. Likewise, an oscilloscope's housing and electronics must be connected to earth ground. When we connect a bench instrument's ground clip to the DIB ground, we form a large ground loop, as shown in Figure 13.39. Ground loops can also be formed when we attach test equipment to other external equipment such as handlers and probers.

Any fluctuating magnetic fields that intersect the area of the ground loop cause currents to flow around the loop. The parasitic resistance of the wires and cables in the loop cause voltage drops, which in turn cause unwanted noise in our grounding system, corrupting whatever measurement we are trying to make. A ground loop often causes the tester's signals to appear terribly corrupted with 60-Hz power hum and other noise components. The test engineer has to realize that these signals are not present in the tester itself. They disappear as soon as we disconnect the bench instrument. Unfortunately, the instrumentation ground loop problem cannot be easily solved without violating company safety rules (e.g., by disconnecting the third prong of the oscilloscope's power cord from earth ground using a "cheater plug"). Ground loops caused by handlers and probers can sometimes be resolved by breaking unnecessary ground connections between the handler or prober and the test head.

13.6 DIB COMPONENTS

13.6.1 DUT Sockets and Contactor Assemblies

The DUT pins and the circuit traces on a DIB must be connected temporarily during test program execution. A hand-test socket or a handler contactor assembly makes the temporary connection.

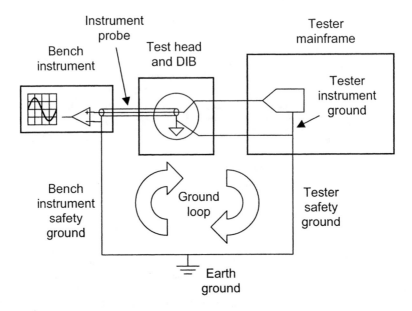

Figure 13.39. Bench instrumentation ground loop.

There are thousands of different socketing and contactor schemes; so we will not try to discuss them in any detail. The most important thing to note is that the metallic contacts of the socket or contactor assembly represent an extra resistance, inductance, and capacitance to ground that will not exist when the DUT is soldered directly to the PCB in the customer's system-level application. Sometimes the parasitic elements are unimportant to a device's operation, but other times, particularly at high frequencies, they can be extremely critical. Occasionally, the test engineer will find that socket or contactor pins are the cause of correlation errors between measurements made on the customer's application and measurements made on the tester.

13.6.2 Contact Pads, Pogo Pins, and Socket Pins

Contact pads are metal pads formed on the outer trace layers of a DIB PCB. They appear on DIB schematics as circles, black dots, or connector bars. These pads allow a relatively reliable, nonabrasive connection between one layer of interface hardware and the next. Two common uses for connector pads are pogo pin connections and DUT socket pin or contactor pin connections (Figure 13.40). A pogo pin is a spring-loaded gold-plated rod that provides a connection between two connector pads. Pogo pins may have blunted ends, pointed ends, or crown-shaped ends, depending on the connection requirements. Pointed or crown-shaped ends tend to dig into the pad surface, providing a reliable, low-resistance connection to the pad. However, the digging action may eventually destroy the pad. Blunted pogo pins are less abrasive to the pads, but are slightly less reliable since they do not dig into the pad surface. Contact pads can also be used to form a connection to DUT socket pins, such as illustrated in Figure 13.40.

Contact pads are usually plated with gold to prevent corrosion and to lower the contact resistance between the pad and the gold-plated pogo pin or socket pin. Without the gold plating, corrosion would lead to higher contact resistance, or even worse, it might result in a complete

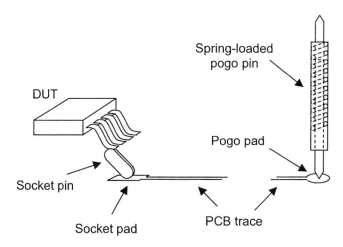

Figure 13.40. Contact pads, pogo pins, and DUT socket pins.

open circuit even when the pogo pin and pad are in physical contact. Since gold and copper are both soft metals, gold-on-copper contact pads can be damaged by repeated connect/disconnect cycles. To increase their hardness, the copper pads are often coated with a nickel alloy before the gold plating is applied.

A connection formed with a contact pad can be treated as an ideal zero-impedance connection in most cases. The connection adds a small amount of resistance, usually on the order of a few tens of milliohms. However, the resistance may change by 50% or more as the pad is connected and reconnected to the pogo pin or socket pin.

Pogo pins and socket pins also add inductance to the signal path. The inductance may be as little as one nanohenry or as large as a few tens of nanohenrys. Socket pins and pogo pins may also introduce pin-to-pin or pin-to-ground parasitic capacitance on the order of a few picofarads. In general, long thin pins add more inductance than short fat ones. The best way to reduce the effects of pogo or socket pin inductance is to return all high-frequency currents through an adjacent pin. This minimizes the area of the loop through which the current must flow, reducing the inductance of the loop. Unfortunately, the shape and size of the current loops are often determined by the pinout of the device; so the test engineer can do little to lower the inductance caused by socket pins. Of course, the best way to minimize parasitic socket pin inductance and capacitance is to choose a socket with very short pins.

13.6.3 Electromechanical Relays

One of the more common DIB components used in mixed-signal testing is the electromechanical relay. The relay is an electromagnetically controlled mechanical switch. Relays allow the DIB circuits to be appropriately reconfigured for each measurement in the test program. As one of the few moving parts on a DIB, relays represent a potential reliability problem in production. Very high-reliability relays must be chosen so that the DIB can operate through hundreds of millions of open/close cycles without failure.

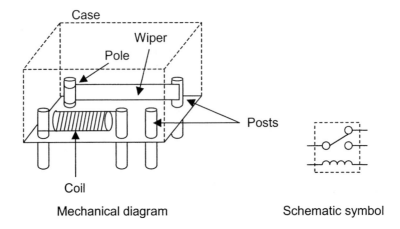

Figure 13.41. Electromechanical relay (single pole, double throw).

The metal contacts in a relay are pulled open or closed using an electromagnetic field generated by a DC current passing through a coil of wire (Figure 13.41). The current is switched on and off under test program control as the test code is executed. As mentioned in Chapter 5, "Tester Hardware," flyback diodes are sometimes added to the DIB in parallel with each relay coil to prevent the coil's inductive kickback voltage from damaging the current-driving electronics located inside the tester.

In conventional relays such as the one in Figure 13.41, the moving armature is called the *wiper*. Since it pivots on its pole, it may eventually wear out and get stuck. A more reliable relay is the reed relay, which uses two springy metal reeds that become magnetized by the coil's electromagnetic field. They are attracted to each other by the induced magnetism. Since the reeds do not swing on any pivot, there are no parts to wear out other than the point of contact between the two reeds. Reed relays are often used on DIBs because of their superior reliability.

Occasionally, a resistive buildup can occur between the contacts of a relay, causing an open circuit. Relay damage can also result from poor DIB design. Care should be taken to avoid passing currents through a relay that exceed its rated current specifications. Damage can also be caused to the wiper and contacts by abrupt changes in the current passing through the contacts as they open and close. The high di/dt current changes can induce large inductive voltage spikes, leading to a spark that welds the contacts together. Care should be taken to avoid discharging capacitors directly through relays without a series resistance to limit the discharge current. Otherwise, the sudden surge in current from the capacitor may weld the relay contacts together.

Relays, like manually activated switches, are available in a variety of configurations. The most common versions are single-pole/single-throw (SPST), single-pole/double-throw (SPDT), double-pole/single-throw (DPST), and double-pole/double-throw (DPDT). The schematic representation of each of these configurations is shown in Figure 13.42.

The parasitic behavior of relays is fairly complicated. They may exhibit a number of possible nonideal characteristics, including series resistance through the wiper and posts (R_{WP}), series inductance through the wiper and contacts (L_W), capacitive coupling between the contacts and ground (C_{PG}), capacitive coupling between the wiper and the coil (C_{WC}), capacitive coupling

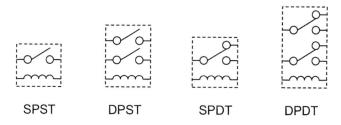

Figure 13.42. Electromechanical relay schematic representations.

from contact to contact, and mutual inductance between the coil and the wiper. A simplified model of a SPDT relay is shown in Figure 13.43. Series resistance is typically only a few hundred milliohms, although the exact value changes from one closure to the next. Series inductance is often fairly high, and may exceed 10 millihenrys. Capacitance values are usually around 1 to 5 picofarads.

Some relays contain a built-in electrostatic shield that can be grounded to prevent capacitive coupling from external signals to the wiper. The shield helps to prevent electromagnetic interference from corrupting the signal passing through the relay. Similarly, controlled-impedance relays contain a shield conductor that can be grounded to form a continuation of a transmission line. This allows high-frequency signals to propagate smoothly through the relay without generating signal reflections. (Cascaded transmission lines were discussed in Section 13.4.2.)

It is good design practice to connect relays so that they are in the most commonly desired position when they are not activated (i.e., when there is no current passing through the coil). For example, if a 1-kΩ resistor is to be connected from a DUT pin to ground during only one test, then it makes sense to connect the relay in the normally open configuration as shown in Figure 13.44(a). Configuring the relay in this manner, the resistor is only connected when the test code sets the relay driver into the nondefault state. This is a minor point, but it tends to save debug time since the test engineer does not have to explicitly add or remove the resistor during each of the remaining tests. Conversely, if the resistor is desired in all but one test, the relay should be connected in the normally closed configuration Figure 13.44(b).

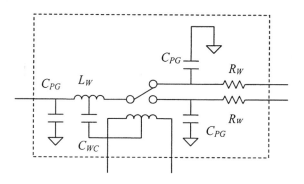

Figure 13.43. Parasitic model of SPDT relay.

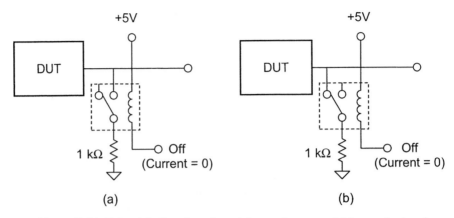

Figure 13.44. Relay default configurations: (a) normally open and (b) normally closed.

13.6.4 Socket Pins

Since relays and active circuits such as op amps are subject to electrical or mechanical failures, they must be replaced from time to time. Although op amps and relays can be soldered directly onto the DIB PCB, replacement is far easier to perform if the relays are mounted in socket pins (Figure 13.45). Socket pins should ideally be used for any component having more than two or three leads.

Surface-mounted components with more than two pins are difficult to unsolder without damaging the board. Therefore, leaded components should be used in conjunction with socket pins whenever possible. Unfortunately, the reduced pin inductance of surface-mount components is sometimes required for very high-frequency testing. Also, the extra capacitance

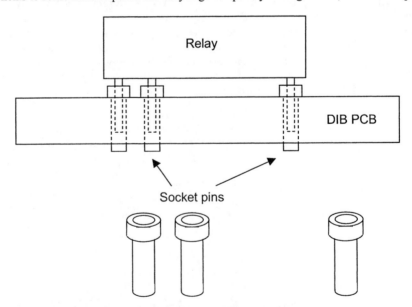

Figure 13.45. Socket pins allow easy repair and maintenance of DIBs.

and inductance of socket pins may make a socketed connection inferior at high frequencies. In such cases, surface-mounted relays, op amps, and other active devices may be the only viable alternative, even though they make the DIB more difficult to repair. Surface-mounted relays and active components should be used only when needed for electrical performance.

13.6.5 Resistors

Resistors are available in a variety of package types, including surface-mount, axial-leaded, and non-axial-leaded varieties (Figure 13.46). They can be constructed using a wide variety of resistive materials, most commonly carbon or metal (e.g., aluminum). Resistors can be constructed either as a solid core of resistive material, a coil of resistive wire, or a thin film of resistive carbon or metal. Carbon film and metal film resistors are constructed by depositing a thin film of the resistive material onto an insulator such as ceramic. The thin film gives a higher resistance than a solid core of the same material.

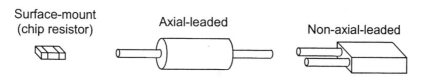

Figure 13.46. Common resistor packages.

The choice of material and package type determines the power dissipation capabilities of the resistor, its accuracy, its stability over time and temperature, and its cost. Power dissipation is basically a function of the resistor's size. Larger resistors can typically dissipate more heat than small resistors. The test engineer should determine the maximum power that must be dissipated by each resistor on a DIB using the equation

$$power = \frac{V^2}{R} \text{ watts} \tag{13.19}$$

where V = maximum RMS voltage dropped across the resistor at any point in the test program. The RMS voltage should include both DC and AC components.

The test engineer should also consider the accuracy requirements of each resistor. In general, metal film and wire-wound resistors are more accurate than carbon resistors. For tolerances of 5% to 20%, solid carbon or carbon film resistors are typically acceptable. Tolerances of 1% generally require metal film resistors. Tolerances below 1% can be attained using a trimmed metal film resistor. Some component vendors are able to provide resistor tolerances of 0.01% or better using a trimmed metal film process. These resistors are not only highly accurate, but are also very stable over time and temperature. Extremely high accuracy can also be achieved using a wire-wound resistor constructed from a coil of wire.

In general, the cost of a resistor is inversely proportional to its accuracy. A 20% carbon resistor is far less expensive than a custom-trimmed 0.001% metal film resistor. However, DIBs are usually constructed for maximum accuracy and reliability. Therefore, the cost of components is a secondary issue in DIB design. It is far more important to get a high test yield than to save a few dollars on DIBs. For this reason, resistors (and most other DIB components for that matter) are chosen based on their performance rather than their cost.

Performance is not only determined by a resistor's accuracy and power-handling capabilities; it is also determined by how closely the resistor can be modeled as a pure resistance. A resistor's material, shape, and size affect its nonideal performance characteristics. Resistors can be modeled to a first approximation as a resistance in series with an inductance. At low frequencies, the inductance may not be at all important and the resistor can be modeled as a pure resistance. At higher frequencies, the series inductance may become a significant reactive element that affects circuit performance.

The most highly inductive resistors are wire-wound varieties, since they are constructed from a long, highly inductive coil of wire. These resistors are therefore used mainly in low-frequency applications requiring high power dissipation and high accuracy. Leaded resistors suffer from a small amount of inductance caused by their wire leads, typically on the order of 1 to 10 nH. However, this inductance is often quite tolerable up to several tens of megahertz.

At higher frequencies, the smaller surface-mount packages typically provide the least series inductance. This is partly due to their lack of wire leads and partly due to the fact that they can be soldered directly onto the DIB without the use of a through-hole or via. Through-holes and vias, add a small amount of series inductance as well as a few picofarads of parasitic capacitance to ground. Beyond the range of a few megahertz, the series inductance of vias, PCB traces, IC bond wires, and socket pins may become more of an issue than the inductance of the resistor itself.

13.6.6 Capacitors

Like resistors, capacitors are also available in axial-leaded, non-axial-leaded, and surface-mount varieties. They can be constructed using a simple configuration of two parallel plates (called *electrodes*) separated by a nonconductive dielectric material, a sandwich of plates, or a pair of rolled foil plates. Capacitor performance is largely dependent on the shape and size of the capacitor as well as the type of dielectric material.

Small values of capacitance can be achieved using the simple two-plate arrangement in which capacitance is given by Eq. (13.9). The value A is the area of one of the plates, D is the distance separating the plates, and $\varepsilon = \varepsilon_r \varepsilon_o$ is the electrical permittivity of the dielectric material separating the plates. Small-value two-plate capacitors are often separated by a thin dielectric film of ceramic, mica, NPO, or even air. These capacitors are limited by physical constraints to a few hundred picofarads.

A larger value of capacitance requires either a larger area A, a smaller distance D, or a larger permittivity ε. Larger area can be achieved by stacking multiple layers of dielectric materials between parallel plates. In effect, this forms a group of parallel capacitors in a small space, leading to a relatively large capacitor value in a compact package. High-value surface-mount capacitors up to 1 µF or more can be fabricated using this type of stacked configuration.

Various types of foil capacitors can be constructed using a pair of long metal foil strips separated by a dielectric film such as mylar, polystyrene, or polypropylene. The long foil strips provide a large area A, but the strips and dielectric film must be rolled up and encapsulated in a tubular package to reduce the physical size of the capacitor.

To achieve dramatically larger values of capacitance, the dielectric film can be replaced by a permeable material such as paper soaked in an electrolytic liquid such as ammonia. The foil/liquid roll is sealed in an airtight package to prevent the liquid from evaporating. Then a small current is passed from one foil plate to the other through the electrolytic liquid. This process deposits an extremely thin layer of insulating material on one foil plate, while the other plate remains in contact with the conducting liquid.

In effect, the liquid becomes one plate, while the extremely thin insulating deposit forms the dielectric. This produces a very small value of D and allows very high values of capacitance in a small package. Such a capacitor is known as an *electrolytic capacitor*. Aluminum electrolytic and tantalum electrolytic capacitors are constructed using foils of aluminum and tantalum, respectively.

Unfortunately, the deposition of the insulting layer in electrolytic capacitors is a reversible process. DC current passing through the capacitor in the wrong direction can lead to destruction of the insulating dielectric film and the two plates of the capacitor can come into direct electrical contact through the electrolytic liquid. The large short circuit that results can lead to a rather destructive explosion. Therefore, electrolytic capacitors are marked with a polarity marker, typically a + or – sign or a black band representing a – sign.

Electrolytic capacitors must never be reverse biased, although two of them can sometimes be connected in a back-to-back series connection to form a nonpolarized electrolytic capacitor. In this configuration, DC current is always blocked in one direction by the insulating film of one of the capacitors, regardless of the polarity of the applied voltage. Thus neither capacitor's insulating layer can be damaged by a reversed DC current.

Capacitors suffer from a number of parasitic elements. The dielectric material can be slightly conductive, giving rise to an effective high-value resistance, R_P, in parallel with the capacitor's plates. This leads to current leakage from one plate to the other whenever the capacitor is charged. The capacitor's leads and plates also contribute series inductance, L_S, and series resistance, R_S. Thus a simple parasitic model of a physical capacitor including dielectric leakage and parasitic series resistance and inductance is shown in Figure 13.47.

The dielectric leakage resistance R_P is usually quite large, exceeding tens of megaohms in many cases. The highest leakage occurs in electrolytic capacitors, while the lowest leakage typically occurs in polystyrene or polypropylene capacitors. Smaller values of capacitance generally exhibit lower leakage currents than large ones simply because they generally have less plate and dielectric area.

A capacitor's series resistance is caused by the metallic leads and plates. At high frequencies, the dielectric material can become lossy enough to cause an additional frequency-dependent series resistance. The series inductance, dielectric losses, and metallic resistance give rise to a

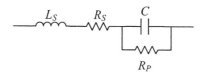

Figure 13.47. Capacitor parasitic model.

lower quality factor, or Q factor, for the capacitor. Capacitors with very good Q factors are required in very high-frequency applications such as microwave and RF systems. Surface-mounted NPO chip capacitors exhibit very low dielectric losses and series resistance, making them ideal for high-frequency applications. Also, surface-mounted ceramic capacitors are fairly well suited to high-frequency applications.

The series inductance of a capacitor can vary widely from a fraction of a nanohenry to hundreds of nanohenrys, depending on the shape and size of the capacitor. In general, electrolytic capacitors exhibit very high series inductance, while surface-mounted chip capacitors exhibit very low series inductance.

13.6.7 Inductors and Ferrite Beads

Inductors are usually built using a length of wire, often looped into a coil to increase the inductance. Larger values of inductance can be achieved by wrapping the wire coil around a magnetic core, such as iron or a ceramic material with magnetic properties. Inductors are available in both surface-mount packages and leaded packages. However, the surface-mounted packages are typically limited to smaller values of inductance.

Inductors are occasionally required as part of a DUT circuit such as a voltage doubler. They may also be used as part of a passive load circuit that must be connected to a DUT output to simulate a speaker coil or similar system-level component. Inductors can also be used in conjunction with capacitors to simulate long transmission lines by building an LC network like the one in Figure 13.25.

Inductors can be modeled as an inductance in series with a resistance. Depending on the magnetic core material chosen, the inductor may also exhibit lossy behavior at high frequencies, resulting in an apparent increase in the series resistance of the coil wire. In fact, a class of inductors called *ferrite beads* are intentionally designed with very lossy core materials to achieve a component with near-zero resistance at low frequencies and higher resistance at higher frequencies. A typical ferrite bead impedance chart is shown in Figure 13.48. As shown, both the inductance $Z_L(f)$ and the series resistance $R(f)$ of the ferrite bead are functions of frequency.

Ferrite beads are useful for blocking high-frequency interference signals. They can reduce AC crosstalk from one circuit to another, while allowing DC current to flow freely. For example, a ferrite bead can be placed in series with a power supply to prevent supply current spikes drawn by one circuit block from disturbing another circuit block, as shown in Figure 13.49. Both the inductance and resistance of the bead are near zero at DC; so power supply current can flow freely through the ferrite beads. Note that the schematic symbol for a ferrite bead is the same as that for an inductor. This is because a ferrite bead is an inductor with a lossy core.

13.6.8 Transformers and Power Splitters

Transformers are sometimes used on DIBs to translate a high-frequency single-ended signal into a differential signal or vice versa. Unfortunately, the transformer's frequency response is highly dependent on the output impedance of the transmitting circuit as well as the input impedance of the receiving circuit. Consequently, the frequency response of the transformer is difficult to accurately calibrate, since it may change from one DUT to the next.

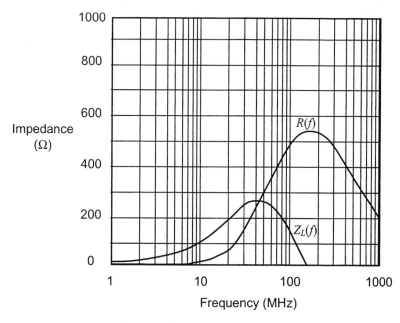

Figure 13.48. Typical ferrite bead impedance chart.

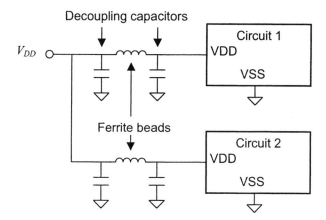

Figure 13.49. Ferrite beads in power supply connections.

Power splitters are controlled impedance transformers that are useful at very high frequencies. They are typically used in RF and microwave systems, but occasionally find use on mixed-signal DIBs. Their useful operation is often limited to a small range of frequencies, whereas transformers are generally able to handle a fairly wide range of frequencies.

Active circuits such as instrumentation amplifiers and other op amp circuits are often superior to transformer circuits, since they can be accurately calibrated. Also, they present a consistent, high-impedance load to the DUT. Transformers and power splitters, by contrast, present a low-impedance inductive or resistive load the DUT. Since many DUTs cannot drive low impedances, transformers and power splitters are often unusable. However, because they can pass frequencies well above those passed by active op amp circuits, they are sometimes the only viable choice. Also, when attached to differential DUT inputs, transformers allow the DUT to set its own common-mode voltage at the differential input. By contrast, the active differential outputs of an op amp single-ended to differential converter forces the common-mode input voltage to a predetermined voltage, which may or may not be acceptable for a particular DUT input.

13.7 COMMON DIB CIRCUITS

13.7.1 Local Relay Connections

One of the simplest DIB circuits is the local relay connection. A relay can be used to temporarily connect two points on the DIB, such as a DUT input and a V_{MID} output. Although it might be possible to achieve this connection using the tester's relay matrix or relays inside one of the tester instruments, a local DIB relay is often a superior connection. Consider the comparison between the relay matrix connection and the local DIB relay connection in Figure 13.50. The DUT's V_{MID} output can be connected to the V_{IN} input to idle the DUT's analog input signal for tests such as idle channel noise and DC channel offset. These two connection schemes are completely equivalent from a circuit schematic point of view.

If we consider the electrical noise susceptibility of the two schemes, we can see why the local relay may be superior in many cases. The traces and wires running between the DUT input and the V_{MID} output in the first design may be several feet long. This makes them very susceptible to

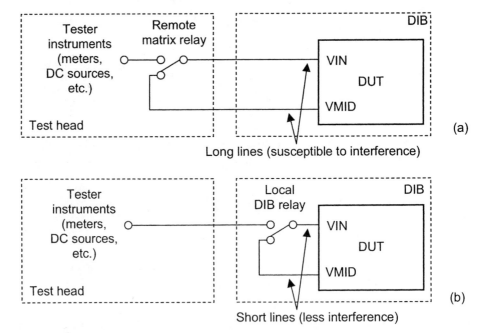

Figure 13.50. (a) Relay matrix V_{MID} connection versus (b) local relay V_{MID} connection.

electromagnetic interference from external sources such as 60-Hz power supply lines, radio stations, and cellular telephone transmissions. The local DIB relay connection is far less susceptible to this type of interference, since the relay is located very close to the DUT.

The long wires of Figure 13.50(a) also introduce another problem. In Section 13.3, we saw how traces and wires can add a significant amount of parasitic series inductance and trace-to-ground capacitance. These undesirable parasitic reactances form a filter between the V_{MID} output and the V_{IN} input. Since V_{MID} is a DUT-generated signal that may contain some noise added to its DC value, the parasitic filter will result in an undesirable AC voltage difference between the V_{MID} output and the V_{IN} input. The AC voltage difference will result in higher noise levels in the DUT's internal circuits. The local relay allows the noise from the V_{MID} output to feed directly into the V_{IN} input, allowing the noise on V_{MID} to be fully cancelled inside the DUT.

Another similar case in which a local relay is useful is in the measurement of AC common-mode rejection ratio (CMRR). This test requires that we apply the same AC signal to both pins of a DUT differential input stage and then measure the output (which ideally should be zero). We can connect the input lines together at the signal source through relays in the tester, as shown in Figure 13.51, or we can connect them together at the DUT using a local DIB relay.

At first glance, there does not appear to be any difference between these schemes. However, due to mismatches in the parasitic resistance, inductance, and capacitance to ground in the signal lines of Figure 13.51(a), the signals reaching the DUT's two inputs may not be perfectly matched. Perfect matching is required in the input signals of an AC CMRR test so that no differential input voltage exists. The local relay guarantees that both inputs receive the same signal, regardless of the effects of the parasitic elements distributed along the wires and traces connecting the tester to the DUT.

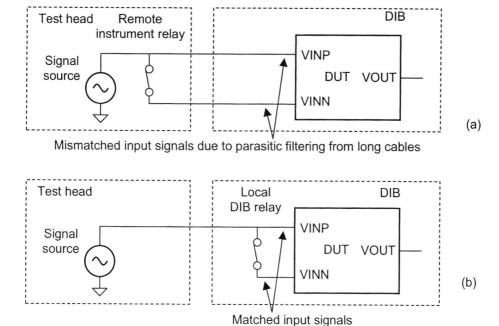

Figure 13.51. (a) CMRR input using remote relay versus (b) CMRR input using local relay.

The concept of using local DIB relays can be taken to extremes, of course. If a test engineer tries to use a local relay for every possible sensitive circuit node, the result will be a cluttered DIB. There is limited space on any DIB; so the test engineer should only use local relays when necessary.

13.7.2 Relay Multiplexers

Another common use for local DIB relays is signal multiplexing or demultiplexing (Figure 13.52). Examples of relay multiplexing include distribution of a tester signal source to multiple DUT inputs and distribution of a DUT output to multiple tester measurement instruments. Examples of relay demultiplexing include distribution of multiple tester instruments to one DUT input and distribution of multiple DUT outputs to one tester instrument.

Most multiplexing and demultiplexing is performed using relays inside the tester. Local DIB multiplexing is required for critical tests (such as the CMRR example in the previous section), or tests in which the tester's multiplexing scheme is not sufficiently flexible to make and break all the necessary connections with its own built-in relays. There are two ways to build a multiplexer or demultiplexer: the parallel configuration and the branching configuration (Figure 13.53). Each has its advantages and disadvantages.

The parallel configuration uses more relays than the branching configuration, but it requires the signal to travel through only one relay rather than many. The shorter signal path is sometimes advantageous when low signal path inductance is desired. On the other hand, the branching configuration can be used with coaxial controlled impedance relays to provide a controlled impedance multiplexer compatible with high-frequency transmission lines. The parallel configuration is not compatible with controlled-impedance transmission lines, since the common connection point of the relays results in a transmission line discontinuity known as a *stub*. (Transmission lines and their properties were discussed in Section 13.4.)

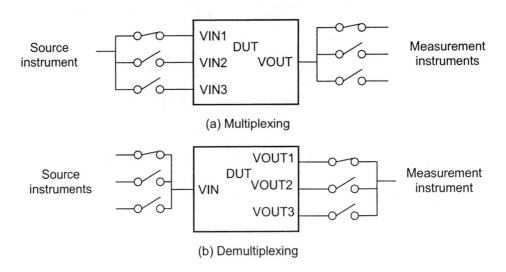

Figure 13.52. (a) DIB relay multiplexing and (b) demultiplexing.

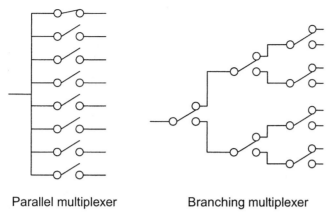

Parallel multiplexer   Branching multiplexer

Figure 13.53. Relay multiplexing schemes.

13.7.3 Selectable Loads

DIB relays are very useful for changing the DUT's electrical environment at various times during test program execution. For example, the distortion of a DUT earphone output may need to be tested while driving each of three different loads. These loads can be attached, one at a time, using relays (Figure 13.54). Distortion can be measured with any of the loads, or with no load at all.

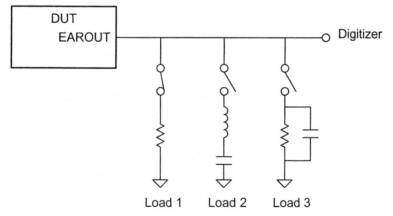

Figure 13.54. Selectable DUT output loads.

13.7.4 Analog Buffers (Voltage Followers)

Sometimes a device output is incapable of driving the parasitic capacitance presented by the traces, cables, and relays leading to a tester instrument. An analog voltage follower with higher capacitive drive capability can be used to buffer the output signal before it is passed to the tester. The primary concerns with analog buffers are offset, signal bandwidth, and added noise from the amplifier. Generally, higher-bandwidth amplifiers generate more noise while low-noise

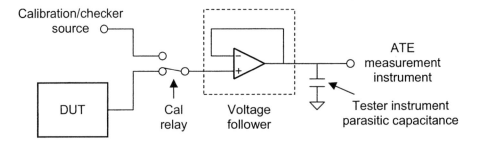

Figure 13.55. Buffer amplifier with calibration relay.

amplifiers have a limited bandwidth. Offset and gain errors can be removed through a focused calibration process (see Chapter 10). However, noise generated from the buffer may or may not be removable through a calibration process. Figure 13.55 shows a simple op amp buffer circuit including a relay for calibration and functional checking of the buffer.

Sometimes, an oscillating DUT amplifier can be stabilized using a small series resistor between the amplifier output and the tester. The resistor restores phase margin to the amplifier, eliminating the need for a buffer amplifier. This saves considerable complexity on the DIB.

13.7.5 Instrumentation Amplifiers

Another type of commonly used op amp circuit is the differential to single-ended converter, also known as the *instrumentation amplifier*. Figure 13.56 shows an instrumentation amplifier constructed using three op amps and four matched resistors. The two voltage followers at its input give the instrumentation amplifier a very high input impedance. Without these voltage followers, the resistors surrounding the third amplifier would present a load impedance of $2R$ to one of the DUT outputs and R to the other output. Of course, if the DUT outputs can drive these resistors without a problem, then the voltage followers are unnecessary. Like the previous analog buffer circuit, this circuit needs calibration relays to allow focused calibrations and checkers using a differential calibration/checker source.

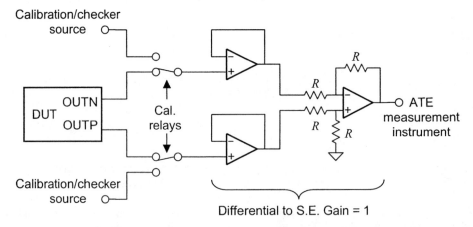

Figure 13.56. Op amp differential to single-ended converter with calibration relays.

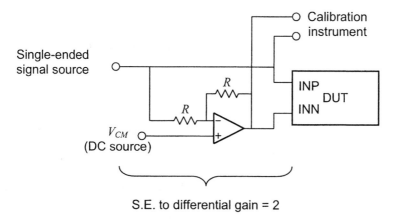

Figure 13.57. Single-ended to differential converter.

The reason this type of circuit is sometimes needed on a DIB is that certain tester instruments are not capable of receiving a differential signal. In such instances, the differential DUT output must be converted to a single-ended signal before the tester can measure it.

A related problem arises with differential inputs that must be driven from an instrument having only a single-ended output. The simple single-ended to differential converter in Figure 13.57 uses an inverter to generate an inverted image of a single-ended input. The differential signal is centered around a common mode voltage, V_{CM}, which is set to the desired voltage level by a tester instrument or by a DUT V_{MID} output. Notice that this single-ended to differential converter has a gain of 2 V/V. Therefore, the signal level of the single-ended source must be attenuated to one-half of the desired differential signal level.

Since the inverter in Figure 13.57 may introduce a significant phase shift at higher frequencies, this circuit is not ideal for high-frequency differential circuits. A somewhat better circuit can be built using a voltage follower in series with the noninverted signal to balance out the phase shift somewhat. Since the amplifiers are not in the same configuration, even the enhanced circuit produces some phase mismatch at very high frequencies. Fortunately, the circuit in Figure 13.57 is good enough for many DUTs having differential inputs.

13.7.6 V_{MID} Reference Adder

Sometimes, a DUT produces a V_{MID} voltage to which all input signals must be referenced. This type of input is commonly used in microphone inputs. Since microphones are basically differential signal generators, any noise or ripple present on the V_{MID} output gets cancelled by the differential input circuits of the DUT. Therefore, an input signal generated by the tester has to fluctuate with any noise and ripple on the V_{MID} signal to simulate the differential nature of a microphone. A V_{MID} adder can be built using the simple op amp circuit in Figure 13.58. The V_{MID} signal is added to the input signal from the tester, simulating the differential nature of a microphone. Since many DUTs are designed with a very weak V_{MID} output driver, a voltage follower is used to buffer the V_{MID} output before it is passed to the op amp adder. Obviously, if the DUT's V_{MID} driver is strong enough to drive a load of $2R$, then this extra buffer is unnecessary.

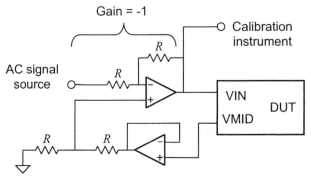

Figure 13.58. V_{MID} reference adder.

The limitations of this circuit are its bandwidth and the small amount of noise generated by the op amps. The gain and offset of this circuit must be calibrated for maximum accuracy. A transformer might be used instead of the active op amp circuit to work around noise and bandwidth limitations, but the transformer's frequency response would have to be carefully calibrated.

13.7.7 Current-to-Voltage and Voltage-to-Current Conversions

Tester instruments do not generally provide a means to directly measure AC currents. We can measure an AC current by dropping it across a resistor, but the parasitic capacitance of the tester instruments sometimes makes this an unacceptable solution. A low-impedance voltage output is a preferable signal for measurement. The circuit in Figure 13.59 can be used to convert current outputs to voltage outputs. The current to voltage translation is defined by Ohm's law ($V = I R$), although the op amp injects a factor of −1 into the equation ($V_{out}/I_{in} = -R$). The DUT sees a virtual ground at its output due to the feedback loop of the op amp. If the DUT is designed to drive its current into a different termination voltage (such as V_{MID}), then the noninverting input of the op amp can be connected to the desired termination voltage instead of ground. The simple I-to-V amplifier in Figure 13.59 is limited by the bandwidth, gain, and offset of the op amp. Its exact offset and voltage-over-current "gain" versus frequency must be calibrated using a focused calibration process.

Tester instruments also do not generally provide a means to force AC currents into the DUT. A transconductance amplifier circuit (Figure 13.60) can be used to make this conversion. In this circuit, the instrumentation amplifier senses the voltage drop across the source resistor, R_S, and feeds that voltage back to the inverting input of an op amp. The op amp adjusts its output voltage until the current forced across R_S generates a voltage drop equal to V_{in}. Of course, this transconductance amp must be calibrated at all frequencies of interest if maximum accuracy is to be achieved.

13.7.8 Power Supply Ripple Circuits

For some reason, mixed-signal testers have never included easily programmed ripple sources compatible with PSRR tests. PSRR tests require that we add a sinusoidal or multitone signal to one or more of the DUT's power supply voltages. This is not as easy as it might seem, since the

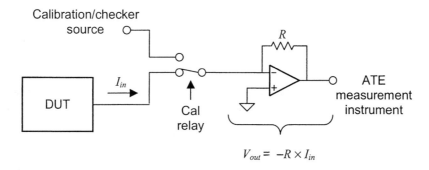

Figure 13.59. Current to voltage converter.

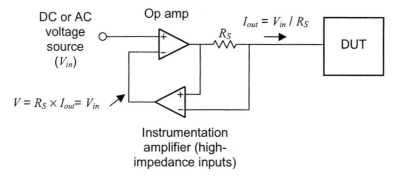

Figure 13.60. Voltage-to-current converter (transconductance amplifier).

output of power supplies include large bypass capacitors specifically designed to dampen ripple on the supply voltage. If the desired ripple cannot be provided by the tester itself, the test engineer can utilize any of a number of DIB circuits.

The simplest ripple injection approach takes advantage of the programmable DUT power supply's Kelvin sense line to force it to ripple its output. This ripple scheme is illustrated in Figure 13.61. The ripple source (a sine wave generator or arbitrary waveform generator) applies an AC signal to the sense line of the Kelvin connection through a resistor.

The power supply is forced to adjust its output to maintain a fixed voltage at its sense line. The power supply thus behaves as an op amp, with the sense line acting as a high-impedance inverting input and the programmed voltage level acting as a DC source at the op amp's noninverting input. The addition of the two resistors in the sense path forms an inverting gain stage, as shown in Figure 13.62. Once the Kelvin-connected DUT source is redrawn in this manner, it is easy to see how an AC signal can be injected into the power supply's output voltage. The output signal is given by

$$v_S = -\frac{R_2}{R_1}(v_R - V_P) \tag{13.20}$$

where v_S = DUT power supply voltage (including its normal DC value and the injected AC ripple), v_R = signal from the ripple source, and V_P = programmed DC voltage level set by the test program.

The values of R_2 and R_1 are chosen to give an attenuation, rather than a gain. This allows very small ripple voltages to be applied to the power supply using a fairly large ripple source amplitude. (Remember from Chapter 4 that large tester signal amplitudes are desirable because they are less susceptible to noise.) Values of R_1 = 10 kΩ and R_2 = 1 kΩ are commonly used, giving the ripple circuit a gain of 1/10.

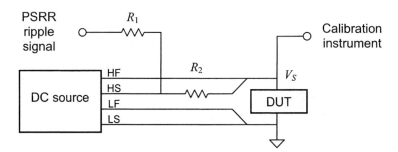

Figure 13.61. Kelvin sense ripple injection circuit.

If the circuit in Figure 13.62 is used, the ripple signal must include a DC offset equal to V_P. Otherwise, the ripple signal will introduce a DC offset into the power supply voltage. To use a ripple signal with no DC offset, a DC blocking capacitor can be added in series with the ripple source. Of course, this turns the ripple circuit into a first order high pass filter, as shown in Figure 13.63.

The cutoff frequency F_c of this filter is given by

$$F_c = \frac{1}{2\pi R_1 C} \text{ Hz} \tag{13.21}$$

Note that the circuit in Figure 13.63 includes a calibration path. This connection is absolutely necessary, since the frequency response of the Kelvin ripple circuit is not known. Each

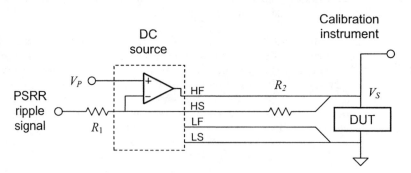

Figure 13.62. Equivalent op amp inverting gain stage.

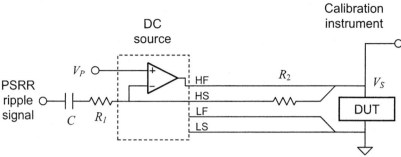

Figure 13.63. Equivalent ripple circuit with DC blocking capacitor.

frequency in the injected signal must be calibrated during a focused calibration process to achieve acceptable accuracy in the injected power supply ripple signal. If the DUT needs a large DIB decoupling capacitor on the rippled power, the capacitor must be removed from the circuit temporarily (with a relay) to prevent it from damping the ripple signal. The Kelvin ripple circuit has one major drawback. It is impossible to ripple most power supplies at a frequency higher than a few kilohertz. At higher frequencies, a different approach must be taken.

One possibility is to use a DIB ripple buffer which, can provide the DC plus AC signals needed to drive the DUT during the supply ripple tests. This circuit is simply an op amp adder circuit with a high-current buffer amplifier connected to its output (Figure 13.64). Again, a calibration path is added to improve the circuit's accuracy. This supply ripple circuit can be inserted into the power supply line using a SPDT relay, as shown.

Note that the large decoupling capacitor is automatically removed when the relay is thrown to the ripple circuit output. The large decoupling capacitor, typically a 10-µF electrolytic variety, provides relatively low-frequency currents to the DUT. Its own series inductance is many times that of the relay; so we can safely connect and disconnect it using the relay. A smaller decoupling capacitor is often needed to provide higher-frequency currents to the DUT. It must be located very close to the DUT to minimize series inductance between the capacitor and the

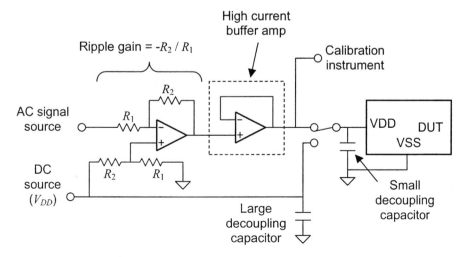

Figure 13.64. Buffer amp ripple injection circuit.

DUT power pin. Since the ripple circuit relay would provide too much series inductance, the small capacitor must be located next to the DUT. Therefore, the small high-frequency capacitor cannot be removed by the relay during the power supply ripple tests.

Exercises

13.10. When placed in a particular test mode, a DUT routes a weak internal circuit node to an analog test pin for measurement. The signal at the test node contains frequencies from DC to 100 kHz. Its output impedance is 10 kΩ (purely resistive). We wish to measure the signal using a digitizer having a distributed cable capacitance of 350 pF plus a lumped input capacitance of 50 pF. Can this signal node be measured directly, or will a buffer amplifier be needed on the DIB?

Ans. At relatively low frequencies, the DUT output impedance forms a low-pass filter with the digitizer input capacitance. The 3-dB cutoff frequency of the filter is 3.98 kHz, preventing accurate measurement at 100 kHz. A local buffer amplifier, such as the one in Figure 13.55, will be needed on the DIB.

13.11. The Kelvin PSRR ripple circuit illustrated in Figure 13.61 is constructed using the values $R_2 = 1$ kΩ and $R_1 = 10$ kΩ. The power supply is programmed to 3.3 V DC. Describe a signal, $v(t)$, at the input to R_1 that would produce a DUT power supply ripple of 75 mV RMS at 1 kHz. (Assume no errors due to circuit bandwidth, component mismatch, etc.)

Ans. $v(t) = 3.3 \text{ V} - \sqrt{2} \times 75 \text{ mV} \times \dfrac{10 \text{ k}\Omega}{1 \text{ k}\Omega} \sin(2\pi f t) = 3.3 \text{ V} - 1.06 \text{ V} \times \sin(2\pi \times 1000 t)$.

13.12. Repeat Exercise 13.11 using the PSRR buffer circuit illustrated in Figure 13.64. (Assume V_{DD} is set to 3.3 V DC.)

Ans. $v(t) = -1.06 \text{ V} \times \sin(2\pi \times 1000 t)$.

13.8 COMMON DIB MISTAKES

13.8.1 Poor Power Supply and Ground Layout

One of the most common sources of noise injection in mixed-signal DIBs is poor power and ground layout. The best way to avoid problems with power distribution and grounding is to use as many planes as needed, without regard to DIB cost. Although each layer in a multilayer DIB adds fabrication cost, the expense is fairly negligible compared to the production yield loss due to poor DIB performance. Therefore, a good DIB might include one or two layers dedicated to digital transmission line ground and noisy current returns, one layer dedicated to analog current returns, one layer dedicated to low-current analog ground (quiet ground serving the purpose of a zero-volt reference layer), and at least one layer dedicated to split power planes. Thus a ten-layer DIB may contain five or six layers dedicated to power and ground distribution.

13.8.2 Crosstalk

Another common problem on mixed-signal DIBs is crosstalk, especially between digital and analog signal lines. The digital-to-analog crosstalk problem can be dramatically reduced by placing analog signals on a separate PCB layer from digital signals, with an analog ground plane between the two signal layers. Analog-to-analog crosstalk can be minimized by simply realizing which signals are most susceptible (i.e., which signals have the highest impedance) and preventing high-frequency, high-amplitude signals from passing nearby. Also, the sensitive high-impedance nodes should be as short as possible to avoid crosstalk and coupling of external noise sources such as radio waves.

One of the most common sensitive nodes on a mixed-signal DUT is its current reference input. This input is typically tied to V_{DD} or ground through a very high-impedance bias resistor. The node between the bias resistor and the DUT is an extremely sensitive one. Noise injected into this node will translate directly into noise throughout the DUT. Therefore, current bias nodes should always be kept extremely short, preferably surrounded by a shield ring.

Another type of sensitive node is the DUT reference voltage. Reference voltages are typically driven by a low-impedance tester source, and are therefore less susceptible to crosstalk than high-impedance bias nodes. However, any noise injected into the reference voltage will translate directly into noise in the DUT circuits. Therefore, reference voltage nodes should also be laid out as if they were extremely vulnerable to crosstalk.

13.8.3 Transmission Line Discontinuities

Small discontinuities in transmission lines can lead to glitches on the rising and falling edges of very fast digital signals. Such glitches can sometimes lead to timing errors or double-clocked logic in the DUT. The discontinuities are caused by lumped capacitance or inductance at transition points along the transmission line. For example, a lumped capacitance and/or inductance exists whenever a digital signal trace is routed between layers through a via or other through hole. It is best to avoid routing digital signals from one layer to another, unless absolutely necessary.

13.8.4 Resistive Drops in Circuit Traces

As we saw in Example 13.1, even relatively short traces may have a series resistance of several hundred milliohms. If we try to force current through such a trace, we will get a voltage drop due to the parasitic resistance of the trace. Sometimes these voltage drops are unimportant, but other times they can lead to errors nearly as large as the parameter we are trying to measure. The test engineer should always consider the effects of series resistance on each trace on the DIB. If the series resistance is serious enough to cause a problem, the trace can either be made wider, or a sensing circuit such as a Kelvin connection can be used to compensate for the resistive drops in the PCB trace.

13.8.5 Tester Instrument Parasitics

The various cables and wires that connect a DUT to a tester's instruments can present a significant capacitive load to the DUT's pins. Often, the loading is high enough to cause gain errors, phase shifts, or even DUT circuit oscillations. It is very important for a test engineer to ask the design engineer responsible for each DUT circuit what its capacitive drive capabilities will be.

If the output impedance of the DUT is incompatible with the tester's load capacitance, then a voltage follower will probably be needed on the DIB buffer the DUT's output. Even if the DUT is designed with a low output impedance, the load capacitance of some tester instruments may cause it to break into oscillations. The test engineer and design engineer should determine whether the unbuffered tester inputs might cause any DUT oscillations. If so, analog buffers must be added to the DIB design.

13.8.6 Oscillations in Active Circuits

Operational amplifiers used in buffer amplifiers or other DIB circuits may break into oscillations if they are not laid out properly. For example, the inverting and noninverting inputs to an op amp are extremely sensitive to parasitic capacitance. If these PCB traces are laid out so that they are more than a few tenths of an inch long, the amplifier will often break into oscillations. This problem is commonly seen in the nulling amplifier circuits described in Chapter 3, "DC Measurements."

Another source of oscillations is poor power supply and decoupling capacitor layout. If the decoupling capacitors attached to an amplifier's power supplies are not positioned very close to the amplifier's power pins and ground plane, then the amplifier may break into oscillations. The oscillations are due to the extra parasitic inductance of the connecting PCB traces. This is especially true of high-bandwidth operational amplifiers. Decoupling capacitors in general should always be placed very close to their supply pin.

Oscillating amplifiers pose a particularly tricky problem. When measuring a DC offset using an oscillating buffer amplifier, the tester's DC voltmeter ignores the oscillation. Instead, it measures the average voltage level at the oscillating amplifier's output, which may or may not have any relation to the DUT signal to be buffered. Thus significant DC errors can be introduced by the oscillating amplifier.

13.8.7 Poor DIB Component Placement and PCB Layout

If the test engineer gets nothing else out of this chapter, at least one fact should come across loud and clear. The physical layout of the DIB is extremely critical to mixed-signal DUT performance. We have seen many cases throughout this chapter in which a short PCB trace is the ideal interconnection. Short traces have less parasitic resistance, inductance, and capacitance than long traces. The best way to achieve short PCB traces is to arrange the DIB components in a way that allows short traces, especially in critical nodes.

Component placement, power and ground schemes, trace layout, and other physical decisions must be made with knowledge of the DUT and DIB circuits and their required performance. This fact makes it very difficult for automatic routing software to lay out mixed-signal DIBs. In fact, it is very difficult to get a good DIB layout from a manual process, unless the test engineer sits with the PCB designer as the critical components and traces are placed on the DIB. This fact escapes many novice test engineers, who literally throw the DIB schematic into the PCB designer's lap and walk away, assuming all will turn out well.

13.9 SUMMARY

A good DIB is one of the most critical elements in a successful mixed-signal test solution. Without good DIB performance, the DUT may be unable to meet its specifications, regardless of the quality of the test code. Many things lead to good mixed-signal DIB design, including proper component selection and placement, proper power and ground layout, proper PCB stackup, and proper attention to parasitic components related to PCB traces and DIB components.

A good schematic design is essential as well as a good DIB layout. If the test engineer forgets to provide for an important connection between the tester and the DUT, then all the clever software routines in the world will not make the DIB useable. Also, if the test engineer does not provide for all the necessary hardware hooks to calibrate the DIB circuits, then it will be equally useless because it will not provide the necessary accuracy.

Often, the only way to produce a good DIB schematic is to have a complete test plan and data sheet to begin with. If the design specifications, DUT pinout, and test list are constantly changing, then it will be impossible to design a good DIB. Also, if the test engineer and design engineers do not work together closely, the DIB and DUT will often be incompatible with one another. It is critical for the test engineer to review his test plan and DIB design with the design engineers before the DIB is laid out and fabricated. Otherwise, the DIB may turn out to be an expensive but useless piece of test hardware, proving once again that concurrent engineering is critical to mixed-signal product development.

APPENDIX A.13.1

Review of *RC* and *RL* Filters

Since much of this chapter was devoted to evaluating the interactions between parasitic reactances and DUT impedances, a quick summary of simple one-pole filters is in order. The following equations describe the cutoff frequency, gain, and phase shift of the four most common types of *RC* and *RL* filters. (Some of these equations can be found elsewhere in this chapter, but they are repeated here for completeness.)

The four first-order *RC* and *RL* filters are shown in Figure 13.65. The ground connection in each filter can be replaced by any DC voltage without affecting the gain or phase response of the filter.

3-dB Cutoff Frequency

RC high-pass or *RC* low-pass filter

$$F_C = \frac{1}{2\pi RC} \text{ Hz} \qquad (13.22)$$

RL high-pass or *RL* low-pass filter

$$F_C = \frac{R}{2\pi L} \text{ Hz} \qquad (13.23)$$

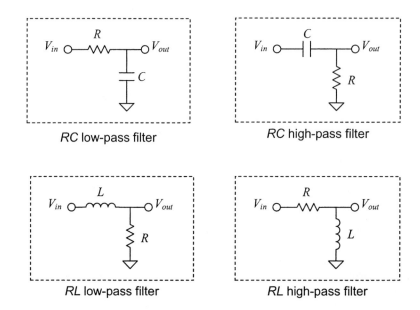

Figure 13.65. First-order *RC* and *RL* filter configurations.

Frequency Response

RC low-pass or *RL* low-pass gain versus frequency

$$|H(f)| = \left[\frac{1}{\sqrt{1+\left(\frac{f}{F_C}\right)^2}}\right] \text{ V/V} \qquad (13.24)$$

RC high-pass or *RL* high-pass gain versus frequency

$$|H(f)| = \left[\frac{\frac{f}{F_C}}{\sqrt{1+\left(\frac{f}{F_C}\right)^2}}\right] \text{ V/V} \qquad (13.25)$$

RC low-pass or *RL* low-pass phase versus frequency

$$\angle H(f) = -\frac{180}{\pi}\tan^{-1}\left(\frac{f}{F_C}\right) \text{ degrees} \qquad (13.26)$$

RC high-pass or *RL* high-pass phase versus frequency:

$$\angle H(f) = \frac{180}{\pi}\tan^{-1}\left(\frac{F_C}{f}\right) \text{ degrees} \tag{13.27}$$

(phase shift = 90 degrees at $f = 0$)

Problems

13.1. Calculate the parasitic resistance of a 13-in. PCB trace having a width of 20 mils and a thickness of 1 mil. A pair of these traces are used as the high-force and low-force lines of a Kelvin-connected voltage regulator located on the DIB. The regulator feeds a 3.3-V DC signal to a 5-Ω load resistance. How much current will flow through the four Kelvin lines? What will be the differential voltage between the high-force and low-force output of the voltage regulator, measured at the regulator side of the PCB traces?

13.2. Using Eq. (13.3), calculate the parasitic inductance per unit length and total inductance of a 3-in. stripline trace having a width of 50 mils and a spacing of 8 mils to the current-return ground plane. If this trace feeds a 5-MHz, 625-mV RMS sinusoidal signal to a 100-Ω load resistance, what will be the error of the RMS voltage at the load as a percentage of the source voltage? (Assume zero trace resistance and zero signal source impedance.) Compare your answers with those obtained using the refined inductance estimate of Figure 13.10.

13.3. Using Eq. (13.10), calculate the parasitic capacitance per unit length and total capacitance of a 3-cm-long, 35-mil-wide stripline trace with a spacing of 10 mils to the ground plane, fabricated on a Teflon® PCB. Compare your answer with that obtained using the refined capacitance estimate of Figure 13.16.

13.4. The stripline trace in Problem 13.3 is connected to the 50-kΩ source impedance of a DUT output. What are the gain and phase shift of the DUT output signal as a function of frequency, compared to its unloaded output (i.e., if it were not connected to the trace)? (Use the estimate of capacitance from Figure 13.16.)

13.5. Using Eq. (13.13), calculate the parasitic capacitance per unit length and total capacitance between two 8-in.-long, 12-mil-wide coplanar traces separated by a spacing of 20 mils, fabricated on an FR4 PCB. Compare your answer with that obtained using the refined capacitance estimate of Figure 13.18.

13.6. Using Eq. (13.10), calculate the parasitic capacitance of an 11.5-in.-long, 15-mil-wide stripline trace with a spacing of 12 mils to the ground plane, fabricated on an FR4 PCB. This trace feeds a 300-kHz, 1-V RMS sinusoidal signal from a DUT output having a 75-kΩ output resistance to an unterminated coaxial cable (tester instrument input) having a distributed capacitance of 35 pF. How much will the distributed capacitance of the trace and coaxial cable capacitance attenuate the DUT signal? (Express your answer as a voltage gain in decibels.) Would a DIB buffer amplifier be required for this output? Compare your answer with that obtained using the refined capacitance estimate of Figure 13.16.

13.7. An 8-in. stripline trace is fabricated on an FR4 PCB with a width of 24 mils. It is separated from its ground plane by a spacing of 16 mils. Using the refined estimates of Figures 13.10 and 13.16, calculate the stripline's parasitic capacitance per meter and inductance per meter. What is the stripline's characteristic impedance? If we want to lower the characteristic impedance, would we make the layer spacing from trace to ground larger or smaller? If we were constrained to a layer spacing of 16 mils, would we make the trace wider or smaller? If we increased the length of the trace to 16 in., what would happen to the characteristic impedance?

13.8. What is the velocity of a signal propagating along the stipline of Problem 13.7? Express your answer in m/s and in a percentage of the speed of light. What is the stipline's propagation delay? What is its distributed capacitance?

13.9. What is the wavelength, in meters, of a 20-MHz sine wave travelling along the FR4 stripline in Problem 13.7? Can we treat the parasitic reactances of the stripline as lumped elements, or do we have to treat the stripline as a transmission line at this frequency?

13.10. A 1.200-GHz sinusoidal signal is transmitted from a DUT output to tester digitizer along a 20-cm 50-Ω terminated coaxial cable having a signal velocity of $0.8c$. What is the wavelength of the transmitted signal as a percentage of the cable length? What is the distributed capacitance of this coaxial cable? What is the phase shift, in degrees, between the DUT output and the digitizer input? (Your answer may exceed 360 degrees.)

13.11. A DUT output signal under test contains frequencies from DC to 44 kHz. The DUT output impedance is guaranteed to fall between 50 and 100 Ω (purely resistive). We wish to measure the signal using a digitizer having a distributed cable capacitance of 120 pF plus a lumped input capacitance of 5 pF. Can this signal node be measured directly, or will a buffer amplifier be needed on the DIB?

13.12. The in-phase and quadrature outputs (IOUT and QOUT) of a cellular telephone DUT are produced by two supposedly identical amplifier circuits. According the the data sheet, the resistive output impedance of each amplifier circuit is specified at 75 Ω (min) to 125 Ω (max). We wish to measure the phase mismatch between these two outputs when a 70-kHz sine wave is driven from each output. Assuming a perfectly matched capacitive load of 300 pF from each of two digitizers, what is the worst-case phase mismatch caused by the tester loading? How could we eliminate the tester-induced phase error problem?

13.13. The Kelvin PSRR ripple circuit illustrated in Figure 13.61 is constructed using the values $R_2 = 1$ kΩ and $R_1 = 4.7$ kΩ. The power supply is programmed to 5.0 V DC. Describe a signal, $v(t)$, at the input to R_1 that would produce a DUT power supply ripple of 100 mV peak-to-peak at 2.4 kHz. (Assume no errors due to circuit bandwidth, component mismatch, etc.)

13.14. Repeat Problem 13.13 using the PSRR buffer circuit illustrated in Figure 13.64. (Assume V_{DD} is set to 5.0 V DC.)

References

1. John D. Kraus, Keith R. Carver, *Electromagnetics*, Second Edition, p. 85, McGraw Hill, New York, NY, 1973, ISBN: 0070353964, p. 85

2. Howard W. Johnson, Graham Martin, *High Speed Digital Design: A Handbook of Black Magic*, Prentice Hall, Englewood Cliffs, NJ, April 1993, ISBN: 0133957241
3. Brian Wadell, *Transmission Line Modelling Handbook (Artech House Microwave Library)*, Artech House, June 1991, ISBN: 0890064369
4. Henry W. Ott, *Noise Reduction Techniques in Electronic Systems*, Second Edition, John Wiley & Sons, New York, NY, March 1998, ISBN: 0471850683

CHAPTER **14**

Design for Test (DfT)

14.1 OVERVIEW

14.1.1 What Is DfT?

Design-for-test (DfT, also DFT[*]) is a major topic of interest in the automated testing field. Any design methodology or circuit that results in a more easily or thoroughly testable product can be categorized as DfT. DfT, when properly implemented, can offer lower production costs and higher product quality. Extensive literature exists on the subject of DfT,[1-3] though much of it pertains to purely digital circuits. Since so much digital DfT literature is already available elsewhere, this chapter will concentrate mostly on analog and mixed-signal DfT. Nevertheless, some of the more common digital concepts will be reviewed to give the reader a basic overview of digital DfT.

There are many types of DfT. Some DfT approaches are highly structured, using industry-defined standards. Other approaches are totally ad hoc, invented by the design engineer or test engineer to solve a specific test problem on a particular device or category of devices. Some DfT concepts are based on built-in circuits that allow easier or more complete testing. Other methodologies, such as increasing design margin to reduce test cost, are equally cost effective but may not be recognized as DfT concepts. In the end, the choice of DfT approach depends very much on the specifics of the device under test (DUT) and the demands placed on it by its system-level application.

In the past, design engineers were sometimes reluctant to add testability features to a device, since DfT added design cycle time, die area, and/or power consumption. Their reluctance was often reinforced by managers who judged design engineering performance based mainly on these criteria rather than the overall cost effectiveness, marketability, and quality of the finished product. Fortunately, the attitude has changed in recent years from reluctance to enthusiasm as design engineers, managers, and customers have embraced the competitive advantages of DfT. Now DfT is seen as a major technological differentiator that can reduce production costs, enhance quality control, and even provide customers with value-added testability features for use in their system-level products.

---

[*] This text discusses both design-for-test and the discrete Fourier transform. For clarity, the lower case notation (DfT) is used throughout this text when referring to design-for-test while the upper case notation (DFT) is used when referring to the discrete Fourier transform.

Although we have only devoted a single chapter to the topic of DfT, this subject is of extreme importance to the semiconductor industry. Our light treatment of the topic is due to the broad, introductory nature of this text and to the fact that this book is targeted toward test engineers rather than design engineers. While test engineers are not typically expected to implement the DfT concepts suggested in this chapter, they are expected to participate in the DfT planning phase of new product development.

Test engineers also need to understand the types of DfT they may encounter when developing a test program for a new device. Our intention is to provide the test engineering professional a cursory introduction to digital and mixed-signal DfT rather than providing a design engineer with detailed knowledge to implement digital and mixed-signal DfT circuits.

14.1.2 Built-In Self-Test

Built-in self-test (BIST) circuits allow the DUT to evaluate its own quality without elaborate automated test equipment (ATE) support. Although BIST and DfT are often treated as if they were separate concepts, BIST is actually a type of DfT. Digital BIST circuits usually return a simple pass/fail bit or a multibit "signature" that allows the ATE tester to evaluate the quality of the device with a very simple (i.e., low cost) test. A BIST circuit may require little more than a power supply and a master clock from the tester. Since the DUT tests itself using BIST, a much less expensive ATE tester can be used. The limited tester resources required by BIST and the ability to perform parallel testing of multiple circuits on the DUT are key advantages of BIST-based testing methodologies.

Unfortunately, analog BIST technology has lagged behind digital BIST because of difficulties in guaranteeing the accuracy of signals generated and measured on-chip. Many books and technical papers have been written on digital BIST techniques,[4-6] but fewer have been written about analog and mixed-signal BIST.[7-10]

14.1.3 Differences between Digital DfT and Analog DfT

DfT for purely digital designs has been extensively utilized for many years. Using a variety of software tools and industry standards, a digital design engineer can follow a well-defined path toward a testable design. Software tools can automatically insert the necessary DfT circuits into a digital design. The same tools can automatically generate the digital patterns to test the design.

Mixed-signal DfT is much less standardized because the testing requirements and failure mechanisms for the analog circuits in a mixed-signal device are often not particularly well understood or well defined. Digital circuits, for example, can be separated into subcircuits using a divide-and-conquer approach. The subcircuits are fairly independent from one another except for race conditions and other timing problems. To a large extent, these timing problems can be avoided using additional automated software tools. As a result, we can test the subsections of a digital circuit to guarantee the operation of the whole.

Mixed-signal circuits can also be subdivided, but the quality of the whole is seldom guaranteed by the quality of the parts. Analog circuits are frequently prone to obscure crosstalk problems and other subtle interactions between circuit blocks. The divide-and-conquer approach is necessary for characterization and diagnosis of mixed-signal devices, but it may not be sufficient to guarantee the system-level specifications.

14.1.4 Why Should We Use DfT?

DfT circuits and methodologies offer a tremendous advantage in the marketplace. An IC designed without attention to testability may work perfectly well in the customer's application, but a competitor's IC may win in the marketplace because of superior DfT features. The advantages of DfT include lower testing costs, higher product and process quality, ease in design diagnostics and characterization, ease in test program development, and enhanced diagnostic capabilities in the customer's system-level application. Let us look at some examples of each of these advantages.

14.2 ADVANTAGES OF DfT

14.2.1 Lower Cost of Test

Perhaps the most visible advantage of DfT is that it can lead to lower testing costs. Consider the power-down logic block in Figure 14.1. This logic block controls the power-down status of all the circuit blocks in the device under test (DUT). The digital logic block accepts five digital inputs from a variety of external device pins and internal control register bits. It uses combinational logic to map the thirty-two possible input states to one of three valid power modes (normal mode, power-down mode, and standby mode). Each of the three power-down modes is expected to produce a unique combination of supply currents, I_{DDA} and I_{DDD}. The three power modes are determined by two power mode control lines, PWRMODE0 and PWRMODE1. The test engineer is required to verify the truth table of the power-down logic block, as well as to measure the I_{DDA} and I_{DDD} power supply currents in all three power modes.

From the design engineer's perspective, this circuit may appear to be well designed. The truth table for mapping the five input signals into three power-down states works perfectly with a minimum number of gates. The analog circuits power down as expected in each of the three modes. What could be wrong with this design?

The problem is that the two outputs of the power-down logic block can only be observed by measuring the power supply current drawn by the DUT. Because the DUT requires power supply decoupling capacitors, the transition from one power state to another may require settling time as the decoupling capacitors charge and discharge. This is especially true when switching to the ultra-low-current power-down mode, since current discharges very slowly from the

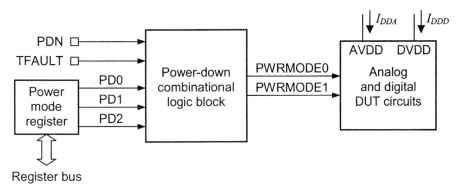

Figure 14.1. Power-down logic block without DfT.

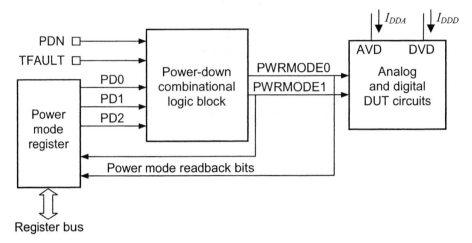

Figure 14.2. Power-down logic block with readback DfT.

decoupling capacitors in this mode. In addition to the decoupling capacitors and other DIB circuits, the DUT and ATE meter may also require settling time. Due to the various settling time requirements, each measurement of I_{DDA} and I_{DDD} might take five milliseconds to complete. The total test time for this logic block could therefore take as much as 320 ms (32 input state combinations times 2 supply currents times 5 ms). While 320 ms may sound like a reasonable test time, it is an eternity in production testing. Obviously, this is a ridiculous way to test a simple digital logic block. The problem is that the output bits of the power-down logic block can only be observed by making time-consuming analog measurements of I_{DDA} and I_{DDD}.

One possible solution to this problem is a very simple DfT circuit allowing the ATE tester to directly observe the power control bits of the logic block (Figure 14.2). This DfT circuit is implemented as a simple two-bit readback function using two unused bits in the existing power mode register (or two bits in a dedicated test register if there are no unused power mode register bits). Alternatively, the bits could be read back through a scan chain DfT structure, which we will discuss later. Using the readback DfT approach, all gates in the combinational logic block can be verified in less than a millisecond using a very fast digital pattern. After the digital logic has been verified, each of the three power supply current combinations only needs to be measured once. This DfT-based approach leads to a total test time of about 30 ms (3 input states times 2 supply current measurements times 5 ms) as opposed to the 320 ms of test time required for the non-DfT version of this design. The lower test cost resulting from the lower test time easily justifies the few logic gates required to add this type of readback DfT capability.

Another way in which DfT can reduce testing costs is by reducing the requirements of the ATE tester. A test that requires a 50-MHz digital pattern will cost more than a test operating at 25 MHz, because an ATE tester capable of running at high frequencies is generally more expensive than one that is only capable of lower frequencies. If the IC design engineer can find a way to test high-frequency signals using low-frequency stimulus and measurement hardware, the test cost savings can be substantial. The same comment applies to digital channel count. A device that requires a 64-channel tester will be much less expensive to test than a device that requires 256 channels. If the design engineer can find ways to reduce digital channel count (using multiplexed I/O pins, for example), then the test cost can be reduced significantly.

14.2.2 Increased Fault Coverage and Improved Process Control

Economic considerations are only one of the advantages of DfT. Another advantage is increased fault coverage. Fault coverage is defined as the percentage of possible failure modes that can be detected by a given test or series of tests. Therefore, increased fault coverage reduces the probability that defective devices will be shipped to the customer. While the economic advantage of lower test time is fairly easy to calculate, the economic advantage of happy customers is much harder to quantify financially. Many forms of DfT are designed to allow increased test coverage, with the understanding that the smiling face of a satisfied customer is well worth slightly higher test time and silicon cost.

I_{DDQ} testing is one such DfT methodology allowing detection of noncatastrophic defects in digital logic. An I_{DDQ} test configures all the gates in a CMOS device into a static digital state and then measures the tiny current leaking from power to ground. Excessive I_{DDQ} current indicates one or more resistive defects between power and ground that may or may not be detectable as a catastrophic failure in the operation of the DUT. When the I_{DDQ} tests suddenly begin rejecting many dies, it often indicates that the wafer fabrication process has gone awry for some reason, producing resistive shorts between circuit nodes. I_{DDQ} testing therefore allows the semiconductor wafer fab to monitor its process to detect and correct problems quickly.

14.2.3 Diagnostics and Characterization

When a design is first released to production, the new product undergoes a characterization process to determine whether or not it meets the customer's requirements. The first-pass design often has problems that must be corrected before a final version of the design can be released to production. To produce a production-worthy design in a short timeframe, we must be able to characterize the IC's performance and diagnose internal circuit problems very quickly. Lack of proper DfT observability circuits can make the diagnostic process extremely difficult or even impossible.

As an example of the diagnostic capabilities of DfT, consider a mixed-signal ADC channel (Figure 14.3). It may include several components, including an input amplifier, a programmable gain amplifier (PGA), a low-pass antialiasing filter, and the ADC itself. If a significant percentage of production devices fail the ADC channel's signal-to-distortion test, then the design needs to be corrected. Without DfT, it can be difficult to determine which of the four circuit blocks is introducing the distortion.

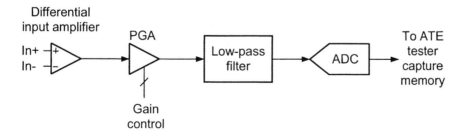

Figure 14.3. ADC channel without DfT.

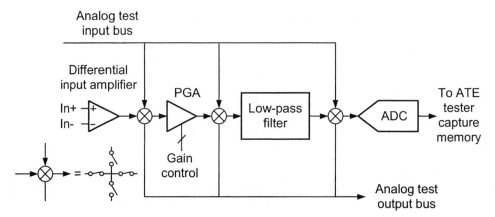

Figure 14.4. ADC channel with analog test access bus.

DfT can provide the necessary diagnostic capabilities to resolve problems like this. By providing test access points to each input and output node in the signal path, the test engineer can test each section of the circuit independently (Figure 14.4). The CMOS switch matrix at each circuit node is capable of three modes. First, it can pass the signal from one block to the next for normal operation. Second, it can disconnect the output from one block and connect the input of the following block to the analog test input bus. The tester can then inject a test signal into the input of the circuit under test. Finally, the switch matrix allows observation of the output of the circuit under test through the analog test output bus.

The defective circuit can be isolated quickly by injecting signals into each block and observing the block's output. Once the defective circuit has been isolated, it can be redesigned to correct the problem. Observability and controllability test modes such as this are a major cornerstone of mixed-signal DfT. Similar observability/controllability circuits can be employed in digital circuits to accelerate the diagnosis of circuit problems. The CMOS switches are replaced by digital multiplexer circuits or scan cells to achieve the same results.

14.2.4 Ease of Test Program Development

Another advantage offered by DfT is easier test program or test hardware development. A number of design decisions can potentially make the test engineer's life unnecessarily difficult. For example, a digital circuit that has no reset capability will come up in an unknown state after the DUT is first powered up. In many cases, the test engineer must set the device into a known state before a particular test can be executed. Since the device cannot be reset to a known state, the test engineer has to start clocking the DUT until it reaches the desired state before the test can proceed.

Many testers include a "match mode," which allows the digital pattern to loop until a certain device state is reached. Match mode search loops are often complicated by pipelining issues in the ATE tester's digital pattern sequencer, which makes the test engineer's task more difficult. It is usually possible to test a device with incomplete reset circuitry, but it requires quite a bit of extra effort on the test engineer's part. Furthermore, the match mode search process can sometimes lead to excessive test times.

Since match mode solves the reset problem without extra design effort and with little extra test cost in most cases, reset capabilities may be overlooked in the testability planning phase. However, in cases where the reset capability comes at little extra design effort or silicon area, the reduced test development difficulty can lead to a shorter product cycle time. This is one of those gray areas where the test engineer and design engineer need to negotiate to determine an appropriate balance of design effort and silicon area versus test effort and test time.

On a related topic, lack of digital resets causes other problems that the design engineer may find more important. First, software simulations of digital circuits are more reliable when the models of the device can be reset to a known state at the beginning of a test sequence. Second, ATE test patterns cannot be generated automatically from design simulations if the device lacks a reset capability on all state-based circuits (digital flip-flops, registers, and state machines, etc.). Compatibility with automated test vector generation tools is probably the most important reason resets should be provided for all digital circuits.

14.2.5 System-Level Diagnostics

A final advantage of DfT is that it often allows the customer to incorporate DfT into the end application (cellular phone, graphics card, etc.) more easily. Examples of IC-level DfT that are geared toward end application DfT include the IEEE 1149.1 and IEEE 1149.4 boundary scan standards. These standards, which allow chip-to-chip and circuit-to-circuit testing, will be reviewed in more detail later in this chapter. Customers may also request custom DfT test modes to allow easier integration of board-level DfT features.

14.2.6 Economics of DfT

Considering all the advantages of DfT, there must surely be drawbacks. One of the biggest disadvantages of DfT is increased circuit complexity and the resulting increase in silicon die area. Ultimately, the tradeoff between the economic disadvantage of DfT and the various advantages must be evaluated by the design and test team. The most straightforward way to justify the increase in die cost is to evaluate each proposed DfT structure and determine how much it will save in production test costs per die. If the increased manufacturing cost of the extra die area is less than the decreased production test cost, then DfT expense can be justified.

Unfortunately, some advantages of DfT, such as enhanced diagnostic capabilities, are much harder to quantify. The advantage of enhanced diagnostics often leads to lower time to market, which is difficult to translate into an exact dollars-and-cents value. However, it is generally accepted that lower time to market results in much higher profit margins over the life of the product.[11] Since there are fewer competitors early in a differentiated product cycle, there is less pricing pressure on the product. Consequently, the highest profit margins are typically attained early in the product cycle, after the high costs associated with the initial learning curve have been resolved (Figure 14.5).

After a brief period of limited competition, profit margins begin to decline as more competitors fight for the same market share. Thus a delay in time to market will usually result in substantially lower profit margin over the product's shortened life span. A delay may also result in lost business if a competitor's solution is designed into the customer's system. It is difficult to quantify exactly how much profit is lost as a result of delayed time to market, although the amount is known to be quite high. It is even more difficult to estimate how much the time to

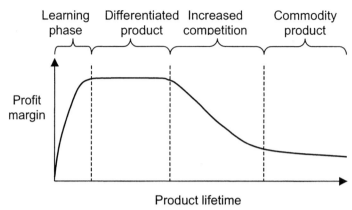

Figure 14.5. Profit margin over a product's market life.

market will be reduced by a particular DfT approach. For these reasons, financial calculations are not a particularly reliable way to evaluate the value of DfT for lower time to market. Diagnostic DfT is more of a religion than a science. The true believers are the engineers and managers who have suffered through never-ending design revisions, brought on by lack of adequate diagnostic capabilities.

14.3 DIGITAL SCAN

14.3.1 Scan Basics

Scan circuits allow a digital block to be isolated from surrounding circuits for the purpose of testing. Scan circuits facilitate a divide-and-conquer approach to testing that is exceptionally well suited to digital circuits. The scan circuits allow the normal inputs of the subcircuit under test to be replaced by tester-injected digital vectors. The tester injects the vectors into the subcircuit under test using a series of flip-flops called a *scan chain*.

A scan chain acts as both a serial-in, parallel-out (SIPO) shift register and a parallel-in, serial-out (PISO) shift register. The SIPO register allows a parallel stimulus vector to be shifted into the DUT through a dedicated serial input pin. A series of parallel stimulus vectors can be used to exercise the internal DUT circuits or to verify interconnections from one IC to another on a finished printed circuit board. Scan chains would be somewhat useless if they did not allow observation of a circuit's response to the stimulus vectors; so the PISO readback capability is provided for capturing the output response from the circuit under test.

There are several different types of scan, including boundary scan (IEEE Std. 1149.1), full scan, and partial scan. Boundary scan is primarily directed toward the board-level chip-to-chip interconnection testing problem, although it is also extensible for testing internal circuits as well. While 1149.1 boundary scan can be used to test internal circuits, full scan and partial scan methodologies are more commonly used for this purpose. These simpler forms of scan are somewhat more efficient in their use of extra circuitry than the more elaborate boundary scan architecture. Let us look at each of these methodologies very briefly to understand the advantages of scan over less-structured digital testing approaches.

14.3.2 IEEE Std. 1149.1 Standard Test Access Port and Boundary Scan

The IEEE Std. 1149.1 test access port and boundary scan standard[12] was developed by a consortium of industry participants from Europe and North America. The consortium, known as the Joint Test Action Group (JTAG, pronounced jay-tag), developed the standard to allow many different IC vendors to design chips compatible with a consistent board-level testing architecture. As a result, IEEE Std. 1149.1 is often referred to as JTAG 1149.1 boundary scan. JTAG 1149.1-compliant devices allow a system-level developer to test chip-to-chip interconnects on a finished printed circuit board. In addition, the system developer can reuse the production test vectors for each of the individual JTAG 1149.1 ICs to perform system-level diagnostics. In an 1149.1-based system, the test structures from multiple ICs can be tied together in a daisy chain configuration so that the entire system can be accessed through a single JTAG 1149.1 interface port.

Figure 14.6 illustrates a JTAG 1149.1 scan implementation compatible with chip-to-chip scan testing. Each shift-and-load element in the chain is called a *scan cell*. Figure 14.7 shows a scan cell from the JTAG 1149.1 boundary scan standard. The Mode signal is used to select normal mode or test mode. In normal mode, the Signal in data are passed directly to Signal out, bypassing the scan cell altogether. In test mode, Signal out gets its data from the test stimulus register.

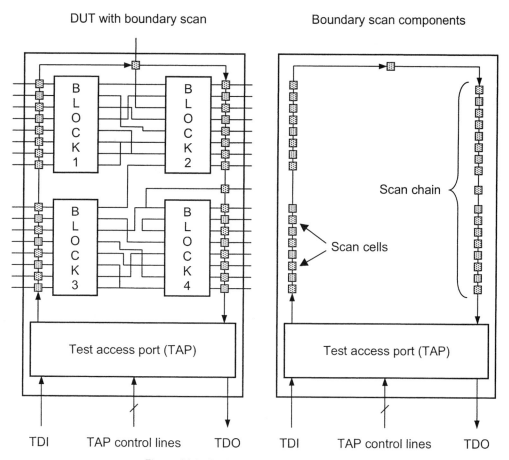

Figure 14.6. Basic boundary scan architecture.

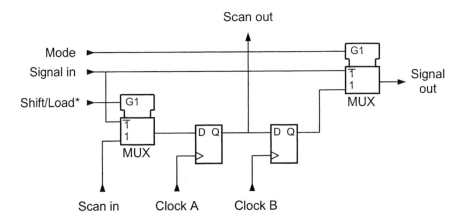

Figure 14.7. IEEE Std. 1149.1 boundary scan cell (reproduced with permission from IEEE).

A load signal updates all the scan cell outputs in a scan chain simultaneously after all the serial bits of the test vector have been shifted into position. Similarly, a response vector from the circuit under test can be loaded into the scan chain by a second load signal after the stimulus vector has been updated. While the next stimulus vector is shifted into the scan chain, the response vector is simultaneously shifted out to a dedicated serial output pin. Therefore, it is possible for each shift and load cycle to apply one parallel stimulus vector to the circuit under test and then read back one parallel response vector. Chip-to-chip interconnects can be verified by applying vectors from one IC's boundary scan circuits and reading back the response from another IC's scan circuits (Figure 14.8).

The advantage of boundary scan chains is that a multibit test stimulus vector can be applied to any number of DUT circuit inputs using only a few device pins. Using JTAG 1149.1, the interface can be made consistent from one device type to another, allowing system-level test and diagnostics. Standardization allows a much more automated test generation process.

A typical scan interface requires only four or five signals, which are connected to the ATE tester through dedicated test pins. In the JTAG 1149.1 standard, the serial scan interface uses a block of control logic called a *test access port* (TAP). The JTAG 1149.1 scan architecture uses four signals: TCK (clock), TMS (test mode select), TDI (test data in), and TDO (test data out). The JTAG 1149.1 TAP controls the clocks, load signals, and mode control signals of the scan cell in Figure 14.7 using a state machine. The test-mode control signal, TMS, controls the operation of the state machine as it is clocked by the TCK signal. The state machine flow diagram for the 1149.1 TAP controller is shown in Figure 14.9. The state machine generates the appropriate load and shift signals as needed. The state machine allows a sophisticated level of operation while using a minimum number of test-specific device pins. For example, the state machine can be used to initiate more advanced operations, such as initiating a BIST operation.

The state machine starts in a test reset mode that disables all 1149.1 scan circuits, allowing normal operation of the device. The state of the TMS pin is used to direct the state machine through its various operational states. The first of these is the run-test idle state, which enables the test mode of the scan cells, taking them out of their normal (i.e., bypass) mode. In this state, the scan cells apply the parallel test stimulus vectors to the circuits under test. From the run-test idle state, the flow diagram continues to either a data register (DR) path or the instruction

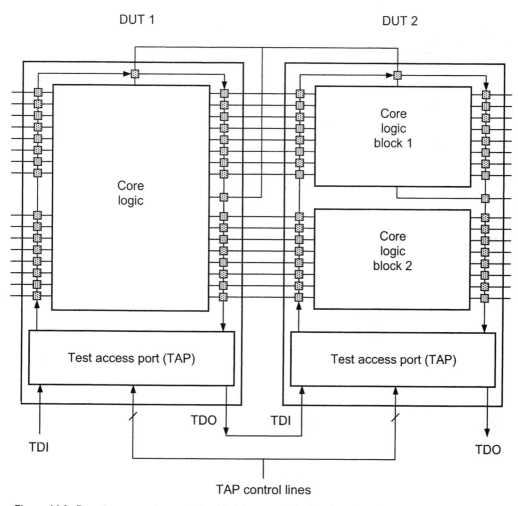

Figure 14.8. Boundary scan allows chip-to-chip interconnect testing through a single board-level interface.

register (IR) path. These are identical except that the DR flow path shifts data into and out of the scan paths, while the IR flow path shifts data into and out of instruction registers.

Instruction registers can be used to initiate BIST operations, to implement proprietary scan chain addressing schemes, or to perform any of a myriad of standard or ad hoc testing operations. The inclusion of instruction registers allows the 1149.1 standard to enable much more powerful operations than a simpler data-only scan architecture could provide. The IR capability is a major forte of the 1149.1 standard, which should allow the standard to grow with advances in BIST and DfT for many years to come.

14.3.3 Full Scan and Partial Scan

As previously mentioned, testing of logic blocks internal to the DUT can be accomplished using a non-JTAG scan methodology called *full scan*. The full-scan methodology breaks complex

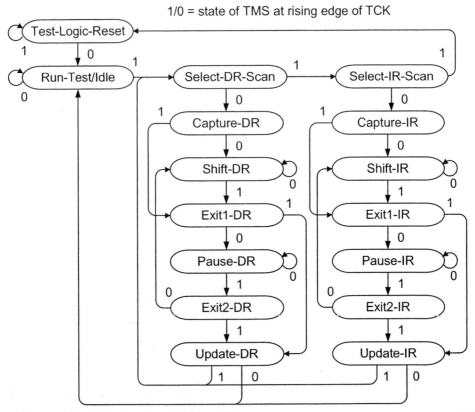

Figure 14.9. IEEE Std. 1149.1 TAP controller state diagram (reproduced with permission from IEEE).

circuits into small, easily tested blocks of simple combinational (i.e., nonclocked) logic. In a full-scan design, each clocked circuit element in the design (flip-flop or latch) serves a dual role. In normal operation, the scan mode is disabled and each flip-flop or latch behaves the same as its nonscannable counterpart. In scan mode, a multiplexer replaces the normal data input, D, of the clocked element with a scan input, SD. A buffered version of the flip-flop's Q output is passed out of the cell as a scan output, SQ. Figure 14.10(a) shows a scannable D flip-flop.

Connecting the scan output from one flip-flop to the scan input of the next, a scan chain can be formed. When scan mode is enabled, we can shift data into the first flip flop in a scan chain through a device pin. Shifting a series of data bits into the scan chain, we can preset all the flip-flops in the design into a desired state. This applies a test vector to all the combinational logic attached to the flip-flop outputs. The response from the combinational logic can then be latched into the flip-flops using a single clock cycle in normal mode. Then the captured response can be shifted out to a scan output pin using the scan chain.

Using the flip-flops of the scan chain, we can apply a series of parallel test vectors to all the combinational logic blocks in a design and read the response from the blocks using a very simple serial interface. Figure 14.10(b) shows a trivial example of a 4-bit scannable state machine. Using the flip-flops in scan mode, we can test the combinational logic without having to cycle the state machine through all its possible states.

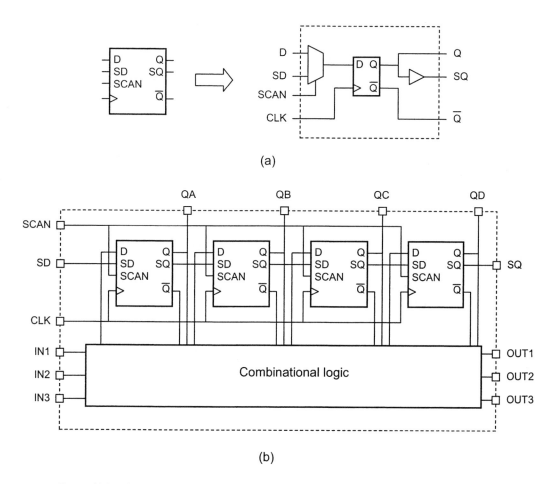

Figure 14.10. Scan circuits (a) scannable D flip-flop and (b) scannable 4-bit state machine.

Fortunately, we do not have to write scan vectors manually. Because test vectors for simple combinational logic blocks can be easily calculated, we can automatically generate high fault-coverage test vectors for a fully scannable device using an automated software tool. It is also worth noting that a scannable design can be generated automatically using a design methodology called *design synthesis*. In design synthesis, a high-level hardware description language such as VHDL (VLSI Hardware Description Language) can be used to describe the desired circuit functionality. A software translator can then synthesize not only the netlist for a scannable version of the circuit, but also the test vectors to guarantee the circuit's quality.

Since a scannable flip-flop or latch requires more circuit area than a nonscannable equivalent, full scan adds about 10-15 percent to the area of a typical digital circuit. To reduce this area overhead, a modified version of full scan is sometimes used. The modified methodology, called *almost full scan,* prunes some of the scan circuits out of a full scan design. The result is a more area-efficient scannable design that can still be tested with fault coverage equivalent to the full-scan version of the circuit.

Another scan methodology, called *partial scan*, is similar to full scan, except that the scannable cells are added to a nonscan design until a desired level of fault coverage is attained. Using a partial scan methodology, we start with no scan capability and work our way toward a full-scan design instead of starting with a full-scan design and working our way toward a design with no scan. As a result, the tools and methodologies for full scan and partial scan are different from one another.

A full treatment of scan circuits and scan methodologies are beyond the scope of this textbook. Although our coverage of scan has been very light, the subject is of extreme importance to design and test engineers. Fortunately, many books and papers have been written on the subject of scan. Rather than duplicating this information here, the reader is referred to the existing literature[13,14] for more in-depth information.

14.4 DIGITAL BIST

14.4.1 Pseudorandom BILBO Circuits

BILBO stands for built-in logic block observation. BILBO circuits are a form of BIST consisting of three parts: a pseudorandom data generator, a signature analyzer such as a cyclic redundancy checker (CRC), and a controller to synchronize the generator and analyzer. The pseudorandom data generator produces digital stimulus to be applied to the circuit under test, and the signature analyzer performs one of several mathematical operations (such as a check sum or CRC) to verify that the digital logic produced the correct sequence of outputs.

An example pseudorandom data generator is shown in Figure 14.11. This circuit is also known as a *linear feedback shift register* (LFSR). The pseudorandom generator circuit in Figure 14.11 sequences through all values from 1 to 511 in a pseudorandom sequence before repeating. (If initialized to all zeros, it will hang up in the all-zero state; so it must be initialized to a nonzero value.)

The pseudorandom values can be passed through a digital circuit under test (Figure 14.12). The output of the digital circuit can be verified using a CRC circuit or other signature analyzer. A very simple example would be a simple checksum circuit, consisting of an 8-bit adder with no carry. During a BILBO BIST operation, all digital blocks are preset to a known state. A BILBO controller circuit then starts clocking pseudorandom data patterns through the circuit under test

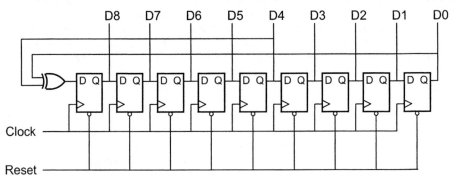

Figure 14.11. LFSR pseudorandom number generator.

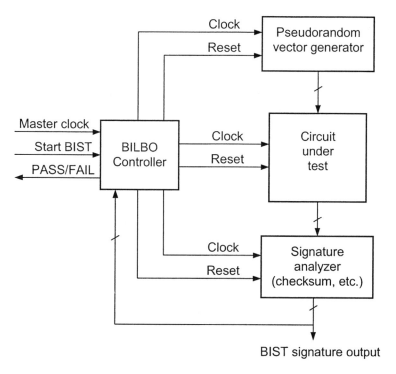

Figure 14.12. BIST testing of a circuit under test using BILBO.

and into the signature analyzer. After a fixed number of clock cycles, the BILBO controller stops the process and the output of the signature analyzer is compared against its expected value. The comparison can be performed by the ATE tester or by the DUT itself. A defective circuit under test is highly likely to produce an incorrect signature.

One advantage of BILBO test circuits is that many of them can operate in parallel, saving test time. Another important advantage is that the circuit under test can be tested at its full digital clock rate without passing high-speed digital signals from the DUT into the ATE tester. This can allow very high-speed digital circuits to be tested on a slower (i.e., less expensive) tester. The tester only needs to supply a high-frequency clock for the BILBO circuit to operate at full speed.

14.4.2 Memory BIST

Memory BIST circuits are similar in nature to pseudorandom BILBO circuits, except that the data patterns are not generated by a pseudorandom algorithm. Instead, the bits of the memory are loaded with specific patterns, such as checkerboard and inverse checkerboard patterns (Figure 14.13), walking ones and zeros (Figure 14.14), and other standard patterns. (Actually, the checkerboard and walking bit patterns are becoming obsolete. We have used them as an example due to their simplicity.) The patterns in Figures 14.13 and 14.14 represent the bits written into an 8x8 RAM array and then verified by a readback operation. Memory testers are designed to generate memory test patterns algorithmically (on-the-fly) rather than storing the repetitive test patterns in deep vector memory in the tester.

```
10101010  11001100  11110000        01010101  00110011  00001111
01010101  11001100  11110000        10101010  00110011  00001111
10101010  00110011  11110000        01010101  11001100  00001111
01010101  00110011  11110000        10101010  11001100  00001111
10101010  11001100  00001111        01010101  00110011  11110000
01010101  11001100  00001111        10101010  00110011  11110000
10101010  00110011  00001111        01010101  11001100  11110000
01010101  00110011  00001111        10101010  11001100  11110000
```

        (a) Checkerboard patterns   (b) Inverse checkerboard patterns

Figure 14.13. Checkerboard and inverse checkerboard RAM test patterns.

```
00000100  00000010  00000001        11111011  11111101  11111110
00000100  00000010  00000001        11111011  11111101  11111110
00000100  00000010  00000001        11111011  11111101  11111110
00000100  00000010  00000001        11111011  11111101  11111110
00000100  00000010  00000001        11111011  11111101  11111110
00000100  00000010  00000001        11111011  11111101  11111110
00000100  00000010  00000001        11111011  11111101  11111110
00000100  00000010  00000001        11111011  11111101  11111110
```

        (a) Walking ones   (b) Walking zeros

Figure 14.14. Walking ones and zeros RAM test patterns.

Rather than using an automated tester, BIST can be used to test memory circuits. On-chip BIST circuits can generate memory test patterns with a minimum number of gates. Alternatively, the patterns can be generated by a general-purpose microcontroller if one happens to be included as part of the DUT. Again, the advantages of memory BIST include the ability to test circuits at full speed with minimal tester support and the ability to test the memory in parallel with other circuits to save test time. Memory testing is a topic unto itself, and will not be covered in detail in this book. Several good books and technical papers have been written on the subject of memory testing and memory BIST.[15,16,17]

14.4.3 Microcode BIST

If a DUT includes a microprocessor or microcontroller, it can be programmed to test itself using test-specific microcode instructions. For example, the microprocessor's arithmetic logic unit (ALU) can be verified by performing a series of mathematical operations, such as additions, subtractions, bit-wise ANDs, bit-wise XORs, etc. The result of each operation is compared by the microprocessor against expected values. Microcode-based testing can also be performed on RAM blocks, I/O ports, and mixed-signal blocks such as ADCs and DACs.

The BIST instructions can be either hard coded into the microprocessor's ROM section, or it can be downloaded by the tester into the microprocessor's program RAM. The advantage of RAM-based BIST is that the BIST instructions do not occupy valuable program ROM space. However, RAM-based microcode BIST requires a longer test time, since the BIST instructions must be downloaded into the program RAM before the BIST testing can be performed. As always, tradeoffs between test time and silicon area must be considered.

14.5 DIGITAL DfT FOR MIXED-SIGNAL CIRCUITS

14.5.1 Partitioning

Highly complex digital circuits benefit from a structured testing approach, preferably using automated software tools to generate DfT structures and test vectors. However, simpler digital circuits such as those in many low-complexity mixed signal devices can be tested quite well without structured scan DfT techniques, saving the overhead of the structured approaches. While structured approaches with automated tools have become the rule rather than the exception as mixed-signal devices have become more complex, it is worth reviewing some of the common ad hoc DfT techniques that have been used on simpler circuits. It is interesting to note that the various structured approaches are based on many of the same concepts as the ad hoc methods that we will review in this section. For example, most structured testing approaches are based on the concept of circuit partitioning to reduce test time and increase circuit observability.

Digital circuits are particularly well suited to a divide-and-conquer approach to testing. In general it is possible to partition a complex digital circuit into pieces and test the pieces separately to guarantee the functionality of the whole. This approach gives us several advantages. First, the test time for exercising a complex circuit can be reduced significantly for certain types of circuits. Second, a complex circuit with many feedback paths may be difficult to force into each of the necessary logic states to guarantee good fault coverage. Partitioning allows the feedback paths of such a circuit to be broken, resulting in many simple circuits rather than one complex one. Finally, partitioning allows automated test generation software to produce test patterns for the simpler circuits that result from a divide-and-conquer approach. Clearly, full scan is the ultimate form of circuit partitioning, since it breaks a digital DUT into many simple subcircuits for quick and thorough testing.

As a very simple example of the test time reduction advantage of partitioning, consider the long divider chain illustrated in Figure 14.15. This circuit divides a 16.777-MHz input clock by 2^{24} to produce a 1-Hz clock. Testing this circuit in a straightforward manner requires that the divider chain step through each of its states, which would take 1 s of test time. Obviously, this is unacceptable. The simplest solution to this problem is to break the divider into three divide-by-256 sections using a test mode (Figure 14.16). Each divider can be tested separately, guaranteeing the operation of the whole. The three separate tests can be executed in a fraction of the test time required to test the divider chain as a whole.

The multiplexers in Figure 14.16 can be controlled using bits in a test register. During normal mode the two multiplexers at the divider stage inputs are configured to pass the output from the previous stage into the following divider stage input. When configured into the test mode, the two input multiplexers replace the previous stage's output with the 16.777-MHz clock

Figure 14.15. Divider chain without DfT.

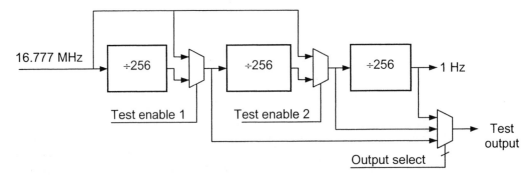

Figure 14.16. Partitioning of a digital divider chan for testability.

to speed up the countdown process for the following stage. Any of the three divider stage outputs can be selected for observation using the third multiplexer, which is controlled by another test-mode setting. The inputs and outputs of test multiplexers such as the ones in Figure 14.16 can be controlled and observed using register-based test modes and readback bits or they can be controlled using a multiplexed digital stimulus/observation bus connected to external device pins. Direct test access of internal circuit nodes through external pins is commonly referred to as parallel module testing, or PMT.

14.5.2 Digital Resets and Presets

One of the simplest but most critical DfT requirements in a digital design is the ability to reset or preset all register and flip-flop circuits into a known state before application of test vectors. In theory, this is usually unnecessary, since most testers can clock the device until it reaches the desired state. In practice, though, resets and presets allow faster testing and much easier test program development.

Consider the 4-bit ripple counter circuit in Figure 14.17. If the counter has no preset capability, it may come up in any of 16 states when power is applied to the DUT. Most testers have the ability to observe the four output bits of the counter and clock it until a desired output state is reached. The capability to search for a particular data pattern from the DUT is called *match mode*. Using match mode, a tester applies clocks to the counter until it sees a desired state, such as 0000, at the DUT output. The match mode search process takes extra test time, which is not significant for the four-bit counter example. But if the circuit is more complex, it may take a very long time for the tester to find the desired state.

Complicating matters is the fact that most testers have a significant pipeline delay that keeps them from immediately sensing the match condition. By the time a match is detected by the

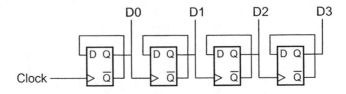

Figure 14.17. Four-bit counter without reset/preset.

tester's pattern controller, the digital pattern may be many test vectors past the point where the match occurred. The test engineer has to keep the pipelining in mind as the match mode code is developed. In short, match mode is a workaround developed by tester companies to compensate for digital designs with poor testability. A well-designed circuit with proper resets has no need for match mode. Using a reset, the four-bit counter can simply be reset to 0000 before testing its 16 states.

Resets and presets also allow accurate simulations of the digital design that can then be converted directly to the tester's format to produce an automatically generated test program. Simulation software seldom intentionally introduces random states at power up; so a simulation may appear to produce a predictable output pattern from the DUT even without resets. Since the actual nonresettable DUT does not behave predictably on power up, the simulations will not necessarily match the real DUT. Digital patterns that are automatically generated for such a design will be useless, since they are based on a simulation that starts in a nonrandom state.

14.5.3 Device-Driven Timing

Device-driven timing is another problematic issue for test engineers. In theory, it seems that a tester should be able to synchronize its digital pattern to a clock source or data strobe from the DUT. In practice, it is impractical to build a general-purpose tester with enough local circuitry to immediately respond to DUT outputs in real time. The pin card electronics in a typical ATE tester are located several inches if not several feet from the DUT because of mechanical constraints. The ATE tester's digital pattern generator and formatting circuits may be even farther away from the DUT. These are often placed inside the tester's mainframe cabinet, several feet away from the pin card drivers and comparators.

The delays caused by these paths are compounded by the pipelined architecture in the tester's high-speed digital pattern generator. It is common to see a pipeline depth of 60 or more digital pattern vectors between driven data from the tester and received data from the DUT. The tester's software compensates for the pipeline delay between the driven data in a pattern and the expect (compare) data in a pattern so that the test engineer normally does not need to worry about the pipeline delay. Match mode is one of the few instances where pipeline delay is not compensated by the tester software. Figure 14.18 shows a series of vectors with both driven and expected data. In reality, the drive data leave the pattern memory many cycles before the expect pass/fail result from the same vector arrives back at the pattern generator. The tester software takes this pipeline into account for all operations except match mode. This explains why the match mode has trouble immediately responding to a pattern from the DUT. By the time the tester senses the match, the pattern generator may have sent out dozens of additional cycles of drive data.

Because pipeline delay prevents the tester from immediately responding to device outputs, it is impossible for the tester to wait until it sees a sync pulse (data-ready signal) from the DUT to start clocking data into its capture memory. By the time the pattern generator sees the sync pulse, it is too late to begin generating the necessary signals to shift the data out of the DUT. If timing is not critical, it may be possible to use match mode to learn where the next sync pulse is going to occur and then start clocking data out at that time whether a sync pulse occurs or not.

A subtler problem occurs when the device produces a sync pulse or clock whose timing relative to input or output data is very critical. In this case, it is very difficult to apply the input data or capture the output data at just the right time. The tester would have to learn exactly where the sync pulse or clock edge occurs relative to the digital pattern's bit rate, then adjust its

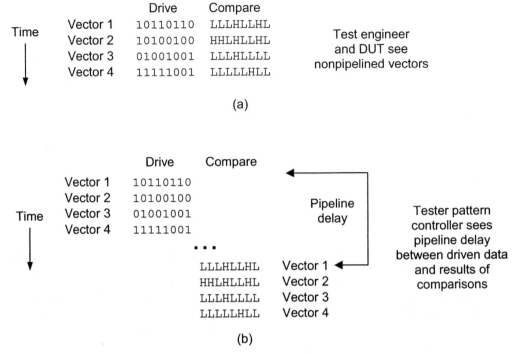

Figure 14.18. Pipeline delay in digital vectors.

own timing vernier circuits on the fly to find the exact position of the valid data. Testers are not able to do this easily or cost effectively.

It is much easier to tell the tester to accept data 2 ns after a rising edge of one of its own signals than to tell the tester to accept data 2 ns after the DUT suddenly decides to toggle a data-ready signal. Whenever possible, the DUT should be designed to provide data or accept data at a time specified by a tester signal. In cases where the DUT must define timing to meet system-level requirements, a separate test mode can be added that switches the DUT into a slave mode rather than a master mode; so that the ATE tester can define when events should occur.

As an example of DfT to allow ATE-driven timing, consider an on-chip crystal oscillator that normally generates the master clock for a device. Oscillators, PLLs, and other master clock generating circuits should always include a bypass mode as shown in Figure 14.19 for testing purposes. Sometimes the same effect can be achieved by simply driving one side of the crystal with a digital clock. This is only acceptable if the DUT pin is truly a digital input, allowing the master clock to be stopped or single stepped as needed. AC-coupled input clocks are unacceptable, since they do not allow the DUT master clock to be reliably halted for static testing of I_{DDQ}, low-speed digital patterns, etc.

The bypass mode allows the tester to control the timing of all digital events in the DUT rather than trying to let the DUT drive all timing. Allowing the tester to drive master clock timing is absolutely crucial in DSP-based mixed-signal testing, since the tester must have complete control of all sampling frequencies in the DUT to achieve coherent sampling (see Chapter 6, "Sampling Theory").

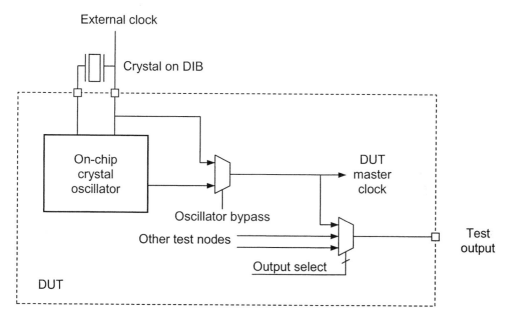

Figure 14.19. Oscillator bypass DfT.

In addition to the bypass mode, the output of the on-chip clock generation circuitry must be observable, either directly or indirectly, to guarantee its functionality in the normal operational mode. Observability can be achieved by either using another test access point, as shown in Figure 14.19, or by observing a digital output whose operation depends on proper operation of the on-chip clock generator.

14.5.4 Lengthy Preambles

Another problem with complex digital circuits such as plug-and-play multimedia devices is that they cannot be enabled without a lengthy digital setup procedure, called a *preamble*. In a poorly designed device, the preamble must be executed every time a new measurement is performed. Since hundreds of AC channel tests may be performed on a stereo audio IC, the preamble must be executed hundreds of times, leading to needless test overhead. Whenever possible, a test mode should be provided to put the device directly into a test-ready state.

14.6 MIXED-SIGNAL BOUNDARY SCAN AND BIST

14.6.1 Mixed-Signal Boundary Scan (IEEE Std. 1149.4)

The IEEE Std. 1149.4 mixed-signal boundary scan standard[13] was developed by many companies and academic institutions around the world. IEEE 1149.4 is built upon the 1149.1 digital boundary scan standard. As an analog complement to the 1149.1 boundary scan for digital circuits, the 1149.4 standard allows chip-to-chip interconnect testing of analog signals. Optionally, it allows testing of internal circuit nodes. The 1149.4 standard provides a consistent interface for analog and mixed-signal tests for those signals that can tolerate the loading, series

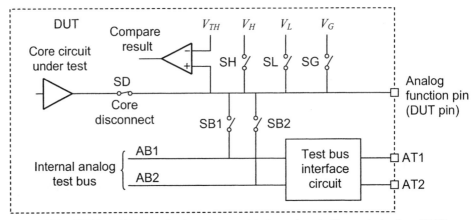

Figure 14.20. IEEE Std. 1149.4 analog boundary module (reproduced with permission from IEEE).

resistance, and crosstalk issues inherent in the physical implementation of the standard in the target IC process (e.g., CMOS).

The 1149.4 mixed-signal boundary scan standard is compliant with the 1149.1 digital TAP and boundary scan architecture. The major difference between 1149.4 and 1149.1 is that the 1149.4 standard includes some new test pins and analog switches for exercising nondigital circuits. Figure 14.20 shows the analog boundary module (ABM) for the 1149.4 standard. The ABM provides standardized access to analog input and output signals at the external device pins.

The 1149.4 standard allows a simple chip-to-chip interconnect verification scheme similar to that used in traditional digital boundary scan. A pair of switches at each analog input and output pin of the IC allows the pin's normal (analog) signal to be replaced by digital signal levels, V_H and V_L. V_H and V_L would typically be connected to V_{DD} and digital ground. In effect, the analog input or output becomes a simple digital driver. The interconnect between ICs can be tested by forcing either V_H or V_L from the pin and then checking the status of a receiver at the other end of the interconnection. The receiver, also part of the 1149.4 standard is an analog comparator tied to an 1149.1 digital boundary scan cell. It compares the incoming voltage against a threshold voltage, V_{TH}. In addition to the two logic level connections, the analog pin can also be connected to a quality ground, V_G (typically analog ground).

Although the 1149.4 standard is primarily targeted for chip-to-chip interconnect testing, it does include optional extensions for internal analog signal testing. For this purpose, the 1149.4 standard uses a pair of analog test buses, similar in nature to the analog test input and output buses in Figure 14.4. The analog switches are controlled by shifting control bits into the 1149.1 TAP, allowing a standardized method of setting up analog stimulus and measurement interconnects. The analog buses can be used for a variety of purposes, including internal testing as well as external (chip-to-chip) interconnect testing. As an example of external testing, engineers at Hewlett Packard and Ford Motor Company developed a method to use this structure to verify the interconnects between ICs and networks of passive components such as resistors, capacitors, and inductors.[18] This method forces DC or AC current through one test bus while measuring the voltage response through the other bus.

It should be noted that the switches defined by the 1149.4 standard do not necessarily need to be physical switches. For example, if the output of a particular circuit can be set to a high-impedance state, then it does not need to be disconnected using a switch. Similarly, if a circuit's output can be set to force a high level and a low level under 1149.1 digital control, then separate V_{DD} and ground switches are not needed. The switches defined by the standard are therefore behavioral in nature, rather than physical requirements. The advantage of eliminating switches when possible is twofold. First, the series impedance and/or capacitive loading of a CMOS transmission gate or other switching structure is not introduced into the signal path. Second, the silicon required to implement the 1149.4 standard can be minimized if the number of switches can be minimized.

The 1149.4 standard cannot be employed blindly to test internal signals without consideration of the effects of the standard on the analog circuits to be measured. Actually, it is not the standard that is the problem; it is the practical implementation of the standard using CMOS or other types of analog switches. Signal crosstalk, capacitive loading, and increased noise and distortion are possible problems that may occur when using CMOS switches in sensitive analog circuits. In some cases, the design engineer might need to use T-switch configurations (see Section 14.7.3) to minimize signal crosstalk and injected noise, though the 1149.4 standard does not specify the physical embodiment of the switches. The issues of crosstalk, noise injection, and loading are identical to those in the more general ad hoc mixed-signal test bus configurations, which have been used successfully for many years. The problems are not insurmountable; they simply require the design engineer to evaluate which nodes can and cannot tolerate the potential imperfections introduced by the analog switches. The potential problems and some common solutions will be discussed in a later section on ad hoc mixed-signal test busses.

Like the 1149.1 standard, the 1149.4 standard carries more overhead than the traditional ad hoc methods. But like the 1149.1 standard, the extra baggage is well justified by the tremendous enhancement in standardization of test access. For the same reasons outlined in the 1149.1 section, the overhead will eventually be much less of a problem as processing geometries shrink.

14.6.2 Analog and Mixed-Signal BIST

The IC industry has recently begun to apply BIST concepts to traditional specification-oriented parameters on mixed-signal circuits. Some of the more promising concepts have been presented at the International Test Conference in recent years,[19-21] although many of these are designed for very focused test applications. The mixed-signal BIST designer faces some challenging problems that are not faced by digital BIST designers. Let us look at some of the more common challenges encountered when implementing mixed-signal BIST.

First, the more obvious implementations of analog BIST sometimes lack robust traceability to central standards such as those maintained by the NIST (the National Institute of Standards and Technology). It is usually easy to let the device wiggle its analog signals to see if it is basically functional or whether it is completely defective. Unfortunately, many parameters are very close to the specification limits, even on a good device. The use of uncalibrated on-chip analog stimulus and measurement circuits throws doubt into the accuracy of measurements, since there is a question about the quality of the signals generated and measured on a given DUT. Thus the analog BIST designer must define a calibration strategy for the analog circuits of the analog BIST structure.

For example, let us say we try to use an on-chip ADC to test the amplitude of a sine wave generated by a DAC (Figure 14.21). How do we know that the ADC gain on a given DUT is not in error by –0.5 dB, canceling out a +0.5 dB amplitude error in its DAC? One possible solution

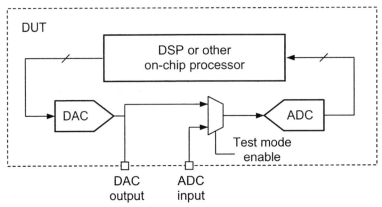

Figure 14.21. ADC/DAC loopback BIST.

is to provide a calibration signal of a known amplitude to the ADC and let the DUT calibrate itself. Certain parameters such as distortion do not tend to cancel, but are instead additive. These can be tested fairly effectively using ADC/DAC BIST without extra calibration.

Another issue with the DAC-and-ADC-based BIST is that the on-chip instrumentation is often inferior to the types of programmable equipment available on ATE equipment. These digitizers have programmable antialiasing filters and other features that allow a much more thorough evaluation of AC signals than can be achieved by an on-chip ADC. In the previous example, how do we know that the ADC output does not contain aliased signal components from the DAC's images? For that matter how can we measure DAC images if the ADC samples at the same rate as the DAC? The Nyquist criterion becomes a problem.

Finally, the circuit overhead to implement a complete suite of production analog tests using BIST is often overwhelming, unless most of the circuits are already present in the design. In the previous ADC/DAC example, some kind of processor would be necessary to provide sine wave samples to the DAC and collect samples from the ADC. A useful BIST operation would then require the processor to perform an FFT on the results, evaluating *signal-to-noise* ratio, fundamental amplitude, distortion components, etc. To truly perform on-chip BIST in this manner requires a fairly powerful processor such as a digital signal processor (DSP). If no processor exists on-chip, there is no straightforward way to let the device test itself for these types of parameters.

Despite this rather gloomy analysis of this particular BIST structure, there is a bright side to analog BIST. We have to keep in mind that one goal of BIST is to allow a customer to perform field diagnostics in the end equipment. For field testing, verification of basic functionality is perfectly adequate in many cases, giving analog BIST a very powerful advantage in system-level testing.

Another promising solution to the challenges faced in analog BIST is the use of defect-oriented testing (DOT). The problems outlined in the example are based on the assumption that we wish to perform traditional specification-oriented testing (SPOT). In other words, the problems are based on the fact that we are trying to measure system-level parameters such as gain, distortion, and noise that are very close to specification limits. However, if design margins and processing controls are maintained so that we do not need to measure these parameters with such extreme accuracy, then we can begin to take a defect-oriented approach to mixed signal testing. Inductive fault modelling is one example of defect-oriented testing that has been researched heavily in recent years.[22]

In defect-oriented testing, we try to detect the cause of the failing parameter rather than the symptom. For example, we might measure the variation in a resistor value that ultimately results in frequency-response errors rather than trying to measure frequency response. Resistance is far easier to measure than frequency response; so a much simpler BIST circuit might be used to detect this type of fault. Of course, this is a very simplistic example, but it serves to illustrate the thought process behind DOT. Detailed coverage of DOT is beyond the scope of this introductory text.

14.7 Ad Hoc Mixed-Signal DfT

14.7.1 Common Concepts

Besides the IEEE 1149.4 analog boundary scan standard, there are few standardized approaches to mixed-signal DfT. Most of the more useful mixed-signal DfT concepts have been developed in an ad hoc manner, as needed for a particular application. These are very specific to the exact type of circuit under test. Some of the more common concepts are presented in the sections that follow.

14.7.2 Accessibility of Analog Signals

Accessibility of critical analog signals is one of the most important mixed-signal DfT concepts. DC voltage references, bias current generators, and other critical analog circuits should be accessible to the ATE tester, both for signal measurement and for insertion of signals from the tester. Let us look at a fictional cellular phone voice-band interface device to see how lack of analog test capability can hinder device debug and characterization. Remember that debug and characterization of the DUT is one of the more time-consuming tasks that frequently delays the release of a new product to market.

Figure 14.22 shows a portion of a cellular telephone voice-band interface device that converts received digital voice samples into an audio signal for the telephone's earpiece. This section of the device consists of a 16-bit DAC, an anti-imaging filter, a programmable gain amplifier (volume control), a power amplifier to drive the speaker, and a DC reference that sets the full-scale range of the DAC. The specifications for this particular device call for a gain error of 0 dB plus or minus 0.05 dB at 1 kHz, with a load of 32 Ω at the power amplifier output. A channel with 0-dB gain error is defined in the data sheet as one that produces 1 V RMS at the amplifier output while the DAC input is supplied with a digitized sine wave that is 3 dB below full scale. In addition to the gain error specification, the channel must have a harmonic distortion level at least 85 dB below the 1-kHz test tone.

The test engineer measures this device and discovers that the signal-to-distortion ratio is 75 dB, which fails by 10 dB. Also, the gain is –0.5 dB, failing by -0.45 dB. After the design engineers verify this result using bench equipment, they try to figure out which block is causing the distortion and gain error. Since there is no way to observe any voltages other than the power amplifier output, they can either try to find the cause of the problem by running more elaborate simulations, or they can try to probe internal circuit nodes using tiny whisker probes.

574 *An Introduction to Mixed-Signal IC Test and Measurement*

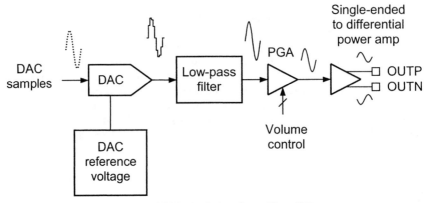

Figure 14.22. Audio interface without DfT.

Meanwhile, a competitor has designed a pin-compatible equivalent device with the same architecture, but this device includes DfT for analog signal observability (Figure 14.23). By a curious coincidence, this device has exactly the same problems as those discovered by the first group of engineers. Since there are multiple test points in the circuit, the test engineer is able to measure the output of the DAC directly and discovers that it is actually about 0.2 dB too high. The distortion is absent at the input to the power amplifier, proving that the power amplifier is probably introducing the distortion. After further investigation, it is discovered that the power amplifier is introducing a gain error of –0.7 dB, which explains why the total channel gain error is –0.5 dB (0.2 dB – 0.7 dB). Since the power amplifier has been shown to be the cause of distortion, the other design engineers concentrate on other tasks while the power amp designer corrects the distortion and gain problems.

Next, the DAC designer asks the test engineer to measure the DAC reference voltage level using the test bus. The level is 0.19 dB too high, which explains most of the 0.2-dB DAC gain

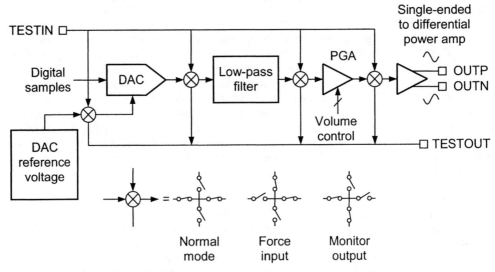

Figure 14.23. Audio interface with analog test accessibility.

error. The DC reference design engineer then discovers a parasitic resistance that explains the gain reference error. After a design adjustment to correct the parasitic resistance and power amplifier transistor sizes, the second pass design works perfectly. Since the second company's product is first to market, the DfT-enabled design wins in the marketplace over the non-DfT design.

We might be tempted to simply sprinkle test pads throughout the design to achieve the same diagnostic capabilities as analog observability DfT. Test pads require less silicon area than T-switches and an analog test bus snaking its way around a die. While test pads are generally a good idea, they can only be accessed using whisker probes that must be positioned by hand under a microscope. This is a time-consuming process that obviously cannot be applied to thousands of units during normal production. It is critical to be able to collect large amounts of characterization using an ATE tester to find correlation between errors in the various internal signals of the DUT.

It is important to realize that breaking a mixed-signal device into sections and testing the pieces is a necessary but insufficient means of guaranteeing system specifications like gain and distortion. Unlike digital circuits, analog circuits do not always behave as a sum of the individual parts; so a divide-and-conquer approach will not always allow system-level specifications to be ignored. Certainly to a first degree, a mixed-signal system behaves as a sum of its parts. Unfortunately, there are many subtle interactions between the various analog and digital circuit blocks that may make the whole behave differently than the individual pieces would indicate. The test engineer should always be prepared to measure both the system-level performance of the whole device as well as the performance of the individual circuit blocks.

14.7.3 Analog Test Buses, T-Switches, and Bypass Modes

Once a design and test team decide to add analog observability DfT into a new product, the exact method of test point insertion must be chosen. One of the more common ways to provide access to internal analog signals is through analog test buses, such as the ones in Figure 14.20 and Figure 14.23. Using one or more analog test buses, the ATE test program can gain access to internal nodes by opening and closing the appropriate transmission gates or other switching structures. The 1149.4 standard provides for just such internal test access through the AB1 and AB2 analog bus lines. In the 1149.4 standard, the appropriate switches are opened and closed using digital control bits injected through the 1149.1 test access port. In ad hoc architectures, the switches may be closed by any of a variety of means, including test-specific register bits or negative logic levels on normally positive digital input pins (V_{DD} and ground for digital signals, negative voltage to enable test modes). The possibilities for switch control are virtually endless. The more important considerations for this type of test access architecture is the nature of the switches themselves.

One of the biggest problems introduced by analog test buses is the danger of crosstalk between all the observed nodes. Figure 14.24(a) shows a test bus capable of accessing three internal DUT nodes. The most common CMOS structure for implementing an analog transmission gate is a back-to-back P-channel and N-channel CMOS transistor pair. As shown in Figure 14.24(b), an inverter provides complementary control signals to the two transistors so that they are both on or off at the same time. The reason that both an n-channel transistor and a p-channel transistor is required is that the pair allow a larger range of voltages to pass through the transmission gate. If only a P-channel transistor were used then signals near V_{DD} would not

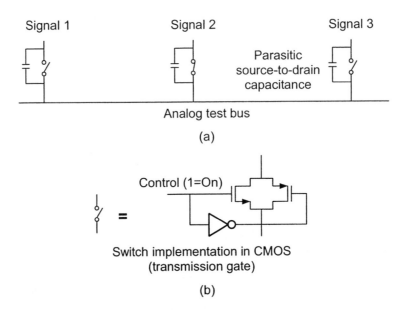

Figure 14.24. Analog test bus DfT: (a) capacitive coupling between signals caused by parasitic capacitance and (b) analog switch implemented using a CMOS transmission gate.

pass due to the transfer characteristics of the P-channel transistor. Likewise, signals near ground would not pass through an N-channel transistor alone.

At first glance, the circuit in Figure 14.24 does not appear to have any problems. But if we realize that the switches are implemented as CMOS transmission gates with parasitic capacitance from drain to source, we see a problem. There is an AC signal path from each signal to the others through the drain-to-source capacitance of each transmission gate. The capacitive coupling path exists even when the transmission gates are all turned off during normal mode. The circuit may suffer from crosstalk problems or it may even break into oscillations due to feedback from one node to a previous node. At low frequencies and low output impedances, this may not pose a problem. Nevertheless, it is a risky approach.

One common solution to the crosstalk problem is the use of a T-switch configuration (Figure 14.25). A T-switch consists of three switches; two in series and one providing a ground to the midpoint of the series connected switches. The grounding switch is closed any time the series switches are opened, thereby shunting any potential crosstalk signals to ground. The only problem is that the resulting switch always presents a small capacitive load to ground on the node to be sensed. Fortunately, in many cases the extra load capacitance is entirely acceptable. A T-switch implemented in CMOS is shown in Figure 14.26.

Another possible means of implementing analog test accessibility in CMOS circuits is to simply power down a circuit block to provide a bypass path through its circuitry. In this type of scheme, each circuit block can be isolated by powering down some or all of the other blocks in the signal path. The advantage of this approach is that it adds no extra loading or crosstalk paths to the circuit in normal operation. Unfortunately, some circuits can be bypassed by disconnecting their power while others cannot.

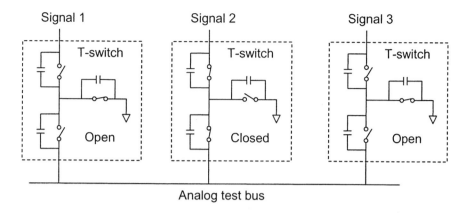

Figure 14.25. T-switches prevent crosstalk between signals.

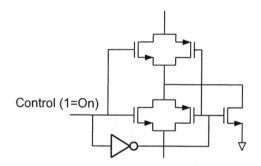

Figure 14.26. CMOS T-switch implementation.

14.7.4 Separation of Analog and Digital Blocks

One of the most important forms of mixed-signal DfT is the separation of analog and digital circuit blocks using test modes. The power supply current example of Figure 14.2 is a prime example of this form of DfT. The separation of analog and digital circuits provides several advantages including lower test time and better control over analog circuits so they can be more easily characterized. Let us look at another example of this form of DfT: a digitally controlled automatic gain control (AGC).

A purely analog AGC (Figure 14.27) includes a variable gain amplifier whose gain is automatically adjusted until its output reaches a desired peak amplitude. The peak output level is typically sensed with a peak detector and window comparator to determine whether the amplifier's gain is too high or too low. AGC circuits are commonly found in microphone amplifier circuits. The purpose of the AGC is to maintain a constant voice signal level no matter how loudly or softly the person is speaking. An analog AGC uses the output of the window comparator to adjust a control voltage, which in turn adjusts the gain of the amplifier.

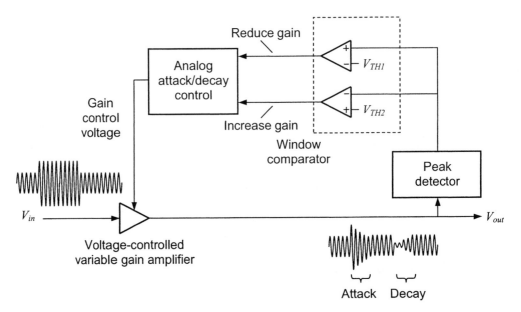

Figure 14.27. Analog automatic gain control (AGC).

A digitally controlled AGC (Figure 14.28) is similar in function to an analog AGC, but it uses digital logic in the feedback path to adjust the amplifier's gain. The digital logic adjusts the gain of a programmable gain amplifier (PGA) up or down depending on the outputs from the window comparator. Since the feedback path is implemented using digital logic, many of the characteristics of the AGC can be adjusted with values written to control registers. Characteristics such as attack time and decay time can be controlled by adjusting the digital logic's response to the window comparator's output. For example, if the decay time is set to be very long, the digital logic would increase the volume of the PGA very slowly after the microphone signal amplitude decreases.

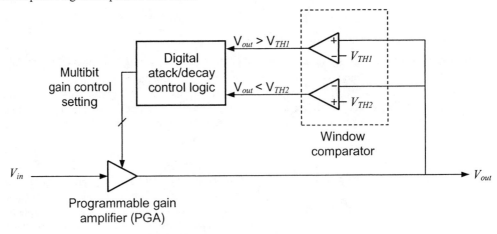

Figure 14.28. Digitally controlled AGC.

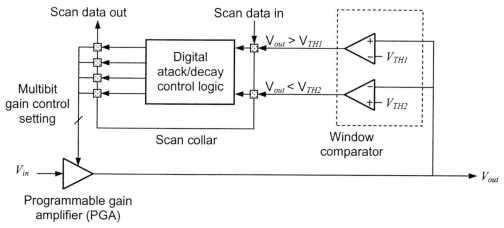

Figure 14.29. Digitally controlled AGC with scan-based DfT.

The digital logic in such a circuit can provide hundreds of combinations of attack and decay time. It would be extremely inefficient to measure all the combinations of attack and decay using a variable input signal and observing the gain changes in the PGA to make sure the digital logic increases and decreases volume levels at the correct times. Instead, the digital logic should be separated from the analog circuits to break the AGC into three parts: a PGA, a window comparator, and a digital logic block. One way to break the circuit into pieces is to insert boundary scan cells between the digital logic and the analog cells, as shown in Figure 14.29. The scan cells form what is known as a *scan collar*. The scan collar allows a convenient method to inject the digital stimulus to the feedback logic and to measure its response, as well as to stimulate and measure the digital interfaces to the analog cells. Of course, full scan of the digital logic is an even better approach. Ad hoc register/MUX based isolation circuits are also usable.

Once the circuits are broken into pieces, the digital logic can be tested using traditional digital methods to guarantee its quality. The remainder of the AGC is a simple PGA and a window comparator, each of which is very easy to test in isolation. The DfT circuits must of course allow the PGA gain to be set directly without the feedback logic, and it must allow direct analog and digital access to the window comparator. Characterization may prove that the AGC does not behave as a sum of pieces; so the test engineer should also be prepared to perform standard AGC tests, such as measuring the output signal envelope (gain vs. time) with a stepped input (Figure 14.30). This is one case where the sum of the individual pieces will most likely reflect the operation of the whole, because the analog circuits are so simple in nature. A divide-and-conquer approach often works well for circuits that are primarily digital with simple analog interfaces.

14.7.5 Loopback Modes

Another common mixed-signal DfT approach is the use of digital or analog loopback modes. Figure 14.31 shows the transmit and receive channels from a cellular telephone voice-band interface circuit. This example shows both analog loopback and digital loopback. In analog loopback, the device is placed into a mode that loops the analog DAC channel output (earphone signal) back into the ADC channel input (microphone signal).

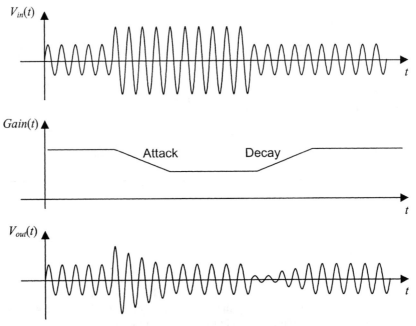

Figure 14.30. AGC attack/decay gain envelope.

In analog loopback mode, it is possible to transmit digitized signals such as sine waves into the DAC channel's digital input and capture digitized samples from the ADC channel's digital output. All circuits in the DAC and ADC channels must work properly for the captured signal to appear at the correct signal amplitude and with low noise and distortion. Loopback mode thus provides a quick and inexpensive way to perform a gross functional test involving most DUT circuits. Any major defect in any of the channel circuits can be detected this way. Analog loopback mode is often used by the original equipment manufacturer (OEM) customer to implement system-level diagnostics that allow the end product (cellular telephone, modem, etc.) to test its own functionality. Loopback DfT thus allows system-level BIST.

Digital loopback mode is similar to analog loopback mode, except that the digital output from the ADC channel is looped back to the digital input of the DAC channel. The test stimulus is a sine wave or other analog signal. The output can be captured and analyzed using a digitizer or other analog measurement instrument.

14.7.6 Precharging Circuits and AC Coupling Shorts

One of the simplest and most effective ways to reduce test time is to provide shorting paths to reduce long *RC* charging times. Such precharging circuits are useful in a number of different circuit configurations. DC reference decoupling capacitors and DC blocking capacitors are two good examples where shorting paths can be used to bring the DUT to a quiescent state quickly.

Figure 14.32 shows a DC V_{MID} reference circuit with a 1-μF decoupling capacitor located on the device interface board (DIB). This circuit includes a test mode that temporarily reconfigures the buffer amplifier connections to quickly precharge the decoupling capacitor. Except for the offset of the amplifier, the precharge voltage will exactly match the voltage at the midpoint of

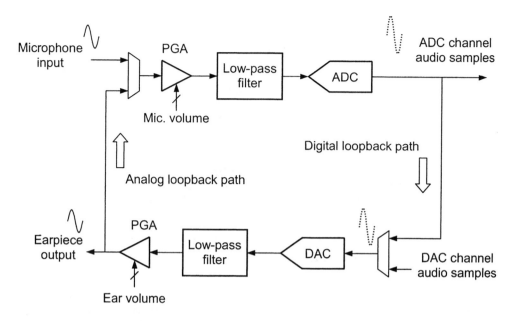

Figure 14.31. Analog and digital loopback paths.

the two resistors in the voltage divider, even if the resistors are mismatched. Once the circuit is returned to the normal mode, the decoupling capacitor only has to charge to compensate for the few millivolts of buffer amplifier offset.

Reduction of settling time can also be useful in DC blocking circuits such as the *RC* high-pass filter configuration illustrated in Figure 14.33. One possible way to reduce settling time is to temporarily short the far end of the capacitor to ground (or V_{MID}, depending on the circuit configuration). Yet another possible solution to this problem is to provide a complete bypass of the blocking capacitor during testing, eliminating the *RC* time constant altogether. However, this technique assumes that there will be no clipping problems or impedance matching problems between the two circuits with the *RC* high-pass filter stage removed.

Sometimes, these precharging techniques can be implemented on the ATE device interface board. In these cases, on-chip DfT may be unnecessary. Other times, the nodes are not accessible from the external pins of the DUT and a special test mode is required. As usual, the reduction in test time must be balanced against the added silicon area to determine the cost effectiveness of this type of DfT approach.

14.7.7 On-Chip Sampling Circuits

As DUT signals extend into the megahertz range and beyond, it may become difficult to get the signal under test into the ATE tester instruments without corrupting the signal. Most ATE tester instruments are connected to the DUT through 50-Ω cables with 50-Ω termination resistors at the far end. To the extent that a DUT cannot drive the transmission lines and other reactive loads presented by the tester, a special test structure may be required.

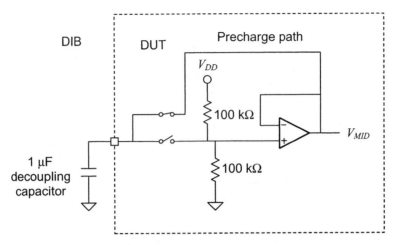

Figure 14.32. Decoupling capacitor precharge circuit.

One such test structure is an undersampling strobed comparator, located on-chip.[23,24] Such comparators are used as a high-bandwidth local front end to an undersampling ADC. The timing circuits, successive approximation register (SAR), and SAR DAC are located off-chip to save silicon cost. Since the comparator of the ADC is located on-chip, it is capable of sensing very high-frequency signals. This is an example of a DfT structure that does not necessarily provide lower cost, but may provide the capability to measure signals that otherwise could not be tested in a production environment.

Another similar structure is an on-chip sample-and-hold circuit. Such a circuit can be used in an undersampling mode to down-convert very high-frequency signals on the DUT to lower frequency signals that can easily be measured by an ATE tester. The only drawback to sample-

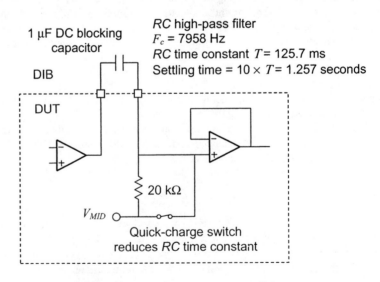

Figure 14.33. Quick-charge switches reduce settling time.

and-hold circuits is that they draw instantaneous currents from the signal under test while charging, thereby introducing current spikes into the signal under test. An undersampling comparator does not suffer from this problem.

Yet another approach to the high-speed signal measurement problem is the use of an on-chip flash ADC, assuming one already exists in the design. This technique is particularly effective in circuits such as hard disk drive PRML channels in which a high-bandwidth ADC is already present on the DUT. The only caveat to this approach and the others presented in this section is that the frequency response of the on-chip sensing element needs to be calibrated on a device-by-device basis for maximum absolute accuracy.

14.7.8 PLL Testability Circuits

The phase-locked loop (PLL) can be one of the more difficult mixed-signal circuit blocks to test, depending on the tightness of the specification limits. Some of the more common specification requirements are settling (lock) time, jitter, center frequency, and frequency range. Lock time and center frequency measurements can be made easier by breaking the analog feedback path of the PLL using a test mode (Figure 14.34). By inserting a midscale voltage into the voltage-controlled oscillator (VCO), it is possible to directly measure the center frequency. Similarly, the maximum and minimum frequencies can be measured by forcing full-scale and minus full-scale voltages into the VCO and observing the VCO's output frequency.

Settling time can be measured by observing the VCO's input voltage during normal operating mode with the analog feedback path enabled for normal operation. Settling time is typically measured by applying a reference frequency that abruptly changes from one frequency to another. The VCO input voltage settles at roughly the same rate that the output frequency

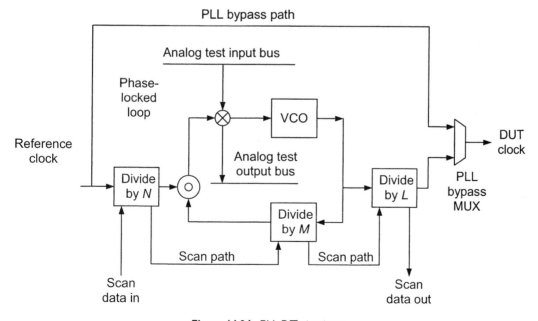

Figure 14.34. PLL DfT structures.

settles. Since it is easier to digitize a time-varying voltage than to measure a time-varying frequency on most testers, the VCO input voltage represents a useful test node.

The jitter measurements specified in most PLL circuits can be difficult to measure, especially when they approach the picosecond range. DfT circuits have proven useful in the measurement of PLL jitter and other specified parameters.

PLLs are often used as part of a frequency multiplier circuit, which includes a digital divider in the PLL feedback path. If the PLL circuit also includes one or more digital divider blocks, these should be isolated from the analog portions of the PLL to facilitate thorough testing of the divider logic without performing frequency measurements. Scan-based isolation techniques are well suited to this task. Since frequency measurements are usually much more time consuming on ATE testers than simple digital pattern tests, this type of testability can be useful in reducing test time.

As mentioned in Section 14.5.3, it is critical that PLLs provide a bypass mode so that the tester can inject clock signals into the DUT without the using the PLL. This allows the tester to drive the master clock or other clock input to the DUT in a direct manner.

One final PLL testability signal that has proven useful is the LOCK signal, which tells the tester that the PLL believes it has stabilized to a final frequency. This signal is often needed by the system-level application as well. The LOCK signal allows the system microcontroller to verify that PLL frequency lock has occurred.

14.7.9 DAC and ADC Converters

One of the most time-consuming tests for ADC and DAC converters is all-codes linearity testing. Integral nonlinearity (INL) and differential nonlinearity (DNL) are tests that require each code to be measured with a high degree of accuracy. The large number of measurements associated with a brute force all-codes test such as this leads to very long test times. Each bit of resolution requires twice as much test time; so a 12-bit converter takes at least 16 times as long to test as an 8-bit converter.

Also contributing to long test times for ADCs and DACs is slow conversion speed. A dual-slope ADC that takes 100 ms per conversion cannot be economically tested for INL and DNL (see Section 12.5.2). Dual-slope ADCs should only be used when absolutely necessary. If a faster successive approximation ADC can be used, it will lead to much lower test times even though its high speed may not be needed by the system-level application.

One DfT technique that leads to lower converter test time is segmentation of the DAC or ADC. This is one of those "easier said than done" suggestions. If a 12-bit DAC can be segmented into two 6-bit DACs, the test time will be significantly lower (see Section 11.3.7). This approach works very nicely as long as the performance of the 12-bit DAC can be reconstructed as a weighted sum of the two 6-bit DACs. The real difficulty lies in superposition errors. The fine DAC (lower 6 bits) may not behave the same when the upper 6-bits are all zero as it does when all upper bits are all ones. But if the lower 6 bits can be made to produce the same voltage curve regardless of the upper bits, then superposition allows for a segmented test approach that is far faster than the 12-bit all-codes test. The design engineer who can find a way to design a robust converter such as this can reduce testing costs significantly.

14.7.10 Oscillation BIST

One of the more recent DfT concepts is the use of oscillation modes to measure the performance of certain types of circuits.[25] Using oscillation BIST, or OBIST, the elements of the circuit to be tested are configured into an oscillator using a test mode. The frequency of the circuit's oscillation is then measured to determine whether the circuit elements are performing as expected.

OBIST is well suited for monitoring process shifts. For example, the oscillation frequency typically correlates closely to the speed of the transistors in the circuit. The technique is especially well suited for cases where high speed is more important than very exacting analog performance. Examples of circuits that might benefit from OBIST include high-speed digital transceivers and high-speed op amps.

14.7.11 Physical Test Pads

Often, a design engineer adds small test pads to the top level of metal on an IC. These pads are connected to critical circuit nodes. The test pads can be connected to bench equipment or ATE tester instruments through tiny microprobes or "whisker probes." Unfortunately, the connections can only be made if the die is not sealed in a plastic or ceramic package. An opening must be printed and etched in the protective overcoat (PO) layer of the die to allow the microprobes to touch the metal pads. There may be hundreds of critical nodes in a circuit that only need to be probed for the purpose of design debug. Since these nodes do not need to be measured during production testing, they may not justify the expense of full-blown analog test bus access. Though test pad DfT has little to do with automated testing, such test pads can be invaluable to device debug.

Another use of physical test pads is to make the device easier to probe with electron beam (e-beam) probers. e-Beam probers are basically scanning electron microscopes that can see differences in voltage on the die (see Chapter 1). Voltages on the top level of metal are far easier to see than voltages underneath layers of oxide. Oxide effectively forms a dielectric layer, producing an AC coupling effect. For this reason, test pads on the top level of metal (with openings in the PO) allow easier use of e-beam probers. If a design lacks a top-level metal test pad, a focused ion beam (FIB) machine can be used to drill holes into the buried metal lines and build up test pads. Unfortunately, the FIB process often damages the circuits of interest, defeating the purpose of e-beam probing.

14.8 SUBTLE FORMS OF ANALOG DfT

14.8.1 Robust Circuits

The DAC segmentation example in the previous section brings up a very interesting and profound DfT issue. Not all forms of DfT involve extra circuit elements. For instance, DfT can sometimes be achieved by simply making design choices that lead to more robust designs having tighter statistical distributions centered between the upper and lower test limits. As we have seen throughout this textbook, well-centered parameters having tight statistical distributions can be tested on lower-cost testers because they do not require extreme accuracy. Test costs can sometimes be reduced by orders of magnitude by simply improving the existing design using extra silicon area or by choosing a more robust circuit architecture. This and other forms of subtle DfT are often overlooked by design engineers and test engineers simply because they do not have an obvious connection to DfT. Subtle forms of analog DfT can be even more important to a product's success than the highly touted extra-transistor circuits of traditional DfT.

14.8.2 Design Margin as DfT

One of the primary causes of long test time in many devices is poor design margin. If a device under test is required to perform at a signal-to-distortion ratio of 80 dB and it actually performs at 80.1 dB, then technically it is a shippable product. The problem is that the test engineer cannot reliably measure the actual 80.1-dB value without averaging thousands of samples. If instead the design engineer uses up a little more silicon area to produce a design with an average reading of 85 dB, then the test engineer's margin of error would be much larger. Since high levels of measurement accuracy result in long test times on expensive testers, the cost of tight design margins is incredibly high.

When measuring a noisy signal (i.e., any signal in the real world), the test time required to make a production-worthy measurement increases exponentially as the design margin tightens. The general rule of thumb for averaging noisy signals is that it takes four times as many samples or averaged measurements to produce a reading that is twice as repeatable. Therefore, doubling the design margin will reduce sample collection time by a factor of four; quadrupling design margin reduces collection time by sixteen, etc. Design engineers sometimes make margin decisions based purely on silicon area without realizing the devastating impact it can have on test time. The test engineer should be prepared to discuss the test time impact of critical parameters that may be designed too close to the specified limits.

14.8.3 Avoiding Overspecification

One method of avoiding tight design margins is to simply question the need for tight specifications in the first place. Occasionally tight specifications are listed in a data sheet without any system-level need. This sometimes occurs when a second-generation device is designed based on a previous data sheet. The tight specification may not be needed in the second-generation design, but the tight specification is carried over because nobody took the time to reevaluate its importance. Though loosened limits are not really a DfT technique, the design engineer should be prepared to make note of those parameters that will be the most difficult to design with acceptable margin.

14.8.4 Predictability of Failure Mechanisms

Another way that designs can be made more robust from a testability standpoint is the use of circuits whose failure mechanisms are very simple. A successive approximation ADC that has very good superposition characteristics will only fail if one or more of its binary-weighted circuit elements are too large or small. If the test engineer can measure only the binary-weighted elements using the major carrier method (see Section 11.3.6), then test time can be dramatically reduced. If, on the other hand, the weight of the LSB depends on the value of the other bits, superposition will not hold. The major carrier technique cannot be used when superposition is not a valid assumption. Instead, a lengthy (i.e., expensive) all-codes test may be the only production-worthy test solution. Whenever possible, a design engineer should consider the use of circuits with simple, predictable failure mechanisms, even if they require more silicon area.

14.8.5 Conversion of Analog Functions to Digital

The AGC in Figure 14.29 is a good example of the use of digital circuits in place of analog circuits. Another good example is the use of a digital filter in place of a switched capacitor filter. Digital circuits are generally much more production-worthy than analog circuits. Digital circuits tolerate much more variation in process without any degradation in performance. The design margins so important in analog circuits are not applicable to digital circuits (with the exception of critical timing and race conditions). Using digital circuits, the time-consuming DSP-based tests required for analog circuits can be replaced by pass/fail digital patterns. Also, digital circuits can benefit from test synthesis and automatic test pattern generation (ATPG) tools to automate the test development process.

14.8.6 Reduced Tester Performance Requirements

Another powerful test cost reduction DfT approach is the reduction of ATE tester requirements. In general a low-frequency tester is much less expensive than one capable of testing high-frequency digital patterns and analog signals. For example a tester with 100-MHz digital pattern capabilities will be much more expensive than one capable of only 25-MHz patterns. If features can be added to the DUT to reduce the demands on the tester, then test costs may be reduced substantially.

Similar comments can be made about the accuracy requirements of the target tester. If a design can be produced that only requires a 60-dB measurement capability, then a variety of low-cost testers can be used to make the measurement. If 110-dB measurement capabilities are required, then a much more expensive tester may be required. Again, this falls into the "easier said than done" category of DfT, but it is something the design and test engineers should keep in mind.

14.8.7 Avoidance of Trim Requirements

Trimming of analog values such as DC offsets, AC gains, and PLL center frequencies can take a huge amount of test time. Whether laser trimming or fuse trimming is involved, the time required to search for the appropriate trim condition can be extremely long. If a trimming circuit is used to compensate for a poorly designed circuit or an out-of-control process, then it may be far more economical to find the root cause of the error than to use the ATE tester to correct the defect.

14.9 I_{DDQ}

14.9.1 Digital I_{DDQ}

As mentioned in the introductory section, I_{DDQ} testing is a DfT methodology allowing detection of noncatastrophic defects in digital logic. An I_{DDQ} test sets all the CMOS devices into a static digital state and then measures the tiny current leaking from power to ground. Excessive I_{DDQ} current indicates a resistive defect, which may or may not be catastrophic (Figure 14.35). A device that fails an I_{DDQ} test might otherwise appear to be perfectly good based on its response to digital pattern tests. Without I_{DDQ} testing, the subtle, noncatastrophic defect would be passed on

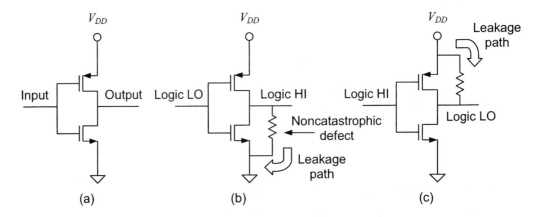

Figure 14.35. CMOS inverter leakage paths detected with I_{DDQ} tests: (a) nondefective inverter, (b) inverter with leakage path from output to ground, and (c) inverter with leakage path from output to V_{DD}.

to the customer. The customer might not see a problem with the device until weeks or months after the finished product left the factory floor.

Since I_{DDQ} testing can only be performed on a device that has been designed properly, I_{DDQ} is considered a DfT methodology rather than simply a quality assurance test technique. The digital vectors necessary to set each P/N transistor pair into each of its two states can be generated automatically, especially if the design is fully scannable. Several off-the-shelf software tools are capable of producing the I_{DDQ} test vectors automatically.

14.9.2 Analog and Mixed-Signal I_{DDQ}

I_{DDQ} testing of analog and mixed-signal circuits is not nearly as well developed as digital I_{DDQ} testing. Part of the reason analog I_{DDQ} testing has not been widely used in the past is that many analog circuits cannot be set into a quiescent mode that draws no current. Whereas the complementary transistors in a digital gate are never supposed to be on at the same time during a static state, the complementary transistors of an analog circuit are configured in their linear region. Both transistors at the output of an op amp are on at the same time; so the quiescent current of an amplifier may be much higher than the leakage currents that one might like to detect.

This is not to say that analog circuits cannot be designed for I_{DDQ} testing. It just means that the design engineer has to avoid any design topology that cannot be set to draw little or no current from power to ground. Amplifiers must include a power-down mode that shuts off the quiescent current of the amplifier without disconnecting it from its power supplies. Circuits like resistor dividers from power to ground are not allowed in an I_{DDQ} testable design unless the resistors can be disconnected. Perhaps the best way to implement resistors with such a disconnection mode is to use P-channel or N-channel CMOS transistors as resistors. The gate can be set to V_{DD} or ground to shut the resistor off or turn it on.

Because the techniques to set each analog block into a zero current quiescent mode are somewhat ad hoc in nature, there are no commercially available automated tools to generate the necessary I_{DDQ} test vectors for analog and mixed-signal circuits. The test engineer has to generate the appropriate test conditions on a case-by-case basis.

I_{DDQ} testing is an entire topic unto itself. Only a brief overview has been presented in this chapter. The reader is encouraged to refer to the existing literature on the subject of I_{DDQ} testing.[26,27]

14.10 Summary

This chapter has attempted to introduce all the major areas of DfT for mixed-signal circuits, at least in a very cursory fashion. However, trying to learn everything there is to know about DfT is like taking a sip from a fire hose. Obviously there are many sides to this seemingly straightforward topic, and we have only looked at some of the possibilities. Many of the dramatic mixed-signal testing breakthroughs in the future will probably be based on totally new concepts in DfT that we have not even discussed here. Hopefully, this brief summary of digital and mixed-signal DfT will inspire the design engineer and test engineer to think of new ways to improve the testability of their mixed-signal circuits.

Often the design engineers or test engineers get confused as to the purpose of a particular DfT technique. The design engineer may be thinking of ways to evaluate internal circuit nodes to allow easier design debug, while the test engineer may be primarily concerned with test time reduction. There are at least five different objectives of DfT, including lower test cost, higher product and process quality, ease in design diagnostics and characterization, ease in test program development, and enhanced diagnostic capabilities in the customer's system-level application. The design engineers and test engineers should keep in mind what purpose or purposes each DfT structure or technique serves.

The task of evaluating the DfT considerations for a given design can be overwhelming. The design engineers may get carried away adding hundreds of test modes that the test engineer finds useless, while forgetting an incredibly important test mode that costs millions of dollars over the life of the product. To help the new test engineer and the design engineer remember all the aspects of DfT presented in this chapter, a DfT checklist has been compiled for use early in the design phase of a new product. The checklist on the following pages is undoubtedly incomplete, but it should serve as a good starting point.

APPENDIX A.14.1

DfT Checklist

General Questions
1. Has a digital scan methodology been considered, and if not, why not?
2. Has 1149.1 boundary scan been considered? Does the customer require it?
3. Has 1149.4 mixed-signal boundary scan been considered?

4. Is the digital portion of the design compatible with I_{DDQ} testing?
5. Can all analog circuits be set to draw zero current for analog I_{DDQ} testing?

Digital Logic
1. Consider the use of pseudorandom BILBO circuits in data channels, especially high-speed data channels.
2. Consider RAM BIST to allow the DUT to test its own memory while other tests are being performed, or to allow long memory tests to be performed on an inexpensive tester.
3. Consider use of on-chip microprocessors or DSPs to perform BIST testing operations. Test code can be hard coded in ROM or downloaded into program RAM.
4. Partition long divider chains and other complex circuits using scan or ad hoc test modes to allow fewer test vectors (lower test time).
5. Break digital feedback paths to allow the tester to take control of all inputs to a digital block. This will save test time and prevent untestable circuits.
6. Always provide a reset or preset for every digital block. Provide hardware or software resets rather than just power-on reset (POR). POR requires long test time. Reset ALL state machines, dividers, etc. so that the tester can put the device into a known state before each test.
7. Do not use device-driven timing, or at least provide a tester-driven timing mode. This is especially important on DAC and ADC interfaces where digitized samples are to be applied to the device. Do not assume the tester can watch the DUT to know when to perform an action on a nanosecond's notice.
8. Provide bypass modes for all clock generators (PLLs, crystal oscillators, etc.). The tester must be able to drive the DUT's master clock to achieve coherent DSP-based testing. Also, automatically generated digital pattern cannot be used if the tester cannot control the master clock directly.
9. Never force the tester to apply a clock to the DUT through an AC coupled path. Example: Overdriving a crystal oscillator through an on-chip capacitor is not generally acceptable. A DC coupled path is required to allow automated test vectors to be used.
10. If lengthy digital preambles are required by the end application (i.e. plug and play preamble), then provide a quick means of getting through the preamble to save test time.

Analog and Mixed-Signal Circuits
1. Provide test modes to check digital functionality with digital test vectors rather than analog measurements. Use scan chains or ad hoc test modes to implement the circuit separation. Examples: power-down logic with direct access to digital outputs (no supply current measurements), digitally controlled AGC with scan chain splitting up the analog and digital portions.
2. Insert analog test access points at each analog or mixed-signal subcircuit's input and output. The test point should allow the tester to take control of each subcircuit's input and measure its output independent of the other subcircuits.
3. Always provide analog test point access to DC references.
4. Use T-switches when connecting multiple analog signals to a common test bus.
5. Consider the use of ADCs and DACs to test each other in a BIST configuration. Use on-chip microprocessors if possible to allow a complete self-contained test.

6. Use loopback modes on DAC/ADC channels to allow a full circuit test with one simple measurement. Analog loopback connects the DAC channel analog output to the ADC channel input. Digital loopback connects the ADC channel digital output to the DAC channel digital input.

7. Look for circuits with long settling times (>10 ms). Provide precharge capabilities to quickly put the device into a settled state. Example: DC reference decoupling capacitor precharge, RC high-pass filter precharge.

8. High-frequency analog signals can be sampled with on-chip ADCs, on-chip strobed comparators, or high-bandwidth sample-and-hold circuits. Undersampling is a powerful technique for reducing the frequency of signals that must be passed to the ATE tester.

9. For PLLs, allow the tester to force a voltage into the VCO or to measure the voltage at the VCO's input. Provide a bypass mode so the tester can get past the PLL. Allow separation of the PLL into its separate digital and analog components, assuming this will not destroy its performance in normal modes.

10. Segment DACs and ADCs so they can be tested in pieces, if possible. This is especially important on converters with high resolution (>10 bits).

11. Consider use of oscillation DfT to test the high-frequency performance of subcircuits.

12. Design physical test pads into the top-level metal for signals that do not need to be measured in production but which might prove useful in debugging defective devices.

13. Always try to give adequate design margin to avoid performance near specification limits. Tight limits require accurate, repeatable measurements that are time-consuming. Every factor of two in test repeatability requires four times the data collection time. This is one of the most important DfT considerations!

14. Use robust circuits that do not need to be tested thoroughly, even if more silicon area is required.

15. Avoid allowing the customer or marketing to push specifications past reasonable limits. Again, tight specs lead to very long test times, not to mention reduced yields.

16. Try to design circuits that will have predictable failure mechanisms. Example: ADCs and DACs with excellent bit-weight superposition. This allows smaller numbers of simple (faster) tests.

17. Consider converting analog functions into digital functions, even if more silicon area is required. Digital circuits are far easier and faster to test than analog functions. Yields are also higher on digital circuits.

18. Try to find ways to reduce the demands on the ATE tester, especially with respect to high frequencies. Low-frequency testers are much less expensive.

19. Avoid analog trim if possible. Consider adding extra silicon area if necessary to achieve higher accuracy without production trimming. Production trimming is a major test time consumer.

Problems

14.1. List five advantages of DfT.

14.2. Why are robust circuits more testable than circuits whose performance is near failure?

14.3. Why does product cycle time affect profit margins?

14.4. What is the primary purpose of JTAG 1149.1 boundary scan? What is the purpose of the JTAG TAP controller? How can we force the JTAG TAP controller into its reset state regardless of its initial state? What is the maximum number of TCK clock cycles required to accomplish the reset?

14.5. If a digital circuit cannot be reset, what is the likely effect on test time? What is the likely effect on test development cycle time?

14.6. A DSL modem board is fabricated using 5 JTAG 1149.1 compatible devices. Sketch the interconnection of the TAP controllers for this board. How many JTAG interface signals must be connected to the board-level automated tester?

14.7. A design contains 325 flip-flops. Assuming we use a full-scan design methodology with a single scan chain, how many flip-flops would be in the scan chain? If we exercise the scan clock at 10 MHz, how fast can we supply arbitrary parallel test vectors to the DUT circuits under test? How would multiple scan chains affect our maximum test vector rate?

14.8. Gate oxide integrity (GOI) failures result in small leakage currents between the gate of a CMOS transistor and its channel. How might we quickly detect GOI failures in a complex digital design (sketch a transistor-level diagram showing a detected failure)?

14.9. The amplifier of Figure 14.32 has 10 mV of DC offset. The nominal value of V_{MID} is 1.50 V. Assuming a CMOS transmission gate resistance of 2 kΩ, how long does it take for the decoupling capacitor to charge from 0 V to its final value ± 1 mV? After precharging the decoupling capacitor and reversing the state of the two switches, how long should we wait for the V_{MID} voltage to restabilize to within 1 mV of its final value? (Assume the settling time of the amplifier is negligible). Compare the total settling time of the circuit in Figure 14.32 with one lacking DfT precharging switches.

14.10. Using MATLAB or similar programming language, simulate the data sequence produced by the LFSR pseudorandom number generator in Figure 14.11. Starting with an initial preset value (seed) of 1, produce a listing of the values produced by the random number generator. How many values are produced before the sequence repeats?

14.11. Using MATLAB or similar programming language, apply the LFSR output values of the previous problem to a 4-bit adder (circuit under test) with carry output. Apply bits D7 through D4 of the LFSR to the A input of the adder and bits D3-D0 to the B input. Apply the output of the adder, including the carry bit, to an 8-bit checksum adder (an 8-bit running sum adder without carry). Reset the 8-bit checksum adder to 0 and apply the pseudorandom sequence to the 4-bit adder and checksum adder through one cycle of the random number sequence. What value does the checksum adder produce? Simulate a defect by forcing the 4-bit adder D1 output to logic LO. What value (signature) is produced by the checksum adder? What type of circuit would we add to a BILBO controller to evaluate the functionality of the adder based on the checksum signature value?

14.12. Design a scannable 4-bit binary counter with synchronous active high reset and ripple carry output using the scannable flip-flop illustrated in Figure 14.10(a). Sketch the CLK, SCAN, and SD inputs required to shift the vector 1101 to the QD-QA outputs, latch the response of the combinational logic, and shift out the response (reset the counter to 0000 before performing the scan sequence). Sketch the output at SQ as the scan sequence is performed.

14.13. A 1-kΩ resistor is connected between an output of DUT 1 and an input of DUT 2. Both DUTs are IEEE Std. 1149.4 compatible. Propose a test methodology to verify the connectivity of the resistor to both DUT pins and to verify its value using the AB1 and AB2 analog measurement buses of the 1149.4 analog boundary module. Would your test methodology detect a short to ground on either side of the resistor?

14.14. Propose a calibration procedure for the ADC/DAC BIST scheme of Figure 14.21, assuming we wish to measure absolute gain and frequency response of the DAC and the ADC at the following frequencies: 1, 5, 9, and 13 kHz. You may use only one external signal as a calibration standard, although you may assume it is error-free.

14.15. In the circuit illustrated in Figure 14.23, the DAC reference voltage is found to be defective, making the channel unusable. We still want to characterize the performance of the remaining circuit blocks while the DC reference block is redesigned. How can we accomplish this characterization using the existing DfT structure?

14.16. Would the following clock generator device be well-suited to ATE testing? Why or why not? How might it be improved using DfT (sketch your DfT proposal)?

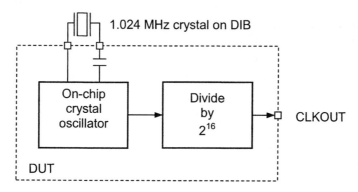

References

1. Miron Abramovici, Melvin A. Breuer, Arthur D. Friedman, *Digital Systems Testing and Testable Design*, Revised Printing, IEEE Press, New York, NY, January, 1998, ISBN: 0780310624

2. E. B. Eichelberger, E. Lindbloom, J. A. Waicukauski, T. W. Williams, *Structured Logic Testing*, (Prentice Hall Series in Computer Engineering), E. J. McCluskey, Ed., Prentice Hall, Englewood Cliffs, NJ, 1991

3. Francis C. Wang, *Digital Circuit Testing: A Guide to DfT, ATVG, and Other Techniques*, Academic Press, New York, NY, August 1991, ISBN: 0127345809

4. P. H. Bardell, W. H. McAnney, J. Savir, *Built-in Test for VLSI, Pseudorandom Techniques*, John Wiley & Sons, New York, NY, 1987

5. L. Avra, E. J. McCluskey, *Synthesizing for Scan Dependence in Built-in Self-Testable Designs*, Proc. International Test Conference, 1993, pp. 734-43

6. S. Hellebrand et al., *Generation of Vector Patterns through Reseeding of Multiple-Polynomial Linear Feedback Shift Registers*, Proc. International Test Conference, 1992, pp. 120-29

7. E. Hawrysh, G. Roberts, *An Integration of Memory-Based Analog Signal Generation into Current DfT Architectures*, Proc. International Test Conference, 1996, pp. 528-537

8. S. Sunter, N. Nagi, *A Simplified Polynomial-Fitting Algorithm for DAC and ADC BIST*, Proc. International Test Conference, 1997, pp. 389-95

9. M. Slamani, B. Kaminska, G. Quesnel, *An Integrated Approach for Analog Circuit Testing with a Minimum Number of Detected Parameters*, Proc. International Test Conference, 1994, pp. 631-40

10. Gordon W. Roberts, Albert K. Lu, *Analog Signal Generation for Built-In-Self-Test of Mixed-Signal Integrated Circuits*, Kluwer Academic Publishers, Boston, MA, April 1995, ISBN: 0792395646

11. C. Dislis, J. H. Dick, I. D. Dear, A. P. Ambler, *Test Economics and Design for Testability*, 1st edition, Prentice Hall, Englewood Cliffs, NJ, January 1995, ISBN: 0131089943

12. Colin M. Maunder, Ed., *IEEE Std 1149.1-1993a, Standard Test Access Port and Boundary-Scan Architecture*, IEEE Standards Board, New York, NY, October 1993, ISBN:1559373504

13. Brian Wilkins, Ed., *IEEE Std 1149.4-1999, IEEE Standard for a Mixed-Signal Test Bus*, IEEE Standards Board, New York, NY, March 2000, ISBN: 0738117552

14. Miron Abramovici, Melvin A. Breuer, Arthur D. Friedman, *Digital Systems Testing and Testable Design*, Revised Printing, IEEE Press, New York, NY, January, 1998, ISBN: 0780310624

15. A. J. van de Goor, *Testing Semiconductor Memories: Theory and Practice*, John Wiley and Sons, Chichester, UK, 1991

16. P. Mazumder, K. Chakraborty, *Testing and Testable Design of High-Density Random-Access Memories (Frontiers in Electronic Testing)*, Kluwer Academic Publishers, Boston, MA, June 1996, ISBN: 0792397827

17. A. J. van de Goor, *The Implementation of Pseudorandom Memory Tests on Commercial Memory Testers*, Proc. International Test Conference, 1997, pp. 226-35

18. K. Parker, J. McDermid, S. Oresjo, *Structure and Metrology for an Analog Testability Bus*, Proc. International Test Conference, 1993, pp. 309-17

19. G. Devarayanadurg, P. Goteti, M. Soma, *Hierarchy based Statistical Fault Simulation of Mixed-Signal ICs*, Proc. International Test Conference, 1996, pp. 521-27

20. R. Voorakaranam et al., *Hierarchical Specification-Driven Analog Fault Modeling for Efficient Fault Simulation and Diagnosis*, Proc. International Test Conference, 1997, pp. 903-12

21. C. Y. Pan, K. T. Cheng, *Fault Macromodeling for Analog/Mixed-Signal Circuits*, Proc. International Test Conference, 1997, pp. 913-22

22. M. Soma et al., *Analog and Mixed-Signal Test*, B. Vinnakota, Ed., Prentice Hall, Englewood Cliffs, NJ, April 1998, ISBN: 0137863101

23. K. Lofstrum, *Early Capture for Boundary Scan Timing Measurements*, Proc. International Test Conference, 1996, pp. 417-22

24. M. Burns, *Undersampling Digitizer with a Sampling Circuit Positioned on an Integrated Circuit*, U.S. Patent No. 5,578,935

25. K. Arabi, B. Kaminska, *Oscillation-Test Strategy for Analog and Mixed-Signal Integrated Circuits*, Proc. 14th VLSI Test Symposium, 1996

26. Sreejit Chakravarty, Paul J. Thadikaran, *Introduction to IDDQ Testing*, Kluwer Academic Publishers, Boston, MA, May 1997, ISBN: 0792399455

27. Y. K. Malaiya, *Bridging Faults and Iddq Testing* (IEEE Computer Society Press Technology Series), R. Rajsuman, Ed., IEEE Computer Society Press, Washington, DC, October 1992, ISBN: 0818632151

CHAPTER 15

Data Analysis

15.1 INTRODUCTION TO DATA ANALYSIS

15.1.1 The Role of Data Analysis in Test and Product Engineering

Data analysis is the process by which we examine test results and draw conclusions from them. Using data analysis, we can evaluate DUT design weaknesses, identify DIB and tester repeatability and correlation problems, improve test efficiency, and expose test program bugs. As mentioned in Chapter 2, debugging is one of the main activities associated with mixed-signal test and product engineering. Debugging activities account for about 20% of the average workweek. Consequently, data analysis plays a very large part in the overall test and product engineering task.

In addition to supporting the silicon and test program debugging task, many data analysis tools are designed to help improve the silicon fabrication process itself. The fabrication process can be improved through statistical data analysis of production test results. A methodology called *statistical process control* (SPC) formalizes the steps by which this improvement is achieved. In this chapter we will examine various data visualization tools, study the statistics that describe repeatability and process variations, and introduce the topic of statistical process control. Although our treatment of SPC is very brief and incomplete, we do not mean to treat it as an unimportant subject. SPC is a powerful tool for continuous improvement that helps reduce defects and lower testing costs.

15.1.2 Visualizing Test Results

Many types of data visualization tools have been developed to help us make sense of the reams of test data that are generated by a mixed-signal test program. In Chapter 4, "Measurement Accuracy," we saw two common data analysis tools: the datalog and the histogram. In this chapter we will review these tools as well as several others including the lot summary, shmoo plot, and wafer map.

Statistical analysis software packages also offer a host of visualization tools. These include scatter plots, control charts, and C_p and C_{pk} Pareto charts. Some or all these tools may be included in the ATE tester's operating system. The rest can be added using data analysis software specifically tailored for statistical process control. Before we discuss SPC or statistics in general, let us examine some common data visualization tools available on most mixed-signal ATE testers.

15.2 DATA VISUALIZATION TOOLS

15.2.1 Datalogs (Data Lists)

A datalog, or data list, is a concise listing of test results generated by a test program. Datalogs are the primary means by which test engineers evaluate the quality of a tested device. The format of a datalog typically includes a test category, test description, minimum and maximum test limits, and a measured result. The exact format of datalogs varies from one tester type to another, but datalogs all convey similar information.

A short datalog from a Teradyne Catalyst tester is shown in Figure 15.1. Each line of the datalog contains a shorthand description of the test. For example, "DAC Gain Error" is the name of test number 5000. The gain error test is part of the S_VDAC_SNR test group and is executed during the T_VDAC_SNR test routine. The minimum and maximum limits for the test are also listed. Using test number 5000 as an example, the lower limit of DAC Gain Error is –1.00 dB, the upper limit is +1.00 dB, and the measured value for this DUT is –0.13 dB.

The datalog displays an easily recognizable fail flag beside each value that falls outside the test limits. For instance, test 7004 in Figure 15.1 shows a failure in which the measured value is 1.23 LSBs. Since the upper limit is 0.9 LSBs, this test fails. In this particular example, the failure is flagged with an (F) symbol. Hardware and software alarms from the tester also result in a datalog alarm flag, such as (A). Alarms can occur for a variety of reasons, including mathematical divisions by zero and power supply currents that exceed programmed limits. When alarms are generated, the test program halts (unless instructed by the test engineer to ignore alarms). The tester assumes that the DUT is defective and treats the alarm as a failure.

```
Sequencer:   S_continuity
   1000 Neg PPMU Cont        Failing Pins:   0
Sequencer:   S_VDAC_SNR
   5000 DAC Gain Error    T_VDAC_SNR   -1.00  dB  <      -0.13  dB    <    1.00  dB
   5001 DAC S/2nd         T_VDAC_SNR    60.0  dB  <=      63.4  dB
   5002 DAC S/3rd         T_VDAC_SNR    60.0  dB  <=      63.6  dB
   5003 DAC S/THD         T_VDAC_SNR    60.00 dB  <=      60.48 dB
   5004 DAC S/N           T_VDAC_SNR    55.0  dB  <=      70.8  dB
   5005 DAC S/N+THD       T_VDAC_SNR    55.0  dB  <=      60.1  dB
Sequencer:   S_UDAC_SNR
   6000 DAC Gain Error    T_UDAC_SNR   -1.00  dB  <      -0.10  dB    <    1.00  dB
   6001 DAC S/2nd         T_UDAC_SNR    60.0  dB  <=      86.2  dB
   6002 DAC S/3rd         T_UDAC_SNR    60.0  dB  <=      63.5  dB
   6003 DAC S/THD         T_UDAC_SNR    60.00 dB  <=      63.43 dB
   6004 DAC S/N           T_UDAC_SNR    55.0  dB  <=      61.3  dB
   6005 DAC S/N+THD       T_UDAC_SNR    55.0  dB  <=      59.2  dB
Sequencer:   S_UDAC_Linearity
   7000 DAC POS ERR       T_UDAC_Lin  -100.0  mV  <        7.2  mV    <  100.0  mV
   7001 DAC NEG ERR       T_UDAC_Lin  -100.0  mV  <        3.4  mV    <  100.0  mV
   7002 DAC POS INL       T_UDAC_Lin   -0.90  lsb <        0.84 lsb   <    0.90 lsb
   7003 DAC NEG INL       T_UDAC_Lin   -0.90  lsb <       -0.84 lsb   <    0.90 lsb
   7004 DAC POS DNL       T_UDAC_Lin   -0.90  lsb <        1.23 lsb (F)<   0.90 lsb
   7005 DAC NEG DNL       T_UDAC_Lin   -0.90  lsb <       -0.83 lsb   <    0.90 lsb
   7006 DAC LSB SIZE      T_UDAC_Lin    0.00  mV  <        1.95 mV    <  100.00 mV
   7007 DAC Offset V      T_UDAC_Lin  -100.0  mV  <        0.0  mV    <  100.0  mV
   7008 Max Code Width    T_UDAC_Lin    0.00  lsb <        1.23 lsb   <    1.50 lsb
   7009 Min Code Width    T_UDAC_Lin    0.00  lsb <        0.17 lsb   <    1.50 lsb
Bin:   10
```

Figure 15.1. Example datalog from a Teradyne Catalyst tester.

Because the device in Figure 15.1 fails test 7004, it is categorized into bin 10 as displayed at the bottom of the datalog. Bin 1 usually represents a good device, while other bins usually represent various categories of failures and alarms. Sometimes there are multiple grades of shippable devices, which are separated into different passing bins. For example, a certain percentage of 500-MHz microprocessors may fail at 500 MHz, but may operate perfectly well at 400-MHz. The 400-MHz processors might be sorted into bin 2 and shipped at a lower cost, while the higher grade 500 MHz processors are sorted into bin 1 to be sold at full price.

15.2.2 Lot Summaries

Lot summaries are generated after all devices in a given production lot have been tested. A lot summary lists a variety of information about the production lot, including the lot number, product number, operator number, etc. It also lists the yield loss and cumulative yield associated with each of the specified test bins. The overall lot yield is defined as the ratio of the total number of good devices divided by the total number of devices tested:

$$\text{yield} = \frac{\text{total good devices}}{\text{total devices tested}} \quad (15.1)$$

The lot yield is listed in the lot summary, but it does not tell us everything we need to know. If a particular lot exhibits a poor yield, we want to know *why* its yield was low. We want to know what category or categories of tests dominated the failures so we can look into the problem to determine its cause. For this reason, lot summaries also list test categories and what percentage of devices failed each category. Figure 15.2 shows a simplified lot summary, including yields for a variety of test categories.

A lot summary can help us identify which failures are most common to a particular type of DUT. This allows us to focus our attention on the areas of the design, the process, and the test program that might be causing the most failures in production. The lot summary in Figure 15.2 shows that our highest yield loss is due to the RECV channel AC tests. We might think that our XMIT channel has no problems, because it causes only a 0.30% yield loss.

However, we have to be careful in making such judgements based on data collected during production. We have to remember that once a DUT fails any test, the tester immediately rejects it and moves on to the next device. After all, there is no point in continuing to test a DUT once it has been disqualified for shipment to the customer.

Since the test program halts after the first DUT failure, the earlier tests will tend to cause more yield loss than later ones, simply because fewer DUTs proceed to the later tests. The earlier failures mask any failures that would have occurred in later tests. For example, any or all the devices that failed the RECV channel tests in Figure 15.2 might also have failed the XMIT channel tests if given the chance. Therefore, during the device characterization phase we may want to instruct the tester to collect data from all tests whether the DUT passes or not. Of course, the extra testing leads to a longer average test time; so we do not want to perform continue-on-fail testing in production unless necessary.

We can sometimes improve our overall production throughput by moving the more commonly failed tests toward the beginning of the test program. Average test time is reduced by the rearrangement because we do not waste time performing tests that seldom fail only to lose

```
Lot Number:  122336
Device Number: TLC1701FN
Operator Number: 42
Test Program: F779302.load

Devices Tested: 10233
Passing Devices: 9392
Test Yield: 91.78%
```

Bin#	Test Category	Devices Tested	Failures	Yield Loss	Cum. Yield
7	Continuity	10233	176	1.72%	98.28%
2	Supply Currents	10057	82	0.80%	97.48%
3	Digital Patterns	9975	107	1.05%	96.43%
4	RECV Channel AC	9868	445	4.35%	92.08%
5	XMIT Channel AC	9423	31	0.30%	91.78%

Figure 15.2. Simplified lot summary.

yield to subsequent tests that often fail. Once again, we have to remember that the order of tests may affect the lot summary output. Therefore, whenever we wish to reorder our test program based on yield loss, we would prefer to use lot summaries in which the tester does not halt on the first failure. That way, we know which tests are truly the ones that catch the largest number of defective DUTs.

When rearranging test programs based on yield loss, we also have to consider the test time that each test consumes. For example, the RECV channel tests in Figure 15.2 may take 800 ms, while the digital pattern tests only takes 50 ms. The digital pattern test is more efficient at identifying failing DUTs since it takes so little test time. Therefore, it might not make sense to move the longer RECV test to the beginning of the program, even though it catches more defective DUTs than the digital pattern test. Clearly, test program reordering is not a simple matter of moving the tests having the highest yield loss to the beginning of the test program.

15.2.3 Wafer Maps

A wafer map (Figure 15.3) displays the location of failing die on each probed wafer in a production lot. Unlike lot summaries, which only show the number of devices that fail each test category, wafer maps show the physical distribution of each failure category. This information can be very useful in locating areas of the wafer where a particular problem is most prevalent. For example, the continuity failures are most severe at the upper edge of the wafer illustrated in Figure 15.3. Therefore, we might examine the bond pad quality along the upper edge of the wafer to see if we can find out why the continuity test fails most often in this area. Also, the RECV channel failures are most severe near the center of the wafer. This kind of ring-like pattern often indicates a processing problem, such as uneven diffusion of dopants into the semiconductor surface or photomask misalignment. If all steps of the fabrication process are within allowable tolerances, consistent patterns such as this may indicate that the device is simply too sensitive to normal process variations and therefore needs to be redesigned. Wafer maps are a powerful data analysis tool, allowing yield enhancement through a cooperative effort between the design, test, product, and process engineers.

Naturally, it is dangerous to draw too many conclusions from a single wafer map. We need to examine many wafer maps to find patterns of consistent failure distribution. For this reason, some of the more sophisticated wafer mapping tools allow us to overlay multiple wafer maps on top of one another, revealing consistency in failure distributions. From these composite failure maps, we can draw more meaningful conclusions about consistent processing problems, design weaknesses, and test hardware problems.

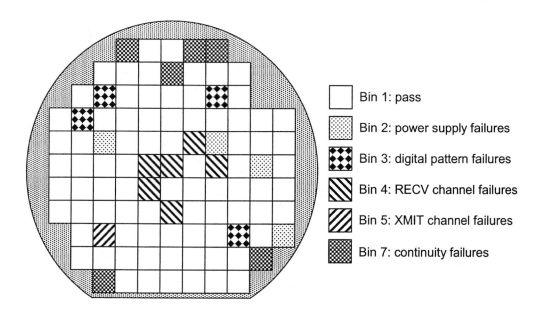

Figure 15.3. Wafer map.

15.2.4 Shmoo Plots

Shmoo plots were among the earliest computer-generated graphic displays used in semiconductor manufacturing.[1] A shmoo plot is a graph of test results as a function of test conditions. For example, some of the earliest shmoo plots displayed pass/fail test results for PMOS memory ICs as a function of V_{DD} and V_{SS}. The origins of the name "shmoo plot" are not known for certain. According to legend, some of the early plots reminded the engineers of a shmoo, a squash-shaped cartoon character from Al Capp's comic strip "Li'l Abner." Although few shmoo plots are actually shmoo-shaped, the name has remained with us.

The graph in Figure 15.4 is called a *functional shmoo plot*, since it only shows which test conditions produce a passing (functional) or failing (nonfunctional) test result. This type of plot is commonly used to characterize purely digital devices, since digital test programs primarily produce functional pass/fail results from the digital pattern tests. Measured values such as supply current and distortion cannot be displayed using a functional shmoo plot.

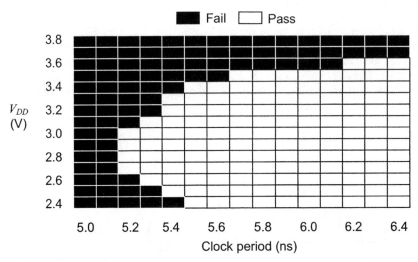

Figure 15.4. Functional shmoo plot.

Analog and mixed-signal measurements often require a different type of graph, called a *parametric shmoo plot*. Analog and mixed-signal test programs produce many parametric values, such as gain error and signal-to-distortion ratio. Parametric shmoo plots, such as the those shown in Figures 15.5 and 15.6, can be used to display analog measurement results at each combination of test conditions rather than merely displaying a simple pass/fail result. Naturally, we always have the option of comparing the analog measurements against test limits, producing pass/fail test results compatible with a simple functional shmoo plot.

Parametric shmoo plots give the test engineer a more complete picture of the performance of mixed-signal DUTs under the specified range of test conditions. This information can tell the engineering team where the device is most susceptible to failure. Assume, for example, that the DUT of Figure 15.5 needs to pass a minimum S/THD specification of 75 dB. The shmoo plot in Figure 15.5 tells us that this DUT is close to failure at about 40°C and it is somewhat marginal at 30°C if our V_{DD} supply voltage is near either end of the allowable range. Once the device weaknesses are understood, then the device design, fabrication process, and test program can be improved to maximize production yield. Also, shmoo plots can help us identify worst-case test conditions. For example, we may choose to perform the S/THD test at both low and high V_{DD} based on the worst-case test conditions indicated by the shmoo plot in Figure 15.5.

Shmoo plots can be generated, at least in principle, using data collected through manual adjustment of test conditions. However, such a process would be extremely tedious. For example, if we wanted to plot pass/fail results for 10 values of V_{DD} combined with 10 values of master clock period, then we would have to run the test 100 times under 100 different test conditions, adjusting the test conditions by hand each time. Clearly, software automation in the tester is required if the shmoo data collection process is to be a practical one. For this reason, modern ATE tester operating systems often include built-in shmoo plotting tools. These tools not only display the shmoo plots themselves, but they also provide automated adjustment of test conditions and automated collection of test results under each permutation of test conditions.

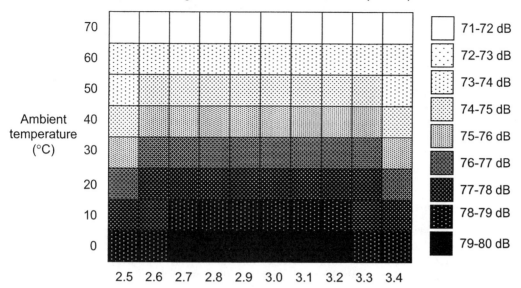

Figure 15.5. Two-dimensional parametric shmoo plot.

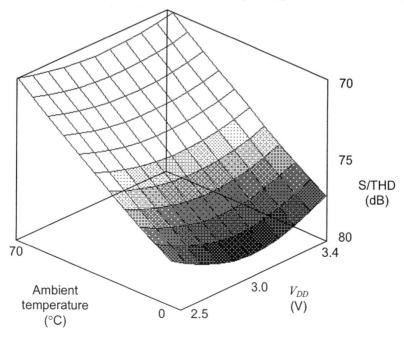

Figure 15.6. Three-dimensional parametric shmoo plot.

The shmoo plots illustrated in Figures 15.4–15.6 only represent a few of the many types of shmoo plots that can potentially be created. For example, we could certainly imagine a 3D shmoo plot showing pass/fail results for combinations of three test conditions instead of two. It is important to note that any of the many factors affecting DUT performance can be used as shmoo plot test conditions.

Common examples of shmoo test conditions include power supply voltage, master clock frequency (or period), setup and hold times, ambient temperature, I_{OL} or I_{OH} load current, etc. However, we are free to plot any measured values or pass/fail results as a function of any combination of test conditions. This flexibility makes the shmoo plot a very powerful characterization and diagnostic tool whose usefulness is limited only by the ingenuity and skill of the test or product engineer.

15.2.5 Histograms

In Chapter 4, we saw how a single DUT tested multiple times produces fluctuating measurement results due to the additive effects of random noise. For example, a DAC gain error test may show slight repeatability errors if we execute the test program repeatedly, as shown in Figure 15.7. (In this example, only the results from test 5000 have been enabled for display.)

We can view the repeatability of a group of measurements using a visualization tool called a *histogram*. A histogram corresponding to the DAC gain error example is shown in Figure 15.8. It shows a plot of the distribution of measured values as well as a listing of several key statistical values. The plot is divided into a number of vertical histogram cells, each indicating the percentage of values falling within the cell's upper and lower thresholds. For example, approximately 5% of the DAC gain error measurements in this example fell between –0.137 and –0.136 dB.

The histogram is a very useful graphical tool that helps us visualize the repeatability of measurements. If the measurement repeatability is good, the distribution should be closely packed, as the example in Figure 15.8 shows. But if repeatability is poor, then the histogram spreads out into a larger range of values. Although histograms are extremely useful for analyzing

```
5000 DAC Gain Error    T_VDAC_SNR   -1.000 dB <  -0.127 dB   < 1.000 dB
5000 DAC Gain Error    T_VDAC_SNR   -1.000 dB <  -0.129 dB   < 1.000 dB
5000 DAC Gain Error    T_VDAC_SNR   -1.000 dB <  -0.125 dB   < 1.000 dB
5000 DAC Gain Error    T_VDAC_SNR   -1.000 dB <  -0.131 dB   < 1.000 dB
5000 DAC Gain Error    T_VDAC_SNR   -1.000 dB <  -0.129 dB   < 1.000 dB
5000 DAC Gain Error    T_VDAC_SNR   -1.000 dB <  -0.128 dB   < 1.000 dB
5000 DAC Gain Error    T_VDAC_SNR   -1.000 dB <  -0.132 dB   < 1.000 dB
5000 DAC Gain Error    T_VDAC_SNR   -1.000 dB <  -0.130 dB   < 1.000 dB
5000 DAC Gain Error    T_VDAC_SNR   -1.000 dB <  -0.134 dB   < 1.000 dB
5000 DAC Gain Error    T_VDAC_SNR   -1.000 dB <  -0.131 dB   < 1.000 dB
```

Figure 15.7. Repeated test executions result in fluctuating measurements.

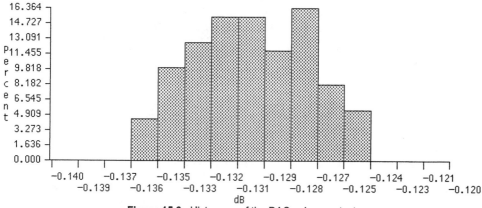

Figure 15.8. Histogram of the DAC gain error test.

measurement stability, repeatability studies are not the only use for histograms. They are also used to look at distributions of measurements collected from many DUTs to determine the extent of variability from one device to another. Excessive DUT-to-DUT variability indicates a fabrication process that is out of control or a device design that is too susceptible to normal process variations.

In addition to the numerical results and a plot of the distribution of measured values, the example histogram in Figure 15.8 displays a number of other useful values. For example, the population size is listed beside the heading "Total Results=." It indicates how many times the measurement was repeated. In the case of a DUT-to-DUT variability study, the "Total Results=" value would correspond to the number of DUTs tested rather than the number of measurement repetitions on the same DUT. In either case, the larger the population of results, the more trustworthy a histogram becomes. A histogram with fewer than 50 results is statistically questionable because of the limited sample size. Ideally a histogram should contain results from at least 100 devices (or 100 repeated test executions in the case of a single-DUT repeatability study).

In Chapters 4 and 12, we touched lightly on the subject of statistical distributions and probability. In this chapter we will examine these topics in a little more detail than in previous chapters. Unfortunately, a full treatment of statistics, probability theory, and random variables is beyond the scope of this book. We can only briefly review some of the topics that are most relevant to analog and mixed-signal testing. For a more in-depth presentation of statistics, including the derivation of fundamental equations and properties of statistics, the reader should refer to a book on the subject of statistics and probability theory.[2-4]

Exercises

15.1. A 5-mV signal is measured with a meter ten times resulting in the following sequence of readings: 5 mV, 6 mV, 9 mV, 8 mV, 4 mV, 7 mV, 5 mV, 7 mV, 8 mV, 11 mV. What is the mean value? What is the standard deviation?

Ans. 7 mV, 2.108 mV.

15.2. What are the mean and standard deviation of a set of samples of a coherent sine wave having a DC offset of 5 V and a peak-to-peak amplitude of 1.0 V?

Ans. $\mu = 5.0$ V, $\sigma = 354$ mV.

15.3 STATISTICAL ANALYSIS

15.3.1 Mean (Average) and Standard Deviation (Variance)

The frequency or distribution of data described by a histogram characterizes a sample set in great detail. Often, we look for simpler measures that describe the statistical features of the sample set. The two most important measures are the arithmetic mean μ and the standard deviation σ.

The arithmetic mean or simply the mean μ is a measure of the central tendency, or location, of the data in the sample set. The mean value of a sample set denoted by $x(n)$, $n = 0, 1, 2, ..., N-1$, is defined as

$$\mu = \frac{1}{N} \sum_{n=0}^{N-1} x(n) \tag{15.2}$$

For the DAC gain error example shown in Figure 15.8, the mean value from 110 measurements is -0.1300 dB.

The standard deviation σ, on the other hand, is a measure of the dispersion or uncertainty of the measured quantity about the mean value, μ. If the values tend to be concentrated near the mean, the standard deviation is small. If the values tend to be distributed far from the mean, the standard deviation is large. Standard deviation is defined as

$$\sigma = \sqrt{\frac{1}{N} \sum_{n=0}^{N-1} \left[x(n) - \mu\right]^2} \tag{15.3}$$

Standard deviation and mean are expressed in identical units. In our DAC gain error example, the standard deviation was found to be 0.0029 dB. Another expression for essentially the same quantity is the variance or mean square deviation. It is simply equal to the square of the standard deviation, that is, variance = σ^2.

Often, the statistics of a sample set are used to estimate the statistics of a larger group or population from which the samples were derived. Provided the sample size is greater than 30, approximation errors are insignificant. Throughout this chapter, we will make no distinction between the statistics of a sample set and those of the population, as it will be assumed that sample size is much larger than 30.

There is an interesting relationship between a sampled signal's DC offset and RMS voltage and the statistics of its samples. Assuming all frequency components of the sample set are coherent, the mean of the signal samples is equal to the signal's DC offset. Less obvious is the fact that the standard deviation of the samples is equal to the signal's RMS value, excluding the DC offset. The RMS of a sample set is calculated as the square root of the mean of the squares of the samples

$$RMS = \sqrt{\frac{1}{N}\sum_{n=0}^{N-1}\left[x(n)\right]^2} \quad (15.4)$$

If the value of μ in Eq. (15.3) is zero (i.e., if the sample set has no DC component), then Eq. (15.3) becomes identical to the RMS calculation in Eq. (15.4). Thus we can calculate the standard deviation of the samples of a coherent signal by calculating the RMS of the signal after subtracting the average value of the sample set (i.e., the DC offset).

15.3.2 Probabilites and Probability Density Functions

The histogram in Figure 15.8 exhibits a feature common to many analog and mixed-signal measurements. The distribution of values has a shape similar to a bell. The bell curve (also called a *normal distribution* or *Gaussian distribution*) is a common one in the study of statistics. According to the central limit theorem,[5] the distribution of a set of random variables, each of which is equal to a summation of a large number ($N > 30$) of statistically independent random values trends toward a Gaussian distribution. As N becomes very large, the distribution of the random variables becomes Gaussian, whether or not the individual random values themselves exhibit a Gaussian distribution. The variations in a typical mixed-signal measurement are caused by a summation of many different random sources of noise and crosstalk in both the device and the tester instruments. As a result, many mixed-signal measurements exhibit the common Gaussian distribution.

Figure 15.9 shows the histogram count from Figure 15.8 superimposed on a plot of the corresponding Gaussian probability density function (pdf). The pdf is a function that defines the probability that a randomly chosen sample X from the statistical population will fall near a particular value. In a Gaussian distribution, the most likely value of X is near the mean value, μ. Thus the pdf has a peak at $x = \mu$.

Notice that the height of the histogram cells only approximates the shape of the true Gaussian curve. If we collect thousands of test results instead of the 110 used in this example, the height

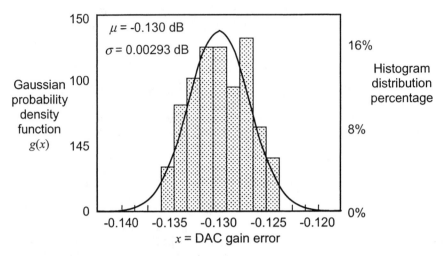

Figure 15.9. Continuous normal (Gaussian) distribution for DAC gain example.

of the actual histogram cells should more closely approach the shape of the probability density function. This is the nature of statistical concepts. Actual measurements only approach the theoretical ideal when large sample sets are considered.

The bell-shaped probability density function $g(x)$ for any Gaussian distribution having a mean μ and standard deviation σ is given by the equation

$$g(x) = \frac{1}{\sigma\sqrt{2\pi}} e^{\frac{-(x-\mu)^2}{2\sigma^2}} \tag{15.5}$$

Since the shape of the histogram in Figure 15.9 approximates the shape of the Gaussian pdf, it is easy to assume that we can calculate the expected percentage of histogram counts at a value by simply plugging the value of into Eq. (15.5). However, a pdf represents the probability *density*, rather than the probability itself. We have to perform an integration on the pdf to calculate probabilities and expected histogram counts.

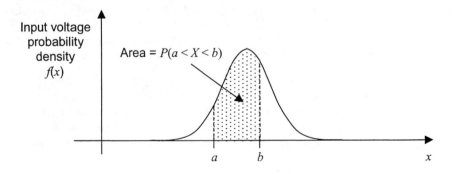

Figure 15.10. The probability over the range a to b is the area under the pdf, $f(x)$, in that interval.

The probability that a randomly selected value in a population will fall between the values a and b is equal to the area under the pdf curve bounded by $x = a$ and $x = b$, as shown in Figure 15.10. Stating this more precisely, for any probability density function $f(x)$, the probability P that a randomly selected value X will fall between the values a and b is given by

$$P(a < X < b) = \int_a^b f(x)\, dx \tag{15.6}$$

The value of P must fall between 0 (0% probability) and 1 (100% probability). In the case where $a = -\infty$ and $b = \infty$, the value of P must equal 1, since there is a 100% probability that a randomly chosen value will be a number between $-\infty$ and $+\infty$. Consequently, the total area underneath any pdf must always be equal to 1.

As $f(x)$ is assumed continuous, the probability that a random variable X is *exactly* equal to any particular value is zero. In such case we can replace either or both of the signs $<$ in Eq. (15.6) by $\leq$, allowing us to write

$$P(a \leq X \leq b) = \int_a^b f(x)\, dx \tag{15.7}$$

Incorporating the equality in the probability expression is therefore left as a matter of choice.

The probability that a Gaussian distributed randomly variable X will fall between the values of a and b can be derived from Eq. (15.6) by substituting Eq. (15.5) to obtain

$$P(a \leq X \leq b) = \int_a^b g(x)\, dx = \int_a^b \frac{1}{\sigma\sqrt{2\pi}} e^{\frac{-(x-\mu)^2}{2\sigma^2}}\, dx \tag{15.8}$$

Unfortunately, Eq. (15.8) cannot be solved in closed form. However, it can easily be solved using numerical integration methods. For instance, in our Gaussian DAC gain example, let us say we want to predict what percentage of measured results should fall into the seventh histogram cell. From the histogram, we see that there are 15 evenly spaced cells between –0.140 and –0.120 dB. The seventh cell represents all values falling between the values

$$a = -0.140 \text{ dB} + 7\, \frac{\left[-0.120 \text{ dB} - (-0.140 \text{ dB})\right]}{15} = -0.132 \text{ dB}$$

and

$$b = -0.140 \text{ dB} + 8\, \frac{\left[-0.120 \text{ dB} - (-0.140 \text{ dB})\right]}{15} = -0.1307 \text{ dB}$$

We can calculate the probability that a randomly selected DAC gain error measurement X will fall between -0.1320 and -0.1307 using the equation

$$P(-0.132 < X < -0.1307) = \int_{-0.132}^{-0.1307} \frac{1}{\sigma\sqrt{2\pi}} e^{\frac{-(x-\mu)}{2\sigma^2}}\, dx$$

where $\mu = -0.130$ dB and $\sigma = 0.00293$ dB. Using mathematical analysis software, the value of $P(-0.1320 < X < -0.1307)$ is found to be 0.163, or 16.3%. Theoretically, then, we should see 16.3% of the 110 DAC gain error measurements fall between –0.1307 and –0.1320. Indeed the seventh histogram cell shows that approximately 15.5% of the measurements fall between these values.

This example is fairly typical of applied statistics. Actual distributions never exactly match a true Gaussian distribution. Notice, for example, that the ninth and tenth histogram cells in Figure 15.9 are badly out of line with the ideal Gaussian curve. Also, notice that there are no values outside the fourth and twelfth cells. An ideal Gaussian distribution would extend to infinity in both directions. In other words, if one is willing to wait billions of years, one should eventually see an answer of +200 dB in the DAC gain error example. In reality, of course, the answer in the DAC gain example will never stray more than a few tenths of a decibel away from the average reading of –0.130 dB, since the actual distribution is only near-Gaussian.

Nevertheless, statistical analysis predicts actual results well enough to be very useful in analyzing test repeatability and manufacturing process stability. The comparison between ideal results and actual results is close enough to allow some general statements. First, the standard deviation of a near-Gaussian distribution is roughly equal to one sixth of the total variation from the minimum observed value to the maximum observed value

$$\sigma \approx \frac{1}{6}(\text{max value} - \text{min value}) \tag{15.9}$$

In the DAC gain distribution example, the standard deviation is 0.00293 dB. Therefore, we would expect to see values ranging from approximately –0.139 to –0.121 dB. These values are displayed in the example histogram in Figure 15.8 beside the labels "Mean –3 sigma" and "Mean +3 sigma." The actual minimum and maximum values are also listed. They range from –0.136 to –0.125 dB, which agrees fairly well with the ideal values of $\mu \pm 3\sigma$.

At this point we should note a common misuse of statistical analysis. We have used as our example a gain measurement, expressed in decibels. Since the decibel is based on a logarithmic transformation, we should actually use the equivalent V/V measurements to calculate statistical quantities such as mean and standard deviation. For example, the average of three-decibel values, 0, -20, and -40 dB is 20 dB. However, the true average of these values as calculated using V/V is given by

$$\text{average gain} = \frac{1 + \frac{1}{10} + \frac{1}{100}}{3} = 0.37 \text{ V/V} = -8.64 \text{ dB}$$

A similar discrepancy arises in the calculation of standard deviation. Therefore, a Gaussian-distributed sample set converted into decibel form is no longer Gaussian. Nevertheless, we often use the nonlinearized statistical calculations from a histogram to evaluate parameters expressed in decibel units as a time-saving shortcut. The discrepancy between linear and logarithmic calculations of mean and standard deviation become negligible as the range of decibel values decreases. For example, the range of values in the histogram of Figure 15.8 is quite small; so the errors in mean and standard deviation are minor. The reader should be careful when performing statistical analysis of decibel values ranging over several decibels.

15.3.3 The Standard Gaussian Cumulative Distribution Function $\Phi(z)$

Computing probabilities involving Gaussian distributions is complicated by the fact that numerical integration methods must be used to solve the definite integrals involved. Fortunately, a simple change of variable substitution can be used to convert the integral equations into one involving a Gaussian distribution with zero mean and unity standard deviation. This then enables a set of tables or approximations that require numerical evaluation to be used to solve the probabilities associated with an arbitrary Gaussian distribution. Hence, the test engineer can completely avoid the need for numerical integration routines. Let us consider how this is done.

The probability that a randomly selected value X will be less than a particular value x can be calculated directly from Eq. (15.6). We set $a = -\infty$, $b = x$, and write

$$F(x) = P(X < x) = P(-\infty < X < x) = \int_{-\infty}^{x} f(y) \, dy \tag{15.10}$$

This integral is central to probability theory and is given a special name called the *cumulative distribution function* $F(x)$. Here we view $F(x)$ as an ordinary function of the variable x. The probability that X lies in the range a to b can then be expressed in terms of the difference of $F(x)$ evaluated at $x = a$ and $x = b$ according to

$$P(a < X < b) = F(b) - F(a) \tag{15.11}$$

In the case of a Gaussian distribution, $F(x)$ is equal to

$$F(x) = \int_{-\infty}^{x} \frac{1}{\sigma\sqrt{2\pi}} e^{\frac{-(y-\mu)^2}{2\sigma^2}} \, dy \tag{15.12}$$

If we consider the simple change of variable $z = (y - \mu)/\sigma$, Eq. (15.12) can be rewritten as

$$F(x) = \int_{-\infty}^{\left(\frac{x-\mu}{\sigma}\right)} \frac{1}{\sqrt{2\pi}} e^{\frac{-z^2}{2}} \, dz \tag{15.13}$$

Except for the presence of μ and σ in the upper integration limit, the integration kernel no longer depends on these two values. Alternatively, one can view $F(x)$ in Eq. (15.13) as the cdf of a Gaussian distribution having zero mean and unity standard deviation. By tabulating a single function, say

$$\Phi(z) = \frac{1}{\sqrt{2\pi}} \int_{-\infty}^{z} e^{\frac{-u^2}{2}} \, du \tag{15.14}$$

we can write $F(x)$ as

$$F(x) = \Phi\left(\frac{x-\mu}{\sigma}\right) \tag{15.15}$$

In other words, to determine the value of a particular cdf involving a Gaussian random variable with mean μ and standard deviation σ at a particular point, say, x, we simply normalized x by subtracting the mean value followed by a division by σ, that is, $z = (y-\mu)/\sigma$, and compute $\Phi(z)$.

The function $\Phi(z)$ is known as the *standard* Gaussian cdf. The variable z is known as the *standardized point of reference*. Traditionally, $\Phi(z)$ has been evaluated by looking up tables that list $\Phi(z)$ for different values of z. A short tabulation of $\Phi(z)$ is provided in Table 15.1. Rows of $\Phi(z)$ are interleaved between rows of z. A plot of $\Phi(z)$ versus z is also provided in Figure 15.11. For reference, we see from this plot that $\Phi(-\infty) = 0$ and $\Phi(\infty) = 1$. Also evident is the antisymmetry about the point $(0, 0.5)$, giving rise to the relation $\Phi(-z) = 1 - \Phi(z)$.

More recently, the following expression[6] has been found to give reasonably good accuracy for $\Phi(z)$

$$\Phi(z) \approx \begin{cases} \left[1 - \left(\dfrac{1}{(1-\alpha)z + \alpha\sqrt{z^2 + \beta}}\right)\right] \dfrac{1}{\sqrt{2\pi}} e^{\frac{-z^2}{2}} & 0 \le z \le \infty \\ \left(\dfrac{1}{(\alpha-1)z + \alpha\sqrt{z^2 + \beta}}\right) \dfrac{1}{\sqrt{2\pi}} e^{\frac{-z^2}{2}} & -\infty < z < 0 \end{cases} \quad (15.16)$$

where $\alpha = 1/\pi$ and $\beta = 2\pi$. Equation (15.16) is generally very useful when we require a standardized value that is not contained in Table 15.1, or any other Gaussian cdf table for that matter. To illustrate the application of the standard cdf, the probability that a Gaussian distributed random variable X with mean μ and standard deviation σ lies in the range a to b is written as

$$P(a < X < b) = \Phi\left(\frac{b-\mu}{\sigma}\right) - \Phi\left(\frac{a-\mu}{\sigma}\right) \quad (15.17)$$

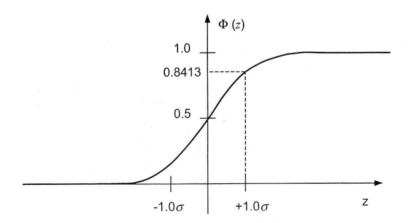

Figure 15.11. Standard Gaussian cumulative distribution function.

Table 15.1. Gaussian Cumulative Distribution Function Values, $\Phi(z)$

z	-3.0	-2.9	-2.8	-2.7	-2.6	-2.5	-2.4	-2.3	-2.2	-2.1
$\Phi(z)$	0.0013	0.0019	0.0026	0.0035	0.0047	0.0062	0.0082	0.0107	0.0139	0.0179
z	-2.0	-1.9	-1.8	-1.7	-1.6	-1.5	-1.4	-1.3	-1.2	-1.1
$\Phi(z)$	0.0228	0.0287	0.0359	0.0446	0.0548	0.0668	0.0808	0.0968	0.1151	0.1357
z	-1.0	-0.9	-0.8	-0.7	-0.6	-0.5	-0.4	-0.3	-0.2	-0.1
$\Phi(z)$	0.1587	0.1841	0.2119	0.2420	0.2743	0.3085	0.3446	0.3821	0.4207	0.4602
z	0.0	0.1	0.2	0.3	0.4	0.5	0.6	0.7	0.8	0.9
$\Phi(z)$	0.5000	0.5398	0.5793	0.6179	0.6554	0.6915	0.7257	0.7580	0.7881	0.8159
z	1.0	1.1	1.2	1.3	1.4	1.5	1.6	1.7	1.8	1.9
$\Phi(z)$	0.8413	0.8643	0.8849	0.9032	0.9192	0.9332	0.9452	0.9554	0.9641	0.9713
z	2.0	2.1	2.2	2.3	2.4	2.5	2.6	2.7	2.8	2.9
$\Phi(z)$	0.9772	0.9821	0.9861	0.9893	0.9918	0.9938	0.9953	0.9965	0.9974	0.9981

Equation (15.17) is a direct result of substituting Eq. (15.15) into (15.11). Equation (15.17) can be used to generate certain rules of thumb when dealing with Gaussian random variables. The probability that a random variable will fall within:

1σ of its mean is

$$P(\mu-\sigma < X < \mu+\sigma) = \Phi\left(\frac{\mu+\sigma-\mu}{\sigma}\right) - \Phi\left(\frac{\mu-\sigma-\mu}{\sigma}\right) = \Phi(1) - \Phi(-1) = 0.6826$$

2σ of its mean is

$$P(\mu-2\sigma < X < \mu+2\sigma) = \Phi\left(\frac{\mu+2\sigma-\mu}{\sigma}\right) - \Phi\left(\frac{\mu-2\sigma-\mu}{\sigma}\right) = \Phi(2) - \Phi(-2) = 0.9544$$

3σ of its mean is

$$P(\mu-3\sigma < X < \mu+3\sigma) = \Phi\left(\frac{\mu+3\sigma-\mu}{\sigma}\right) - \Phi\left(\frac{\mu-3\sigma-\mu}{\sigma}\right) = \Phi(3) - \Phi(-3) = 0.9974$$

It is also instructive to look at several limiting cases associated with Eq. (15.17), specifically when $a = -\infty$ and b is an arbitrary value, we find

$$P(X < b) = P(-\infty < X < b) = \Phi\left(\frac{b-\mu}{\sigma}\right) - \Phi\left(\frac{-\infty-\mu}{\sigma}\right) = \Phi\left(\frac{b-\mu}{\sigma}\right) \quad (15.18)$$

Conversely, with a arbitrary and $b = \infty$, we get

$$P(a < X) = P(a < X < \infty) = \Phi\left(\frac{\infty - \mu}{\sigma}\right) - \Phi\left(\frac{a - \mu}{\sigma}\right) = 1 - \Phi\left(\frac{a - \mu}{\sigma}\right) \qquad (15.19)$$

Of course, as previously mentioned, with $a = -\infty$ and $b = \infty$, $P(-\infty < X < \infty) = 1$.

Notice that the probability that X will fall outside the range $\mu \pm 3\sigma$ is extremely small, that is, $P(X < \mu - 3\sigma) + P(\mu + 3\sigma < X) = \Phi(-3) + 1 - \Phi(3) = 0.0026$. As a result, one would not expect many measurement results to fall very far beyond $\mu \pm 3.0\sigma$.

To summarize, the steps involved in calculating a probability involving a Gaussian random variable are as follows:

1. Estimate the mean μ and standard deviation σ of the random variable from the sample set using

$$\mu = \frac{1}{N}\sum_{n=0}^{N-1} x(n) \quad \text{and} \quad \sigma = \sqrt{\frac{1}{N}\sum_{n=0}^{N-1}\left[x(n) - \mu\right]^2}$$

2. Determine the probability interval limits, a and b, and write a probability expression in terms of the standard Gaussian cumulative distribution function $\Phi(z)$

$$P(a < X < b) = \Phi\left(\frac{b - \mu}{\sigma}\right) - \Phi\left(\frac{a - \mu}{\sigma}\right)$$

3. Evaluate $\Phi(z)$ through a table lookup (Table 15.1) or by using a numerical approximation [Eq. (15.16)].

The following example will illustrate this procedure.

Example 15.1

A DC offset measurement is repeated many times, resulting in a series of values having an average of 257 mV. The measurements exhibit a standard deviation of 27 mV. What is the probability that any single measurement will return a value larger than 245 mV?

Solution:

If X is used to denote the Gaussian random variable, then we want to know: $P(245 \text{ mV} < X)$. Comparing the probability limits with the expression listed in Eq. (15.19), we can state $a = 245$ mV and $b = \infty$. Further, since $\mu = 257$ mV and $\sigma = 27$ mV, we can write

$$P(245 \text{ mV} < X) = 1 - \Phi\left(\frac{245 \text{ mV} - 257 \text{ mV}}{27 \text{ mV}}\right) = 1 - \Phi(-0.44)$$

Referring to Table 15.1, we see that the value Φ(-0.44) is not listed. However, it lies somewhere between 0.3085 and 0.3446; so one can either make a crude midpoint interpolation of 0.33 or, alternatively, one can use the approximation for Φ(z) given in Eq. (15.16) and write

$$\Phi(-0.44) \approx \left[\frac{1}{\left(\frac{1}{\pi}-1\right)(-0.44) + \left(\frac{1}{\pi}\right)\sqrt{(-0.44)^2 + 2\pi}}\right] \frac{1}{\sqrt{2\pi}} e^{\frac{-(-0.44)^2}{2}} = 0.3262$$

We shall select the latter value of 0.3262 and state that the probability that the measurement will be *greater* than 245 mV is equal to 1-0.3262, or 0.6738. Consequently, there is a 67.38% chance that any individual measurement will exceed 245 mV.

Exercises

15.3. For the following specified values of z, compute the value of Φ(z):

(a) $z = -2.5$; (b) $z = +0.34$; (c) $z = +4.3$.

Ans. (a) 0.0062, (b) 0.6369, and (c) 1.0.

15.4. Calculate the following probabilities associated with a Gaussian-distributed random variable having the stated mean and standard deviation values: (a) $P(0 < X < 15$ mV) when $\mu = 0$, $\sigma = 12$ mV; (b) $P(-20$ mV $< X < 30$ mV) when $\mu = 10$ mV, $\sigma = 10$ mV; (c) $P(-1.5$ V $< X < 1.5$ V) when $\mu = 0$, $\sigma = 1$ V.

Ans. (a) 0.3953, (b) 0.9760, and (c) 0.8673.

15.5. A series of op amp gain measurements is found to have an average value of 62 dB and a standard deviation of 1.4 dB. What is the probability that any single measurement will return a value less than 60 dB? What is the probability that any single measurement will return a value greater than 60 dB?

Ans. $P(X < 60$ dB$) = 0.0760$; $P(X > 60$ dB$) = 0.9240$.

15.3.4 Non-Gaussian Distributions

It is fairly common to encounter distributions that are not Gaussian. Two common deviations from the familiar bell shape are bimodal distributions such as that shown in Figure 15.12, and distributions containing outliers as shown in Figure 15.13. When evaluating measurement repeatability on a single DUT, these distributions are a warning sign that the test results are not sufficiently repeatable. When evaluating process stability (consistency from DUT to DUT), these plots may indicate a weak design or a process that needs to be improved.

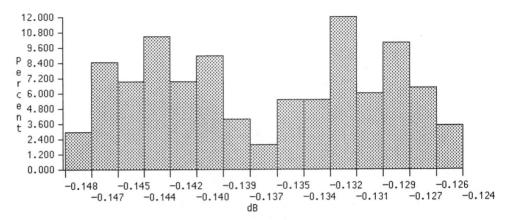

Figure 15.12. Bimodal distribution.

There are many other non-Gaussian distributions, one of which is the uniform distribution. The uniform distribution shown in Figure 15.14 is described by the pdf equation

$$f(x) = \begin{cases} \dfrac{1}{B-A}, & A \leq x \leq B \\ 0, & \text{elsewhere} \end{cases} \qquad (15.20)$$

The probability that a uniformly distributed random variable X will fall in the interval a to b, where $A \leq a < b \leq B$, is obtained by substituting Eq. (15.20) into (15.6) to get

$$P(a < X < b) = \int_a^b f(x)\, dx = \int_a^b \frac{1}{B-A}\, dx = \frac{b-a}{B-A} \qquad (15.21)$$

The mean and standard deviation of an ideal uniform distribution are given by

$$\mu = \frac{A+B}{2} \qquad (15.22)$$

$$\sigma = \frac{(B-A)}{\sqrt{12}} \qquad (15.23)$$

Uniform distributions occur in at least two instances in mixed-signal test engineering. The first instance is random number generators found in various programming languages, such as the *rand* function in MATLAB and the *random* function in C. This type of function returns a randomly chosen number between a minimum and maximum value (typically 0 and 1, respectively). The numbers are supposed to be uniformly distributed between the minimum and maximum values. There should be an equal probability of choosing any particular number, and therefore a histogram of the resulting population should be perfectly flat. One measure of the quality of a random number generator is the degree to which it can produce a perfectly uniform distribution of values.

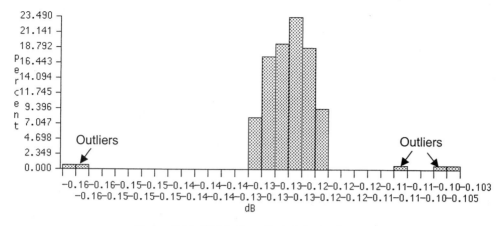

Figure 15.13. Distribution with statistical outliers.

The second instance in which we commonly encounter a uniform distribution is the errors associated with the quantization process of an analog-to-digital converter (ADC). It is often assumed that statistical nature of these errors is uniformly distributed between $-1/2$ LSB (least significant bit) and $+1/2$ LSB. This condition is typically met in practice with an input signal that is sufficiently random. From Eq. (15.22), we see that the average error is equal to zero. Using (15.23) and, assuming that the standard deviation and RMS value are equivalent, we expect that the ADC will generate $1/\sqrt{12}$ LSB of RMS noise when it quantizes a signal. In the case of a full-scale sinusoidal input having a peak of 2^{N-1} LSBs (or an RMS of $2^{N-1}/\sqrt{2}$), the signal-to-noise ratio (SNR) at the output of the ADC is

$$SNR = 20 \, \log\left(\frac{\text{signal RMS}}{\text{noise RMS}}\right)$$
$$= 20 \, \log\left(\frac{2^{N-1}/\sqrt{2} \text{ LSB}}{1/\sqrt{12} \text{ LSB}}\right) \quad (15.24)$$
$$= 20 \, \log_2\left(\sqrt{6} \times 2^{N-1}\right)$$

Simplifying further, we see that the SNR depends linearly on the number of bits, N, according to

$$SNR = \frac{20 \log_2\left(\sqrt{6} \times 2^{N-1}\right)}{\log_2(10)} = 1.761 \text{ dB} + 6.02 \text{ dB} \times N \quad (15.25)$$

In the situation where we know the SNR, we can deduce from Eq. (15.25) that the equivalent number of bits for the ADC, that is, $ENOB = N$, is

$$ENOB = \frac{SNR - 1.761 \text{ dB}}{6.02 \text{ dB}} \quad (15.26)$$

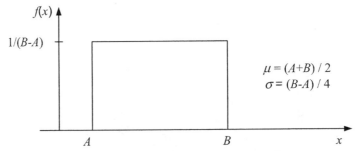

Figure 15.14. Uniform distribution pdf.

15.3.5 Guardbanding and Gaussian Statistics

Guardbanding is an important technique for dealing with the uncertainty of each individual measurement in a test program. If a particular measurement is known to be accurate and repeatable with a worst-case uncertainty of $\pm\varepsilon$, then the final test limits should be tightened by ε to make sure no bad devices are shipped to the customer. In other words,

$$\begin{aligned}\text{guardbanded upper test limit} &= \text{upper specification limit} - \varepsilon \\ \text{guardbanded lower test limit} &= \text{lower specification limit} + \varepsilon\end{aligned} \quad (15.27)$$

If the data sheet limit for the offset of a buffer output is −100 mV minimum, 100 mV maximum, and an uncertainty of ±10 mV exists in the measurement, the test program limits should be set to −90 mV minimum and 90 mV maximum. This way, if the device output is 101 mV and the error in its measurement is −10 mV, the resulting reading of 91 mV will cause a failure as required. Unfortunately, a reading of 91 mV may also represent a device with a 81 mV output we would prefer to ship rather than disqualify. We would like to set the guardbands to zero so that good devices are not thrown away, but zero guardbands would require zero measurement error.

In practice, we need to set ε equal to 3 to 6 times the standard deviation of the measurement to account for measurement variability. This is illustrated in Figure 15.15. This diagram shows a marginal device with an average (true) reading equal to the upper specification limit. The upper

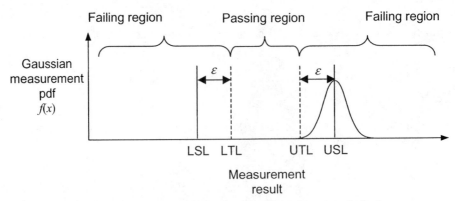

Figure 15.15. Guardbanded measurement with Gaussian distribution.

and lower specification limits (USL and LSL, respectively) have each been tightened by $\varepsilon = 3\sigma$. The tightened upper and lower test limits (UTL and LTL, respectively) reject marginal devices such as this, regardless of the magnitude of the measurement error. A more stringent guardband value of $\varepsilon = 6\sigma$ gives us an extremely low probability of passing a defective device, but this is sometimes too large a guardband to allow a manufacturable yield.

Example 15.2

If our specification limits in Example 15.2 were 250 mV plus or minus 50 mV, where would we have to set our 6σ guardbanded upper and lower test limits?

Solution:

The value of σ is equal to 27 mV; so the width of the 6σ guardbands would have to be equal to 162 mV. The upper test limit would be 300 mV − 162 mV, and the lower test limit would be 200 mV + 162 mV. Clearly there is a problem with the repeatability of this test, since the lower guardbanded test limit is higher than the upper guardbanded test limit! Averaging would have to be used to reduce the standard deviation.

If a device is well designed and a particular measurement is sufficiently repeatable, then there will be few failures resulting from that measurement. But if the distribution of measurements from a production lot is skewed so that the average measurement is close to one of the test limits, then production yields are likely to fall. In other words, more good devices will fall within the guardband region and be disqualified. Obviously, a measurement with poor accuracy or poor repeatability will just exacerbate the problem.

The only way the test engineer can minimize the required guardbands is to improve the repeatability and accuracy of the test, but this requires longer test times. At some point, the test time cost of a more repeatable measurement outweighs the cost of throwing away a few good devices. Thus there are inherent tradeoffs between repeatability, test time, guardbands, and production yield.

The standard deviation of a test result calculated as the average of N values from a statistical population is given by

$$\sigma_{ave} = \frac{\sigma}{\sqrt{N}} \qquad (15.28)$$

So, for example, if we want to reduce the value of a measurement's standard deviation σ by a factor of two, we have to average a measurement four times. This gives rise to an unfortunate exponential tradeoff between test time and repeatability.

We can use Gaussian statistical analysis to predict the effects of nonrepeatability on yield. This allows us to make our measurements repeatable enough to give acceptable yield without wasting time making measurements that are *too* repeatable. It also allows us to recognize the situations where the average device performance or tester performance is simply too close to failure for economical production.

Example 15.3

How many times would we have to average the DC measurement in Example 15.1 to achieve 6σ guardbands of 10 mV? If each measurement takes 5 ms, what would be the total test time for the averaged measurement?

Solution:

The value of σ_{ave} must be equal to 10 mV divided by 6 to achieve 6σ guardbands. Rearranging Eq. (15.28), we see that N must be equal to

$$N = \left(\frac{\sigma}{\sigma_{ave}}\right)^2 = \left(\frac{27 \text{ mV}}{10 \text{ mV}/6}\right)^2 = 262 \text{ measurements}$$

The total test time would be equal to 262 times 5 ms, or 1.31 s. This is clearly unacceptable for production testing of a DC offset. The 27-mV standard deviation must be reduced through an improvement in the DIB hardware or the DUT design.

Exercises

15.6. A series of AC RMS measurements reveal an average value of 1.25 V and a standard deviation of 35 mV. If our specification limits were 1.2 V plus or minus 150 mV, where would we have to set our 3σ guardbanded upper and lower test limits? If 6σ guardbands are desired, how many times would we have to average the measurement to achieve guardbands of 40 mV?

Ans. 3σ guardbanded test limits are: 1.15 and 1.245 V. $N = 28$.

15.3.6 Effects of Measurement Variability on Test Yield

Consider the case of a measurement result having measurement variability caused by additive Gaussian noise. This test has a lower test limit (LTL) and an upper test limit (UTL). If the true measurement result is exactly between the two test limits, and the repeatability error never exceeds ±1/2 (UTL-LTL), then the test will always produce a passing result. The repeatability error never gets large enough to push the total measurement result across either of the test limits. This situation is depicted in Figure 15.16, where the pdf plot is shown.

On the other hand, if the average measurement is exactly equal to either the LTL or the UTL, then the test results will be unstable. Even a tiny amount of repeatability error will cause the test to randomly toggle between a passing and failing result when the test program is repeatedly executed. Assuming the statistical distribution of the repeatability errors is symmetrical, as in the case of the Gaussian pdf, the test will produce an equal number of failures and passing

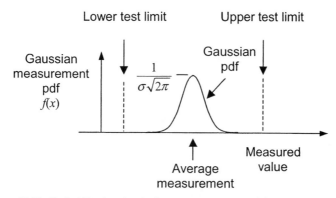

Figure 15.16. Probability density plot for measurement result between two test limits.

results. This is illustrated by the pdf diagram shown in Figure 15.17. The area under the pdf is equally split between the passing region and the failing region; so we would expect 50% of the test results to pass and 50% to fail.

For measurements whose average value is close to but not equal to either test limit, the analysis gets a little more complicated. Consider an average measurement μ that is δ_1 units below the upper test limit as shown in Figure 15.18.

Any time the repeatability error exceeds δ_1 the test will fail. In effect, the measurement noise causes an erroneous failure. The probability that the measurement error will *not* exceed δ_1 and cause a failure is equal to the area underneath the portion of the pdf that is less than the UTL. This area is equal to the integral of the pdf from minus infinity to the UTL of the measurement results. In other words, the probability that a measurement will not fail the upper test limit as adopted from Eq. (15.17) is

$$P(X < UTL) = \Phi\left(\frac{UTL - \mu}{\sigma}\right) \tag{15.29}$$

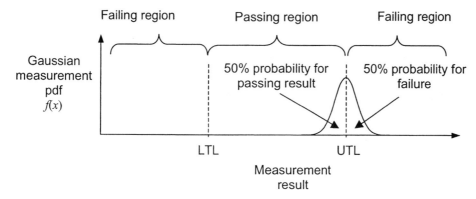

Figure 15.17. Probability density plot for nonrepeatable measurement centered at the UTL.

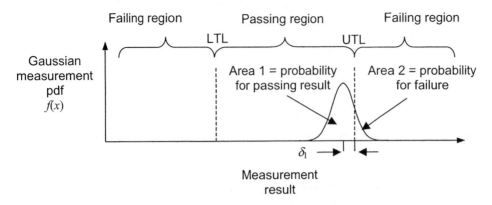

Figure 15.18. Probability density plot for average reading, μ, slightly below UTL by δ_1.

Conversely, the probability of a failing result due to the upper test limit is

$$P(UTL < X) = 1 - \Phi\left(\frac{UTL - \mu}{\sigma}\right) \qquad (15.30)$$

Similar equations apply to the lower test limit

$$P(LTL < X) = 1 - \Phi\left(\frac{LTL - \mu}{\sigma}\right) \qquad (15.31)$$

and

$$P(X < LTL) = \Phi\left(\frac{LTL - \mu}{\sigma}\right) \qquad (15.32)$$

If the distribution of measurement values becomes very large relative to the test limits, then we have to consider the area in both failing regions as shown in Figure 15.19. Clearly, if the true measurement result μ is near either test limit, or if the standard deviation σ is large, the test program has a much higher chance of rejecting a good DUT.

Considering both UTL failures and LTL failures, the probability of a passing result given this type of measurement repeatability according to Eq. (15.17) is

$$P(LTL < X < UTL) = \Phi\left(\frac{UTL - \mu}{\sigma}\right) - \Phi\left(\frac{LTL - \mu}{\sigma}\right) \qquad (15.33)$$

The probability of a failing result due to measurement variability is

$$P(X < LTL \text{ or } UTL < X) = P(X < LTL) + P(UTL < X)$$
$$= 1 + \Phi\left(\frac{LTL - \mu}{\sigma}\right) - \Phi\left(\frac{UTL - \mu}{\sigma}\right) \qquad (15.34)$$

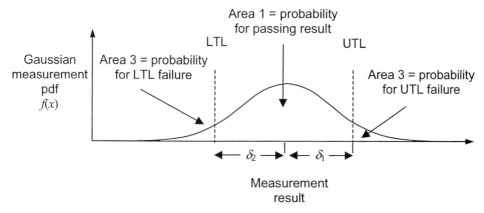

Figure 15.19. Probability density with large standard deviation.

Example 15.4

What is the probability that the nonaveraged offset measurement in Example 15.1 will fail on any given test program execution? Assume an upper test limit of 300 mV and a lower test limit of 200 mV.

Solution:

The probability that the test will lie outside the test limits of 200 and 300 mV is obtained by substituting the test limits into Eq. (15.34),

$$P(X < 200 \text{ mV}) + P(300 \text{ mV} < X) = 1 + \Phi\left(\frac{200 \text{ mV} - 257 \text{ mV}}{27 \text{ mV}}\right) - \Phi\left(\frac{300 \text{ mV} - 257 \text{ mV}}{27 \text{ mV}}\right)$$
$$= 1 + \Phi(-2.11) - \Phi(1.59)$$

Using Table 15.1, we estimate the cdf values as

$$P(X < 200 \text{ mV}) + P(300 \text{ mV} < X) \cong 1 + 0.0179 - 0.9452 = 0.0727$$

Here we see that there is a 7.27% chance of failure, even though the true DC offset value is known to be within acceptable limits.

15.3.7 Effects of Reproducibilty and Process Variation on Yield

Measured DUT parameters vary for a number of reasons. The factors affecting DUT parameter variation include measurement repeatability, measurement reproducibility, and the stability of the process used to manufacture the DUT. So far we have examined only the effects of measurement repeatability on yield, but the equations in the previous sections describing yield loss due to measurement variability are equally applicable to the total variability of DUT parameters.

> **Exercises**
>
> **15.7.** An AC gain measurement is repeated many times, resulting in a series of values having an average of 0.985 V/V. The measurements exhibit a standard deviation of 0.2 V/V. What is the probability that the gain measurement will fail on any given test program execution? Assume an upper test limit of 1.2 V/V and a lower test limit of 0.98 V/V.
>
> **Ans.** 0.4894.

Inaccuracies due to poor tester-to-tester correlation, day-to-day correlation, or DIB-to-DIB correlation appear as reproducibility errors. Reproducibility errors add to the yield loss caused by repeatability errors. To accurately predict yield loss caused by tester inaccuracy, we have to include both repeatability errors and reproducibility errors. If we collect averaged measurements using multiple testers, multiple DIBs, and repeat the measurements over multiple days, we can calculate the mean and standard deviation of the reproducibility errors for each test. We can then combine the standard deviations due to repeatability and reproducibility using the equation

$$\sigma_{tester} = \sqrt{\left(\sigma_{repeatability}\right)^2 + \left(\sigma_{reproducibility}\right)^2} \tag{15.35}$$

Yield loss due to total tester variability can then be calculated using the equations from the previous sections, substituting the value of σ_{tester} in place of σ.

The variability of the actual DUT performance from DUT to DUT and from lot to lot also contributes to yield loss. Thus the overall variability can be described using an overall standard deviation, calculated using an equation similar to Eq. (15.35)

$$\sigma_{total} = \sqrt{\left(\sigma_{repeatability}\right)^2 + \left(\sigma_{reproducibility}\right)^2 + \left(\sigma_{process}\right)^2} \tag{15.36}$$

Since σ_{total} ultimately determines our overall production yield, it should be made as small as possible to minimize yield loss. The test engineer must try to minimize the first two standard deviations. The design engineer and process engineer should try to reduce the third.

Example 15.5

A six-month yield study finds that the total standard deviation of a particular DC offset measurement is 37 mV across multiple lots, multiple testers, multiple DIB boards, etc. The standard deviation of the measurement repeatability is found to be 15 mV, while the standard deviation of the reproducibility is found to be 7 mV. What is the standard deviation of the actual DUT-to-DUT offset variability, excluding tester repeatability errors and reproducibility errors? If we could test this device using perfectly accurate, repeatable test equipment, what would be the total yield loss due to this parameter, assuming an average value of 2.430 V and test limits of 2.5 V $\pm$ 100 mV?

Solution:

Rearranging Eq. (15.36), we write

$$\sigma_{process} = \sqrt{(\sigma_{total})^2 - (\sigma_{repeatability})^2 - (\sigma_{reproducibility})^2}$$
$$= \sqrt{(37 \text{ mV})^2 - (15 \text{ mV})^2 - (7 \text{ mV})^2}$$
$$= 33 \text{ mV}$$

Thus, even if we could test every device with perfect accuracy and no repeatability errors, we would see a DUT-to-DUT variability of $\sigma = 33$ mV. The value of μ is equal to 2.430 V; so our overall yield loss for this measurement is found by substituting the above values into Eq. (15.34) as

$$P(X < 2.4 \text{ V}) + P(2.6 \text{ V} < X) = 1 + \Phi\left(\frac{2.4 \text{ V} - 2.43 \text{ V}}{33 \text{ mV}}\right) - \Phi\left(\frac{2.6 \text{ V} - 2.43 \text{ V}}{33 \text{ mV}}\right)$$
$$= 1 + \Phi(-0.91) - \Phi(5.15)$$

From Table 15.1, $\Phi(-0.91) \cong \Phi(-0.9) = 0.1841$, and we estimate $\Phi(5.15) \cong 1$; hence

$$P(X < 2.4 \text{ V}) + P(2.6 \text{ V} < X) \cong 1 + 0.1841 - 1 = 0.1841$$

We would therefore expect an 18% yield loss due to this one parameter, due to the fact that the DUT-to-DUT variability is too high to tolerate an average value that is only 30 mV from the lower test limit. Repeatability and reproducibility errors would only worsen the yield loss; so this device would probably not be economically viable. The design or process would have to be modified to achieve an average DC offset value closer to 2.5 V.

The probability that a particular device will pass all tests in a test program is equal to the product of the passing probabilities of each individual test. In other words, if the values P_1, P_2, P_3, ..., P_n represent the probabilities that a particular DUT will pass each of the n individual tests in a test program, then the probability that the DUT will pass all tests is equal to

$$P(\text{DUT passes all tests}) = P_1 \times P_2 \times P_3 \times \cdots \times P_n \quad (15.37)$$

Equation (15.37) is of particular significance, because it dictates that each of the individual tests must have a very high yield if the overall production yield is to be high. For example, if each of the 200 tests has a 2% chance of failure, then each test has only a 98% chance of passing. The yield will therefore be $(0.98)^{200}$, or 1.7%! Clearly, a 1.7% yield is completely unacceptable. The problem in this simple example is not that the yield of any one test is low, but that so many tests produce a small amount of yield loss.

Example 15.6

A particular test program performs 857 tests, most of which cause little or no yield loss. Five measurements account for most of the yield loss. Using a lot summary and a continue-on-fail test process, the yield loss due to each measurement is found to be:

Test #1: 1%, Test #2: 5%, Test #3: 2.3%, Test #4: 7%, Test #5: 1.5%

All other tests combined 0.5%

What is the overall yield of this lot of material?

Solution:

The probability of passing each test is equal to 1 minus the yield loss produced by that test. The values of P_1, P_2, P_3, ..., P_5 are therefore

$$P_1 = 99\%, \quad P_2 = 95\%, \quad P_3 = 97.7\%, \quad P_4 = 93\%, \quad P_5 = 98.5\%$$

If we consider all other tests to be a sixth test having a yield loss of 0.5%, we get a sixth probability

$$P_6 = 99.5\%$$

Using Eq. (15.37) we write

$$P(\text{DUT passes all tests}) = 0.99 \times 0.95 \times 0.977 \times 0.93 \times 0.985 \times 0.995 = 0.8375$$

Thus we expect an overall test yield of 83.75%.

Because the yield of each individual test must be very high, a methodology called *statistical process control* (SPC) has been adopted by many companies. The goal of SPC is to minimize the total variability (i.e., to try to make $\sigma_{total} = 0$) and to center the average test result between the upper and lower test limits [i.e. to try to make $\mu = (UTL+LTL)/2$]. Centering and narrowing the measurement distribution leads to higher production yield, since it minimizes the area of the Gaussian pdfs that extend into the failing regions as depicted in Figure 15.20. In the next

Exercises

15.8. A particular test program performs 600 tests, most of which cause little or no yield loss. Four measurements account for most of the yield loss. The yield loss due to each measurement is found to be: Test #1: 1.5%, Test #2: 4%, Test #3: 5.3%, Test #4: 2%. All other tests combined 5%. What is the overall yield loss of this lot of material?

Ans. Yield loss = 16.63%.

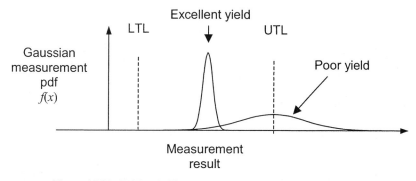

Figure 15.20. DUT-to-DUT mean and standard deviation determine yield.

section, we will briefly examine the SPC methodology to see how it can help improve the quality of the manufacturing process, the quality of the test equipment and software, and most important the quality of the devices shipped to the customer.

15.4 Statistical Process Control

15.4.1 Goals of SPC

Statistical process control (SPC) is a structured methodology for continuous process improvement. SPC is a subset of total quality control (TQC), a methodology promoted by the renowned quality expert, Joseph Juran.[5] SPC can be applied to the semiconductor manufacturing process to monitor the consistency and quality of integrated circuits.

SPC provides a means of identifying device parameters that exhibit excessive variations over time. It does not identify the root cause of the variations, but it tells us when to look for problems. Once an unstable parameter has been identified using SPC, the engineering and manufacturing team searches for the root cause of the instability. Hopefully, the excessive variations can be reduced or eliminated through a design modification or through an improvement in one of the many manufacturing steps. By improving the stability of each tested parameter, the manufacturing process is brought under control, enhancing the inherent quality of the product.

A higher level of inherent quality leads to higher yields and less demanding test requirements. If we can verify that a parameter almost never fails, then we may be able to stop testing that parameter on a DUT-by-DUT basis. Instead, we can monitor the parameter periodically to verify that its statistical distribution remains tightly packed and centered between the test limits. We also need to verify that the mean and standard deviation of the parameter do not fluctuate wildly from lot to lot as shown in the four rightmost columns of Figure 15.21.[*] Once the stability of the distributions has been verified, the parameter might only be measured for every tenth device or

---

[*] The authors acknowledge the efforts of the Texas Instruments SPC Guidelines Steering Team, whose document "Statistical Process Control Guidelines, The Commitment of Texas Instruments to Continuous Improvement Through SPC" served as a guide for several of the diagrams in this section.

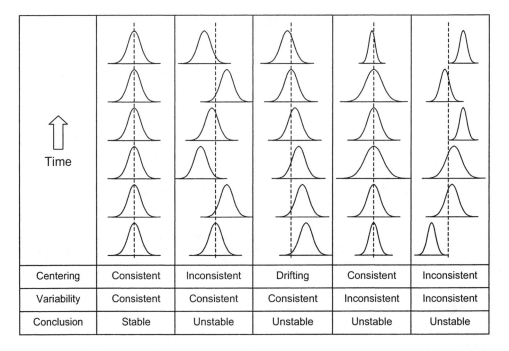

Figure 15.21. Process stability conclusions.

every hundredth device in production. If the mean and standard deviation of the limited sample set stays within tolerable limits, then we can be confident that the manufacturing process itself is stable. SPC thus allows statistical sampling of highly stable parameters, dramatically reducing testing costs.

15.4.2 Six-Sigma Quality

If successful, the SPC process results in an extremely small percentage of parametric test failures. The ultimate goal of SPC is to achieve six-sigma quality standards for each specified device parameter. A parameter is said to meet six-sigma quality standards if its standard deviation is no greater than 1/12 of the difference between the upper and lower specification limits *and* the center of its statistical distribution is no more than 1.5σ away from the center of the upper and lower test limits. These criteria are illustrated in Figure 15.22.

Six-sigma quality standards result in a failure rate of less than 3.4 defective parts per million (dppm). Therefore, the chance of an untested device failing a six-sigma parameter is extremely low. This is the reason we can often eliminate DUT-by-DUT testing of six-sigma parameters.

15.4.3 Process Capability, C_p, and C_{pk}

Process capability is the inherent variation of the process used to manufacture a product. Process capability is defined as the $\pm 3\sigma$ variation of a parameter around its mean value. For example, if

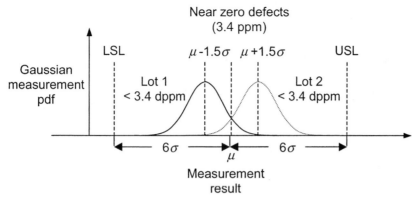

Figure 15.22. Six-sigma quality standards lead to low defect rates (< 3.4 defective parts per million).

a given parameter exhibits a 10-mV standard deviation from DUT to DUT over a period of time, then the process capability for this parameter is defined as 60 mV.

The centering and variation of a parameter are defined using two process stability metrics, C_p and C_{pk}. The process potential index, C_p, is the ratio between the range of passing values and the process capability

$$C_p = \frac{USL - LSL}{6\sigma} \tag{15.38}$$

C_p indicates how tightly the statistical distribution of measurements is packed, relative to the range of passing values. A very large C_p value indicates a process that is stable enough to give high yield and high quality, while a C_p less than 2 indicates a process stability problem. It is impossible to achieve six-sigma quality with a C_p less than 2, even if the parameter is perfectly centered. For this reason, six-sigma quality standards dictate that all measured parameters must maintain a C_p of 2 or greater in production.

The process capability index, C_{pk}, measures the process capability with respect to centering between specification limits

$$C_{pk} = C_p(1-k) \tag{15.39}$$

where

$$k = \frac{|T - \mu|}{0.5(USL - LSL)} \tag{15.40}$$

and

T = specification target (ideal measured value), μ = average measured value.

For one-sided specifications, such as a signal-to-distortion ratio test, we only have an upper or lower specification limit. Therefore, we have to use slightly different calculations for C_p and C_{pk}.

In the case of only the upper specification limit being defined, we use

$$C_{pk} = C_p = \frac{USL - \mu}{3\sigma} \tag{15.41}$$

Alternatively, with only the lower specification limit defined, we use

$$C_{pk} = C_p = \frac{\mu - LSL}{3\sigma} \tag{15.42}$$

The value of C_{pk} must be 1.5 or greater to achieve six-sigma quality standards as shown in Figure 15.22.

Example 15.7

The values of an AC gain measurement are collected from a large sample of the DUTs in a production lot. The average reading is 0.991 V/V and the upper and lower specification limits are 1.050 and 0.950 V/V, respectively. The standard deviation is found to be 0.0023 V/V. What is the process capability and the values of C_p and C_{pk} for this lot? Does this lot meet six-sigma quality standards?

Solution:

The process capability is equal to 6 sigma, or 0.0138 V/V. The values of C_p and C_{pk} are given by Eqns. (15.38), (15.39), and (15.40):

$$C_p = \frac{USL - LSL}{6\sigma} = \frac{1.050 - 0.950}{0.0138} = 7.245$$

$$k = \frac{|T - \mu|}{0.5(USL - LSL)} = \frac{|1 - 0.991|}{0.5(1.050 - 0.950)} = 0.18$$

$$C_{pk} = C_p(1-k) = 5.94$$

This parameter meets six-sigma quality requirements, since the values of C_p is greater than 2 and C_{pk} is greater than 1.5.

15.4.4 Gauge Repeatability and Reproducibility

As mentioned previously in this chapter, a measured parameter's variation is partially due to variations in the materials and the process used to fabricate the device and partially due to the tester's repeatability errors and reproducibility errors. In the language of SPC, the tester is known as a gauge. Before we can apply SPC to a manufacturing process, we first need to verify the accuracy, repeatability, and reproducibility of the gauge. Once the quality of the testing process has been established, the test data collected during production can be continuously monitored to verify a stable manufacturing process.

Gauge repeatability and reproducibility, denoted GRR, is evaluated using a metric called *measurement* C_p. We collect repeatability data from a single DUT using multiple testers and different DIBs over a period of days or weeks. The composite sample set represents the combination of tester repeatabilty errors and reproducibility errors [as described by Eq. (15.35)]. Using the composite mean and standard deviation, we calculate the measurement C_p using Eq. (15.38). The gauge repeatability and reproducibility percentage (precision-to-tolerance ratio) is defined as

$$\%GRR = \frac{100}{\text{measurement } C_p} \qquad (15.43)$$

The general criteria for acceptance of gauge repeatability and reproducibility are listed in Table 15.2.

Table 15.2. %GRR Acceptance Criteria

Measurement C_p	%GRR	Rating
1	100	Unacceptable
3	33	Unacceptable
5	20	Marginal
10	10	Acceptable
50	2	Good
100	1	Excellent

15.4.5 Pareto Charts

A Pareto chart is a graph of values in ascending or descending order of importance. Pareto charts help us identify the most significant factors in a sea of data. For example, we may wish to concentrate our process improvement efforts on the ten parameters that have the lowest C_{pk} values. We can plot the value of C_{pk} for every parameter in a test program, starting with the lowest and progressing toward the highest as shown in Figure 15.23. If we have hundreds of tests, this technique allows us to quickly isolate the tests having the worst centering and variability.

15.4.6 Scatter Plots

Once it has been determined that a problem exists, it is often useful to investigate suspected cause-and-effect relationships. The scatter plot is a very useful tool for this purpose. The example scatter plot in Figure 15.24 displays the correlation between transistor threshold voltage

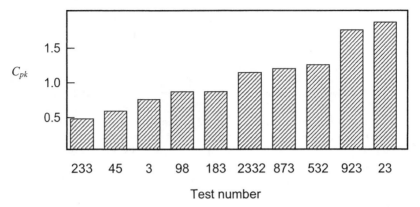

Figure 15.23 C_{pk} Pareto chart showing the ten least stable parameters in a test program.

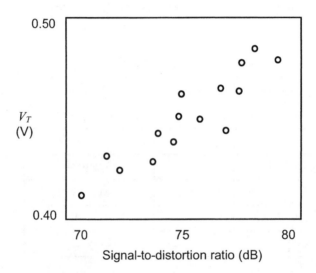

Figure 15.24. Example scatter plot of V_T versus distortion.

and distortion. Each point in the plot represents the threshold voltage of the transistors in a particular lot and the average value of a particular distortion measurement for that lot. If all the points in a scatter plot form a line, then there is a strong correlation between the factors. If they are randomly placed throughout the chart, then there is no correlation. As the example scatter plot shows, the threshold voltage and distortion exhibit a fairly strong correlation. The engineering team would then know that the distortion parameter might be stabilized by stabilizing the transistor threshold voltage.

Scatter plots are also useful for finding correlations between parameters on a single device. For example, we might find that a particular DC reference offset correlates strongly with gain error, giving us a hint as to as to where to start looking for the cause of excessively unstable gain error measurements.

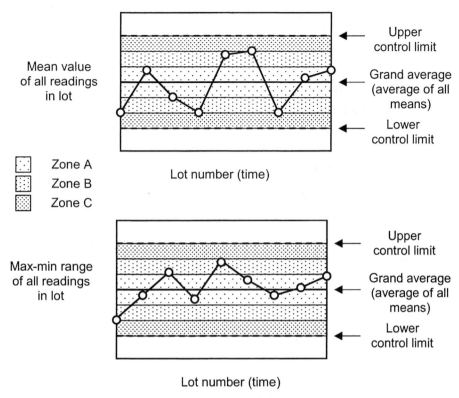

Figure 15.25. (a) X-bar control chart and (b) range control chart.

15.4.7 Control Charts

In addition to monitoring the C_p and C_{pk} of critical parameters, we can also monitor the stability of a process using control charts. A control chart is a graph of parameter stability over time. An effective SPC implementation depends in large part on selecting the appropriate critical parameters to monitor and then choosing an appropriate set of control charts. Control charts are the mechanism by which we determine when the quality metric of interest is drifting out of control.

For example, we may choose to monitor the mean and range (range = maximum reading minus minimum reading) of a particular parameter for each production lot. We can track the fluctuations in these mean and range values over time, creating an X-bar control chart and a range control chart. We then define upper and lower control limits for each chart, as shown in Figure 15.25.

The average of all the points on a control chart is the centerline, while the upper and lower control limits determine when the process has gone out of control. The space between the upper and lower limit can be divided into zones so that more sophisticated control rules can be applied. For example, the Western Electric (WECO) run rules divide the control region into six zones, as shown in Figure 15.25. Instability is defined as any one of the following conditions:

1. One or more points fall beyond the control limits.
2. Two points out of three successive points fall on the same side of the centerline in Zone A or beyond.
3. Four points out of five successive points fall on the same side of the centerline in Zone B or beyond.
4. Eight successive points fall on one side of the centerline.
5. Seven consecutive points show increasing or decreasing values, indicating a systematic drift toward failure.

The rules defining an out-of-control process are not chiseled in stone. To some extent, we are free to choose our own set of quality metrics and control chart rules for determining what "out of control" really means. The metrics and rules must be derived based on experience, best practices, and common sense. We can begin by basing our quality standards on six-sigma criteria, WECO run rules, and other accepted best practices. In the end, though, SPC is best viewed as a methodology for continuous improvement rather than a strictly defined goal to be achieved, celebrated, and then forgotten.

15.5 SUMMARY

There are literally hundreds if not thousands of ways to view and process data gathered during the production testing process. In this chapter, we have examined only a few of the more common data displays, such as the datalog, wafer map, scatter plot, and histogram. Using statistical analysis, we can predict the effects of a parameter's variation on the overall test yield of a product. We can also use statistical analysis to evaluate the repeatability and reproducibility of the measurement equipment itself.

Statistical process control allows us not only to evaluate the quality of the process, including the test and measurement equipment, but it tells us when the manufacturing process is not stable. We can then work to fix or improve the manufacturing process to bring it back under control. We have really only scratched the surface of SPC and TQC in this chapter. Although every test engineer may not necessarily get involved in SPC directly, it is important to understand the basic concepts. The limited coverage of this topic is only intended as an introduction to the subject rather than a complete tutorial. For a comprehensive treatment of these subjects, the reader is encouraged to refer to books devoted to TQC and Six Sigma.[7-9]

Problems

15.1. The thickness of printed circuit boards is an important characteristic. A sample of eight boards had the following thickness (in millimeters): 1.60, 1.55, 1.65, 1.57, 1.55, 1.62, 1.52, and 1.67. Calculate the sample mean, sample variance and sample standard deviation. What are the units of measurement for each statistic?

15.2. An electronics company manufacturers power supplies for a personal computer. They produce several hundred power supplies each shift, and each unit is subjected to a 12-h burn-in test. The number of units failing during this 12-h test each shift is shown in the following table.

3	6	4	7	6	7	6	8	4	9
4	7	8	2	1	4	5	7	6	4
2	9	4	6	4	8	4	14	15	13
5	10	10	9	13	7	3	13	4	3
6	14	14	10	12	3	2	12	7	6
10	13	8	7	10	6	8	5	5	5
5	10	12	9	2	7	10	4	3	10
4	9	4	16	5	8	9	6	2	6
3	8	5	11	7	4	11	5	6	7
11	10	14	13	10	12	8	11	10	9
9	3	2	3	4	6	14	7	14	10
2	2	8	13	2	17	6	2	8	7
7	4	6	3	2	5	4	8	12	9
8	6	10	7	6	10	4	10	4	16
4	4	8	3	4	8	8	7	9	3
2	10	6	2	10	9	7	4	2	6

Construct a histogram of these data and comment on the properties of the data.

Find the sample mean, sample variance, and sample standard deviation.

Suppose the data points listed were all multiplied by a factor of 100. How would the sample mean, sample variance, and sample standard deviation be affected?

15.3. An electronics company manufacturers analog-to-digital converters for the sound card in a personal computer. An important parameter of device operation is the signal-to-noise-plus-distortion ratio (SNDR). Below is a sample of the measured SNDR (in decibels).

88.5	87.7	83.4	86.7	91.5	88.6	89.0	96.1	93.3	91.8	92.3	90.4
100.3	95.6	93.3	94.7	91.1	91.0	90.1	93.0	88.7	89.9	89.8	89.6
94.2	87.8	89.9	88.3	87.6	84.3	87.4	88.4	88.9	91.2	89.3	94.4
86.7	88.2	90.8	88.3	98.8	94.2	92.7	91.8	91.6	90.4	91.1	92.6
92.7	93.2	91.0	93.4	88.5	90.1	89.8	90.6	91.1	90.4	89.3	89.7
89.2	88.3	85.3	87.9	88.6	90.9	90.3	91.6	90.5	93.7	92.7	92.2
87.6	84.3	87.4	88.4	88.9	91.2	92.2	91.2	91.0	92.2	90.0	90.7

Construct a histogram of these data.

Find the sample mean, sample variance and sample standard deviation.

From the data provided, what is the sample yield if a good device is one with an SNDR greater than 90 dB?

How does the sample standard deviation compare with that predicted by Eq. (15.9). What conclusion can you therefore draw about the properties of the data?

Calculate the mean, variance, and standard deviation using linear values (V/V) instead of dB. Do the answers agree with those obtained using decibel values?

15.4. Construct a functional shmoo plot from the following set of test data. Assume that the SNR lower test limit is 70 dB.

TEMP	SNR (dB)								
	2.5 V	2.6 V	2.7 V	2.8 V	2.9 V	3.0 V	3.1 V	3.2 V	3.3 V
0° C	66.8	66.7	66.4	68.2	69.1	69.4	69.1	68.1	65.0
10° C	67.5	68.1	71.0	73.2	74.2	75.1	75.1	74.6	74.9
20° C	69.4	73.2	74.1	74.5	74.1	74.1	75.2	73.2	73.2
30° C	72.2	72.6	73.3	73.7	73.9	74.3	74.8	75.4	73.6
40° C	72.2	72.4	73.4	73.5	73.4	75.1	74.1	73.4	73.1
50° C	69.1	71.0	70.4	70.4	72.0	73.8	74.1	74.6	73.2
60° C	67.2	68.1	69.1	69.5	69.8	70.5	71.6	72.6	72.8
70° C	66.8	68.9	69.0	69.8	70.0	70.5	71.1	72.2	71.2
80° C	66.1	68.5	69.1	69.5	71.1	71.1	71.5	71.2	70.8
90° C	65.2	68.4	68.4	68.9	70.0	70.1	70.1	70.0	68.2

15.5. Repeat Problem 15.4 but this time create a parametric shmoo plot. Divide the SNR into the following classes: 64-66 dB, 66-68 dB, 68-70 dB, 70-72 dB, 72-74 dB, 74-76 dB and 76-78 dB.

15.6. A random variable X has the probability density function

$$f(x) = \begin{cases} ce^{-3x} & x > 0 \\ 0 & x \leq 0 \end{cases}$$

(a) Find the value of the constant c.

(b) Find $P(1 < X < 2)$.

(c) Find $P(X \leq 3)$.

(d) Find $P(X < 1)$.

(e) Find the cumulative distribution function $F(x)$.

15.7. Compare the tabulated results for $\Phi(z)$ listed in Table 15.1 with those generated by Eq. (15.16). Provide a plot of the two curves. What is the worst-case error?

15.8. For the following specified values of z, estimate the value of $\Phi(z)$ using Eq. (15.16) and compare the results with those obtained from Table 15.1. Use linear interpolation where necessary.

(a) $z = -3.0$, (b) $z = -1.9$, (c) $z = -0.56$, (d) $z = -0.24$, (e) $z = 0.0$,
(f) $z = -0.09$, (g) $z = +0.17$, (h) $z = +3.0$, (i) $z = +5.0$, (j) $z = -5.0$.

15.9. Using Eq. (15.16), show that the following relationships are true.
$P(X < b) = 1 - P(X > b)$
$P(a < X < b) = 1 - P(X < a) - P(X > b)$
$P(|X| > c) = 1 - P(-c < X < c)$

15.10. Calculate the following probabilities associated with a Gaussian-distributed random variable having the following mean and standard deviation values:
(a) $P(0 < X < 30 \text{ mV})$ when $\mu = 0$, $\sigma = 10$ mV
(b) $P(-30 \text{ mV} < X < 30 \text{ mV})$ when $\mu = 1$ V, $\sigma = 10$ mV
(c) $P(-1.5 \text{ V} < X < 1.4 \text{ V})$ when $\mu = 0$, $\sigma = 1$ V
(d) $P(-300 \text{ mV} < X < -100 \text{ mV})$ when $\mu = -250$ mV, $\sigma = 50$ mV
(e) $P(X < 250 \text{ mV})$ when $\mu = 100$ mV, $\sigma = 100$ mV
(f) $P(-200 \text{ mV} > X)$ when $\mu = -75$ mV, $\sigma = 150$ mV
(g) $P(|X| < 30 \text{ mV})$ when $\mu = 0$, $\sigma = 10$ mV
(h) $P(|X| > 30 \text{ mV})$ when $\mu = 0$, $\sigma = 10$ mV

15.11. In each of the following equations, find the value of z that makes the probability statement true. Assume a Gaussian distributed random variable with zero mean and unity standard deviation.
(a) $\Phi(z) = 0.9452$
(b) $P(Z < z) = 0.7881$
(c) $P(Z < z) = 0.2119$
(d) $P(Z > z) = 0.2119$
(e) $P(|Z| < z) = 0.5762$

15.12. In each of the following equations, find the value of x that makes the probability statement true. Assume a Gaussian distributed random variable with $\mu = -1$ V and $\sigma = 100$ mV.
(a) $P(X < x) = 0.7881$
(b) $P(X < x) = 0.2119$
(c) $P(X > x) = 0.2119$
(d) $P(|X| < x) = 0.3830$

15.13. It has been observed that a certain measurement is a Gaussian-distributed random variable of which 25% are less than 20 mV and 10% are greater than 70 mV. What are the mean and standard deviation of the measurements?

15.14. If X is a uniformly distributed random variable over the interval (0, 100), what is the probability that the number lies between 23 and 33? What are the mean and standard deviation associated with this random variable?

15.15. It has been observed that a certain measurement is a uniformly distributed random variable of which 25% are less than 20 mV and 10% are greater than 70 mV. What are the mean and standard deviation of the measurements?

15.16. A DC offset measurement is repeated many times, resulting in a series of values having an average of -110 mV. The measurements exhibit a standard deviation of 51 mV. What is the probability that any single measurement will return a positve value? What is the probability that any single measurement will return a value less than -200 mV? Provide sketches of the pdf, label critical points, and highlight the areas under the pdf that corresponds to the probabilities of interest.

15.17. A series of AC gain measurement is found to have an average value of 10.3 V/V and a variance of 0.1 $(V/V)^2$. What is the probability that any single measurement will return a value less than 9.8 V/V? What is the probability that any single measurement will lie between 10.0 V/V and 10.5 V/V? Provide sketches of the pdf, label critical points and highlight the areas under the pdf that correspond to the probabilities of interest.

15.18. A noise measurement is repeated many times, resulting in a series of values having an average RMS value of 105 µV. The measurements exhibit a standard deviation of 21 µV RMS. What is the probability that any single noise measurement will return an RMS value larger than 140 µV? What is the probability that any single measurement will return an RMS value less than 70 µV? What is the probability that any single measurement will return an RMS value between 70 and 140 µV? Provide sketches of the pdf, label critical points, and highlight the areas under the pdf that correspond to the probabilities of interest.

15.19. A series of DC offset measurements reveal an average value of 10 mV and a standard deviation of 11 mV. If our specification limits were 0 mV plus or minus 50 mV, where would we have to set our 3σ guardbanded upper and lower test limits? If 6σ guardbands are desired, how many times would we have to average the measurement to achieve guardbands of 20 mV?

15.20. A DC offset measurement is repeated many times, resulting in a series of values having an average of -100 mV. The measurements exhibit a standard deviation of 38 mV. What is the probability that the offset measurement will fail on any given test program execution? Assume an upper test limit of 0 mV and a lower test limit of -150 mV. Provide a sketch of the pdf, label critical points, and highlight the area under the pdf that corresponds to the probability of interest.

15.21. An AC gain measurement is repeated many times, resulting in a series of values having an average of 0.99 V/V. The measurements exhibit a standard deviation of 0.2 V/V. What is the probability that the gain measurement will fail on any given test program execution? Assume an upper test limit of 1.2 V/V and a lower test limit of 0.98 V/V. Provide a sketch of the pdf, label critical points, and highlight the area under the pdf that corresponds to the probability of interest.

15.22. The standard deviation of a measurement repeatability is found to be 12 mV, while the standard deviation of the reproducibility is found to be 8 mV. Determine the standard deviation of the tester's variability. If process variation contributes an additional 10 mV of uncertainty to the measurement, what is the total standard deviation of the overall measurement?

15.23. An extensive study of yield finds that the total standard deviation of a particular DC offset measurement is 25 mV across multiple lots, multiple testers, multiple DIB boards, etc. The standard deviation of the measurement repeatability is found to be 19 mV, while the standard deviation of the reproducibility is found to be 11 mV. What is the standard deviation of the actual DUT-to-DUT offset variability, excluding tester repeatability errors and reproducibility errors? If we could test this device using perfectly accurate, repeatable test equipment, what would be the total yield loss due to this parameter, assuming an average value of 2.235 V and test limits of 2.25 V ± 100 mV.

15.24. A particular test program performs 1000 tests, most of which cause little or no yield loss. Seven measurements account for most of the yield loss. The yield loss due to each measurement is found to be: Test #1: 1.1%, Test #2: 6%, Test #3: 3.3%, Test #4: 8%, Test #5: 2%, Test #6: 2%, Test #7: 3%, all other tests; 1%. What is the overall yield of this lot of material?

15.25. The values of an AC noise measurement are collected from a large sample of the DUTs in a production lot. The average RMS reading is 0.12 mV and the upper and lower RMS specification limits are 0.15 and 0.10 mV, respectively. The standard deviation is found to be 0.015 mV. What is the process capability and the values of C_p and C_{pk} for this lot? Does this lot meet six-sigma quality standards?

References

1. Keith Baker, Jos van Beers, *Shmoo Plotting: The Black Art of IC Testing*, Proc. International Test Conference, 1996, pp. 932, 933

2. Athanasios Papoulis, *Probability, Random Variables, and Stochastic Processes*, 3rd edition, McGraw Hill, December 1991, ISBN: 0070484775

3. Julius S. Bendat, Allan G. Piersol, *Random Data: Analysis and Measurement Procedures*, John Wiley & Sons, 605 Third Avenue, New York, NY, April 1986, ISBN: 0471040002

4. George R. Cooper, Clare D. McGillem, *Probabilistic Methods of Signal and System Analysis*, 3rd edition, Oxford University Press, New York, NY, 1999, ISBN: 0195123549

5. Mark J. Kiemele, Stephen R. Schmidt, Ronald J. Berdine, *Basic Statistics, Tools for Continuous Improvement*, Fourth Edition, Air Academy Press, 1155 Kelly Johnson Blvd., Suite 105, Colorado Springs, CO 80920, 1997, ISBN: 1880156067, pp. 9-71.

6. P.O. Börjesson, C. E. W. Sundberg, *Simple Approximations of the Error Function Q(x) for Communication Applications*, IEEE Trans. On Communications, March 1979, pp. 639-43.

7. J. M. Juran (Editor), A. Blanford Godfrey, *Juran's Quality Handbook*, 5th Edition, January 1999, McGraw Hill, New York, NY, ISBN: 007034003X

8. Thomas Pyzdek, *The Complete Guide to Six Sigma*, Quality Publishing, Tucson, AZ, 1999, ISBN: 0385494378

9. Forrest W. Breyfogle, *Implementing Six Sigma: Smarter Solutions Using Statistical Methods*, 2nd edition, June 7, 1999, John Wiley & Sons, New York, NY, ISBN: 0471296597

CHAPTER 16

Test Economics

16.1 PROFITABILITY FACTORS

16.1.1 What Is Meant by Test Economics?

In simplistic terms, profitability is the difference between the revenues generated by a company's products and the costs associated with developing, manufacturing, and selling them. The test engineer has direct or indirect influence over the revenues, the development costs, and the manufacturing costs of semiconductor products. In this chapter, we will examine the role of test engineering in each of these areas. The direct and indirect influences of testing and test development on profitability is a subject we will call *test economics*.

It may not be obvious at first, but test economics is a subject that extends well beyond day-to-day production costs. Of course, we will examine direct testing costs and their obvious effect on overall manufacturing expenses. But we will also examine other less obvious aspects of test economics. For example, the test engineer's debugging skills have a direct and profound effect on time to market and yield enhancement. Both of these factors are of extreme importance to profitability.

In this chapter we will examine the debugging process and attempt to formulate some common sense techniques and ideas that can help the beginning test engineer debug common problems. Finally, we will examine some of the emerging trends that show promise in improving various issues related to test economics. These trends range from software tools that allow more rapid test debug and development to methodology shifts that are changing the very nature of mixed-signal testing.

16.1.2 Time to Market

At first, it might not seem that a test engineer has any control over the revenue portion of the profitability equation. After all, the free market determines the selling price of most products, and the sales and marketing department is responsible for pricing and selling the product. However, the test engineer plays an important role in revenue generation. In Chapter 14, "Design for Test," we examined the importance of time to market on profitability. Late product introduction can result in missed opportunities if a competitor's product gains market share. Conversely, early product introduction leads to higher initial selling prices because of limited competition. Therefore, time to market is perhaps as important to a company's profitability as direct manufacturing costs. The test engineer certainly has influence (good or bad) over time to market, and therefore has influence over the selling price and total revenues.

16.1.3 Testing Costs

The test engineer has a very obvious connection to manufacturing costs. Many years ago, production testing represented a fairly small portion of the cost of manufacturing a semiconductor device. Today, testing often represents a painfully large percentage of the total production cost. This is especially true of mixed-signal semiconductor devices. Shrinking IC geometries and ever-increasing performance requirements are the primary driving forces behind this trend.

Every time fabrication geometries shrink by a factor of two, we can build approximately four times as many circuits onto a given area of silicon. The fabrication costs do not quadruple and in fact may decline over time. However, the complexity of the circuit and the functions it performs certainly do quadruple. Circuit complexity is directly related to testing costs. If a circuit has to perform four times as many functions as a previous generation device, then its testing time tends to quadruple as well. Since the cost of printing four times as much circuitry does not quadruple, the testing costs grow faster than the fabrication costs. This is a primary reason that testing costs have become a greater percentage of total manufacturing costs.

Increasing circuit complexity and performance requirements give rise to another major mixed-signal testing problem. As more circuits are added to the same die, crosstalk problems and other interaction mechanisms grow geometrically. For example, a device with four mixed-signal circuit blocks has twelve possible block-to-block interaction categories (i.e., block A to block B, B to A, A to C, etc.), any of which can lead to marginal device performance. By contrast, a circuit block with eight mixed-signal circuit blocks has 56 possible block-to-block interaction categories! As we saw in Chapter 15, "Data Analysis," marginal device performance requires longer test times (and thus higher testing costs) because of increased averaging and accuracy requirements. One of the reasons that purely digital circuits have not suffered quite as severe an explosion in testing costs is that digital circuit blocks do not tend to interact with one another. The digital blocks can be broken into pieces and tested simultaneously using built-in self-test (BIST). There is little or no concern that the digital circuits will interfere with one another through unexpected coupling mechanisms.

More demanding DUT performance requirements also drive up the cost of testing, while reducing yield. A 100-dB signal-to-noise ratio (SNR) test is far more costly to implement than a 60-dB SNR test, simply because less averaging is required to achieve an acceptable level of repeatability at 60 dB. Also, high-performance devices that push the capability of the fabrication process often have very poor design margin. In Chapter 15, we studied the importance of good design margin to yield. When testing high-performance DUTs, the test engineer must often compensate for poor design margin through an averaging process, to increase repeatability. The absolute accuracy of each marginal measurement must also be very high. Extreme levels of accuracy are usually more expensive to attain than less extreme levels of accuracy. Finally, the cost of test equipment is directly proportional (perhaps even exponentially proportional) to performance requirements. Clearly, higher DUT performance requirements drive up the cost of mixed-signal testing in many ways.

16.1.4 Yield Enhancement

Yield enhancement is another important area of test economics. A good test engineer or product engineer can play a critical role in increasing the overall yield of a product. Identifying and resolving problems in device design, fabrication process, test hardware, or test software can enhance production yields. As we have stressed many times in this text, the entire engineering team is responsible for the yield enhancement task. The yield enhancement team should include members from test engineering, product engineering, process engineering, systems engineering, and design engineering.

16.2 DIRECT TESTING COSTS

16.2.1 Cost Models

Determining the exact cost of production testing is not as simple as one might think. In Chapter 1, "Overview of Mixed-Signal Testing," we presented a very simplistic model of testing cost. This simplified model gave the cost of test as the product of cost per second times test time

$$\text{cost of test} = \text{test cost per second} \times \text{test time} \qquad (16.1)$$

Equation (16.1) is useful in that it shows how important test time is to the cost of test, but it is somewhat deceptive. There are many factors that affect cost of test, including tester depreciation, handler index time, tester down time, tester idle time, etc. These complex factors are often imprecisely lumped together in the test cost per second value. More realistic cost models can become fairly difficult to develop and maintain in a dynamic production environment.

There are many cases in which the correct decision from a total cost of test viewpoint is not immediately obvious. For example, one might assume that the tester with the lowest purchase price always achieves the lowest cost of test. This is a common misconception, but it is often completely wrong. The problem is that we have to consider the total factory throughput rather than just considering tester purchase price.

16.2.2 Cost of Test versus Cost of Tester

A tester's throughput and yield are often more important to profitability than its purchase price. Tester throughput is defined as the average number of passing DUTs that can be tested in a given amount of time.

$$\text{throughput} = \frac{\text{number of passing DUTs}}{\text{time to test all DUTs}} \qquad (16.2)$$

Very expensive testers can often be justified because they allow a higher throughput than slower, less expensive testers. For example, an inexpensive mixed-signal tester may have a very slow data movement or DSP process compared to a more expensive tester. A less efficient hardware architecture or operating system can slow down each of the many tests in a typical mixed-signal program, perhaps doubling or tripling the test time.

It might seem from Eq. (16.1) that a tester costing one-tenth the price of a faster tester would be justified, even at twice the test time of the expensive tester. However, the numerator in Eq. (16.1) is test cost per second rather than tester depreciation expense per second. Test cost per second is much more complex than the depreciation of the tester's purchase price over its lifetime. The total cost per second of test time certainly includes a large amount of tester depreciation, but it also includes the cost of factory facilities, handlers and probers, factory personnel, equipment maintenance, electricity, and general corporate expenses. These other expenses are largely independent of the cost of the tester itself. However, fixed expenses are somewhat dependent on the *size* of the tester, since we can pack more testers into a given factory floor space if the testers are small.

If we show the total cost per second associated with a fully utilized production tester, we can begin to understand how tester throughput is far more important than tester purchase price. Figure 16.1 shows a pie chart detailing the test cost per second for an expensive mixed-signal tester.[*] Now consider a mixed-signal tester costing one-third the price of the expensive tester (Figure 16.2). Since tester depreciation expenses are directly proportional to the tester's purchase price, they will drop by 2/3. However, assuming the other expenses are independent of tester purchase price, the total test cost per second in this example only drops to

$$\frac{\text{inexpensive tester test cost per second}}{\text{expensive tester test cost per second}} = \frac{(40/3)+40+15+5}{40+40+15+5} = 73.3\%$$

Thus, in this example, the testing cost per second associated with the inexpensive tester is actually only 26.7% lower than that of the expensive tester. To achieve equivalent testing costs, the test time of the lower-priced tester must be no greater than 1/0.733, or 136% of the test time of the expensive tester. Clearly, then, a low-cost tester must not only be inexpensive, but it must have almost the same throughput as its expensive counterpart.

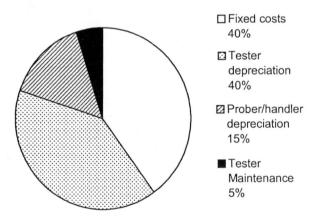

Figure 16.1. Expensive mixed-signal tester total cost per second.

---

[*] Actual cost models and cost data are highly proprietary information. The cost models and examples given in this chapter are intentionally skewed from actual data. They are presented for instructional purposes only, as a means of illustrating some of the factors that go into a calculation of the cost of test.

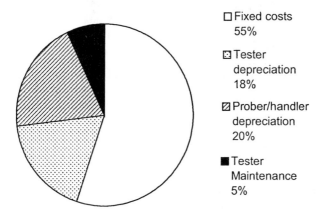

Figure 16.2. Inexpensive mixed-signal tester total cost per second.

An even more subtle issue relates to test yield. If the number of passing DUTs is reduced by a lower-quality tester, then the numerator in Eq. (16.2) drops, reducing throughput. Thus an inexpensive tester must also maintain the same test yield as a higher-cost tester to achieve equivalent cost effectiveness.

Another aspect of tester cost is flexibility and ease of use. This affects time to market, because an inexpensive tester will typically have an inexpensive, less developed operating system that may extend test development time. For all these reasons, mixed-signal testers have grown in price over the years to allow fast, accurate measurements with very high throughput.

16.2.3 Throughput

Like test cost per second, throughput is not a simple calculation. We might at first assume that the denominator in Eq. (16.2) is approximately equal to the test time per DUT multiplied by the total number of DUTs tested. In reality, there are many factors that affect the average test time per DUT. These factors include handler or prober index time, multihead versus single-head testing, multisite versus single site testing, equipment down time, and tester idle time.

Of course, the most important factor in throughput is the test time for a passing DUT. This time tends to dominate the average test time per DUT. Test time reduction is one of the main contributions a test or product engineer can make towards increased profitability. Unfortunately, test time reduction is a very tester-specific topic. The details of each tester's usage are so unique to that tester that we cannot effectively teach test time reduction techniques in this text. Coherent DSP-based testing is one of the common test time reduction techniques found on all mixed-signal testers. Other obvious examples of test time reduction are the elimination of unnecessary settling time and the use of simultaneous testing of multiple circuit blocks. Beyond these general categories of test time reduction, the test engineer has to study the architectural quirks of any particular tester to determine ways to streamline the testing process on that tester. The tester vendors are a very good source of test time reduction techniques, since it is in their best interests to help reduce the effective cost of their tester.

Handler or prober index time is the amount of time it takes to remove one DUT from the tester and replace it with the next DUT. Index times for probers are typically on the order of

100 ms, but handler index times can be very long. Robotic (pick-and-place) handlers are especially slow, since they are designed for very flexible operation. The index time of a robotic handler is typically 1 s or more.

Because index times can be so lengthy, testers are often equipped with two test heads. The heads are not designed for simultaneous use. Rather, the mainframe tester instruments are multiplexed between the heads so that testing can proceed on one head while the handler or prober is indexing on the other head (Figure 16.3). In effect, the index time of the handler or prober is hidden. Of course, the extra test head adds expense to the tester, but the added cost is justified by the increased throughput.

Another technique that can increase throughput is multisite testing. A tester having multisite capabilities can test multiple DUTs at the same time. Each DUT tested in parallel is called a *site*. The advantage of multisite testing is obvious. If four devices can be tested at the same time, then a tester's throughput goes up by a factor of four. This assumes a fully parallel test capability, of course. Some testers are only capable of semiparallel testing, in which portions of the test program are performed in parallel while other portions are performed serially, one DUT at a time.

The serial portions of a semiparallel multisite test program are required because certain tester resources may be limited. For example, if a tester has only one AWG and digitizer, then quad-site parallel testing of AC parameters is not possible. However, the digital patterns might be fully compatible with quad-site testing. The effect of semiparallel multisite testing is that the full effect of multisite testing is not achieved. Instead of a fourfold increase in throughput, we might only get a doubling of throughput. Nevertheless, this advantage is almost always enough to justify the extra expense of a multisite tester. Referring again to Figures 16.1 and 16.2, an inexpensive tester lacking multisite capabilities is often at a tremendous disadvantage compared to an expensive multisite tester.

Down time and idle time are two final factors to consider in cost of test. Down time is defined as any time the tester cannot be used for production testing. This time includes time required for maintenance, repair, calibration, and changeover time from one DUT type to another. Idle time is any time the tester is available for production, but is not testing devices. Idle time can result from poor production planning or from a temporary lack of demand for devices that can be tested on a particular tester. Both down time and idle time increase the effective cost of a tester.

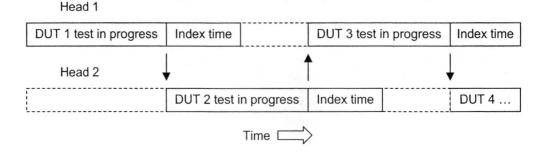

Figure 16.3. Multihead testing hides index time.

Tester depreciation is a somewhat ambiguous quantity, since testers are fully depreciated in five to seven years. This makes them "free" as far as their depreciation is concerned. Rather than trying to calculate the depreciation for each tester individually, we typically group all similar testers together, using an average depreciation cost per second in our tester cost calculations. This figure includes the amortized purchase price plus the cost associated with having otherwise investable capital tied up in the equipment during the depreciation period. In this chapter, we will simply lump all the costs associated with the purchasing price of a piece of equipment into a single quantity called *depreciation*.

Taking all these factors into account, we can propose a fairly complete model for testing costs. This formula should not be taken as an absolute truth, but rather as a starting point for calculating the approximate test cost per second. The exact cost model will vary significantly from one company to the next

$$\text{test cost per second} = (D_T + D_H + C_F)\left(\frac{T_{TEST} + T_{INDEX}}{T_{TEST}}\right)\left(\frac{T_{PROD} + T_{DOWN} + T_{IDLE}}{T_{PROD}}\right) \quad (16.3)$$

where

D_T = depreciation of the tester (cents per second)

D_H = depreciation of the handler/prober (cents per second)

C_F = tester's share of fixed costs (cents per second)

T_{TEST} = test time (seconds)

T_{INDEX} = index time (seconds)

T_{PROD} = average hours per week tester is used for production

T_{DOWN} = average hours per week tester is down

T_{IDLE} = average hours per week tester is idle

Example 16.1

During a particular month, a tester has an average down time of 12 h/wk. During that same time it has an average idle time of 16 h/wk, leaving a total of 140 h/wk of useful production time. The depreciation cost for this type of tester is 1 cent per second. A particular DUT requires a robotic handler, which carries a depreciation cost of 0.5 cents per second. Handler index time is 1 s, but dual head testing (requiring two handlers) effectively masks this time. Single-site test time is 5 s and the test yield is 95%. This tester's share of the test floor's fixed costs is 1 cent per second. What is the overall cost per second of this tester/handler combination? How does the test cost compare with single-head testing? By what percentage could we increase the cost of the tester (i.e., the tester depreciation cost) if we wanted to perform quad-site testing at the same cost per second? (Assume the handlers are already capable of quad-site handling.) Compare the test cost of single-site testing versus quad-site testing assuming fully depreciated equipment.

Solution:

Using Eq. (16.3), we can calculate the effective test cost per second as follows:

$$\text{test cost per second} = \left(1\cent/s + 2\times 0.5\cent/s + 1\cent/s\right)\left(\frac{5\,s+0\,s}{5\,s}\right)\left(\frac{140\,h+12\,h+16\,h}{140\,h}\right) = 3.6\cent/s$$

The total testing cost for this DUT would be 3.6 cents per second times 5 seconds, or 18 cents per DUT. With a yield of 95%, the test cost per *passing* DUT increases to 18 / 0.95 = 18.95 cents. The cost of test per second using a single handler is:

$$\text{test cost per second} = \left(1\cent/s + 0.5\cent/s + 1\cent/s\right)\left(\frac{5\,s+1\,s}{5\,s}\right)\left(\frac{140\,h+12\,h+16\,h}{140\,h}\right) = 4.32\cent/s$$

However, the test time has increased from 5 s per DUT to 6 s. Thus the total test cost per passing DUT for single-head testing is 6 s x 4.32 cents per second/0.95 = 27.28 cents. Clearly, dual-head testing is more economical in this example.

Multisite testing gives us an even more dramatic cost reduction. Quad-site full-parallel testing would reduce the effective test time to 1.25 s. Solving for the tester depreciation that gives less than 18 cents test cost at a test time of 1.25 s

$$\left(D_T + 2\times 0.5\cent/s + 1\cent/s\right)\left(\frac{1.25\,s+0\,s}{1.25\,s}\right)\left(\frac{140\,h+12\,h+16\,h}{140\,h}\right)1.25\,s\ <\ 18\cent$$

Rearranging to solve for the maximum tolerable tester depreciation, D_T

$$D_T < \left[\frac{18\cent}{1.25\,s}\times\left(\frac{140\,h}{140\,h+12\,h+16\,h}\right)\right] - 2\times 0.5\cent/s - 1\cent/s\ =\ 10\cent/s$$

Assuming this cost model is correct, we see that we can afford a tester costing as much as $(10\cent/s)\div(1\cent/s)$, or 1000% more than a single-site tester, as long as we can achieve fully parallel quad-site testing. Clearly, multisite capability is a must for economical production testing.

Repeating the exercise for fully-depreciated equipment, the test cost of dual-head, single-site testing is

$$\text{test cost per DUT} = \left(0\cent/s + 2\times 0\cent/s + 1\cent/s\right)\left(\frac{5\,s+0\,s}{5\,s}\right)\left(\frac{140\,h+12\,h+16\,h}{140\,h}\right)5\,s = 6\cent$$

Using depreciated equipment, the cost for single-head, single-site testing is still greater than for dual-head testing

$$\text{test cost per DUT} = \left(0\cent/s + 0\cent/s + 1\cent/s\right)\left(\frac{5\,s+1\,s}{5\,s}\right)\left(\frac{140\,h+12\,h+16\,h}{140\,h}\right)(5\,s+1\,s) = 8.64\cent$$

Finally, the cost for fully-depreciated, quad-site, dual-head testing is dramatically lower

$$\text{test cost} = \left(0\cent/s + 0\cent/s + 1\cent/s\right)\left(\frac{1.25\text{ s} + 0\text{ s}}{1.25\text{ s}}\right)\left(\frac{140\text{ h} + 12\text{ h} + 16\text{ h}}{140\text{ h}}\right)1.25s = 1.5\cent$$

Since a production floor typically contains a mixture of depreciated and nondepreciated equipment, the true cost of testing lies somewhere between these extremes.

16.3 DEBUGGING SKILLS

16.3.1 Sources of Error

Murphy's law states that anything that can go wrong will go wrong. A mixed-signal test program, hardware, tester, and DUT provide many opportunities for something to go wrong. As a result, test engineers spend much of their time debugging hardware and software errors. Good debugging skills are vital to a mixed-signal test engineer's effectiveness. Debugging is a process that is difficult to lay out in a formalized, step-by-step methodology. Nevertheless, in this section we will try to outline some of the common debugging rules and techniques.

One of the most difficult aspects of mixed-signal test program debugging is that it is not simply an issue of defective software routines. Most of the particularly difficult, time-consuming problems are eventually found to be defects in hardware rather than defects in the test code. Problems can be caused by defective tester hardware, poor DIB design, or poor DIB layout. Most important, problems can be caused by the DUT itself. Many times, a problem that appears to be a test program bug is actually a defect in the DUT. Rapid identification and debug of DUT design flaws allows a much shorter time to market. Much of the test engineer's task is to identify DUT flaws while at the same time debugging the test software and hardware.

Test program development is always far easier if a known good DUT is used in the debugging process. Unfortunately, we seldom have such a luxury. We have to assume that anything that can be defective is defective, to paraphrase Murphy's law. Anytime a problem arises, we have to suspect the test software, the DIB, the tester, and, of course, the DUT.

16.3.2 The Scientific Method

In high school science classes, we all learned the five-step scientific method, which can be used in the investigation of any problem:

       1. State the problem

       2. Form a hypothesis

       3. Design experiments to test the hypothesis

       4. Test the hypothesis

       5. Draw conclusions

We can easily apply the scientific method to test program debugging. We usually have no problem stating the problem (the DAC will not work, the DUT explodes when power is applied, etc.). We next have to come up with a list of possible causes, which are our hypotheses. Next, we have to design experiments that will rule out each of the hypotheses in a logical order. Then we conduct experiments to find out which of the possible causes is giving us the problem. Finally, the conclusions are drawn (the bug is fixed, the DIB needs to be modified, the DUT needs to be redesigned, etc.)

It seems that steps 2 and 3 are the hardest ones for most test engineers. It is easy to state problems, draw conclusions, and perform experiments, once we know what experiments we need to perform. Forming a hypothesis, on the other hand, requires a little imagination and a lot of experience. When we have a problem, we have to imagine all the things that could possibly cause it. When our experience with that type of problem is limited, we have a lack of past history to draw upon. It is often necessary to ask a co-worker who has dealt with this type of issue in the past to help form the hypotheses.

Fortunately, designing experiments is not as difficult as forming a hypothesis out of thin air. The biggest problem most test engineers have with this step is limiting the number of experimental variables. We cannot change five things at once and expect to gain any meaningful experimental results. For example, if we change some test code, drop a different DUT into the test socket, hook up a scope probe to a circuit node that was previously unloaded, and then try to draw conclusions, we are asking for trouble. We do not know which of the variables might have had an effect on the quantity we were trying to measure.

Clearly, if we do not control our experimental variables, we cannot make any progress in debugging. If we modify a routine extensively and it does not work anymore, then we should reload the previous version of the program and change it one line at a time until it stops working. This allows us to find the offending portion of the new code. If the routine has not been changed and it worked yesterday but not today, then we have to ask what variables changed.

Did we use a different DIB board or a different tester? Perhaps we should try using the previous DIB board or even the previous tester. If that clears up the problem, then we can have a logical place to start looking for the cause of the problem. Did we use a different DUT yesterday? Maybe we should try the original DUT to see if it still works. Returning to a known good state and then working our way back toward the bad state is a very effective technique for isolating bugs. Using this methodology, we can eliminate multiple experimental variables and reintroduce them one at a time.

Let us look at a common example of error caused by uncontrolled experimental variables in a mixed-signal measurement. Assume that we wish to correlate the absolute gain of an analog channel that was measured using our test program with the gain that was measured using a spectrum analyzer. If we make the measurement of the gain using the tester and then we connect the spectrum analyzer probe to the output of the channel to measure its signal level, we have changed an experimental variable.

The output of the DUT is unloaded during the first measurement, but it is loaded by the parasitic capacitance of the spectrum analyzer during the second measurement. If the loading changes the output signal level, then we will generate a correlation error. The correct method is to maintain a constant load on the DUT output. We should leave the spectrum analyzer probe connected to the DUT output during the tester measurement so that the test conditions are identical during both measurements.

In practice, we seldom actually write down a formal document outlining our scientific method. Debugging methodology is a mental process that we just take for granted once we have developed it. However, when all else fails and we are out of ideas, we can always fall back to the formal scientific method.

16.3.3 Practical Debugging Skills

Effective test program debugging is often a matter of breaking a problem into pieces and examining the pieces in a logical order. For example, if we have a continuity test that is failing, we have to imagine all the things that could possibly have gone wrong. We might list a set of hypotheses as follows:

1. The DUT is not in the socket
2. The DUT is defective
3. One or more DUT pins are bent
4. The socket is not properly seated on the DIB board
5. There is a short between pins on the DIB board
6. The DIB board is not properly seated on the test head
7. The DUT power supplies are not connected to the DUT and set to 0V
8. The tester's pin card electronics have gone bad
9. The continuity test code has a bug

Next, we decide which of these problems is most likely. We usually base this on experience and common sense. If our continuity test worked yesterday and not today, and we have not changed the code, then it is very unlikely that our test code is defective. We usually try a new DUT as the first and easiest hardware experiment. If several DUTs all show the same failure, then we probably have a DIB or tester hardware problem.

Once we have decided our continuity failure is most likely a hardware defect, then we can attack this problem in a logical manner. Using a breakpoint in the test program, we can simply trap the program at the continuity test routine where it forces current across the DUT's protection diodes. Then we look for the expected 0.6-V diode drop at various points along the signal path using a hand-held voltmeter.

If there is a short to ground anywhere along the signal path, we will see 0 V relative to ground and -5 V relative to the 5 V power supply. If there is a break in the circuit along the signal path, then we will see the clamp voltage of the current source up to the break in the circuit, and then we will see a high impedance from that point on. A high impedance will appear as 0 V relative to ground and will also appear as 0 V relative to 5 V.

If we see a short, we can pull the DIB off the tester and look for stray wires, solder blobs, and other common causes of short circuits. If we determine that we have an open circuit, we can examine all the interconnections and make sure they are making proper contact. Eventually, by making assumptions about what might be going wrong and performing experiments to verify which assumption is correct, we find the cause of the problem.

16.3.4 Importance of Bench Instrumentation

Many engineers think test program debugging is primarily a software engineering task. As the previous example illustrates, the problem is often not software related at all. Unfortunately, some test engineers are more comfortable with software than with hardware. As a result, they will sometimes stare at a computer screen for hours, trying to find the cause of their problems using the various tools and displays of the tester's operating system.

In the continuity example, let us say that the problem with the continuity test was a stray wire that had fallen into the test head by accident, causing a short circuit. There is no tool in any tester's operating system that would have identified a problem such as this. Yet it is fairly common to find an inexperienced test engineer writing and rewriting test code to try to fix such a problem, *even though the code worked perfectly well the day before*.

The simple fact is that most test problems are far easier to debug using a bench instrument such as a voltmeter, oscilloscope, or spectrum analyzer. It is impossible to debug mixed-signal test programs efficiently without observing test signals with bench instruments. The test engineer must not avoid these tools because it takes time to learn how to use them or because it takes time to set them up beside the tester.

Another important use of non-ATE instrumentation is bench correlation. The design engineers often set up a completely separate non-ATE test fixture to perform measurements on the DUT. These bench setups are both a blessing and a curse to the test engineer. They can be a blessing because they can confirm problems that the test engineer believes to exist in the DUT. They can also prove that a DUT is passing when the test program says it is not. Either way, this keeps the test engineer from wasting test time trying to get a defective DUT to function or trying to prove that a functional DUT is defective.

On the other hand, the bench equipment frequently gets different answers than the ATE tester. This causes a great deal of extra effort, as the two measurements must be brought into agreement. Sometimes the test engineer has to figure out what is wrong with the bench setup, but other times, the bench setup shows a weakness in the ATE test program. Bench correlation may seem like a lot of extra work, but it serves to validate the mixed-signal test program. Any analog measurement performed with a single piece of test equipment is highly suspect. Correlation between two independent measurement techniques is a necessary step to prove that the measurements are correct.

16.3.5 Test Program Structure

One of the most effective ways to shorten the time spent debugging test problems is to structure the test program so that it is "debuggable." There are several general guidelines that help to make a test program easier to debug. Some of these relate to code structure, while others relate to the ease of observation of signals using external measurement equipment.

One simple example of debuggable code structure is extensive use of comments in the program. Adequate code commenting is one of those obvious good practices that we sometimes fail to follow because we are in a hurry to get the code working. However, we invariably pay a heavy price when someone else has to figure out what we have done or when we ourselves pick up the code after two years and cannot figure out what we were thinking at the time that we wrote it.

Another example of good mixed-signal test code structure is avoidance of over-procedurized code. It is extremely frustrating to deal with a piece of test software that drops through eight levels of subroutines just to make a simple DC measurement. The argument for heavily procedurized test code is that extensive procedurization makes the code modular and reusable. However, if taken to an extreme, modularization just makes a simple test complicated, increasing debug time.

With the exception of a few waveform analysis subroutines and waveform creation algorithms, test programs consist of a straightforward, linear sequence of tester instrument setup instructions to be followed step by step. The potential time-to-market advantage won by reusability of general-purpose routines is sometimes lost in the confusion caused by the contorted program flow and extensive variables needed to direct the operation of overly general-purpose subroutines. The test engineer should be careful to use general purpose procedures where they make sense and straight-line code where it makes sense.

Digital patterns can be structured so that they, too, aid the debugging process. Mixed-signal sampling loops should always be written so that they can either stop after one unit test period (UTP) or loop indefinitely. Infinite looping should happen automatically when the test engineer traps at a designated debug point in the test program. Without the ability to drop into an infinite signal loop, it is impossible to use oscilloscopes and spectrum analyzers to debug problems and perform correlation studies. Any signal that is generated for only a few tens of milliseconds will be difficult or impossible to measure using bench equipment.

Finally, it is a good idea to avoid mystery constants in a test program. For example, if we see a line of code that reads

PeakSineVoltage = 0.501782 V;

then we have no idea where the 0.501782 V came from. Instead, we should show the derivation of the constant in the code.

/* -9 dB Sine Wave Peak Voltage, 0 dB Ref = 1.0 V RMS */
PeakSineVoltage = 1.0V * SQRT2 * pow(10,-9dB/20dB);

We can more easily see that this is the peak amplitude of a –9 dB waveform whose 0-dB reference is 1.0 V RMS. In the process of debugging a DUT, we are often asked by the design team to make measurements under a variety of test conditions. Explicit computations help us do this very quickly. For instance, if we wish to raise the sine wave amplitude in this example program to –6 dB, we can do so by simply changing the 9 to 6 without hunting for a calculator. The time it takes for the tester computer to perform the calculation on the fly is insignificant compared to the overall test time. Therefore, the extra test time caused by the computation is no reason to avoid explicit calculations such as this.

16.3.6 Common Bugs and Techniques to Find Them

Almost every mixed-signal test engineer eventually encounters a number of common bugs. The following is a list of common problems and debugging suggestions. Hopefully, these hints will help the new test engineer at least get started in the right direction when one of these problems occurs.

The test program worked yesterday, but it does not work today.

First, be sure the code really is the same as it was yesterday. Sometimes, a corrective change was made at the end of the previous day and it was not saved. Other times, the test was just barely passing on the previous day and a slight drift in the tester's performance causes the DUT to fail. If the code is definitely the same code that previously passed, then something in the hardware setup has probably changed. Usually it is a simple continuity failure or a short circuit on the DIB. Make sure the DUT is the same one that passed before. Also, try using the same DIB that worked before and if possible, use the same tester that worked before. If this fixes the problem *do not just move on – find the source of the problem on the defective equipment*. Make sure the continuity test is working properly to verify that the DIB is properly connected to the tester. If the tester uses dual heads, make sure the test program is loaded on the correct test head. Finally, make sure the tester passes all its checkers.

When debugging a new test, the DUT is completely nonfunctional.

Make sure the DUT is powered up. Do not simply call up a debug display on the tester computer, but observe the voltage at the DUT pins to make sure they are receiving the correct voltages and signals. Using an oscilloscope, verify that the analog and digital signals are arriving at the DUT as expected. Then verify that the DUT is producing the expected analog and digital outputs. If signals are not arriving at the DUT, then turn to the test code and software debug tools to find out why the signal is absent. If the signals are arriving as expected, but the DUT is not producing the expected output, then make sure the program is correctly setting up the DUT's internal control registers and other test conditions specified in the data sheet.

Try loosening the digital timing and setting the digital logic levels of the tester so that the DUT is not stressed to its test limits. If the DUT produces the correct output signal and the signal makes it all the way back to the tester inputs, then turn to the test code and debug displays to find out why the tester is not receiving the correct data. Do not just observe the output of a mathematical routine such as an FFT, look at the raw data from the DUT to make sure it matches the expected signal. If all else fails, get the design engineer to help determine what is wrong. Sometimes, the problem is caused by a last-minute design change that did not appear in the data sheet. The design engineer can quickly identify this type of error, and many others.

The FFT output shows excessive spikes that cause a failure in signal-to-noise ratio.

If the spikes are confined to individual FFT spectral bins, then they are probably not coming from an external source such as 60-Hz power lines or cellular telephones. They are coming from inside the DUT or perhaps from the tester. If a spike is located at a frequency that is a submultiple of the DUT master clock, then it may be caused by digital-to-analog crosstalk. Ask the design engineer if there are any internal or external DUT clocks or other signals operating at that frequency. If not, then try changing the frequency of the test tone. If the spike shifts to a different spectral bin, then it is either caused by distortion, imaging, aliasing, or mixing of the test tone with another frequency. Move the tone several times and see what pattern appears between its movement and the movement of the spike. Try to figure out which of the distortion, imaging, aliasing, or mixing mechanisms are possible and then design experiments to try to isolate which mechanism is responsible for each spike. Once a possible mechanism has been identified, see if the designer can identify a weakness in the DUT or in the test fixture that might be causing the problem. If the same spikes appear on a bench setup, then they are probably not caused by the DIB. If they are absent, then the DIB and tester are highly suspect.

The DUT works, but noise levels are too high.

Excessive noise levels are perhaps the most common mixed-signal testing problem, and they are among of the most difficult to debug. The problem is that noise can come from a large variety of sources. Because noise is such a small signal, it is often very difficult to observe with bench equipment. Fortunately, noise usually couples into a DUT through a number of common mechanisms.

The most common mechanism is direct injection of noise into the high-impedance voltage reference inputs, bias current inputs, or V_{MID} inputs. These nodes should always be carefully shielded from external noise sources, as discussed in Chapter 15, "DIB Design." Another common coupling mechanism is noise coupling directly into the analog inputs or outputs. Sometimes the tester's signals are simply not clean enough to allow the DUT to pass. In these cases, special test techniques such as DIB filters must be added to clean up the signal before it is sourced or measured by the tester. A third common coupling mechanism is power supply or ground noise. Proper power and ground layout and good decoupling practices can help to reduce this problem.

The process of purposely aggravating a circuit to see if it is sensitive can be very effective. Sensitive nodes can often be located by purposely injecting noise into the circuit using the body as a radio antenna and the finger as a signal source. This is a peculiar but effective way to debug noise problems which some have called "tactile engineering" or "the laying of hands upon the DIB." However odd it may seem, the practice of purposely trying to make a problem worse to discover the circuit's weakness is often more effective than trying to figure out how to make it better. Once we know where a circuit is vulnerable, we can try to figure out ways to protect it from unintentional sources of aggravation.

I have been trying to debug a problem for two weeks and I am not getting anywhere.

Ask someone for help. Do not stare at the same problem for hours or days without asking someone else to look at it. If a problem has persisted for more than two or three days, it is time to attack it from a new angle. Often the problem can only be seen from the fresh perspective of a second set of eyes. In particular, be sure to keep the design engineers abreast of the problems encountered in debugging the DUT and test program. Sometimes, they possess critical but undocumented information about the DUT that is key to solving the test problem. Without their assistance it is sometimes impossible to debug test problems.

16.4 EMERGING TRENDS

16.4.1 Test Language Standards

One of the factors that drives up the cost of mixed-signal test equipment and extends test development is the lack of a common software environment for mixed-signal testing. One of the recent developments in test engineering is the emergence of a generic test language called STIL (standard test interface language).[1-3] There have been several attempts in the past to create a standardized test language that is tester-independent. One example was the ATLAS test language developed for military applications. Initially, STIL was primarily targeted at purely digital devices, but analog and mixed-signal test commands are being added to the language.

In theory, a standard test language could reduce testing costs in a number of ways. First, it might allow tester software/hardware interchangeability similar to that of the Microsoft/Intel PC. ATE hardware vendors could concentrate on the production of very low-cost, high-performance hardware compatible with the standard language. Independent ATE software companies could concentrate on the standard language and operating system without being caught up in the hardware development headaches associated with a mixed-signal tester. Ideally, if all testers were compatible with a common computing platform, a semiconductor manufacturer could purchase whatever tester represented the best value at any given time. The huge manual effort required to convert test programs from one tester platform to another would be a "thing of the past" as futurists and inventors are fond of saying. Many experienced mixed-signal test engineers are skeptical about the idea of "effortless" transfer from one tester platform to another, since Murphy's law is especially prevalent in mixed-signal test program development. Nevertheless, the concept of a unified test language is a highly attractive one.

Another advantage promised by a standard language is reduced training costs. Multiple tester platforms require multiple training, which offers no inherent financial advantage to a semiconductor vendor. Not only does the engineer lose productive time to the extra training sessions, but his or her level of expertise is diluted by the reduction of time spent on any particular type of tester. The only advantage that might come with multiple tester platforms is that the semiconductor vendor encourages competition between the ATE vendors, hopefully reducing tester purchase price. However, the free market already encourages competition; so any price reductions are typically short-lived. Obviously, it would be attractive to be able to train all test and product engineers on a single tester platform, without locking the company into a single vendor. A truly standard, flexible mixed-signal test language and operating system would allow multiple ATE vendors without requiring multiple training and diluted expertise.

16.4.2 Test Simulation

Another interesting development in recent years has been the advent of test simulation.[4-6] Although still in its infancy, test simulation promises us the ability to debug our test programs and DIB designs before we have received actual silicon DUTs. Modern tester operating systems allow us to perform a great deal of test code debugging even without test simulation. Using a standalone workstation rather than an expensive ATE tester, we can debug much of our test program errors off-line. However, we cannot fully debug our programs without the DUT and its corresponding DIB. We can only verify that the tester will produce the DUT input signals that we tell it to produce. We would prefer to verify that these input signals are the appropriate ones, that the DUT will react correctly to them, and that our test program will capture and analyze the DUT responses correctly. Test simulation provides a closed-loop simulation process that allows us to verify the test code before the DUT or DIB have been fabricated.

Test simulation links the tester's off-line simulation software with a software model of the device. Typically, the device model is developed by the design engineers for the purpose of design verification (Figure 16.4). Device models may be developed using SPICE, VHDL, or other software modeling languages. The tester simulator generates DUT stimulus signals based on the test program. It passes these signals, called *events*, through models of the tester and DIB to the DUT model. The DUT model produces simulated responses to the test stimuli. The DUT responses are passed back to the tester's simulator, which continues as if the signals had been captured from a physical DUT. Using test simulation, both the test program and the DUT design can be verified at the same time.

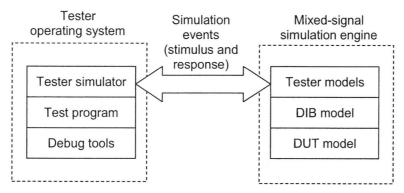

Figure 16.4. Test simulation signal flow (diagram courtesy Teradyne, Inc.)

Without test simulation, we are forced to wait until a physical DUT has been fabricated. We often discover basic flaws in our DIB or our test approach that force us to discard code or hardware. The resulting delay in test program development extends the time to market, reducing profit margins or even allowing a competitor's device to win market share. Figure 16.5 illustrates the difference between a traditional new product development flow and one based on test simulation.

Using test simulation, a portion of the test program debugging effort can be performed before the silicon is fabricated. The test program is developed in parallel with the device design, reducing the overall cycle time of the new product. An added benefit of the test simulation approach is that flaws in the design can be identified by the test program before the design is released to fabrication. These flaws can be corrected in a timely manner, saving further cycle time.

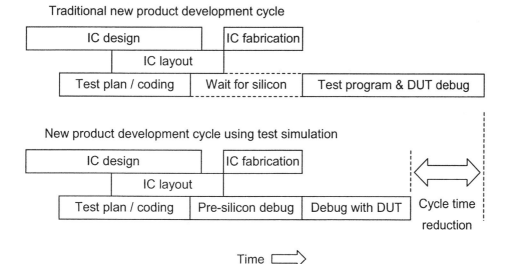

Figure 16.5. New product development using test simulation (diagram courtesy Teradyne, Inc.).

16.4.3 Noncoherent Sampling

In Chapter 6, "Sampling Theory," the importance of coherent sampling was emphasized. The exact relationships between the signal frequencies, digital pattern rates, and the tester's various sampling rates lead to a maze of restrictive sampling criteria. A recent trend is the elimination of coherence as a fundamental requirement for DSP-based testing.

Noncoherent sampling can be utilized using a number of mathematical routines. In Chapter 7, "DSP-Based Testing," we examined the use of windows and their inherent disadvantages. Recently, mathematical resampling routines have been used to change the effective sampling rate of the tester's digitizer and capture memory. This allows a fully accurate, coherent measurement based on noncoherent test tones. Unlike windowing techniques, these resampling algorithms do not discard useful signal information, and they do not allow energy from adjacent spectral bins to bleed into one another.

The relaxed coherence requirements allow test cost savings in two ways. First, the clock generation circuits of the tester do not have to be as flexible as they would need to be in a coherent tester. Less demanding clocking requirements can help reduce the cost of the tester hardware. The second advantage of noncoherent testing is that it reduces the difficulty associated with the complicated calculation of coherent sampling systems. This in turn may reduce the test development cycle time, since it allows the test engineer to concentrate on the signals themselves rather than the methods used to generate and analyze them.

16.4.4 Built-In Self-Test

In Chapter 14, "Design for Test," the subject of built-in self-test (BIST) was introduced briefly. To be properly classified as built-in self-test, the measurement circuit must perform the intended measurement of quality and return a pass/fail result using only minimal externally generated signals such as power supply voltages and digital clocks. As mentioned, BIST is commonly used in digital circuits but has proven more difficult to introduce into high-performance mixed-signal circuits. A number of interesting BIST circuits have been proposed at the annual International Test Conference. Among these are a number of sigma-delta noise-shaping BIST concepts.[7-9] Some of the other interesting BIST concepts are listed in the references at the end of Chapter 14.

To maintain high accuracy standards, analog and mixed-signal BIST circuits should ideally be traceable to the National Institute of Standards and Technology (NIST), or its non-U.S. equivalent. However, we saw in Chapter 15, "Data Analysis," that wide design margins can lead to reduced accuracy requirements in the production testing process. Therefore, in many cases the promise of low-cost analog and mixed-signal testing through BIST is tightly coupled to the improvement of design margins to allow less stringent testing. Otherwise, the BIST scheme must incorporate a well-engineered and robust means of standards traceability. An alternative is to use BIST circuits that are inherently accurate by design, or circuits that derive their accuracy from external sources, such as frequency generators operating at calibrated frequencies.

16.4.5 Defect-Oriented Testing

Defect-oriented testing (DOT) is yet another cost-saving test methodology that shows great promise.[10-13] Most of this book has been devoted to the traditional testing approach, known as *specification-oriented testing* (SPOT). However, over the past decade or so many companies have begun to explore and adopt a methodology called *defect-oriented testing*. DOT is based on the assumption that we can understand and predict some or all a circuit's major failure mechanisms through mathematical analysis and software modeling, and that these failure mechanisms can be made to exhibit themselves in the form of a limited set of measured parameters. The reduced set of measured parameters leads to a lower test cost while providing more thorough coverage of failure mechanisms. For example, if we can design a special test mode that converts a 100-ms signal-to-noise ratio test into a 5-ms DC offset measurement, then obviously we can save a great deal of test cost.

In Chapter 14, "Design for Test," we examined the role of I_{DDQ} testing in the improvement of product quality. I_{DDQ} testing is a very simple type of DOT that efficiently tests for the presence of undesirable current leakage paths in an improperly fabricated IC. Another very simple example of defect-oriented testing is the major carrier method presented in Chapter 11 for the testing of DAC and ADC INL and DNL. The major carrier method eliminates the redundancy inherent in all-codes linearity testing. It allows us to predict the shape of the overall transfer curve of a converter by measuring only a selected set of converter levels.

This particular DOT methodology is based on an assumption about the DUT's robustness that must be verified. In order for the major carrier technique to work, the failure mechanism of the converter must be simple; that is, superposition must hold. If we can verify that the converter exhibits low superposition errors, then we can assume that any output level is a simple summation of the converter's bit weights. Testing every code is a waste of time if superposition holds, since the same bit weight errors show up repeatedly in the complete transfer curve. Defect-oriented testing such as the major carrier method seeks to reduce testing redundancy, complexity, and expense.

The I_{DDQ} and the major carrier examples are very trivial and do not represent the current state of the art in DOT. True DOT involves an analysis of failure mechanisms through simulation and/or collection of emprical data to build models of failure mechanisms, called *fault models*. These models are then used to predict failures that would normally require lengthy, expensive production tests. For example, the major carrier method relies on a fault model that assumes all linearity failures are due to mismatched components in the binary-weighted components of a converter circuit. However, this fault model is far too simplistic to work in all cases. Defect-oriented testing allows a more thorough modeling process to predict failures more reliably than that obtained by a simple major carrier technique.

Unfortunately, defect oriented testing is a very complex subject that is too advanced for an introductory text such as this. Like many of the advanced subjects introduced in this text (e.g., digital scan testing, BIST, etc.,) we have presented DOT only to introduce the terminology to the reader and to encourage further study.

16.5 SUMMARY

In this chapter we have seen a variety of factors affecting test economics. We have seen how testing costs are based not only on the purchase price of the tester, but on the tester's overall throughput and yield. We have discussed the importance of time to market on revenues and profit margins, and we have seen how a test engineer's debugging skills can reduce the cycle time of a new semiconductor product. Finally we have examined some of the emerging trends that are changing the manner in which we develop and utilize test programs for the production of mixed-signal ICs.

Throughout this book, we have seen that robust designs are the key to lowering testing costs. The importance of design and process quality is summed up by the statement, "Testing adds no value to the product." Taken out of context, this statement may be offensive to test engineers, but it emphasizes the importance of good designs. Wide design margins allow lower testing costs, less averaging, less expensive test equipment, BIST, and DOT. Most important, though, robust designs result in high-yielding, high-quality products that do not need to be tested in the first place. Realistic or not, this is our common goal as engineers.

The future of mixed-signal testing is difficult to predict. Perhaps we will eventually discover ways to test devices in production using only simple BIST circuits and DOT methodologies combined with traditional statistical process control methodologies. Perhaps we will just keep performing specification-oriented testing for hundreds of years. More likely, the future of mixed-signal testing will involve a continuing evolution in which the various tried-and-true techniques are combined with new methodologies as appropriate.

At the 1999 International Test Conference, a different view of the future of IC testing was proposed at one of the panel sessions. Toward the end of the session, a gentleman from the audience asked a somewhat rhetorical but sobering question: "Imagine it is the year 2100 and we have developed integrated circuits that are one cubic centimeter in size, contain trillions and trillions of transistors, and operate at a master clock rate of 1 terahertz. How on earth are we going to test such a device?" One of the other participants offered perhaps the only feasible solution to such a difficult problem:

"We'll just ask it how it feels!"

Problems

16.1. A production lot of 5000 units yields 92% and consumes a tester for 8 h, including changeover time, calibration time, etc. What is the throughput of the device on this tester, measured in passing DUTs per hour?

16.2. During a particular month, a tester has an average down time of 9 h/wk. During that same time it has an average idle time of 5 h/wk. The average depreciation cost for this type of tester is 0.89 cents per second. A particular DUT requires a handler having a depreciation cost of 0.7 cents per second. Handler index time is 450 ms. Single-site test time is 3.8 s. This tester's share of the fixed costs is 1.4 cents per second. What is the overall cost per second of this tester/handler combination?

16.3. By what maximum percentage could we increase the cost of the tester in Problem 16.2 (i.e., the tester depreciation cost) if we were able to perform fully parallel dual-site testing on the more expensive tester? (Assume the chosen handler is already capable of dual-site handling.)

16.4. Compare the cost per DUT of single-head, single-site testing with dual-head, single-site testing in Problem 16.2, assuming the tester and handler are already capable of dual-head testing.

16.5. Compare the cost per DUT of single-head, single-site testing with dual-head, dual-site testing in Problem 16.2, assuming the tester and handler are already capable of dual-site and dual-head testing.

16.6. In a DUT having 4 circuit blocks, there are 12 possible categories of block-to-block interactions (i.e., block A to block B, block B to block A, block A to block C, etc.). In a DUT having 8 circuit blocks, there are 56 interaction categories. Derive the general equation describing the number of categories of block-to-block interactions as a function of the number of blocks, N. Plot the number of interaction categories as a function of the number of blocks. Limit your plot to the range $N = 2$ to 10.

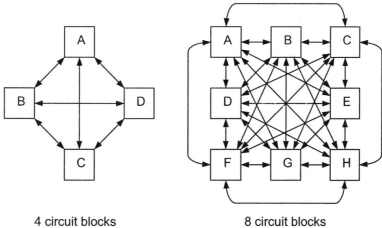

4 circuit blocks
12 interaction categories

8 circuit blocks
56 interaction categories

16.7. Assume that 30% of the possible interaction categories in Problem 16.6 result in a failure mechanism that must be tested in production. Furthermore, assume that each extra test requires an additional 150 ms of test time. Plot the additional cost of block-to-block interaction testing as a function of the number of circuit blocks. Assume a simple cost model of 2.5 cents per second. Limit your plot to the range $N = 2$ to 10.

16.8. Assume that each circuit block in Problem 16.6 consumes 5×10^6 μm^2 of silicon area and that the cost of fabricating a 200-mm-diameter wafer is $1500 (this number is intentionally skewed from reality). Plot the fabrication cost of each DUT as a function of the number of circuit blocks, N. Use the simplistic assumption that the wafer is perfectly circular and that there is 15% noncircuit area for bond pads, interconnections, scribe lines, unusable wafer area, etc.

16.9. Assume that each circuit block in Problem 16.6 takes 500 ms of test time, excluding the extra time required for block-to-block interaction testing. Plot the total test cost as a function of the number of circuit blocks. Again, use a simple model of 2.5 cents per second of test time. Plot the ratio of total test cost versus fabrication cost as a function of the number of circuit blocks, N.

16.10. Explain the relevance of Problems 16.6 to the cost of test versus DUT complexity. Explain the relevance of Problems 16.6–16.9 to the cost of testing as a percentage of total manufacturing costs as devices become more complex.

16.11. An IC incorporates a clever BIST circuit that eliminates the need for the AWG and digitizer in a $1.5M tester. As a result, the tester cost drops by $100,000, which is the cost of the digitizer/AWG pair. The BIST circuit occupies 50,000 μm^2 of silicon area. Using the wafer cost model of Problem 16.8, how much expense does the BIST circuit add to each DUT? Using a single-site, dual-head test setup, how much is the test cost reduced if the cost of the original $1.5M tester depreciation is 1 cent per second, the handler depreciation is 0.4 cents per second and the fixed costs are 1 cent per second? (Assume the same down time and idle time as Problem 16.2 and that test time remains at 5 s with or without the BIST circuit). Is the cost of the BIST circuit justified from a production economics standpoint (i.e., is the added silicon cost less than the test cost savings)? If not, list at least two other reasons we might implement this BIST circuit.

16.12. The BIST circuit from Problem 16.11 eliminates 500 ms of test time from the 5-s test program. The BIST circuit occupies the same area and costs the same as the BIST scheme in Problem 16.11. Given a 1 cent per second tester depreciation cost, a 0.4 cents per second handler depreciation cost, and a 1 cent per second fixed cost, does this BIST scheme make sense from a production economics standpoint?

References

1. *Standard Test Interface Language (STIL) for Digital Test Vectors*, IEEE P1450 (working standard)

2. Tony Taylor, Gregory A. Mastron, *Standard Test Interface Language (STIL) – A New Language for Patterns and Waveforms*, Proceedings of the International Test Conference, 1996, p. 565

3. Tony Taylor, *Standard Test Interface Language (STIL), Extending the Standard*, Proc. International Test Conference, 1998, pp. 962-70

4. Tom Austin, Nash Khouzam, Jean Q. Xia, *Faster Mixed-Signal Development Using CAD to Model IC, Package, and Test Systems*, Proceedings of the First International Conference on Electronics, Circuits & Systems, 1994, pp. 216-22

5. Scott Bullock, *Report on Pilot Project Sucessfully Implementing a Design-to-Test Methodology*, Proc. International Test Conference, 1995, pp. 771-80

6. Craig Force, Tom Austin, *Testing the Design: The Evolution of Test Simulation*, Proc. International Test Conference, 1998, pp. 612-21

7. Xavier Haruie, Gordon Roberts, *Arbitrary-Precision Signal Generation for Bandlimited Mixed-Signal Testing*, Proc. International Test Conference, 1995, pp. 78-86

8. E. Hawrysh, Gordon Roberts, *An Integration of Memory-Based Analog Signal Generation into Current DfT Architectures*, Proc. International Test Conference, 1996, pp. 528-37

9. Benoit R. Veillette, Gordon Roberts, *Stimulus Generation for Built-In Self-Test of Charge-Pump Phase-Locked Loops*, Proc. International Test Conference, 1998, pp. 698-707

10. Anne Meixner and Wojciech Maly, *Fault Modeling for the Testing of Mixed Integrated Circuits*, Proc. International Test Conference, 1991, pp. 564-72

11. M. M. A. va Rosmalen, Keith Baker, E. M. J. G. Bruls, J. A. G. Jess, *Parameter Monitoring: Advantages and Pitfalls*, Proc. International Test Conference, 1993, pp. 115-24

12. M. Soma et al. *Analog and Mixed-Signal Test*, B. Vinnakota, Ed. Prentice Hall, Englewood Cliffs, NJ, April 1998, ISBN: 0137863101

13. Y. Xing, *Defect-Oriented Testing of Mixed-Signal ICs: Some Industrial Experience*, Proc. International Test Conference, 1998, pp. 678-87

Answers to Selected Problems

Chapter 1
1.1. Operational amplifiers, active filters, comparators, voltage regulators, analog mixers, analog switches, and transistors. **1.2.** Comparators, analog switches, PGAs, AGCs, ADCs, DACs, PLLs, and switched capacitor filters. **1.3.** Programmable gain amplifier (PGA). **1.4.** Digital-to-analog converter (DAC). **1.5.** Analog-to-digital converter (ADC). **1.6.** Digital signal processor (DSP). **1.7.** Automatic gain control (AGC). **1.8.** Cellular telephone microphone gain (volume) control. **1.9.** Short between metal traces (i.e. blocked etch). Bonus answer: Can also cause a gap in the trace, if using a negative process. **1.10.** A clean room eliminates particles in the air, thus reducing the particulate defects such as those in Figure 1-8. **1.11.** Wafer probe testing, bond wire attachment into lead frames, plastic encapsulation (injection molding), lead trimming, final testing. **1.12.** Two devices do not represent a good statistical sample from which to draw conclusions. Need data from hundreds or thousands of devices. **1.13.** Mainframe, test head, user computer, system computer, DIB. **1.14.** Provides a temporary electrical interface between the ATE tester and the DUT. Also provides DUT-specific circuits such as load circuits and buffer amplifiers. **1.15.** Wafer prober. **1.16.** Allows design engineers and test engineers to agree upon an appropriate set of tests. Also serves as test program documentation. **1.17.** Time to market, accuracy/repeatability/correlation, electromechanical fixturing, economics of prodution testing (test economics). **1.18.** We have to test a total of 5,555,555 devices to get 5 million good ones since we have a 90% yield (yield = ratio of good devices to total devices tested). For the good devices, the time saved is 1.5 s times 5 million devices, or 7.5 million seconds of reduced test time per year. Multiplying this by 3 cents per second, we get a total savings of $225,000 per year in reduced testing costs for the good devices. Multiplying the 0.5-s test time reduction for the bad devices by 555,555 devices per year, we see an additional savings of 3 cents per second times 0.5 s times 555,555 devices, or $8333 per year. Thus the total test cost savings is $233,333 per year. **1.19.** The profit margin is 20%. Therefore, we would have to ship $233,333/20% = 1.166665 million dollars worth of additional product to equal the extra profit offered by the reduced test time. Thus, we have to ship approximately 650,000 extra devices at $1.80 per unit to get the same incremental profit as we get from the 1.5-s test time reduction for this device. Obviously, reducing test time can have as high an impact on profits as selling and shipping millions of extra devices!

Chapter 2
2.1. (a) It serves as a design specification, helps test and product engineers define the test plan, helps the customer use the device in the end application, and serves as the formal communication channel between engineering personnel. (b) Feature summary and description, principles of operation, absolute maximum ratings, electrical specifications, timing diagrams, application information, characterization data, circuit schematic, die layout. (c) Electical specifications. **2.2.** A test list is a written list of tests and test procedures that will be used to verify the quality of a given device in production. No. No. **2.3.** No. **2.4.** 100 Ω, 13 pF. **2.5.** $V_O = V_I \times (D/256)$ where V_O = output voltage, V_I = fixed input voltage, D = digital input code, converted to decimal. **2.6.** Both signals low. The DAC output remains constant, since the chip select signal is high (disabled). **2.7.** No signal is connected to pin 16. Pin 9 is data pin DB4. Pin 18 is connected to

V_{DD}. **2.8.** 5 mW. No, it dissipates 5 V × 1.5 mA = 7.5 mW, which is too high. **2.9.** 20 kHz. No, this is typical data generated through characterization. It does not apply to each device. **2.10.** No. There are infinite permutations of supply voltage, input signal levels, output loading conditions, etc.; so we have to choose a subset of tests. **2.11.** Binning. **2.12.** The DIB checker code verifies that the DIB board is not defective. **2.13.** Improves the accuracy of tester instruments, compensates for errors introduced by DIB circuits such as op amps.

Chapter 3

3.1. 610 mV, 6.1% **3.2.** 540 mV, 10.8% **3.3.** 8.85 V **3.4.** 5.6 V **3.5.** 540 Ω **3.6.** 1 kΩ **3.7.** 105 Ω **3.8.** 200 Ω **3.9.** -10.48 V (input), 5 V (output) **3.10.** 0.950 V **3.11.** 0.5 or 50% **3.12.** -100 mV (OUTP), 200 mV (OUTN), -300 mV (diff.), 50 mV (c.m.) **3.13.** -627.5 mV **3.14.** 16.5 V/V, 24.35 dB **3.15.** 8 V, 11.5 V, 3.5 V/V, 10.88 dB **3.16.** 2.85 V, 8.65 V, 2.9 V/V, 9.25 dB **3.17.** 15390 V/V, 83.75 dB **3.18.** -674 µV **3.19.** 3.001 V **3.20.** PSS = 40 mV/V, -27.95 dB, PSRR = 4.08 mV/V, -47.78 dB **3.21.** 51.98 µV/V, -85.68 dB **3.22.** 56.23 µV **3.23.** 2.506 V **3.24.** 2.02 V **3.25.** 2.59 V **3.26.** 2.49 V **3.27.** *1st iteration*: point1 = (-1 V,-7.75 V) point2 = (+1 V, 4.25 V), estimated zero crossing at 291.7 mV. *2nd itertaion*: point1 = (+1 V, 4.25V), point2 = (291.7 mV, 228.7 mV), estimated zero crossing at 327.84 mV. *3rd iteration*: point1 = (291.7 mV, 228.7 mV), point2 = (327.84 mV, 6.08 mV), estimated zero crossing at 328.83 mV – converged with 3 iterations. Input offset voltage is approximately -328.83 mV. A binary search would have taken 10 iterations to achieve a 1-mV convergence.

Chapter 4

4.1. 55.6 mV, 600 µV. **4.2.** ±305 µV, ±0.0061 %. **4.3.** 17 bits, We would set the input range to ±125 mV to achieve the desired resolution. **4.4.** The input could have been anywhere between 322 mV and 324 mV. **4.5.** 1.08 V/V, 41 mV. $v_{calibrated}$ = ($v_{measured}$ - 41 mV)/1.08. **4.6.** 1.258 V. **4.7.** 0.970 V/V, 0.894 V/V, 0.800 V/V. We would request 515 mV, 559 mV, and 625 mV. **4.8.** 1.155 V/V, -135 mV. **4.9.** 1.961 V, uncalibrated measurement error = 169 mV. **4.10.** 500 nV RMS, 1.07 µV RMS. **4.11.** 62.1 ms. The second filter requires less settling time (only 13.7 ms.) **4.12.** We need to reduce the 6σ spread by a factor of 5. Since σ is proportional to the noise at the meter input, we need to reduce the noise by a factor of 5. Therefore, we need to reduce the cutoff frequency of the filter by a factor of 5^2, or 25. Thus, the *RC* time constant (i.e. settling time) would have to increase by a factor of 25. **4.13.** The less repeatable measurement costs us $7.50 per every 100 units tested (6 good devices discarded times $1.25). The cost per device of the less repeatable measurement is therefore 7.5 cents per unit tested. To eliminate this cost, we have to add 250 ms of test time, which costs us 3.5 cents/second times 250 ms of extra test time, or 0.875 cents. In this case, the slower measurement is more economical.

Chapter 5

5.1. System computers, DC sources, DC meters, relay control lines, relay matrix lines, time measurement hardware, arbitrary waveform generators, waveform digitizers, clocking and synchronization sources, and a digital subsystem for generating and evaluating digital patterns and signals. **5.2.** It improves DC measurement repeatability by removing high frequency noise from the signal under test. **5.3.** A PGA placed before the meter's ADC allows proper ranging of the instrument to minimize the effects of the ADC's quantization error. **5.4.** Using the two-measurement approach, we obtain two readings on the ±5 V-range that have errors of ±5 mV, giving a total differential error of ±10 mV. Using the meter's sample-and-difference mode, the 1-V range can be selected, giving a worst-case error of ±1 mV. **5.5.** Using the two-pass (normal) measurement mode, we have to set the meter to the 5-V range to accommodate the 3.5-V common-mode offset. This gives an error of 0.01% of 5 V, or 0.5 mV. The worst-case error is twice this amount, or 1 mV, since we are subtracting two measurements. Using the sample-and-difference mode, the meter range can be set to the 1-V range. The single measurement has a worst-case error of 0.01% of 1 V, or 0.1 mV. The sample-and-difference mode is much more accurate. **5.6.** Kelvin connections compensate for the *IR* voltage drops in the high force line and current return line of a DC source. **5.7.** A local relay can provide a low-noise ground to the DUT inputs. (Alternate answer, can provide local connections to load circuits, buffer amplifiers, and other sensitive DIB circuits). **5.8.** Flyback diodes prevent inductive kickback caused by the relay coil from damaging the drive circuitry. **5.9.** A digital pattern generates digital 1/0 waveforms and high/low comparisons. A digital signal contains waveform information such as samples of a sine wave. **5.10.** The number of vectors in the frame loop and the frequency of the digital vectors in a sampling frame determine the sampling frequency of the mixed-signal circuit (DAC or ADC, for example). **5.11.** Source memory stores digital signal samples and supplies them to a mixed-signal circuit such as a DAC. **5.12.** Capture memory captures and stores digital signal samples from a DUT circuit such as an ADC. **5.13.** The DAC samples are sourced at a frequency equal to 6 MHz / 600 = 10 kHz. The total time to supply all 256 samples to the DAC is equal to 256×(1 / 10 kHz) = 25.6 ms. **5.14.** X's are required to place the driver into a HIZ state so that data from the DUT can be read from SDATA. Otherwise the driver would conflict with the output of the SDATA serial interface. **5.15.** Formatting and timing information are combined with one/zero information to reduce the amount of digital pattern memory required to produce a particular digital waveform. It also reduces the required vector (bit cell) rate for complex patterns (see Figure 5.11.) Finally, it gives us better control of edge placement. **5.16.** The NRZ waveform could be produced using clocked digital logic operating at 2 MHz. The RZ waveform would require clocked digital logic operating at a period of 100 ns, or 10 MHz. If the stop time for the RZ formatted waveform had to be delayed to 901 ns, we would need digital logic operating at a clock rate of 1 GHz. **5.17.** The CW source and RMS voltmeter are only able to measure a single frequency during each measurement, leading to long test times compared to DSP-based testing. Also, the RMS voltmeter can't distinguish between the DUT's signal and its distortion and noise. **5.18.** The low-pass filter is used to reconstruct, or smooth, the stepped waveform from the AWG's DAC output. **5.19.** The PGA sets the measurement range, reducing the effects of the digitizer's ADC quantization error. **5.20.** Distributed DSP processing reduces test time by splitting the processing task among several processors that perform the mathematical operations in parallel.

Chapter 6

6.1. The frequency of sine wave, f_o, UTP, and fundamental frequency F_f are: (a) 1/32 Hz, 32 s, 1/32 Hz (b) 13/64 Hz, 64 s, 1/64 Hz (c) 5/64 Hz, 64 s, 1/64 Hz (d) 31/64 Hz, 64 s, 1/64 Hz (e) 63/128 Hz, 128 s, 1/128 Hz (f) 5/64 Hz, 64 s, 1/64 Hz. **6.3.** The frequency of sine wave, f_o, UTP, and fundamental frequency F_f are: (a) 250 Hz, 1/250 s, 250 Hz (b) 1650 Hz, 1/125 s, 125 Hz (c) 625 Hz, 1/125 s, 125 Hz (d) 3875 Hz, 1/125 s, 125 Hz (e) 3937.5 Hz, 2/125 s, 125/2 Hz (f) 625 Hz, 1/250 s, 250 Hz. **6.7.** With M=5, the frequency of the analog reconstructed signal is 1.25 kHz. With M=25, the output signal frequency becomes 1.75 kHz. **6.8.** 1 LSB = 732.6 µV; Quantization noise = 211.5 µV RMS. **6.9.** 3.29 mV RMS. **6.10.** 261.9 mV and 309.5 mV. **6.11.** 22.2 µV. **6.12.** 33 ns. **6.13.** 353 ps. **6.14.** 142.9 MHz. **6.15.** 5.2 bits (ADC), 2.7 bits (DAC). **6.16.** 1.79 mV **6.17.** 25.5 mV. **6.18.** *6-bit DAC*: ideal values: mean value = 0, RMS value = 9.2 mV; simulation values: mean value = 1.43×10^{-8}, RMS value = 8.8 mV. *8-bit DAC*: ideal values: mean value = 0, RMS value = 2.2 mV; simulation values: mean value = 1.46×10^{-18}, RMS value = 2.4 mV. Simulatin results are in close agreement with theory. **6.19.** 19.5 Hz, 51.2 ms. **6.20.** 16 cycles. 15.8 cycles. The signal that completes 16 cycles in one

UTP will be coherent. **6.21.** 62.5 Hz, 125 Hz, 187.5 Hz, 250 Hz, 312.5 Hz, 375 Hz, 437.5 Hz, 550 Hz. **6.25.** Peak-to-RMS = 3.39 when the following phases (radians) are used:
4.9342, 2.7349, 4.4638, 5.9738, 4.4438, 0.8677, 2.0356, 4.9567, 1.8078,
5.0128, 1.3495, 2.4216, 5.8606, 6.1796, 4.7200, 3.8936, 0.3495, 2.1271,
1.5063, 3.3569, 4.2536, 1.3820, 4.1650, 0.7499, 5.3641, 1.0551, 3.8608,
3.9997, 4.8341, 3.4298, 2.5608, 1.3350, 3.9429, 5.2267, 4.7921, 4.7961,
4.4067, 5.1394, 1.1798, 1.2371, 1.3297, 4.8141, 5.7427, 1.4506, 2.1806,
4.3915, 0.1061, 3.5092, 3.8089, 2.8280, 0.1911, 6.2151, 0.2577, 0.2993,
0.7163, 1.9450, 4.4592, 0.7747, 3.2199, 0.6044, 4.3580, 1.7283, 2.9038,
4.0410, 0.2353, 5.6214, 6.0695, 1.2226, 0.9672, 1.3684, 3.4856, 5.5875,
1.9998, 2.8378, 2.8407, 3.6993, 4.4819, 4.4830, 1.6024, 0.7223, 0.6823,
0.7283, 2.2069, 4.5689, 2.3992, 4.4281, 2.2526, 0.0921, 1.0373, 5.4354,
3.5057, 0.2174, 1.0509, 1.0709, 0.6066, 4.7496, 1.5292, 4.7927, 3.3856,
5.9554.
6.26. Spectral bins: 23, 47, 67 and 71 where $N=1024$. A prunning table reveals there is no spectral overlap. **6.27.** 195.3 kHz. **6.28.** 2.98 Hz to 51.2 GHz.

Chapter 7
7.1.

(a) $v(t) = \dfrac{20}{\pi}\left[\sin(2\pi\ 1000\ t) + \dfrac{1}{3}\sin(6\pi\ 1000\ t) + \dfrac{1}{5}\sin(10\pi\ 1000\ t) + \cdots\right]$

(b) $v(t) = \dfrac{20}{\pi}\left[\cos(2\pi\ 1000\ t) - \dfrac{1}{3}\cos(6\pi\ 1000\ t) + \dfrac{1}{5}\cos(10\pi\ 1000\ t) - \cdots\right]$

(c) $v(t) = -\dfrac{10}{\pi}\left[\sin(2\pi\ 1000\ t) + \dfrac{1}{2}\sin(4\pi\ 1000\ t) + \dfrac{1}{3}\sin(6\pi\ 1000\ t) + \cdots\right]$

(d) $v(t) = 2.5 + \dfrac{20}{\pi^2}\left[\cos(2\pi\ 1000\ t) + \dfrac{1}{3^2}\cos(6\pi\ 1000\ t) + \dfrac{1}{5^2}\cos(10\pi\ 1000\ t) + \cdots\right]$

7.2.

$x(t) = \dfrac{2}{\pi}\left[\sin(2\pi t) - \dfrac{1}{2}\sin(4\pi t) + \dfrac{1}{3}\sin(6\pi t) - \cdots\right]$

7.3.

$x(t) = \dfrac{2}{\pi}\left[\cos\left(2\pi t - \dfrac{\pi}{2}\right) - \dfrac{1}{2}\cos\left(4\pi t + \dfrac{\pi}{2}\right) + \dfrac{1}{3}\cos\left(6\pi t - \dfrac{\pi}{2}\right) - \cdots\right]$

7.4.

$x[n] = 0.6155\cos\left[\left(\dfrac{2\pi}{10}\right)n - \dfrac{\pi}{2}\right] + 0.2753\cos\left[2\left(\dfrac{2\pi}{10}\right)n + \dfrac{\pi}{2}\right]$

$+ 0.1453\cos\left[3\left(\dfrac{2\pi}{10}\right)n - \dfrac{\pi}{2}\right] + 0.0650\cos\left[4\left(\dfrac{2\pi}{10}\right)n + \dfrac{\pi}{2}\right]$

7.5.

$x[n] = 0.3667 + 0.2167\cos\left[\left(\dfrac{2\pi}{6}\right)n\right] - 0.8372\sin\left[\left(\dfrac{2\pi}{6}\right)n\right] - 1.8167\cos\left[2\left(\dfrac{2\pi}{6}\right)n\right]$;

$+ 0.8949\sin\left[2\left(\dfrac{2\pi}{6}\right)n\right] + 1.333\cos\left[3\left(\dfrac{2\pi}{6}\right)n\right]$

$$x[n] = 0.3667+0.8647 \cos\left[\left(\frac{2\pi}{6}\right)n+1.3175\right] + 2.0251 \cos\left[2\left(\frac{2\pi}{6}\right)n-2.6839\right]$$

$$+1.333 \cos\left[3\left(\frac{2\pi}{6}\right)n\right]$$

7.6.
$$x[n] = 1.4458-1.3125 \cos\left[\left(\frac{2\pi}{6}\right)n\right]-0.5413 \sin\left[\left(\frac{2\pi}{6}\right)n\right]+0.5542 \cos\left[2\left(\frac{2\pi}{6}\right)n\right];$$

$$+1.9991 \sin\left[2\left(\frac{2\pi}{6}\right)n\right]-0.6875 \cos\left[3\left(\frac{2\pi}{6}\right)n\right]$$

$$x[n] = 1.4458+1.4197 \cos\left[\left(\frac{2\pi}{6}\right)n+2.7504\right]+2.0745 \cos\left[2\left(\frac{2\pi}{6}\right)n+1.3004\right]$$

$$+0.6875 \cos\left[3\left(\frac{2\pi}{6}\right)n-3.1416\right]$$

7.10.
$$x[n] = 1.0+0.4 \cos\left[\left(\frac{2\pi}{8}\right)n-\frac{\pi}{4}\right]+0.6 \cos\left[3\left(\frac{2\pi}{8}\right)n+\frac{\pi}{3}\right]$$

7.11.
$$x[n] = 1.0+5.2345 \cos\left[2\left(\frac{2\pi}{8}\right)n-0.8124\right]+1.5811 \cos\left[3\left(\frac{2\pi}{8}\right)n+0.3218\right]$$

7.15.
(a) $$x[n] = 0.25+(0.25-j0.05)e^{j\left(\frac{2\pi}{8}\right)n}+1.05e^{j2\left(\frac{2\pi}{8}\right)n}+(-0.45+j0.05)e^{j3\left(\frac{2\pi}{8}\right)n}$$
$$+(-0.45-j0.05)e^{j5\left(\frac{2\pi}{8}\right)n}+1.05e^{j6\left(\frac{2\pi}{8}\right)n}+(0.25+j0.05)e^{j7\left(\frac{2\pi}{8}\right)n}$$

(c) same as part (a).

7.16.
(a) $$x[n] = 0.8944 \cos\left[\left(\frac{2\pi}{10}\right)n-1.1071\right]+0.7071 \cos\left[3\left(\frac{2\pi}{10}\right)n+0.7854\right]$$

(c) same as part (a).

7.17. $c_{rms}(0)$=152.3815, $c_{rms}(1)$=203.2356, $c_{rms}(2)$=108.7383, $c_{rms}(3)$=79.3807, $c_{rms}(4)$=67.6312, $c_{rms}(5)$=32.1734, $\phi(0)$=0, $\phi(1)$=−1.3585, $\phi(2)$=−1.9765, $\phi(3)$=−2.3970, $\phi(4)$=−2.7756, $\phi(5)$=3.1416. **7.18.** (a) For NOI=64, Amp = 9.978900240726147×10^{-1}, (b) For NOI=512, Amp = 9.978482396853521×10^{-1}, (c) For NOI=1024, Amp = 9.986680525528827×10^{-1}, (d) For NOI=8192, Amp = 9.998648337894499×10^{-1} (Amplitude is being estimated over approximately the same bandwidth.)
7.20. (a) Rectagular, ε = 1.000 , (b) Blackman, ε = 0.5497 , (c) Kaiser, β = 10, ε = 0.5292.
7.21. (a) For NOI=64, Amp = 9.999998605676857×10^{-1}, (b) For NOI=512, Amp = 9.999999991009749×10^{-1}, (c) For NOI=1024, Amp = 9.999999994688535×10^{-1}, (d) For NOI=8192, Amp = 9.999999999999813×10^{-1}. (Amplitude is being estimated over approximately the same bandwidth.)
7.22. (a) For NOI=64, Amp = 9.999999965277837×10^{-1}, (b) For NOI=512, Amp = 9.999999999777280×10^{-1}, (c) For NOI=1024, Amp = 9.999999999983369×10^{-1}, (d) For

NOI=8192, Amp = 9.999999999398433×10^{-1}. (Amplitude is being estimated over approximately the same bandwidth.)

7.23. Frequency resolution with 128 samples in the observation interval = 7.8125 kHz; with 8192 samples in the observation interval, frequency resolution = 0.1221 kHz.

7.24. For NOI=64: Amplitude estimate using Blackman window over bins 1 to 7 = 0.99999628. Amplitude estimate using rectangular window over bins 1 to 7 = 0.9778153735077986; An improvement of 1.0266 times. For NOI=8192: Amplitude estimate using Blackman window over bins 1 to 7 = 0.99999982579. Amplitude estimate using rectangular window over bins 1 to 7 = 0.995466430944; An improvement of 1.00455 times. **7.29.** RMS = 0.9487 **7.30.** RMS = 0.2372 **7.31.** (a) x = [1.9500 1.2399 -1.6500 -0.7399 2.7500 -0.7399 -2.0500 1.2399], (b) x = [1.5828 0.8204 1.8025 0.8447 0.4172 1.1796 0.1975 1.1553], (c) x = [0.9000 0.1638 0.7738 1.3356 -0.1744 -0.9000 -0.1638 -0.7738 -1.3356 0.1744]. **7.33.** Original time resolution =1.0 µs, effective time resolution = 0.1667 µs

7.34.

$$x[n] = 0.01 + 0.1069 \cos\left[\left(\frac{2\pi}{16}\right)n + 1.2566\right] + 5.6215 \times 10^{-4} \cos\left[5\left(\frac{2\pi}{16}\right)n + 3.6197\right]$$

7.35. RMS noise = 7.0711×10^{-6}

Chapter 8

8.1. (a) 6.019 dBV (b) -12.83 dBV (c) (differential) -20.0 dBV, (single-ended) -26.02 dBV (d) 0.4950 V. **8.2.** Amp(V)=0.7140 V; phase=8.0501 degrees; RMS=0.5049 V; Amp(dBV)= –5.9359 dBV. **8.3.** Amp(V)=0.7071 V; phase=–135 degrees; RMS=0.5 V; Amp(dBV)= –6.0206 dBV.

8.4. (a) $G = 0.99$, (b) $G = 2 + 0.2 V_{in} + 0.03 V_{in}^2$, (c) $G = a_1 + 2a_2 V_{in} + 3a_3 V_{in}^2 + \cdots + Na_N V_{in}^{N-1}$,

(d) $G = \dfrac{4}{\sqrt{1+V_{in}^2}}$

8.5. (a) When V_{in}=2.5 V, G_{error} = –0.3886 dB; When V_{in}=4.0 V, G_{error} = –0.6303 dB.
(b) G_{error} = –0.1586 dB, 0 dB, 0.1942 dB, 0.3700 dB, 0.6221 dB.

8.6. F_t = 9.4062 kHz; G = 2.1 V/V = 6.4444 dB. **8.7.** 0.484 kHz; 0.828 kHz; 0.844 kHz; 8.234 kHz; 8.5 kHz; 11.703 kHz; 15.797 kHz. **8.8.** RMS value of multi-tone signal = 0.1114 V RMS. **8.9.** Amplitude of each tone = 0.3651 V peak **8.10.** Absolute gains: 0.7060 V/V; 0.6700 V/V; 0.6280 V/V; 0.6620 V/V; 0.6980 V/V; 0.5140 V/V; 0.3160 V/V; 0.1620 V/V; relative gains: 0 dB; -0.4546 dB; -1.0169 dB; -0.5589 dB; -0.0990 dB; -2.7568 dB; -6.9824 dB; -12.7858 dB. **8.11.** At 19,698 Hz the gain is –1 dB.

8.12. (a) Input RMS (bin 0 to bin 8) = 0.7071; 0.7071; 0.7071; 0.7071; 0.7071; 0.7071; 0.7071; 0.7071; 0.7071; output RMS (bin 0 to bin 8) = 0.7071; 0.6894; 0.6676; 0.7018; 0.6675; 0.0045; 0.0219; 0.0161; 0. (b) Phase shift of input (bin 0 to bin 8) = 16.0697; 90.6131; 160.3493; 39.3309; 135.0567; 56.8661; -57.5130; -76.5905; 221.1078; phase shift of output (bin 0 to bin 8) = 16.0697; 120.1443; 220.9370; 146.4316; 29.6011; -86.3670; 3.5186; 0.1508; 0. (c) Absolute gain (bin 0 to bin 8) = 1 V/V; 0.9750 V/V; 0.9441 V/V; 0.9925 V/V; 0.9440 V/V; 0.0063 V/V; 0.0310 V/V; 0.0228 V/V; 0 V/V. (d) Relative gain (bin 0 to bin 8) = 1.0593 V/V; 1.0328 V/V; 1 V/V; 1.0514 V/V; 1 V/V; 0.0067 V/V; 0.0328 V/V; 0.0242 V/V; 0 V/V. (e) Phase difference (bin 0 to bin 8) = 0 degrees; 29.5312 degrees; 60.5877 degrees; 107.1007 degrees; -105.4556 degrees; -143.2332 degrees; 61.0316 degrees; 76.7413 degrees; 0. (f) Group delay (adjacent tones) = 0.0820 ms; 0.0863 ms; 0.1292 ms; -0.5904 ms; -0.1049 ms; 0.5674 ms; 0.0436 ms; Group delay distortion (adjacent tones) = 0.0935 ms; 0.0978 ms; 0.1407ms; -0.5789 ms; -0.0934 ms; 0.5789 ms; 0.0552 ms.

8.13. (a) S/2nd harmonic = 34.88 dB, (b) S/3rd harmonic = 54.8 dB dB, (c) S/THD = 34.84 dB, (d) S/THD+N = 12.91 dB.

8.14. $S/3^{rd} = 20 \log_{10}\left(\dfrac{4a_1}{a_3 A^2}\right)$

8.15. 2nd order intermodulation frequencies: 0.8 kHz, 1.9 kHz, 3.4 kHz, 4 kHz, 4.5 kHz, 6.6 kHz; 3rd order intermodulation frequencies: 0.5 kHz, 0.6 kHz, 2.7 kHz, 2.9 kHz, 4.7 kHz, 5.1 kHz, 5.5 kHz, 5.8 kHz, 7.7 kHz, 7.9 kHz, 9.3 kHz, 11.9 kHz.

8.16. Third-order intermodulation products ($2\omega_1+\omega_2$ and $\omega_1+2\omega_2$) are: $\dfrac{3}{4}a_3 A_1^2 A_2$ and $\dfrac{3}{4}a_3 A_1 A_2^2$.

8.17. (a) Noise = 51.3 µV RMS, (b) S/N = 82.78 dB, (c) S_n=1.147 µV /$\sqrt{\text{Hz}}$.

8.18. S_n= 8.5856 µV /$\sqrt{\text{Hz}}$. **8.19.** -46.1979 dBm **8.20.** 73rd spectral bin corresponds to 1140.6 Hz; G = 0.2783 dB. **8.22.** (a) S/2nd = 37.05 dB, (b) S/3rd = 54.06 dB, (c) S/THD = 36.96 dB, (d) Total noise = 0.5139 V RMS, (e) SNR = –0.1866 dB, (f) SNDR = –0.3219 dB. **8.23.** (a) S/IMD$_2$ = 40.00 dB, (b) S/IMD$_3$ = 53.45 dB.

Chapter 9

9.1. ADC: $F_{S\text{-}ADC}$ = 32 kHz, N_{ADC} = 2048; DAC: $F_{S\text{-}DAC}$ = 32 kHz, N_{DAC} = 512; AWG: $F_{S\text{-}AWG}$ = 64 kHz, N_{AWG} = 4096; digitizer: $F_{S\text{-}DIG}$ = 128 kHz, N_{DIG} = 2048. **9.2.** $F_{S\text{-}DAC}$ = 17,875 Hz and $F_{S\text{-}DIG}$ = 32,832 Hz. (Small value of n imposes a large change in the sampling frequencies.) **9.3.** $F_{S\text{-}ADC}$= 43,008 Hz and $F_{S\text{-}AWG}$ = 125,640 Hz. **9.4.** $F_{S\text{-}AWG}$ = 64 kHz, N_{ADC} = 256. **9.5.** A 1-V RMS, 55-kHz signal sampled at 24 kHz will have the following six lowest frequency components: 7 kHz, 17 kHz, 31 kHz, 41 kHz, 55 kHz and 65 kHz. The amplitudes are all equal at 1 V RMS. **9.6.** A 1-V RMS, 63-kHz signal sampled at 24 kHz will have the following six lowest frequency components: 9 kHz, 15 kHz, 33 kHz, 39 kHz, 57 kHz and 63 kHz. The amplitudes are all equal at 1 V RMS. **9.7.** A 1-V RMS, two-tone multitone signal consisting of frequencies 55 kHz and 63 kHz sampled at 24 kHz will not have any frequency components that overlap. (See the spectra provided in the solutions for Problems 9.5 and 9.6.) If the test frequencies are changed to 55 kHz and 65 kHz, both tones will create a frequency component at 7 kHz. **9.8.** Test tone amplitude = 749.8 mV; gain factor = 1.33 V/V. **9.9.** In-band tone amplitude = 0.221 V, 1st image = 0.158 V, 2nd image = 0.065 V, 3rd image = 0.058 V. **9.10.** In-band tone amplitude = 1.765 V, 1st image = 0.0260 V. **9.12.** DAC ideal gain = 1.466 mV/bit. **9.13.** 2's complement DAC: V_{MID} = 3.0159 V, V_{DAC} (D = 1001001) = 2.2857 V; sign/magnitude DAC: V_{MID} = 3.0 V, V_{DAC} (D = 1001001) = 2.7097 V. **9.15.** Intrinsic error for M=1 is 0.0866 LSB; intrinsic error for M=8 is 0.2426 LSB **9.16.** Actual output amplitude = 0.5730 V. **9.17.** With N=1024, M=191, A=0.5, P=0: $G_{intrinsic}$ = 1.2752 V/V; sample-and-hold effect, $G_{sin(x)/x}$ = 0.944 V/V. **9.18.** With N=512, M=127, A=0.5, P=0: $\Delta G_{intrinsic}$ = 0.2694 V/V. **9.19.** G_{DAC} = 13.51 mV/bit and ΔG_{DAC} = 1.459 V/V = 3.28 dB. **9.20.** G_{ADC} = 203.5 bits/V and ΔG_{ADC} = 0.9946 V/V = –0.0469 dB. **9.21.** G_{filter}(79 kHz) / G_{filter}(15 kHz) = -53.5 dB **9.22.** The spurious tone will alias down to 3 kHz. **9.23.** Spectral bin for the 15th harmonic = 367, frequency of 15th harmonic= 5.7344 kHz; spectral bin for the 23rd harmonic = 393, frequency of 23rd harmonic= 6.1406 kHz. **9.24.** With M_1=191 and M_2=205, 1st–order intermodulation distortion: $2M_1$=130 (2031.25 Hz), $3M_1$=61 (953.1250 Hz), $2M_2$=102 (1593.750 Hz), $3M_2$=103 (1609.3750 Hz), 2nd– order intermodulation distortion: M_1+M_2 = 116 (1812.50 Hz), M_1-M_2 = 14 (218.750 Hz); 3rd– order intermodulation distortion: $2M_1+M_2$ = 75 (1171.8750 Hz), $2M_1-M_2$ = 177 (2765.6250 Hz), M_1+2M_2 = 89 (1390.6250 Hz), $2M_2-M_1$ = 219 (3421.8750 Hz). No spectral component overlap. **9.25.** G_{ADC} = 52.33 bits/V; PSRR = 2.150 V/V = 6.65 dB. **9.26.** V_{noise} = 2.30057 mV RMS; SNR = 49.75 dB; ENOB = 7.97 bits.

Chapter 10

10.1. $G_{COMP} = -1.0682$; $\text{offset}_{COMP} = -0.0058$; $V_{CALIBRATED} = -(V_{DIG}+0.0058)/1.0682$.
10.2. Three stages: $G_{COMP} = G_1 \times G_2 \times G_3$; $\text{offset}_{COMP} = O_1 \times G_2 \times G_3 + O_2 \times G_3 + O_3$; four stages: $G_{COMP} = G_1 \times G_2 \times G_3 \times G_4$; $\text{offset}_{COMP} = O_1 \times G_2 \times G_3 \times G_4 + O_2 \times G_3 \times G_4 + O_3 \times G_4 + \times O_4$; N stages: $G_{COMP} = G_1 \times G_2 \times G_3 \times G_4 \times \cdots \times G_N$; $\text{offset}_{COMP} = O_1 \times G_2 \times G_3 \times G_4 \times \cdots \times G_N + O_2 \times G_3 \times G_4 \times \cdots \times G_N + O_3 \times G_4 \times \cdots \times G_N + \cdots + \times O_N$. **10.3.** $G_{DIG} = 1.0960$; $G_{AWG}(2\text{kHz}) = 0.8868$; $G_{AWG}(3\text{kHz}) = 0.9305$; $G_{AWG}(4\text{kHz}) = 0.9580$; $G_{FILTER}(2\text{kHz}) = 0.9691$; $G_{FILTER}(3\text{kHz}) = 0.9903$; $G_{FILTER}(4\text{kHz}) = 0.940$
10.4. $V_{AWG} = 2.2321$ V; $V_{DUT} = 1.5597$ V **10.5.** $G_{DUT}(1\text{kHz}) = 52.5$ bits/V; $G_{DUT}(2\text{kHz}) = 52.4482$ bits/V; $G_{DUT}(3\text{kHz}) = 51.9775$ bits/V. **10.6.** $G_{DUT}(1\text{kHz}) = 0.9027$ V/V; $G_{DUT}(2\text{kHz}) = 1$ V/V; $G_{DUT}(3\text{kHz}) = 1.0942$ V/V **10.7.** $\phi_{DUT} = -70$ degrees **10.8.** $\phi_{DUT} = -66$ degrees
10.9. SNR = 18.9 dB; SNR(calibrated) = 302.8 dB; SNR(with noise) = 62.9 dB

Chapter 11

11.2. $V_{FSR} = 5.5050$ V - 0.0465 V = 5.4585 V **11.3.** All the possible output levels: (in volts):
0.000, 0.067, 0.133, 0.200, 0.267, 0.333, 0.400, 0.467,
0.533, 0.600, 0.667, 0.733, 0.800, 0.867, 0.933, 1.000
The input code 0 corresponds to an output level of 0 V
11.4. All the possible output levels: (in volts)
2.000, 2.032, 2.065, 2.097, 2.129, 2.161, 2.194, 2.226
2.258, 2.290, 2.323, 2.355, 2.387, 2.419, 2.452, 2.484
2.516, 2.548, 2.581, 2.613, 2.645, 2.677, 2.710, 2.742
2.774, 2.806, 2.839, 2.871, 2.903, 2.935, 2.968, 3.000
The input code 0 corresponds to an output level of 2.516V
11.5. (a) A best-fit line through the actual points provides: Gain = 0.3654 V/V; Gain Error = –8.6418%; Offset= –0.0017 V; Offset Error= –0.0017 V.
(b) LSB Step size using endpoints = 0.3639 V.
(c) absolute error transfer curve:

Ideal Output (Volts)	Actual Output, S (Volts)	Output Change, S' (Volts)	Output Change, S' (LSB)
0.0	0.0465	0.0465	0.1162
0.4	0.3255	-0.0745	-0.1863
0.8	0.7166	-0.0834	-0.2085
1.2	1.0422	-0.1578	-0.3945
1.6	1.5298	-0.0702	-0.1755
2.0	1.8236	-0.1764	-0.4410
2.4	2.1693	-0.2307	-0.5768
2.8	2.5637	-0.2363	-0.5908
3.2	2.8727	-0.3273	-0.8183
3.6	3.3443	-0.2557	-0.6393
4.0	3.6416	-0.3584	-0.8960
4.4	4.0480	-0.3520	-0.8800
4.8	4.3929	-0.4071	-1.0178

5.2	4.7059	-0.4941	-1.2353
5.6	5.0968	-0.5032	-1.2580
6.0	5.505	-0.4950	-1.2375

(d) Yes, the DAC output is monotonic.
(e) DNL curve using best-fit method (beginning with first transition):
-0.2365, 0.0702, -0.1090, 0.3343, -0.1960, -0.0540, 0.0793, -0.1544, 0.2905, -0.1864, 0.1121, -0.0562, -0.1435, 0.0697, 0.1170.
This DAC passes the ±1/2 LSB specification for DNL.
(f) DNL curve using the endpoint method (beginning with first transition):
-0.2333, 0.0747, -0.1052, 0.3399, -0.1926, -0.0500, 0.0838, -0.1509, 0.2960, -0.1830, 0.1168, -0.0522, -0.1399, 0.0742, 0.1217.
This DAC passes the ±1/2 LSB specification for DNL.
(g) INL curve using the best-fit method (beginning with first code):
0.1320, -0.1046, -0.0343, -0.1433, 0.1910, -0.0050, -0.0590, 0.0202, -0.1342, 0.1563, -0.0301, 0.0820, 0.0258, -0.1177, -0.0480, 0.0690.

11.6. (a) A best-fit line through the actual points provides: Gain = 0.1193 V/V; Gain Error = –10.4933%; Offset = –0.9615 V; Offset Error = –0.0068 V
(b) LSB Step size using endpoints = 0.1189 V.
(c) absolute error transfer curve:

Ideal Output (Volts)	Actual Output, S (Volts)	Output Change, S' (Volts)	Output Change, S' (LSB)
-1.0667	-0.9738	0.0929	0.6965
-0.9333	-0.8806	0.0527	0.3955
-0.8000	-0.6878	0.1122	0.8415
-0.6667	-0.6515	0.0152	0.1137
-0.5333	-0.3942	0.1391	1.0435
-0.4000	-0.3914	0.0086	0.0645
-0.2667	-0.2497	0.0170	0.1272
-0.1333	-0.1208	0.0125	0.0940
0.0000	-0.0576	-0.0576	-0.4320
0.1333	0.1512	0.0179	0.1340
0.2667	0.2290	-0.0377	-0.2825
0.4000	0.4460	0.0460	0.3450
0.5333	0.4335	-0.0998	-0.7488
0.6667	0.5999	-0.0668	-0.5007
0.8000	0.6743	-0.1257	-0.9428
0.9333	0.8102	-0.1231	-0.9235

(d) No, the DAC output is not monotonic as the transition from 0.4460 to 0.4335 is negative.
(e) DNL curve using best-fit line (beginning with first transition):
-0.2191, 0.6155, -0.6958, 1.156, -0.9765, 0.1873, 0.0801, -0.4704, 0.7496, -0.3481, 0.8183, -1.1047, 0.3943, -0.3766, 0.1387,
This DAC does not pass the ±1/2 LSB specification for DNL.
(f) DNL curve using the endpoint method (beginning with first transition):
-0.2164, 0.6211, -0.6948, 1.1634, -0.9765, 0.1914, 0.0838, -0.4686, 0.7556, -0.3459, 0.8246, -1.1051, 0.3991, -0.3744, 0.1427.
This DAC does not pass the ±1/2 LSB specification for DNL.
(g) INL curve using the best-fit method (beginning with first code):
-0.1029, -0.3219, 0.2936, -0.4022, 0.7537, -0.2228, -0.0354, 0.0446, -0.4258, 0.3238, -0.0243, 0.7940, -0.3107, 0.0836, -0.2930, -0.1543.

11.7. As the best-fit linearity method is used, Integration constant, C = Offset Error/LSB= (-0.4919+0.5045)/0.0631=0.1997 INL Values:
0.1997, 0.1182, -0.0174, -0.1307, -0.125, -0.1032, 0.0276, -0.0085, -0.1035, 0.0101, -0.1532, 0.0569, 0.1081, 0.12, 0.0494, -0.0425
This DAC passes the 1/2 LSB specification for INL

11.8. DNL Values:
0.0814, -0.2957, 0.0467, 0.0057, 0.0217, 0.1308, -0.0361, -0.095, 0.1135, -0.1633, 0.2102, 0.0512, 0.0119, -0.0706, -0.0061
This DAC passes the 1/2 LSB specification for DNL.

11.10. DAC output = $D_0W_0 + D_1W_1 + D_2W_2 + D_3W_3 + D_4W_4 + 100$ mV where D_4, D_3, D_2, D_1, D_0 varies from 00000 to 11111 and W_0=0.1939, W_1=0.3272, W_2=0.6519, W_3=1.3046, W_4=2.6121. Reconstructed output, beginning with code 00000 is therefore:
0.1, 0.2939, 0.4272, 0.6211, 0.7519, 0.9458, 1.0791, 1.273, 1.4046, 1.5985, 1.7318, 1.9257, 2.0565, 2.2504, 2.3837, 2.5776, 2.7121, 2.906, 3.0393, 3.2332, 3.364, 3.5579, 3.6912, 3.8851, 4.0167, 4.2106, 4.3439, 4.5378, 4.6686, 4.8625, 4.9958, 5.1897

11.11. DAC output = $D_0W_0 + D_1W_1 + D_2W_2 + \overline{D}_3W_3 + 500$ mV where D_3, D_2, D_1, D_0 varies from 1000 to 0111 and W_0=0.1049, W_1=0.2082, W_2=0.4129, W_3=0.8276. Reconstructed output, beginning with code 1000 is therefore:
0.5, 0.6049, 0.7082, 0.8131, 0.9129, 1.0178, 1.1211, 1.226, 1.3276, 1.4325, 1.5358, 1.6407, 1.7405, 1.8454, 1.9487, 2.0536

11.13. An examination on the two sets of output levels reveals that they both share the same binary weights and DC base: W_0=0.0552, W_1=0.0655, W_2=0.0655, W_3=0.0615 and DC base=0.0064. However, only the second set of output levels coincide with the reconstructed DC voltage ramp. That means only the second DAC can be tested by the major carrier test technique as the first DAC has a superposition error. **11.14.** No (DAC output voltage at t=10ns > 1.05V); Overshoot=45%; Rise time = 1.5ns; Glitch Energy = 0.5 × 3 ns × 0.45 V + 0.5 × 3.5 ns × -0.2 V + 0.5 × 3.5 ns × 0.1 V = 0.5 ns-V (triangle approximation). **11.15.** 1% Settling Time=24 ns; Rise-time=2.5 ns. **11.16.** High Speed DACs: Resistive Divider DACs, Binary-Weighted DACs. Low Speed DACs: PWM DACs, Sigma Delta DACs.

Chapter 12

12.1. $P(V < 40$ mV$)$=0.7881; $P(V > 10$ mV$)$=0.4207; $P(-10$ mV $< V < 40$ mV$)$=0.3674
12.2. $P(V < 40$ mV$)$=0.7257; $P(V > 10$ mV$)$=0.5; $P(-10$ mV $< V < 40$ mV$)$=0.3811.
12.3. 41.1 mV. **12.4.** P(code=325)=0.2496; 300 out of 400 codes are expected to be produced for code 324 leaving the remaining 100 for code 325. **12.5.** 21.4 mV **12.6.** 79 out of 500 codes are expected to be code 115, 269 codes are expected to be code 116, and 152 codes are expected to be code 117. **12.7.** Average LSB size = 0.2067 V. Code width (V) beginning with

code 0: undefined, 0.2126, 0.1417, 0.2126, 0.2480, 0.2480, 0.1771, undefined. Code edge location (V): 0.0100, 0.2226, 0.3643, 0.5769, 0.8249, 1.0729, 1.2500.
12.8. Average LSB size = 0.3204 V. Code width (V) beginning with code -8: 0.3381, 0.2930, 0.2705, 0.2254, 0.2705, 0.2705, 0.3155, 0.3155, 0.2930, 0.3381, 0.4282, 0.3606, 0.3155, 0.4508, undefined. Code edge location (V): 0.0750, 0.4131 , 0.7061, 0.9765, 1.2019, 1.4723, 1.7428, 2.0583, 2.3738, 2.6668, 3.0049, 3.4331, 3.7937, 4.1092, 4.5600.
12.9. Average LSB size = 0.5155 V. Accuracy = 96.7 mV
12.10. Average LSB size = 0.1222 V. Code width (V) beginning with code 0: undefined, 0.0888, 0.1028, 0.1158, 0.1251, 0.1246, 0.1355, 0.1300, 0.1296, 0.1342, 0.1293, 0.1218, 0.1403, 0.1266, 0.1251, undefined. Code edge location (V): 0.0140, 0.1028, 0.2056, 0.3215, 0.4466, 0.5713, 0.7068, 0.8368, 0.9665, 1.1007, 1.2300, 1.3517, 1.4921, 1.6187, 1.7438.
12.11. Endpoint DNL (LSBs): 0.0606, -0.0101, -0.1515, -0.2222, -0.1515, -0.1515, -0.0101, -0.0101, -0.0808, 0.0606, 0.1313, 0.1313, -0.0101, 0.4141. Endpoint INL (LSBs): 0.0, 0.0606, 0.0505, -0.1010, -0.3232, -0.4747, -0.6263, -0.6364, -0.6465, -0.7273, -0.6667, -0.5354, -0.4040, -0.4141, 0.0.
12.12. Endpoint DNL (LSBs): 0.0553, -0.0854, -0.1558, -0.2965, -0.1558, -0.1558, -0.0151, -0.0151, -0.0854, 0.0553, 0.3367, 0.1256, -0.0151, 0.4070. Endpoint INL (LSBs): 0.0, 0.0553, -0.0302, -0.1859, -0.4824, -0.6382, -0.7940, -0.8090, -0.8241, -0.9095, -0.8543, -0.5176, -0.3920, -0.4070, 0.0.
12.13. Endpoint DNL (LSBs): 0.1250, -0.0625, 0.1250, -0.2500, 0.1250, -0.0625.
Endpoint INL (LSBs): 0, 0.1250, 0.0625, 0.1875, -0.0625, 0.0625, 0.
12.14. High speed: flash ADC. Medium speed: Successive approximation ADC, semi-flash ADC. Low speed: integrating (dual- or single-slope) ADC, PDM (sigma-delta) ADC.

Chapter 13

13.1. 195 mΩ, HF:635 mA, LF: 635 mA, HS 0 mA, LS 0 mA, V_{drop} = 124 mV per line, V_{OUT} = 3.3V + 2 × 124 mV = 3.548 V. **13.2.** 201 nH/m, 46 nH total, 0.01%; 150 nH/m, 35 nH total, 0.006%. **13.3.** 83.7 pF/m, 2.5 pF total; 126 pF/m, 3.8 pF total.

$$|H(f)| = \left[\frac{1}{\sqrt{1+\left(\frac{f}{F_C}\right)^2}}\right] \text{ V/V} \quad \text{and} \quad \angle H(f) = -\frac{180}{\pi}\tan^{-1}\left(\frac{f}{F_C}\right) \text{ degrees}$$

$$\text{where} \quad F_C = \frac{1}{2\pi RC} \text{ Hz} = \frac{1}{2\pi \times 50 \times 10^3 \times 3.8 \times 10^{-12}}$$

13.4. 45 pF/m, 9.2 pF total; 50 pF/m, 10.3 pF total. **13.5.** 50 pF/m, 14.7 pF + 35 pF = 49.7 pF total, gain = 0.141 = −17.02 dB, buffer amplifier is needed. Using refined estimate: 110 pF/m, 32.4 pF + 35 pF = 67.4 pF total, gain = 0.104 = −19.63 dB, buffer amplifier is needed. **13.6.** 35 pF/m, 400 nH/m, Z_o = 58 Ω; To lower the characteristic impedance, we would make the spacing smaller or widen the trace. Changing the length has no effect on characteristic impedance in an ideal transmission line. **13.7.** v_{signal} =1.44 × 10^8 m/s = 0.481 c, T_d = 2.1 ns, 36.9 pF. **13.8.** L = 7.22 m, or 281 in., much longer than the stipline length. We can use a lumped element model for this line. **13.9.** 100 % (cable length = one wavelegth), 16.7 pF, 360 degrees. **13.10.** F_c = 12.7 MHz. We should not need a buffer amplifier unless the output becomes unstable when loaded with 125 pF. **13.11.** ±0.38 degrees; We can reduce this error using a pair of buffer amplifiers placed near the DUT to isolate the DUT outputs from the tester capacitance. Of course we need to calibrate the phase mismatch between the DIB buffer amplifiers.

13.12.
$$v(t) = 5.0 \text{ V} - \frac{100 \text{ mV}}{2} \times \frac{4.7 \text{ k}\Omega}{1 \text{ k}\Omega} \sin(2\pi ft) = 5.0 \text{ V} - 235 \text{ mV} \times \sin(2\pi \times 2400\ t)$$

13.13.
$$v(t) = -235 \text{ mV} \times \sin(2\pi \times 2400\ t)$$

Chapter 14

14.1. Lower cost of test, increased fault coverage / improved process control, diagnostics and characterization, ease of test program development, and system-level diagnostics. **14.2.** Robust circuits can tolerate more measurement error without failing test limits. Since measurement accuracy generally comes at the expense of longer test time (averaging, for example), robust circuits with wide design margins can be tested more economically. **14.3.** Delayed time to market results in lower unit prices due to more competition. Lower unit prices lead to smaller profit margins. **14.4.** The IEEE Std. 1149.1 test interface is primarily designed for board-level, chip-to-chip interconnect testing. The TAP controller provides a standard, consistent interface to the scan circuits of the 1149.1 boundary scan circuits. Hold TMS at logic 1 while clocking TCK. No more than five TCK clock cycles are required. **14.5.** Test time will increase due to the tester's match mode search process. Test development time will likely increase due to match mode code development. **14.6.** Only four lines are required: TMS, TCK, TDI, and TDO **14.7.** 325 flip flops (all flip flops are in the scan chain). 10 MHz / (325+1). (we need one clock cycle to capture circuit response between parallel vectors). Multiple scan chains allow parallel testing, which will reduce the time require to scan data in and out. **14.8.** IDDQ testing. **14.9.** Using Eq. 4.12, t_s = ln(1 mV/1.5 V) × 2 kΩ × 1µF = 15 ms. After precharging, R increases to 50 kΩ (two 100 kΩ resistors in parallel). To recover from the worst-case op-amp offset of 10 mV takes an additional settling time of t_s = ln(1 mV / 10 mV) × 50 kΩ × 1µF = 115 ms for a total settling time of 130 ms. (Notice that the op-amp's offset causes most of the settling time). Without DfT, the circuit would normally take t_s = ln(1 mV / 1.5 V) × 50 kΩ × 1µF = 366 ms. **14.13.** Force 1 V at the AB1 line and connect DUT 1's signal pin to the AB1 line. Connect DUT2's signal line to the internal ground connection through the ABM ground switch. Measure the voltages at the two signal pins using AB2 and subtract to get V_{drop}. Measure the current, I, supplied into AB1. Use $V_{drop} = IR$ to calculate the value of R. This method would not catch shorts to ground, but repeating the process from DUT 2 to DUT 1 will catch shorts. **14.14.** A reference multitone at a known amplitude at 1, 5, 9, and 13 kHz is applied to the input of the ADC. The DSP measures the ADC signal level and calculates the ADC gain. Then the test mux is switched to connect the DAC output to the ADC. The DAC is set to produce a multitone at the same frequencies. Its output is measured using the ADC. The ADC gain error at each frequency is removed through an on-chip calibration routine. This gives the DAC gain at each frequency. **14.15.** Place the first switch network into the "Force input" mode and apply a reference voltage to the DAC using the TESTIN analog bus. **14.16.** No, the circuit has a very long divide time. Also, it can only be clocked through a coupling capacitor. To improve the design, provide a bypass mode around the divider to allow observation of the oscillator output frequency in it's normal mode of operation. Also, provide a bypass path to clock the digital divider directly so it can be driven without the timing shift produced by the capacitor. Finally, use partitioning to split the divider into subcircuits that will count faster, or better yet, use a full-scan methodology for the divider to allow very fast testing.

Chapter 15

15.1. $(\mu, \sigma, \sigma^2) = (1.5913$ mm, 0.0528 mm, 0.0028 mm$^2)$

15.2. (a) $(\mu, \sigma, \sigma^2) = (6.9333, 3.5363, 12.5056)$, (b) scale data by 100 gives $(\mu, \sigma, \sigma^2) = (693.3333, 353.6325, 1.2506e+05)$

15.3. Working with decibels: $(\mu, \sigma, \sigma^2) = (90.6872, 2.8122, 7.9084)$; Yield = 58.97 %; Working with V/V: $(\mu, \sigma, \sigma^2) = (3.6172e+04, 1.3618e+04, 1.8545e+08) = (91.1675$ dB, 82.6824 dB, 38.3483 dB). Clearly, the statistics do not agree, only the mean value seems to converge to similar values. Since the estimate of standard deviation, (est, equals 2.8167, the distribution is near-gaussian.

15.6. (a) $c=3$, (b) $P(1 < X < 2) = 0.0473$, (c) $P(X < 3) = 0.9999$, (d) $P(X < 1) = 0.9502$, (e) $F(x) = 1 - e^{-3x}$

15.8.

Problem	z	Probability Eqn. (15.16)	Probability (Table 15.1)	Error (%)
(a)	-3	0.0013	0.0013	3.5098
(b)	-1.9	0.0286	0.0287	-0.4435
(c)	-0.56	0.2844	0.288	-1.2692
(d)	-0.24	0.4016	0.4053	-0.907
(e)	0	0.5	0.5	0
(f)	-0.09	0.4621	0.4642	-0.4403
(g)	0.17	0.5705	0.5675	0.5409
(h)	3	0.9987	0.9987	-0.004
(i)	5	1	1	0
(j)	-5	0	0	100

15.10. (a) $P(0 < X < 30$ mV$) = 0.4987$, (b) $P(-30$ mV $< X < 30$ mV$) = 0$, (c) $P(-1.5$ V $< X < 1.4$ V$) = 0.8535$, (d) $P(-300$ mV $< X < -100$ mV$) = 0.8430$, (e) $P(X < 250$ mV$) = 0.9337$, (f) $P(-200$ mV $> X) = 0.2001$, (g) $P(|X| < 30$ mV$) = 0.9973$, (h) $P(|X| > 30$ mV$) = 0.0027$.

15.11. (a) $z = 1.5900$, (b) $z = 0.7900$, (c) $z = -0.7900$, (d) $z = 0.7900$, (e) $z = -0.7900$.

15.12. (a) $P(X < x) = 0.7881$ when $x = -0.9210$, (b) $P(X < x) = 0.2119$ when $x = -1.0790$, (c) $P(X > x) = 0.2119$ when $x = -0.9210$, (d) $P(|X| < x) = 0.3830$ when $x = -1.0490$.

15.13. $(\mu, \sigma) = (37.2$ mV, 25.8 mV$)$ **15.14.** $P(23 < X < 33) = 0.1$ and $(\mu, \sigma) = (50, 28.87)$

15.15. $A = 0.76923$ mV and $B = 77.7$ mV.

15.16. $P(X > 0) = 0.0154$; $P(X < -200$ mV$) = 0.0386$.

15.17. $P(G < 9.8) = 0.0565$; $P(10.0 < G < 10.5) = 0.5699$

15.18. $P(N < 70$ (V)$) = 0.0475$; $P(N > 140$ (V)$) = 0.0475$; $P(70$ (V) $< N < 140$ (V)$) = 0.9050$

15.19. 3σ guardbanded limits: $LTL = -17$ mV, $UTL = +17$ mV. 6-sigma guardbanded limits: $LTL = +16$ mV, $UTL = -16$ mV. Number of measurements to average$=11$.

15.20. P(Failure)$=0.0976$. **15.21.** P(Failure)$=0.6242$. **15.22.** $\sigma_{tester}=14.4$ mV; $\sigma_{total}=17.5$ mV.

15.23. $\sigma_{process}=12.0$ mV; Yield loss$=0$ (no loss). **15.24.** Test yield $= 76.28\%$. **15.25.** Process capability $= 90$ µV; $C_p=0.5556$, $C_{pk}=0.4444$. Since $C_p < 2$ and $C_{pk} < 1.5$, this lot does not meet six sigma quality standards.

Chapter 16

16.1. 575 passing DUTs / h **16.2.** 3.648¢/s **16.3.** Following the development in Example 16.1 (using a single-head, single-site cost model) The total cost to test a DUT is 3.648¢/s × 3.8 s = 13.863¢. We can increase the tester's depreciation cost to 3.307¢/s if we can drop the test time in half using dual-site testing. Since depreciation is directly proportional to purchase price, this represents a 271.5% increase in tester purchase price. **16.4.** Using an index time of 0 (hidden by dual-head testing), the test cost drops to 12.395¢, compared to 13.863¢, more than 10% lower. Note that if two single-head handlers are required, the economics do not make sense in this example. **16.5.** Test costs drop to 6.612¢ per DUT in this case; a 52% reduction in test costs.
16.6.

$$N_{CATEGORIES} = 2\sum_{i=1}^{N-1} i = N(N-1)$$

16.10. (a) As mixed-signal devices become more complex, the additional block-to-block failure mechanisms drives up the testing costs. (b) Mixed-signal testing costs tend to increase faster than the cost of the added silicon. Thus test costs continue to represent a larger percentage of total manufacturing costs. **16.11.** Using the cost model of Eq. (16.3), we get a cost savings of 0.14¢. However, the cost of the added BIST circuit in silicon area is equal to 0.28¢, so this BIST circuit does not pay for itself in lowered testing costs. However, this cost is not unreasonable. If the customer gains diagnosability in the field, or if more device defects can be detected, then this BIST scheme is probably worthwhile. **16.12.** In this case, the BIST circuit saves 1.31¢ and only costs 0.28¢ in silicon area. The net gain is about 1¢. In a device that ships 10 million units per year, this BIST circuit would add $100,000 of profit per year!

Index

A

A/D (*see* analog-to-digital converter)
A580 tester, 114, 115
ABM (*see* analog boundary module)
absolute error,
 definition 410-412
 in ADCs 479
 in DACs 403, 405, 435
 MATLAB example 438
absolute gain,
 definition, 256
 in analog channels, 256-262
 in sampled channels, 351-355
 MATLAB example 438
absolute level,
 definition, 251
 in analog channels, 251
 in sampled channels, 351
absolute linearity, 413
absolute maximum ratings, 24, 31
absolute phase shift, 360, 396
accuracy
 calibrations, 41, 93-103, 369
 definition, 87-92
 tester accuracy, 103-106
accuracy standards, 369-371, 658
ADC (*see* analog-to-digital converter)
ADC testing
 aperture jitter, 472, 479, 480
 conversion time, 470
 linear ramp histogram, 456
 missing codes, 470
 servo method, 455-457
 sparkling, 472
AGC (*see* automatic gain control)
Agilent Technologies, 11
AGND (*see* analog ground)
A-law, 339, 343, 344, 346
alias tones, 265, 320, 333
aliasing
 definition, 160
 use in undersampling, 333
all-codes testing, 420, 431, 435
almost full scan, 561
ammeters, 83, 136
amplifier
 differential, 66, 75, 76, 84
 differential to single-ended, 125, 249, 256, 534

instrumentation, 125, 249, 256, 534
nulling, 68-70, 74, 75, 542
programmable gain, 2-4, 249
 in ATE instruments, 125, 140
 in AGC circuits, 578, 579
 testing PGAs, 260-264
sample-and-hold, 315, 320
 in SAR ADCs, 473
 in ATE instruments, 126, 140
 sin(x)/x rolloff, 336, 338
 testing S/H amps, 352, 356
 single-ended to differential, 62, 63, 249, 535
analog boundary module, 570
analog channels, 249
analog ground, 60, 517, 518, 540
analog loopback, 319, 579-581
analog multiplexer, 2, 5
analog switch, 1, 2, 570, 576
analog test bus, 570, 575, 576, 585
analog-to-digital converters
 ADC applications, 1-3, 479, 480
 architectures, 473-478
 histogram testing, 456-467
 INL and DNL, 468
 missing codes, 470
 monotonicity, 469
 statistical behavior, 448-454
 testing 447-480
antenna
 in cellular telephones, 3, 316
 parasitic antennae, 50, 130, 655
antialiasing, 160, 265
 (*see also* filters, antialiasing)
arbitrary waveform generator,
 architecture, 139-141
 calibrating, 97-100, 382-394
 synchronization, 141, 142, 320
array processing, 144, 190
array processors, 133, 144, 322
ATE (*see* automated test equipment)
ATLAS test language, 655
audio reconstruction, 436
automated test equipment
 definition, 1, 11-13
 instrumentation, 123-145
 vendors, 11
automatic gain control, 4, 577-580
averaging
 improving repeatability, 111, 293, 619, 620

N-point running averager, 111
testing costs of averaging, 642
AWG (*see* arbitrary waveform generator)

B

ball grid array package, 13
base stations, 3-5, 315
base-band interface, 3-5, 123, 316
bell curve, 117, 607
bench equipment
 correlation to ATE 18, 91
 calibration and accuracy, 369
 importance to debugging, 652
 windowing, 241
best-fit linearity 407-409, 413
best-straight-line linearity, 413
BGA (*see* ball grid array package)
BILBO (*see* built-in logic block observability)
bimodal distributions, 118, 615
binary searches, 79, 455, 473
binary-weighted DACs, 430
binning, 38, 40
BIST (*see* built-in self-test)
bond pad, 10, 11, 600
bond wire, 7, 10, 526
boundary scan, 555-559,
 mixed-signal , 569
buffer amplifier (voltage follower),
 in DUTs 249, 429
 on DIBs 4, 13, 252, 533-535
 calibration, 41, 97, 100-102, 370, 376, 379, 380
built-in logic block observability, 562, 563, 590, 592
built-in self-test,
 definition, 550
 digital, 558, 562-565
 economics, 658
 mixed-signal, 571-573
butterfly network, 218
bypass capacitors, 52, 53

C

calibration
 AC amplitudes, 382
 avoiding calibration, 397
 AWGs and digitizers, 382
 cal factors (*see* calibration factors)

calibration equation, 121
calibration paths, 95, 101, 538
cancellation of errors, 374
distortion, 396
focused, 41, 96-99, 369-376
gain, 378
gain and phase matching, 397
hardware, 93, 369, 370
low-level AC signals, 389
noise, 397
offset, 378
periodic system calibration, 96
phase shifts, 373, 392, 396
premeasurement of input signals, 375
reference source, 93, 96, 369
residual errors, 371, 384, 390
software, 41, 93, 97, 370
source and meas. paths, 373
system 369-371, 374-376
calibration factors
definition, 95, 347-381
composite, 374
calibration interval, 375
calibration laboratory, 93, 369
calibration source, 96, 128
capacitance
distributed, 507, 514, 545
parasitic, 6, 48, 61, 101, 496-499, 502-504, 514, 521, 526, 542, 576
capacitance per unit length, 497
capacitors
dielectric materials, 526
matching, 252
package types, 526
capture memory, 132, 322-328
carrier tray, 11
Catalyst tester, 11, 12
cdf (*see* cumulative distribution function)
cellular telephones, 3-5, 315-318
center code testing, 454
central limit theorem, 117, 294, 607
characteristic impedance, 506-512
characteristic pulse shape, 153-158
characterization, 11, 31, 42, 599
checker programs, 41, 96
chemical vapor deposition, 5
chipset, 5
chords, 344-346
circuits
analog (linear), 1-4, 249, 550, 587
digital, 1-4, 550, 587
load circuits, 13
mixed-signal, 1-3, 315
protection, 46, 82
clamp voltages, 48
clipping, 173, 252

clock
master clock, 182-184, 321-323
reference clock, 24, 27, 183, 184
clock and data feedthrough, 293
clocking and synchronization, 123, 141, 181-184, 320, 326, 658
CMOS fabrication, 5, 6
CMRR (*see* common-mode rejection ratio)
coaxial cable, 13, 136, 502, 506, 509-512
code edges, 344-346, 363, 453-455, 459-461
codecs, 320, 326, 344
coherence, 171, 172
common-mode rejection ratio
AC CMRR, 285-290, 362
DC CMRR, 72-76
companded encoding format,
companding, 343, 344
comparators, 1, 77-80, 135, 473-477, 577-579, 582
compare data, 39, 137
component footprints, 487
component shifts, 91
concurrent engineering, 16, 17, 27
conductivity, 489, 490
configuration board, 485
contact pads, 520, 521
contact testing (*see* continuity)
contactor assembly, 13, 45, 483
continuity, 38, 45-51, 651, 654
continuous time, 148, 149
continuous wave source, 139
control charts, 597, 633, 634
conversion time, 425, 470, 475
convolution, 153, 156, 236, 453
Cooley, J. W., 216
Cooley-Tukey fast Fourier transform, 217
coplanar shielding, 503
correlation, 18, 91, 624
cosine/sine pairs, 218
cost
handler, 643-645
tester, 239, 321, 643-648
testing, 9, 19, 549
cost models, 643, 644
C_p, and C_{pk}, 597, 628-631
CRC (*see* cyclical reduduncy check)
crosstalk, 179, 289-293, 361, 498-500, 502-504, 570, 575-577, 654
definition, 289
cumulative distribution function, 449-452
cumulative yield, 599
current meters, 83, 136
current sources, 47-50, 57-59

cutoff frequency (3-dB), 106, 267, 494, 495, 498, 543, 544
cyclical redundancy check, 562
cynate ester PCBs, 489, 490

D

D/A (*see* digital-to-analog converter)
DAC (*see* digital-to-analog converter)
data analysis, 144, 597
data compression, 4, 132, 136, 343
data modulation, 316, 436
data sheets, 23-35
ambiguities, 23, 30, 52, 468
data transceivers, 123
datalogs, 114-116, 597-599
DC references, 55, 435
DC sources, 127, 128
debugging, 17, 42, 649-657
debugging skills, 641, 649-655
decibels
0 dB reference level, 612
conversion from V/V, 65
dBm, 254
dBm0, dBrn, dBrn0, dBrnC0, 296
dBV, 253
decimation-in-time and -freq., 218
decision levels (*see* code edges)
defect-oriented testing, 572, 658-660
depreciation, 643-645
design engineering, 16, 17
design for test
advantages, 551-555
definition, 549
diagnostic capabilities, 553
design margin, 120, 273, 586, 658
design synthesis, 561
deskewing, 513
device description, 24, 27, 37
device ground sense, 128, 516, 517
device interface boards, 12, 45, 483, 519, 530
device under test, 9
device zero, 128
DfT (*see* design for test)
DFT (*see* discrete Fourier transform)
DGND (*see* digital ground)
DGS (*see* device ground sense)
diagnostics,
DIB (*see* device interface board)
DIB circuits, 530-540
dielectric leakage, 527
differential gain and phase, 399, 400, 437
differential impedance, 59

differential nonlinearity
  ADCs, 468, 469
  DACs, 412-420
  definition, 412
differential voltage, 67, 74, 75
digital audio channels, 318
digital drivers,
  pin card drivers, 12, 135, 514
  output testing, 82, 570
digital ground, 517
digital loopback, 320, 579-581
digital patterns, 30, 39, 131-139
digital pattern generators, 137, 183, 567
digital pattern loop, 132-134, 179-181, 328
digital signal processing, 190-241
  definition, 190
  example operations, 191
digital signal processors, 2, 144, 315-317
digital signals, 131, 190
digital subsystem, 124, 131-139
digital vectors, 131
digital waveforms, 136-138
digital-to-analog converters
  applications, 435-437
  DAC architectures, 428-435
  DAC-to-DAC skew, 426
  DNL, 412
  INL, 416
diode bridge, 82, 136
DIP (*see* dual inline package)
discrete Fourier transform, 216-218
discrete time, 149
discrete waveforms, 315
discrete-time aperiodic signals, 213
discrete-time Fourier series, 198-204, 211-216
distortion
  asymmetrical, 280
  harmonic, 175-178, 280-283, 360
  intermodulation, 175-178, 283, 284, 360
  symmetrical, 280
distributed capacitance, 507, 514
distributed processing, 144
distributed-element model, 505
dividers (digital) 3, 27, 566, 583
DNL (*see* differential nonlinearity)
doping, 5
DOT (*see* defect-oriented testing)
down time, 46, 643-647
drive data, 39, 131, 137, 567
driven guards, 503
dropout voltage, 53, 54
DSP (*see* digital signal processing)
DSP-based testing, 139, 189

DTFS (*see* discrete-time Fourier series)
dual inline package, 13
dual-slope ADCs, 474, 475, 584
dual-head testing, 19, 648, 649
DUT (*see* device under test)
duty cycle,
DVM (*see* digital voltmeter)
dynamic loads, 135
DZ (*see* device zero)

E

earpiece, 4, 316, 573
earth ground, 60, 519
e-beam prober (*see* electron beam prober)
edge code testing, 454
EEPROM (*see* electrically erasable programmable read-only memory)
effective number of bits, 363, 617
electical characteristics, 27-30, 37
electrical permitivity, 10, 497, 526
  of free space, 497
electrical specification, 27-30, 37
electrically-erasable probrammable read-only memory, 55, 56
Electroglas, 13
electrolytic capacitors, 527, 528
electromagnetic compatibility, 293, 515
electromagnetic interference, 18, 293, 503, 515, 523, 531
electron beam prober, 14, 15, 585
electrostatic discharge, 7, 46-48
electrostatic shields, 486, 502, 503, 515
EMC (*see* electromagnetic compatibility)
EMI (*see* electromagnetic interference)
encoding formats
  A-law, 343, 344, 346
  mu-law (μ-law), 343- 345
  one's complement, 341
  sign/magnitude, 341, 342
  two's complement, 340
  unsigned binary, 339
ENOB (*see* equivalent number of bits)
endpoint linearity, 413
equivalent number of bits, 363, 617
error band, 425-428, 445
ESD (*see* electrostatic discharge)
ESD protection, 46, 47
etching, 5
expect data, 39, 137

F

fall time, 135, 141-143, 426, 427
family board, 12, 484
fast binning, 40
fast Fourier transform, 216-220
fault coverage, 553, 561, 562, 565
fault models, 573, 659
feature summary, 24
Federal Communications Commission, 293, 298
feedback loops, 65, 68, 127, 182, 536
ferrite beads, 528
FFT (*see* fast Fourier transform)
FIB (*see* focused ion beam)
fiberglass PCBs, 488, 497, 499, 504
filaments, 50
filters
  ANSI C-weighting, 300
  antialiasing, 160, 183, 333, 360, 382-385, 387, 400, 553, 572
  anti-imaging, 99
  A-weighting, 239, 300
  band-pass, 147, 239, 266
  brick-wall, 240, 335
  Butterworth, 238, 305
  C-message, 296, 300, 303, 312
  discrete-time, 111, 235
  high-pass, 266, 293, 498, 581
  low-pass, 4, 105, 106, 125, 139-141, 160
  notch, 147
  N-point running averager, 111
  pass-band, 336
  psophometric, 300
  Q factor and settling time, 266
  RC and *RL* first-order, 543
  RC high-pass, 400, 498, 543-545
  RC low-pass, 106-108, 543-545
  reconstruction, 4, 155, 350, 403, 434
  RL high-pass, 543-545
  RL low-pass, 495, 504, 543-545
  stop-band, 336
  switched-capacitor, 249, 315, 320, 323, 351, 352, 355, 587
  weighting, 239, 296, 300
final test, 10, 384, 618
flash ADCs, 472, 476, 477, 583
flash memory, 55
flyback diodes, 130, 522
flying adders, 181-183
focused calibration (*see also* calibration, focused), 97
focused ion beam, 6, 7, 15, 585
forced-temperature system, 15, 16
formatters, 136

Fourier analysis, 191
Fourier integral, 192
Fourier series, 192-195, 198, 200
Fourier spectral bins, 173
   spectral bin selection, 175
Fourier transform, 215
Fourier, Jean Baptiste Joseph, 192
FR4 PCBs (*see* fiberglass PCBs)
frame (*see* sampling frame)
frequency denormalization, 210
frequency domain, 98, 99, 191
frequency leakage, 216, 224, 226, 228, 230
frequency resolution, 172, 177, 216
frequency response, 265-278, 356-360
   problems testing, 267
frequency synthesizers, 181-184
frequency-domain filtering, 234-241
fringe effects, 497
full scan, 556, 559-562
full-scale range, 406-409
full-scale voltage (*see* maximum full-scale voltage)
functional block diagram, 31, 33
functional shmoo plot, 601, 602,
fundamental frequency, 171
   mismatched, 330
fundamental tone, 280
fuse blowing, 9, 55

G

gain,
   absolute, 36, 256, 259, 271, 351, 438
   AC, 147, 256, 351, 587
   calibration, 372-374, 378
   closed-loop, 65
   common-mode, 72, 73
   DC, 65, 72, 265, 378-383, 385
   differential (*see also* differential gain and phase), 66, 67, 286-289
   open-loop, 68-71
gain error, 256-261, 351-355
   definition, 256
   intrinsic gain error 347-355
gain matching, 397
gain tracking, 258-260, 355, 389
gauge repeatability and reproducibility, 631
Gauss, C. F., 217
Gaussian distribution, 117, 118, 607-611, 618
Gauss-Jordan elimination, 95
GBD (*see* guaranteed by design)
Gerber plots, 487
Gibb's phenomenon, 195

glitch energy (glitch impulse), 427, 428
go/no-go testing, 27, 82, 103
ground loops, 518-520
ground planes, 492-494, 515-518
grounding, 514-520
group delay and group delay distortion, 278-280, 360
GRR (*see* gauge repeatability and reproducibility)
guaranteed by design, 30, 432, 435
guardbanding, 113, 119, 618-620

H

handlers
   depreciation cost, 643-645
   gravity-fed, 13
   pick-and-place (robotic), 13, 14, 646
hand-test socket, 483, 519
Hanning window, 226-229
hard disk drive, 3, 192, 279, 315, 318, 479
HDD (*see* hard disk drive)
harmonic distortion (*see* distortion, harmonic)
Hewlett Packard, 570
histograms
   data analysis 115-119, 604, 605
   linear ramp testing of ADCs, 456
   sinusoidal histogram testing of ADCs, 462-466,
humidity, 90, 91
hysteresis
   in comparators, 78-80,
   in ADC linearity testing, 461, 474

I

IC (*see* integrated circuits)
ICN (*see* idle channel noise)
I_{DDQ} (quiescent I_{DD}) testing
   analog I_{DDQ}, 588
   digital I_{DDQ}, 85, 553, 568, 587-590, 659
idle channel noise
   definition, 294
   in analog channels, 294-296
   in sampled channels, 363
idle time, 643-647
IEEE Std. 1149.1, 555, 557-560
IEEE Std. 1149.2, 555, 569-570
IFFT (*see* inverse FFT)
I_{IH} (*see* input high current)
I_{IL} (*see* input low current)
images, 160, 329, 334, 335, 336
   out-of-band, 356

impedance
   DC differential impedance, 59
   DC input impedance, 56-58
   DC output impedance, 58
   of transmission lines, 506-512
impulse function, 150
incomplete etching, 7
index time, 19, 646
inductance
   distributed, 505, 506
   parasitic, 18, 491-496, 514, 542
inductance per unit length, 492
infant mortality, 50
initialization run, 322
inking, 10, 40
INL (*see* integral nonlinearity)
input high current, 50, 82
input high voltage, 51, 82, 135
input impedance,
input low current, 58, 82
input low voltage, 51, 82, 135
input regulation, 53
input rejection, 54
instrument ranging, 103-105, 125-127, 372-375, 393-395
integral nonlinearity,
   ADCs, 468, 469
   DACs, 416-419
   definition, 416
integrated circuits, 5-8
intermodulation distortion (*see* distortion, intermodulation)
interpolation
   using convolution, 153-158
   using the inverse FFT, 233
   logarithmic, 300
intrinsic gain error, 347-355
intrinsic parameters, 403
inverse FFT, 231-240
   definition, 230
ion implantation, 5
I_{OSH} (*see* output short circuit current, high)
I_{OSL} (*see* output short circuit current, low)

J

jitter
   aperture jitter, 166-168, 472, 479, 480
   sampling jitter, 166-170
jitter-induced noise,
Joint Test Action Group, 557, 558
JTAG (*see* Joint Test Action Group)
jump discontinuities, 195, 199, 216, 274

Index **681**

K

Kelvin connections, 127, 516-518, 537, 538

L

laser trimming, 9, 56, 587
LCC (*see* leadless chip carrier)
lead frames, 10
leadless chip carrier package, 483
leakage testing, 50-52
least significant bit, 115, 162-165, 325, 340-342, 409-418
least-squated-error line,
LFSR (*see* linear feedback shift register)
line regulation, 53, 54
linear feedback shift register, 562
linear regression, 407
linear searches, 80, 81
load boards, 370
load current, 53, 491, 492, 604
load regulation, 53
loading conditions, 11, 54, 252, 321
logarithmic interpolation, 300
loopback modes, 319, 320, 579-581,
lot summaries, 599-600
lower specification limit, 113, 618, 619, 628-630
lower test limit, 113, 618, 619, 628-630
LSB (*see* least significant bit)
LSB-first, 325
LSB step size, 409
LSL (*see* lower specification limit)
LTL (*see* lower test limit)
LTX Corporation, 11
lumped-element model, 504-508

M

magnetic permeability, 492
  of free space, 492
magnitude spectrum, 193, 206, 207, 209
Mahoney, Matthew, 241, 344, 465
major carrier method, 420-423, 586, 659
manipulator (test head), 12
masking
  digital failures, 135
  mask limits, 266-271
  crosstalk tones, 361
match mode, 554, 555, 566, 567
MATLAB, 156-158, 221-230
  model of an analog channel, 304
maximum full-scale voltage, 405

mean, 89, 116, 117
mechanical docking, 18
memory BIST, 563
microcode instructions
  in digital patterns, 321, 325
  in microcode BIST, 564
midscale voltage, 405
minimum full-scale voltage, 405
missing codes, 470
mixer, analog, 2, 317
modeling
  DUT, 42, 656
  fault models, 573, 659
  linear models, 95, 101
  MATLAB model of an analog channel, 304
  sampling process, 151
  test setup, 101
  tester instruments, 93, 94
  testing costs, 643, 644
  transmission lines, 504-508
modems, 4, 293, 483, 580, 592
modulation
  data 4, 316, 436
  sigma-delta, 432, 433
monotonicity, 412-414, 469
most significant bit, 325, 340-342
mother board, 484
MSB (*see* most significant bit)
MSB-first, 325
mu-law (μ-law), 339, 343-345
multihead testing, 19, 645-649
multilayer PCBs, 488, 489, 519
multimeter, 125, 128
multisite testing, 20, 645-649
multitone leveling, 99
multitone signals, 98-100, 173-179
  definition, 98
  amplitude setting, 266
mutually prime bins, 175, 178

N

National Bureau of Standards, 369
National Institute of Standards and Technology, 93, 128, 369, 571, 658
NBS (*see* National Bureau of Standards)
near-Gaussian distribution, 118, 610
netlists, 487, 561
networking hardware, 144
new product development, 24, 657
NIST (*see* National Institute of Standards and Technology)
noise
  1/f, 293
  electrical, 18, 105, 135, 389, 530
  idle channel, 294-296, 343, 363

pink, 294
removal of noise using the inverse FFT, 233
shot, 161
thermal, 88, 293
white, 294, 479
noise shaping, 433, 434, 477, 658
noise spectral density, 107, 109, 112
noise weighting, 239, 300
non-Gaussian distributions, 118, 615
normal distribution, 117, 118, 607-611, 618
normalization, 195, 410, 464, 465
normalized frequency, 195, 305
N-point running averager, 111
nulling amplifiers, 68-70, 74, 75, 542
Nyquist, 140, 159-161
Nyquist criterion, 140, 159
Nyquist frequency, 159, 333-336
Nyquist interval, 159, 209

O

OBIST (*see* oscillation BIST)
observability, 16, 553, 554, 562, 565, 569, 574, 575
observation interval, 216, 224-226
offline simulation, 17, 38, 656
offset
  calibration, 376, 378
  common-mode, 62, 63
  differential, 62, 63, 126
  input offset voltage, 64, 377
  input-referred, 61, 68, 70
  output offset voltage, 61-64
  single-ended,
op-amps, 60-76
orthogonal basis functions, 203
oscillation, 542, 585
oscillation BIST, 585
oscillators, 182, 568, 569, 583, 585
oscilloscopes, 93, 98, 652
oscilloscope probes, 93, 519, 520
outliers, 118, 119, 472, 615-617
output high voltage, 82, 135
output impedance, 58, 59, 136, 435
output low voltage, 82, 135
output no-load voltage, 53
output short circuit current, high, 82, 83
output short circuit current, low, 82, 83
overshoot, 426, 514
overshoot suppression, 136
overspecification of device parameters, 586

P

parallel continuity testing, 48, 49
parallel leakage testing, 51
parallel testing, 48-51, 550, 646
parallel traces, 494, 496, 497, 500
parametric shifts, 50
parametric shmoo plot, 602, 603
parametric testing, 27, 38
parasitic capacitance, 6, 48, 61, 101, 496-499, 502-504, 514, 521, 526, 542, 576
parasitic elements, 490
parasitic inductance, 18, 491-496, 514, 542
parasitic loading, 50, 61
parasitic resistance, 128, 486, 490, 491, 519, 531, 541, 542
Pareto charts, 631, 632
partial scan, 556, 562
partial transfer curves, 419
particulate, 7, 8, 50
partitioning circuits for testability, 565, 566
pass/fail test results, 27, 40, 41, 131, 550, 601
passive loads, 130, 528, 533
patterns (*see* digital patterns)
PCB (*see* printed circuit board)
pdf (*see* probability density function)
PDM (*see* pulse density modulation)
peak-to-RMS ratio, 173, 174, 175
performance board, 12
performance verification, 96, 97
period measurements, 141
periodic extension, 214-217
per-pin measurements, 48, 50, 51, 136
PGA (*see* programmable gain amplifiers)
phase error, 392, 397, 399, 400, 496
phase matching, 397
phase measurement, 274, 392-394, 396, 397, 399
phase response,
  in analog channels, 273, 274
  in sampled channels, 359, 360
phase spectrum, 193, 194, 206-209
phase-locked loop, 170, 181-183, 583, 584
photoelectric effect, 56
photolithography, 5, 19
photomask, 7, 20, 600
photoresist, 5
PIB (*see* probe interface board)
pin card electronics, 82, 134-137
pipeline delay, 566-568

plastic encapsulation, 56
PLL (*see* phase-locked loop)
PO (*see* protective overcoat)
pogo pins, 13, 45, 520, 521
polar notation, 231, 273, 274
polysilicon, 5
polysilicon fuses, 55, 56
POR (*see* power-on reset)
power consumption, 51,
power planes, 492, 493, 517, 540
power splitters, 529
power supply currents, 51-53
power supply rejection, 287, 362
power supply rejection ratio
  DC PSRR, 72
  AC PSRR, 287-290, 362
power supply sensitivity, 71, 72, 405, 410
power-down modes, 52, 551
precharging, 580-582
precision, 87
precision-to-tolerance ratio, 631
prime bins, 175, 178, 354
prime numbers, 175, 331, 332, 354
primitive frequency, 171
primitive period, 171
principles of operation, 24, 26, 27
printed circuit boards
  computer-aided design, 487
  materials, 489
  netlists, 487
  stackups, 488, 489, 492
  trace parasitics, 490-497
probability density function, 448, 449, 453, 607-609
probe card, 13, 484
probe interface boards, 13, 484
  probe tips, 13, 484
probers, 13, 14, 645-647
process capability, 628-630
process potential, 629
process variations, 120, 597, 600
product engineering, 8, 9, 16, 17, 597
production down time, 102, 646
production lot, 599, 600
production test program, 11, 42, 43
production worthiness, 92, 96, 97, 486, 553, 586, 587
profit margin, 17, 555, 556, 641
programmable gain amplifiers, 1-4, 2-4, 249
  in ATE instruments, 125, 140
  in AGC circuits, 578, 579
  testing PGAs, 260-264
propagation delay
  measuring, 141
  of transmission lines, 507, 513
protection diodes, 46-48
protective overcoat, 10, 585

prototype, 11, 16, 486
pseudorandom number generators
  in LFSR circuits, 562
  selecting multitone phases, 174
PSRR (*see* power supply rejection ratio)
PSS (*see* power supply sensitivity)
pulse density modulation, 433, 477, 478
pulse-width modulation, 431-435
PWM (*see* pulse-width modulation)
PWM DACs, 431

Q

QFP (*see* quad flat pack package)
QGND (*see* quiet ground)
quad flat pack package, 11, 483
quanta, 338
quantization errors, 89, 103, 125-127, 161-164, 175, 344, 351
quiet ground, 518, 540

R

R&M (*see* repair and maintenance)
radio frequency, 3, 4, 316, 317
ramp searches, 79, 80
random errors, 88
random variables, 117
recalibration 370-376
receive channel, 315
receive memory, 132
recommended operating conditions, 28, 31
reconstruction
  effects in sampled channels, 335
  sampling and reconstruciton, 148, 152-160
rectangular window, 226-228
reflections in transmission lines, 508, 512
relative error, 62
relative gains, 265
relative permittivity, 497, 499
relay drivers, 130
relay matrix, 128-130
relays
  failure, 522
  failure due to current, 46
  failures caused by age, 103
  inductive kickback, 130
  local DIB relays, 130, 530-532
  pole, 522
  types of relays, 522, 523
  useful DIB circuits, 530
  wiper, 46, 522, 523
repair and maintenance, 96 524
repeatability, 18, 89, 92,
  improving, 109-120

reproducibility, 92, 623-625
resistive divider DACs, 428
resistive ladder DACs, 430
resistors
    matching, 68, 74, 75, 256, 534
    package types, 525
    tolerance, 525
resolution, 88
ringing, 136, 428, 508, 511, 514
ripple rejection, 53, 54
ripple sources, 536-538
rise time, 135, 141-143, 426, 427
RMS (*see* root mean square)
robust design, 269, 585
root mean square, 607
root spectral density, 295

S

S/H (*see* sample and hold)
S/N+THD (*see* signal to noise plus total harmonic distortion)
S/THD (*see* signal to total harmonic distortion)
safety ground, 519, 520
sample-and-difference, 125-127
sample-and-hold (S/H), 315, 320
    in SAR ADCs, 473
    in ATE instruments, 126, 140
    $\sin(x)/x$ rolloff, 336, 338
    testing S/H amps, 352, 356
sampled channels, 315
sampling, 148, 149
sampling frame, 131-134, 141, 182, 321-324
    definition, 131
sampling frequency, 149, 316, 320, 354,
    maximum sampling freq. 470
sampling loops, 131-134, 141, 321-324
sampling period, 149
sampling rate, 149
    constraints in sampled channels, 320-323
sampling theorem, 159, 160
sampling theory, 147
SAR (*see* successive approximation register)
scan testing, 556-562
scan cell, 557, 558
scan collar, 579
scanning electron microscope, 5, 14
scatter plots, 631, 632, 634
Schlumberger Test Equipment, Inc., 7, 11, 14
scientific method, 649
SCR (*see* silicon-controlled rectifiers)

second harmonic distortion, 176, 178, 280, 361
selected-code testing, 419-424
SEM (*see* scanning electon microscope)
semiflash ADCs, 476
send memory, 132
sequencer, 40, 115
serial continuity testing, 48
serial leakage testing, 50
serial testing, 48, 50, 51
settling time
    clock sources, 183
    DACs, 424-426
    DfT precharging, 581, 582
    filter testing, 266
    RC filters, 106, 107
Shannon sampling theorem, 159
shmoo plots, 601
sifting property, 150, 151
sigma-delta ADCs, 477
sigma-delta DACs, 433
sign/magnitude format, 341-343
signal to noise and distortion, 281, 282, 296
signal to noise plus distortion ratio, 635
signal to noise plus total harmonic distortion, 115
signal to total harmonic distortion, 115
signal velocity, 506-508
signal-to-noise ratio, 296, 297, 363
silicon-controlled rectifiers, 46, 47
silicon dioxide, 5,6
silkscreens, 489
simulation,
    analog channel, 305-307
    test code, 42
    test simulation, 656, 657
$\sin(x)/x$ rolloff, 155, 320, 335-337, 354-356
SINAD (*see* signal to noise and distortion)
single-slope ADCs, 474
sink currents, 136
site, 20, 645, 646
six sigma, 628-639
skew
    channel-to-channel, 513
    DAC-to-DAC, 426
small outline IC package, 483
SNDR (*see* signal to noise plus distortion ratio)
SNR (*see* signal-to-noise ratio)
socket (*see* hand-test socket)
socket pins, 524
SOIC (*see* small outline IC package)
solder mask, 489

sorting, 14, 19
source currents, 136
source memory, 132, 322-328
source termination, 511
sparkling, 472, 473
SPC (*see* statistical process control)
specification limits, 113, 618, 619, 628-630
specification-oriented testing, 658
spectral bin selection, 175-178
spectral coefficients, 192
spectral density, 107-112, 294, 295
spectral leveling, 98
spectral lines, 98, 99
spectrum analyzer, 98
split planes, 517
SPOT (*see* specification-oriented testing)
spurious free dynamic range, 298
spurs, 298, 299
stability, 90
stackups (PCBs), 488, 489, 492
staircase waveform, 154, 157
standard deviation, 116-118, 606-612
standard Guassian cdf, 612
standard test interface language, 655
standards agency, 369
star ground, 516, 518
statistical analysis, 606
statistical outliers, 118, 119, 472, 615-617
statistical populations, 116, 605, 607, 619
statistical process control, 627
step response, 106
step searches, 79, 80, 82, 455, 474
STIL (*see* standard test interface language)
striplines, 492-494, 545, 546
strobes and windows, 137
successive approximation, 79, 473-480, 582-586
successive approximation ADCs, 473, 476
successive approximation register, 473, 582
superposition
    in PGAs, 261-264
    major carrier testing, 420, 423, 424, 586, 659
supply pins, 47, 48, 410
swap block, 12, 484, 485
synchronization, 123, 141, 181-184, 320, 326, 658
systematic errors, 88, 91
system-level applications, 278, 315, 435, 479
systems engineering, 8, 9, 16

T

TAP (*see* test access port)
TBD (*see* to be determined)
TDR (*see* time-domain reflectometry)
Teflon® PCBs, 489, 490, 497-499
temperature drifts and shifts 90, 97,
Teradyne, Inc., 11, 12, 19, 21, 114, 115, 598
termination of transmission lines, 508, 511
test access port, 557, 558
test conditions, 30
test economics, 641
test engineering, 8, 9, 16
test head, 11, 12, 18, 19
test limits, 113, 618, 619, 628-630
test list, 11, 31
test modes, 554-558, 573-585
test pads, 585
test plan, 31-37
test programs, 38-42
test sequence, 40
test simulation, 656
test time, 19, 642, 645
test tone selection, 178
test tones, 175-181
testability, 16, 549
tester mainframe, 11, 12
Texas Instruments, Inc., 24, 42, 627
thermal chamber, 14-16
thermal stabilization, 93
third harmonic distortion, 176, 178, 280, 361
threshold voltage, 78, 80, 82, 136
threshold voltage error, 78
throughput, 19, 645
time denormalization, 210
time domain, 98, 99, 191
time measurement systems, 141-143, 471
time set, 138
time to market, 17, 555, 641
time-domain aliasing, 216
time-domain filtering, 234
time-domain reflectometry, 513
timing and formatting, 136
timing-on-the-fly, 131
to be determined, 38
total quality control, 627, 634
TQC (*see* total quality control)
trace resistance, 490, 491
traceability to standards, 93, 96, 128, 369, 571, 658

trace-to-trace crosstalk, 502, 503
transfer characteristics, 60, 164, 265
transfer functions, 108, 191, 265
transformers, 528
transmission lines
 characteristic impedance, 506
 discontinuities, 541
 distributed-element model, 505
 lumped-element model, 504
 signal velocity, 506-508
 stubs, 532
transmission parameters, 403
transmit channel, 315
trigonometric form of the DTFS, 237
trigonometric Fourier series,
trimmable references, 55
T-switches, 576, 577
Tukey, J., 216
TYP (*see* typical specifications)
typical specifications, 30, 31

U

underetching, 7
undersampling, 582
undershoot, 426, 514
uniform distribution, 163, 616-618
unit impulse, 151
unit impulse train, 151
unit test period, 171, 172
unsigned binary format, 339-341
upper specification limit, 113, 618, 619, 628-630
upper test limit, 113, 618, 619, 628-630
user supplies, 128
USL (*see* upper specification limit)
UTL (*see* upper test limit)
UTP (*see* unit test period)

V

variance, 606-612
VHDL (*see* VLSI hardware description language)
vias
 in ICs, 7, 8
 in PCBs, 488, 489, 526, 541
video, 436, 480
 NTSC, 399, 400, 436, 438, 480
video DAC, 405, 436, 437
video palette, 123
V_{IH} (*see* input high voltage)

V_{IL} (*see* input low voltage)
VLSI hardware description language, 561, 656
vocoding, 4
V_{OH} (*see* output high voltage)
voice-band interface, 3-5, 316-318
V_{OL} (*see* output low voltage)
voltage references, 55, 56, 128
voltage regulators, 52-55
voltage sources, 125, 128
voltmeters
 AC, 97, 251
 DC, 125
 differential, 67, 125
 DVM (digital voltmeter), 88
 RMS, 139, 147, 252, 382-390
volume control, 2, 4, 260

W

wafer, 5, 7, 10, 13, 600
watchdog timer, 5
waveform digitizers, 140-146, 251, 382-390
waveforms
 continous, 140, 148, 350
 digital, 136, 138
 discrete, 315
wavelet transforms, 424
WECO run rules, 634
whisker probes, 585
windowing, 216, 226-229
windows,
 Blackman, 246
 Hanning, 226-229
 Kaiser, 246
 rectangular, 226-228, 246
workstation, 11, 12, 17, 144, 656

Y

yield
 cumulative, 599
 definition, 118
 effects of process variation, 623
 yield loss, 540, 599, 600, 623-626

Z

z-domain transfer function, 235, 238